Coastal Vulnerability and Mitigation Strategies

Coastal Vulnerability and Mitigation Strategies: From Monitoring to Applied Research

Editors

Diego Vicinanza
Pasquale Contestabile

MDPI • Basel • Beijing • Wuhan • Barcelona • Belgrade • Manchester • Tokyo • Cluj • Tianjin

Editors
Diego Vicinanza
Università degli Studi della Campania "Luigi Vanvitelli"
Italy

Pasquale Contestabile
University of Campania Luigi Vanvitelli
Italy

Editorial Office
MDPI
St. Alban-Anlage 66
4052 Basel, Switzerland

This is a reprint of articles from the Special Issue published online in the open access journal *Water* (ISSN 2073-4441) (available at: https://www.mdpi.com/journal/water/special_issues/coastal_vulnerability).

For citation purposes, cite each article independently as indicated on the article page online and as indicated below:

LastName, A.A.; LastName, B.B.; LastName, C.C. Article Title. *Journal Name* **Year**, *Article Number*, Page Range.

ISBN 978-3-03943-386-5 (Hbk)
ISBN 978-3-03943-387-2 (PDF)

Cover image courtesy of Diego Vicinanza.

Contents

About the Editors

Diego Vicinanza (Ph.D., 1997) is currently Associate Professor at Università degli Studi della Campania "Luigi Vanvitelli"—Department of Engineering (Italy). He received his National Scientific Qualification as Full Professor in Italian University in 2017. His main research interests are in Coastal Engineering. His recent research has focused on two topics: innovative technologies for the production of marine renewable energy and their integration with traditional breakwaters; solutions/strategies for the mitigation of coastal risk from flooding and erosion as a function of climate change. Research on innovative solutions for marine environmental restoration and bathing represent the currently ongoing topics of investigation. He is Research Fellow of the Stazione Zoologica Anton Dohrn and Italian National Council for Research (CNR-INM). He is member of CoNISMa (Italian National Inter-University Consortium for Marine Sciences) and part of the Directive Committee and of the Administrative Council. He was/is involved in numerous national/international competitive research projects. e is author/co-author of 197 papers (60 international journal papers with impact factor), SCOPUS h-index 25, 1767 citations, and 90 documents. A complete list of his national and international research projects and papers can be accessed at ORCID (http://orcid.org/0000-0002-2033-1902).

Pasquale Contestabile (Ph.D., 2012) is Assistant Professor in Coastal Structures and Special Hydraulic Systems at Università degli Studi della Campania Luigi Vanvitelli—Department of Engineering (Italy). He got the National Scientific Qualification as Associate Professor in Italian University (2018). He has over 12 years of experience in Coastal Engineering, which includes consultancy and expert witness services in Maritime and Environmental Engineering. He is author/co-author of 60 papers (SCOPUS h-index 15). He has served as Project Manager for the Italian Ministry of University and Research within the National Operational Programme for "Research and Competitiveness" (2007–2013) related to innovative coastal structures. He has a strong record of national and international collaboration on research regarding topical issues in coastal systems. Recent and current relevant research projects include life-extension strategy and eco-sustainable reuse of coastal and offshore structures, the use of natural structural biomaterials to contrast coastal erosion, and cost-effective solutions for ecological restoration of coastal areas.

water

MDPI

Editorial

Coastal Vulnerability and Mitigation Strategies: From Monitoring to Applied Research

Pasquale Contestabile [1,2] and Diego Vicinanza [1,3,*]

1 Department of Engineering, University of Campania "Luigi Vanvitelli", 81031 Aversa, Italy; pasquale.contestabile@unicampania.it
2 CoNISMa National Inter-University Consortium of Marine Sciences, Piazzale Flaminio 9, 00196 Roma, Italy
3 Stazione Zoologica Anton Dohrn, Villa Comunale, 80121 Napoli, Italy
* Correspondence: diego.vicinanza@unicampania.it; Tel.: +39-328-482-0770

Received: 28 August 2020; Accepted: 11 September 2020; Published: 17 September 2020

Abstract: This paper intends to offer the readers an overview of the Special Issue on Coastal Vulnerability and Mitigation Strategies: From Monitoring to Applied Research. The main focus of this Special Issue is to provide the state-of-the-art and the recent research updates on the sustainable management strategies for protecting vulnerable coastal areas. Based on 28 contributions from authors from 17 different countries (Australia, China, Ecuador, Germany, Greece, India, Italy, Mexico, The Netherlands, New Zealand, Poland, Spain, Sri Lanka, Taiwan, United Arab Emirates, UK, USA), an ensemble of interdisciplinary articles has been collected, emphasizing the importance of tackling technical and scientific problems at different scales and from different point of views.

Keywords: coastal vulnerability; coastal defense; coastal monitoring; wave climate; coastal morphodynamic; coastal management; coastal ecosystem

---

1. Introduction

Coastal management in the 21st century will require us to face multiple issues including climate change and impacts of sea level rise. Conservation of coastal systems and ecosystems requires multidisciplinary inputs as well as integrated studies and approaches.

In view of this, the following research topics deserve greater attention to speed up the development of suitable coastal management strategies:

(a). relationship between coastal ecosystems and hydrodynamics;
(b). climate change effect on coastal areas;
(c). coastal morphodynamics;
(d). coastal vulnerability;
(e). integrated coastal management.

2. Contributions

This Special Issue provides food for thought on each of these topics.

2.1. Relationship between Coastal Ecosystem and Hydrodynamics

Coastal lagoons and river deltas are complex environments where hydrology and coastal dynamics work together for the ecosystem functioning. Management and environmental policies of such coastal areas are extremely difficult because of continuous conflicts between conservation and development. Modelling is crucial for supporting the analysis of management scenarios. So many research efforts have been made to understand the relationships of coastal hydrodynamics with biotic and abiotic elements of the ecosystems.

Christia et al. [1] developed an integrated environmental assessment methodology on Western Greece coastal lagoons (Rodia, Tsoukalio and Logarou—Amvrakikos Gulf, Kleisova—Messolonghi-Aitoliko, Araxos) demonstrating the link between macrophyte assemblages and abiotic factors typical of coastal lagoon systems. Their results emphasize the crucial impact of the sea water intrusion on the relative abundance and distribution of macrophyte species, as described in other Mediterranean coastal lagoons. The proposed methodology is broadly applicable, since it is based on important parameters affecting coastal lagoon ecosystems, and the provided links between macrophyte assemblages and abiotic factors are of critical importance to improve environmental policies.

Tran Anh et al. [2] combined different models to simulate the hydrodynamics and salinity distributions in the Hau (Bassac) River estuary of the Mekong Delta, southern Vietnam. A combination of 1D and 2D hydrodynamic models were calibrated and applied to simulate future hydrological changes under multiple scenarios of upstream inflow changes, climate change and sea level rise for the 2036–2065 period. The model simulations indicate that a combination of upstream discharge reductions, rainfall changes and rising sea level will substantially exacerbate salinity intrusion.

Interaction between mangrove vegetation and hydrodynamics plays an important role in many coastal tropical and sub-tropical intertidal environments, including coastal protection. Coastal vegetation is effective in dissipating incident wave energy during storm conditions, which offers valuable protection to coastal communities.

Montgomery et al. [3] explored the influence of channelization on mangrove flood attenuation comparing high water events in two contrasting New Zealand mangrove forests. The degree of channelization and, therefore, the capacity of mangroves to reduce flooding depends on the elevation of the vegetation. Observations from sites with the same vegetation type suggest that mangrove properties are important to long wave dissipation only if water transport through the vegetation is the dominant mechanism of fluid transport.

Tan et al. [4] investigated wave propagation and turbulence characteristics through vegetation with different stiffness by means of a physical model in a laboratory wave flume. The results showed different patterns in wave propagation turbulence intensity in different canopies; such knowledge may support the selecting of vegetation species with suitable stiffness for coastal protection purposes.

Yao et al.'s [5] technical note provides a practical set-up to derive both time-varying and period-averaged vegetation drag coefficients following the direct measuring method. Standard force sensors are applied to compose four synchronized force–velocity measuring systems in the current experiment. The newly-developed synchronized force–velocity measuring systems and the automatic realignment algorithm offers information for future experiments on vegetation–wave interactions for better understanding and prediction of vegetation-induced wave dissipation.

Tripepi et al. [6] investigated hydrodynamic forces induced by tsunami-like solitary waves on a horizontal cylinder placed on a horizontal seabed by means of 2D laboratory experiments. An overall good agreement found between analytical solutions and laboratory tests has led, in conjunction with the measurement of experimental forces, to the calibration of the hydrodynamic coefficients in the Morison and transverse equations.

2.2. *Coastal Climate*

Extreme sea and weather events (in terms of storm waves, tsunamis, sea level rise, air temperature, wind and atmospheric precipitation) in coastal areas have highlighted the destructive effects that can occur from hazards of marine origin. Many geomorphological and coastal engineering scenarios require robust estimates of wave climate and design wave height with a certain return period and incorrect estimates can have dramatic effects on the flood risk analysis or on the structural design of maritime structures.

Dentale et al. [7] proposed a procedure based on integrating significant wave height time series generated by model chains with those recorded by wave buoys in the same area (North Atlantic Spanish Coast, South Mediterranean Italian coasts and Gulf of Mexico) in order to provide better estimates of extreme values. A general procedure is provided to improve the reliability of model data for the extreme values analysis; such a procedure can also be used to evaluate the suitability of a given model data archive to the estimation of the probability of extreme sea states.

Molina et al. [8] analyzed a 35-year wave climate dataset concerning four positions equally spaced along the Mediterranean coast of Andalusia (south of Spain). A total of 2961 storm events were recorded and classified as a function of their associated energy flux. In particular, nine stormy years, i.e., years with a high cumulative energy, were recorded.

Hamza et al. [9] investigated the wave climate offshore Saadiyat island situated in the Gulf within the Emirate of Abu Dhabi. They compared the measured ADCP data and propagation results of the NOAA offshore wave dataset by means of the Simulating WAves Nearshore (SWAN) numerical model, the NOAA and ECMWF wave datasets at the closest grid point in shallow water conditions, and the SPM '84 hindcasting method with the NOAA wind dataset used as input have been carried out. They showed that the SPM '84 hindcasting method might be very accurate in shallow water conditions.

Contestabile et al. [10] carried out a multi-comparison between wave propagation model data and direct measurements at Bagnoli-Coroglio bay (central Tyrrhenian Sea, Italy). A non-conventional triple-collocation-based calibration of a wave propagation model is described. GPS-buoy, ADCP data and model virtual numerical points allowed an implicit reciprocal validation of the different data source. The results suggest that numerical model calibration based on short term wave buoy measurements can be easily applied in different areas where detailed wave data are not available.

Tylkowski et al. [11] determined the threshold values for extreme sea and weather events on the Polish Baltic coast. The threshold values presented can be used to forecast changes in climatic and hydrological conditions (maximum and average daily air temperature, daily sum of atmospheric precipitation and maximum and average sea level) in the Baltic coastal zone.

Hydrometeorological conditions especially favorable to the intensification of aeolian processes are the main determinants of geomorphological changes in the coastal zone. Hojan et al. [12] presented the temporal and spatial variability of hydrometeorological conditions conducive aeolian processes on the Southern Baltic coastal zone in Poland. In periods between storms, coastal wind is seen to decrease the balance of beach sediments and lowers the beach area.

2.3. Coastal Morphodynamic

Extreme storms may significantly affect the coastal environment, especially in terms of erosion and sediment transport. They can provoke disastrous consequences such as sediment transport beyond the surf zone to unusual depths [13]. The swash zone is that part of the beach alternately covered and exposed by uprush and backwash. It is characterized by strong and unsteady flows, high turbulence levels, large sediment transport rates and rapid morphological change, and it represents arguably the most dynamic region of the nearshore [14].

Riefolo et al. [15] analyzed experimental data from large scale wave flume under the project SUSCO (Swash zone response under grouping Storm Conditions) founded by EC Programme HYDRALAB III. The authors highlighted the effects of wave grouping and long-wave short-wave combination regimes on low frequency wave generations and clarified their influence on morphodynamics. Some evidence of the influence of low frequency waves on runup and transport patterns are shown. In particular, the generation and evolution of secondary bedforms are consistent with energy transferred between the standing wave modes.

Saponieri et al. [16] analyzed short term morphodynamic response of a beach nourishment protected by a standard and an innovative beach defense system. The 2D physical small-scale models were carried out to study a rubble-mound detached submerged breakwater and a Beach Drainage System deployed together. The Beach Drainage System influenced swash zone hydrodynamics and morphodynamics in the presence of the submerged breakwater while a reversal of the prevalent direction of the net sediment transport seaward was reported offshore from the sheltered region.

Medellín et al. [17] studied the shoreline resistance and resilience associated to a transient disturbance (a temporary groin) by means of field observations and numerical modelling. The study site is a micro-tidal sea-breeze dominated beach located in the northern Yucatan Peninsula. A new one-line numerical model of beach evolution is calibrated with the field surveys, reproducing both the sediment impoundment and subsequent beach recovery after the structure removal. Results suggest that beach resistance associated to the presence of a structure decreases with increasing alongshore sediment transport potential, whereas resilience after structure removal is positively correlated with the alongshore diffusivity.

Torres-Freyermuth et al. [18] investigated beach morphodynamics behind low-crested detached breakwaters build on a micro-tidal sea-breeze-dominated beach located in the northern Yucatán Peninsula in the Gulf of Mexico. Three study sites were monitored trough beach survey (RTK-DGPS), UAV flights, freeboard elevation and breakwater length, waves and sea level measurements. Observations suggest the high sensitiveness of beach morphodynamics to breakwater transmissivity.

2.4. Coastal Vulnerability

Coastal vulnerability is a spatial concept that identifies people and places that are susceptible to disturbances resulting from coastal hazards. Hazards in the coastal environment, such as coastal storms, erosion and inundation, pose significant threats to coastal physical, economic, and social systems. [19].

The Coastal Vulnerability Index (CVI) is a popular index in literature to assess the coastal vulnerability of climate change.

Pantusa et al. [20] presenting a case study proposed a CVI formulation to make it suitable for the Mediterranean coasts. The new formulation considers ten variables divided into three typological groups: geological, physical process and vegetation. For the case study presented in this work, the most influential variables in determining CVI are dune width and geomorphology. The transects presenting a very high vulnerability were characterized by sandy and narrow beaches (without dunes and vegetation) and by the absence of Posidonia oceanica.

Kantamaneni et al. [21] reviewed the existing coastal vulnerability assessment studies along the coastal Andhra Pradesh region in India with the aim to mitigate the existing shortcomings in the assessment techniques used previously in that area. Their study shows that very little was done so far in the area to assess the overall coastal vulnerability, with only a few of the CVI parameters being accounted for and based on relatively low-resolution data. So, this study significantly improved the assessment.

Garcia-Ayllon [22] presented an innovative methodology for analyzing the coastal vulnerability based on the GIS evaluation of the spatial statistical correlation of long-term anthropic impacts and the distribution of current risks. The geo-statistical analysis carried out for the Mar Menor Mediterranean lagoon reveals that the urbanization processes being developed in the last decades have generated imbalances. The proposed approach seems promising to better understand the relationship between territorial transformations on the coast and the current coastal vulnerability of this area.

Favaretto et al. [23] presented a novel 2D model for the inland flood propagation and an approach for the assessment of coastal flooding vulnerability. Hazard maps of two stretches of the Venetian littoral (Northern Adriatic Sea, Italy) were produced, showing the probability of failure in each point of the coast for a given inland inundation level.

Gaeta et al. [24] implemented a coupled wave-2D hydrodynamic simulation by means of the open-source TELEMAC suite in the coastal area of the River Reno mouth (eastern coast of Northern Italy). Past (1971–2000) and future climate change (2071–2100) scenarios showed that flooding hazards and changes in littoral hydrodynamics at the selected site are nowadays already significant, especially during extreme events and are expected to further increase in the future. The highest contribution to the coastal vulnerability of the studied beach is due to the relative rise of sea level, especially when this is combined with extreme sea storms.

2.5. Coastal Management

Coastal management in the coastal zone includes nature conservation, recreational activity, habitat and species restoration and coastal defense (protection from coastal pollution, coastal erosion and flooding).

Ruol at al. [25] described the recent Coastal Plan of the Veneto Region (Italy) proposing erosion mitigation criteria. The authors provided practical guidelines on how to interpret coastal monitoring analysis, select when, where and what mitigation measures should be adopted, and suggest a methodology for assigning a priority level to any action. The criterion used takes into account erosive tendency, existing coastal flooding hazards, coast value, environmental relevance, tourist pressure, urbanization level, the presence of production activities and cultural heritage.

Coastal management criteria based on the coastal risk map for the Yunlin Coast (Taiwan) was drawn by Huang et al. [26]. The results showed vulnerability and potential hazards and proposed design criteria for coastal defense and land use for the various kinds of risks faced. The safety of the present coastal defenses and land use was assessed, and coastal protection measures for hazard prevention were proposed based on the generated risk map. The coastal hazards are constituted mainly by storm surge and ground subsidence and, therefore, an implementation of both engineered and non-engineered control measures is proposed.

Giardino et al. [27] presented an implementation and application of the Bayesian belief network (BBN) for coastal erosion management at the regional scale along the entire Holland coast. The effects of different sand nourishment designs on two pre-identified coastal indicators (i.e., dune foot and momentary coastline position) were assessed at 604 cross-shore transects and spanning a period of over 50 years. BBN provides a very powerful tool to bridge the existing gap between the needs of coastal managers and the currently available data and numerical models.

Coastal pollution is also a significant element of vulnerability. The assessment of pollution sources is critical for support management and if needed remediation actions.

Giglioli et al. [28] analyzed the contaminants' concentrations (i.e., heavy metals and hydrocarbons) in seabed sediments through a statistical multivariate approach in a post-industrial area, Bagnoli (Gulf of Naples, Southern Italy). The main contamination source was found related to anthropogenic activities but, concerning Arsenic and other metals, it was reported that the existence of multiple anthropogenic and geogenic sources might originate from the volcanic rocks present in the area.

Mestanza et al. [29] presented an historical analysis (17th to 19th centuries) of the shore protection works performed at Callao (Peru) as defense from storm waves and tsunamis. The analyses put in evidence that most of the physical processes of coastal dynamics and shore protection were qualitatively understood. The main difference with the modern approach is that new materials to build structures as physical and numerical models to design them are now available. A strategic retreat as the most sustainable solution with respect to the forecasted sea level rise and increased storminess, was even considered.

Pranzini et al. [30] analyzed a long time series (from 1878 to 2017) of data on shorelines and shore protection structures along the Northern Tuscany coast. The presented case studies allow identification as to how shore protection structures are designed to counteract beach erosion. This study shows how sediment bypassing could have been implemented at most important harbors and how softer solutions might have been adopted starting from stable sectors and moving towards eroding areas. However, the authors point out that such strategies would have required a long-term strategy, which in many cases is incompatible with the lifetime of political decision-makers.

Cioffi et al. [31] developed a methodological modeling approach to assess the reliability of hydraulic infrastructures in controlling risks of flooding in a lagoon area in the south of Italy. This zone has an elevation equal to or lower than the mean sea level. The modeling study shows that the carrying capacity of the hydraulic network downstream of the pumping system is insufficient to cope with future sea level rise and intensification of rainfall.

3. Conclusions

The research contents examined in this Special Issue highlight that several stressors affect the coastal zones, determining new challenges to minimize coastal vulnerability through mitigation strategies.

Most papers give us greater insights and open new frontiers to handle risks in coastal zones. Innovation in management tools for coastal managers and applied research in the topic of coastal zone management have been proposed. Several studies have thoroughly investigated specific hydrodynamic and morphodynamic processes in costal zones. An overview of mitigation strategies against flooding and erosion, also in the perspective of climate change effects, is also provided. Moreover, the Special Issue is completed by several contributions concerning ecological coastal defense and innovative monitoring techniques.

The ensemble of interdisciplinary articles collected in this Special Issue emphasizes the importance of tackling technical and scientific problems at different scales and from different points of view.

Author Contributions: D.V. and P.C. carried out the conceptualization, they wrote the original draft and performed the final review. All authors have read and agreed to the published version of the manuscript.

Funding: This research received no external funding.

Conflicts of Interest: The authors declare no conflict of interest.

References

1. Christia, C.; Giordani, G.; Papastergiadou, E. Environmental Variability and Macrophyte Assemblages in Coastal Lagoon Types of Western Greece (Mediterranean Sea). *Water* **2018**, *10*, 151. [CrossRef]
2. Tran Anh, D.; Hoang, L.P.; Bui, M.D.; Rutschmann, P. Simulating Future Flows and Salinity Intrusion Using Combined One- and Two-Dimensional Hydrodynamic Modelling-The Case of Hau River, Vietnamese Mekong Delta. *Water* **2018**, *10*, 897. [CrossRef]
3. Montgomery, J.M.; Bryan, K.R.; Horstman, E.M.; Mullarney, J.C. Attenuation of Tides and Surges by Mangroves: Contrasting Case Studies from New Zealand. *Water* **2018**, *10*, 1119. [CrossRef]
4. Tan, C.; Huang, B.; Liu, D.; Qiu, J.; Chen, H.; Li, Y.; Hu, Z. Effect of Mimic Vegetation with Different Stiffness on Regular Wave Propagation and Turbulence. *Water* **2019**, *11*, 109. [CrossRef]
5. Yao, P.; Chen, H.; Huang, B.; Tan, C.; Hu, Z.; Ren, L.; Yang, Q. Applying a New Force-Velocity Synchronizing Algorithm to Derive Drag Coefficients of Rigid Vegetation in Oscillatory Flows. *Water* **2018**, *10*, 906. [CrossRef]
6. Tripepi, G.; Aristodemo, F.; Veltri, P. On-Bottom Stability Analysis of Cylinders under Tsunami-Like Solitary Waves. *Water* **2018**, *10*, 487. [CrossRef]
7. Dentale, F.; Furcolo, P.; Pugliese Carratelli, E.; Reale, F.; Contestabile, P.; Tomasicchio, G.R. Extreme Wave Analysis by Integrating Model and Wave Buoy Data. *Water* **2018**, *10*, 373. [CrossRef]
8. Molina, R.; Manno, G.; Lo Re, C.; Anfuso, G.; Ciraolo, G. Storm Energy Flux Characterization along the Mediterranean Coast of Andalusia (Spain). *Water* **2019**, *11*, 509. [CrossRef]
9. Hamza, W.; Lusito, L.; Ligorio, F.; Tomasicchio, G.R.; D'Alessandro, F. Wave Climate at Shallow Waters along the Abu Dhabi Coast. *Water* **2018**, *10*, 985. [CrossRef]
10. Contestabile, P.; Conversano, F.; Centurioni, L.; Golia, U.M.; Musco, L.; Danovaro, R.; Vicinanza, D. Multi-Collocation-Based Estimation of Wave Climate in a Non-Tidal Bay: The Case Study of Bagnoli-Coroglio Bay (Tyrrhenian Sea). *Water* **2020**, *12*, 1936. [CrossRef]
11. Tylkowski, J.; Hojan, M. Threshold Values of Extreme Hydrometeorological Events on the Polish Baltic Coast. *Water* **2018**, *10*, 1337. [CrossRef]
12. Hojan, M.; Tylkowski, J.; Rurek, M. Hydrometeorological Conditions for the Occurrence of Aeolian Processes on the Southern Baltic Coast in Poland. *Water* **2018**, *10*, 1745. [CrossRef]
13. Budillon, F.; Vicinanza, D.; Ferrante, V.; Iorio, M. Sediment transport and deposition during extreme sea storm events at the Salerno Bay (Tyrrhenian Sea): Comparison of field data with numerical model results. *Nat. Hazards Earth Syst. Sci.* **2006**, *6*, 839–852. [CrossRef]
14. Puleo, J.A.; Beach, R.A.; Holman, R.A.; Allen, J.S. Swash zone sediment suspension and transport and the importance of bore-generated turbulence. *J. Geophys. Res.* **2000**, *105*, 17021–17044. [CrossRef]

15. Riefolo, L.; Contestabile, P.; Dentale, F.; Benassai, G. Low Frequency Waves Detected in a Large Wave Flume under Irregular Waves with Different Grouping Factor and Combination of Regular Waves. *Water* **2018**, *10*, 228. [CrossRef]
16. Saponieri, A.; Valentini, N.; Di Risio, M.; Pasquali, D.; Damiani, L. Laboratory Investigation on the Evolution of a Sandy Beach Nourishment Protected by a Mixed Soft-Hard System. *Water* **2018**, *10*, 1171. [CrossRef]
17. Medellín, G.; Torres-Freyermuth, A.; Tomasicchio, G.R.; Francone, A.; Tereszkiewicz, P.A.; Lusito, L.; Palemón-Arcos, L.; López, J. Field and Numerical Study of Resistance and Resilience on a Sea Breeze Dominated Beach in Yucatan (Mexico). *Water* **2018**, *10*, 1806. [CrossRef]
18. Torres-Freyermuth, A.; Medellín, G.; Mendoza, E.T.; Ojeda, E.; Salles, P. Morphodynamic Response to Low-Crested Detached Breakwaters on a Sea Breeze-Dominated Coast. *Water* **2019**, *11*, 635. [CrossRef]
19. Bevacqua, A.; Yu, D.; Zhang, Y. Coastal vulnerability: Evolving concepts in understanding vulnerable people and places. *Environ. Sci. Policy* **2018**, *82*, 19–29. [CrossRef]
20. Pantusa, D.; D'Alessandro, F.; Riefolo, L.; Principato, F.; Tomasicchio, G.R. Application of a Coastal Vulnerability Index. A Case Study along the Apulian Coastline, Italy. *Water* **2018**, *10*, 1218. [CrossRef]
21. Kantamaneni, K.; Sudha Rani, N.; Rice, L.; Sur, K.; Thayaparan, M.; Kulatunga, U.; Rege, R.; Yenneti, K.; Campos, L.C. A Systematic Review of Coastal Vulnerability Assessment Studies along Andhra Pradesh, India: A Critical Evaluation of Data Gathering, Risk Levels and Mitigation Strategies. *Water* **2019**, *11*, 393. [CrossRef]
22. Garcia-Ayllon, S. Long-Term GIS Analysis of Seaside Impacts Associated to Infrastructures and Urbanization and Spatial Correlation with Coastal Vulnerability in a Mediterranean Area. *Water* **2018**, *10*, 1642. [CrossRef]
23. Favaretto, C.; Martinelli, L.; Ruol, P. Coastal Flooding Hazard Due to Overflow Using a Level II Method: Application to the Venetian Littoral. *Water* **2019**, *11*, 134. [CrossRef]
24. Gaeta, M.G.; Bonaldo, D.; Samaras, A.G.; Carniel, S.; Archetti, R. Coupled Wave-2D Hydrodynamics Modeling at the Reno River Mouth (Italy) under Climate Change Scenarios. *Water* **2018**, *10*, 1380. [CrossRef]
25. Ruol, P.; Martinelli, L.; Favaretto, C. Vulnerability Analysis of the Venetian Littoral and Adopted Mitigation Strategy. *Water* **2018**, *10*, 984. [CrossRef]
26. Huang, W.-P.; Hsu, J.-C.; Chen, C.-S.; Ye, C.-J. The Study of the Coastal Management Criteria Based on Risk Assessmeant: A Case Study on Yunlin Coast, Taiwan. *Water* **2018**, *10*, 988. [CrossRef]
27. Giardino, A.; Diamantidou, E.; Pearson, S.; Santinelli, G.; Den Heijer, K. A Regional Application of Bayesian Modeling for Coastal Erosion and Sand Nourishment Management. *Water* **2019**, *11*, 61. [CrossRef]
28. Giglioli, S.; Colombo, L.; Contestabile, P.; Musco, L.; Armineto, G.; Somma, R.; Vicinanza, D.; Azzellino, A. Source apportionment assessment of marine sediment contamination in a post-industrial area (Bagnoli, Naples). *Water* **2020**, *12*, 2181. [CrossRef]
29. Mestanza, C.; Piccardi, M.; Pranzini, E. Coastal Erosion Management at Callao (Peru) in the 17th and 18th Centuries: The First Groin Field in South America? *Water* **2018**, *10*, 891. [CrossRef]
30. Pranzini, E.; Anfuso, G.; Cinelli, I.; Piccardi, M.; Vitale, G. Shore Protection Structures Increase and Evolution on the Northern Tuscany Coast (Italy): Influence of Tourism Industry. *Water* **2018**, *10*, 1647. [CrossRef]
31. Cioffi, F.; De Bonis Trapella, A.; Conticello, F.R. Efficiency Assessment of Existing Pumping/Hydraulic Network Systems to Mitigate Flooding in Low-Lying Coastal Regions under Different Scenarios of Sea Level Rise: The Mazzocchio Area Study Case. *Water* **2018**, *10*, 820. [CrossRef]

water

MDPI

Article

Environmental Variability and Macrophyte Assemblages in Coastal Lagoon Types of Western Greece (Mediterranean Sea)

Chrysoula Christia [1], Gianmarco Giordani [2] and Eva Papastergiadou [2,*]

[1] Department of Biology, University of Patras, 26504 Patras, Greece; xchristia@gmail.com
[2] Department of Chemistry, Life Sciences and Environmental Sustainability, University of Parma, Parco Area delle Scienze, 11A, I-43124 Parma, Italy; gm.giordani@gmail.com
* Correspondence: evapap@upatras.gr; Tel.: +30-261-096-9245

Received: 21 December 2017; Accepted: 30 January 2018; Published: 3 February 2018

Abstract: Coastal lagoon types of Western Greece were allocated to a spectrum of meso to polyhaline chocked lagoons; poly to euhaline restricted lagoons; and euhaline restricted lagoons along the Ionian Sea coast. This diversity comprises wide ranges of physical, chemical and environmental parameters in a seasonal and annual scale, which explains the variability in the distribution of benthic macrophytes. Four different macrophyte assemblages were distinguished, characterized by annual or perennial species. Extensive statistical analysis showed that salinity and nitrate concentrations had a great impact on the composition and distribution of macrophyte assemblages into lagoon types that also changed their abundance on a seasonal and annual scale. During the monitoring period, an important salinity shift in a chocked lagoon might cause the gradual loss of *Zostera noltii* and its replacement by *Ruppia cirrhosa*. Restricted lagoons were characterized by higher species diversity, while the other three identified macrophyte assemblages were dominated by the angiosperms *Ruppia cirrhosa* and *Cymodocea nodosa*. This integrated study of coastal lagoons is likely to be broadly applicable, since it was based on important parameters affecting such ecosystems, and the provided links between macrophyte assemblages and abiotic factors are of critical importance to improve management and environmental policies.

Keywords: brackish lagoon types; benthic macrophytes; salinity; succession; univariate variables; Greece

---

1. Introduction

Coastal lagoons are dynamic ecosystems characterized by shallow waters isolated from the open sea by the presence of coastal barriers. Therefore, they represent an ecotone between marine, fresh-water, and terrestrial ecosystems showing some typical characteristics of all these types [1]. These characteristics often result in considerable seasonal changes of environmental variables (e.g., temperature, salinity) and large fluctuations in chemical parameters with consequences to many resident species [2–4].

Coastal lagoons are often sub-divided into choked, restricted, leaky [1] and even open [5] with respect to the characteristics of their hydrodynamic exchange properties with the adjacent open sea. The WFD/2000/60/EC does not include an explicit definition of lagoons, but the definition of transitional waters (TW) specifies a salinity gradient and significant freshwater inputs [6]. Several criteria have been used to define the typology of transitional waters such as salinity, substrate type, formation, isolation, size, morphology, etc. [7,8]. Recently, a classification approach of coastal lagoons of Western Greece was conducted by Christia et al. [9] who revealed four different types, based on criteria defined by the system B of WFD 2000/60/EE and other descriptors indicated as either obligatory or optional. Classification was based solely on abiotic parameters in order to avoid circular reasoning due to biological variation [10–12].

Mediterranean coastal lagoons are generally shallow with tidal ranges below 0.5 m [13]. The extreme meteorological conditions (high temperatures and low precipitation in summer) observed in the last decades in the Mediterranean basin, foster high seasonal and annual variations in physical and chemical parameters, making these ecosystems highly vulnerable to climate change [14]. This tendency will probably continue owing to the global climate changes, leading to the degradation and loss of critical habitats, the increase of eutrophication phenomena and associated algal blooms. Global induced changes lead inevitably to a chain of effects on the ecosystem structure, especially in the submerged macrophytes assemblages. Submerged macrophytes, composed of angiosperms and macroalgae, are important primary producers in coastal lagoons, and many species are considered as ecosystem engineers by creating habitats for aquatic organisms [15]. A coastal lagoon is typically dominated by few submerged macrophytes genera with great plasticity in resource exploitation and adaptation to salinity regimes and other structuring abiotic parameters [16]. During the past 150 to 300 years, eutrophication, habitat modifications, water level and salinity fluctuations have led to a massive decrease of angiosperms and other submerged macrophytes in temperate estuarine and coastal ecosystems in Europe and North America [17,18]. The development of type-specific lagoon management plans and the implementation of proactive adaptation measures became necessary [19].

The recovery of benthic macrophytes is one of the targets of the WFD/2000/60/EC and has led to policy decisions aiming, directly or indirectly, to improve the status of coastal ecosystems [6,20]. Submerged macrophytes have morphological, physiological and ecological adaptations to confront environmental shifts [21]. Benthic macrophytes have a strong influence on the physical and chemical structure of aquatic ecosystems [22,23], forming extensive [15], highly productive [24] and spatio-temporally patchy habitats [25].

A comprehensive presentation of macrophyte distribution in the Mediterranean lagoons and their dynamics based on long-term datasets is necessary in order to depict the high temporal variability of these environments [26,27]. Nevertheless, long-term studies supporting the spatiotemporal dynamics of macrophyte assemblages in lagoons of Greece are scarce in literature [28,29]. Until now, the monitoring of biological quality elements was focused on phytoplankton, benthic invertebrates, zooplankton and fishes [30–33], while benthic macrophytes were monitored only in few lagoons in Northern Greece [34] and in southern and western Greece [2,35]. The knowledge of the ecology of these macrophytes is of prime importance both for the understanding of the ecosystem functioning and for more applied aspects. Macrophytes can be used as ecological indicators of environmental health and ecological status [36,37], as they respond to water nutrients at the community level regarding species diversity (Shannon index), structure and abundance [38]. In the Mediterranean region, three euryhaline species—*Z. noltii*, *Z. marina*, and *C. nodosa*—are present [35]. These species not only provide the physical habitat for a rich fauna but also play a fundamental role in biogeochemical processes contributing to lagoons water quality [39]. This knowledge is crucial to further recommend management and restoration measures.

In this paper, the hypothesis that physical, chemical and environmental parameters of water column have played significant roles in the distribution of macrophyte assemblages was investigated in the identified lagoon types. In this context, the composition of each macrophyte assemblage on a seasonal and annual scale was examined in each lagoon type and correlated with key role parameters such as salinity and nitrogen compounds concentrations as derived by the multivariate analysis. In addition, the species that contributed more to the dissimilarity among lagoon types were identified and the seasonal evolution of their abundance was investigated following the spatial and temporal variations of number of species, species richness, Evenness and Shannon diversity in each lagoon.

2. Materials and Methods

2.1. Study Area

The current study was based on the typological framework of coastal lagoons of Western Greece (Ionian Sea), as derived by Christia et al. [9]. According to them, the investigated area is classified into three different lagoon types based on hydromorphological characteristics (Figure 1): (a) Lagoon Type I includes large, chocked lagoons with meso to polyhaline waters as Rodia which belongs to the natural complex system of Amvrakikos Gulf; (b) Type II consists of large, shallow, restricted lagoons with poly to euhaline salinity regimes and higher sea water exchanges. This type includes Tsoukalio and Logarou lagoons (Amvarkikos Gulf) and Kleisova lagoon that belongs to the lagoonal complex system of Messolonghi-Aitoliko; (c) Type III includes small, shallow and restricted lagoons with euhaline salinity regime and medium seawater intrusion (Araxos lagoon). Detailed information is reported in Christia and Papastergiadou [2] and Christia et al. [9,35]. According to Christia et al. [9] the typological classification of lagoons also revealed a fourth type which includes Kaiafas but this has been omitted from the current research due to its peculiar environmental characteristics: small, deep, mesohaline lagoon with a wide barrier and a unique macrophyte assemblage composed by *Potamogeton pectinatus* and *Chara hispida* f. *corfuensis*. For that reason, it was tested as a case study by Christia et al. [40].

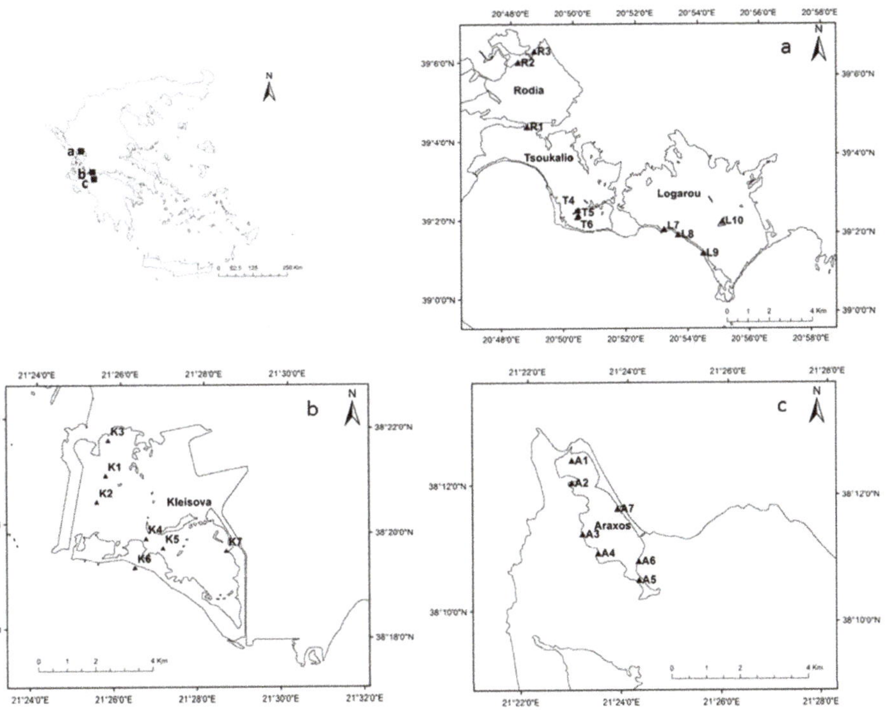

Figure 1. Maps and sampling stations of the investigated Western Greece coastal lagoons: (**a**) Rodia (Type I); Tsoukalio and Logarou (Type II)-Amvrakikos Gulf; (**b**) Kleisova (Type II)-Messolonghi-Aitoliko lagoonal complex; (**c**) Araxos (Type III).

2.2. Sampling Design of Water Quality and Aquatic Macrophytes

Samplings were carried out seasonally (spring, summer, autumn) between 2005 and 2007 in 24 stations of the five studied coastal lagoons of Western Greece. Sampling stations were homogeneously distributed, covering the spatial heterogeneity of each particular lagoonal environment (Figure 1). Depth, transparency, temperature, salinity, dissolved oxygen (DO) and pH were directly measured in situ using portable equipment (Secchi disk, WTW multi 340i/SET, Wissenschaftlich—Technische Werkstätten, Dr- Karl-Slevogt—Straße 1, 82362, Weilheim, Germany). Discrete surface water samples were collected in 1 L polyethylene bottles and preserved at 4 °C for laboratory analysis of the following nutrients: NH_4^+-N, NO_2^--N, NO_3^--N, PO_4^{-3}-P. Water samples for dissolved nutrients analyses were filtered using 0.45 μm pore size filters and immediately frozen ($T = -20$ °C) until analysis, while Chlorophyll-a (Chl-*a*; μg/L) extraction was conducted in 90% acetone for 24 h. All concentrations were measured according to American Public Health Association (APHA) [41]. For total phosphorus (TP), water samples were collected before filtering. Dissolved inorganic nitrogen (DIN) was calculated as the sum of the inorganic nitrogen forms.

The macrophyte sampling campaigns were generally carried out in each station during spring, summer and early autumn in order to evaluate the presence and abundance of species during the whole growth period. Macrophyte compositional and abundance data were measured from a sampling plot of 10 m × 10 m. In each plot, three samples were randomly scraped from the bottom, in a water depth range of 1 to 3 m, on an area of 2 m × 2 m [42]. Plant species abundance was visually scored on a 5-level percentage coverage abundance scale (1 ≤ 20%; 2 = 21–40%; 3 = 31–60%; 4 = 61–80%; 5 = 81–100%). Macrophyte specimens were placed in a plastic bag and transported to the laboratory for identification. The samples were rinsed with water to remove sediments, identified at species level and then fixed in 2% formalin.

2.3. Statistical Analysis

All environmental parameters were log (x + 1) transformed in order to make them closer to normal distribution. Multivariate analysis of variance was applied to investigate the differences of these parameters between sampling periods and macrophyte assemblages. A two-way ANOVA (analysis of variance) test was performed to assess which parameters differed significantly between seasons and years. A factorial ANOVA with interactions between seasons and years was run. Interactions were specified by joining the variables with asterisks, e.g. seasons*years. An LSD test (SPSS V.15) [43] provided direct comparisons between two means from two individual groups (Table S1) in order to address which variables differed significantly among lagoon types.

A detrended correspondence analysis (DCA) was conducted with the CANOCO 4.5 software [44] to explore the different macrophyte assemblages occurring in different coastal lagoon types. A correspondence analysis (CA) was also tested but, due to the presence of an arch effect, the DCA was finally chosen. DCA was performed using the percentage coverage data of the species found in each lagoon type. All data were log (x + 1) to avoid the down weighting of rare species with values approaching to zero. In order to meet criticism rose against DCA on the wedge effect a Multidimentional Scaling (MDS) plot in PRIMER (6.0) [45] was also run.

A redundancy analysis (RDA) takes explanatory variables into account, which allows a direct modeling of the cause-effect relationship between species data and environmental parameters. Explanatory variables were selected using the threshold of $p < 0.5$ of the Monte Carlo permutation test and the threshold of <20 of inflation factors (VIF) [43]. RDA results are displayed by an ordination diagram which reflects the distribution of macrophytes species along coastal lagoon types with different environmental parameters [44].

The structure of macrophyte assemblages was inspected by calculating the total number of species (S), Margalef's species richness (d), diversity index of Shannon (H) and Pielou evenness (J') with PRIMER (6.0). These indices were calculated for each lagoon and their variations were tested with three-way ANOVA on a seasonal, annual and spatial scale (SPSS V.15).

The contribution of individual macrophyte species to the dissimilarity between lagoon types on an annual scale was tested with the similarity percentages (SIMPER) analysis. The zero-adjusted Bray–Curtis coefficient was used to modulate the erratic behavior of Bray-Curtis for near-denuded assemblages in the sampling sites [46]. For this analysis a 90% cut off of the cumulative percentage was applied for taxa with low contributions. Moreover, pairwise Analysis of Similarity (ANOSIM) comparisons applied between all groups, using 10,000 simulations in each case. This analysis was carried out to test the null hypothesis that there were no differences in the composition of macrophyte species among different lagoon types. Both analyses, SIMPER and ANOSIM were based on the Bray–Curtis dissimilarity index and were conducted using the PRIMER (6.0) statistical software. Values were square-root transformed before the analyses; in this way, each species contributed fairly evenly to each analysis [45].

3. Results

3.1. Environmental Change and Water Quality Characteristics

The analysis of variance showed significant variations of environmental parameters, both on seasonal and annual scale, in the three studied lagoon types of Western Greece (Table 1). Water temperature in coastal lagoon types did not show significant differences and followed the typical pattern which is generally characterized by highest values during the dry period (summer). Water depth played an important role not only to the classification of lagoon types [10] but also to the variability of nutrient concentrations. The higher mean depth value was found in lagoon Type I (1.06 m) where the predominant forms of nitrogen were NO_3-N and NH_4-N. During the wet period (spring) they accounted for 396 µg/L and 186 µg/L, respectively, showing significant difference among all lagoon types (Figure 2; Table 1). During the monitoring period the higher concentration of TP was measured in lagoon Type II (156.1 µg/L) in autumn (Figure 2). The concentration of TP showed significant variations between seasons and years (Table 1).

Salinity varied significantly among lagoons and played pivotal role in the classification of lagoon types. On a seasonal scale, it followed a marked similar pattern in all lagoon types with higher values recorded during the dry period. Restricted lagoons showed typically marine conditions, while chocked lagoons are strongly influenced by freshwater inputs. Therefore, lagoon Type III showed the higher mean salinity (40.5‰), while the lower value was recorded in Type I (14.1‰). Low Chl-*a* concentrations were common in all lagoon types during the monitoring period. The highest value (3.7 µg/L) was measured during spring in lagoon Type III. The significance of interactions between the two factors of season, year and season*year was also considered. More specifically, the interaction between season and year indicated significant ($p < 0.05$) effects for temperature, pH, DO, nutrients of N and P, alkalinity and Chl-*a* (Table 1).

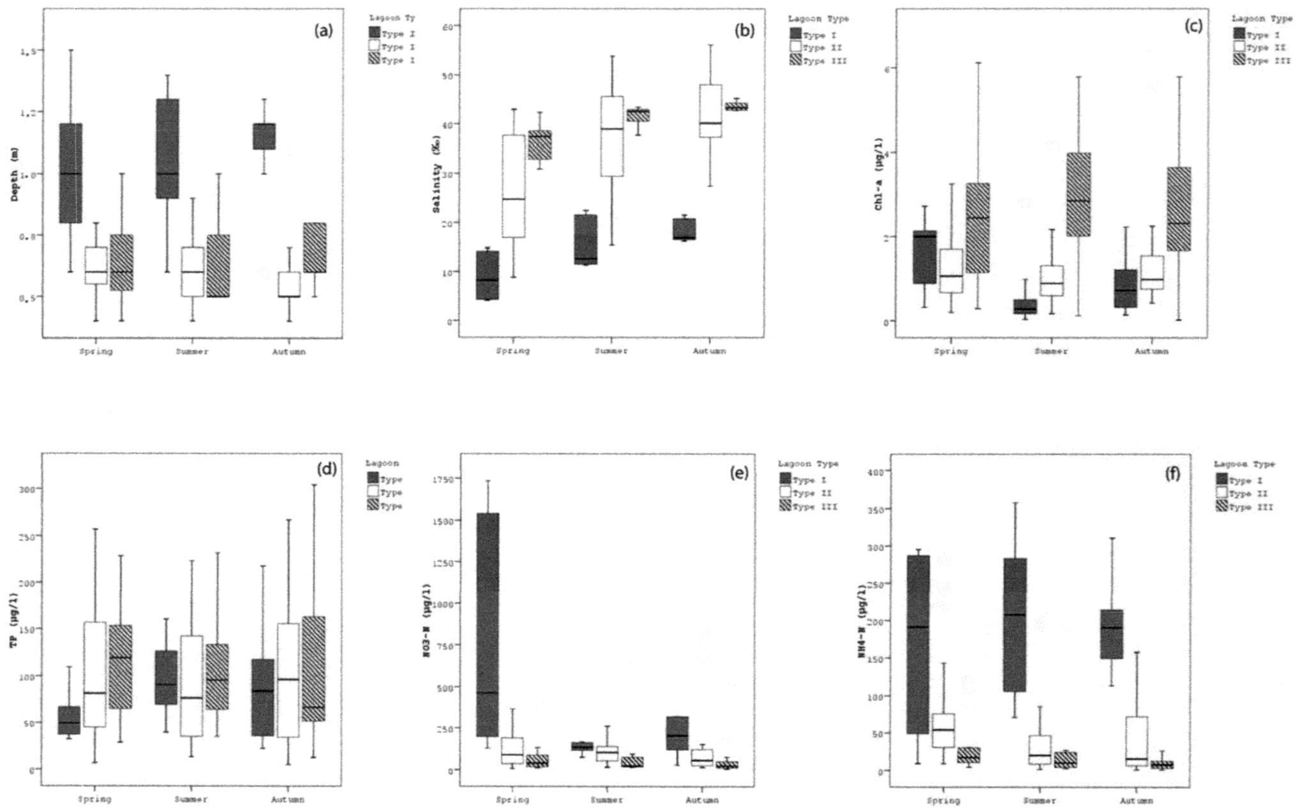

Figure 2. Seasonal variability of: (**a**) water depth (m); (**b**) salinity (‰); (**c**) Chl-*a* (µg/L); (**d**) TP (µg/L); (**e**) NO_3-N (µg/L); (**f**) NH_4-N (µg/L) concentrations in the three lagoon types of Western Greece during the whole sampling period.

Table 1. Analysis of variance with the effects of the factors of season, year and their interactions (Season*Year) on the environmental parameters of three lagoon types of Western Greece.

Environmental Parameters	Type I (n = 27)				Type II (n = 111)				Type III (n = 50)				Interaction	
	Season		Year		Season		Year		Season		Year		Season*Year	
	F	Sig	F	Sig	F	Sig	F	Sig	F	Sig	F	Sig	F	Sig
Depth (m)	411.7	0.000	2.2	ns	0.1	ns	1.2	ns	0.2	ns	1.2	ns	0.5	ns
Transparency (m)	301.7	0.000	0.1	ns	0.2	ns	0.0	ns	0.1	ns	0.0	ns	0.2	ns
Temperature (°C)	32,021.2	0.000	38.1	0.000	652.8	0.000	33.6	0.000	476.8	0.000	39.2	0.000	27.1	0.000
pH	249,734.3	0.000	32.9	0.000	22.3	0.000	25.2	0.000	9.6	0.002	27.1	0.000	33.9	0.000
DO (mg/L)	2308.2	0.000	19.6	0.000	43.1	0.000	7.1	0.006	26.1	0.000	10.9	0.001	37.8	0.000
Salinity (‰)	818,318.8	0.000	12,391.2	0.000	21,547.4	0.000	6788.5	0.000	19,105.6	0.000	7108.8	0.000	1414.0	0.000
PO_4-P (μg/L)	434.8	0.000	9.9	0.001	14.4	0.000	10.0	0.001	8.1	0.003	8.6	0.003	7.4	0.001
TP (μg/L)	78.5	0.000	0.6	ns	2.3	ns	0.7	ns	1.4	ns	0.5	ns	5.1	0.007
NO_2-N (μg/L)	39.4	0.000	11.9	0.001	22.4	0.000	6.2	0.010	29.3	0.000	6.9	0.006	7.2	0.001
NO_3-N (μg/L)	57.3	0.000	10.8	0.001	8.0	0.004	6.6	0.007	2.2	ns	6.6	0.007	8.0	0.001
NH_4-N (μg/L)	192.7	0.000	17.9	0.000	11.5	0.001	5.2	0.017	21.0	0.000	5.6	0.013	10.4	0.000
Alkalinity (mg/L)	132.4	0.000	16.9	0.000	14.8	0.000	11.0	0.001	8.5	0.003	10.6	0.001	6.4	0.002
Chl-*a* (μg/L)	1547.0	0.000	20.3	0.000	8.2	0.003	17.8	0.000	2.0	ns	19.5	0.000	18.2	0.000

3.2. Macrophyte 'Assemblages' in the three Lagoon Types of Western Greece

A total of 38 macrophytic taxa [35] were recorded in the three studied lagoon types: three angiosperm species (*Z. noltii, R. cirrhosa, C. nodosa*), one Charophyte (*Lamprothamnium papulosum*) and 34 macroalgae (Rhodophytes, Chlorophytes, Ochrophytes). In the lagoon Type I, 8 macrophyte species were identified; 25% belonged to the Magnoliophyta phylum, 62.5% to Chlorophyta and 12.5% to Rhodophyta and Charophyta (Figure 3). In Lagoon Type II, 25 species were recorded, 48% belonged to Rhodophyta, 24% to Chlorophyta, 12% to Ochrophyta and Magnoliophyta and only 4% to Charophyta. Finally, in lagoon Type III, 19 species were found, 36.8% accounted for Rhodophyta, 42.1% for Chlorophyta, 10.1% for Ochrophyta and Magnoliophyta, while no Charophyte species were observed.

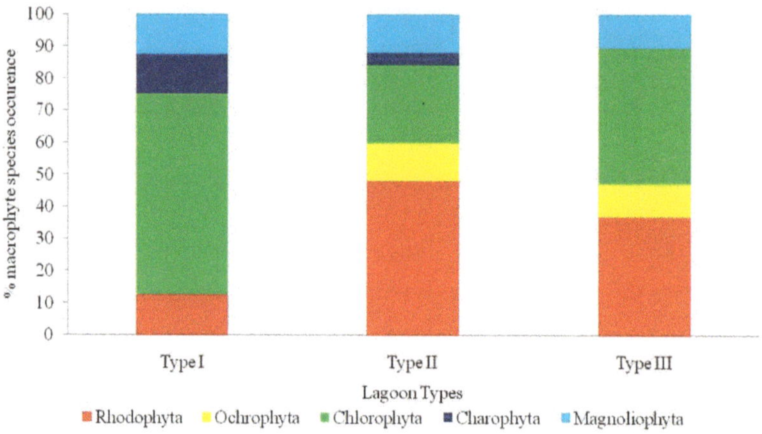

Figure 3. Percentage of macrophyte species occurrence in the three different lagoon types of Western Greece.

DCA analysis revealed four macrophyte assemblages (Figure 4). The first two DCA axes accounted for 84.73% (DCA axis 1: 59.36%; DCA axis 2: 25.37%) of the total variance (Figure 3, Table S2). The angiosperm *Z. noltii* and the charophyte *L. papulosum* are positioned along the left part of the ordination plot, forming the macrophyte assemblage i which is associated with *G. bursa pastoris* and *Cl. glomerata*. According to the results of DCA axes the species of the assemblage i (Table 2) are typical of coastal lagoons of Type I. They seem to prefer mesohaline, deep, high transparent waters with occasional high nitrate concentrations. In the middle part of the ordination plot, the angiosperm *R. cirrhosa* coexisted with *Ac. acetabulum, Gr. longissima, U. rigida* and *Ch. linum* forming the macrophyte assemblage ii. These species are well established in coastal lagoon Type II, showing high adaptability to high salinity shifts and shallow water depths.

The macrophyte assemblage iii (Table 2) is common in both lagoon Types II and III. It is established to the right part of the plot and characterized by the dominance of the angiosperm *C. nodosa* and the epiphyte species *C. diaphanum* and *Ch. capillaris.* Finally, across the left bottom part of the ordination plot, the marine species of *C. barbata, Al. corrallinum, A. nayadiformis, Gr. gracilis, V. aegagropila* and *L. obtusa* are dispersed forming the macrophyte assemblage iv. These marine species were found only in the lagoon Type II, in shallow, euhaline and low nutrient waters, especially in the sampling stations adjacent to the marine inlet channels of the lagoons (Table 2). The identified macrophyte assemblages i, ii and iii are occupied by fast growing opportunistic species such as the green algae *Chaetomorpha* and *Cladophora,* mainly during the dry period (summer).

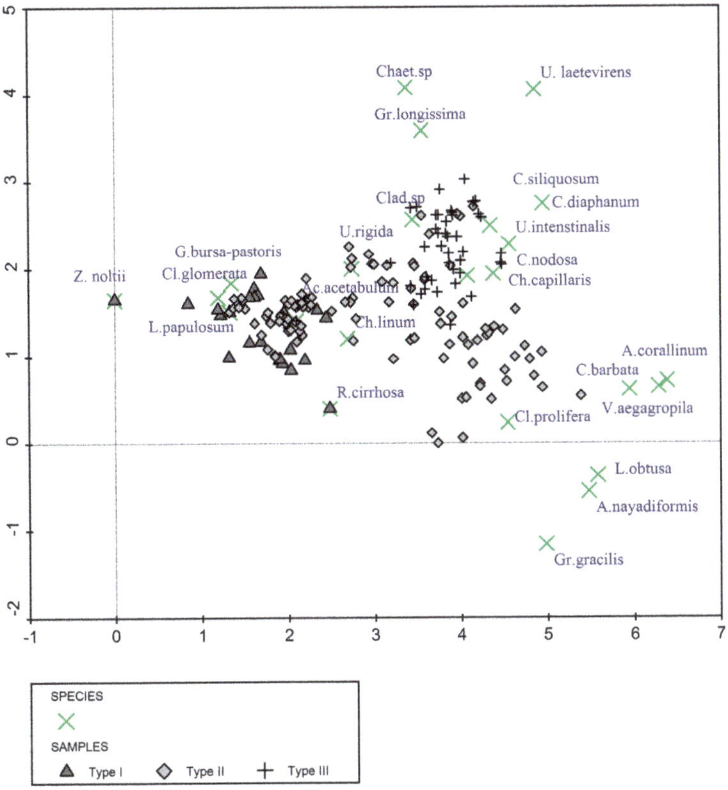

Figure 4. Species ordination obtained by the detrended correspondence analysis (DCA) conducted with species coverages data (%) in the three different lagoon types of Western Greece.

Table 2. List of the species belonging to the four macrophyte assemblages (i–iv) identified after the detrended correspondenceaAnalysis (DCA) in the three lagoon types of Western Greece.

Lagoon Type	Macrophyte Assemblages
Type I	i. *Zostera noltii-Lamprothamnium papulosum-Gracilaria bursa pastoris-Cladophora glomerata* ii. *Ruppia cirrhosa-Acetabularia acetabulum-Gracilariopsis longissima-Ulva rigida*
Type II	ii. *Ruppia cirrhosa-Acetabularia acetabulum-Gracilariopsis longissima-Ulva rigida* iii. *Cymodocea nodosa-Chondria capillaris-Ceramium siliquosum-Ulva species* iv. *Cystoseira barbata, Alsidium corrallinum, Acanthophora nayadiformis, Gracilaria gracilis and Valonia aegagropila*
Type III	iii. *Cymodocea nodosa-Chondria capillaris-Ceramium siliquosum-Ulva species*

The MDS analysis (Figure 5) gave a potentially useful two-dimensional picture of the studied lagoons with no real prospect of a misleading interpretation (stress = 0.16). The pattern in the species ordination was confirmed by the correlations of DCA axes. DCA axis 1 is positively correlated with salinity, while a negative relation is shown with nitrogen forms and depth. However, the DCA axis 2 is negatively related with transparency and ammonium concentrations, while a positive correlation was found for Chl-*a* (Table S2).

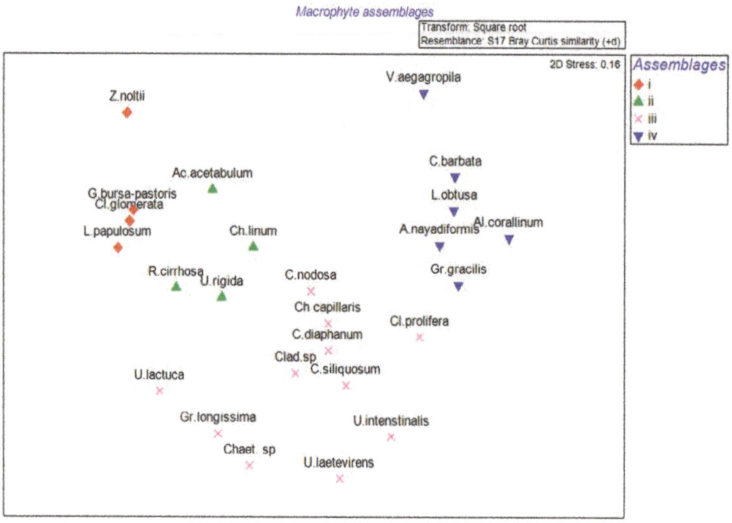

Figure 5. Results of Multidimentional Scaling (MDS) analysis of macrophyte assemblages identified in the three different lagoon types of Western Greece.

3.2.1. Relationship between Macrophytes and Environment

The first two axes of the redundancy analysis accounted for 81% of the total variance (Table S3). Axis 1 (66.9%) explained the higher percentage of the total variance, while axis 2 explained the 14.08%. The ordination diagram of the redundancy analysis (RDA) (Figure 6) with environmental parameters and macrophyte species shows the distribution of macrophyte species and the position of coastal lagoon types in an approximate way. The first axis is highly positively correlated with salinity and DIN concentration, while axis 2 is highly positively correlated with DO, transparency and Chl-*a*. Following the RDA analysis, the clustering allows the classification of the macrophyte species into lagoon types according to physical and chemical parameters. Thus, sampling stations of lagoon Type I are positioned to the upper left section of the plot, the samplings of lagoon Type II are mainly dispersed in the bottom part, while samplings of lagoon Type III are clustered to the right section of the plot. The angiosperm species *Z. noltii*, *R. cirrhosa* and the charophyte *L. papulosum*, positioned to the upper left part of axis 1, have a relatively large distribution span in waters with medium salinity, high transparency and high concentrations of total inorganic nitrogen as mainly found in lagoon Type I. Species located at the right part of Axis 1 are mainly found in poly to euhaline waters. The angiosperm species *C. nodosa,* as well as the macrophyte species *C. siliquosum*, *C. capillaris*, *A. nayadiformis* and *C. barbata* are typical of coastal lagoons classified in Type III. In the center of the diagram, along axis 1, the macrophyte species *Gr. bursa-pastoris*, *Ac. acetabulum* and *R. cirrhosa*, which belongs to lagoon Type II, show their preference to high salinity, lower nutrients concentrations and high marine water exchanges. The second axis reflected the gradient of photosynthetic activity with taxa located to the lower part showing higher adaptability to lower transparency, DO and Chl-*a* waters. Macrophyte species positioned in the upper part of the diagram were mainly present in sampling stations with high transparency and higher Chl-*a*, DO and PO_4-P concentrations.

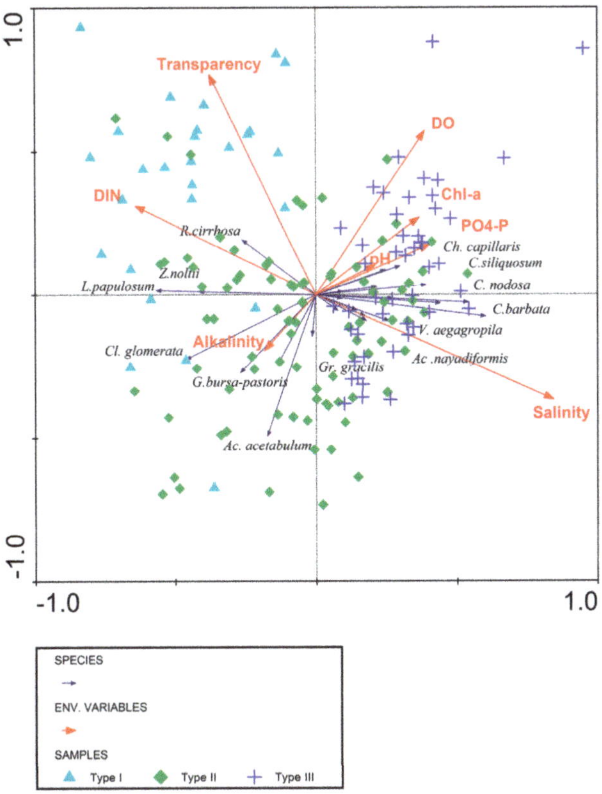

Figure 6. Graph plot of the redundancy analysis (RDA) conducted between environmental parameters and submerged macrophytes coverages in the three different lagoon types of Western Greece.

3.2.2. Comparisons of Macrophyte Assemblages among Lagoon Types

Following the results of similarity percentage (SIMPER) analysis, macrophyte assemblages of Type I differ significantly from Types II and III (Tables 3 and 4). The highest average dissimilarity of Type I was recorded during spring (96.44) and autumn (96.9) especially with Type III. The main species contributed to this difference were the angiosperms *Z. noltii*, *C. nodosa* and the charophyte *L. papulosum*. Lower average dissimilarity (83.6) was observed between lagoon Type I and II, due to the presence of more common species. At the opposite, the angiosperms *R. cirrhosa*, *Z. noltii* and the Rhodophyte *Gr. longissima* contributed to the differences between these lagoon types.

During the monitoring period an interesting shift in distribution and abundance of the angiosperm *Z. noltii* was noticed in lagoon Type I related to changes in salinity regime. The gradual loss of *Z. noltii* and its replacement by *R. cirrhosa* in Type I was recorded through years 2005 to 2007 (Table 3, Figure S1), while the mean average abundance of *R. cirrhosa* and *C. nodosa* followed an increasing trend (Table 3). The average abundance of *R. cirrhosa* was null in 2005, 36.6% in 2006 and 40.8% in 2007 while salinity increased from 8.2‰ to 4.2‰ and to 22.4‰ in the same years. The mean average abundance of *R. cirrhosa* (Table 4) was highest in summer (31.7) when salinity rose from 10.7‰ (2006) to 17.6‰ (2007). In Type II lagoons, *R. cirrhosa* showed high average abundance in spring and followed an increasing trend from 2005 (6.4) to 2007 (25.4).

Table 3. SIMPER (similarity percentage) analysis results: average dissimilarity (%) within the three lagoon types of Western Greece during the three years 2005, 2006 and 2007. The macrophyte species primarily responsible for differences between macrophyte assemblages and its contribution within the lagoon type are also reported.

Average Dissimilarity (%) between Lagoon Types	**2005**			**2006**			**2007**		
	Av. Dis. (I–II)= 77.1 **Av. Dis. (I–III) = 95.9** **Av. Dis. (II–III) = 83.8**			**Av. Dis. (I–II) = 77.7** **Av. Dis. (I–III) = 95.2** **Av. Dis. (II–III) = 81.6**			**Av. Dis. (I-II) = 77.8** **Av. Dis. (I-III) = 97.6** **Av. Dis. (II-III) = 83.0**		
	Type I	**Type II**	**Type III**	**Type I**	**Type II**	**Type III**	**Type I**	**Type II**	**Type III**
Species	**Av. Abund.**	**Av. Abund.**	**Av. Abund.**	**Av. Abund.**	**Av. Abund.**	**Av. Abund.**	**Av. Abund.**	**Av. Abund.**	**Av. Abund.**
Acanthophora nayadiformis	0	0	0	0	3	0	0	6.1	0
Acetabularia acetabulum	0	14.3	0	0	4.2	0.2	0	7.1	0.5
Ceramium diaphanum	0	0	0	0	1.7	3.4	0	0.9	2.3
Chaetomorpha linum	6.6	5.6	1.5	1.8	2.9	1.6	0.7	3.4	0.9
Chondria capillaris	0	0	6	0	2.1	5.6	0	2.8	6.1
Cladophora glomerata	10	6.5	0	0.4	2.1	0	2.7	1.7	0
Cladophora prolifera	0	0	0	0	0	0	0	2.9	4.1
Cladophora sp.	0	0	0	0	2.8	3.2	0	0	0
Cymodocea nodosa	0	15.9	44.4	0	17.2	27.0	0	18.6	34.9
Cystoseira barbata	0	0	0	0	2.3	0	0	0	0
Gracilaria bursa-pastoris	0	6.4	0	0	0	0	0	4.4	3.3
Gracilariopsis longissima	0	0	4.5	0.8	4.1	16.1	0	0	9.3
Lamprothamnium papulosum	25.9	20.2	0	18.9	6.8	0	4.2	8.4	6.6
Ruppia cirrhosa	0	6.4	1.4	36.3	18.9	1.2	40.8	25.4	19.4
Ulva lactuca	0	0	12.1	6	0.4	9.7	8.4	0.4	4.6
Ulva laetevirens	0	0	0	0	0.5	2.2	0	0	0
Ulva rigida	0	5.9	2	0	2.4	3.0	0	0	1.8
Zostera noltii	27.8	4.5	0	2.3	0	0	2.8	0	0

Table 4. SIMPER (similarity percentage) analysis results: average dissimilarity (%) within the three lagoon types of Western Greece during spring, summer and autumn. The macrophyte species primarily responsible for the differences between macrophyte assemblages and its contribution within the lagoon type are also reported.

Average Dissimilarity (%) between Lagoon Types	**Spring**			**Summer**			**Autumn**		
	Av. Dis. (I–II) = 83.6 Av. Dis. (I–III) = 96.44 Av. Dis. (II–III) = 87.6			**Av. Dis. (I–II) = 82.4 Av. Dis. (I–III) = 95.9 Av. Dis. (II–III) = 80.1**			**Av. Dis. (I-II) = 81.2 Av. Dis. (I-III) = 96.9 Av. Dis. (II-III) = 83.2**		
	Type I	**Type II**	**Type III**	**Type I**	**Type II**	**Type III**	**Type I**	**Type II**	**Type III**
Species	**Av. Abund.**	**Av. Abund.**	**Av. Abund.**	**Av. Abund.**	**Av. Abund.**	**Av. Abund.**	**Av. Abund.**	**Av. Abund.**	**Av. Abund.**
Acanthophora nayadiformis	0	2.4	0	0	4.7	0	0	3.1	0
Acetabularia acetabulum	0	8.4	0	0	8.0	1.3	0	6.9	0
Ceramium diaphanum	0	0	0	0	1.5	4	0	1.3	2.4
Chaetomorpha linum	1.5	2.8	1.5	6.2	3.3	0.8	1.6	5.0	1.5
Chaetomorpha sp.	0	0	2.4	0	0	0	0	0	0
Chondria capillaries	0	1.4	3.6	0	2.9	5.1	0	1.8	8.4
Cladophora glomerata	2.6	1.7	0	5.1	3.6	0	5.9	3.6	0
Cladophora prolifera	0	1.4	3.3	0	1	2.3	0	1.4	1
Cladophora sp.	0	1.5	2.1	0	0.6	2.3	0	0	0
Cymodocea nodosa	0	13.2	24.2	0	20.9	35.5	0	17.9	37.6
Cystoseira barbata	0	2.1	0	0	0	0	0	0	0
Gracilaria bursa-pastoris	0	3.1	0	0	3.3	0	0	4.1	0
Gracilariopsis longissima	0	0.5	18.9	0	2.7	8.9	1	1.7	10.4
Lamprothamnium papulosum	4.1	9.3	0	17	10.4	0	19.6	12.2	0
Ruppia cirrhosa	22.5	23.9	0.4	31.7	15.8	0.5	29.7	15.6	0.3
Ulva lactuca	3	0.4	9.6	9.2	0.7	5.2	5.7	0.4	10.2
Ulva rigida	0	2.9	1.9	0	3.7	2.3	0	2.3	2.4
Zostera noltii	16.3	0.46	0	12	1.7	0	6.4	1.1	0

3.3. Univariate Variables of Diversity Indices and Environmental Variables

The list of diversity indices applied to each lagoon type during the study period point out significant differences on seasonal and annual scale (Figure 7a–e). Coastal lagoons of Type II (Tsoukalio and Kleisova) and Type III (Araxos) showed highest Shannon and species richness values. For example Kleisova and Araxos showed the higher species richness in spring 2007 (3.5) and summer 2005 (2.8), respectively. Figure 7e shows the clear relationship of salinity with univariate variables. Thus, coastal lagoons with higher salinity and higher seawater exchange have higher values of diversity indices. Lower Shannon values were found in the highly confined lagoon (Rodia), which is more influenced by freshwater inputs.

The differences of univariate variables among lagoon types and the conceptual linear regression analysis between univariate variables showed that seasons and stations played significant role to the results of the variables. No significant interaction was obtained between the factors season*year*station (Table S4). Number of species, species richness and Shannon differed significantly between seasons and stations, but no significant temporal variations were observed with the only exception of the Shannon index (H) in lagoon Type I.

4. Discussion

The results derived from the monitoring of the five selected coastal lagoons of Western Greece are representative of several Mediterranean ecosystems. These lagoons belong to three different lagoon types and show the typical gradient of environmental conditions observed in many transitional water ecosystems due to the mixing of freshwater, seawater and human impacts [47]. The observed salinity followed a seasonal trend, typical of all Mediterranean lagoons, with higher values in the dry period and in the restricted lagoon types. Salinity was the main variable driving the distribution of submerged macrophytes in these coastal lagoons, explaining more than 71% of the variance; DIN can explain 19%. Highly confined lagoons with lower salinity such as Rodia (Type I) show high concentrations of nitrogen compounds (NO_3-N and NH_4-N) during the wet period (spring, autumn). Freshwater inputs and agricultural runoff from the adjacent drainage channels [2] affect the nutrient concentrations which increase with habitat isolation [48]. The highest NO_3-N and DIN concentrations observed in chocked lagoon revealed the inverse relationship of salinity with nitrates, which is the most abundant nitrogen compound in these coastal lagoons [48]. Less confined and euryhaline lagoons (Type II and III) had higher TP concentrations probably due to high salinity values and associated high sulfate reduction rates [49]. The dissimilatory reduction of sulfate is very abundant in marine waters and produces sulfide ions that precipitate ferrous iron. With the removal of iron, phosphate can be released from the sediment to the water mass [50]. Even if nutrient loads range was similar to other transitional water ecosystems in the Mediterranean region [51,52] a buffering capacity or a feedback mechanism of submerged macrophytes in stabilizing phosphorus is expected in lagoons with high habitat isolation [48]. The dense mats of charophyte species such as *L. papulosum*, typical of such lagoon types, can prevent sediment resuspension and mitigate the phosphorus binding capacity of particulate matter [53,54].

Changes in water and sediment quality subsequently lead to changes in the macrophyte community composition and vice versa [9,40,55]. High light penetration and high water transparency can be the result of low phytoplankton densities (as indicated by low Chl-*a* concentrations) but can also be promoted by the distribution of macrophyte assemblages as the dominance of angiosperms that diminishes the resuspension of the sediments [56]. The highest Chl-*a* mean values recorded in the lagoon Type III were supported by the runoff of adjacent agricultural lands.

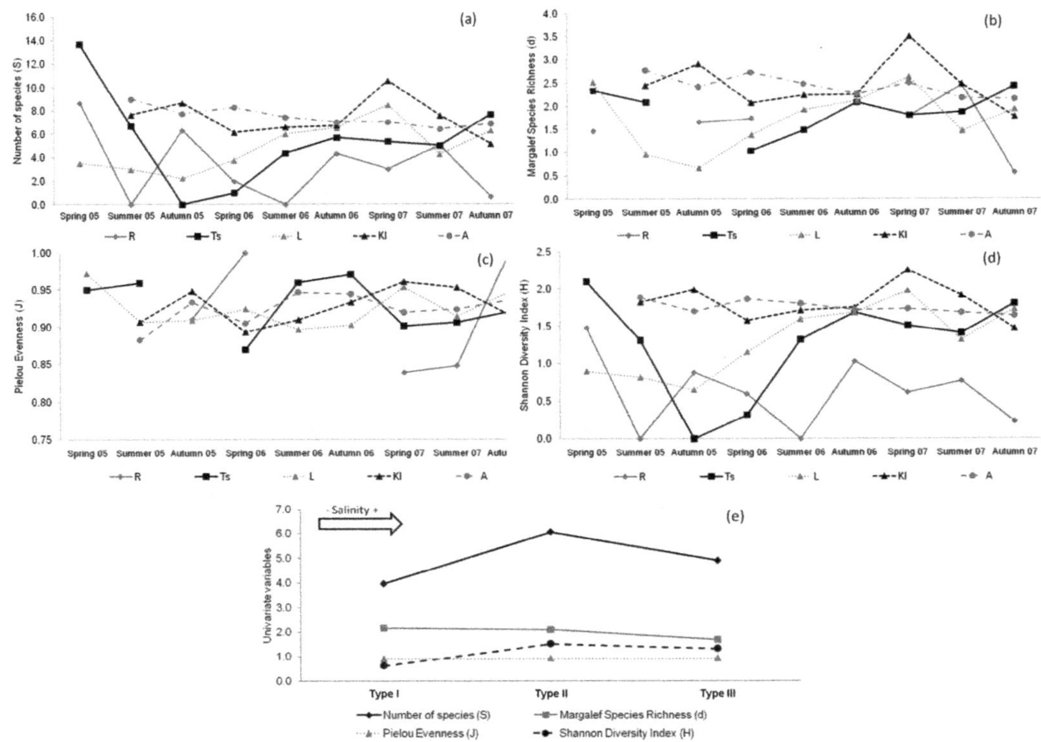

Figure 7. (**a**–**e**) The seasonal pattern of univariate variables of: (**a**) Number of species (S); (**b**) Margalef Species Richness (d); (**c**) Pielou Evenness (J); (**d**) Shannon Diversity Index (H) calculated from spring 2005 to autumn 2007 in each lagoon and (**e**) mean values of the univariate variables during the whole monitoring period in lagoon types of Western Greece.

The results also supported the hypothesis that physical and chemical parameters of the water column may determine the composition and distribution of macrophyte assemblages. In coastal ecosystems with low seawater inflow, several specialist species may tolerate severe environmental conditions and potentially develop large populations in a wide range of salinity gradients. Lagoon specialists are better adapted to high environmental variability, most likely afforded by a degree of genetic plasticity [57]. However, in extreme salinity conditions, a drop of species richness is expected [58]. From the seven species of phanerogams that have been signaled in the Mediterranean [59], three of them: *Z. noltii*, *R. cirrhosa* and *C. nodosa* have been formed extensive meadows in the studied areas [35] and support four macrophyte assemblages. The structure and composition of these four macrophyte assemblages distinguished in the coastal lagoons of Western Greece was determined by the abiotic gradients [1] and the degree of isolation by the sea [7]. Due to the higher variability of abiotic gradients (transparency and salinity) and the hydrological regime in lagoons, diversity is generally lower than in more stable and marine environments [60,61]. The detrended correspondence analysis showed that lagoon type, salinity and nutrient concentrations played relatively important roles on species distribution and succession [62]. The results also show that chocked and more isolated lagoons (Type I) with lower salinity values had lower Shannon index diversity. The seasonal variations of diversity (H) reflected the seasonally high abundance of a few dominant species, such as *Z. noltii*, *R. cirrhosa* and *C. nodosa*.

Macrophyte assemblages formed by *R. cirrhosa* and *C. nodosa* and accompanied with the opportunistic species of *Ulva* and various Rhodophyceae of the genus *Gracilaria* and *Gracilariopsis* were found in the lagoons Types II and III characterized by high salinity values (>30‰). Also, in the lagoons with high sea water exchange and heterogeneous physical and chemical characteristics [63], the diversity indices (number of species, species richness and Shannon) are higher than those observed in the isolated lagoons or with little exchanges with the sea [64].

R. cirrhosa presents high ranges of habitability both in terms of salinity and inorganic nitrogen concentrations and can be found from oligotrophic to hypertrophic environments [65,66]. The higher densities of *R. cirrhosa* were observed in spring and summer, while from late summer to autumn, the senescence of the plants associated with intense grazing and the development of opportunistic species and epiphytes, may limit the growth of this phanerogam in the Mediterranean lagoons [67,68]. *C. nodosa* appears to be more vulnerable to salinity changes and was found to colonize areas of the lagoons more affected by marine intrusions [15,69]. In the lagoon Type III, characterized by high salinity values induced by low freshwater inflows and high influence of the sea water, *C. nodosa* can dominate or can be a competitor of *R. cirrhosa* [70]. The abundance of *C. nodosa* observed from summer to autumn is one of the highest among other Mediterranean lagoons [24,71].

The angiosperm *Z. noltii*, recorded in the lagoon Type I, forms the macrophyte assemblage i with the charophyte *L. papulosum*., This assemblage is typical of lagoons with low salinity, high transparency and high concentrations of total inorganic nitrogen. *Z. noltii* is a relatively small and fast-growing species having a high tolerance to changes of environmental conditions, such as light irradiance, temperature and nutrient concentrations [72]. It can be established on a wide range of substrata [70] and form mixed meadows with *R. cirrhosa* and *C. nodosa* in areas where salinity fluctuate as estuaries and coastal lagoons [73,74]. Based on its field distribution *Z. noltii* is classified as euryhaline species [73]. The growth and survival of *Z. noltii* are both significantly affected by water salinity [75]. In the current study, *Z. noltii* was found at salinities lower than 20‰, while in other Mediterranean lagoons such as Mar Menor the species was found at higher salinities (42‰ to 47‰). The average abundance of *Z. noltii* can be reduced by 50% at salinities lower than 10–20‰, whereas high rates of leaf production were found when salinity ranges from 20 to 31‰ [76]. In our study, *Z. noltii* was found only in the confined lagoon type I. In the marshes of Rodia, large freshwater inputs from Louros River in June and July 2003 and March 2004 combined with high precipitation rates, increased the water level of the lagoon and probably contributed to the reduction of *Z. noltii* [2]. However, *Z. noltii* was recorded at low average abundances in lagoon Type II and absent in the lagoon Type III. These differences in salinity tolerance could be explained by individuals' adaptation to different and variable local conditions that occur naturally in their habitats, as it is the case of other widespread species [74].

Transparency and Chl-*a* concentrations may affect the composition and distribution of macrophytes as indicated by several studies [9,77]. With low nutrient levels and clear water conditions, such as those typical of an oligotrophic state, *Zostera* spp. and the aquatic plants of the *Ruppia* genus are the dominant macrophytes taxa of the lagoon [9,39]. The seasonal salinity fluctuations and especially the increase of the gradient registered in 2007 resulted in the deterioration of *Z. noltii* abundance and its replacement by the angiosperm *R. cirrhosa* [78]. The degradation of submerged phanerogam meadows is generally indicated by a reduction of water transparency and the consequent decrement of the depth limits for all macrophytes growth. Moreover, a gradual loss of plant communities containing charophytes can be also observed [42]. Also, high nutrient concentrations can lead to damages of submerged meadows, losses of diversity and increments of angiosperms mortality [76]. However, increase in nutrient availability enhances the development of fast growing macroalgae and epiphytic communities that shade aquatic angiosperms and may affect their abundance [67]. Typical green algae, such as *Chaetomorpha* spp., *Cladophora* spp. and *Ulva* spp. display enhanced growth in euryhaline environments and its abundance is favored by the confinement with the sea [51]. Since macrophytes are typically adapted to euryhaline waters, drastic variations in salinity may be an important local factor contributing to the species losses observed not only in Western Greece lagoons, but also in the Baltic Sea and in the Catalan area [18,42].

In less confined lagoon types, macrophyte species typical of marine environments were recorded. The angiosperms *C. nodosa* and *R. cirrhosa*, associated with several epiphytes or opportunistic species, were forming dense mats [15,77]. Blooms of *Ulvaceae*, *Cladophoraceae* and *Gracilariaceae* could decrease the abundance of these angiosperms and restrict their distribution to areas close to the sea inlets [39]. Both species were adapted to polyhaline waters ranging from 27‰ to 43‰ mean salinity values [76] but, in accordance with the ordination analysis, they differ in their responses to nutrient concentrations. *R. cirrhosa* (as *Z. noltii*) is more adapted to high DIN concentrations and is abundant in spring where nutrient concentrations are at the maximum [79]. *C. nodosa*, on the other hand, prevails in sampling sites with lower nitrate or ammonia concentrations and high salinity values. In Kleisova and Araxos lagoons, even if *C. nodosa* formed mixed meadows with *R. cirrhosa*, it was the species with the highest average abundance.

Finally, the presence of marine species such as *C. barbata*, *Al. corrallinum*, *A. nayadiformis*, *Gr. gracilis* and *L. obtusa* are common in lagoons representing slow-growing, sun-adapted perennial to annual macroalgae favoured in pristine and moderately degraded environments [80,81]. Stands of *C. barbata* could be found together with *C. nodosa* and *R. cirrhosa* [81]. The species of the genus *Cystoseira* are usually the dominant element of the benthic vegetation on unpolluted hard substratum and the *Cystoseira* algal community is considered as the final stage (climax) in a succession of photophilic algal communities [82]. The species *C. barbata* is an important element of upper infralittoral benthic vegetation in semi enclosed bays and even in small fishing ports [83]. According to Montesanto and Panayiotides [80] species of genus *Cystoseira* could be considered as indicator species of unpolluted waters, with the exception of *C. barbata* which seems to be tolerant of moderate eutrophication conditions.

A decline of benthic angiosperms was referenced on a worldwide scale during last decades [84]. Climate change, induced land cover/use changes, eutrophication and hydrological alteration are the main threats of benthic macrophytes in transitional water ecosystems [84]. The high temperatures predicted for the Mediterranean area through the end of 21st century will significantly impact the biodiversity of coastal lagoons [85]. Conservation actions such as the improvement of water quality by the reduction of pollution sources, water drainage and habitat modifications are needed to preserve macrophyte species. The knowledge of spatial variability and the temporal changes in macrobenthic assemblages of coastal lagoons can be highly relevant namely for the establishment of monitoring programs and develop national conservation strategies for transitional water ecosystems.

5. Conclusions

Our findings support the identification of macrophyte assemblages distinguished in three different lagoon types of Western Greece. These are composed by species that are common in coastal environments and are able to form populations capable to acclimate to the particular environmental conditions of these ecosystems. The four macrophyte assemblages are characterized by the presence of the angiosperms *Z. noltii*, *R. cirrhosa*, *C. nodosa* and the charophyte *L. papulosum*. In these lagoons, the adaptations and the replacement of macrophyte species are more likely to occur during the wet period (spring) by taking advantage of the more favorable environmental conditions rather than in the extreme conditions typical of summer. Submerged macrophytes have to cope with large and frequent changes in their environment by means of morphological, physiological and life-cycle adaptations. Our findings support the crucial impact of sea water intrusion to the relative abundance and distribution of macrophyte species, as has occurred in other Mediterranean coastal lagoons. Furthermore, the shifts in salinity regime may introduce alterations in the abundance and distribution of the angiosperm *Z. noltii* especially in chocked lagoons. Due to the important structuring effects of macrophytes in shallow ecosystems, gaining insights into the connections between macrophytes structuring and environmental conditions is of critical importance to improve management and environmental policies.

Supplementary Materials: The following are available online at http://www.mdpi.com/2073-4441/10/2/151/s1. Table S1: Mean values and standard deviation of environmental parameters observed into the different macrophytic assemblages (i–iv), as well as the results of One Way ANOVA and LSD test. Table S2: Summary on variable correlations to DCA axes based on species coverage (%). Table S3: Intra-relationships of Correlation coefficients between the environmental variables and the principal component axes of Redundant Direct Analysis in three lagoon types. Table S4: Results of mixed analysis of variance in three different coastal lagoon types of Western Greece showing the effects of the factors Season, Year, Station and their interactions (Season*Year*Station) on the univariate variables: (i) Number of species (S), (ii) Margalef Species Richness (d), (iii) Pielou Evenness (J), (iv) Shannon Diversity Index (H). Figure S1: MDS analysis based on the seasonal variation of abundances of the angiosperms *Zostera noltii*, *Cymodocea nodosa* and *Ruppia cirrhosa* into the three different lagoon types of Western Greece.

Acknowledgments: The authors would like to thank all the staff of the Patras Laboratory of Ecology involved in the monitoring program and the Fishermen Cooperatives of Amvrakikos Gulf lagoons, Messolonghi, and Araxos lagoons for their help during field samplings. This study received funding from the Research committee of Patras University and from the Grants of Plant Ecology division. We greatly appreciate Paraskevi Manolaki, for the linguistic revision of the manuscript and Aikaterini Kostara, for her help on the preparation of figures. Many thanks to Pierluigi Viaroli, Department of Life Sciences, University of Parma, for his hospitality and for his helpful comments and constructive criticism. Finally, we are grateful to the Editors Diego Vicinanza and Pasquale Contestabile, as well as the two anonymous reviewers for the careful revision and valuable critical comments on the manuscript.

Author Contributions: All the authors have extensively contributed to the work presented in this paper; Chrysoula Christia for data collection and paper writing, Gianmarco Giordani, for data analysis and paper revision; Eva Papastergiadou for experimental data collection conception, research supervision and paper revision.

Conflicts of Interest: There is no conflict of interest to declare.

References

1. Kjerfve, B. *Coastal Lagoons Processes*; Elsevier Oceanography Series 60; Elsevier Science Publishers: Amsterdam, The Netherlands, 1994.
2. Christia, C.; Papastergiadou, E. Spatial and temporal variations of aquatic macrophytes and water quality in six coastal lagoons of Western Greece. *Belg. J. Bot.* **2007**, *140*, 39–50.
3. Rodriguez-Climent, S.; Caiola, N.; Ibanez, C. Salinity as the main factor structuring small-bodied fish assemblages in hydrologically altered Mediterranean coastal lagoons. *Sci. Mar.* **2013**, *77*, 37–45.
4. Pulina, S.; Brutemark, A.; Suikkanen, S.; Padedda, B.M.; Grubisic, L.M.; Satta, C.T.; Caddeo, T.; Farina, P.; Sechi, N.; Lugliè, A. Effects of warming on a Mediterranean phytoplankton community. *Web Ecol.* **2016**, *16*, 89–92. [CrossRef]
5. Bird, E.C.F. Physical setting and geomorphology of coastal lagoons. In *Coastal Lagoon Processes*; Kjerfve, B., Ed.; Elsevier: Amsterdam, The Netherlands, 1994; Chapter 2; pp. 9–40.

6. Directive, E.C.W.F. Water Framework Directive 2000/60/EC of the European Parliament and of the Council establishing a framework for the Community action in the field of water policy. *Off. J. Eur. Communities* **2000**, *43*, 1–72.
7. Guelorget, O.; Perthuisot, P. *Le Domaine Paralique: Expressions Geologiques, Biologiques et Economiques du Confinement*; Travaux du Laboratoire de Geologie de l'Ecole Normale Superiere: Paris, France, 1983; Volume 16, p. 136.
8. Basset, A.; Sabbeta, L.; Fonnesu, A.; Mouillot, D.; Do Chi, T.; Viaroli, P.; Giardani, G.; Reizopoulou, S.; Abbiati, M.; Carrada, G.C. Typology in Mediterranean transitional waters: New challenges and perspectives. *Aquat. Conserv. Mar. Freshw. Ecosyst.* **2006**, *16*, 441–455. [CrossRef]
9. Christia, C.; Giordani, G.; Papastergiadou, E. Assessment of ecological quality of coastal lagoons with a combination of phytobenthic and water quality indices. *Mar. Pol. Bul.* **2014**, *86*, 411–423. [CrossRef] [PubMed]
10. Umgiesser, G.; Ferrarin, C.; Cucco, A.; De Pascalis, F.; Bellafiore, D.; Ghezzo, M.; Bajo, M. Comparative hydrodynamics of 10 Mediterranean lagoons by means of numerical modeling. *J. Geophys. Res.* **2014**, *119*, 2212–22226. [CrossRef]
11. Perez-Ruzafa, A.; Garcıa-Charton, J.A.; Barcala, E.; Marcos, C. Changes in benthic fish assemblages as a consequence of coastal works in a coastal lagoon: The Mar Menor (Spain, Western Mediterranean). *Mar. Poll. Bull.* **2006**, *53*, 107–120. [CrossRef] [PubMed]
12. Battaglia, B. Final resolution of the symposium on the classification of brackish waters. *Archo Oceanogr. Limnol.* **1959**, *11*, 243–248.
13. Tagliapietra, D.; Volpi Ghirardini, A. Notes on coastal lagoon typology in the light of the EU Water Framework Directive: Italy as a case study. *Aquat. Conserv. Mar. Freshw. Ecosyst.* **2006**, *16*, 457–467. [CrossRef]
14. De Wit, R. Challenges for applying vulnerability assessments in coastal lagoons. *Transit. Wat. Bull.* **2011**, *5*, 32–41.
15. Agostini, S.; Marchand, B.; Pergent, G. Temporal and spatial changes of seagrass meadows in a Mediterranean coastal lagoon. *Oceanol. Acta* **2003**, *25*, 297–302. [CrossRef]
16. Adams, W.J.; Kimerle, R.A.; Barnett, R.A., Jr. Sediment quality and aquatic life assessment. *Environ. Sci. Technol.* **1992**, *26*, 1865–1875. [CrossRef]
17. Lotze, H.K.; Lenihan, H.S.; Bourque, B.J.; Bradbury, R.H.; Cooke, R.G.; Kay, M.C.; Kidwell, S.M.; Kirby, M.X.; Peterson, C.H.; Jackson, J.B.C. Depletion, degradation and recovery potential of estuaries and coastal seas. *Science* **2006**, *312*, 1806–1809. [CrossRef] [PubMed]
18. Chappuis, E.; Gacia, E.; Ballesteros, E. Changes in aquatic macrophyte flora over the last century in Catalan water bodies (NE Spain). *Aquat. Bot.* **2011**, *95*, 268–277. [CrossRef]
19. Chapman, P.M. Management of coastal lagoons under climate change. *Estuar. Coast. Shelf Sci.* **2012**, *110*, 32–35. [CrossRef]
20. Duarte, C.M.; Borja, A.; Carstensen, J.; Elliott, M.; Krause-Jensen, D.; Marbà, N. Paradigms in the Recovery of Estuarine and Coastal Ecosystems. *Estuar. Coasts* **2015**, *38*, 1202–1212. [CrossRef]
21. Littler, M.M.; Littler, D.S. A relative dominance moel for biotic reefs. In Proceedings of the Joint Meeting of the Atlantic Reef Committee Society of Reef Studies, Miami, FL, USA, 26–28 October 1984.
22. Fodge, J.D.; Thomas, G.L.; Pauley, G.B. Effects of canopy formation by floating and submergent aquatic macrophytes on the water quality of two shallow Pacific Northwest lakes. *Aquat. Bot.* **1990**, *38*, 231–248. [CrossRef]
23. Moore, K.A. Influence of seagrasses on water quality in shallow regions of the lower Chesapeake Bay. *J. Coast. Res.* **2004**, *45*, 162–178. [CrossRef]
24. Malea, P.; Kevrekidis, Th.; Mogias, A. Annual versus perennial growth cycle in *Ruppia maritima* L: Temporal variation in population characteristics in Mediterranean lagoons (Monolimni and Drana Lagoons, Northern Aegean Sea). *Bot. Mar.* **2004**, *47*, 357–366. [CrossRef]
25. Verhoeven, J.T.A. The ecology of *Ruppia*-dominated communities in Western Europe. III. Aspects of production, consumption and decomposition. *Aquat. Bot.* **1980**, *8*, 209–253. [CrossRef]
26. Comín, F.A.; Menéndez, M.; Herrera, J.A. Spatial and temporal scales for monitoring coastal aquatic ecosystems. *Aquat. Conserv. Mar. Freshw. Ecosyst.* **2004**, *14* (Suppl. S1), 5–17. [CrossRef]

27. Viaroli, P.; Bartoli, M.; Azzoni, R.; Giordani, G.; Mucchino, C.; Naldi, M.; Nizzoli, D.; Tajé, L. Nutrient and iron limitation to *Ulva* blooms in a eutrophic coastal lagoon (Sacca di Goro, Italy). *Hydrobiologia* **2005**, *550*, 57–71. [CrossRef]
28. Haritonidis, S.; Malea, P. Bioaccumulation of metals by the green alga *Ulva rigida* from Thermaikos Gulf, Greece. *Environ. Pollut.* **1999**, *104*, 365–372. [CrossRef]
29. Orfanidis, S.; Panayotidis, P.; Stamatis, N. An insight to the Ecological Evaluation Index (EEI). *Ecol. Ind.* **2003**, *3*, 27–33. [CrossRef]
30. Simboura, N.; Zenetos, A. Benthic indicators to use in ecological quality clas-sification of Mediterranean soft bottom marine ecosystems, including a new biotic index. *Med. Mar. Sci.* **2002**, *3*, 77–111. [CrossRef]
31. Nicolaidou, A.; Reizopoulou, S.; Koutsoubas, D.; Orfanidis, S.; Kevrekidis, T. Biological components of Greek lagoonal ecosystems: An overview. *Mediterr. Mar. Sci.* **2005**, *6*, 31–50. [CrossRef]
32. Katselis, G.; Koukou, K.; Dimitriou, E.; Koutsikopoulos, C. Short- term seaward fish migration in the Messolonghi-Etoliko lagoons (Western Greek coast) in relation to climatic variables and the lunar cycle. *Estuar. Coast. Shelf Sci.* **2007**, *73*, 571–582. [CrossRef]
33. Vasileiadou, K.; Pavloudi, C.; Kalantzi, I.; Apostolaki, E.T.; Chatzigeorgiou, G.; Chatzinikolaou, E.; Pafilis, E.; Papageorgiou, N.; Fanini, L.; Konstas, S.; et al. Environmental variability and heavy metal concentrations from five lagoons in the Ionian Sea (Amvrakikos Gulf, W. Greece). *Biodiv. Data J.* **2016**, *4*, e8233. [CrossRef] [PubMed]
34. Orfanidis, S.; Panayotidis, P.; Stamatis, N. Ecological evaluation of transitional and coastal waters: A marine benthic macrophytes-based model. *Mediterr. Mar. Sci.* **2001**, *2*, 45–65. [CrossRef]
35. Christia, C.; Tziortzis, I.; Fyttis, G.; Kashta, L.; Papastergiadou, E. A survey of the benthic aquatic flora of transitional water systems of Greece and Cyprus (Mediterranean Sea). *Bot. Mar.* **2011**, *54*, 169–178. [CrossRef]
36. Sfriso, A.; Facca, C.; Ghetti, P.F. Validation of the Macrophyte Quality Index (MaQI) set up to assess the ecological status of Italian marine Transitional environments. *Hydrobiologia* **2009**, *617*, 117–141. [CrossRef]
37. Le Fur, I.; De Wit, R.; Plus, M.; Oheix, J.; Simier, M.; Ouisse, V. Submerged benthic macrophytes in Mediterranean lagoons: Distribution patterns in relation to water chemistry and depth. *Hydrobiologia* **2017**. [CrossRef]
38. Sondergaard, M.; Johansson, L.S.; Lauridsen, T.L.; Jorgensen, T.B.; Liboriussen, L.; Jeppensen, E. Submerged macrophytes as indicators of the ecological quality of lakes. *Freshw. Biol.* **2010**, *55*, 893–908. [CrossRef]
39. Viaroli, P.; Bartoli, M.; Giordani, G.; Naldi, M.; Orfanidis, S.; Zaldivar, J.M. Community shifts, alternative stable states, biogeochemical controls and feedbacks in eutrophic coastal lagoons: A brief overview. *Aquat. Conserv. Mar. Freshw. Ecosyst.* **2008**, *18*, 105–117. [CrossRef]
40. Christia, C.; Papastergiadou, E.; Papatheodorou, G.; Geraga, M.; Papadakis, E. Seasonal and spatial variations of water quality, substrate and aquatic macrophytes based on side scan sonar, in an eastern Mediterranean lagoon (Kaiafas, Ionian Sea). *Environ. Earth Sci.* **2014**, *71*, 3543–3558. [CrossRef]
41. AHPA. *Standard Methods for the Examination of Water and Waste Water*, 18th ed.; American Public Health Association: New York, NY, USA, 1989.
42. Selig, U.; Schubert, M.; Eggert, A.; Steinhardt, T.; Sagert, S.; Schubert, H. The influence of sediments on soft bottom vegetation in inner coastal waters of Mecklenburg-Vorpommern (Germany). *Estuar. Coast. Shelf Sci.* **2007**, *71*, 241–249. [CrossRef]
43. SPSS, Inc. *SPSS v. 15.0 for Windows*; SPSS, Inc.: Chicago, IL, USA, 2006.
44. Ter Braak, C.J.F.; Šmilauer, P. *CANOCO Reference Manual and CanoDraw for Windows User's Guide: Software for Canonical Community Ordination (Version 4.5)*; Microcomputer Power: Ithaca, NY, USA, 2002.
45. Clarke, K.R.; Gorley, R.N. *PRIMER v6: User Manual/Tutorial*; PRIMER-E: Plymouth, UK, 2006.
46. Clarke, K.R.; Somerfield, P.J.; Chapman, M.G. On resemblance measures for ecological studies, including taxonomic dissimilarities and a zero-adjusted Bray–Curtis coefficient for denuded assemblages. *J. Exp. Mar. Biol. Ecol.* **2006**, *330*, 55–80. [CrossRef]
47. Ferrarin, C.; Umgiesser, G. Hydrodynamic modelling of a coastal lagoon: The Cabras lagoon in Sardinia, Italy. *Ecol. Mod.* **2005**, *188*, 340–357. [CrossRef]
48. Rosqvist, K.; Mattila, J.; Sandström, A.; Snickars, M.; Westerbom, M. Regime shifts in vegetation composition of Baltic Sea coastal lagoons. *Aquat. Bot.* **2010**, *93*, 39–46. [CrossRef]

49. Barker, P.; Leng, M.J.; Gasse, F.; Huang, Y. Century-tomillennial scale climatic variability in Lake Malawi reveales by isotope records. *Earth Planet. Sc. Lett.* **2007**, *261*, 93–103. [CrossRef]
50. Blomqvist, S.; Gunnars, A.; Elmgren, R. Why the limiting nutrient differs between temperate coastal seas and freshwater lakes: A matter of salt. *Limnol. Oceanogr.* **2004**, *49*, 2236–2241. [CrossRef]
51. Carvalho, S.; Pereira, P.; Pereira, F.; de Pablo, H.; Vale, C.; Gaspar, M.B. Factors structuring temporal and spatial dynamics of macrobenthic communities in a eutrophic coastal lagoon (Obidos lagoon, Porugal). *Mar. Environ. Res.* **2011**, *71*, 97–110. [CrossRef] [PubMed]
52. Acquavita, A.; Aleffi, I.F.; Benci, C.; Bettoso, N.; Crevatin, E.; Milani, L.; Tamberlich, F.; Toniatti, L.; Barbieri, P.; Licen, S.; et al. Annual characterization of the nutrients and trophic state in a Mediterranean coastal lagoon: The Marano and Grado Lagoon (northern Adriatic Sea). *Reg. Stud. Mar. Sci.* **2015**, *2*, 132–144. [CrossRef]
53. Van den Berg, M.S.; Coops, H.; Meijer, M.; Simons, J. Clear water associated with a dense *Chara* vegetation in the shallow and turbid lake Veluwenmeer. In *The Structuring Role of Submerged Macrophytes in Lakes*; Jeppesen, E., Søndergaard, M., Søndergaard, M., Christoffersen, K., Eds.; Springer: New York, NY, USA, 1998.
54. Blindow, I.; Hargeby, A.; Andersson, G. Seasonal changes of mechanisms maintaining clear water in a shallow lake with abundant *Chara* vegetation. *Aquat. Bot.* **2002**, *72*, 315–334. [CrossRef]
55. Del Pozo, R.; Fernandez-Alaez, C.; Fernadez-Alaez, M. The relative importance of natural and anthropogenic effects on community composition of aquatic macrophytes in Mediterranean ponds. *Mar. Freshw. Res.* **2011**, *62*, 101–109.
56. Guidelines for Impact Assessment on Seagrass Meadows. Available online: http://www.rac-spa.org/sites/default/files/doc_vegetation/ld_etude_impact_en.pdf (accessed on 3 February 2018).
57. Bamber, R.N.; Batten, S.D.; Sheader, M.; Bridgwater, N.D. On the ecology of brackish water lagoons in Great Britain. *Aquat. Conserv.* **1992**, *2*, 65–94. [CrossRef]
58. Cognetti, G. Colonization of stressed coastal environment. *Mar. Pol. Bul.* **1992**, *24*, 12–14. [CrossRef]
59. Green, E.P.; Short, F.T. *World Atlas of Seagrasses*; UNEP, WCMC: Berkeley, CA, USA, 2003.
60. Reizopoulou, S.; Nicolaidou, A. Benthic diversity of coastal brackish-water lagoons in western Greece. *Aquat. Conserv.* **2004**, *14*, 93–102. [CrossRef]
61. Rodriguez-Gallego, L.; Meerhoff, E.; Clemente, J.M.; Conde, D. Can ephemeral proliferations of submerged macrophytes influence zoobenthos and water quality in coastal lagoons? *Hydrobiologia* **2010**, *646*, 253–269. [CrossRef]
62. Perez-Ruzafa, A.; Marcos, C.; Perez-Ruzafa, I.M.; Perez-Marcos, M. Coastal lagoons: "transitional ecosystems" between transitional and coastal waters. *J. Coast Conserv.* **2011**, *15*, 369–392. [CrossRef]
63. Millet, B.; Guelorget, O. Spatial and seasonal variability in the relationships between benthic communities and physical environment in a lagoon ecosystem. *Mar. Ecol. Prog. Ser.* **1994**, *198*, 161–174. [CrossRef]
64. Salas, F.; Neto, J.M.; Borja, A.; Marques, J.C. Evaluation of the applicability of a marine biotic index to characterize the status of estuarine ecosystems: The case of Mondego estuary (Portugal). *Ecol. Ind.* **2004**, *4*, 215–225. [CrossRef]
65. Azzoni, R.; Giordani, G.; Bartoli, M.; Welsh, D.T.; Viaroli, P. Iron, sulphur and phosphorus cycling in the rhizosphere sediments of a eutrophic Ruppia cirrhosa meadow (Valle Smarlacca, Italy). *J. Sea Res.* **2001**, *45*, 15–26. [CrossRef]
66. Zaldivar, J.M.; Cardoso, A.C.; Viaroli, P.; Newton, A.; De Wit, R.; Ibanez, C.; Reizopoulou, S.; Somma, F.; Razinkovas, A.; Basset, A.; et al. Eutrophication in transitional waters: An overview. *Transit. Waters Monogr.* **2008**, *1*, 1–78.
67. Menéndez, M. Net production of *Ruppia cirrhosa* in Ebro Delta. *Aquat. Bot.* **2002**, *73*, 107–113. [CrossRef]
68. Mannino, A.M.; Sara, G. The effect of *Ruppia cirrhosa* features on macroalgae and suspended matter in a Mediterranean shallow system. *Mar. Ecol.* **2006**, *27*, 350–360. [CrossRef]
69. Warwick, R.M.; Clarke, K.R. Relationship between body-size, species abundance and diversity in marine benthic assemblages: Facts or artefacts? *J. Exper. Mar. Biol. Ecol.* **1996**, *202*, 63–71. [CrossRef]
70. Charpentier, A.; Grillas, P.; Lescuyer, F.; Coulet, E.; Auby, I. Spatio-temporal dynamics of a *Zostera noltii* dominated community over a period of fluctuating salinity in a shallow lagoon, Southern France. *Estuar. Coast. Shelf Sci.* **2005**, *64*, 307–315. [CrossRef]

71. Signorini, A.; Massini, G.; Migliore, G.; Tosoni, M.; Varrone, C.; Izzo, G. Sediment biogeochemical differences in two pristine Mediterranean coastal lagoons (in Italy) characterized by different phanerogam dominance—A comparative approach. *Aquat. Conserv. Mar. Freshw. Ecosyst.* **2008**, *18*, 27–44. [CrossRef]
72. Pergent-Martini, C.; Pasqualini, V.; Ferrat, L.; Pergent, G.; Fernandez, C. Seasonal dynamics of *Zostera noltii* Hornem. In two Mediterranean lagoons. *Hydrobiologia* **2005**, *543*, 233–243. [CrossRef]
73. Den Hartog, C. *The Sea-Grasses of The World;* North-Holland Publishing Company: Amsterdam, The Netherlands, 1970.
74. Vermaat, J.E.; Verhagen, F.C.A.; Lindenburg, D. Contrasting responses in two populations of Zostera noltii Hornem. To experimental photoperiod manipulation at two salinities. *Aquat. Bot.* **2000**, *67*, 179–189. [CrossRef]
75. Torquemada, Y.F.; Lizaso, J.L.S. Responses of two Mediterranean seagrasses to experimental changes in salinity. *Hydrobiologia* **2011**, *669*, 21–33. [CrossRef]
76. Prado, P.; Caiola, N.; Ibáñez, C. Spatio-temporal patterns of submerged macrophytes in three hydrologically altered mediterranean coastal lagoons. *Estuar. Coasts* **2013**, *36*, 414–429. [CrossRef]
77. Marba, N.; Holmer, M.; Gacia, E.; Barron, C. Seagrass beds and coastal biogeochemistry. In *Seagrasses: Biology, Ecology and Conservation*; Larkum, A.W.D., Orth, R.J., Duarte, C.M., Eds.; Springer: Dordrecht, The Netherlands, 2006; pp. 135–157.
78. Fernández-Torquemada, Y.; Sánchez-Lizaso, J.L. Responses of two Mediterranean seagrasses to experimental changes in salinity. *Hydrobiologia* **2011**, *669*, 21–33. [CrossRef]
79. Menéndez, M.; Peñuelas, J. Sesonal photosynthetic and respiratory responses of *Ruppia cirrhosa* (Petagna) Grande to changes in light and temperature. *Arch. Hydrobiol.* **1993**, *129*, 221–230.
80. Montesanto, B.; Panayotidis, P. The *Cystoseira* spp. Communities from the Aegean Sea (NE Mediterranean). *Med. Mar. Sci.* **2001**, *2*, 57–67. [CrossRef]
81. Falace, A.; Curiel, D.; Sfriso, A. Study of the macrophyte assemblages and application of phytobenthic indices to assess the Ecological Status of the Marano-Grado lagoon (Italy). *Mar. Ecol.* **2009**, *30*, 480–494. [CrossRef]
82. Peres, J.M.; Picard, J. Nouveau manuel de bionomie benthique de la mer Méditerranée. *Rec. Trav. St. Mar. Endoume* **1964**, *31*, 5–137.
83. Giaccone, G.; Bruni, A. *Le Cistoseire e la Vegetazione Sommersa del Mediterraneo*; Atti dell' Instituto Veneto de Scienze: Lett. ed Arti, Venezia, Italy, 1973; pp. 59–103.
84. Waycott, M.; Duarte, C.M.; Carruthers, T.J.B.; Orth, R.J.; Dennison, W.C.; Olyarnik, S.; Calladine, A.; Fourqurean, J.W.; Heck, K.L., Jr.; Hughes, A.R.; et al. Accelerating loss of seagrasses across the globe threatens coastal ecosystems. *Proc. Natl. Acad. Sci. USA* **2009**, *106*, 12377–12381. [CrossRef] [PubMed]
85. Aral, M.M.; Chang, B. Spatial Variation of Sea Level Rise at Atlantic and Mediterranean Coastline of Europe. *Water* **2017**, *9*, 522. [CrossRef]

water

MDPI

Article

Simulating Future Flows and Salinity Intrusion Using Combined One- and Two-Dimensional Hydrodynamic Modelling—The Case of Hau River, Vietnamese Mekong Delta

Duong Tran Anh [1,*], Long Phi Hoang [2], Minh Duc Bui [1] and Peter Rutschmann [1]

1 Institute of Hydraulic and Water Resources Engineering, Technische Universität München, Arcisstrasse 21, D-80333 München, Germany; bui@tum.de (M.D.B.); peter.rutschmann@tum.de (P.R.)
2 Water Systems and Global Change Group, Wageningen University, P.O. Box 47, 6700 AA Wageningen, The Netherlands; long.hoang@wur.nl
* Correspondence: tran.duong@tum.de; Tel.: +49-176-477-225-52

Received: 14 May 2018; Accepted: 3 July 2018; Published: 6 July 2018

Abstract: Salinity intrusion in the Vietnamese Mekong Delta (VMD) has been exacerbated significantly in recent years by the changing upstream inflows, sea level rise resulting from climate change, and socioeconomic development activities. Despite significant damage to agricultural production and freshwater supplies, quantitative assessments of future flows and salinization remain limited due to lack of observation data and modelling tools to represent a highly complex hydraulic network. In this study, we combine 1D-MIKE 11 and 2D-MIKE 21 hydrodynamic models to simulate future flows, water level and salinity intrusion in the Hau River—one main river branch in the Mekong Delta. Future hydrological changes are simulated under multiple scenarios of upstream inflow changes, climate change and sea level rise for the 2036–2065 period. We first use the 1D-MIKE 11 to simulate the flow regime throughout the whole VMD using upstream discharges, outlet water levels and rainfall data as boundary conditions. Output from this step is then used to force the 2D-MIKE 21 model to estimate flow velocity, water level and salinity concentration in the Hau River, focusing on the salinization-prone section between Can Tho, Dinh An, and Tran De estuaries. Simulation results show that salinization will increase substantially, characterized by (1) higher salinity intrusion length under spring tide from 6.78% to 7.97%, and 8.62% to 10.89% under neap tide; and (2) progression of the salinity isohalines towards the upper Mekong Delta, from 3.29 km to 3.92 km for 1 practical salinity unit (PSU) under spring tide, and 4.36 km to 4.65 km for 1 PSU concentration under neap tide. Additionally, we found that salinity intrusion will make it more difficult to re-establish the freshwater condition in the estuary in the future. In particular, the flushing time required to replace saltwater with freshwater at the estuaries tends to increase to between 7.27 h for maximum discharge of 4500 m^3/s and 58.95 h for discharge of 400 m^3/s under the most extreme scenario. Increasing salinization along the Hau River will have important consequences for crop production, freshwater supplies and freshwater ecosystems, therefore requiring timely adaptation responses.

Keywords: salinity intrusion; MIKE modelling; climate change; sea level rise; Mekong Delta

---

1. Introduction

The prospective for water resources management in the Vietnamese Mekong Delta (VMD) is very challenging in the 21st century due to multiple issues including increasing water use, climate change impacts and sea level rise [1,2]. These changes will likely have considerable impacts on economic development and the livelihoods of people living in the delta [3]. In this context, the Mekong Delta is considered to be one of the most vulnerable catchment areas throughout the basin because of the cumulative impacts of upstream development activities. The Mekong Delta has long been identified as a hotspot in terms of vulnerability to sea level rise and climate change, as well as to hydropower dam construction in the upper basin [4,5]. As a result of these factors, the Mekong Delta experiences increasing salinity intrusion, which represents a critical challenge for water resources management and agriculture production.

The VMD plays a crucial role in the economy of Vietnam. The delta contributes approximately 27% of the country's GDP, supports 16 million inhabitants (nearly 22% of Vietnam's total population), and provides about 50% of the annual rice production [6,7]. Vietnam is also one of the most vulnerable countries to climate change and sea-level rise [8]. Currently, the VMD is experiencing rapid development in terms of population growth and infrastructure development [9,10], which create substantial pressure on water resources [11,12]. Moreover, hydropower development and water withdrawal in upstream countries cause changes in the hydraulic regime and salinization in the VMD [5,13]. All these changes pose critical challenges for sustaining water resources for economic development. In view of these challenges, it is important to assess future hydrological changes, especially salinization, in order to support decision making and planning for sustainable delta management.

Salinity intrusion is a natural phenomenon occurring in the lands, estuaries, and aquifers adjacent to the sea. Salinization in deltas and estuaries varies considerably depending on the tidal regimes, river flows and topography [14]. Tidal dynamics cause turbulent mixing via the transportation of saltwater, as well as tidal trapping and tidal transport via the interaction between tide and terrain [15]. The tide is a powerful source of mixed fresh and saltwater, that also plays an important role in saltwater intrusion. Recent studies also identified a wide range of factors affecting salinity intrusion, including river flows, topography, morphology, river bed slope, wind velocity and direction, and water temperature [16,17].

Salinization in the Mekong Delta under climate change and sea level rise has been addressed in a number of recent studies. Several studies have shown that a combination of climate change, sea level rise and upstream inflow changes will increase salinization, resulting in important consequences for water supplies and agriculture production [2,8,18–20]. For instance, Wassmann et al., Le et al., and Dinh et al. [21–24] investigated the effects of sea level rise on water levels in the VMD. Khang et al. [20] demonstrated the relationship between salinity intrusion and sea level rise and river flow change in the VMD during the dry season. Most studies covered the whole VMD using the 1D-MIKE 11 hydrodynamic model and, therefore, did not pay particular attention to salinity intrusion and hydraulic regime in the estuaries.

Distributed hydrological models are frequently applied in water resources assessments at the Mekong basin level, but little attention has been paid to the hydraulic regimes in the estuaries due to highly complex river–estuary–ocean interactions. The existing modelling frameworks are not well-designed to cope with the complex hydrodynamic relationship between river flow, tides and human activities. To address this challenge, we propose a combination of hydrodynamic models in order to capture the hydrodynamics of the Mekong Delta's estuary sufficiently. Therefore, the objective of this study is to combine 1D-MIKE 11 and 2D-MIKE 21 hydrodynamic models to investigate the hydraulic regimes and salinity intrusion in a specific river section, namely the Hau River, stretching between Can Tho to Tran De and Dinh An estuaries. To achieve this objective, a 1D model was first applied to simulate the entire hydrodynamic of the VMD with different scenarios of upstream discharge and downscaled precipitation from five general circulation models (GCMs) with two representative pathway scenarios (RCP) from Coupled Model Intercomparison Project Phase 5 (CMIP5). We employed statistical downscaling and bias-correction methods to prepare climate change input data, as shown in detail in Duong et al. [25]. The outputs of downscaled daily precipitation for 2036–2065 were used to run a rainfall-runoff model to simulate runoff for 1D hydrodynamic simulations. Finally, we used the results at Can Tho and branch discharge along the Hau River from MIKE 11 simulation as a boundary condition to force the 2D hydrodynamic model to simulate river–estuary–ocean interactions. As such, our modelling framework allows evaluation of the combined impacts of changing upstream discharges, the variabilities of precipitation and sea level rise from climate change on salinity intrusion in Hau River, VMD.

2. Study Area

The Vietnamese Mekong Delta (VMD) is located at the end of the Lower Mekong Basin. It is a large, relatively flat and low-lying area, with an elevation of approximately 0.5–1.0 m above mean sea level. The delta's hydraulic regimes are complicated, with two major distributaries—the Tien (Mekong River) and Hau River (Bassac River)—which drain out to the South China Sea through eight estuaries (Figure 1). The structures comprise 7000 km of main canals, 4000 km of secondary on-farm canal systems, 193 spills, 409 reservoirs, 528 junctions, 29 sluices, 749 floodplains and more than 20,000 km of dykes to prevent early floods [26,27]. The mean annual flow is 15,000 m^3/s, the maximum discharge is about 60,000 m^3/s in the flood season, and the minimum discharge is around 2000 m^3/s in the dry season [28,29]. The climate of the VMD has tropical monsoon characteristics, with two separate seasons per year. The rainy season normally lasts from May to October, whereas the dry season lasts from December to March [30]. The mean annual precipitation is 1500 mm for the entire VMD and the range varies from 1600 mm to 2400 mm/year. The total precipitation in the rainy season contributes to nearly 90% of the annual precipitation [31]. The average discharge in the dry season fluctuates greatly, from 1700 m^3/s to 6000 m^3/s between January and May, and leads to water shortages for irrigating about 1.5 million hectares of irrigated crops [27]. The Mekong Delta experiences two types of tidal regime, namely the semi-diurnal and diurnal tides, affecting the hydraulic condition in estuaries of the East and West seas, respectively [32]. The salinity intrusion in the VMD—and in all the river networks—is substantial in recent years, particularly during the dry season, during which about 2.1 million hectares in the Mekong Delta suffered from salinity intrusion [21,27].

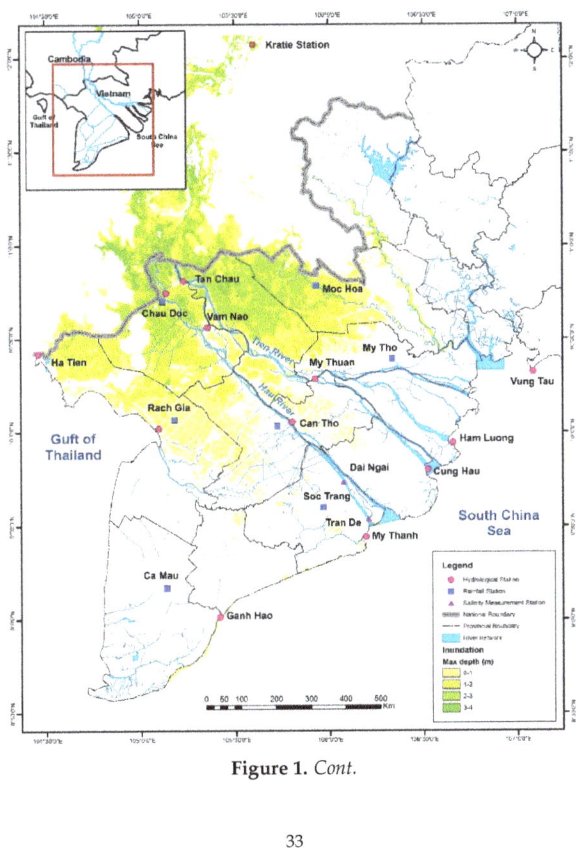

Figure 1. *Cont.*

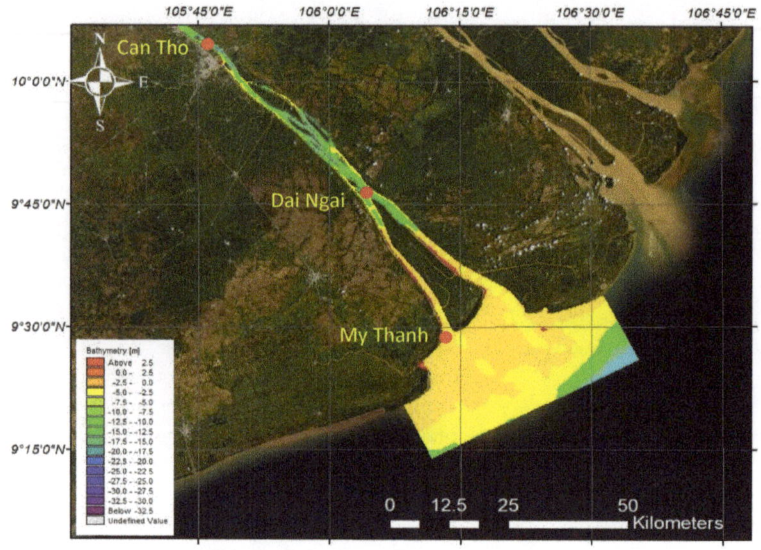

Figure 1. The study area for the whole Vietnamese Mekong Delta (VMD) (top panel) and 2D study domain (bottom panel).

3. Methodology and Model Setup

3.1. Hydrological and Hydraulic Data

The hydrological input data for the MIKE 11 model includes river discharges at the delta inlet at Kratie (Cambodia), outlet water levels, precipitation, evapotranspiration and water demands for agriculture, industry and domestic sectors in the entire VMD. The hourly Kratie discharge and hourly water levels at 10 major stations, including Vung Tau, Vam Kenh, Binh Dai, An Thuan, Ben Trai, My Thanh, Ganh Hao, Song Doc, Rach Gia and Xeo Ro, were measured from 2009 to 2011 and all input data is summarized in Table 1. More details about these data can be found in Duong et al. [25].

The daily precipitation scenarios for boundary conditions in the MIKE 11 model were generated from the CMIP5 using five GCMs: ACCESS 1.0, CCSM4, CSIRO-Mk 3.6, HadGEM and MPI-ESM-LR, with two RCP4.5 and RCP8.5 scenarios (Table 2). The bilinear interpolation was applied to downscale climate data to $0.5^{\circ} \times 0.5^{\circ}$ grid before applying bias correction. Future precipitation change scenarios were generated using three bias correction methods, including linear scaling [33], local intensity scaling [34] and distribution mapping [35,36]. The motivations for selecting the GCMs and RCP scenarios were explained in detail in [25,37].

The hydraulic data for the MIKE 21 model consists of hourly discharge at Can Tho station, offshore tidal levels and salinity concentration at Tran De and Dai Ngai stations, in the years 2011 and 2010 for calibrating and validating the models, respectively. The tidal levels in the years 2010 and 2011 were derived from the Global Tidal Model in the MIKE Zero Toolbox and were calibrated with tidal levels data at My Thanh station, provided by the Institute of Coastal and Offshore Engineering (http://www.icoe.org.vn/index.php?pid=551). Furthermore, the branch discharges along the Hau River were simulated from the MIKE 11 model, for five main intakes, including Mang Thit, Rach Mop, Cau Quan, Nga Bay and Dai Ngai. The future discharges at Can Tho were simulated from MIKE 11 with four scenarios of upstream discharges at Kratie and precipitation in the VMD.

Regarding the wind data for the MIKE 21 simulation, there are two dominant wind directions; the wind blows south-west in the wet season and northeast in the dry season, with stronger speeds as a result of the monsoon in the South China Sea. The average north-east wind speed varied between

3 and 6 m/s, and the maximum speed was approximately 15–20 m/s towards the east, blowing from the sea. The predominant wind directions were calculated and are shown in Table 3.

Table 1. Main input data for the model simulations.

Model	Input Data	Observed	Projection
MIKE 11	Precipitation	Daily precipitation from 7 stations (locations in Figure 1) Period: 1978–2011 Source: SRHMC	Projected precipitation based on downscaled and bias-corrected GCM simulations. Period: 2036–2065 Source: CMIP5
	River discharge at Kratie	Daily discharge data Period: 2010–2011 Source: MRC + SIWRR	NA
	River discharge in the Mekong Delta	Daily discharges and water levels at 10 stations (locations in Figure 1) Period: 2005–2011 Source: SIWRR	NA
	Sea water level and tides	Period: 2005–2011 Source: SIWRR	
	Hydraulic network	Period: 2005–2011 Source: SIWRR	
MIKE 21	Discharges at Can Tho	Hourly discharge Cali. Period 2011 Vali. Period 2010 Source: SRHMC + DHI	Future Period 2036–2065
	Tidal magnitudes	Period 2010–2011 Source: SRHMC + DHI	Future Period 2036–2065
	Branches discharges	Five main channels/rivers connecting Hau river	
		Period 2010–2011	

SRHMC: Southern Regional Hydro-Meteorological Center, SIWRR: Southern Institute of Water Resources Research, CMIP5: Coupled Model Intercomparison Project Phase 5, DHI: Danish Hydraulic Institute, GCM: General Circulation Model, MRC: Mekong River Commission.

Table 2. Description of the GCMs used in this study.

General Circulation Model (GCM)	Abbreviation	Country	Spatial Resolution
ACCESS 1.0	ACCESS	Australia	1.25° × 1.875°
CCSM4	CCSM4	NCAR/USA	0.94° ×1.25°
CSIRO-Mk 3.6	CSIRO	Australia	1.875° × 1.875°
HadGEM2-ES	HadGEM2	Hadley/UK	1.875° × 1.25°
MPI-ESM-LR	MPI	Germany	1.875° × 1.875°

Table 3. Direction of wind in Vung Tau station, Southern Vietnam (SIWRR, 2011).

Month	Jan.	Feb.	Mar.	Apr.	May	June	July	Aug.	Sept.	Oct.	Nov.	Dec.
Dominant direction	NE	NE	E	SE	SE	SW	SW	SW	SW	NW	ENE	NE

Note: E = east; ENE = east-north-east; NE = north-east; NW = north-west; SE = south-east; and SW = south-west.

Salinity data was used for calibrating and validating the MIKE 21 model for the years 2011 and 2010, respectively. The salinity concentration in the Hau River changed significantly during the dry season and reached its highest value between March and May (see Figure 2). The Dinh An and Tran De estuaries are characterized by an irregular semi-diurnal tide, which has a strong effect due to tidal oscillations [38,39]. Similar to salinity, the discharge at Can Tho shows considerable fluctuation within one day, and the diurnal differences are evident. The salinity and tidal levels showed phase differences, with the maximum salinity at the Dai Ngai station occurring during the spring tide. The maximum salinity in 2011 at the Tran De station was about 23.0 PSU (practical salinity unit; 1 PSU equals 1‰)

and the minimum value was 0.5 PSU [40]. The salinity concentration at Dai Ngai station was smaller than Tran De because this station was located about 31 km further inland. In 2011, the maximum value of salinity level in Dai Ngai reached 11.0 PSU, and minimum values varied between 0.0 and 0.5 PSU.

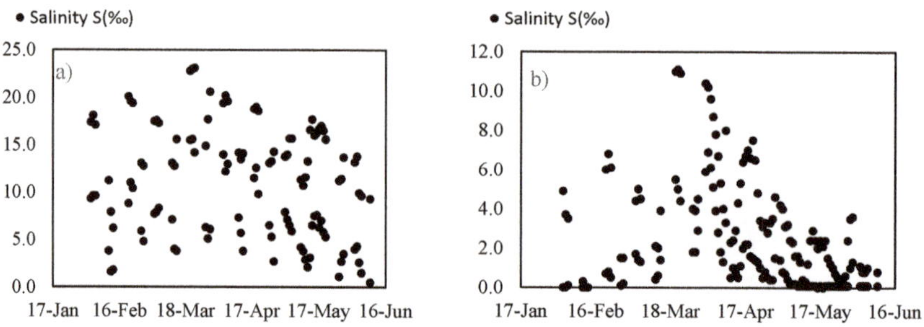

Figure 2. The salinity concentration at Tran De (**a**) and Dai Ngai (**b**) stations in the dry season of 2011.

3.2. *Model Schematization*

The detailed bathymetry in numerical modelling plays a critical role in achieving accurate hydrodynamic simulations. In the present study, the bottom topography data in the coastal area and the Hau River estuary were obtained from the DHI Vietnam (Danish Hydraulic Institute), the Department of Transportation in Ho Chi Minh City, and the Southern Institute of Water Resources Research (SIWRR). The modelling domain in the horizontal plane covers 100×110 km, including the main river stream and the coastal sea. The greatest depth in the study area is 20 m below mean sea level, near the south-east corner of the modelling domain in the coastal sea. It is important to consider the model stability by satisfying the Courant–Friedrich–Levy number: the selection of time step (Δt) and grid step (Δx) is crucial in order to balance the trade-off between numerical stability and computational time. The simulation grid cell selection for the study area is flexible mesh (FM) or unstructured mesh, where triangular cells of bathymetry are used to optimize the simulation, with small sizes in river domain and larger sizes in offshore settings. The triangular element sizes are about 80–100 m in the river and 800–1000 m offshore, with the total triangular being 35,000 elements and 28,000 nodes. Bathymetry was constructed with the MIKE zero tool and obtained from an updated digitizing map in 2011 from the Southern Institute for Water Resource Planning and DHI Vietnam. The bathymetry and grid resolution is presented in Figure 3.

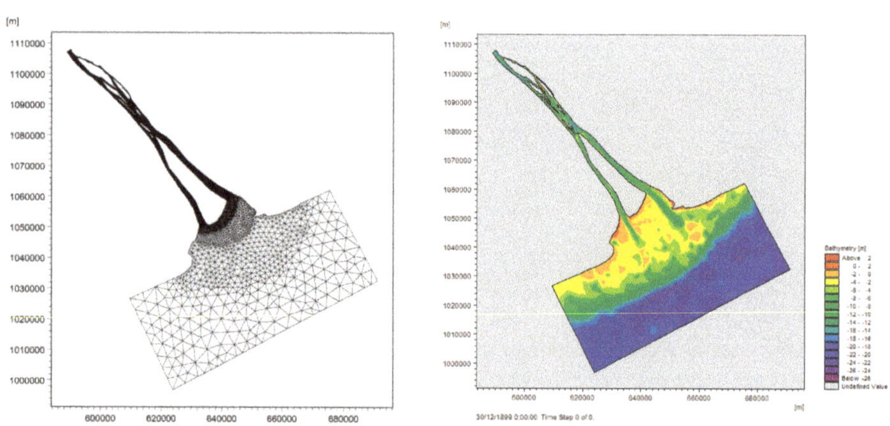

Figure 3. Bathymetry (**left** panel) and computational flexible grid mesh (**right** panel).

3.3. MIKE 11 Hydrodynamic Model

The MIKE 11 is an unsteady 1-dimensional hydrodynamic model, which is based on one-dimensional equations and solves the vertically integrated equations regarding conservation of continuity and momentum. The solution of continuity and momentum equations are employed as an implicit finite difference scheme with a 6-point Abbott scheme [41]. The main governing equations are Saint-Venant equations [42]. In order to effectively simulate floodplain and field areas, and to connect main channels and rivers, we employed the quasi-2D modelling approach developed by Dung et al. [43]. The rice fields and the floodplain are considered to be artificial channels or "virtual channels", having wide cross sections and were extracted from the digital elevation model (DEM). The calibration process of these channels and crest of dikes was explained in detail in [43,44].

3.4. MIKE 21 Hydrodynamic Model

The MIKE 21 is a dynamic modelling system applicable for coastal and estuarine environments. The MIKE 21 comprises several modules, including the hydrodynamic module, advection–dispersion module, spectral wave module, and transport module. This model is based on the numerical solution of the two dimensional incompressible Reynolds-averaged Navier–Stokes equations with the assumptions of Boussinesq and hydrostatic pressure. It comprises continuity, momentum, temperature, salinity and density equations [45].

This study applies MIKE 21 with two modules, namely hydrodynamic (HD) and transport modules (TR), using a flexible mesh. The mutual interaction between the flow, wind, and velocity and sea tide is considerable. The wind is specified as a spatially constant value for the entire domain and temporally variable values. The schematic presentation of the modelling framework is shown in Figure 4.

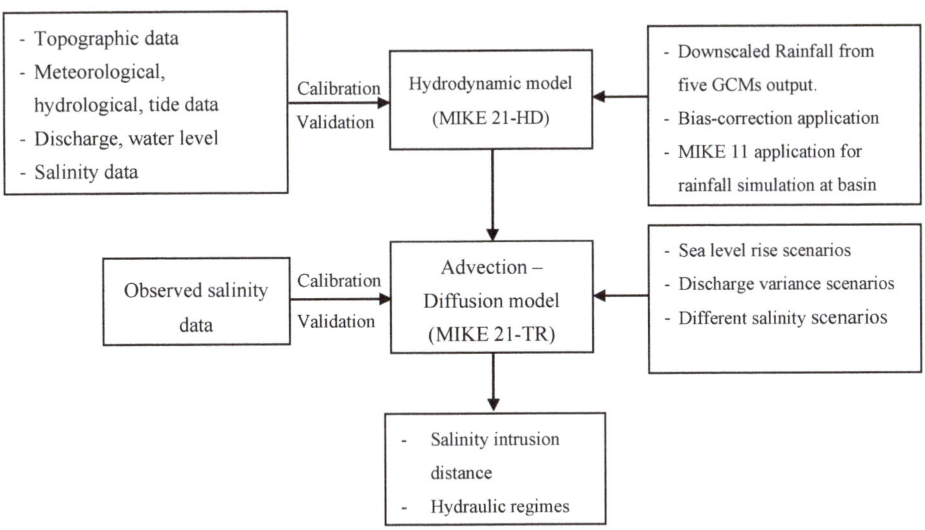

Figure 4. Schematic of the modelling framework for this study.

The MIKE 21 FM is based on a flexible mesh approach. The spatial discretization of the primitive equation is performed using a cell-centered finite volume method. The spatial domain is discretized by subdivision of the continuum into a non-overlapping cell. The mesh is divided by triangles or quadrilateral elements, as described in Figure 3.

3.5. Model Parameterization

The model parameters to be defined for the HD model are the roughness coefficient (n), or Manning number, and for the advection-diffusion model, the horizontal dispersion coefficient (D_h), eddy viscosity, Courant-Friedrich-Levy (CFL) and Smagorinsky coefficient. In order to realistically reproduce the physical phenomena of river mechanics, the Manning number is defined as a function of water depth (h) and bed river type, and can be calculated based on depth and drag coefficient [46,47]. In the MIKE 21 manual, the Manning number (M) was defined via the drag coefficient in Equation (1) with g as gravity constant.

$$C_d = \frac{g}{\left(Mh^{\frac{1}{6}}\right)^2} \tag{1}$$

The C_d can be defined through total water depth (h) and empirical drag coefficient (at 1 m above the bed) for different bottom types (from mud to gravel) by Equation (2):

$$C_d = \left(\frac{1}{0.32h}\right)^{1/7} \cdot C_{100} \tag{2}$$

C_{100} ranges between 0.022 ÷ 0.0047, and a table of full empirical drag coefficients can be found in Soulsby [47]. The Manning coefficients were available for each grid and varied from estuary to upstream river in the entire domain. The range of the Manning values varied from 20 to 40 $m^{1/3}/s$, and depend on grainsize (rippled sand-gravels) and water depth (shallow to deep water) during model calibration. We adjusted the Manning coefficient through the revision of C_{100} by defining the type of riverbed and seabed. The eddy viscosity was defined in a Smagorinsky formulation and was adjusted during calibration with a range 0.25–0.27, with an initial value of 0.28.

3.6. Boundary Conditions

For the delta-wide modelling with MIKE 11, we proposed four scenarios to cover future changes of upstream discharges, sea level rise and in-delta precipitation changes, and used these scenarios as boundary conditions for our modelling exercises (Table 4). More detailed information about rationales and designs of the scenarios can be found in Duong et al. [25]. Results from the Mike 11 modelling, particularly the discharge data at Can Tho, were then used as boundary conditions for the salinity intrusion simulation using MIKE 21. The changes in upstream discharges at Kratie are selected with a range of +10% to −20% relative to baseline discharge in 2011, based on literature review on projected future flow changes of Lauri et al. [48] and Hoang et al. [37]. Notably, both studies [37,48] did not consider the other anthropogenic factors such as irrigated land expansion, urbanization, and inter-basin water transfer. In addition, other studies [5,49] show different ranges of future hydrological changes due to differences in GCM outputs and climate change scenarios selection. Therefore, we extended the range of inflow changes at Kratie to capture possible future hydrological alterations in the upper Mekong Delta. For the climate change scenarios, we selected two RCPs for precipitation, namely RCP4.5 and RCP8.5.

Our sea level rise scenarios were obtained from the scenarios provided by the Ministry of Natural Resources and Environment. We selected an average of the predicted sea level rise, resulting in an increasing 23 cm between 2030 and 2040 and 35 cm between 2050 and 2065 [50]. For the MIKE 21 simulation, the hourly discharges at the Can Tho station were taken from the MIKE 11 simulation with four distinct time series of flows. The predicted sea levels were calculated linearly based on the averaged multi-annual tidal magnitudes from the global tidal model [51]. The seawater density was assumed to be constant and the salinity level at sea was predicted to remain at 35.0 PSU in the future condition. There are two approaches to obtaining the sea level rise condition. One is to impose an instantaneous elevation of sea level at coastal sea boundaries by applying linear adjustment of the mean sea level from the long observed sea level; the other is to run the hydrodynamic model over a long period to obtain predicted sea level. In this study, we obtained the sea level rise projection from the Vietnamese Ministry of Natural Resources and Environment (MONRE) and applied linear adjustment for predicted sea level rise during the 2036–2065 period.

The boundary condition for the MIKE 21 modelling includes the upstream hourly discharge at Can Tho station and tidal magnitude at downstream boundaries in the coast (Figure 1). To obtain salinity boundary conditions under future sea level rise and to reduce the influence of the boundary conditions on the in-delta modelling, a larger offshore domain that covers Hau estuary and adjacent shelf area was used. Secondly, five river branches connecting to the Hau River include Mang Thit, Rach Mop, Cau Quan, Nga Bay and Dai Ngai, were simulated as the sources and sinks. Thirdly, wind speeds and directions data were collected at the Vung Tau hydrological station and applied for the entire domain. Finally, time series of water level at My Thanh, Dai Ngai and Tran De stations were measured during 2010–2011 for model calibration and validation (Table 2). For salinity at downstream boundaries, we set up the salinity concentration as 35.0 PSU and assumed the salinity in upstream to be 0.1 PSU.

Table 4. Four scenarios of changing of discharges, sea level rise and precipitation.

Scenarios	Sea Level Rise (cm)	Changes Upstream Discharge	Precipitation Scenarios
Scen. 1	23	+10%	RCP 4.5, 8.5
Scen. 2	23	−10%	RCP 4.5, 8.5
Scen. 3	35	−15%	RCP 4.5, 8.5
Scen. 4	35	−20%	RCP 4.5, 8.5

3.7. Computation of Flushing Time

The flushing time can be determined by the freshwater fraction approach [14,52], which can be determined from the salinity distributions. This technique provides an estimation of the time scale over which contaminants and/or other material (saltwater in this study) released in the estuary are removed from the system. Using the freshwater fraction method, the flushing time (T_f) in an estuary can be expressed as Equation (3):

$$T_f = \frac{F}{Q} = \frac{\int_{vol} f \cdot d(V)}{Q} \quad (3)$$

where F is the accumulated freshwater volume in the estuary, which can be calculated by integrating the freshwater volume $d(V)$ in all the sub-divided model grids over a period of time. In estuaries with unsteady river flow and tidal variations, F and Q are the approximate average freshwater volume and average freshwater input, respectively, over several tidal cycles for a period of time. The term "f" is the freshwater content or the freshwater fraction, which is described by Equation (4):

$$f = \frac{S_o - S}{S_o} \quad (4)$$

where S_o is the salinity in the ocean and S is the salinity at the study location.

4. Model Calibration and Validation

4.1. Model Calibration

The MIKE 21 FM-HD and TR models were first calibrated with the observed averaged water level and salinity concentration at the middle cross section of the Hau River's estuaries. The calibration period was from January to December 2011 for hydrodynamic simulation, and six months from January to June 2011 for salinity simulation. The water levels measured at My Thanh and Dai Ngai were used to calibrate the model. In the inlet of the model at Can Tho station, initial salinity values were set at 0.1 PSU (‰). In the outlet of model (offshore), the salinity concentration was set at 35 PSU.

Several parameters of the HD and TR modules of MIKE 21 were adjusted to improve model performance. The calibration process aims to match simulated results and observed data, including water level and salinity concentration in different locations, by changing the Manning number and CFL in MIKE 21 HD and TR. In the simulations we applied a value of 0.8 for CFL-number and Smagorinsky coefficient within the range 0.25 and 0.27. Model calibration results have achieved the realistic results and indicated that the selected parameters were reasonable, as shown in Figure 5. To compare the observed and simulated water levels and salinity concentrations, the latter were taken at the center point of the cross section, as mean values for comparing with observed values. The performance indices include mean absolute error (MAE), root mean square error (RMSE) and correlation coefficients (R). The comparison between modelling results and observed salinity data also reveals a strong correlation and an excellent prediction, with the R-value higher than 0.8 (Table 5).

Figure 6 compares the observed and simulated salinity concentrations at the Tran De and Dai Ngai stations. The simulated salinity time series compared favorably with the discrete salinity measurements at the two aforementioned stations. Overall, the model reflected the large dynamic variation of salinity between 0 and 30 PSU over a tidal cycle, with decreasing values of mean salinity as freshwater discharge increased.

Table 5. Statistical performance of calibration and validation at My Thanh and Dai Ngai stations.

	Calibration (2011)			Validation (2010)		
Stations	**Root Mean Square Error (RMSE) (m)**	**Mean Absolute Error (MAE) (m)**	**R**	**RMSE**	**MAE**	**R**
My Thanh	0.29	0.23	0.923	0.31	0.26	0.913
Dai Ngai	0.13	0.11	0.997	0.16	0.14	0.967

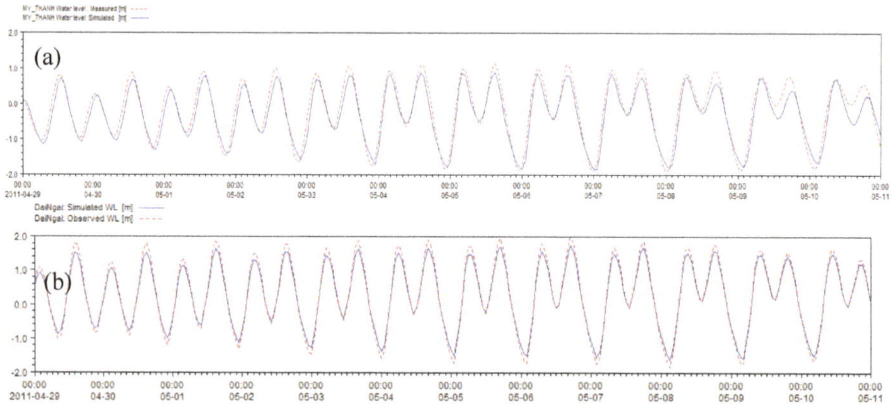

Figure 5. Calibration of water level at My Thanh (**a**) and Dai Ngai station (**b**).

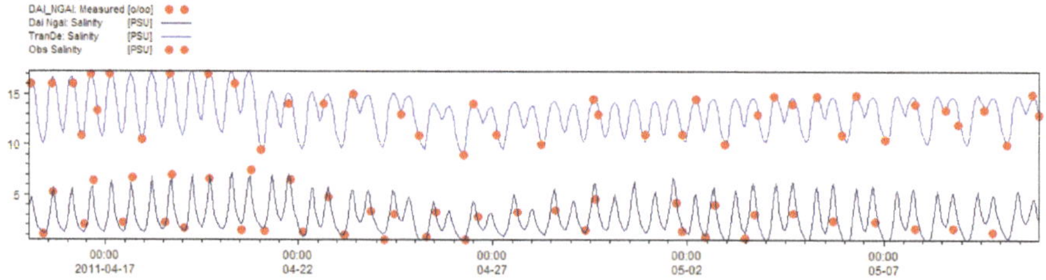

Figure 6. Calibration of salinity concentration at Tran De and Dai Ngai stations.

4.2. Model Validation

The MIKE 21 model was further validated with the calibrated parameters, focusing on water levels and salinity concentration. The time series data of hourly discharge at the Can Tho station in 2010 was used as upstream boundary conditions to drive the model simulations. The hourly tidal values taken from a global tide model were adopted as a forcing function at the coastal sea boundaries. The hourly tidal data and daily freshwater discharges were collected from SIWRR and the Southern Regional Hydro-Meteorological Center (SRHMC). The water level and salinity concentrations from Tran De and Dai Ngai stations were employed to evaluate the model. The comparison of observed data and simulated results for water level and salinity concentration is verified to check the model's performance.

Figure 7 compares the simulated water levels and observed data with the time series at My Thanh and Dai Ngai during the period between 29 April and 11 May 2010. In general, the modelling results show realistically simulated water level variations. The comparison demonstrates the model's capability to reproduce the water levels, even under large variations of daily freshwater influx from the upstream Can Tho station. To compare the observed and simulated water levels and salinity concentrations, the simulated water level and salinity were taken at the center point of the cross section, as mean values for comparing with observed values. Overall, the model satisfactorily simulated the water level at My Thanh and Dai Ngai on the Hau River. The calibrated model parameters were, therefore, adopted for our modelling exercises and scenario analyses.

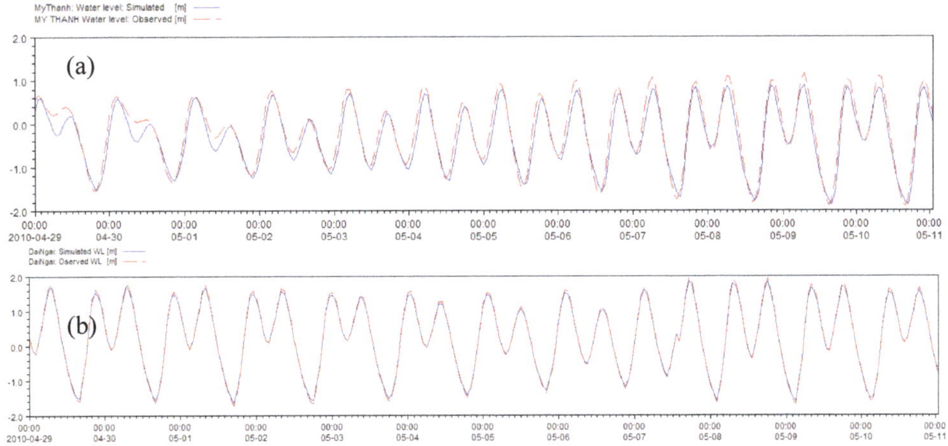

Figure 7. Validation of water level at My Thanh (**a**) and Dai Ngai (**b**) stations.

5. Results and Discussion

5.1. Changes in Salinity Intrusion

Simulating the spatial variations of salinity in the estuary and further upstream of the Hau River shows the detailed changes in the salinity dynamics under different discharges, sea level rise (SLR) and rainfall scenarios. The discharge at Can Tho varied significantly throughout the year, with a maximum discharge less than 20,000 m^3/s in the wet season and a minimum discharge of $-15{,}000$ m^3/s (Figure 8). The inverse flow direction in the dry season is caused by the tidal flow. Discharges at Can Tho were influenced not only by the upstream flow from Tan Chau, Vam Nao, but also by tidal regimes from the East Sea. The tidal feature in South China Sea is semidiurnal asymmetry; the peak spring tide reaches 3.0 m between December and January every year and reaches its lowest levels from June to August with a variability of around 0.5 m.

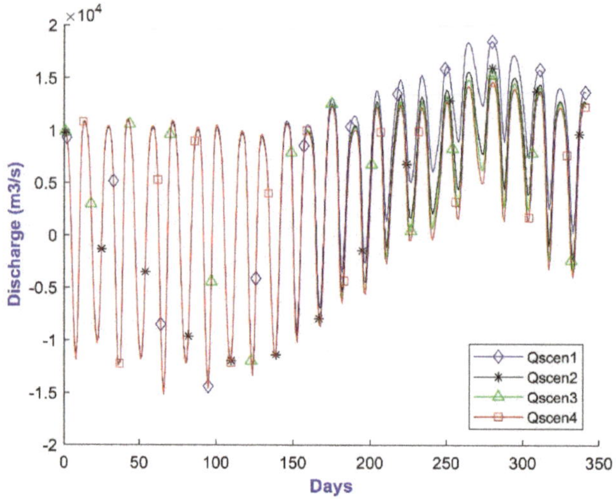

Figure 8. River discharges at Can Tho simulated from the MIKE 11 model for four scenarios.

In order to quantify the spatial variations in the salinity concentrations along the river, the minimum and maximum difference in salinity concentrations between the baseline and four scenarios was calculated and is presented in Table 6. Figures 9 and 10 presents the spatial distributions of maximum and average salinity levels along the Hau River estuary under the four scenarios, respectively. The result shows radical changes in the salinity levels across the entire estuary. Compared to the reference isohaline of 4.0 PSU, the salinity profile with 4.0 PSU moves farther upstream by 48.55, 49.13, 49.16 and 49.18 km from Scenario 1 to Scenario 4, respectively (Table 7). The relative changes compared to the baseline are 3.29, 3.87, 3.90, and 3.92 km for the four scenarios, respectively (Table 8). Scenario 4 shows the farthest salinity intrusion, which is explained by strong upstream discharge reduction and substantial sea level rise of 35 cm (Figure 11). Similarly, Scenario 1 shows weaker salinity intrusion from the sea compared to other scenarios. This is explained by slight increases in the upstream discharge and the sea level rise of 25 cm. Figure 12 presents the salinity distribution under four scenarios at different points along the river: from the river mouth (My Thanh station) to upstream (Can Tho station) in spring tide (Figure 12a) and neap tide simulations (Figure 12b). The mean values are taken at the middle point of the cross-section. Figure 9 shows the difference in salinity between the four scenarios and the distance of salinity intrusion from the river mouth.

Here we discuss our results for salinization modelling in view of the relevant studies on this topic. Smajgl et al. [53] studied the effects of a wide range of driving factors on salinization, including land use changes, sea level rise of 30 cm, development of all proposed upstream reservoir and irrigation, and an increasing number of dry years. The results stated that the isohaline of 4.0 PSU is relocated throughout the Hau River, with distance from the river mouth reaching approximately 70 km. Considerable differences in the 4.0 PSU isohaline between this study and Smajgl et al. [53] can be attributed to differences between the scenario setups and boundary conditions in the two studies. Nguyen and Savenije [32] tested analytical solutions of salinity simulation in the Mekong estuaries based on observed data in 2005 for the Co Chien, Cung Hau and Hau estuaries. The results showed that salinity intrusion distances from the river mouth were 41 km and 23 km for spring tide and neap tide during the dry season in 2005, respectively. Furthermore, the finding by Trieu and Phong [54] concluded that the salinity intrusion of 1.0 PSU in Hau River was approximately 55–60 km for the dry season of 2010. This projection is higher than our results for all scenarios. All in all, results from this study show a strong dependency of salinization on upstream inflow and the changing tidal dynamics under sea level rise. Such a mechanism is also commonly reported in large river deltas of the world, including the Bangladesh Delta [55,56], and the Dutch Delta [57].

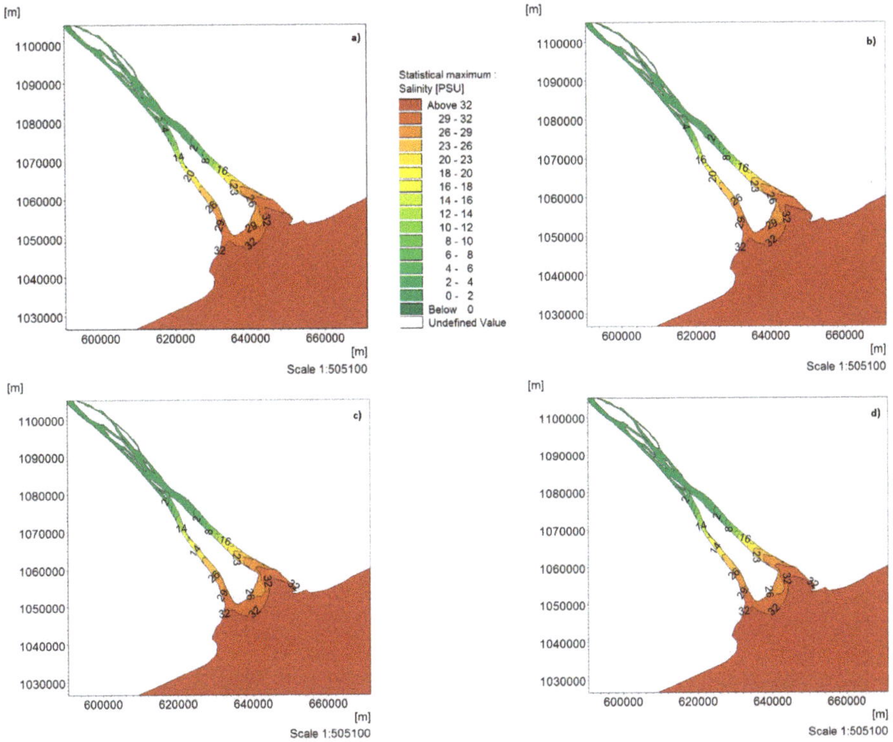

Figure 9. Spatial distribution of maximum salinity level across the modelling domain under Scenario 1 (**a**), Scenario 2 (**b**), Scenario 3 (**c**) and Scenario 4 (**d**).

Table 6. Maximum and minimum salinity concentration at Dai Ngai and Tran De stations.

Stations	Maximum and Minimum Salinity (Practical Salinity Units (PSU))							
	Scenario 1		Scenario 2		Scenario 3		Scenario 4	
	Maximum	Minimum	Maximum	Minimum	Maximum	Minimum	Maximum	Minimum
Dai Ngai (1st June)	5.44	0.076	5.74	0.071	5.83	0.066	5.91	0.054
Tran De (1st June)	29.41	25.02	30.39	25.31	30.27	25.20	30.31	25.44

Table 7. Salinity intrusion length for thresholds of 1.0 PSU and 4.0 PSU for four scenarios, under spring tide and neap tide.

Scenarios	Distance from the Mouth (km)			
	Spring Tide		Neap Tide	
	1.0 PSU	4.0 PSU	1.0 PSU	4.0 PSU
Baseline	45.26	41.46	25.49	20.13
Scenario 1	48.55	43.58	29.85	22.03
Scenario 2	49.13	44.05	30.44	23.01
Scenario 3	49.16	44.17	30.07	22.48
Scenario 4	49.18	44.27	30.14	22.59

Table 8. The relative change of salinity intrusion length for four scenarios compared to baseline.

Scenarios	Relative Changes of Saline Intrusion (km)			
	Spring Tide		Neap Tide	
	1.0 PSU	4.0 PSU	1.0 PSU	4.0 PSU
Scenario 1	3.29	2.12	4.36	1.90
Scenario 2	3.87	2.59	4.95	2.88
Scenario 3	3.90	2.71	4.58	2.35
Scenario 4	3.92	2.81	4.65	2.46

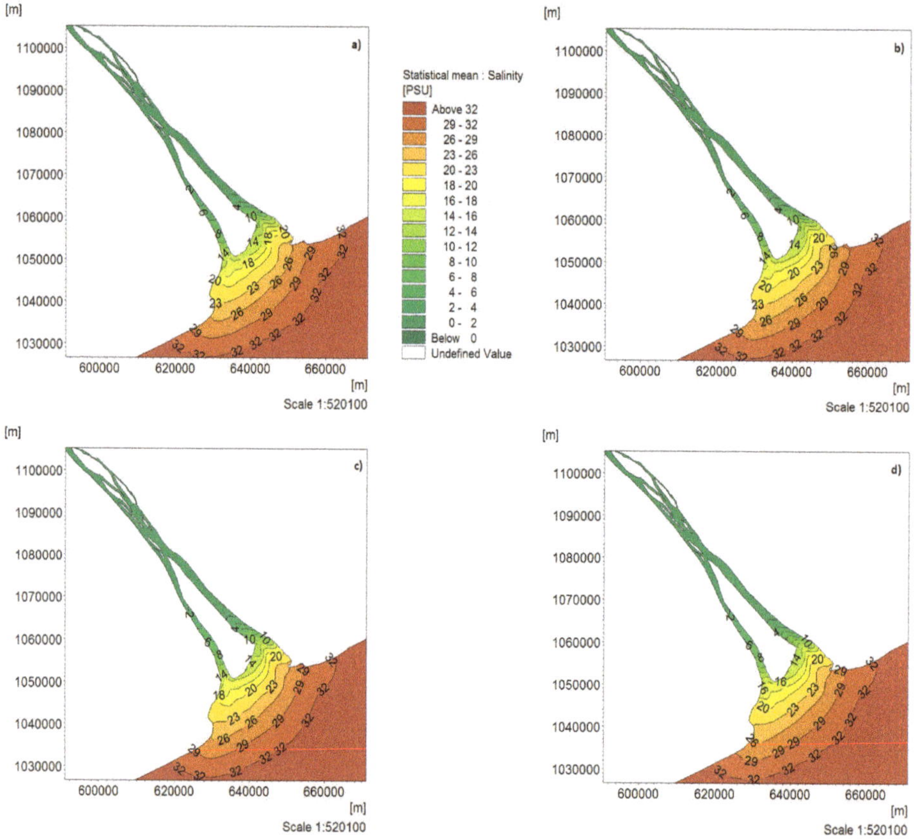

Figure 10. Spatial distribution of average salinity concentration across the modelling domain under Scenario 1 (**a**), Scenario 2 (**b**), Scenario 3 (**c**) and Scenario 4 (**d**).

5.2. Changes in Salinity Intrusion Length

We further assessed the changes in salinity intrusion length under the four future scenarios. The salinity intrusion length is defined as the distance from the estuary mouth to the location upstream with mean salinity at the cross section. There are three types of intrusion length: intrusion length at low water slack, intrusion length at high water slack, and tidal average intrusion length, which is considered to be an average of low and high [58–60]. The salinity intrusion length is defined in this study as the distance from the Hau River mouth to the upstream limit location where the bottom salinity level drops down to a certain threshold, e.g., 1 PSU or 4 PSU. Figure 12 presents salinity distribution along the river for four scenarios, under spring tide and neap tide.

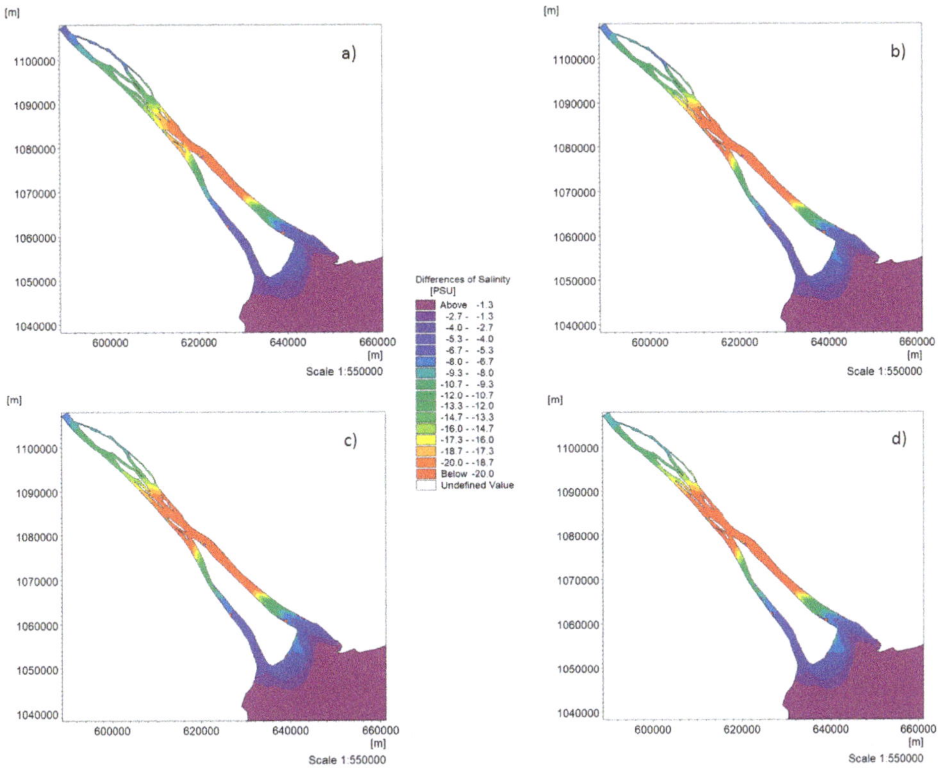

Figure 11. Differences of maximum salinity concentration between future scenarios and baseline, for Scenario 1 (**a**), Scenario 2 (**b**), Scenario 3 (**c**) and Scenario 4 (**d**).

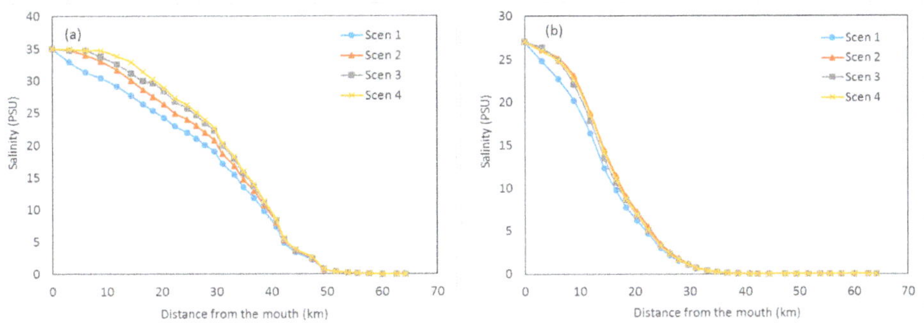

(**a**) (**b**)

Figure 12. Salinity distribution at center line of cross section along the river from the mouth to upstream during spring tide (**a**) and neap tide (**b**).

Table 7 shows that the salinity intrusion lengths under spring tide are larger than those under neap tide. By comparing the baseline to the four scenarios, it can be seen that the intrusion lengths for all scenarios are consistently farther than that of the baseline. In Scenario 1, river discharge increases by 10% at Kratie, and therefore the inflow at Can Tho is also slightly higher than the other scenarios. This explains the difference in salinity intrusion length between the baseline and Scenario 1. For Scenarios 2, 3 and 4, (with sea level rises of 23 cm, 35 cm and 35 cm, respectively) the differences in salinity intrusion lengths between the baseline and these three scenarios are relatively small. In essence, the results show that upstream discharge changes do not substantially affect salinity intrusion lengths, in contrast to the more dominant impacts of tidal regime.

Under sea level rise scenarios, salinity intrusion length in the Hau River would increase between 3.29 and 3.92 km (for 1.0 PSU salinity level) compared to the baseline in the spring tide condition. During neap tide, this length ranges between 4.36 and 4.65 km for the 1.0 PSU salinity level (Table 8). Scenario 4 shows the longest salinity intrusion length for spring and neap tide (Table 7) for both 1.0 PSU and 4.0 PSU isohaline. Such variation in the magnitude of salinity intrusion length is caused by changes in the stratification of the estuary. As the sea level rises, the river depth increases and the horizontal gradient of salinity changes, resulting in increased estuarine circulation. The model simulations indicate that the location of 4 PSU isohaline would migrate up from 43.58 to 44.27 km if sea level rises from 23 to 35 cm and the upstream discharges change from +10% to −20% at Kratie. Our findings about increasing salinity intrusion length are in line with observed data, which confirms that when river discharge reduces in the Hau River, saline water moves upstream to Can Tho [54,61]. Any increase in the magnitude or duration of salinity intrusion at the location of the irrigation water intake would likely affect crop yield and aquaculture in Mekong Delta.

5.3. Changes in Flushing Time

To examine the effect of river discharge changes on the temporal dynamics of salinity intrusion in the Hau River, we calculate flushing time with varying discharge conditions for baseline and four future scenarios. Flushing time is defined as the time required to replace the existing saltwater in the estuary using inflows at certain discharge levels [14,62]. To calculate the flushing time in the estuary, different freshwater discharges were selected based on maximum and minimum discharges from the dry season in 2011, with a range of 400 to 4500 m^3/s. The simulated flushing times for the baseline and the four scenarios were calculated and are presented for different discharge values in Figure 13.

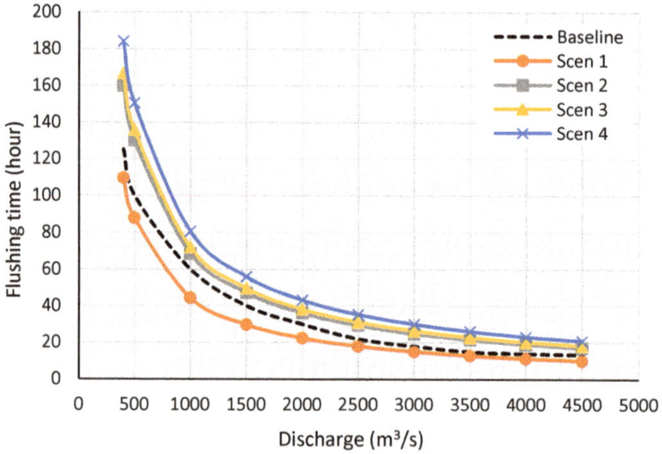

Figure 13. Estimated flushing time under varying discharge levels (Q, m^3/s) for four scenarios.

The flushing time varies considerably under all considered scenarios, ranging between 10 h (at discharge 4500 m^3/s) and 109 h (at discharge 400 m^3/s). For the most extreme scenario (i.e., Scenario 4), flushing time varies within a range from 20.7 h to 182 h. The results also indicate that, for high flow, the flushing time under Scenario 1 is lower than under baseline conditions, while Scenarios 2, 3 and

4 consistently exhibit longer flushing time than the baseline. Longer flushing time caused by reducing discharges means that it would require a higher volume of freshwater to push saltwater back to the river mouth in the future. Reductions in river discharge also cause the salinity boundary to move further upstream (Figure 11). Salinity deceases substantially when river discharge increases, and as a result the salinity intrusion boundary migrates farther downstream. As freshwater discharge increases from 400 to 4500 m^3/s with different sea levels, the flushing time reduces significantly from 87.72 h to 10 h for Scenario 1, and 150 h to 20.7 h for Scenario 4.

5.4. Uncertainties, Limitations and Further Research

The hydraulic models, including 1D-MIKE 11 and 2D-MIKE 21 HD and TR used in this study, entail limitations and assumptions. These limitations and assumptions exist in both the data and modelling approach. The major data used in this study were predicted river discharge, rainfall scenarios under climate change, and sea level rise. The climate change and sea level rise scenarios used in this study were obtained from different sources, such as GCMs outputs, SRHMC and SIWRR, and therefore our results might be subjected to errors caused by data inconsistencies. However, we implemented quality checks on our input data and controlled for inconsistencies. Additionally, we acknowledge that climate change is a non-stationary and dynamic process; while the projected salinity intrusion from upstream discharge changes, rainfall and sea level rise alterations were treated in this study as part of a steady-state system. That means we assume that the relationship between these driving factors are also valid in the future and did not include their interrelations in the simulations. This assumption could lead to biases in future discharge and rainfall estimates, which in turn result in uncertainties in the simulated salinity intrusion. However, we think that the hydraulic and salinity dynamics in the Hau River of the Mekong Delta are primarily driven by changes in the considered driving factors and their interrelations play marginal roles. Therefore, the developed modelling approaches are useful and applicable for our modelling purposes.

Furthermore, we employed a two-step modelling procedure (see Section 3—Methodology and Model setup) for assessing changes in salinity intrusion in the Mekong Delta. A 1D model was applied to simulate delta-wide hydrodynamic conditions (i.e., discharge and water level), thereby providing boundary conditions for the more detailed hydrodynamic and salinity simulation with the 2D-MIKE 21 model. While the current data availability and computational capacity does not allow for detailed 2D simulation for the whole Mekong Delta domain, we suggest gradually increasing the domain size to cover larger areas in future studies. Last but not least, while our calibrated parameter set yields realistic simulation results for the current state of the delta system (i.e., 2010–2011), we did not consider changing delta settings and its hydraulic characteristics in the future. Therefore, we suggest focusing on this topic in future studies.

6. Conclusions

This study combined 1D-MIKE 11 and 2D-MIKE 21 hydrodynamic (HD) and salt transport models (TR) to simulate the hydrodynamics and salinity distributions in the Hau (Bassac) River estuary of the Mekong Delta, southern Vietnam. The model was calibrated and verified using observational water level, salinity distribution and tidal data from 2010 to 2011. The model simulation results agree well with the observed data during calibration and validation periods. The calibrated model was used to simulate future salinity intrusion, focusing on the potential impacts of upstream discharges, rainfall variabilities and sea level rise on salinity intrusion length and saltwater flushing time. The simulation results indicate that a combination of upstream discharge reductions, rainfall changes and rising sea level will substantially exacerbate salinity intrusion in the Mekong Delta. By the 2036–2065 period, salinity intrusion moves farther upstream by between 4855 and 4918 m, depending on the scenario considered. As a result, the flushing time required to re-establish freshwater conditions in Hau River will increase in the future. While flushing time under baseline ranges between 125 h and 13.5 h, it tends to increase in the future, ranging between 180 h and 10 h. Increasing salinization in the future will make it more difficult to re-establish the freshwater condition in the estuary. In particular, the flushing time required to replace saltwater with freshwater at the estuaries tends to increase between 7.27 h for maximum discharge of 4500 m^3/s and 58.95 h for discharge of 400 m^3/s under the most extreme scenario. Increasing salinity intrusion under upstream discharge changes, rainfall alterations and sea level rise, will likely have serious consequences for crop production, freshwater supplies and freshwater ecosystems, therefore requiring timely adaptation responses.

Author Contributions: D.T.A. designed the study, processed and analyzed the data, developed the models, interpreted the results and wrote the paper. L.P.H. provided data, assisted in the data analyses and drafting the manuscript. The study has been carried out under the supervision of M.D.B. and P.R., who contributed to the model development stage with theoretical considerations and practical guidance, assisted in the interpretations and integration of the results and helped in preparation of this paper with proof reading and corrections.

Funding: This research was funded by the CUOMO Foundation and the IPCC Scholarship Program, grant number IPCC2013-2017.

Acknowledgments: The contents of this paper are solely the liability of the authors and under no circumstances may be considered as a reflection of the position of the Cuomo Foundation and/or the IPCC. The German Research Foundation (DFG) and the Technical University of Munich (TUM) supported this work in the framework of the Open Access Publishing Program. We greatly appreciate their help. Wholehearted thanks are given to DHI Vietnam, Hanoi water resource University, base 2, Southern Regional Hydro-meteorological Center, National meteorological center for providing all necessary data. Special thanks to Dang Quang Thanh and Nguyen Van Song on for valuable support and comments to complete this paper. We thank the editor and three anonymous reviewers for their constructive comments, which helped us to improve the manuscript.

Conflicts of Interest: The authors declare no potential conflict of interest.

References

1. Gupta, A.D. *Challenges and Opportunity for Integrated Water Resources Management in Mekong River Basin*; Asian Institute of Technology: Bangkok, Thailand, 2005.
2. World Bank. The Impact of Sea Level Rise on Developing Countries: A Comparative Analysis. World Bank Policy Research Working Paper 4136. 2007. Available online: http://go.worldbank.org/775APZH5K0 (accessed on 15 March 2017).
3. Todd, M.C.; Taylor, R.G.; Osborn, T.J.; Kingston, D.G.; Arnell, N.W.; Gosling, S.N. Uncertainty in climate change impacts on basin-scale freshwater resources—Preface to the special issue: The QUEST-GSI methodology and synthesis of results. *Hydrol. Earth Syst. Sci.* **2011**, *15*, 1035–1046. [CrossRef]
4. Hoanh, C.T.; Jirayoot, K.; Lacombe, G.; Srinetr, V. *Impacts of Climate Change and Development on Mekong Flow Regime, First Assessment—2009*; MRC Technical Paper No. 29; Mekong River Commission: Vientiane, Laos, 2010.
5. Kingston, D.G.; Thompson, J.R.; Kite, G. Uncertainty in climate change projection of discharge for the Mekong River Basin. *Hydrol. Earth Syst. Sci.* **2011**, *15*, 1459–1471. [CrossRef]
6. Dasgupta, S.; Laplante, B.; Meisner, C.; Wheeler, D.; Yan, J. *The Impact of Sea Level Rise on Developing Countries: A Comparative Analysis*; World Bank Policy Research Working; The World Bank: Washington, DC, USA, 2007; p. 4136.
7. ICEM. *Strategic Environmental Assessment of Hydropower on the Mekong Mainstream—Final Report*; Mekong River Commission: Vientiane, Laos, 2010.
8. IPCC. *Climate Change 2007: Synthesis Report Contribution of Working Groups I, II and III to the Fourth Assessment Report of the Intergovernmental Panel on Climate Change*; IPCC: Geneva, Switzerland, 2007.
9. Varis, O.; Kummu, M.; Salmivaara, A. Ten major rivers in monsoon Asia-Pacific: An assessment of vulnerability. *Appl. Geogr.* **2012**, *32*, 441–454. [CrossRef]
10. Keskinen, M.; Chinvanno, S.; Kummu, M.; Nuorteva, P.; Snidvongs, A.; Varis, O.; Västilä, K. Climate change and water resources in the Lower Mekong River Basin: Putting adaptation into the context. *J. Water Clim. Chang.* **2010**, *1*, 103–117. [CrossRef]
11. Lebel, L.; Garden, P.; Imamura, M. The politics of scale, position, and place in the governance of water resources in the Mekong Region. *Ecol. Soc.* **2005**, *10*. Available online: http://www.ecologyandsociety.org/vol10/iss2/art18/ (accessed on 10 February 2016). [CrossRef]
12. Piman, T.; Lennaerts, T.; Southalack, P. Assessment of hydrological changes in the lower Mekong Basin from Basin-Wide development scenarios. *Hydrol. Process.* **2013**, *27*, 2115–2125. [CrossRef]

13. Västilä, K.; Kummu, M.; Sangmanee, C.; Chinvanno, S. Modelling climate change impacts on the flood pulse in the Lower Mekong floodplains. *J. Water Clim. Chang.* **2010**, *1*, 67–86. [CrossRef]
14. Dyer, K.R. *Estuaries: A Physical Introduction*, 2nd ed.; John Wiley and Sons Ltd.: Chichester, UK, 1997; 195p.
15. Fei, Y.; Xu, L.; Du, P.; Guan, Q.; Kang, X.; Xiao, W. Analysis of characteristics of time-frequency correlations between runoff, tidal range and salinity in the Changjiang Estuary. *Acta Oceanol. Sin.* **2011**, *30*, 84–93. [CrossRef]
16. Pritchard, D.W. Estuarine circulation patterns. *Proc. Am. Soc. Civ. Eng.* **1955**, *81*, 717.
17. Cameron, W.M.; Pritchard, D.W. Estuaries, in the Sea: Ideas and Observations on Progress in the Study of the Seas. In *The Composition of Sea-Water, Comparative and Descriptive Oceanography*; Hill, M.N., Ed.; Wiley-Interscience: New York, NY, USA, 1963; Volume 2, pp. 306–324.
18. Nijssen, B.; O'Donnell, G.M.; Hamlet, A.F.; Lettenmaier, D.P. Hydrologic Sensitivity of Global Rivers to Climate Change. *Clim. Chang.* **2001**, *50*, 143–175. [CrossRef]
19. Hoanh, C.T.; Tuong, T.P.; Gallop, K.M.; Gowing, J.W.; Kam, S.P.; Khiem, N.T.; Phong, N.D. Livelihood impacts of water policy changes: Evidence from a coastal area of the Mekong River Delta. *J. Water Policy* **2003**, *5*, 475–488. [CrossRef]
20. Khang, D.K.; Kotera, A.; Sakamoto, T.; Yokozawa, M. Sensitivity of Salinity Intrusion to Sea Level Rise and River Flow Change in Vietnamese Mekong Delta-Impacts on Availability of Irrigation Water for Rice Cropping. *J. Agric. Meteorol.* **2008**, *64*, 167–176. [CrossRef]
21. Wassmann, R.; Hien, N.X.; Hoanh, C.T.; Tuong, T.P. Sea level rise affecting the Vietnamese Mekong Delta: Water elevation in the flood season and implications for rice production. *Clim. Chang.* **2004**, *66*, 89–107. [CrossRef]
22. Le, T.V.H.; Nguyen, H.N.; Wolanski, E.; Tran, T.C.; Haruyama, S. The combined impact on the flooding in Vietnam's Mekong River delta of local man-made structures, sea level rise, and dams upstream in the river catchment. *Estuar. Coast. Shelf Sci.* **2007**, *71*, 110–116. [CrossRef]
23. Dinh, N.Q.; Balica, S.; Popescu, I.; Jonoski, A. Climate change impact on flood hazard, vulnerability and risk of the Long Xuyen Quadrangle in the Mekong Delta. *Int. J. River Basin Manag.* **2012**, *10*, 103–120. [CrossRef]
24. Dat., T.Q.; Kanchit, L.; Thares, S.; Trung, N.H. Modeling the influence of river discharge and sea level rise on salinity intrusion in Mekong delta. In Proceeding of the Environment Asia International Conference, Bangkok, Thailand, 22–25 March 2011.
25. Duong, T.A.; Hoang, P.L.; Bui, M.D.; Rutschmann, P. Modeling seasonal flows alterations in the Vietnamese Mekong Delta under upstream discharge changes, rainfall changes and sea level rise. *J. River Basin Manag.* **2018**, in press.
26. Van, P.D.T.; Popescu, I.; van Griensven, A.; Solomatine, D.P.; Trung, N.H.; Green, A. A study of the climate change impacts on fluvial flood propagation in the Vietnamese Mekong Delta. *Hydrol. Earth Syst. Sci.* **2012**, *16*, 4637–4649. [CrossRef]
27. Tuan, L.A.; Hoanh, C.T.; Miller, F.; Sinh, B.T. Floods and salinity management in the Mekong delta, Vietnam. In *Challenges to Sustainable Development in the Mekong Delta: Regional and National Policy Issues and Research Needs*; Be, T.T., Sinh, B.T., Miller, F., Eds.; The Sustainable Mekong Research Network: Bangkok, Thailand, 2007; pp. 15–68.
28. MRC. *Overview of the Hydrology of the Mekong Basin, Tech. Rep*; Mekong River Commission: Vientiane, Laos, 2005.
29. Trung, N.H.; Tuan, L.A.; Trieu, T.T.; Bastakoti, R.; Lebel, L. Multi-level governance and adaptation to floods in the Mekong delta. In *Governing the Mekong. Engaging in the Politics of Knowledge Centre*; Daniel, R., Lebel, L., Manorom, K., Eds.; Strategic Information and Research Development: Petaling Jaya, Malaysia, 2013; pp. 111–126.
30. De, N.N. *Farmers, Agriculture and Rural Development in the Mekong Delta of Vietnam*; Education Publishing House: Hanoi, Vietnam, 2005; pp. 50, 51, 58.
31. Thoi, N.H.; Gupta, A.D. Assessment of Water Resources and Salinity Intrusion in the Mekong Delta. *Water Int.* **2001**, *26*, 86–95. [CrossRef]
32. Nguyen, A.D.; Savenije, H.H. Salt intrusion in multi-channel estuaries: A case study in the Mekong Delta, Vietnam. *Hydrol. Earth Syst. Sci.* **2006**, *10*, 743–754. [CrossRef]
33. Lenderink, G.; Buishand, A.; Deursen, W.V. Estimates of future discharges of the river Rhine using two scenario methodologies: Direct versus delta approach. *Hydrol. Earth Syst. Sci.* **2007**, *11*, 1145–1159. [CrossRef]

34. Schmidli, J.; Frei, C.; Vidale, P.L. Downscaling from GCM precipitation: A benchmark for dynamical and statistical downscaling methods. *Int. J. Clim.* **2006**, *26*, 679–689. [CrossRef]
35. Sennikovs, J.; Bethers, U. Statistical downscaling method of regional climate model results for hydrological modelling. In Proceedings of the 18th World IMACS/MODSIM Congress, Cairns, Australia, 13–17 July 2009; Available online: http://mssanz.org.au/modsim09 (accessed on 20 July 2017).
36. Block, P.J.; Souza Filho, F.A.; Sun, L.; Kwon, H.H. A Streamflow Forecasting Framework using Multiple Climate and Hydrological Models. *J. Am. Water Resour. Assoc.* **2009**, *45*, 828–843. [CrossRef]
37. Hoang, L.P.; Lauri, H.; Kummu, M.; Koponen, J.; van Vliet, M.T.H.; Supit, I.; Leemans, R.; Kabat, P.; Ludwig, F. Mekong River flow and hydrological extremes under climate change. *Hydrol. Earth Syst. Sci.* **2016**, *20*, 3027–3041. [CrossRef]
38. Nguyen, C.T. Processes and Factors Controlling and Affecting the Retreat of Mangrove Shorelines in South Vietnam. Ph.D. Thesis, Kiel University, Kiel, Germany, 2012.
39. Vinh, V.D.; Ouillon, S.; Van Thao, N.; Ngoc Tien, N. Numerical Simulations of Suspended Sediment Dynamics Due to Seasonal Forcing in the Mekong Coastal Area. *Water* **2016**, *8*, 255. [CrossRef]
40. MARD. The Drought and Salinity Intrusion in the Mekong River Delta of Vietnam. Assessment Report. 2016. Available online: https://ccafs.cgiar.org/publications/ (accessed on 15 March 2018).
41. Abbott, M.B.; Ionescu, F. On the numerical computation of nearly-horizontal flows. *J. Hydraul. Res.* **1967**, *5*, 97–117. [CrossRef]
42. Shooshtari, M.M. *Principles of Flow in Open Channels*; Shahid Chamran University Press: Ahvaz, Iran, 2008; Volume 15, pp. 643–745.
43. Dung, N.V.; Merz, B.; B´ardossy, A.; Thang, T.D.; Apel, H. Multi-objective automatic calibration of hydrodynamic models utilizing inundation maps and gauge data. *Hydrol. Earth Syst. Sci.* **2011**, *15*, 1339–1354. [CrossRef]
44. Manh, N.V.; Dung, N.V.; Hung, N.N.; Merz, B.; Apel, H. Large-scale suspended sediment transport and sediment deposition in the Mekong Delta. *Hydrol. Earth Syst. Sci.* **2014**, *18*, 3033–3053. [CrossRef]
45. DHI. *Mike 21 Flow Model FM Hydrodynamic Module Scientific Documentation*; DHI Water and Environment: Hørsholm, Denmark, 2007.
46. Lambkin, D. *A Review of the Bed Roughness Variable in MIKE 21 FLOW MODEL FM, Hydrodynamic (HD) and Sediment Transport (ST) Modules*; University of Southampton: Southampton, UK, 2007.
47. Soulsby, R.L. *Dynamics of Marine Sand: A Manual for Practical Application*; Thomas Telford: London, UK, 1997; ISBN 0-7277-2584-X.
48. Lauri, H.; de Moel, H.; Ward, P.J.; Räsänen, T.A.; Keskinen, M.; Kummu, M. Future changes in Mekong River hydrology: Impact of climate change and reservoir operation on discharge. *Hydrol. Earth Syst. Sci.* **2012**, *16*, 4603–4619. [CrossRef]
49. Thomson, A.; Calvin, K.; Smith, S.; Kyle, G.P.; Volke, A.; Patel, P.; Delgado-Arias, S.; Bond-Lamberty, B.; Wise, M.; Clarke, L.; et al. RCP4.5: A pathway for stabilization of radiative forcing by 2100. *Clim. Chang.* **2011**, *109*, 77–94. [CrossRef]
50. MONRE. *Climate Changes and Sea Level Rise Scenarios for Vietnam*; Ministry Natural Resources Environment (Viet Nam): Hanoi, Vietnam, 2009; p. 34.
51. Andersen, O.B. The ANDERSEN-2006 Global Empirical Ocean Tide Model. 2006. Available online: http://ebookbrowse.com/andersen-ag06-tide-pdf-d351826746 (accessed on 20 May 2018).
52. Lauff, G.H. *Lyered Sediments of Tidal Flats, Beaches; Shelf Bottoms of the North Sea Estuaries*; Publication 83; American Association for the Advancement of Science: Washington, DC, USA, 1967.
53. Smajgl, A.; Toan, T.Q.; Nhan, D.K.; Ward, J.; Trung, N.H.; Tri, L.Q.; Tri, V.P.D.; Vu, P.T. Responding to rising sea levels in the Mekong Delta. *Nat. Clim. Chang.* **2015**, *5*, 167–174. [CrossRef]
54. Trieu, T.T.N.; Phong, N.T. The impact of climate change on salinity intrusion and Pangasius (Pangasianodon Hypophthalmus) farming in the Mekong Delta, Vietnam. *Aquac. Int.* **2014**, *23*, 523. [CrossRef]
55. Bhuiyan, M.J.A.N.; Dutta, D. Assessing impacts of sea level rise on river salinity in the Gorai river network, Bangladesh. *Estuar. Coast. Shelf Sci.* **2012**, *96*, 219–227. [CrossRef]
56. Nobi, N.; Gupta, A.D. Simulation of regional flow and salinity intrusion in an integrated stream-aquifer system in coastal region: Southwest region of Bangladesh. *Groundwater* **1997**, *35*, 786–796. [CrossRef]
57. Oude Essink, G.H.P.; Van Baaren, E.S.; De Louw, P.G. Effects of climate change on coastal groundwater systems: A modeling study in The Netherlands. *Water Resour. Res.* **2010**, *46*, 76. [CrossRef]

58. Van den Burgh, P. *Ontwikkeling van een Methode voor het Voorspellen van Zoutverdelingen in Estuaria, Kanalen en Zeeen*; Rijkwaterstaat: Assen, The Netherlands, 1972; pp. 10–72. (In Dutch)
59. Savenije, H.H.G. Predictive model for salt intrusion in estuaries. *J. Hydrol.* **1993**, *148*, 203–218. [CrossRef]
60. Prandle, D. Saline intrusion in partially mixed estuaries. *Estuar. Coast. Shelf Sci.* **2004**, *59*, 385–397. [CrossRef]
61. Takagi, H.; Tsurudome, C.; Thao, N.D.; Anh, L.T.; Ty, T.V.; Tri, P.V.D. Ocean tide modelling for urban flood risk assessment in the Mekong Delta. *Hydrol. Res. Lett.* **2016**, *10*, 21–26. [CrossRef]
62. Williams, B.L. Flushing time calculations for the Upper Waitemata Harbour, New Zealand. *N. Z. J. Mar. Freshw. Res.* **1986**, *20*, 455–465. [CrossRef]

water

MDPI

Article

Attenuation of Tides and Surges by Mangroves: Contrasting Case Studies from New Zealand

John M. Montgomery *, Karin R. Bryan, Erik M. Horstman and Julia C. Mullarney

Faculty of Science and Engineering, School of Science, University of Waikato, Private Bag 3105, Hamilton 3240, New Zealand; karin.bryan@waikato.ac.nz (K.R.B.); e.m.horstman@utwente.nl (E.M.H.); julia.mullarney@waikato.ac.nz (J.C.M.)

* Correspondence: jmontgom31@gmail.com; Tel.: +64-021-036-7446

Received: 29 June 2018; Accepted: 17 August 2018; Published: 23 August 2018

Abstract: Mangroves have been suggested as an eco-defense strategy to dissipate tsunamis, storm surges, and king tides. As such, efforts have increased to replant forests along coasts that are vulnerable to flooding. The leafy canopies, stems, and aboveground root structures of mangroves limit water exchange across a forest, reducing flood amplitudes. The attenuation of long waves in mangroves was measured using cross-shore transects of pressure sensors in two contrasting environments in New Zealand, both characterized by mono-specific cultures of grey mangroves (*Avicennia marina*) and approximate cross-shore widths of 1 km. The first site, in the Firth of Thames, was characterized by mangrove trees with heights between 0.5 and 3 m, and pneumatophore roots with an average height of 0.2 m, and no substantial tidal drainage channels. Attenuation was measured during storm surge conditions. In this environment, the tidal and surge currents had no alternative pathway than to be forced into the high-drag mangrove vegetation. Observations showed that much of the dissipation occurred at the seaward fringe of the forest, with an average attenuation rate of 0.24 m/km across the forest width. The second site, in Tauranga harbor, was characterized by shorter mangroves between 0.3 and 1.2 m in height and deeply incised drainage channels. No attenuation of the flood tidal wave across the mangrove forest was measurable. Instead, flow preferentially propagated along the unvegetated low-drag channels, reaching the back of the forest much more efficiently than in the Firth of Thames. Our observations from sites with the same vegetation type suggest that mangrove properties are important to long wave dissipation only if water transport through the vegetation is a dominant mechanism of fluid transport. Therefore, realistic predictions of potential coastal protection should be made prior to extensive replanting efforts.

Keywords: eco-defense; coastal defense; coastal morphodynamics; mangroves; flood attenuation; natural defense

1. Introduction

Mangroves are the dominant species of vegetation in many tropical and sub-tropical intertidal environments. These salt-tolerant trees provide a valuable habitat for a range of animal species, reduce hydrodynamic forces, promote sedimentation, and provide protection from floods [1]. Additionally, mangroves are significantly more efficient than many terrestrial ecosystems at sequestering carbon [2]. Mangroves thrive in the zone between mean sea-level and high water and thus are sensitive to changes in inundation regime. Their zonation and ability to prevent erosion or increase sedimentation may provide a mechanism for mangroves to adapt to sea level rise and alleviate the threat of coastal retreat [3]. Despite the diverse array of valuable services, worldwide mangrove populations are in steep decline, with the loss of over one-quarter of global mangrove cover since 1980 [1,4].

Extreme flooding events are projected to increase with sea level rise [5,6]. Additionally, coastal populations and infrastructure are increasing [7], driving demand for effective coastal protection.

Conventional engineering solutions are often costly and may have a limited lifespan, destroy or fragment sensitive habitat, and have been associated with enhanced erosion [8,9]. Coastal vegetation has been proposed as an alternative to hard engineering solutions. Mangroves can provide coastal protection by reducing storm waves, dissipating currents, and stabilizing sediments [10,11]. Additionally, sedimentation in mangrove forests may provide a mechanism to maintain present coastlines with respect to sea level rise [12].

The reduction in the wave height of short period wind-generated waves due to interaction with mangroves is well established [10,13,14]. Less well established are the protective benefits of mangroves with respect to storm surge [15–17]. Mangroves reduce peak flood levels by limiting fluid exchange across the forest [18]. Dissipation of storm surges through coastal vegetation has previously been quantified as a reduction in peak water level (cm) per distance of flood propagation (km) with values categorized by vegetation type [15–18]. Although providing an easily accessible solution, using fixed dissipation rates over wide-ranging sites may oversimplify flood protection provided by coastal vegetation.

Alongi [12] noted that flood protection provided by coastal vegetation is dependent on vegetation properties, local bathymetry, and storm parameters. At forest-wide scales applicable to coastal inundation issues, obtaining mangrove properties is problematic. Vegetation can be heterogeneously distributed [19], and quantifying the drag-inducing elements (leaves, stems, trunks, and pneumatophores) can be unwieldy. Several different summary statistics are used for large-scale hydrodynamics, including frontal area density, the proportion of volume occupied by the solid canopy, and the blockage factor [20,21]. However, Nepf [21] comments that at reach scales in vegetated rivers, the patch distribution plays a larger role in determining flow resistance than individual plant geometry. Typically, vegetation drag is large relative to bed drag and therefore in heterogeneously vegetated environments, flow is channelized and deflected away from vegetation/high-drag patches [21,22].

The influence of channelization on mangrove flood attenuation is explored through the comparison of high water events in two contrasting New Zealand mangrove forests. The study sites are similar in length, with the forest extending ~1 km in the direction of flood propagation, and both sites are comprised of the same mangrove species, *Avicennia marina* var. *australasica* [23]. The key distinction is that the Tauranga mangrove forest is highly channelized in comparison with the Firth of Thames site.

2. Materials and Methods

2.1. Study Sites

2.1.1. Firth of Thames

The Firth of Thames (FoT) is a ~800 km^2 estuary on New Zealand's North Island (37°12′ S, 175°27′ E) (Figure 1b). The mesotidal estuary has a spring tidal range of 2.8 m and due to a shallow bed slope and plentiful fine-sediment supply, a large intertidal mudflat has developed [24]. The basin is bounded to the east and west by mountain ranges and the Hauraki Plains to the South. A stopbank (visible as a diagonal track on Figure 2a) prevents the inundation of the Hauraki Plains to the south of the Firth. The basin is exposed to moderate waves from the North and subject to a high terrigenous sediment supply from the Waihou and Piako rivers. The southern boundary of the Firth is colonized by a 1 km wide forest of grey mangroves (*Avicennia marina* var. *australasica*).

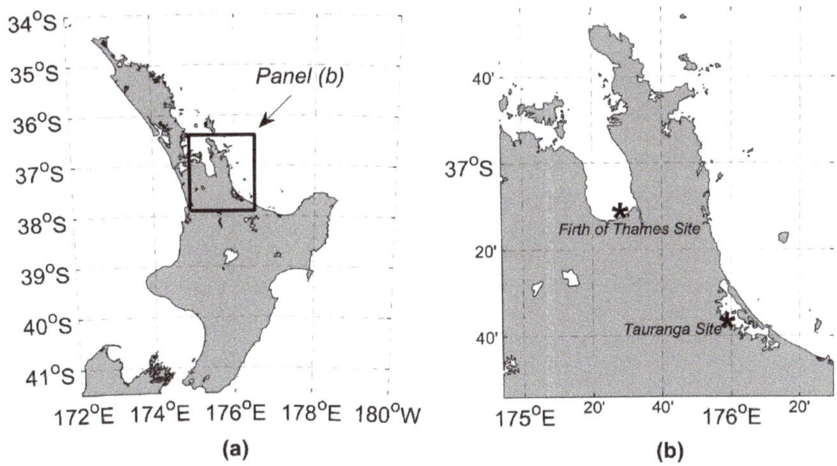

Figure 1. (**a**) The North Island of New Zealand with a panel (**b**) outlined; (**b**) Section of North Island of New Zealand showing the proximity of the Firth of Thames and Tauranga mangrove sites.

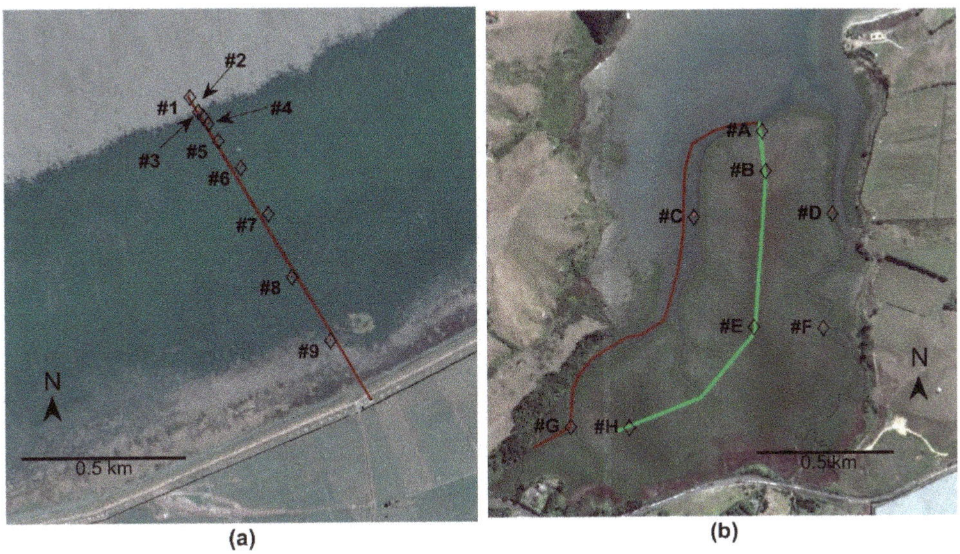

Figure 2. (**a**) The Firth of Thames study site with 9 (#1 seaward–#9 landward) instrument locations noted and bathymetry survey transect (brown). Station #1 is on the unvegetated mudflat, station #2 is in the vegetation fringe, stations #3–#5 are in the gently sloping intertidal, and stations #6–#9 are on the intertidal flat; (**b**) The Tauranga study site with 8 instrument locations noted with channel thalweg bathymetry survey (brown) and central mangrove survey (green). Station #A is on the vegetation fringe on the seaward boundary, stations #C and #D are in the western channel, stations #B, #E, and #H are in the central mangrove forest, stations #D and #F are in the eastern channel.

The cross-shore profile of the vegetated region (Figure 3a) consists of a level mangrove forest ~1.7–1.9 m above mean sea level (MSL) extending ~800 m seaward of the stopbank [24]. The sloping vegetation extends an additional ~100 m seaward to the mudflat. The topography and forest characteristics are relatively homogenous in the longshore direction. The elevation of the seaward fringe of the forest is close to a mean high water neap tide level (0.98 m MSL), so the tidal prism within the forest is relatively small and no substantial creeks have developed (Figure 3a) [25].

Mangrove characteristics vary throughout the forest. Along the forest fringe, trees are characterized by open spreading forms (Figure 4a,b). Within the forest, trees tend to have straight vertical trunks (Figure 4c). Tree height ranges from 0.5 to 3.5 m. Dense pneumatophores, as many as ~500 m^{-2}, emerge from the bed up to 25 cm in height and ~1 cm in diameter (Table 1).

In November 2016, a supermoon and low-pressure event occurred to produce an unusually large flood event in the Firth of Thames (Figures 4d and 5a,b). The water levels reached 2.36 m above MSL, corresponding to an event with a ~10-year return period for the Firth of Thames. The study area was flooded for several tidal cycles prior to the peak high water and remained flooded for several tidal cycles after the peak water level.

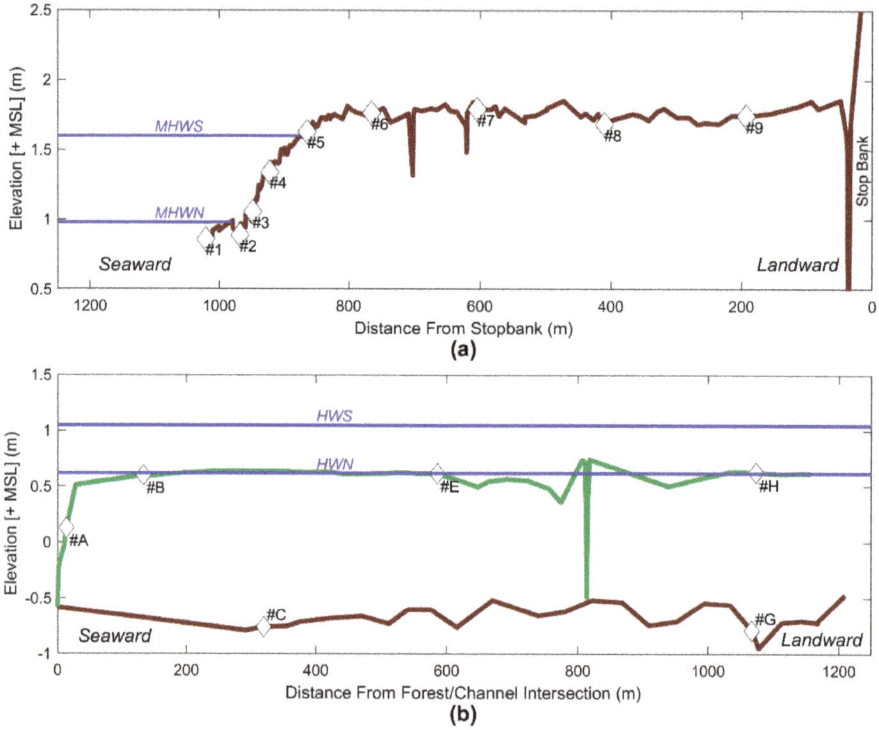

Figure 3. (**a**) The elevation profile along the instrument transect in the Firth of Thames. Mean High Water Spring (MHWS) and Mean High Water Neap (MHWN) are noted. The main forest is higher than normal tidal levels and therefore no drainage channels have been scoured by tidal water flow; (**b**) the Tauranga RTK survey of transect through central mangrove forest (green) and thalweg (brown). High Water Spring (HWS) and High Water Neap (HWN) are marked (blue). The semi-diurnal tidally-driven flow through the forest is responsible for channelization at the site.

Figure 4. The images of Firth of the Thames study site. (**a**) Forest fringe at low tide; (**b**) Fringe at mid tide, trees are characterized with open spreading branches; (**c**) Interior mangrove forest with two researchers for scale. Trees are tall with vertical trunks; (**d**) Two researchers in the mangrove forest during the flood event.

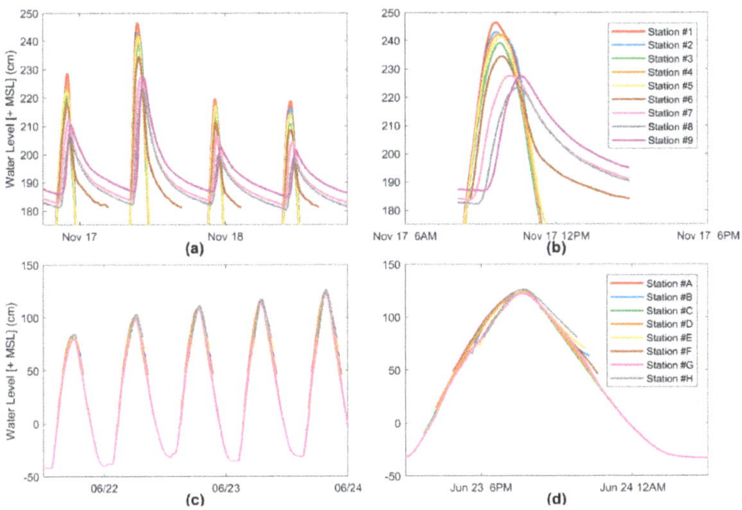

Figure 5. (**a**) The Firth of Thames water level at each instrument station for five consecutive spring tidal cycles. The upper intertidal flat (stations #6–#9) did not fully drain for several tidal cycles; (**b**) Firth of Thames water level during maximum inundation event; (**c**) Tauranga water level for five consecutive spring tidal cycles. Note that only station #G was submerged at low tide; (**d**) Tauranga water level during the largest tidal cycle.

2.1.2. Tauranga Site

Tauranga harbor is a 200 km^2 barrier-enclosed lagoon on the North Island of New Zealand (37°39′ S, 176° E) (Figure 1). The mesotidal estuary has an average spring tidal range of 1.62 m and neap range of 1.24 m [26]. Due to the complexity of the estuary, exact tidal ranges are location-dependent [27].

The shallow lagoon, with an average depth of 3 m at low tide, has extensive intertidal areas that make up nearly 2/3 of the estuary area [28]. The estuary has two entrances and is comprised of many sub-estuarine basins. The mangroves in Tauranga have expanded rapidly, from 13 hectares in the 1940s to 168 hectares in 1999 [29]. The mangroves in Tauranga are at the southern boundary of their latitudinal range, which causes the forests to be less productive and the trees to be shorter [23]. The focus of the presented work is a basin north of Pahoia (Figure 1) that nearly drains at low tide.

The Pahoia field site is comprised of a ~1 km long intertidal mangrove forest that occupies ~2/3 of the basin surface area (Figure 2b). Two unvegetated steep-sided channels, along the eastern and western sides of the forest, maintain a near-uniform depth throughout the study site and dominate water flow into the area (Figures 3b and 6b). The western channel bifurcates around a ~300–400 m wide central mangrove platform. The vegetated regions are at the same elevation as high-water neap tidal levels and are approximately flat. A small creek drains into the western channel and further divides the central mangrove forest. The significant tidal prism in the forest is likely responsible for creating the channel network [3].

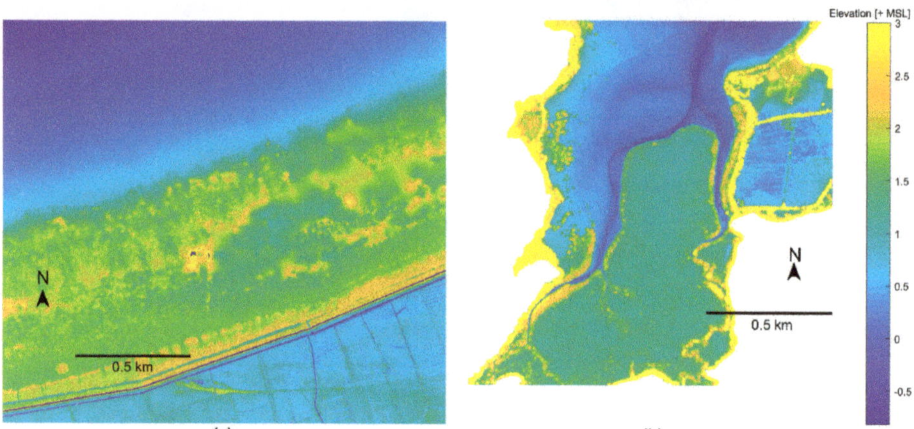

(a) (b)

Figure 6. (**a**) The Firth of Thames LiDAR devoid of a channel network. Patchy higher elevations likely indicate a vegetation canopy; (**b**) Tauranga LiDAR data. Deep, incised channels and a level vegetated intertidal characterize the site. High elevation along channels displays a dense mangrove canopy. The color bar shows an elevation scale for both subplots.

The forest is comprised of small shrub-like grey mangroves less than 1.2 m in height (average 0.41 m). Individual trees have complex geometry (Figure 7a) and present a low dense canopy (Figure 7b). The pneumatophore density averages 75 per square meter, with individual pencil-roots of similar dimension to the pneumatophores in the Firth of Thames (Table 2).

Figure 7. The images of the Tauranga study site. (**a**) Example mangrove tree. Vegetation is characterized by a complex trunk structure and a low canopy height; (**b**) Mangrove forest at mid tide; (**c**) Mangrove lined channel with steep densely vegetated banks; (**d**) Weather station recording barometric pressure and wind speed during high spring tide with canopy almost submerged.

Typical spring tides nearly fully submerge the Pahoia mangrove forest (Figure 7d). Figure 5c,d displays data from a series of spring tides in June 2017. The peak water level reached ~1.25 m above MSL. Note that the study site nearly drained at low tide, leaving most of the instruments exposed.

2.2. Field Data Collection

Field observations were collected during two experiments conducted in November 2016 in the southern Firth of Thames and in April 2017 in the Pahoia sub-estuary of Tauranga Harbor. In each case, arrays of water level sensors were deployed, along with RTK-GPS surveying and manual vegetation surveys. Details of surveys and deployments are given below.

2.2.1. Vegetation Survey

Vegetation surveys were conducted to quantify the rigid vegetation at the study sites. The heterogeneous distribution of vegetation, large study area, and diverse mangrove properties necessitated a unique approach to measure vegetation. The survey objective was targeted at the two types of structure that characterize the flow-reducing properties of *Avicennia marina*: (1) The pneumatophores and seedling which comprise small, but dense structures near the seabed and (2) the trees and branches which are much less dense but can form a large blocking mechanism at high tide and surge levels. Flexible leafy canopies were not quantified. Mangrove canopies in the Firth of Thames were at sufficient elevation to not submerge and therefore did not influence flow resistance and were not quantified. The shrubby mangroves at Pahoia were submerged at high tide (Figure 7). Attempts were made to quantify canopies in Tauranga using the structure from motion, but were unsuccessful due to the difficulty of gridding the complex leaf geometry. Therefore, the leaves and small branches will provide an unquantified additional blocking. Visual observations indicated that flows were not sufficiently strong to cause the flexing of the leafy vegetation.

Due to the different growth forms of the mangroves at the two sites, slightly different strategies were employed. In the Firth of Thames, mangrove trees in some areas were nearly impenetrable, 5 m by 5 m quadrats were established along the cross-shore instrument transect. In Pahoia, where the mangroves were generally less than a meter tall, a cross-shore and an along-shore transect were established, and vegetation properties measured every 10 m.

In the Firth of Thames, the number of trees was counted in each quadrat. The height and width of the canopy and the stem diameter at 0.3 m from the seabed were manually measured for 5 trees closest to the pre-defined coordinates within the quadrat. To quantify pneumatophores, five 0.5 by 0.5 m quadrats were selected, one at each corner and one at the center of the large quadrat. Within these small quadrats, all the pneumatophores and seedlings were counted. In addition, the height, top diameter, and bottom diameter of five pneumatophores were measured resulting in a total of 25 pneumatophores measured at each station (Table 3).

In Pahoia, a transect through the central mangrove forest was established. Along the transect, the canopy height was measured at 5 m intervals. Pneumatophore and seedling characteristics were measured in 0.5 by 0.5 m quadrats every 5 m using the same method as that used at the Firth of Thames site. Pneumatophore statistics in Table 4 are based on the measurement of 5 pencil roots.

The vegetation drag is estimated as proportional to vegetation frontal area [30]. The frontal area (a_v) assumes that both the trunks and pneumatophores are rigid emergent vegetation ($a_v = nd$) where n is vegetation element density and d is the average diameter. Note that the contribution of pneumatophores to the total frontal area is reduced when water levels fully submerge the pneumatophores. The direct measurement of trunks in the Firth of Thames was used to generate the frontal area parameter. The complex configuration of plants in Tauranga (Figure 7a) made the measurement of trunks problematic. Therefore, general relationships between trunk density and canopy height, as well as trunk diameter and canopy height, were obtained for the vegetation data in the Firth of Thames. Assuming that the general structural properties of New Zealand *Avicennia* growth forms are the same between sites, these relationships were used to estimate the trunk density and diameter for Tauranga based on local measured canopy height. This method generates values for frontal area density (with uncertainty because of the assumptions of similarities between sites).

2.2.2. Bathymetry

Manual RTK GPS surveys were conducted to obtain bathymetry at both study sites. In the Firth of Thames, an RTK survey of the instrument transect was conducted using a Trimble R8 GNSS system (Figure 3a). In Tauranga, RTK elevations were obtained for the western channel and a transect through the central mangrove forest (Figure 3b) with a Leica GS18 T GNSS system. Additionally, LIDAR data was provided by local government organizations for both sites (Figure 6). All elevation surveys were initiated and terminated with verification of vertical measurement accuracy with a fixed survey mark, the measurement error never exceeded 3 cm.

2.2.3. Water Level

Pressure sensors were deployed at each of the study locations (Figure 2, Tables 1 and 2). The array of water level sensors deployed in the Firth of Thames in November 2016 consisted of a single transect of 9 instruments, extending from the stopbank to the mudflat (Figure 2a): Station 1 was seaward of the vegetation on the mudflat, stations 2–5 were located across the sloping forest region and stations 6–9 were spread across the forest platform (Figure 3a). A man-made channel is located just seaward of the stopbank and is evident in the transect survey (Figure 3a). At PAHOIA in Tauranga, the central mangrove forest and both the east and west channels were instrumented with pressure gauges (Figure 2b). Station #A is at the seaward edge of the mangroves and at the intersection of the two primary channels. Three stations (#B, #E, and #H) were located at increasing distances into the vegetated intertidal platform. An additional two gauges were positioned in each of the two channels.

Pressure sensors were corrected for variations in barometric pressure and for temperature dependence and referenced to mean sea level using survey data. Pressure signals were smoothed using a low-pass filter and converted to the water level using a constant water density of 1025 kg/m^3.

Table 1. The Firth of Thames instrument array. Note that during factory calibration, the Aquadopp depth uncertainty over the range of interest was <5 cm and the Vector uncertainty was <2 cm.

Station	Instrument	Sampling		Temperature Dependence	Depth Measurement Uncertainty
		Regime	Details	Correction Applied	(cm)
#1	Nortek Aquadopp	Burst	2^{12} samples at 8 Hz every 15 min	Y	<10
#2	Nortek Vector	Burst	7.5 min sampling at 16 Hz every 15 min	Y	<5
#3	Nortek Aquadopp	Burst	2^{12} samples at 8 Hz every 15 min	Y	<10
#4	Nortek Aquadopp	Burst	2^{12} samples at 8 Hz every 15 min	Y	<10
#5	RBR Concerto	Continuous	4 Hz	N	<1
#6	RBR Duet	Continuous	8 Hz	N	<1
#7	Solinst Levelogger	Continuous	1/60 Hz	N	<1
#8	Solinst Levelogger	Continuous	1/60 Hz	N	<1
#9	Solinst Levelogger	Continuous	1/60 Hz	N	<1
	Weather Station	Continuous	1/5 min	N	<1

Table 2. The Tauranga instrument array.

Station	Instrument	Sampling		Temperature Dependence	Depth Measurement Uncertainty
		Regime	Details	Correction Applied	(cm)
#A	RBR Solo	Continuous	8 Hz	N	<1
#B	RBR Solo	Continuous	Every 2 min Average 1 min of data sampled at 4 Hz	N	<1
#C	RBR Duet	Continuous	8 Hz	N	<1
#D	RBR Duet	Continuous	8 Hz	N	<1
#E	RBR Solo	Continuous	Every 2 min Average 1 min of data sampled at 4 Hz	N	<1
#F	RBR Solo	Continuous	8 Hz	N	<1
#G	RBR Solo	Continuous	8 Hz	N	<1
#H	RBR Solo	Continuous	8 Hz	N	<1
	Weather Station	Continuous	1/5 min	N	<1

3. Results

Water levels across the Firth of Thames and Pahoia study sites for successive tidal cycles are displayed in Figure 5. The ~1 km wide mangrove forest in the Firth of Thames site reduced the peak water levels and delayed the inundation signal, with the reduction and delay increasing with distance into the forest. The largest inundation wave in the Firth of Thames reached a maximum of 72 cm above the tidal flat at the seaward forest fringe and decayed to 53 cm above the tidal flat at the landward most station. This water height reduction of 19 cm across the 800 m separation between station 1 and station 9 corresponds to a dissipation rate of ~24 cm/km. The temporal delay in peak water between station 1 and 9 is evident in Figure 5 and estimated at 60 $\pm$ 5 min. The average velocity of peak inundation through the Firth of Thames mangrove forest is ~0.2 m/s.

At the Pahoia site, no measurable reduction in water level occurred over the ~1 km separation between instruments. The dissipation of flood levels was less than the uncertainty of the elevation measurements. Additionally, no identifiable temporal delay in peak water occurred over the study site (Figure 5c,d).

The magnitudes of the inundation waves for the two locations are similar with respect to the elevation of the mangrove forest. The inundation waves for both sites range from ~20 to ~70 cm above the average forest elevation (Figure 8). Nonetheless, the elevation of inundation relative to MSL was different between the two sites (Figure 5), with the Firth of Thames ranging from ~220 to ~245 cm above MSL at the seaward boundary of the mangrove forest, while the Tauranga inundation wave varied between 80 and 125 cm above the MSL.

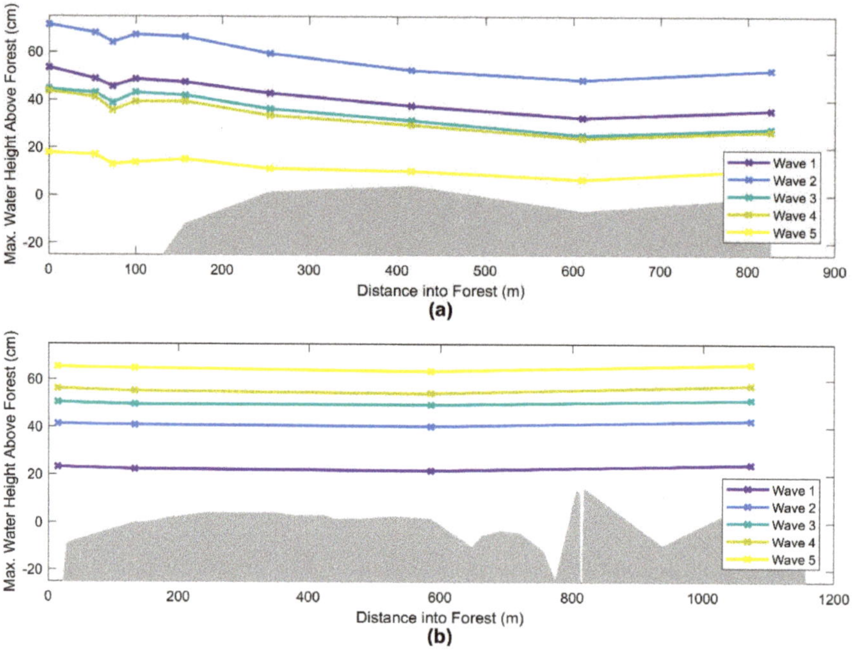

Figure 8. (**a**) The peak water level at each instrument station along the transect in the Firth of Thames for 5 consecutive inundation events; (**b**) the peak water level at each instrument along the central mangrove transect in Tauranga (#A, #B, #E, and #H). Elevations are with respect to average elevation of the forest floor, profiles of which are shown in grey.

Table 3. The Firth of Thames vegetation survey summary. The values given are averages with the standard deviation where appropriate.

Location		Trees				Pneumatophores			
Description	Distance from Fringe	Density	Height	Diameter	Frontal Area	Density	Height	Diameter	Frontal Area
	(m)	(# per m^2)	(m)	(cm)	(m^{-1})	(# per m^2)	(cm)	(cm)	(m^{-1})
Mudflat	−49	0.1	0.6	1.6	0.00	0	0	0	0
Fringe (Front)	0	0.1	2.3	6.4	0.01	126	25.4 (8.5)	0.6 (0.2)	0.76
Fringe (Back)	22	1.0	1.9 (1.1)	4.8 (5.4)	0.05	385	22.9 (9.1)	0.6 (0.2)	2.31
Forest	36	1.0	2.9 (0.3)	5.8 (3.1)	0.06	386	17.0 (6.9)	0.5 (0.2)	1.93
Forest	114	6.4	2.3 (0.2)	3.0 (0.7)	0.19	322	12.5 (5.6)	0.6 (0.2)	1.93
Forest	166	11.4	1.5 (0.3)	2.8 (1.2)	0.32	357	7.7 (4.4)	0.6 (0.2)	2.14
Forest	262	12.2	1.2 (0.1)	1.9 (0.4)	0.23	481	8.3 (4.1)	0.6 (0.1)	2.89
Forest	376	6.4	3.1 (0.1)	3.6 (1.1)	0.23	342	8.7 (5.4)	0.5 (0.1)	1.71
Forest	476	1.4	3.8 (0.3)	4.2 (1.1)	0.06	330	10.5 (5.5)	0.6 (0.1)	1.98
Forest	561	3.6	1.6 (0.2)	3.9 (2.2)	0.14	570	12.0 (6.1)	0.5 (0.1)	2.85
Forest	667	5	2.6 (0.3)	3.8 (0.6)	0.19	482	11.0 (4.7)	0.5 (0.1)	2.41

Table 4. The Tauranga vegetation survey summary. The values given are averages with the standard deviation where appropriate.

Location		Trees		Pneumatophores			
Transect	**Distance from Fringe**	**Canopy Height**	**Frontal Area ***	**Density**	**Height**	**Diameter**	**Frontal Area**
	(m)	(m)	(m^{-1})	(# per m^2)	(cm)	(cm)	(m^{-1})
Along-shore	0	0	0.26	40	1.5 (0.5)	0.3 (0.2)	0.12
Along-shore	5	0.97	0.26	164	1.7 (0.5)	0.3 (0.2)	0.49
Along-shore	15	0.48	0.26	296	1.1 (0.4)	0.4 (0.2)	1.18
Along-shore	25	0.41	0.26	308	1.1 (0.5)	0.5 (0.2)	1.54
Along-shore	35	0.36	0.26	296	0.6 (0.5)	0.4 (0.2)	1.18
Along-shore	45	0.38	0.26	252	1 (0.6)	0.4 (0.2)	1.00
Along-shore	55	0.365	0.26	248	6 (3.7)	0.5 (0.2)	1.24
Along-shore	65	0.34	0.26	356	8.7 (4.2)	0.6 (0.1)	2.14
Along-shore	75	0.37	0.26	196	10.8 (2.8)	0.6 (0.2)	1.18
Along-shore	85	0.29	0.25	244	7 (2.6)	0.6 (0.1)	1.46
Along-shore	95	0.23	0.26	240	5.1 (2.4)	0.6 (0.1)	1.44
Along-shore	105	0.26	0.26	216	10.3 (4.1)	0.5 (0.2)	1.08
Along-shore	115	0.27	0.25	172	4.3 (1.8)	0.6 (0.2)	1.03

Table 4. *Cont.*

Location		Trees		Pneumatophores			
Transect	Distance from Fringe	Canopy Height	Frontal Area *	Density	Height	Diameter	Frontal Area
	(m)	(m)	(m^{-1})	(# per m^2)	(cm)	(cm)	(m^{-1})
Along-shore	125	0.22	0.26	172	5.3 (3.2)	0.5 (0.2)	0.86
Along-shore	135	0.29	0.26	264	6.9 (3.6)	0.5 (0.1)	1.32
Along-shore	145	0.26	0.25	248	7.4 (3.5)	0.5 (0.2)	1.24
Along-shore	155	0.23	0.25	364	6.3 (3.7)	0.6 (0.2)	2.18
Along-shore	165	0.23	0.25	292	6.1 (3.3)	0.6 (0.2)	1.75
Along-shore	175	0.22	0.25	252	6.2 (2.1)	0.7 (0.2)	1.76
Along-shore	185	0.2	0.25	172	6.3 (2.5)	0.7 (0.1)	1.20
Along-shore	195	0.2	0.26	200	13.6 (7.1)	0.6 (0.2)	1.20
Along-shore	205	0.32	0.26	212	5.4 (3.1)	0.5 (0.1)	1.06
Along-shore	215	0.3	0.26	224	11.2 (2.9)	0.6 (0.1)	1.34
Along-shore	225	0.24	0.26	192	7 (6.0)	0.5 (0.1)	0.96
Along-shore	200	0.27	0.26	276	11.7 (2.8)	0.6 (0.2)	1.66
Along-shore	190	0.29	0.26	324	3.7 (2.5)	0.6 (0.2)	1.94
Cross-shore	180	0.33	0.26	344	8.2 (6.3)	0.6 (0.2)	2.06
Cross-shore	170	0.52	0.26	192	11.7 (8.0)	0.6 (0.1)	1.15
Cross-shore	160	0.36	0.26	340	14.2 (7.1)	0.6 (0.1)	2.04
Cross-shore	150	0.35	0.26	356	8.5 (5.2)	0.6 (0.1)	2.14
Cross-shore	140	0.47	0.26	348	12.9 (7.3)	0.6 (0.2)	2.09
Cross-shore	130	0.54	0.26	396	10.2 (3.9)	0.6 (0.2)	2.38
Cross-shore	110	0.59	0.26	404	15.1 (7.0)	0.6 (0.2)	2.42
Cross-shore	100	0.63	0.26	536	12.7 (4.3)	0.6 (0.1)	3.22
Cross-shore	90	0.64	0.26	400	10.7 (2.5)	0.5 (0.1)	2.00
Cross-shore	80	0.71	0.26	552	8.1 (4.0)	0.5 (0.1)	2.76
Cross-shore	70	0.93	0.26	568	7.4 (2.7)	0.5 (0.1)	2.84
Cross-shore	60	0.55	0.26	404	8.8 (4.1)	0.6 (0.2)	2.42
Cross-shore	50	0.56	0.26	464	15.9 (8.6)	0.6 (0.1)	2.78
Cross-shore	40	0.42	0.26	408	7.6 (6.1)	0.5 (0.1)	2.04
Cross-shore	30	0.44	0.26	444	8.9 (4.2)	0.6 (0.1)	2.66
Cross-shore	20	0.72	0.25	308	10.5 (5.0)	0.5 (0.1)	1.54
Cross-shore	10	1.3	0.26	432	5.4 (2.8)	0.5 (0.1)	2.16
Cross-shore	0	0.97	0.26	268	12.8 (5.3)	0.5 (0.1)	1.34

* Note that tree stem diameter and density for frontal area density calculation has been estimated (see text for details).

4. Discussion

Mangroves reduce peak water levels during a flood by limiting the exchange of water through the vegetation [18]. Krauss et al. [15] observed that the presence of channels decreased the efficacy of mangrove flood attenuation from 9.4 cm/km to 4.2 cm/km. Using a combination of observations and numerical simulations, Zhang et al. [16] found that the amplitude of storm surge was reduced at a rate of 40–50 cm/km through mangrove forests and ~20 cm/km through patchy regions consisting of a combination of mangrove islands and open water. Flood level reduction during the series of large inundation events in the non-channelized Firth of Thames averaged 24 cm/km, which agrees with rates previously published for unchannelized forests by Krauss et al. and Zhang et al. [15,16]. In contrast, the channelized New Zealand mangrove site in Tauranga had no measurable reduction of flood amplitude. We statistically tested whether tree height and pneumatophore density, diameter and height differed between sites. Because the raw (and transformed) data did not meet the assumptions of normality (Shapiro-Wilk W test $p < 0.001$) and/or homogeneity of variances (Levene's test $p < 0.01$) required for t-tests of differences between means, we used the non-parametric Mann-Whitney U test (a = 0.05). All statistical analyses were conducted in Statistica version 13.2. The results indicated that tree height, pneumatophore density, and pneumatophore height were all significantly different ($p < 0.001$). As a consequence, the frontal area, and therefore the flow resistance, from trunks was greater in Tauranga than in the Firth of Thames (Tables 3 and 4). Additionally, Chen et al. [31] demonstrate that mangrove canopies only increase the generation of turbulent kinetic energy by about 10% compared to when the flow is below the bottom of the leaf structures. Despite flow engagement of the dense leafy canopy in Tauranga (Figure 7d), the vegetation had no flood mitigating influence due to preferential routing through the channels.

The interaction of water and vegetation is complex and has been investigated at multiple length scales. At small scales (O (mm)), the boundary layers and shear caused by individual stems, roots, and leaves cause turbulent eddies that shed off each individual stem [32]. Turbulence is also generated at the shear layer between the faster moving flow over submerged vegetation, and the damped flow within the canopy [33]. Intermediate scales (O (m)) comprise flow at the canopy or patch scale involving a community of vegetation. Larger length scale interactions (O (km)) occurs at the forest level [34]. To appropriately investigate a process of interest, a reasonable spatial scale, associated conceptual model, and relevant measurements and methods must be identified [21]. Vegetated regions produce high drag with respect to unvegetated areas and flow is diverted to the path of least resistance. In areas described as dense vegetation patches, most flow is directed around the patches and a forest-wide approach is required. In sparse or homogeneously distributed vegetation, smaller-scale resistance dominates and a smaller-scale approach is justified [35].

In mangrove forests, it is not just the vegetation geometry that controls water transport, the intertidal bathymetry, and water level relative to the elevation of the vegetation also play a role [36–38]. Flow through mangrove forests has been categorized into creek flow or sheet flow depending on the primary mechanism of fluid transport. Creek flow dominates in channelized mangroves at low water levels. Sheet flow, the transportation over the vegetated platform through the mangroves, becomes increasingly important with reduced channelization and at increasing water levels [36]. Our results show that the Tauranga mangrove forest is dominated by creek flow, and the density of mangrove vegetation therefore only has minimal contribution to the flow restriction; no evidence of reduced inundation level nor delay in the flood signal exists. In Tauranga, flow resistance is best described by the larger-scale distribution of vegetation and degree of channelization. Conversely, the Firth of Thames mangrove forest is not channelized, and the primary shoreward water transportation mechanism is sheet flow through the vegetation. Here, vegetation properties are important for impeding water exchange across the forest, reducing inundation levels and slowing the flood wave propagation. In the Firth of Thames, the flow resistance relates to the vegetation properties along the one-dimensional cross-shore transect. The cumulative influence of large quantities of individual stems, stalks, and leaves on fluid flows at forest wide scales necessitates simplifying vegetation summary statistics [20]. Several different statistical parameters have been used to describe the influence of vegetation on large-scale flow resistance, including the solid volume fraction, vegetation porosity, and frontal area per bed area [21].

Channelization in mangrove environments develops as the trees grow and create flow resistance and concentrate the flow into channels [3]. However, for initiation of the feedback process that allows the channels to develop, the intertidal platform must be at a sufficiently low elevation with respect to the tidal excursion that currents occur on the vegetated platform. In the case of a very high platform, inundation only occurs at slack water close to high tide, at which time conditions promote sediment deposition. The Tauranga mangrove forest is inundated during normal tidal levels. The drainage channels in the Tauranga study site have likely resulted from scouring by tidally-driven water transport through the forest. The forest elevation in the Firth of Thames is higher than the Mean High Water Spring (MHWS) water levels (Figure 3a) and therefore water is only infrequently transported through the forest and channels cannot develop.

During the study, water depths in the relatively flat mangrove forests in Tauranga and the Firth of Thames were of similar magnitude (Figure 8). The capacity of mangroves to provide coastal flood protection is ultimately related to water transport pathways. Extrapolating from our case study environments, we can expect that lower intertidal areas with channelization will be far less capable of protection than higher intertidal areas with little channelization. Krauss et al. [15] and Zhang et al. (2012) found that the reduction of flood levels along a river corridor was less than through unchannelized vegetation but still provided flood protection. Both previous investigations focused on hurricane-driven storm surges in the south-east United States. The extreme water levels greatly exceeded the capacity of the channel networks, likely resulting in flow pathways through the vegetation and therefore the capacity to mitigate flood levels was apparent but reduced compared to unchannelized locations. Moreover, the sediment regime has been shown to contribute to the development of a profile shape, with muddy profiles often associated with high convex intertidal geometries [39].

5. Conclusions

The influence of mangroves on long wave propagation is strongly dependent on flow routing. In highly channelized mangrove forests, such as our Tauranga case study area, water is preferentially transported via the channels and flood levels are not reduced substantially across the forest. In these cases, the vegetation does not contribute significantly to flow resistance, so specific plant properties are irrelevant with respect to limiting fluid transport. However, we hypothesize that for sufficiently large flood events, in which the conveyance capacity of the channels is exceeded, a proportion of the flow will be forced through the vegetation, which will thus provide an intermediate level of attenuation (still reduced relative to an unchannelized environment). Conversely, in homogeneously vegetated forests without channels, such as the Firth of Thames study site, water is transported through the mangroves and the trees reduce flood levels by limiting fluid exchange through the forest. In these cases, knowledge of vegetation characteristics is essential for the prediction of the rate of flood level reduction.

The degree of channelization and therefore the capacity of mangroves to reduce flooding depends on the elevation of the vegetation. Mangrove forests that occur at relatively low, frequently inundated elevations are subjected to tidal currents that promote channelization, which in turn reduces their capacity to mitigate the flood water level. Higher elevation mangrove forests are inundated only at the peak tide when currents are at a minimum and the sediment regime is depositional. No channel network is created nor maintained, and the capacity of the mangrove forest to reduce flood events is maximized.

Author Contributions: Conceptualization, J.M.M. and K.R.B.; Methodology, J.M.M., K.R.B., E.M.H. and J.C.M.; Formal Analysis, J.M.M.; Data Curation, J.M.M., and E.M.H.; Writing-Original Draft Preparation, J.M.M. & K.R.B.; Writing-Review & Editing, J.M.M., K.R.B., E.M.H. and J.C.M.; Funding Acquisition, K.R.B. and J.C.M.

Funding: This research was funded by the Natural Hazards Platform program (contract C05X0907) AND THE Royal Society Marsden Fund (grant 14-UOW-011).

Acknowledgments: The authors gratefully acknowledge the field help provided by Dean Sandwell, Heide Friedrich, Caitlyn Gillard, Hieu Nguyen, Dave Culliford, and Rex Fairweather. Help with statistical analysis of vegetation was provided by Conrad Pilditch.

Conflicts of Interest: The authors declare no conflict of interest.

References

1. Food and Agriculture Organization (FAO). *The World's Mangroves 1980–2005*; Food and Agriculture Organization of the United Nations: Rome, Italy, 2007; ISBN 978-92-5-105856-5.
2. McLeod, E.; Chmura, G.L.; Bouillon, S.; Salm, R.; Bjork, M.; Duarte, C.M.; Lovelock, C.E.; Schlesinger, W.H.; Silliman, B.R. A blueprint for blue carbon: Toward an improved understanding of the role of vegetated coastal habitats in sequestering CO_2. *Front. Ecol. Environ.* **2011**, *9*, 552–560. [CrossRef]
3. van Maanen, B.; Coco, G.; Bryan, K.R. On the ecogeomorphological feedbacks that control tidal channel network evolution in a sandy mangrove setting. *Proc. R. Soc. A* **2015**, *471*, 20150115. [CrossRef] [PubMed]
4. Giri, C.; Ochieng, E.; Tieszen, L.L.; Zhu, Z.; Singh, A.; Loveland, T.; Masek, J.; Duke, N. Status and distribution of mangrove forests of the world using earth observation satellite data. *Glob. Ecol. Biogeogr.* **2011**, *20*, 154–159. [CrossRef]
5. Wahl, T.; Haigh, I.D.; Nicholls, R.J.; Arns, A.; Dangendorf, S.; Hinkel, J.; Slangen, A.B.A. Understanding extreme sea levels for broad-scale coastal impact and adaptation analysis. *Nat. Commun.* **2017**, *8*, 16075. [CrossRef] [PubMed]
6. Kroeker, K.J.; Reguero, B.G.; Rittelmeyer, P.; Beckd, M.W. Ecosystem Service and Coastal Engineering Tools for Coastal Protection and Risk Reduction. In *Managing Coasts with Natural Solutions*; World Bank: Washington, DC, USA, 2016.
7. Small, C.; Nicholls, R.J. A global analysis of human settlement in coastal zones. *J. Coast. Res.* **2003**, *19*, 584–599.
8. Plant, N.G.; Griggs, G.B. Interactions between nearshore processes and beach morphology near a seawall. *J. Coast. Res.* **1992**, *8*, 183–200.
9. Airoldi, L.; Abbiati, M.; Beck, M.W.; Hawkins, S.J.; Jonsson, P.R.; Martin, D.; Moschellad, P.S.; Sundelofg, A.; Thompsonf, R.C.; Aberg, P. An ecological perspective on the deployment and design of low-crested and other hard coastal defence structures. *Coast. Eng.* **2005**, *52*, 1073–1087. [CrossRef]
10. Guannel, G.; Ruggiero, P.; Faries, J.; Arkema, K.; Pinsky, M.; Gelfenbaum, G.; Guerry, A.; Kim, C.K. Integrated modeling framework to quantify the coastal protection services supplied by vegetation. *J. Geophys. Res. Ocean.* **2015**, *120*, 324–345. [CrossRef]
11. Temmerman, S.; Meire, P.; Bouma, T.J.; Herman, P.; Ysebaert, T.; De Vriend, H.J. Ecosystem-based coastal defence in the face of global change. *Nature* **2013**, *504*, 79–83. [CrossRef] [PubMed]
12. Alongi, D.M. Mangrove forests: Resilience, protection from tsunamis, and responses to global climate change. *Estuar. Coast. Shelf Sci.* **2008**, *76*, 1–13. [CrossRef]
13. Henderson, S.M.; Norris, B.K.; Mullarney, J.C.; Bryan, K.R. Wave-frequency flows within a near-bed vegetation canopy. *Cont. Shelf Res.* **2017**, *147*, 91–101. [CrossRef]
14. Massel, S.; Furukawa, K.; Brinkman, R. Surface wave propagation in mangrove forests. *Fluid Dyn. Res.* **1999**, *24*, 219–249. [CrossRef]
15. Krauss, K.W.; Doyle, T.W.; Doyle, T.J.; Swarzenski, C.M.; From, A.S.; Day, R.H.; Conner, W.H. Water level observations in mangrove swamps during two hurricanes in Florida. *Wetlands* **2009**, *29*, 142–149. [CrossRef]
16. Zhang, K.Q.; Liu, H.Q.; Li, Y.P.; Xu, H.Z.; Shen, J.; Rhome, J.; Smith, T.J. The role of mangroves in attenuating storm surges. *Estuar. Coast. Shelf Sci.* **2012**, *102–103*, 11–23. [CrossRef]
17. U.S. Army Corps of Engineers. *Hurricane Study for Morgan City, Louisiana and Vicinity*; United States Congress Serial Set: Washington, DC, USA, 1965.
18. McIvor, A.; Spencer, T.; Möller, I.; Spalding, M. Storm surge reduction by mangroves. In *Natural Coastal Protection Series: Report 2*; The Nature Conservancy and Wetlands International: Cambridge, UK, 2012.
19. Chen, R.; Twilley, R.R. A gap dynamic model of mangrove forest development along gradients of soil salinity and nutrient resources. *J. Ecol.* **1998**, *86*, 37–51. [CrossRef]

20. Mullarney, J.C.; Henderson, S.M. Flows Within Marine Vegetaion Canopies in Advances. In *Coastal Hydraulics*; Panchange, V., Kaihatu, J., Eds.; World Scientific Publishing: Singapore, 2018; pp. 1–46.
21. Nepf, H. Hydrodynamics of vegetated channels. *J. Hydraul. Res.* **2012**, *50*, 262–279. [CrossRef]
22. Folkard, A.M. Vegetated flows in their environmental context: A review. *Proc. Inst. Civ. Eng.-Eng. Comput. Mech.* **2011**, *164*, 3–24. [CrossRef]
23. Horstman, E.M.; Lundquist, C.J.; Bryan, K.R.; Bulmer, R.H.; Mullarney, J.C.; Stokes, D.J. The dynamics of expanding mangroves in New Zealand. In *Threats to Mangrove Forests*; Springer: New York, NY, USA, 2018; pp. 23–51. [CrossRef]
24. Swales, A.; Bentley, S.J.; Lovelock, C.; Bell, R.G. Sediment processes and mangrove-habitat expansion on a rapidly-prograding muddy coast, New Zealand. In Proceedings of the Coastal Sediments'07, Processing of the Sixth International Symposium on Coastal Engineering and Science of Coastal Sediment Process, New Orleans, LA, USA, 13–17 May 2007.
25. Swales, A.; Bentley, S.J.; Lovelock, C.E. Mangrove-forest evolution in a sediment-rich estuarine system: Opportunists or agents of geomorphic change? *Earth Surf. Proc. Landf.* **2015**, *40*, 1672–1687. [CrossRef]
26. Heath, R. A review of the physical oceanography of the seas around New Zealand—1982. *N. Z. J. Mar. Freshw. Res.* **1985**, *19*, 79–124. [CrossRef]
27. Tay, H.W.; Bryan, K.R.; de Lange, W.P.; Pilditch, C.A. The hydrodynamics of the southern basin of Tauranga Harbour. *N. Z. J. Mar. Freshw. Res.* **2013**, *47*, 249–274. [CrossRef]
28. Healy, T.R.; Cole, R.; de Lange, W. Geomorphology and ecology of New Zealand shallow estuaries and shorelines. In *Estuarine Shores*; John Wiley and Sons: New York, NY, USA, 1996.
29. Park, S.G. *Aspects of Mangrove Distribution and Abundance in Tauranga Harbour*; Environment Bay of Plenty: Whakatane, New Zealand, 2004.
30. Nepf, H. Vegetated flow dynamics. In *The Ecogeomorphology of Tidal Marshes*; American Geophysical Union: Washington, DC, USA, 2004; pp. 137–163.
31. Chen, Y.N.; Li, Y.; Cai, T.L.; Thompson, C.; Li, Y. A comparison of biohydrodynamic interaction within mangrove and saltmarsh boundaries. *Earth Surf. Proc. Landf.* **2016**, *41*, 1967–1979. [CrossRef]
32. Norris, B.K.; Mullarney, J.C.; Bryan, K.R.; Henderson, S.M. The effect of pneumatophore density on turbulence: A field study in a Sonneratia-dominated mangrove forest, Vietnam. *Cont. Shelf Res.* **2017**, *147*, 114–127. [CrossRef]
33. Horstman, E.; Bryan, K.; Mullarney, J.; Pilditch, C.; Eager, C. Are flow-vegetation interactions well represented by mimics? A case study of mangrove pneumatophores. *Adv. Water Resour.* **2018**, *111*, 360–371. [CrossRef]
34. Mullarney, J.C.; Henderson, S.M.; Reyns, J.A.; Norris, B.K.; Bryan, K.R. Spatially varying drag within a wave-exposed mangrove forest and on the adjacent tidal flat. *Cont. Shelf Res.* **2017**, *147*, 102–113. [CrossRef]
35. Green, J.C. Comparison of blockage factors in modelling the resistance of channels containing submerged macrophytes. *River Res. Appl.* **2005**, *21*, 671–686. [CrossRef]
36. Horstman, E.M.; Dohmen-Janssen, C.M.; Bouma, T.J.; Hulscher, S.J.M.H. Tidal-scale flow routing and sedimentation in mangrove forests: Combining field data and numerical modelling. *Geomorphology* **2015**, *228*, 244–262. [CrossRef]
37. Mazda, Y.; Kanazawa, N.; Wolanski, E. Tidal asymmetry in mangrove creeks. *Hydrobiologia* **1995**, *295*, 51–58. [CrossRef]
38. Mei, C.C.; Chan, I.C.; Liu, P.L.F.; Huang, Z.; Zhang, W. Long waves through emergent coastal vegetation. *J. Fluid Mech.* **2011**, *687*, 461–491. [CrossRef]
39. Bryan, K.R.; Nardin, W.; Mullarney, J.C.; Fagherazzi, S. The role of cross-shore tidal dynamics in controlling intertidal sediment exchange in mangroves in Cù Lao Dung, Vietnam. *Cont. Shelf Res.* **2017**, *147*, 128–143. [CrossRef]

water

MDPI

Article

Effect of Mimic Vegetation with Different Stiffness on Regular Wave Propagation and Turbulence

Chao Tan [1], Bensheng Huang [1], Da Liu [1], Jing Qiu [1], Hui Chen [1], Yulong Li [2,*] and Zhan Hu [2]

[1] Guangdong Research Institute of Water Resources and Hydropower, Guangzhou 510630, China; gdsky_tanchao@foxmail.com (C.T.); bensheng@21cn.com (B.H.); gdsky_liuda@foxmail.com (D.L.); gdsky1202@163.com (J.Q.); chhui127@163.com (H.C.)

[2] School of Marine Sciences, Sun Yat-sen University, Guangzhou 518000, China; huzh9@mail.sysu.edu.cn

* Correspondence: liylong7@mail.sysu.edu.cn; Tel.: +86-(0)20-3933-3262

Received: 28 September 2018; Accepted: 29 December 2018; Published: 10 January 2019

Abstract: Flume experiments were performed to test four plant mimics with different stiffness to reveal the effect of plant stiffness on the wave dissipation and turbulence process. The mimics were built of silica gel rod groups, and their bending elastic modulus was measured as a proxy for stiffness. The regular wave velocity distribution, turbulence characteristics, and wave dissipation effect of different groups were studied in a flume experiment. Results show that, when a wave ran through the flexible rod groups, the velocity period changed gradually from unimodal to bimodal, and the secondary wave peak was more apparent in the more flexible mimics. The change in the turbulence intensity in the different rod groups showed that the higher the rod stiffness, the greater the turbulence intensity. With an increase in the bending elastic modulus of a rod group, the wave dissipation coefficient increased. The increase in the wave dissipation coefficient was not linearly correlated with the bending elastic modulus, but it was sensitive within a certain range of the elastic modulus.

Keywords: vegetation stiffness; bending elastic modulus; velocity distribution; turbulence intensity; wave dissipation

1. Introduction

Waves are one of the most important hydrodynamic force in coastal environments [1–3]. The reduction of coastal erosion induced by waves is an important topic for coastal protection and morphological changes [4–6]. Plants, such as mangroves, play an important role in protecting coasts. The planting of forests for wave attenuation in front of seawalls can reduce the arrival of waves, reduce the impact force of waves, and enhance the security of dams. It is known that different plant properties (e.g., density, stiffness, flexibility, arrangement mode, degree of submergence, and other factors) produce different influences on momentum transfer and the turbulence structure in canopy flow [7–12]. These related processes can lead to different sediment deposition patterns, which can influence the coastal morphology. The interesting point is that even the presence of a short, low-biomass seagrass meadow can lower the beach erosion rates compared to shallow unvegetated nearshore reef flats [13–16]. However, wave interactions between plants with varied stiffness have not been fully understood.

To investigate these interactions, Huang et al. [17] designed a physical model with a rigid main trunk and flexible branches and leaves. They then systematically analyzed wave propagation behaviors on a vegetated floodplain, as well as the effect of plant branches and leaves, tree trunks, the width of the beach, the depth of the water on the beach land, and wave elements on the propagation and deformation of the wave. Jiang et al. [18] used a physical model experiment of a wave flume to study the effect of changes in the wave height and wave form. Incident wave height, plant densities, reflection, transmission coefficient, and the wave energy dissipation were investigated. Moller and Spencer discovered that wave height decreased exponentially in a vegetated area [19]. Quartel et al. [20] performed an experiment at the Red River Delta in Vietnam and found that the wave dissipation ability of mangrove areas is five to 7.5 times more than that produced due to bottom friction. Bradley and Houser [21] quantitatively analyzed the effect of the relative movement of flexible seaweed leaves on wave height reduction in a reversing current. Fonseca and Cahalan [22] and Augustin et al. [11] showed that, when the height of seaweed was greater than or close to the water depth, the wave dissipation effect was obvious, and when the plant was submerged, the wave dissipation effect decreased with increased water depth. Tschirky and Hall [23] and Lima et al. [24] performed experiments that indicated that an increase in plant density enhanced the wave dissipation effect. However, Mazda et al. [25] and Horstman et al. [26] found that when the water depth in the mangrove was more than the height of the aerial roots, an increase in water depth reduced the wave dissipation effect, and when the water depth increased to the height of the mangrove leaves, the wave dissipation effect increased. Cruise and Muslesh [27] used a rigid pole to simulate the emerged portion of rigid vegetation and studied the effect of plant diameter and arrangement on water depth and velocity. White and Nepf [28] also studied the plant drag force, flow turbulence, and diffusion with a rigid rod.

Currently, there are many laboratory studies on vegetation under unidirectional currents and/or waves [29–32], field studies on wave dissipation through flexible vegetation [33–36], and turbulent flow through real mangrove roots [37]. However, the wave propagation and turbulence in vegetation with different stiffness is currently less studied [8].Therefore, in this study, the main aim is to study the wave propagation and turbulence characteristics among vegetation with different stiffness. A type of mimics used in this study is completely rigid, and the other three types of mimics are flexible to mimic the stems of macro algaes like *Fucus vesiculosus* and *Fucus serratus* with comparable elastic modulus (0.121–0.585 Gpa) [38]. The bending elastic modulus were measured using plant mimics made of silica gel rods with different stiffness. The regular wave velocity distribution, turbulence characteristics, and wave dissipation effect of different groups were studied to better understand the wave dissipation process through a vegetated field. The knowledge obtained by this study may provide a scientific reference for the planning and design of coastal protection projects.

2. Experimental Setup

2.1. Experimental Design

Experiments were conducted in a laboratory wave flume. The dimensions of the flume were 66 m long, 1.0 m wide, and 1.6 m deep. A piston-type waves paddle installed at one end of the flume was used to generate regular and irregular waves. For simplicity, we only tested regular waves in the current study. An overview of the flume, with its coordinate system and the wave maker, is shown in Figure 1. All of the instruments were deployed in this flume. Details of the flume dimensions and sensor deployments are also shown in Figure 1. The water surface elevation was measured using capacitance-type wave probes with good long-term stability and linear calibration curves. The wave probes were calibrated just prior to conducting the experiments. A SonTek 16-MHz Micro ADV (Acoustic Doppler Velocimeter) (SonTek/Xylem Inc.: San Diego, CA, USA) was used to measure the three-dimensional water velocity. The sampling frequency of the ADV and wave gages is 50 Hz. The data were collected after the waveform stabilized in the front of the mimic vegetation area, and the data collection lasted for 60 s. In order to eliminate the error, each experiment was repeated for three times. Peak velocities were obtained by taking the maximum value of an entire wave period.

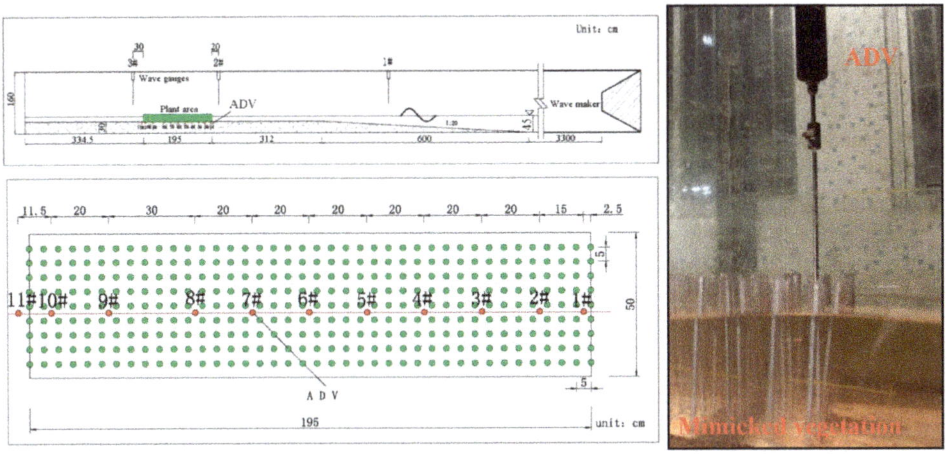

Figure 1. Model configuration and rod arrangement (from left to right, The green solid dots are the rods and the red ones are the ADV (Acoustic Doppler Velocimeter) measuring points).

The vegetation zone was composed of silica gel rods of different stiffness installed on a flat slope. The rods were fixed though the prefabricated holes on the at the slope bed. Wave gauges were installed before and after the vegetation zone to quantitatively measure the wave attenuation. The 3D flow field structure and turbulence characteristics were measured using the ADV. The current velocity was measured at the middle and bottom layers at 10 cm and 2 cm above the bed using the ADV. These two measuring points are regarded as representative heights of the vertical profile, while the information of the whole profile was not obtained.

The wave and flow design included no wave breaking action nor the presence of emergent vegetation. Therefore, the designed water depth of the floodplain was 15 cm, and the corresponding water depth before the wave plate was 45 cm. In addition, the regular wave height was 5 cm, and the wave period was 1.34 s. The resulting wave length was 1.53 m, and the tested water depth (0.15 m) in the vegetation canopy was half water depth. The wave condition is similar to previous lab work with small waves (wave height H $\leq$7 cm) [36,39].

2.2. Experiment Materials

The diameter of plant mimics (d) was 1 cm and mimic height (h_v) was 20 cm, and they were arranged into a rectangle with a total width of 195 cm consisting of 40 rows that were 5 cm apart with columns 5 cm apart (See Figure 1). The height of the mimics was determined to be similar to *F. vesiculosus*. The projecting area was 289.85 cm^2, which is also similar to the field conditions of *F. vesiculosus* [38]. Thus, essentially, this experiment did not involve scaling, as the tested mimics were dynamically similar to *F. vesiculosus* in the field conditions and the tested wave condition was also similar to the real field condition (depth = 0.15 m, wave height = 0.05 m, and period = 1.34 s). In the current experiment, the tested Re (Re = $u \times d/v$, where u is the velocity of the middle and bottom layers) number range was generally between 1000 and 2000.

Eleven measuring point was set along the center-line of the wave flume to minimize the influence of the side-wall. The obtained velocities and turbulence statistics are regarded as the representative measurements of the flume cross section. However, pair ADV measurements in the lateral direction were not conducted in our experiment. Thus, the current study mostly focused on velocity in the streamwise direction, i.e., u. The u velocity is positive when it is in the same direction as wave propagation, and it is negative when it is opposite to wave propagation.

We used a spring scale for the cantilever measurement. 10 rods of the same material were involved in each test and the experimental results are averaged. The elasticity modulus was calculated using the cantilever beam formula, and the stiffnesses of different materials were measured as:

$$\text{Cantilever beam formula}: \ E = FL^3/(3uI) \tag{1}$$

where E is the elasticity modulus (Pa); u is the offset distance (m); F is the transverse tensile force (N); I is the inertia moment, i.e., $I = \pi d^4/64$ for a circle; and L is the rod length (m).

The elastic modulus of the rods is shown in Table 1. Materials 1, 2, 3, and 4 with different stiffnesses are denoted as M1, M2, M3, and M4, respectively (Figure 2). These rods were commercially available. The elastic modulus of M1 was significantly greater than the others, and it was able to keep upright throughout the entire process. Therefore, M1 can be seen as a rigid rod.

Table 1. The bending elastic modulus of the different materials and statistics of the flow velocity peak values.

Material	Application Force (N)	Elasticity Modulus E (Gpa)	Measuring Point of 5#	
			Middle Layer Velocity (cm/s)	Bottom Layer Velocity (cm/s)
M1	30	16.56	22.17	19.95
M2	1.2	0.66	18.10	17.33
M3	0.7	0.39	17.23	16.08
M4	0.2	0.11	15.28	13.54

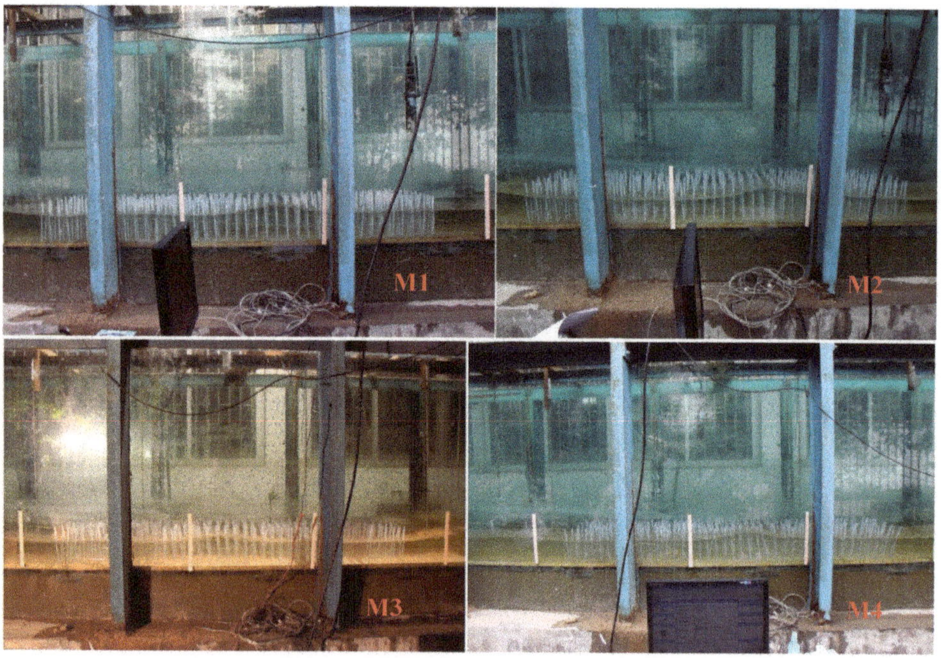

Figure 2. The picture of four materials of the flexible rods bending during the experiment.

2.3. Data Processing

The original data were phase-averaged according to Cox's theory [40]:

$$\mu_a(x,y,z,t_j) = \frac{1}{N}\sum_{n=1}^{N_j} \mu_m(x,y,z,t_n) j = 1,2,\ldots,\frac{T}{\Delta t}. \tag{2}$$

On the basis of $j = \frac{T}{\Delta t}$ points, the original 3D data were divided into three directions and N_j circles. Using the phase average, the velocity of each point in one circle can be obtained, $u_{ia}(i = x, y, z)$, and the fluctuating velocity can be presented as follows:

$$u_i' = u_{im} - u_{ia} i = x, y, z. \quad (3)$$

Turbulence intensity is the root mean square of the fluctuating velocity:

$$\eta = \sqrt{\overline{(u_i')^2}} i = x, y, z. \quad (4)$$

The probability density function of random data means the probability of an instantaneous value being within a specified range. For the turbulent process, the probability of its value, $u(t)$, being in the $(u_0, u_0 + \Delta u)$ can be defined as the following:

$$prob[u_0 < u(t) < u_0 + \Delta u] = \lim_{T \to \infty} \frac{T_s}{T}, \quad (5)$$

where T is the measuring time; and T_s is the sampling time within $(u_0, u_0 + \Delta u)$, $T_s = \sum_{i=1}^{n} \Delta t_i$. A probability density function of velocity measurements was made for each material. If this random process is a normal distribution, then the probability density function can be found using the following equation:

$$f(u_i') = \frac{1}{\sigma_i \sqrt{2\pi}} e^{-\frac{u_i^2}{2\sigma_i^2}}, \quad (6)$$

where f is the probability density; and u_i' is the fluctuating velocity in the i direction.

Reynolds stress is the shear force caused by the momentum exchange of a unit fluid passing through a unit area. The equation is the following:

$$\tau_{ij} = -\rho \overline{\mu_i' \mu_j'}, \quad (7)$$

where when $i = j$, $\sigma = -\rho \overline{\mu_i' \mu_j'}$, and σ is the normal stress; when $i \neq j$, τ_{ij} is the Reynolds shear stress.

3. Results and Discussion

3.1. Velocity Variations in the Mimicked Vegetation Canopy with Different Stiffnesses

3.1.1. Peak Velocities

The peak velocities changed significantly when waves crossed the different rod groups. The more flexible the plant, the smaller the peak the velocity. Table 1 shows the value of the velocity peaks. It was found that with an increase in rod flexibility (from M1 to M4), the velocity peak value in the middle and bottom layer both gradually diminished. Compared to M1, the peak value of the middle and bottom layer of M4 were reduced by 31% and 32%, respectively. The peak velocity value between two rods was increased. It is because rigid rods do not have any swing deformation, which squeezes the passing water and leads to higher velocity. For flexible rods, however, they sway as water passes, and hence do not lead to similar increased velocity. In fact, as the flexibility increases, the averaged velocity reduces (Figure 3).

Flexible rods do not cause contraction of the flow passing the flume section because of the unsynchronized swing. Therefore, the peak wave velocity was small. This is similar to what occurs when a bridge makes a channel narrow and increases the flow velocity.

3.1.2. Phase Averaged Velocity

According to the instantaneous velocity measured using the ADV, phase velocities in the direction of u are shown for different rod groups in Figure 3. Data from measuring point #5 is shown as it is in the center of the vegetation patch and it is representative of the averaged flow condition. The velocity curve shows that with a phase shift, the more flexible the materials are, the lower peak flow is. With a low flow velocity, the differences among the flow velocities of different materials are not obvious.

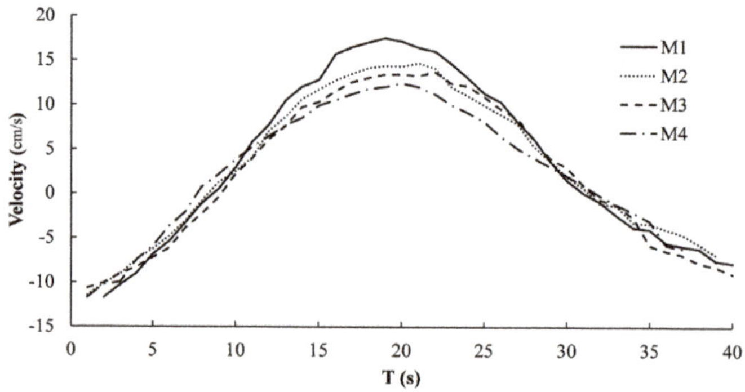

Figure 3. Phase averaged velocities in the u direction of different rod groups (measuring point of #5).

3.1.3. The Secondary Wave Peak in the Flexible Rod Groups

The experimental results show that when waves went through the flexible rod groups, the velocity period changed gradually from unimodal to bimodal, owing to swing in the rod group. The more flexible the rod group, the more obvious the secondary wave peak. Figure 4 shows velocity of the M4 rod group at measuring point #5. The figure shows that both in the middle and the bottom layer, bimodal structures existed during each wave period. The ratio between the secondary wave peak and the main wave peak in the middle layer was 0.49:1, while in the bottom layer it was 0.31:1. This phenomenon indicates that the swing extent of a rod increases as the water surface approaches, and its impact on the secondary peak of the wave velocity also increases.

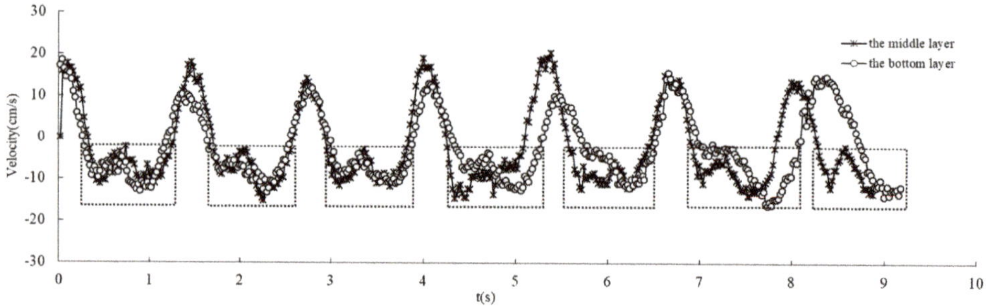

Figure 4. Velocity at measuring point #5 of the material 4 (M4) rod group. The dash boxes show the secondary peaks of the wave velocity in the negative direction.

3.2. *Turbulence Characteristics of the Different Rod Groups*

3.2.1. Turbulence Intensity

The middle layer turbulence intensity distributions in the u direction for the M1 and M4 rod groups are presented in Figure 5. The middle and bottom layer turbulence distributions of the different materials are shown in Figure 6. Spatial changes in the turbulence intensity indicates that the highest value occurs during the period of the wave entrance into the rod group and in the middle of the rod group. Possible explanations include the following: (1) The water's entrance into the rod group means that the wave propagates from one interface to another interface, which can result in intense turbulence; and (2) wave streaming causes intense turbulence when the wave propagates in the middle of the rod group.

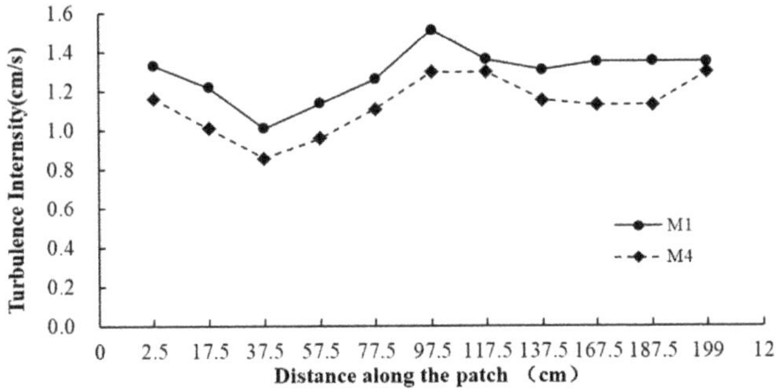

Figure 5. Turbulence intensity in the *u* direction in the material 1 and 4 (M1 and M4) rod groups.

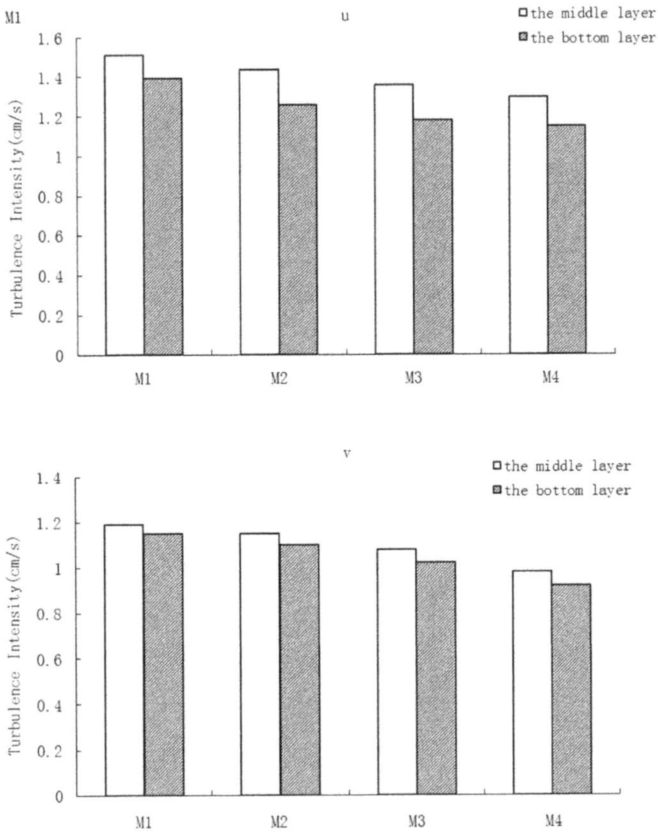

Figure 6. *Cont.*

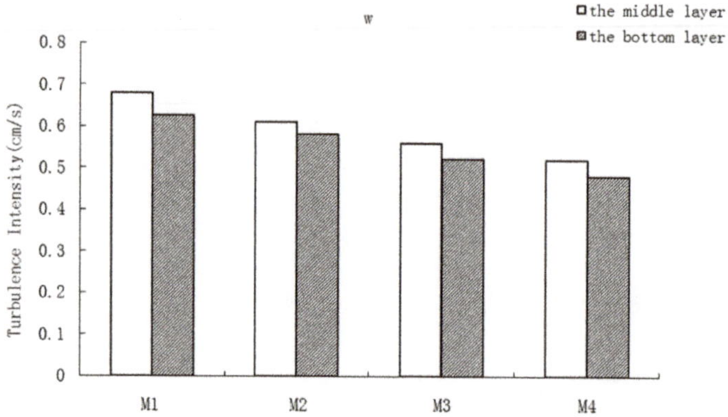

Figure 6. Middle and bottom layer turbulence distributions of the different materials (measuring point #5).

The turbulence intensity changes in different materials indicate that the greater the material stiffness, the stronger the turbulence velocity and intensity. From M1 to M4, the turbulence intensity reduces 5.1%, 5.4%, and 4.4%, respectively, in the *u* direction of measuring point #5. This result is similar to that of a previous experiment by Pujol et al., [29].

The turbulence intensity in different directions shows that the largest was in the u direction followed by the v direction, and the intensity in the w direction was minimal. As far as the vertical distribution, the turbulence intensity in the bottom layer was smaller than in the surface layer. This reflects that turbulence was anisotropic when the vegetation patch was under wavy flows.

3.2.2. Probability Density of the Fluctuating Velocity

If the probability density is a normal distribution, the wider the graph, the larger the velocity deviation. The y axis intercept is the turbulence intensity in Figure 7. Figure 7 shows the probability density distribution of the fluctuating velocity in the *u*, *v*, and *w* directions. Two peak values are found in the probability density function of the *u* direction, and at the same time, the turbulence intensity decreases with an increase in the flexibility of the rod group. While in the *v* direction and *w* direction, the probability density follows a normal distribution, and the differences between different materials are not obvious.

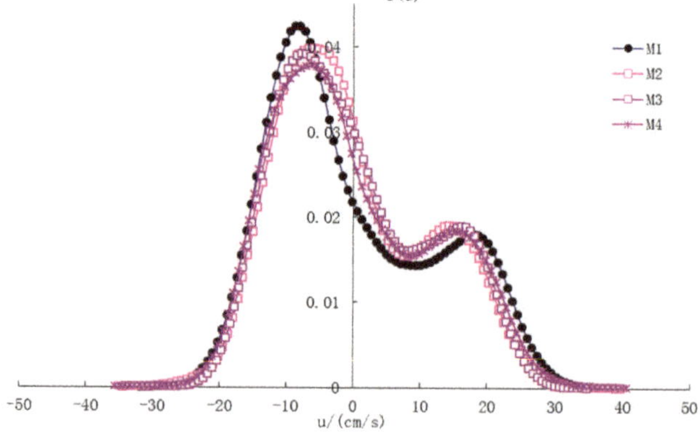

Figure 7. *Cont.*

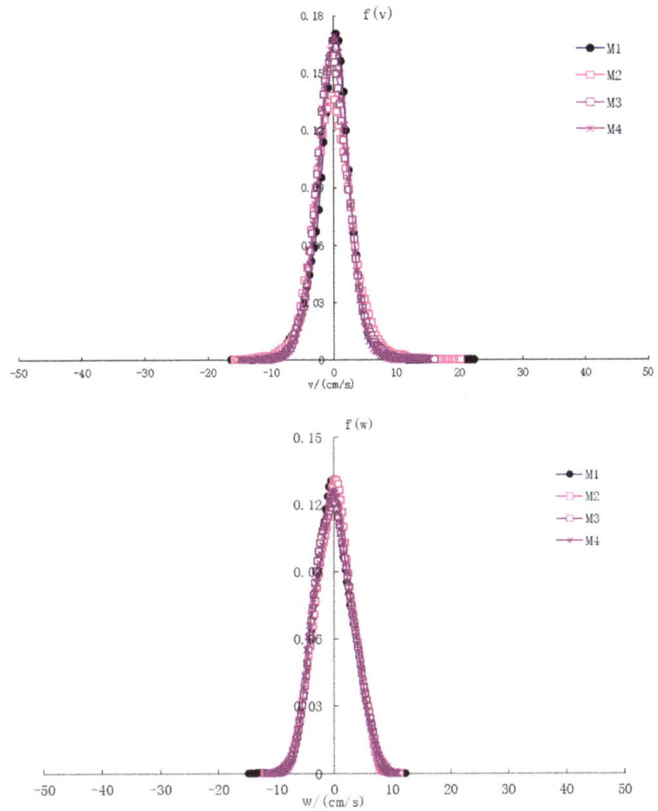

Figure 7. Probability density distribution of the fluctuating velocity in the u, v, and w direction (at measuring point #5).

3.2.3. Reynolds Stress

Reynolds stress is generally greater than the viscous shear force in vegetated flows [41]. Therefore, only near the sidewall is the viscosity term considered, otherwise it is ignored.

Reynolds stress is the result of an uneven flow velocity distribution in a flow field. Therefore, the more uneven the velocity distribution is, the greater the Reynolds stress is, and the stronger the turbulence. Figure 8 shows the change in Reynolds stress for different rod groups, where $R_a = \langle u'v' \rangle$, $R_b = \langle u'w' \rangle$ and $R_c = \langle v'w' \rangle$ (u', v', and w' represent fluctuating velocities in the u, v, and w directions, respectively). Results show that the Reynolds stress decreases with increasing stiffness. The Reynolds stress of the M4 middle layer is only 10% that of M1.

Moreover, the Reynolds stress of the middle layer is larger than the bottom layer, which is similar to Ma's [42] research on the wave turbulence. The middle layer Reynolds stress is about 1.14 times that of the bottom layer in the rigid rod group (M1), and about 1.52 times that of the flexible rod group (M4).

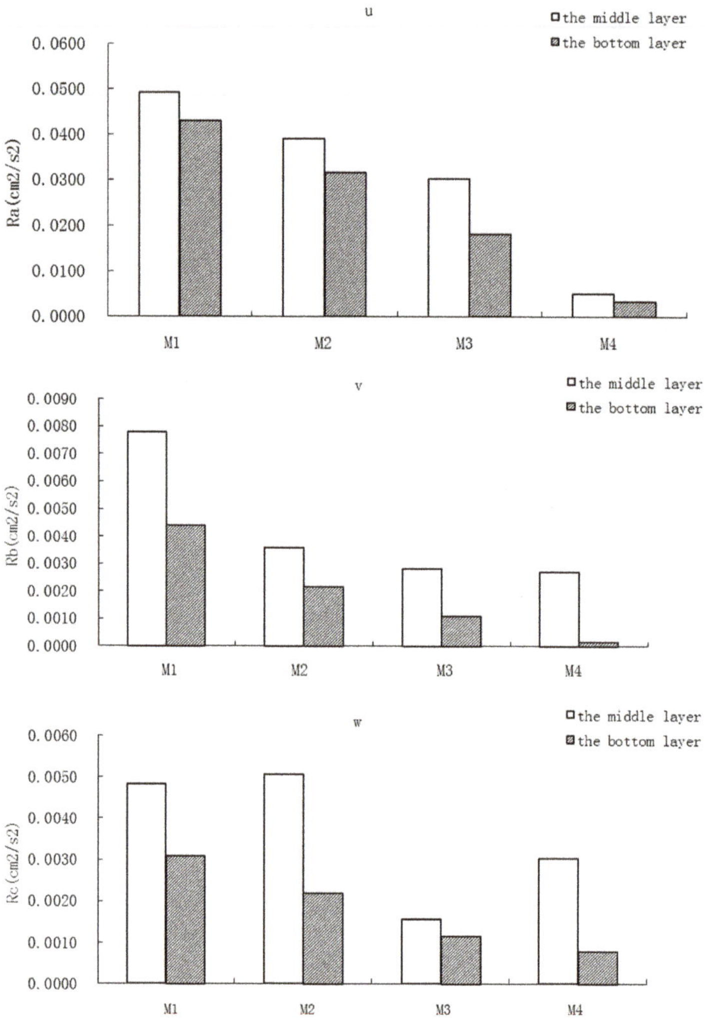

Figure 8. Reynolds stresses for different rod groups (at measuring point #5).

3.2.4. Energy Spectrum Density

The turbulent process can be seen as superposition of simple harmonic waves with different frequencies. Velocity spectra were computed and give the parameters used to compute the power spectra ($E(n)$, $\int_{-\infty}^{\infty} E(t)dn = \frac{1}{2}\overline{u^2}$). The energy spectral density curve represents the distribution of turbulent kinetic energy in a wave band $(w, w + dw)$ in steady time. Time is the inverse of frequency. We set a 95% confidence interval on each of the spectra to eliminate the effect of noise. The high frequency in the energy spectrum represents the quickly changing turbulence, or turbulence on a small time scale [43].

The energy spectral density distributions in the u direction for different material rod groups at measuring point #5 are shown in Figure 9, which is related to the wave energy transmission process. As can be seen from the figure, when a wave propagates in the rod groups, two energy spectral peaks exist, with the main peak value larger than the secondary peak. With a reduction in material rigidity, the main peak value of the wave energy decreases, which means that the wave turbulence intensity is reduced. Furthermore, the secondary peak of M4 is the largest among all the cases. The secondary peaks were related to rod group swing. The more flexible the rod group was, the more obvious the secondary wave peak was.

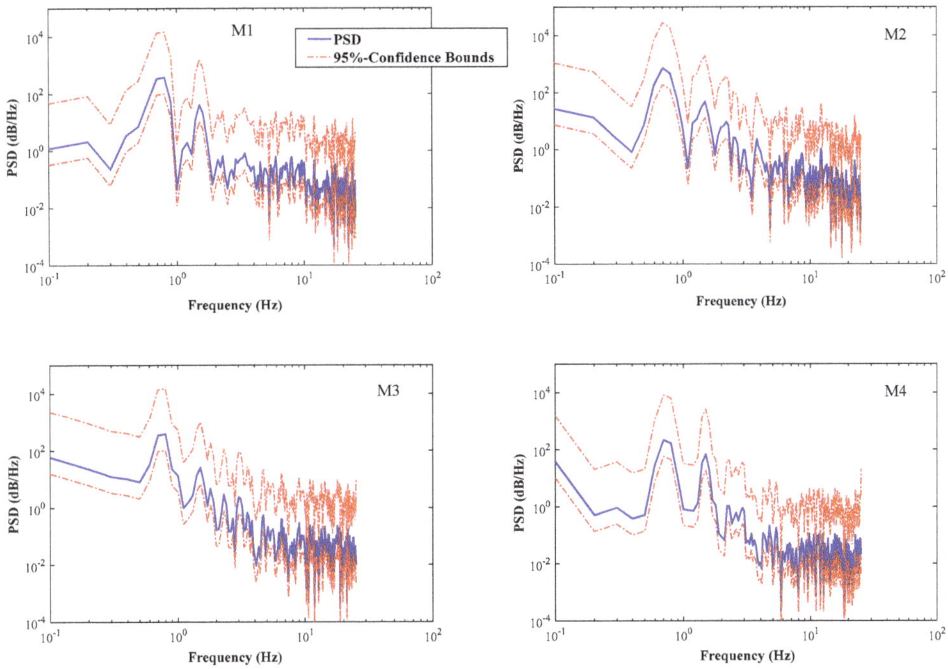

Figure 9. Energy spectral density distributions in the *u* direction for different material rod groups at measuring point #5.

3.3. Wave Dissipation Effect in the Different Rod Groups

The change in wave height before and after the wave moves through the rod groups represent the attenuation of wave energy (see Table 2 and Figure 10). When the bending elastic modulus of the rod group increases from 0.11 GPa to 0.39 Gpa (Compare M4 with M3), the wave dissipation coefficient correspondingly increases from 25.17% to 39.79%. When the flexural elastic modulus increases from 0.39 GPa to 16.56 Gpa (Compare M3 with M1), the wave dissipation coefficient increases from 39.79% to 40.45%, only an increase of 1.66%. M2 and M3 may have happened to be in an area that was insensitive to stiffness, causing the wave dissipation coefficients to fluctuate.

Table 2. Wave height and wave dissipation coefficient before and after the wave moves through the rod groups.

Material	Bending Elastic Modulus E (Gpa)	Wave Height before the Rod Groups (cm)	Wave Height after the Rod Groups (cm)	Wave Dissipation Coefficient
M 1	16.56	5.81	3.46	40.45%
M 2	0.66	5.79	3.56	38.51%
M 3	0.39	5.83	3.51	39.79%
M 4	0.11	5.84	4.37	25.17%

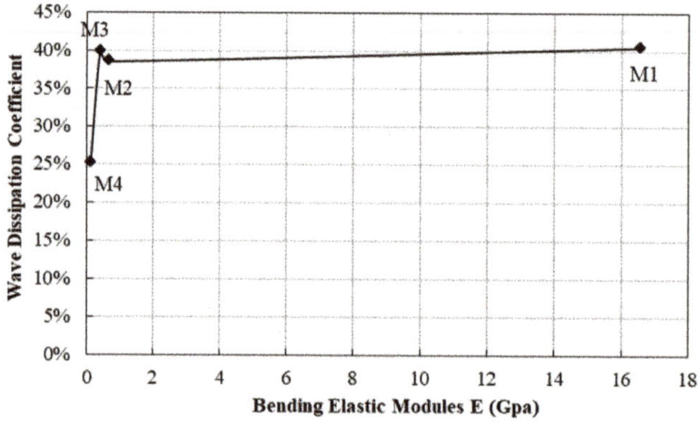

Figure 10. The relation of the Bending Elastic Modules E (Gpa) and the Wave Dissipation Coefficient.

Overall, the wave dissipation coefficient increases with increases in the bending modulus of elasticity in the rod group. More succinctly, the greater the stiffness of the rod group is, the more obvious the energy dissipation effects will be. In addition, the growth of the wave dissipation coefficient is not linear with the bending elastic modulus; but is sensitive within a certain range of the elastic modulus. There is a sharp quick increase in the wave dissipation coefficient. However, when the bending elastic modulus value increases to 0.39 Gpa, the wave dissipation coefficient growth becomes extremely small.

The above behaviors can also be interpreted from the physical phenomenon point of view. M3 and M4 obviously swing more under wave flow. M2 only slightly swings when the wave peak passes. M1 is completely rigid and does not swing. This phenomenon shows that the bending elastic modulus values of M3 and M4 happen to be in the most sensitive ranges for a swing reaction under group wave conditions; that is, in the most sensitive ranges for a change in the wave dissipation coefficient.

Our results showed that when waves ran through flexible vegetation mimics, the velocity period changed gradually from unimodal to bimodal. This phenomenon is likely due to the swaying effect of the flexible vegetation [8,31], as it is more apparent with flexible mimics. The change in the turbulence intensity in the different rod groups showed that the higher the rod stiffness, the greater the turbulence intensity exists. This result is similar to that in Reference [29]. With an increase in the bending elastic modulus of a rod group, the wave dissipation coefficient increased, which is consistent with the previous studies [39,41]. However, the increase in the wave dissipation coefficient was not linearly correlated with the bending elastic modulus. It was more sensitive in a certain range of the elastic modulus than others.

4. Conclusions

The bending elastic modulus was measured using a conceptual plant model that was built of silica gel rod groups of different stiffness. The regular wave velocity distribution, turbulence characteristics, and wave dissipation effect of the different groups were studied. According to the results, the conclusions that follow can be drawn.

(1) When waves went through different material rod groups, the peak velocity of the wave was in decay. The more flexible the rod group, the smaller the peak flow velocity. With a low flow velocity, the differences among the flow velocities of the different materials was not apparent.
(2) When waves go through the flexible rod group, the velocity period gradually changed from unimodal to bimodal. Owing to rod group swing, the more flexible the rod group was, the more obvious the secondary wave peak was. With a reduction in material rigidity, the second peak value of the wave energy decreased, which was related to flow shocks that were caused by

the swing of the flexible rod group. It is expected that, with different wave periods, the swing behavior and the wave energy transmission will be different, which should be further studied.

(3) High turbulence intensity existed in the areas at the front and in the middle of the rod group. This was because when the wave entered the rod group, the wave propagated from one interface to another, resulting in intensified turbulence.

(4) The greater the material stiffness was, the stronger the turbulence velocity and intensity were. The Reynolds stress decreased with increased flexibility. Additionally, the middle layer Reynolds stress was generally larger than that at the bottom layer.

The insights on different patterns in wave propagation turbulence intensity in different canopies may lead to further understanding of the coastal morphological changes with vegetation influence and may assist in selecting vegetation species with suitable stiffness for coastal protection purposes.

Author Contributions: C.T., Y.L., Z.H. conceived and designed the methodology and experiments. C.T., Y.L., B.H., H.C. and D.L. performed the experiments. C.T., J.Q., H.C. and B.H. analyzed the data. All the authors contributed to the writing and editing of the manuscript.

Funding: The authors gratefully acknowledge financial support of the 2018 Guangzhou science and technology project (No. 201806010143), the National Key R&D Program of China (No. 2016YFC0402607), the National Natural Science Foundation of China (No. 51609269).

Acknowledgments: We thank the three anonymous reviewers for their comments and suggestions.

Conflicts of Interest: The authors declare no conflict of interest.

References

1. Green, M.O. Very small waves and associated sediment resuspension on an estuarine intertidal flat. *Estuar. Coast. Shelf Sci.* **2011**, *93*, 449–459. [CrossRef]
2. Green, M.O.; Coco, G. Review of wave-driven sediment resuspension and transport in estuaries. *Rev. Geophys.* **2014**, *52*, 77–117. [CrossRef]
3. Callaghan, D.P.; Leon, J.X.; Saunders, M.I. Wave modelling as a proxy for seagrass ecological modelling: Comparing fetch and process-based predictions for a bay and reef lagoon. *Estuar. Coast. Shelf Sci.* **2015**, *153*, 108–120. [CrossRef]
4. Le Hir, P.; Roberts, W.; Cazaillet, O.; Christie, M.; Bassoullet, P.; Bacher, C. Characterization of intertidal flat hydrodynamics. *Cont. Shelf Res.* **2000**, *20*, 1433–1459. [CrossRef]
5. Stive, M.J.F.; De, S.; Luijendijk, A.P.; Aarninkhof, S.G.J.; Van, G.-M.; Van, T.D.V.; De, V.; Henriquez, M.; Marx, S.; Ranasinghe, R. A new alternative to saving our beaches from sea-level rise: The sand engine. *J. Coast. Res.* **2013**, *29*, 1001–1008. [CrossRef]
6. Temmerman, S.; Meire, P.; Bouma, T.J.; Herman, P.M.; Ysebaert, T.; De Vriend, H.J. Ecosystem-based coastal defence in the face of global change. *Nature* **2013**, *504*, 79–83. [CrossRef]
7. Nepf, H.M. Flow and Transport in Regions with Aquatic Vegetation. *Annu. Rev. Fluid Mech.* **2011**, *44*, 123–142. [CrossRef]
8. Luhar, M.; Nepf, H.M. Wave-induced dynamics of flexible blades. *J. Fluids Struct.* **2016**, *61*, 20–41. [CrossRef]
9. Suzuki, T.; Zijlema, M.; Burger, B.; Meijer, M.C.; Narayan, S. Wave dissipation by vegetation with layer schematization in SWAN. *Coast. Eng.* **2012**, *59*, 64–71. [CrossRef]
10. Cao, H.; Feng, W.; Hu, Z.; Suzuki, T.; Stive, M.J.F. Numerical modeling of vegetation-induced dissipation using an extended mild-slope equation. *Ocean Eng.* **2015**, *110*, 258–269. [CrossRef]
11. Augustin, L.N.; Irish, J.L.; Lynett, P. Laboratory and numerical studies of wave damping by emergent and near-emergent wetland vegetation. *Coast. Eng.* **2009**, *56*, 332–340. [CrossRef]
12. Dalrymple, R.; Kirby, J.; Hwang, P. Wave Diffraction Due to Areas of Energy Dissipation. *J. Waterw. Port Coast. Ocean Eng.* **1984**, *110*, 67–79. [CrossRef]
13. Bos, A.R.; Bouma, T.J.; de Kort, G.L.J.; van Katwijk, M.M. Ecosystem engineering by annual intertidal seagrass beds: Sediment accretion and modification. *Estuar. Coast. Shelf Sci.* **2007**, *74*, 344–348. [CrossRef]

14. Christianen, M.J.A.; van Belzen, J.; Herman, P.M.J.; van Katwijk, M.M.; Lamers, L.P.M.; van Leent, P.J.M.; Bouma, T.J. Low-Canopy Seagrass Beds Still Provide Important Coastal Protection Services. *PLoS ONE* **2013**, *8*, e62413. [CrossRef] [PubMed]
15. Ondiviela, B.; Losada, I.J.; Lara, J.L.; Maza, M.; Galvan, C.; Bouma, T.J.; van Belzen, J. The role of seagrasses in coastal protection in a changing climate. *Coast. Eng.* **2014**, *87*, 158–168. [CrossRef]
16. De Boer, W.F. Seagrass-sediment interactions, positive feedbacks and critical thresholds for occurrence: A review. *Hydrobiologia* **2007**, *591*, 5–24. [CrossRef]
17. Huang, B.S.; Lai, G.W.; Qiu, J.; Lin, S.Z. Hydraulice of compound channel with vegetated flood plains. *J. Hydrodyn. Ser. B* **2002**, *14*, 23–28.
18. Jiang, C.B.; Wang, R.X.; Chen, J.; Chen, J.K.; Zhi-Yuan, W.U. Laboratory investigation on solitary wave transformation through the emergent rigid vegetation. *J. Changsha Univ. Sci. Technol.* **2012**, *9*, 50–55.
19. Möller, I.; Spencer, T. Wave dissipation over macro-tidal saltmarshes: Effects of marsh edge typology and vegetation change. *J. Coast. Res.* **2002**, *36*, 506–521. [CrossRef]
20. Quartel, S.; Kroon, A.; Augustinus, P.G.E.F.; Santen, P.V.; Tri, N.H. Wave attenuation in coastal mangroves in the Red River Delta, Vietnam. *J. Asian Earth Sci.* **2007**, *29*, 576–584. [CrossRef]
21. Bradley, K.; Houser, C. Relative velocity of seagrass blades: Implications for wave attenuation in low-energy environments. *J. Geophys. Res. Earth Surf.* **2009**, *114*, F01004. [CrossRef]
22. Fonseca, M.S.; Cahalan, J.A. A preliminary evaluation of wave attenuation by four species of seagrass. *Estuar. Coast. Shelf Sci.* **1992**, *35*, 565–576. [CrossRef]
23. Tschirky, P.; Hall, K.; Turcke, D. Wave attenuation by emergent wetland vegetation. In Proceedings of the 27th International Conference on Coastal Engineering, Sydney, Australia, 16–21 July 2000. ASCE.865-877.
24. Lima, S.F.; Neves, C.F.; Rosauro, N.M.L. Damping of gravity waves by fields of flexible vegetation. In Proceedings of the 30th International Coastal Engineering Conference, San Diego, CA, USA, 3–8 September 2006; pp. 491–503.
25. Mazda, Y.; Magi, M.; Ikeda, Y.; Kurokawa, T.; Asano, T. Wave reduction in a mangrove forest dominated by *Sonneratia* sp. *Wetl. Ecol. Manag.* **2006**, *14*, 365–378. [CrossRef]
26. Horstman, E.; Dohmen-Janssen, M.; Narra, P.; van den Berg, N.-J.; Siemerink, M.; Balke, T.; Bouma, T.; Hulscher, S. Wave attenuation in mangrove forests; field data obtained in Trang, Thailand. *Coast. Eng. Proc.* **2012**, *1*, 40. [CrossRef]
27. Cruise, J.F.; Musleh, F.A. Functional Relationships of Resistance in Wide Flood Plains with Rigid Unsubmerged Vegetation. *J. Hydraul. Eng.* **2006**, *132*, 163–171. [CrossRef]
28. White, B.; Nepf, H. A vortex-based model of velocity and shear stress in a partially vegetated shallow channel. *Water Resour. Res.* **2008**, *44*, W01412. [CrossRef]
29. Pujol, D.; Serra, T.; Colomer, J.; Casamitjana, X. Flow structure in canopy models dominated by progressive waves. *J. Hydrol.* **2013**, *486*, 281–292. [CrossRef]
30. Lowe, R.J.; Koseff, J.R.; Monismith, S.G. Oscillatory flow through submerged canopies: 1. Velocity structure. *J. Geophys. Res. C Oceans* **2005**, *110*, C10016. [CrossRef]
31. Luhar, M.; Nepf, H.M. Flow-induced reconfiguration of buoyant and flexible aquatic vegetation. *Limnol. Oceanogr.* **2011**, *56*, 2003–2017. [CrossRef]
32. Ghisalberti, M.; Nepf, H. The structure of the shear layer in flows over rigid and flexible canopies. *Environ. Fluid Mech.* **2006**, *6*, 277–301. [CrossRef]
33. Riffe, K.C.; Henderson, S.M.; Mullarney, J.C. Wave dissipation by flexible vegetatio. *Geophys. Res. Lett.* **2011**, *38*, L18607. [CrossRef]
34. Stevens, A.W.; Lacy, J.R. The Influence of Wave Energy and Sediment Transport on Seagrass Distribution. *Estuaries Coasts* **2012**, *35*, 92–108. [CrossRef]
35. Luhar, M.; Infantes, E.; Orfila, A.; Terrados, J.; Nepf, H.M. Field observations of wave-induced streaming through a submerged seagrass (Posidonia oceanica) meadow. *J. Geophys. Res. Oceans* **2013**, *118*, 1955–1968. [CrossRef]
36. Bouma, T.J.; De Vries, M.B.; Low, E.; Peralta, G.; Tánczos, I.C.; Van De Koppel, J.; Herman, P.M.J. Trade-offs related to ecosystem engineering: A case study on stiffness of emerging macrophytes. *Ecology* **2005**, *86*, 2187–2199. [CrossRef]

37. Horstman, E.M.; Bryan, K.R.; Mullarney, J.C.; Pilditch, C.A.; Eager, C.A. Are flow-vegetation interactions well represented by mimics? A case study of mangrove pneumatophores. *Adv. Water Resour.* **2018**, *111*, 360–371. [CrossRef]
38. Paul, M.; Henry, P.-Y.T.; Thomas, R.E. Geometrical and mechanical properties of four species of northern European brown macroalgae. *Coast. Eng.* **2014**, *84*, 73–80. [CrossRef]
39. Bouma, T.J.; De Vries, M.B.; Herman, P.M.J. Comparing ecosystem engineering efficiency of two plant species with contrasting growth strategies. *Ecology* **2010**, *91*, 2696–2704. [CrossRef]
40. Cox, D.T.; Kobayashi, N.; Okayasu, A. *Experimental and Numerical Modeling of Surf Zone Hydrodynamics*; University of Delaware, Center for Applied Coastal Research, Ocean Engineering Laboratory: Newark, DE, USA, 1995; pp. 32–50.
41. Nepf, H.M. Flow Over and Through Biota. In *Treatise on Estuarine and Coastal Science*; Wolanski, E., McLusky, D., Eds.; Academic Press: Waltham, MA, USA, 2011; pp. 267–288. ISBN 978-0-08-087885-0.
42. Ma, D. Wave turbulence and calculation. *J. Tianjin Univ.* **1964**, *14*, 16–19.
43. Cui, C.; Zhang, N.C.; Guo, C.S.; Fang, Z. Impact of water depth variation on simulated freak waves and their time-frequency energy spectrum. *Acta Oceanol. Sin.* **2011**, *33*, 174–179.

water

MDPI

Technical Note

Applying a New Force–Velocity Synchronizing Algorithm to Derive Drag Coefficients of Rigid Vegetation in Oscillatory Flows

Peng Yao [1,2,3], Hui Chen [4], Bensheng Huang [4], Chao Tan [4], Zhan Hu [1,3,*], Lei Ren [3,5,*] and Qingshu Yang [3,5]

1 School of Marine Science, Sun Yat-sen University, Guangzhou 510275, China; yaop3@mail.sysu.edu.cn
2 State Key Laboratory of Hydrology Water Resources and Hydraulic Engineering, Nanjing Hydraulic Research Institute, Nanjing 210029, China
3 Guangdong Province Engineering Research Center for Coasts, Islands and Reefs, Guangzhou 510275, China; yangqsh@mail.sysu.edu.cn
4 Guangdong Research Institute of Water Resources and Hydropower, Guangzhou 510630, China; chhui2@mail2.sysu.edu.cn (H.C.); bensheng@21cn.com (B.H.); gdsky_tanchao@foxmail.com (C.T.)
5 School of Marine Engineering and Technology, Sun Yat-sen University, Guangzhou 510275, China
* Correspondence: huzh9@mail.sysu.edu.cn (Z.H.); renlei7@mail.sysu.edu.cn (L.R.); Tel.: +86-(0)20-3933-3262 (Z.H.)

Received: 24 May 2018; Accepted: 6 July 2018; Published: 9 July 2018

Abstract: Coastal vegetation is effective in dissipating incident wave energy during storm conditions, which offers valuable protection to coastal communities. Determining vegetation drag coefficient (C_D) is of great importance to the quantification of vegetation-induced wave dissipation. Recently, a direct measuring approach has been developed to derive vegetation drag coefficient more accurately compared to the conventional calibration approach. However, as this approach requires perfectly in-phase force and velocity signals, there are two difficulties associated with it. The first difficulty is the availability of a suitable force sensor to compose synchronized force–velocity measuring systems. The second difficulty is related to realigning the obtained timeseries of force and velocity data. This technical note develops a new synchronized force–velocity measuring system by using standard force sensors and an acoustic doppler velocimeter (ADV). This system is applied together with an automatic realignment algorithm to ensure in-phase data for C_D deviation. The algorithm reduces the phase shift between force–velocity signals from ca. 0.26 s to 0.003 s. Both time-varying and period-averaged C_D can be obtained using this method. The derived C_D can be used to accurately reproduce the measured maximum total acting force on vegetation (R^2 = 0.759), which shows the reliability of the automatic alignment algorithm. The newly-developed synchronized force–velocity measuring system and alignment algorithm are expected to be useful in future experiments on vegetation–wave interactions with various hydrodynamic and vegetation settings.

Keywords: drag coefficients; oscillatory flows; force sensors; synchronization; automatic alignment

1. Introduction

Mangroves, saltmarshes, and seagrasses are important coastal ecosystems that are widely distributed in world's coasts [1–3]. The wave-damping capacity of these coastal wetlands has been increasingly recognized [1,4–8]. These coastal wetlands can significantly reduce wave energy even under storm or tsunami conditions [9,10], which provides valuable protection to the coastal communities and properties [11,12]. Over the past decades, the wave heights have a clear increase trend in extreme conditions, together with accelerated sea level rise [13–17]. Therefore, there is a demand for better understanding and predictive ability of vegetation–wave interaction process to reduce coastal flooding risks [4,18–20].

The main impact of vegetation on incident waves is exerting an additional force on water motion [21,22]. This force can be described by the Morison equation, which is composed by drag force (F_D) and inertia force (F_M) [23]. For normal field conditions, the drag force is the dominant force, and most relevant for wave energy dissipation. In the Morison equation, F_D is proportional to the square of impact velocity on vegetation stems. When the velocity scale is determined, the magnitude of F_D varies linearly with vegetation drag coefficients (C_D). In oscillatory (wavy) flows, the C_D values have a large range of variations (i.e., 0.1 to 100) [24]. The C_D values depend on canopy density, hydrodynamic conditions, as well as the morphology of the individual canopy elements. Thus, choosing appropriate C_D values are important for accurate simulation of F_D, and the resultant wave dampening in many modelling studies [18,25–30].

Currently, there are two methods available in determining C_D: the calibration method and the direct measurement method. The calibration method is a convectional method developed in the 1990s [31,32], and has been widely used since [33–35]. It derives C_D by calibrating its values to obtain the best fit between modelled and measured wave height evolution over vegetation fields. The direct measurement method is a new method, which has been developed since the 2010s [36–38]. This method directly applies the Morrison equation and measured in-phase force and velocity data to determine C_D. The main differences between the calibration and the direct measurement method are: (1) the calibration method can only provide period-averaged C_D, but the direct measurement method can derive both period-averaged and time-varying C_D; (2) the direct measurement method can eliminate the potential errors often associated with the calibration method, and lead to C_D–Re (Reynolds number) relations with better fits, which are desirable for model applications [37].

Since the direct measurement method relies on in-phase force and velocity data, there are two difficulties when applying this method. The first difficulty is the availability of suitable force sensors to assemble synchronized force–velocity measuring systems. The force sensors should be waterproofed and durable in wave flumes, where frequent water logging and splashing occur. Additionally, the sensors should be small enough to fit into wave flumes. The second difficulty associated with this method is the data processing technique required to obtain perfectly aligned force–velocity data for C_D derivation [37]. The data alignment is critical for the direct measurement method, as there are time lags between original force data and velocity data signals (ca. 0.2 s), which may lead to large errors in the derived C_D. These time lags may originate from small misalignments between force sensors and velocity measurement [37]. They may also be induced by intrinsic time shifts in instrument recordings. The maximum wave energy dissipation occurs at the peak wave orbital velocity in phase with the peak drag force when C_D values matter the most. Thus, in order to obtain accurate C_D values, it is important to minimize the time lags. Previous studies firstly set an intrinsic time lag between the force and velocity data, and then started iterations to reduce the time lag. Note that this intrinsic time lag varies with different instrument set-ups. This intrinsic time lag needs to be carefully tuned to obtain in-phase data. It is, however, preferable to have an automatic algorithm that can provide generic solutions to the alignment problem.

In this technical note, we have developed (1) a synchronized force–velocity measuring system by using standardized force sensors that can easily fit in wave flumes; (2) and an automatic alignment algorithm to obtain in-phase force–velocity data for C_D derivation. The new force–velocity measuring systems are applied in a flume at four locations in a mimicked mangrove canopy, which were tested with various simulated wave conditions. The automatic alignment algorithm was then applied to reduce the time lags between force–velocity signals. The processed data were subsequently used to derive both time-varying and period-averaged C_D. To evaluate the accuracy of the derived C_D (and also the alignment algorithm), we used the derived C_D to reproduce the total acting force on mimicked vegetation, and compared it with the measurements.

The rest of the technical note is organized as follows: Section 2 introduces the automatic alignment algorithm and the set-up of the synchronized force–velocity measuring system in the wave flume at the Fluid Mechanics Laboratory at Sun Yat-sen University, Zhuhai Campus. Section 3 demonstrates the original and processed force–velocity data, as well as the reproduced force data results. Section 4 discusses the current limitations and provides an outlook for future applications of this method. Finally, Section 5 provides conclusions of the current note.

2. Materials and Methods

2.1. The Direct Measuring Method for C_D Derivation

The force acting on a single stem can be expressed by Morison equation [23] as

$$F = F_D + F_M = \frac{1}{2}\rho C_D h_v b_v U|U| + \frac{\pi}{4}\rho C_M h_v {b_v}^2 \frac{\partial U}{\partial t} \tag{1}$$

where F is the total acting force on vegetation which can be obtained by force measurement. F_D is drag force, and F_M is inertia force. ρ is the density of the fluid. C_D and C_M are the drag and inertia coefficients, respectively. h_v is the height of vegetation in water, and b_v is the diameter of circular cylinder. U is the depth-averaged flow velocity. When linear wave theory is applied, the U varies as a function of sine:

$$U = U_w \sin(\omega t) \tag{2}$$

where U_w is the amplitude of horizontal wave orbital velocity. Following linear wave theory, U_w can be expressed as

$$U_w = \frac{\pi H}{T}\frac{\cos \mathrm{h}[k(h + Z_0)]}{\sin \mathrm{h}(kh)} \tag{3}$$

where H is wave height, T is wave period, k is wave number, h is water depth, and Z_0 is the vertical position of the considered point, which is 0 at the still wave level, and $-h$ at the sea bed. Equation (3) was used to estimate U_w when the velocity measurement is unavailable. C_M is often assumed to be equal to 2 for cylinders (e.g., [39]). To derive the time-varying C_D, we can apply the following equation:

$$C_D = \frac{2F_D}{\rho h_v b_v U|U|} = \frac{2(F - F_M)}{\rho h_v b_v U|U|} \tag{4}$$

where F_M can be derived based on $\frac{\partial U}{\partial t}$ using the timeseries of velocity data, and other parameters (i.e., ρ, b_v, h_v, C_M) in F_M are known. Thus, time-varying C_D can be obtained readily when in-phase force and velocity data is obtained.

Period-averaged C_D is relevant to vegetation-induced wave dissipation. It was not computed as the temporal mean of the time-varying C_D. Time-varying C_D has great variability over one wave period [37]. Specifically, its value is infinite when the velocity is close to zero. However, those C_D values are not relevant for vegetation-induced wave energy dissipation, as dissipation is highest at the velocity peaks. Thus, C_D values at high velocity matter the most. To obtain relevant period-averaged C_D values, the direct measurement method applies the technique of quantifying the power and work done by the acting force (ε) [37]. The time-varying power of F_D and F_M is evaluated as follows:

$$P_D = F_D U \tag{5}$$

$$P_M = F_M U \tag{6}$$

The work done by the total acting force (F) over a wave period (T) is

$$W = \int_0^T FU dt \tag{7}$$

If we substitute F with Equation (1), then we obtain

$$W = W_D + W_M = \frac{1}{T}\int_0^T F_D U dt + \frac{1}{T}\int_0^T F_M U dt = \frac{1}{2T}\int_0^T \rho C_D h_v b_v U^2 |U| dt + \frac{\pi}{4T}\int_0^T \rho C_M h_v b_v{}^2 \frac{\partial U}{\partial t} U dt \quad (8)$$

W_D and W_M is the work done by F_D and F_M over a full period, respectively. As U is a sine function (Equation (2)), the work done by F_M over a full wave period (i.e., second term on the right) is zero. Thus, the work done by F is equal to the work done by F_D:

$$W = \int_0^T F U dt = \frac{1}{2}\int_0^T \rho C_D h_v b_v U^2 |U| dt \quad (9)$$

Finally, the period-averaged C_D can be derived based on the above equation:

$$C_D = \frac{2\int_0^T F_D U dt}{\int_0^T \rho h_v b_v U^2 |U| dt} = \frac{2\int_0^T F U dt}{\int_0^T \rho h_v b_v U^2 |U| dt} \quad (10)$$

The in-phase time series data of total force (F) and velocity (U) can be used directly in Equation (10) to drive period-averaged C_D values. As W is proportional to U^3 (Equation (9)), the integration of FU over a period is largely contributed to by the moments with relatively high velocity, and to a very limited extent, by the moments with low velocity. Thus, deriving period-averaged C_D via the technique of quantifying ε can automatically assign large weight to the moments with high velocities in a wave period, resulting in most relevant C_D values for wave dissipation analysis. To check the validity of the direct measuring method, we used the derived period-averaged C_D values to reproduce the total acting forcing (F_{rep}) using Equation (1), and compare it with the actual measurement. Additionally, another reproduced total force F_{rep}' is included by assuming $C_D = 1$. It is used as a reference for the F_{rep}.

2.2. Synchronized Force–Velocity Measuring System

In-phase force–velocity data are critical to the direct measurement method. To obtain in-phase data, a synchronized force–velocity measuring system was developed, which was composed by a force sensor and an acoustic doppler velocimeter (ADV) (Figure 1a,d). Four measuring systems were deployed in the wave flume at Fluid Mechanics Laboratory at Sun Yat-sen University, Zhuhai Campus (Figure 1a). The wave flume is 20 m long, 0.8 m wide, and 0.6 m deep. A series of capacitance-type wave gauges were installed to monitor wave height changes in the wave flume. The mimicked vegetation canopy was 8 m long, and it was constructed by PVC pipes. The pipes were 0.2 m tall and their diameter was 0.02 m. The mimicked vegetation canopy was built following a stagger pattern with a density of 139 stems/m^2. It was built on top of a false bottom in order to elevate the canopy so that the force sensors can be mounted underneath it. The mimicked vegetation canopy was submerged in water (water depth = 0.25 m), and it was subjected to various wave conditions. The tested wave height varied from 0.03 to 0.09 m, and the tested wave period varied from 0.6 to 1.2 s. A space-averaged C_D can be obtained by taking the mean C_D values of the four measuring spots in each test.

The force sensors selected were model M140 built by Utilcell, Spain (Figure 1b). As the size of this sensors is small, they can be installed at multiple locations in one flume test. The reading of the sensor is in "gram", which can be easily translated into "Newton" by multiplying the acceleration of gravity. The minimum division of the sensor is 3×10^{-3} N, and the maximum measuring load is 30 N. In the current experiment, the measuring frequency of the force sensors is set as 20 Hz. These sensors were chosen also because they are robust and can be easily waterproofed by sealing the cable connection point with glue. Furthermore, the sensor is small (15 cm × 4 cm × 2.5 cm), and can be easily fitted in the flume (Figure 1b). To prevent the sensor being affected by any force acting on itself, an aluminum case was put around the sensor.

For each force sensor, a mimicked vegetation stem (PVC pipe) was firmly screwed to it, so that the force acting on a vegetation stem can be detected (Figure 1b). The PVC pipes that were attached to the force sensor were not different from other pipes in the mimicked vegetation canopy. After the sensor was attached to a PVC pipe, it measured the total acting force on the pipe. For the same force, the reading is constant, regardless of the location of the acting force. This is desirable for our current experiment, in which wave-induced forces acted over the full length of the pipe. Additionally, the sensors can detect the force in both the following and the opposing direction as the wave propagation, which is ideal to measure the force generated by oscillatory flows.

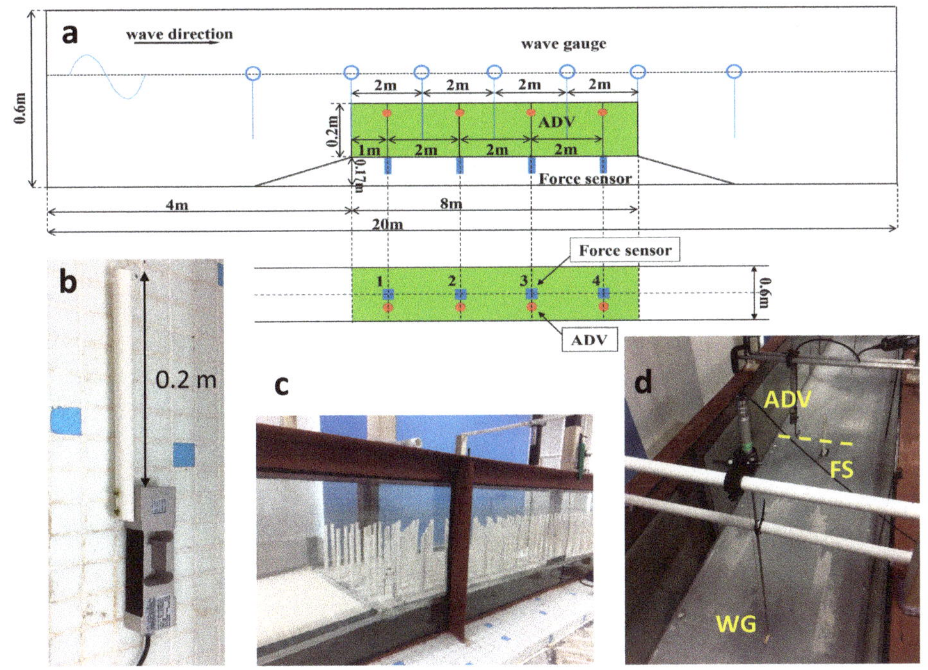

Figure 1. (**a**) Flume experiment set-up. The numbers 1–4 indicate the locations of the in-phase force–velocity measurement; (**b**) A force sensor connected to a PVC pipe; (**c**) Mimicked vegetation canopy constructed by PVC pipes; (**d**) Instrument deployments in the flume without mimicked vegetation. "WG" stands for wave gauge, "ADV" stands for acoustic doppler velocimeter, and "FS" stands for force sensor. The dashed line indicates the ADV and force sensor are placed at the same cross-section to ensure in-phase measurements.

As a first test of the force sensors, we put known weights on one of the sensors attached with PVC pipes. This sensor was held horizontally in this test. The obtained readings were subsequently compared to the known weights. Different weights were put at three different positions on the pipe, i.e., bottom, middle, and tip. The generated forces were in both positive and negative directions. The comparison between the readings and the weights is listed in Table 1. It is clear that for the force measurement in both directions, the relative errors of the sensors are within 1%, compared to the known weights. This first test shows that the force sensors can provide precise measurements.

For the velocity measurement, we used four ADVs. They were deployed at the same cross-section as the force sensors to obtain roughly synchronized signals. The cases wave0312 (3 cm wave height and 1.2 s wave period) and wave0709 (7 cm wave height and 0.9 s wave period) were selected to conduct velocity profile measurements in order to obtain information on velocity structure and depth-averaged in-canopy velocity. The profile was obtained by manually adjusting the vertical measuring location of the ADVs in repeated sequences. For most cases, the velocity measurement was taken at half of the water depth, which was a representative value of the depth-averaged in-canopy velocity. The accuracy of this treatment is acceptable when the submergence ratio (i.e., h/h_v = 1.25) is small [24,37], and it can significantly reduce the labor involved. Two of the ADVs were made by Nortek (*Vectrino* see http://www.nortekusa.com/usa/products/acoustic-doppler-velocimeters/vectrino-1), and the other two were made by SonTek (*MicroADV*, see https://www.sontek.com/argonaut-adv). These four ADVs are common instruments in fluid mechanics labs. Their basis measurement technology is coherent Doppler processing. They measure 3D water velocity of a small cylinder (i.e., within 1 cm^3) that is a few centimeters away from measuring probes in the water. They can measure at frequencies as high as 64 Hz, which are desirable for the direct measuring method. In our experiment, the ADV data acquisition followed their respective user manuals. The measuring frequency was set as 40 Hz in order to accommodate the measuring frequency of the force sensor (i.e., 20 Hz). The obtained data are filtered through a low-pass filter to remove high frequency spikes following a similar method described in Strom and Papanicolaou [40].

Table 1. Comparison between known weights and force sensor reading.

Direction	Known Weights (g)	1st Reading [a] (g)	2nd Reading [b] (g)	3rd Reading [c] (g)	Mean Reading (g)	Absolute Error (g)	Relative Error
+	5 g	5.10	5.00	5.00	5.03	0.03	0.60%
	10 g	10.10	10.10	10.00	10.07	0.07	0.70%
	20 g	20.10	20.00	20.00	20.03	0.03	0.15%
	50 g	49.90	49.90	49.90	49.90	0.10	0.02%
−	5 g	−5.00	−5.00	−5.10	−5.03	−0.03	−0.60%
	10 g	−10.00	−10.10	−10.10	−10.07	−0.07	−0.70%
	20 g	−20.00	−20.00	−20.00	−20.00	0	0%
	50 g	−50.0	−49.9	−49.9	−49.93	0.07	0.14%

[a,b,c] The 1st, 2nd, and 3rd time readings were taking when the weights were put at the bottom, middle, and tip of the testing pipe, respectively.

2.3. Automatic Alignment Algorithm

Although the force and velocity measurements were deployed at the same cross-section of the wave flume, it did not ensure perfectly in-phase data. In fact, small time lags commonly existed between the obtained original force and velocity time series. These time lags were induced by small misalignments between force and velocity measurement, and/or by intrinsic delays of these electronic devices. To reduce the time-lag between velocity and force measurement, the original force and velocity time series should be realigned.

According to the Morison equation [23], velocity (U) and drag force (F_D) should be in phase, which can be used to evaluate the time-lag between velocity and force measurement. A flow chart for the data realignment is shown in Figure 2. The inputs are the timeseries of force (F) and velocity (U). As we can assume that C_M = 2 [39], the inertia force can be calculated based on the velocity. Then, the drag force (F_D) can be computed by subtracting inertia force (F_M) from the total force (F). Subsequently, we can determine the phase shift (Δt) between the velocity and drag force peaks. Lastly, this phase shift (Δt) will be recorded and used to adjust the velocity timeseries, aiming to obtain more in-phase velocity and force data. The obtained new velocity and force data will be used as input in the same loop. This loop continues 30 times, and we chose the minimum phase shift (Δt) and the resultant velocity and force timeseries as outputs for C_D derivation. The automatic alignment algorithm is provided in the Appendix A as a MATLAB script. To verify the 30 loop count criterion, a sensitivity analysis is conducted by changing the loop count to 10, 20, 30, and 50. The resulting phase shifts with those loop counts are subsequently compared.

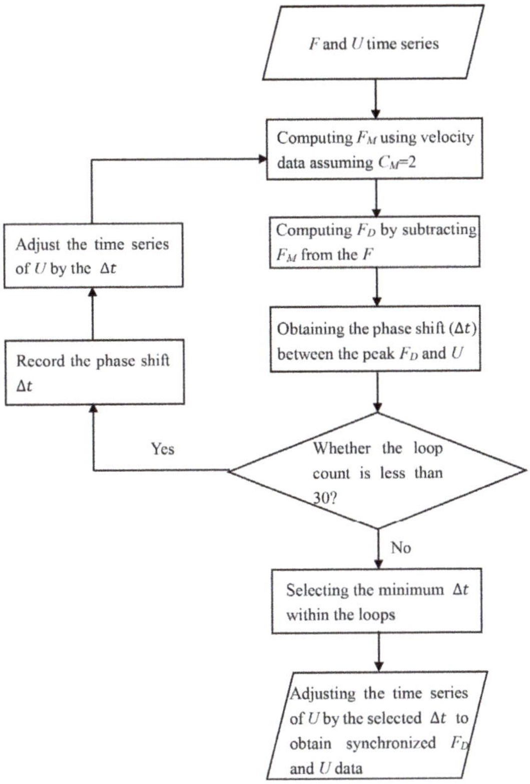

Figure 2. Flow chart to realign velocity and force data signals. The algorithm is provided in the Appendix A as a MATLAB script.

3. Results

3.1. Velocity Profiles in the Vegetation Canopy

Figure 3 shows the velocity profiles were measured at location 3 in case wave0312 (3 cm wave height and 1.2 s wave period) and wave0709 (7 cm wave height and 0.9 s wave period). For the rest of the tested cases, velocity was measured at the half water depth as a proxy of depth-averaged in-canopy velocity. Figure 3a shows that in the case wave0312, the velocity profiles are rather uniform in the vertical direction. The depth-averaged in-canopy velocity amplitude is 0.052 m/s, whereas U_w measured at the half water depth is 0.049 m/s. The difference between these two is small. The velocity profiles in wave0709 have greater vertical gradient, i.e., higher velocity at the top and lower velocity near the bottom. Overall, the difference between U_w measured at the half water depth (0.106 m/s) and the amplitude of depth-averaged in-canopy velocity (0.115 m/s) is small. Therefore, it is acceptable to use U_w measured at the half water depth as a representative value of the depth-averaged in-canopy velocity.

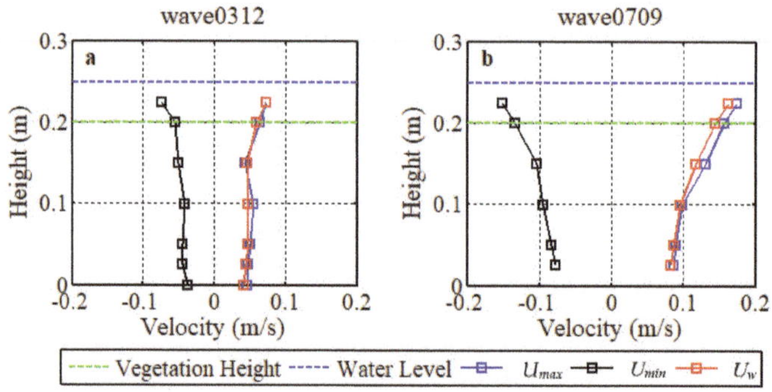

Figure 3. (**a**) Measured velocity profile for case wave0312 with 3 cm wave height and 1.2 s wave period, U_{min} is the highest wave orbital velocity in negative direction (oppose to wave propagation), U_{max} is the highest wave orbital velocity in positive direction (same as wave propagation), and U_w is the amplitude of wave orbital velocity; (**b**) Measured velocity profile for case wave0709 with 7 cm wave height and 0.9 s wave period.

3.2. *Wave Height and Wave Orbital Velocity in the Mimicked Vegetation Canopy*

Reductions of wave height (H) and magnitude of wave orbital velocity (U_w) through mimicked vegetation canopy can be observed in Figure 4. The wave height reduces continuously from the canopy front to the end. The final wave height reduction rate was 55% (Figure 4a). The shown wave orbital velocity is obtained by ADV measurement at location 1–3. The ADV measurement at location 4 failed during the experiment. The shown U_w is obtained by using Equation (3) based on an average wave height between x = 6–8 m in Figure 4a. With the reduced wave height, the magnitude of wave orbital velocity also reduces from 0.155 m/s to 0.095 m/s from the beginning to the end of the canopy. The reduced wave orbital velocity leads to variations in acting force on vegetation stem as well as in vegetation drag coefficient (C_D), which are shown in the following sections.

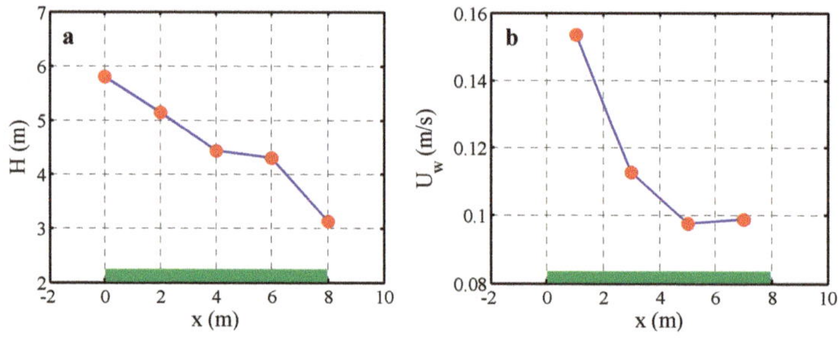

Figure 4. (**a**) Spatial variations of wave height (H) through the vegetation canopy (i.e., x = 0–8 m) indicated as green bars; (**b**) Spatial variations of the magnitude of wave orbital velocity (U_w) through the vegetation canopy. The first 3 U_w data points (x = 0–6 m) are obtained from ADV measurement, whereas the last data point at x = 7 m is obtained by using Equation (3) based on an average wave height between x = 6 m and 8 m in panel (**a**). The shown test case is wave0712 with 7 cm wave height and 1.2 s wave period.

3.3. *Data Alignment and Time-Varying C_D*

The total acting force (F) and velocity measured at four locations in the wave flume are shown in Figure 5. Waves reach these four locations at different moments. The velocity data at the last measurement point were missing due to the failure of the ADV measurement. It is clear that the acting force (F) and velocity reduce as waves pass through mimicked vegetation canopy. It is also apparent that there are time lags in synchronized force–velocity measurements at all locations. However, it should be noted that these seeming time lags (i.e., time shifts between F and U) are not the real time lags (i.e., time shifts between F_D and U), which lead to errors in C_D.

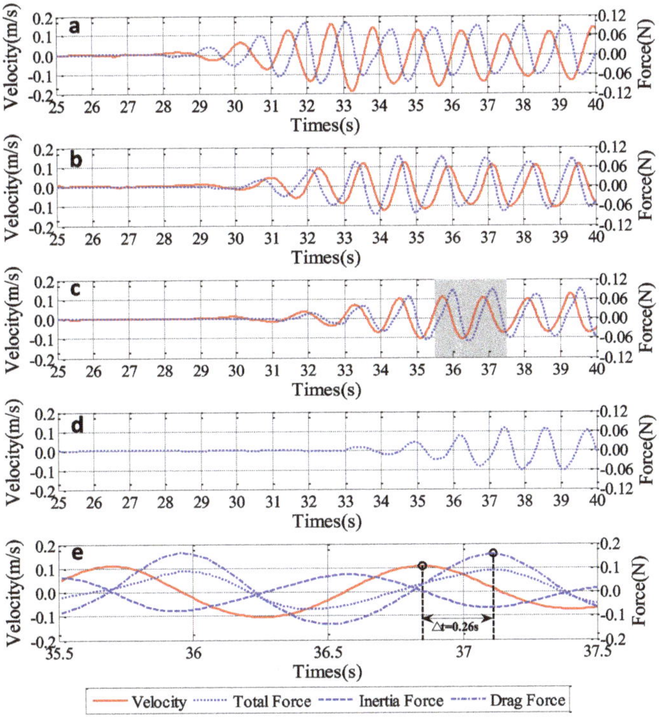

Figure 5. (**a–d**) Raw velocity and total force data measured at four locations (1–4) in the order of the wave propagation in the mimicked vegetation canopy; (**e**) The shaded area in panel (**c**) is blown up in panel (**e**) for detailed analysis, where inertia force and drag force were derived based on non-synchronized data. The time shift (Δt) between the F_D and U is 0.26 s. The shown test case is wave0712 with 7 cm wave height and 1.2 s wave period. The velocity data in panel (**d**) is not included, due to the ADV measurement failure at location 4.

Following the previous study [37], we only tracked the first 2–3 full wave periods after the start-up, but before waves reached the back end of the flume, to avoid possible influence of wave reflection. Figure 5e shows the force and velocity data of the chosen first two wave periods after the start-up. Based on the original data and Equation (1), both F_D and F_M can be estimated as shown in Figure 5e. However, it should be noted that as the total force and velocity data are not in phase. The derived F_D and F_M are the first estimation. The derived peaks of the F_D are even higher than the total force, which is not possible. This result highlights the necessity of obtaining in-phase data. Judging from the peaks between U and estimated F_D peaks, the time lag between those two signals is 0.26 s.

In order to eliminate the time lags between these signals, the realignment algorithm was applied to obtain synchronized velocity and force data (Figure 6a–d). It is clear that after the realignment procedure, the time lag between U and F_D is largely reduced (Figure 6a,b), but it cannot be completely eliminated, as small shifts in signal peaks and troughs still exist. Based on the phase shifts of the two peaks and two troughs (as indicated by the red double arrow lines), the time shift of the realigned

U and F_D is significantly reduced to 0.003 s, which is only 1% of the original time shift before the realignment. Note that this optimized time shift is the mean shift of the tested two wave periods (i.e., shifts of two peaks and two troughs). Apart from the reduced time shift, the magnitude of F_D is also reduced to be lower than the total force (F), which is in line with the original Morison equation. It is noted that the peak drag force before alignment is about twice as large as the peak drag force after the alignment. Thus, if the original F timeseries were used, the derived drag coefficient would be considerably overestimated.

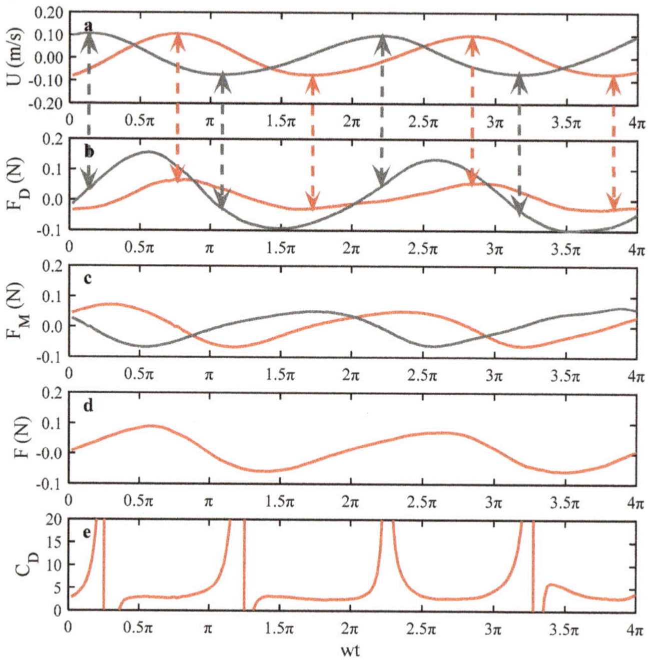

Figure 6. (**a**–**d**) Time-varying U, F_D, F_M, and F data over two wave periods. The gray lines are before realignment, and the red ones are after realignment. The vertical double-arrowed lines in panel (a) and (**b**) indicate the synchronization status before and after the realignment; (**e**) Time-varying C_D derived based on realigned U and F_D data. The shown test case is wave0712 with 7 cm wave height and 1.2 s wave period.

Based on the synchronized velocity and force data and Equation (4), the time-varying C_D can be derived (Figure 6e). The time-varying C_D values vary periodically with the changing velocity signals. The values are small when the velocity is large, but they reach to infinity when the velocity is close to zero. The reason for the unrealistically large C_D values is because the F_D are divided by very small velocity values in Equation (4). These unrealistically large C_D values are not useful in modelling wave dissipation by vegetation, since they are associated with period with very low velocity when the energy dissipation is very limited.

To verify the loop count criterion applied in the realignment algorithm, a sensitivity analysis of the loop is conducted (Figure 7). It is clear that with the increase of loops, the phase shifts can be sufficiently reduced. When 30 loop count is applied, the phase shifts can be reduced to minimum, i.e., 0.003 s for the case 1 and almost zero for the case 2. If the number of loops is increased to 50, the time shifts cannot be further reduced. Thus, the results of sensitivity analysis indicate that applying 30 loop count in the realignment algorithm is a reasonable criterion.

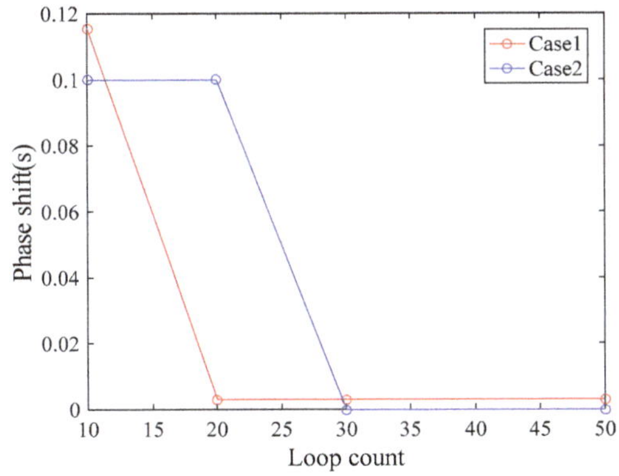

Figure 7. Sensitivity analysis of the number of loops used in the realignment algorithm. Case 1 is the case wave0712 with 7 cm wave height and 1.2 s wave period, and case 2 is the case wave0512 with 5 cm wave height and 1.2 s wave period.

3.4. Deriving Period-Averaged C_D

Based on the realigned force–velocity data, we can also derive period-averaged C_D by quantifying the power and work done, F_D and F_M. It is clear that the time-varying power of F_D is always positive, and its magnitude varies in phase with the velocity magnitude (Figure 8). For ideal sinusoidal velocity signals, P_D at the velocity peaks should be equal to the troughs. However, in our test case (and in real field conditions), the wave orbital velocity is asymmetrical: higher in the positive direction and lower in the negative direction. Thus, P_D is larger near the wave peaks and smaller near the wave trough. The difference between peaks and troughs are much more apparent in P_D compared to the difference in velocity. It is because that P_D is to the third power of velocity. Small asymmetry in velocity will be greatly magnified in P_D. The variation of P_M is different from that of the P_D. It is clear that P_M varies between positive and negative values. The zero-crossings in P_M occur when velocity is zero or when velocity is at its maximum in both directions, i.e., when F_M is zero in Equation (5). Note that there are small fluctuations at the peaks of the P_D. These fluctuations may be induced by the small phase shifts between the F_D and U timeseries.

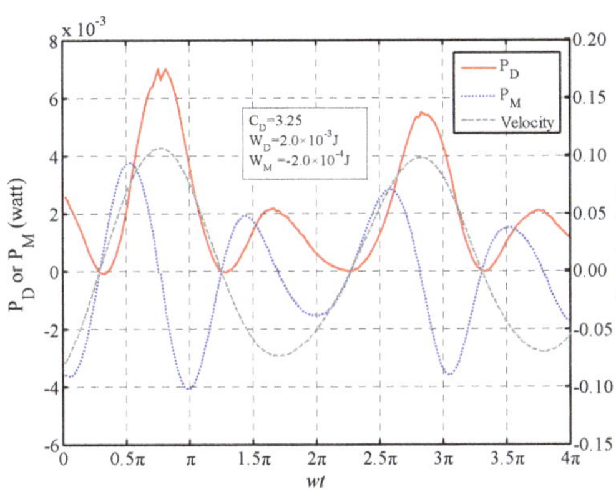

Figure 8. Time-varying U, P_D, and P_M over two wave periods. The shown test case is wave0712, with 7 cm wave height and 1.2 s wave period. The averaged work done by drag force (W_D) and inertia force (W_M) over the shown two periods is 2.0×10^{-3} J and -2.0×10^{-4} J, respectively. The derived period-averaged C_D is 3.25.

By integrating P_D and P_W over one wave period, we can obtain the work done by drag force (W_D) and inertia force (W_M) as shown in Equation (8). The averaged W_D and W_M over the two periods in Figure 8 is 2.0×10^{-3} J and -2.0×10^{-4} J, respectively. In case of ideal sinusoidal velocity signals, W_M should be exactly zero. Due to the asymmetric wave velocity, the W_M is not zero, but the magnitude of W_M is fairly small, i.e., one tenth of the W_D. Since the magnitude of W_M is considerably small compared to W_D, the assumption that W_M can be ignored in the deriving period-averaged C_D (in Equation (8)) is still valid, and the period-averaged C_D in the shown case is derived as 3.25.

Following the same method, period-averaged C_D at three functional measuring locations of all the tested cases are listed in Table 2. It shows that relatively large deviations in C_D values exist among different measuring locations. The C_D values generally increase from locations from 1 to 3. Previous studies have shown that C_D values increase with the reduced velocity (i.e., Reynolds number) [24,37]. The obtained increase of C_D values may be related to the reduction of wave orbital velocity from the front to the end of the vegetation canopy, as shown in Figure 4b Thus, the spatial variation in C_D values is in-line with previous studies. Additionally, it is noted that the cases with larger wave height and wave period (i.e., higher wave orbital velocity) generally lead to smaller spatially averaged C_D, which is also in agreement with previous studies [24,37].

Table 2. Period-averaged C_D in all the tested cases.

Test Number	Wave Height (m)	Wave Period (s)	C_D at Location 1	C_D at Location 2	C_D at Location 3	Space-Mean C_D	Standard Deviation
1	0.03	0.6	5.41	10.02	10.14	8.52	7.28
2	0.03	0.9	3.19	4.88	7.09	5.05	3.82
3	0.03	1.2	3.60	2.79	5.25	3.88	1.56
4	0.05	0.6	7.61	5.83	5.96	6.46	0.98
5	0.05	0.9	3.82	2.36	3.62	3.27	0.62
6	0.05	1.2	2.84	3.03	3.97	3.28	0.37
7	0.07	0.6	3.43	3.01	7.04	4.49	4.90
8	0.07	0.9	1.97	1.83	2.94	2.25	0.37
9	0.07	1.2	1.77	2.77	3.51	2.68	0.76
10	0.09	0.6	3.02	5.76	6.10	4.96	2.86
11	0.09	0.9	1.26	1.79	2.64	1.89	0.49
12	0.09	1.2	1.44	2.54	3.00	2.33	0.64

3.5. Assessing the Derived C_D by Reproducing Acting Force

In order to test the derived period-averaged C_D, we used the derived values to reproduce the total force from the velocity signals using Equation (1). The reproduced total force (F_{rep}) is subsequently compared with the measured actual total force (Figure 9). The reproduced total force (F_{rep}') using $C_D = 1$ is also included as reference. It is clear that F_{rep} is in good agreement with the measured force over the shown two wave periods, although small differences exist between them. Notably, the measured maximum force is well captured in F_{rep} near x = 0.5 π, which is important as the maximum force is critical not only for energy dissipation but also for assessing the stem strength to wave loading. As a comparison, the difference between F_{rep}' and the measurement is large, which shows the validity of using period-averaged C_D to reproduce the total force.

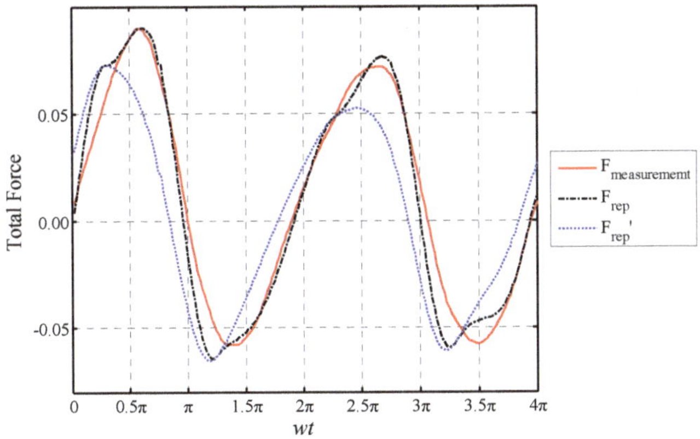

Figure 9. Comparison between reproduced total acting force and measured total force. The red solid line is the measured total force; The black dash line is quantified by using derived period-averaged C_D in Equation (1) (i.e., F_{rep}); The blue dash line is quantified by using $C_D = 1$ in Equation (1) for reference (i.e., F_{rep}'). The shown test case is wave0712, with 7 cm wave height and 1.2 s wave period.

Figure 10 compares the maximum F_{rep} and maximum measured total force obtained at all three functional measuring locations in all test cases. In general, the reproduced maximum force is well in-line with the measurement, as most of the data points are fairly close to the 1:1 reference line. The R^2 value is 0.759 for data of all three functional measuring locations in all test cases. This result indicates the C_D deriving procedure is valid, and the intrinsic errors associated with this procedure are acceptable.

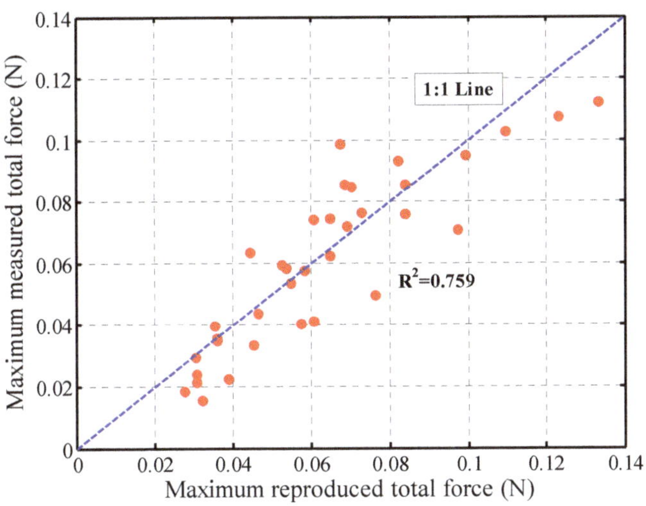

Figure 10. Comparison between the maximum reproduced total acting force (F_{rep}) and the measured maximum total force at three functional measuring locations in all the tested cases.

4. Discussion

4.1. Advantages of the Current Measuring System and Alignment Algorithm

Our results have shown that large spatial variations can exist in the wave particle velocity and C_D (Figure 5 and Table 2). Thus, it is important to have synchronized force–velocity measurement at multiple locations by a number of force–velocity measuring systems. The selected standard force sensors are small enough to be installed at multiple locations in wave flumes. Additionally, these sensors are designed with built-in tapped holes, which facilitate testing various vegetation mimics, e.g., rigid cylinders, flexible stripes, and real vegetation stems.

Our results further show that the realignment process is important to derive both time-varying and period-averaged C_D values (Figures 6 and 7). In our experiment, we aligned the instruments as good as possible (please see Figure 1d of the manuscript), but it is inevitable to have small misalignment to cause the delay. The main source causing the delay may be the inherent difference in instruments' speed of recording and receiving data, as the force and velocity measurements have their separate data acquisition systems. Thus, the automatic synchronizing algorithm is necessary and valuable in the current study. By using this algorithm, the time shift between two signals can be reduced to 0.003 s, which is only 1% of the original time shift before the realignment. The obtained time shifts are believed to be acceptable when comparing to normal wave periods (1–2 s) tested in our lab flume. The time shifts are merely 0.3% to the tested wave period.

The overall good performance between maximum measured force and reproduced force shows the reliability of this alignment algorithm. Importantly, this algorithm can automatically process the force and velocity data. No manual tuning is needed. Hence, it provides a generic solution to the alignment problems in deriving C_D. Furthermore, this algorithm can run very efficiently, which is desirable when processing large data sets from multiple measuring locations. Lastly, this alignment algorithm is applied to process the velocity data from ADVs, but it is worth noticing that this algorithm is also applicable for other velocity measuring technologies, e.g., EMF (electromagnetic flow manufacture meter) and PIV (particle image velocimetry) [41–43].

4.2. Current Limitations and Future Applications

It is noted that the time shifts after the realignment are non-zero, but they are significantly reduced. To further reduce the time shifts, high frequency force and velocity measurements (e.g., >100 Hz) are required to obtain finite time steps for the realignment algorithm. However, it is perhaps not possible to completely eliminate the time shifts for all the tested cases, especially when multiple wave periods are included in the analysis, as the realignment procedure needs to account for time shifts at multiple peaks.

The velocity measurement in the current experiment was conducted by ADV measurement, which is a conventional method in flume experiments. The main limitation of the ADV measurement is that it is a point measurement. To obtain vertical velocity profile, it is required to manually adjust the ADV-measuring locations and repeat the same test conditions for each measuring location. This process is very time-consuming. Thus, we only conducted the velocity profile measurement for two cases, whereas for other cases, the velocity data is taken at the half water depth, which roughly equaled to the mean in-canopy velocity (Figure 3). The same practice is also done in previous study [37]. However, it is possible that the small deviation between the point velocity and depth-averaged in-canopy velocity can lead to errors in the derived C_D values. This may partly explain the difference between the maximum measured force and the reproduced force. In order to improve the velocity measuring accuracy and reduce the labor involved, PIV system can be applied in future experiments. The PIV system can provide detailed velocity information of velocity field [41,42]. By applying such a system, it is also possible to obtain the relative velocity between water motion and the motion of flexible vegetation stems. Thus, the developed technics in the current study can be further applied in flexible vegetation canopies, e.g., saltmarshes and seagrasses, which is interesting to both coastal engineers and ecologists.

5. Conclusions

This technical note provides a practical set-up to derive both time-varying and period-averaged vegetation drag coefficients (C_D) following the direct measuring method [36–38]. Different from previous studies, standard force sensors are applied to compose four synchronized force–velocity measuring systems in the current experiment. These standard force sensors are robust and suitable for flume applications. The composed force–velocity measuring systems can provide synchronized force–velocity measurement. Although one of the ADV instruments failed, the other three ADVs functioned well during the experiment. Importantly, an automatic algorithm was developed to realign the obtained force and velocity signals for direct C_D deviation. This algorithm is expected to be able to accommodate a variety of velocity measuring techniques, providing possibilities to extend current application range. The developed force–velocity measuring systems and the automatic realignment algorithm may assist future experiments on vegetation–wave interactions for better understanding and prediction of vegetation-induced wave dissipation.

Author Contributions: P.Y., Z.H., B.H. and Q.Y. conceived and designed the methodology and experiments. P.Y., H.C., C.T. and L.R. performed the experiments. P.Y., L.R. and B.H. analyzed the data. All the authors contributed to the writing and editing of the manuscript.

Funding: The authors gratefully acknowledge financial support of the National Key R&D Program of China (No. 2016YFC0402607), the National Natural Science Foundation of China (No. 51609269, 51709288), the Joint Research Projects NSFC (No. 51761135022)—NWO (No. ALWSD.2016.026)—EPSRC (No. EP/R024537/1): Sustainable Deltas, and the Open Foundation of State Key Laboratory of Hydrology-Water Resources and Hydraulic Engineering (Grant No. 2016491711), and the 2017 Guangdong Oceanic and Fishery Department scientific research and technological development program.

Acknowledgments: We thank the three anonymous reviewers for their comments and suggestions.

Conflicts of Interest: The authors declare no conflict of interest.

Appendix A. MATLAB Code for Force and Velocity Data Realignment

This appendix provides the MATLAB (R2016a, The MathWorks, Natick, U.S.) script that can be applied to realign force and velocity timeseries for C_D derivations.

```
for i=1:3   % number of Synchronized Force–Velocity Measurement systems
  NN=0;
  D=0;
  DD=0;
  while NN<=30   % number of loops
    NN=NN+1
    Delay(i,NN)=round(abs(DD)+D);
%Adjust the velocity bases on the phase difference, while keep the force is not change
      if i==1
        secADV1=AllADV1new(orpo1−Delay(i,NN):orpo1+8*TT*80−Delay(i,NN),:);
        secADV(:,1)=secADV1(:,2);
      end

      if i==2
        secADV2=AllADV2new(orpo2−Delay(i,NN):orpo2+8*TT*80−Delay(i,NN),:);
        secADV(:,2)=secADV2(:,2);
      end

      if i==3
```

```
            secADV3=AllADV3new(orpo3−Delay(i,NN):orpo3+8*TT*80−Delay(i,NN),:);
            secADV(:,3)=secADV3(:,2);
        end
        % Calculation of inertia force
            Fm(:,i)=2*1000*pi*0.02*0.02*hv*(diff(secADV(:,i))/0.0125)/4;
        % Calculation of drag force
            Fd(:,i)=secCLN(2:end,i)−Fm(:,i)
        %Calculation of in−time Cd
            Cd(:,i)=2*Fd(:,i)./(1000*0.02*hv*secADV(2:end,i).*abs(secADV(2:end,i)));
        %Intercept the peak value of velocity
            [maxE,minE]=peakdet(secADV(:,i),0.3*(max(secADV(:,i))−min(secADV(:,i))));
        %Intercept the peak value of drag force
            [maxFd,minFd]=peakdet(Fd(:,i),0.3*(max(Fd(:,i))−min(Fd(:,i))));

        if maxE(1)<10
            for kk=2:length(maxE)
                maxE(kk−1)=maxE(kk);
            end
         maxE(kk)=NaN;
        end

        if minE(1)<10
            for kk=2:length(minE)
                minE(kk−1)=minE(kk);
            end
         minE(kk)=NaN;
        end

        if maxFd(1)<10
            for kk=2:length(maxFd)
                maxFd(kk−1)=maxFd(kk);
            end
         maxFd(kk)=NaN;
        end

        if minFd(1)<10
            for kk=2:length(minFd)
                minFd(kk−1)=minFd(kk);
            end
         minFd(kk)=NaN;
        end
         loc_E(:,i)=cat(1,maxE(1:2,1),minE(1:2,1));
         loc_Fd(:,i)=cat(1,maxFd(1:2,1),minFd(1:2,1));
         %Phase difference between drag force and velocity
         delay(i,NN)=mean(loc_E(:,i)−loc_Fd(:,i))
         clear maxFd minFd maxE minE

              D=abs(delay(i,NN));
              DD=Delay(i,NN);
    end
end
```

References

1. Bouma, T.J.; van Belzen, J.; Balke, T.; Zhu, Z.; Airoldi, L.; Blight, A.J.; Davies, A.J.; Galvan, C.; Hawkins, S.J.; Hoggart, S.P.G.; et al. Identifying knowledge gaps hampering application of intertidal habitats in coastal protection: Opportunities & steps to take. *Coast. Eng.* **2014**, *87*, 147–157. [CrossRef]
2. Balke, T.; Friess, D.A. Geomorphic knowledge for mangrove restoration: A pan-tropical categorization. *Earth Surf. Process. Landf.* **2016**, *41*, 231–239. [CrossRef]
3. Barbier, E.B. Valuing the storm protection service of estuarine and coastal ecosystems. *Ecosyst. Serv.* **2015**, *11*, 32–38. [CrossRef]
4. Temmerman, S.; Meire, P.; Bouma, T.J.; Herman, P.M.J.; Ysebaert, T.; De Vriend, H.J. Ecosystem-based coastal defence in the face of global change. *Nature* **2013**, *504*, 79–83. [CrossRef] [PubMed]
5. Horstman, E.M.; Dohmen-Janssen, C.M.; Narra, P.M.F.; van den, B.; Siemerink, M.; Hulscher, S.J. Wave attenuation in mangroves: A quantitative approach to field observations. *Coast. Eng.* **2014**, *94*, 47–62. [CrossRef]
6. Zhang, X.; Chua, V.P.; Cheong, H.-F. Hydrodynamics in mangrove prop roots and their physical properties. *J. Hydro-Environ. Res.* **2015**, *9*, 281–294. [CrossRef]
7. Zhang, X.; Chua, V.P.; Cheong, H.-F. Geometrical and material properties of Sonneratia alba mangrove roots. *Trees-Struct. Funct.* **2014**, *29*, 285–297. [CrossRef]
8. Mazda, Y.; Wolanski, E.; King, B.; Sase, A.; Ohtsuka, D.; Magi, M. Drag force due to vegetation in mangrove swamps. *Mangroves Salt Marshes* **1997**, *1*, 193–199. [CrossRef]
9. Alongi, D.M. Mangrove forests: Resilience, protection from tsunamis, and responses to global climate change. *Estuar. Coast. Shelf Sci.* **2008**, *76*, 1–13. [CrossRef]
10. Möller, I.; Kudella, M.; Rupprecht, F.; Spencer, T.; Paul, M.; van Wesenbeeck, B.K.; Wolters, G.; Jensen, K.; Bouma, T.J.; Miranda-Lange, M.; Schimmels, S. Wave attenuation over coastal salt marshes under storm surge conditions. *Nat. Geosci.* **2014**, *7*, 727–731. [CrossRef]
11. Arkema, K.K.; Guannel, G.; Verutes, G.; Wood, S.A.; Guerry, A.; Ruckelshaus, M.; Kareiva, P.; Lacayo, M.; Silver, J.M. Coastal habitats shield people and property from sea-level rise and storms. *Nat. Clim. Chang.* **2013**, *3*, 913–918. [CrossRef]
12. Zhu, Z.; Van, B.; Hong, T.; Kunihiro, T.; Ysebaert, T.; Herman, P.M.J.; Bouma, T.J. Sprouting as a gardening strategy to obtain superior supplementary food: Evidence from a seed-caching marine worm. *Ecology* **2016**, *97*, 3278–3284. [CrossRef] [PubMed]
13. Young, I.R.; Zieger, S.; Babanin, A.V. Global Trends in Wind Speed and Wave Height. *Science* **2011**, *332*, 451–455. [CrossRef] [PubMed]
14. Cheng, Y.; Ezer, T.; Hamlington, B.D. Sea Level Acceleration in the China Seas. *Water* **2016**, *8*, 293. [CrossRef]
15. Lin-Ye, J.; Garcia-Leon, M.; Gracia, V.; Isabel Ortego, M.; Stanica, A.; Sanchez-Arcilla, A. Multivariate Hybrid Modelling of Future Wave-Storms at the Northwestern Black Sea. *Water* **2018**, *10*, 221. [CrossRef]
16. Wolf, J.; Lowe, J.; Howard, T. Climate downscaling: Local mean sea level, surge and wave modelling. In *Broad Scale Coastal Simulation: New Techniques to Understand and Manage Shorelines in the Third Millennium*; Springer: Dordrecht, The Netherlands, 2015; pp. 79–102. ISBN 978-94-007-5258-0.
17. Wolf, J.; Brown, J.M.; Bolaños, R.; Hedges, T.S. Waves in Coastal and Estuarine Waters. In *Treatise on Estuarine and Coastal Science*; Wolanski, E., McLusky, D., Eds.; Academic Press: Waltham, MA, USA, 2011; pp. 171–212. ISBN 978-0-08-087885-0.
18. Van Loon-Steensma, J.M.; Slim, P.A.; Decuyper, M.; Hu, Z. Salt-marsh erosion and restoration in relation to flood protection on the Wadden Sea barrier island Terschelling. *J. Coast. Conserv.* **2014**, *18*, 1–16. [CrossRef]
19. Hu, Z.; Van Belzen, J.; Van Der Wal, D.; Balke, T.; Wang, Z.B.; Stive, M.; Bouma, T.J. Windows of opportunity for salt marsh vegetation establishment on bare tidal flats: The importance of temporal and spatial variability in hydrodynamic forcing. *J. Geophys. Res. Biogeosci.* **2015**, *120*, 1450–1469. [CrossRef]
20. Zhang, M.; Zhang, H.; Zhao, K.; Tang, J.; Qin, H. Evolution of wave and tide over vegetation region in nearshore waters. *Ocean Dyn.* **2017**, *67*, 973–988. [CrossRef]
21. Dalrymple, R.; Kirby, J.; Hwang, P. Wave Diffraction Due to Areas of Energy Dissipation. *J. Waterw. Port Coast. Ocean Eng.* **1984**, *110*, 67–79. [CrossRef]
22. Koch, D.L.; Ladd, A.J.C. Moderate Reynolds number flows through periodic and random arrays of aligned cylinders. *J. Fluid Mech.* **1997**, *349*, 31–66. [CrossRef]

23. Morison, J.R.; O'Brien, M.P.; Johnson, J.W.; Schaaf, S.A. The Force Exerted by Surface Waves on Piles. *J. Pet. Technol.* **1950**, *2*, 149–154. [CrossRef]
24. Nepf, H.M. Flow over and through Biota. In *Treatise on Estuarine and Coastal Science*; Wolanski, E., McLusky, D., Eds.; Academic Press: Waltham, MA, USA, 2011; pp. 267–288. ISBN 978-0-08-087885-0.
25. Augustin, L.N.; Irish, J.L.; Lynett, P. Laboratory and numerical studies of wave damping by emergent and near-emergent wetland vegetation. *Coast. Eng.* **2009**, *56*, 332–340. [CrossRef]
26. Tang, J.; Shen, S.; Wang, H. Numerical model for coastal wave propagation through mild slope zone in the presence of rigid vegetation. *Coast. Eng.* **2015**, *97*, 53–59. [CrossRef]
27. Cao, H.; Feng, W.; Hu, Z.; Suzuki, T.; Stive, M.J.F. Numerical modeling of vegetation-induced dissipation using an extended mild-slope equation. *Ocean Eng.* **2015**, *110*, 258–269. [CrossRef]
28. van Loon-Steensma, J.M.; Hu, Z.; Slim, P.A. Modelled Impact of Vegetation Heterogeneity and Salt-Marsh Zonation on Wave Damping. *J. Coast. Res.* **2016**, *32*, 241–252. [CrossRef]
29. Peruzzo, P.; De Serio, F.; Defina, A.; Mossa, M. Wave Height Attenuation and Flow Resistance Due to Emergent or Near-Emergent Vegetation. *Water* **2018**, *10*, 402. [CrossRef]
30. Tang, J.; Causon, D.; Mingham, C.; Qian, L. Numerical study of vegetation damping effects on solitary wave run-up using the nonlinear shallow water equations. *Coast. Eng.* **2013**, *75*, 21–28. [CrossRef]
31. Kobayashi, N.; Raichle, A.W.; Asano, T. Wave attenuation by vegetation. *J. Waterw. Port Coast. Ocean Eng.-ASCE* **1993**, *119*, 30–48. [CrossRef]
32. Méndez, F.J.; Losada, I.J.; Losada, M.A. Hydrodynamics induced by wind waves in a vegetation field. *J. Geophys. Res. Oceans* **1999**, *104*, 18383–18396. [CrossRef]
33. Ozeren, Y.; Wren, D.G.; Wu, W. Experimental investigation of wave attenuation through model and live vegetation. *J. Waterw. Port Coast. Ocean Eng.* **2014**, *140*. [CrossRef]
34. Anderson, M.E.; Smith, J.M. Wave attenuation by flexible, idealized salt marsh vegetation. *Coast. Eng.* **2014**, *83*, 82–92. [CrossRef]
35. Losada, I.J.; Maza, M.; Lara, J.L. A new formulation for vegetation-induced damping under combined waves and currents. *Coast. Eng.* **2016**, *107*, 1–13. [CrossRef]
36. Infantes, E.; Orfila, A.; Bouma, T.J.; Simarro, G.; Terrados, J. Posidonia oceanica and Cymodocea nodosa seedling tolerance to wave exposure. *Limnol. Oceanogr.* **2011**, *56*, 2223–2232. [CrossRef]
37. Hu, Z.; Suzuki, T.; Zitman, T.; Uijttewaal, W.; Stive, M. Laboratory study on wave dissipation by vegetation in combined current-wave flow. *Coast. Eng.* **2014**, *88*, 131–142. [CrossRef]
38. Maza, M.; Adler, K.; Ramos, D.; Garcia, A.M.; Nepf, H. Velocity and Drag Evolution From the Leading Edge of a Model Mangrove Forest. *J. Geophys. Res. Oceans* **2017**, *122*, 9144–9159. [CrossRef]
39. Dean, R.; Dalrymple, R. *Water Wave Mechanics for Engineers and Scientists*; Advanced Series on Ocean Engineering; World Scientific: Tokyo, Japan, 1991.
40. Strom, K.B.; Papanicolaou, A.N. ADV measurements around a cluster microform in a shallow mountain stream. *J. Hydraul. Eng.-ASCE* **2007**, *133*, 1379–1389. [CrossRef]
41. Persoons, T.; O'Donovan, T.S. High Dynamic Velocity Range Particle Image Velocimetry Using Multiple Pulse Separation Imaging. *Sensors* **2011**, *11*, 1–18. [CrossRef] [PubMed]
42. Martino, R.; Paterson, A.; Piva, M. Double-average mean flow and local turbulence intensity profiles from PIV measurements for an open channel flow with rigid vegetation. *Environ. Fluid Mech.* **2012**, *12*, 45–62. [CrossRef]
43. You, Z.; Chen, Y. The Use of Tactile Sensors and PIV Analysis for Understanding the Bearing Mechanism of Pile Groups. *Sensors* **2018**, *18*, 476. [CrossRef] [PubMed]

water

MDPI

Article

On-Bottom Stability Analysis of Cylinders under Tsunami-Like Solitary Waves

Giuseppe Tripepi, Francesco Aristodemo * and Paolo Veltri

Dipartimento di Ingegneria Civile, Università della Calabria, via Bucci, cubo 42B, 87036 Arcavacata di Rende (CS), Italy; giuseppe.tripepi@unical.it (G.T.); paolo.veltri@unical.it (P.V.)

* Correspondence: francesco.aristodemo@unical.it; Tel.: +39-0984-496554

Received: 28 March 2018; Accepted: 12 April 2018; Published: 16 April 2018

Abstract: A two-dimensional (2D) laboratory investigation on the horizontal and vertical hydrodynamic forces induced by tsunami-like solitary waves on horizontal circular cylinders placed on a rigid sea bed is presented. A series of 30 physical model tests was conducted in the wave channel of the University of Calabria in which a rigid circular cylinder was equipped with 12 pressure transducers placed along its external surface to determine the wave loads, with three wave gauges to record the surface elevation. The observed experimental range was characterized by the prevalence of the inertia component for the horizontal forces and of the lift component for the vertical ones. On the basis of the performance of several time-domain methods, the wave loads and the undisturbed velocity and acceleration derived from the surface elevation of the cylinder section were used to calculate the drag, lift, and horizontal and vertical inertia coefficients in the practical Morison and transverse semi-empirical equations.

Keywords: tsunami-like solitary waves; horizontal cylinders; hydrodynamic loads; experimental tests; Morison and transverse equations; hydrodynamic coefficients

1. Introduction

The occurrence of tsunami events in coastal areas is a source of risk for already-vulnerable marine structures subjected to the action of wind waves and currents. Hence, the stability of marine structures under tsunami action depends on the accurate assessment of the hydrodynamic forces. The reproduction of catastrophic tsunami waves like those that occurred in the Indian Ocean in 2004 and in Japan in 2011 was observed to be dependent upon the magnitude of the specific source, and the resulting shapes of surface elevation can be quite different, leading to a generalized modelling of tsunami waves (e.g., [1,2]). Owing to its robust and suitable approach, the modelling of the leading wave of a tsunami event is usually reproduced by the generation of solitary waves both experimentally and numerically (e.g., [3]). Indeed, when tsunami waves approach the coast, the wave trough disappears and only a positive peak remains.

Different studies have been conducted to analyze the propagation of solitary waves and their interaction with structures such as breakwaters or submerged barriers (e.g., [3,4]), although little attention has been paid to the analysis of hydrodynamic forces in the case of horizontal cylindrical bodies. Preliminary studies describe the general features of breaking and non-breaking solitary wave-induced forces on horizontal cylinders ([5–7]) but without an extensive approach to study this problem in various wave conditions and positions of the cylinder along the depth. In contrast, for horizontal cylinders under the action of currents or regular and random waves, numerous studies have researched this kind of wave–structure interaction process, adopting different degrees of external roughness of the cylinder and of the sea bed (for a comprehensive review see [8,9]). For bottom-mounted cylinders under the above kind of incident flows, values of hydrodynamic coefficients in Morison-type equations (e.g., [10]) were deduced from field tests [11] as well as small- and large-scale laboratory experiments, and for wide ranges of Keulegan–Carpenter (KC) and Reynolds (Re) numbers [12–19]. More complex models as compared to Morison-type ones and dealing with an improved description of flow-cylinder interaction processes have been also developed [20,21]. More recently, Aristodemo et al. [22] performed a laboratory study on non-breaking solitary wave forces with respect to a horizontal cylinder placed at half-water depth that was supported by numerical simulations based on the smoothed particle hydrodynamics technique (e.g., [23–29]). In this case, the effect of the free surface was negligible, i.e., there was no scattering, and that related to the bottom was weak. In this context, the horizontal and vertical force regime was dominated by inertia components and the peaks of the horizontal forces were observed to be between about four and five times higher than the vertical ones.

Here, a new laboratory investigation is presented in the case of a bottom-mounted horizontal cylinder subjected to tsunami-like solitary waves. A set of 30 experimental tests was performed in the wave channel of the University of Calabria. A rigid circular cylinder with longitudinal axis parallel to the cross flume was located at the bed of the flume. The horizontal and vertical loads were deduced from the records of 12 pressure sensors arranged along the external surface of the cylinder. Moreover, three wave gauges were placed in correspondence to the vertical axis of the cylinder and close to it to measure the surface elevation, while an ultrasonic sensor located behind the wavemaker was adopted to measure its displacement. The experiments were conducted at intermediate water depths quite close to shallow ones and for A/d ranging from about 0.08 to 0.18, where A is the wave amplitude and d is the water depth, with KC ranging from about 4 to 7 and $1.83 \times 10^4 < Re < 3.62 \times 10^4$. The resulting force field was characterized by the prevalence of an inertial regime for the horizontal force and of the lift component for the vertical one. It can be observed that, for Re of order of 10^3, the force regime is completely dominated by the inertia components in both directions (no formation of vortex patters). Moreover, for Re of order of 10^5, there is the prevalence of the drag force component in the horizontal direction and of the lift force in the vertical direction (e.g., [9]). The present experimental values of the free stream kinematics at the transversal axis of the cylinder and of the hydrodynamic forces were adopted to calibrate the hydrodynamic coefficients in the Morison [10] and transverse (e.g., [13]) semi-empirical relationships through the application of ordinary and weighted least square approaches.

The contents of the paper are organized in the following manner. The adopted theory to model tsunami-like solitary waves is summarised in Section 2. The experimental investigation in a laboratory wave channel to determine the horizontal and vertical loads induced by solitary waves on a horizontal cylinder placed on a horizontal bottom is illustrated in Section 3. The adopted semi-empirical formulas for a practical evaluation of the wave forces are explained in Section 4. The characteristics of the incident flow field and the hydrodynamic forces are respectively analysed in Sections 5.1 and 5.2. The calibration of the semi-empirical equations through the assessment of the hydrodynamic coefficients and their application to assess the contribution of the force components in the present wave–structure interaction phenomenon are respectively described in Sections 5.3 and 5.4. Finally, conclusions are drawn in Section 6.

2. Tsunami-Like Solitary Waves

Robust and widespread modelling of the leading wave of a tsunami event is given by the solitary wave theory. The time variation of the surface elevation, η, is taken as equal to (e.g., [30]):

$$\eta(t) = Asech^2(\beta ct/2) \tag{1}$$

where β is defined as the outskirts decay coefficient and c represents the wave celerity.

The Rayleigh theory to model a solitary wave is here selected due to the stable evolution of this kind of wave along a plane wave flume [31]. Under this approach, β and c in Equation (1) are respectively determined as:

$$\beta = 2\sqrt{\frac{3A}{4d^2(A+d)}}, \qquad c = \sqrt{g(A+d)} \tag{2}$$

where $\beta = 2k$ and k is the wave number which is considered a finite quantity for engineering purposes even if the wave length, L, of a solitary wave is theoretically taken as equal to infinity. As a result, an apparent wave period is defined as $T = L/c$ and an apparent wave length $L = 2\pi/k$ is then used. The above quantity is determined assuming, at a distance of $L/2$ away from the wave crest, the value of η is reduced to 1% of its maximum value (e.g., [22,32]). Other heuristic methods to define a finite wave length lead to negligible differences in defining a finite time window to analyse the present physical process (e.g., [33]).

Following the Rayleigh theory, the horizontal (u) and vertical (v) velocity values induced by the passage of a solitary wave are calculated as (e.g., [34]):

$$\left\{\begin{array}{l} \frac{u(t)}{c} = B_1\left(\frac{\eta(t)}{A}\right) - \left(\frac{A}{d}\right)^2 B_2\left(\frac{\eta(t)}{A}\right)^2 \\ \frac{v(t)}{c} = B_3\left(\frac{\eta(t)}{d}\right) tanh\left(B_4\frac{ct}{d}\right)\left\{1 + B_5\left[1 - \frac{7\eta(t)}{A} - B_6\left(1 - \frac{3\eta(t)}{A}\right)\right]\right\} \end{array}\right. \tag{3}$$

where z is the vertical coordinate starting from the bed and B_1, B_2, B_3, B_4, B_5, and B_6 are equal to:

$$\begin{array}{l} B_1 = \frac{A}{d} + 3\left(\frac{A}{d}\right)^2\left[\frac{1}{6} - \frac{1}{2}\left(\frac{z}{d}\right)^2\right], \quad B_2 = \frac{7}{4} - \frac{9}{4}\left(\frac{z}{d}\right)^2, \\ B_3 = \sqrt{3\frac{A}{d}}\left(\frac{z}{d}\right), \quad B_4 = -\sqrt{\frac{3A}{4d}}, \quad B_5 = \frac{A}{2d}, \quad B_6 = \left(\frac{z}{d}\right)^2 \end{array} \tag{4}$$

The analytical expressions to calculate the horizontal ($a_H = du/dt$) and vertical ($a_V = dv/dt$) accelerations read as:

$$\left\{\begin{array}{l} a_H(t) = \frac{2\beta c^2\left(\frac{A}{d}\right)^2 B_2 \sinh\left(\frac{\beta c t}{2}\right)}{\cosh^5\left(\frac{\beta c t}{2}\right)} - \frac{\beta c^2 B_1 \sinh\left(\frac{\beta c t}{2}\right)}{\cosh^3\left(\frac{\beta c t}{2}\right)} \\ a_V(t) = \frac{B_3 B_5 c \frac{A}{d} \tanh\left(\frac{B_4 c t}{d}\right)\left(\frac{7\beta c \sinh\left(\frac{\beta c t}{2}\right)}{\cosh^3\left(\frac{\beta c t}{2}\right)} - \frac{3 B_6 \beta c \sinh\left(\frac{\beta c t}{2}\right)}{\cosh^3\left(\frac{\beta c t}{2}\right)}\right)}{\cosh^2\left(\frac{\beta c t}{2}\right)} + \\ \quad - \frac{B_3 \beta c^2 \frac{A}{d} \sinh\left(\frac{\beta c t}{2}\right) \tanh\left(\frac{B_4 c t}{d}\right)\left\{B_5\left[B_6\left(\frac{3}{\cosh^2\left(\frac{\beta c t}{2}\right)} - 1\right) - \frac{7}{\cosh^2\left(\frac{\beta c t}{2}\right)} + 1\right] + 1\right\}}{\cosh^3\left(\frac{\beta c t}{2}\right)} + \\ \quad - \frac{B_3 B_4 c^2 \frac{A}{d}\left[\tanh^2\left(\frac{B_4 c t}{d}\right) - 1\right]\left\{B_5\left[B_6\left(\frac{3}{\cosh^2\left(\frac{\beta c t}{2}\right)} - 1\right) - \frac{7}{\cosh^2\left(\frac{\beta c t}{2}\right)} + 1\right] + 1\right\}}{d \cosh^2\left(\frac{\beta c t}{2}\right)} \end{array}\right. \tag{5}$$

The second-order solution given by Equation (3) leads to a small variation of u and v along z. This is in agreement with the intermediate water depth conditions close to shallow ones as observed through the present laboratory experiments. The free stream velocity field, in conjunction with the acceleration one, will be used in the practical Morison and transverse equations in order to calculate the hydrodynamic coefficients.

A non-linear solution of the horizontal movement for a piston-type paddle, X, able to generate a solitary wave based on the Rayleigh solution is used [31]:

$$X(t) = \frac{2A}{\beta d} tanh\{\beta[ct - X(t)]/2\} \tag{6}$$

3. Experimental Tests

Two-dimensional experimental tests were conducted in the wave flume at the GMI Laboratory of the University of Calabria. The wave channel is 41.0 m long, 1.2 m deep, and 1.0 m wide, with the sidewalls and bottom made of glass. It is equipped with a piston-type wavemaker with a maximum stroke $S = 0.5$ m, and a rubble mound breakwater (slope of 1:4) to dissipate the incoming waves. More specifically, the paddle movement is controlled indirectly by the rotation of a joint of the mechanical chain, which is connected to the paddle. The rotation angle is measured with a resistive encoder that provides a proportional analogue voltage signal. This signal, as well as the set-point signal, is processed by a properly tuned proportional integral derivative (PID) controller. The PID acts in order to minimize the error, i.e., the difference between the set-point and the feedback signals. The output of the PID is connected to the kinematic chain through a hydro-pneumatic actuator. The set-point signal is generated by a Data AcQuision Board (DAQ), thanks to a digital-to-analog converter (DAC) (see, for more details, Tripepi et al. [35]). The longitudinal profile of the experimental layout is highlighted in Figure 1.

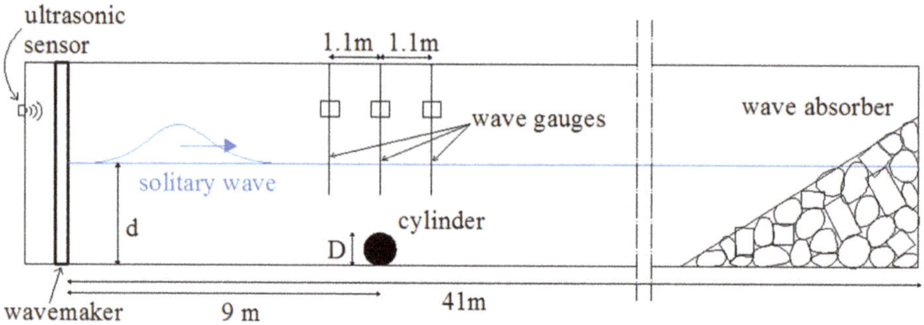

Figure 1. Sketch of the longitudinal profile of the experimental setup.

At about 9 m from the wave paddle, a circular cylinder with diameter $D = 0.127$ m was placed at the bottom flume ($e/D = 0$, where e is the distance between the bottom of the cylinder and the bed) with its longitudinal axis orthogonal to the wave direction (Figure 2a). This physical model was installed in the channel by means of a steel support equipped by a pulley system in order to accurately choose a specific location along the depth. To ensure unwanted displacements of the cylinder, C-shaped PVC supports were used at its edges (see Figure 2b). Moreover, a special glue was adopted to fix the cylinder at the bottom in order to inhibit the passage of water flows below it. Twelve pressure transducers (PDCR1830 model by Druck) were acquired in differential mode due to the Wheatstone-bridge configuration and mounted along the external surface of the cylinder at 30° intervals. Similar to the experimental tests with wind waves and currents performed by Aristodemo et al. [19,20], the transducers were slightly staggered along the longitudinal axis of the central part of the cylinder to avoid the use of a too-large diameter (see Figure 2a). The obtained dynamic pressures, Δp, were determined by subtracting the static pressures from the records of total pressures measured by the transducers. The values of Δp were assumed constant over the influence areas and evaluated as a function of the position of the transducers. Then, the horizontal (F_H) and vertical (F_V,) hydrodynamic forces were deduced as:

$$\begin{cases} F_H(t) = a_1[\Delta p_1(t) + \Delta p_6(t) - \Delta p_7(t) - \Delta p_{12}(t)]+ \\ \quad +a_2[\Delta p_2(t) + \Delta p_5(t) - \Delta p_8(t) - \Delta p_{11}(t)]+ \\ \quad +a_3[\Delta p_3(t) + \Delta p_4(t) - \Delta p_9(t) - \Delta p_{10}(t)] \\ \\ F_V(t) = a_1[\Delta p_3(t) + \Delta p_{10}(t) - \Delta p_4(t) - \Delta p_9(t)]+ \\ \quad +a_2[\Delta p_2(t) + \Delta p_{11}(t) - \Delta p_5(t) - \Delta p_8(t)]+ \\ \quad +a_3[\Delta p_1(t) + \Delta p_{12}(t) - \Delta p_6(t) - \Delta p_7(t)] \end{cases} \tag{7}$$

where the influence areas a_1, a_2 and a_3 are evaluated as:

$$a_1 = \int_{\pi/3}^{\pi/2} \frac{D}{2} cos\beta d\beta, \quad a_2 = \int_{\pi/6}^{\pi/3} \frac{D}{2} cos\beta d\beta, \quad a_3 = \int_{0}^{\pi/6} \frac{D}{2} cos\beta d\beta \tag{8}$$

The reference angle, β, was considered starting from the lower side of the cylinder in clockwise direction (Figure 2c).

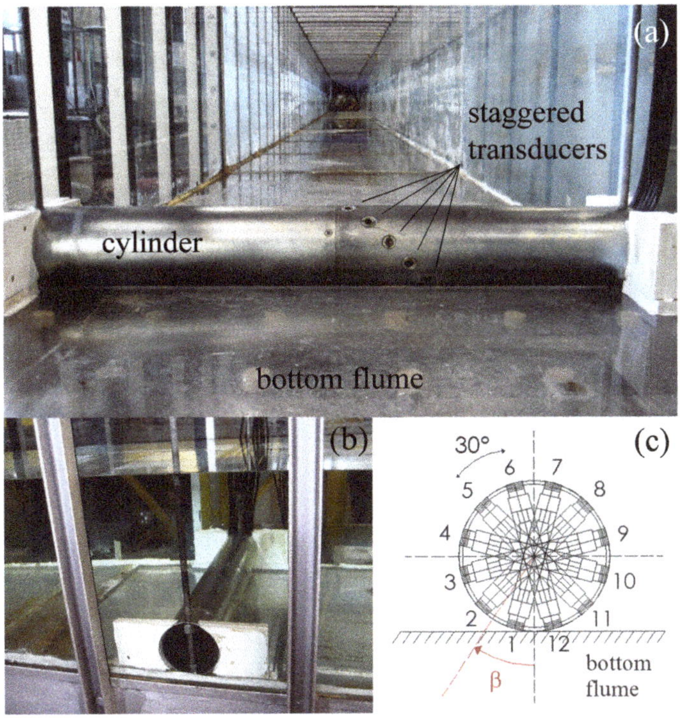

Figure 2. (**a**) Front view from the paddle of the cylinder and the staggered transducer arrangement in the wave channel; (**b**) View of the cylinder placement during the experiments; (**c**) Representative cross section of the transducers around the bottom-mounted cylinder.

A resistive wave gauge by Edif Instruments was located in correspondence to the vertical axis of the cylinder to measure the surface elevation and successively deduce the undisturbed kinematic field at the transversal axis of the cylinder for the application of semi-empirical equations. A further two wave gauges were placed 1.1 m before and after the vertical axis of the considered structure in order to check the value of c obtained by Equation (2) on the basis of the time shifts observed during the propagation of the solitary waves. The wave gauges were acquired in single-ended mode. The sampling frequency (f) of the transducers and gauges was set at 1000 Hz. Both types of instruments were calibrated in static conditions using a water tank equipped with a digital water gauge and a bottom spillway. Measurements were performed every 0.02 m and, for each water level, the signals were acquired for 10 min. All instruments provided a linear calibration function. To verify the correct generation of the solitary wave by applying Equation (6), the horizontal displacement of the piston, X, was measured by an ultrasonic sensor located behind the position at rest of the paddle using f = 50 Hz (see Figure 1).

A set of 30 different experimental tests at increasing A was performed by changing the motion law in the possible range of the stroke S of the present piston-type paddle. Each wave amplitude was generated two times in order to check for repeatability of the experiments. The still water depth, d, of the experimental tests was 0.4 m. Table 1 shows the resulting values of A, T, A/d, $KC = u_{max}T/D$, $Re = u_{max}D/\nu$ and d/L (relative depth), where u_{max} is the maximum value of the free stream horizontal velocity at the transversal axis of the cylinder, and ν is the kinematic viscosity.

Table 1. Characteristics of the experimental tests.

Test Number	A (m)	T (s)	A/d	KC	Re	d/L
1	0.033	3.73	0.083	4.43	1.83×10^4	0.052
2	0.034	3.84	0.085	4.69	1.87×10^4	0.051
3	0.034	3.82	0.085	4.74	1.91×10^4	0.051
4	0.035	3.83	0.088	4.86	1.95×10^4	0.051
5	0.037	3.91	0.093	5.17	2.03×10^4	0.050
6	0.041	3.99	0.103	5.86	2.26×10^4	0.048
7	0.041	3.78	0.103	5.60	2.27×10^4	0.051
8	0.042	3.87	0.105	5.76	2.29×10^4	0.050
9	0.042	4.06	0.105	6.06	2.29×10^4	0.047
10	0.043	3.37	0.108	5.10	2.33×10^4	0.057
11	0.044	3.73	0.110	5.77	2.38×10^4	0.051
12	0.044	3.64	0.110	5.70	2.41×10^4	0.053
13	0.045	3.90	0.113	6.19	2.44×10^4	0.049
14	0.046	3.79	0.115	6.17	2.50×10^4	0.050
15	0.046	3.47	0.115	5.67	2.51×10^4	0.055
16	0.048	4.05	0.120	6.87	2.61×10^4	0.047
17	0.049	3.88	0.123	6.66	2.64×10^4	0.049
18	0.052	3.86	0.130	7.00	2.78×10^4	0.049
19	0.054	3.56	0.135	6.62	2.86×10^4	0.053
20	0.055	3.59	0.138	6.84	2.92×10^4	0.053
21	0.055	3.69	0.138	7.03	2.93×10^4	0.051
22	0.058	3.26	0.145	6.45	3.04×10^4	0.058
23	0.058	3.58	0.145	7.14	3.06×10^4	0.053
24	0.058	3.36	0.145	6.71	3.07×10^4	0.056
25	0.064	3.22	0.160	6.99	3.33×10^4	0.058
26	0.065	3.26	0.163	7.11	3.35×10^4	0.058
27	0.065	3.28	0.163	7.18	3.36×10^4	0.057
28	0.066	3.09	0.165	6.87	3.42×10^4	0.061
29	0.071	3.01	0.178	7.07	3.61×10^4	0.062
30	0.071	2.86	0.178	6.74	3.62×10^4	0.065

It is worth noting that the experimental values of T sometimes highlight a deviation from a decreasing theoretical trend when A increases. This is due to the significant spreading of η around the undisturbed free surface because of the occurrence of the so-called trailing waves [31]. However, the maximum amplitudes of trailing waves are up to 2% of those related to the solitary waves, leading to a slight influence on the final part of the wave loads for which the magnitude is usually low and then irrelevant for stability purposes. The values of the experimental trailing waves were observed to be under the critical threshold suggested by Guizien and Barthélemy [31]. The above features also influence the values of KC and, similarly to the definition of an apparent wave period, it is possible to define an apparent Keulegan–Carpenter number [22]. This parameter was generally linked to the occurrence of vortices around the cylinder and, more generally, used to study the wave-cylinder interaction processes (e.g., [8]). The values of KC will be successively adopted for comparisons with regular wave cases in the literature. It can also be noted that the range of d/L refers to intermediate water depths quite close to shallow ones, allowing for the use of Equation (3) to represent the undisturbed kinematic field at the cylinder location.

4. Semi-Empirical Formulas

For engineering purposes, the use of semi-empirical formulas represents a simple and suitable tool to determine the horizontal and vertical hydrodynamic loads acting on offshore and coastal structures. Owing to their mathematical representation, these formulas require a specific calibration of the hydrodynamic coefficients for their correct application. In the case of in-line loads, the Morison equation [10] is widely adopted for various incident flows and kinds of structures, while the transverse equation (e.g., [13,35]) is adopted to model the vertical forces. It is worth noting that the use of Morison and transverse formulas is here possible since no diffraction effects occur, i.e., the presence of the physical model of the bottom-mounted cylinder does not affect the local free surface, which is considered as a rigid lid [6].

In this context, the in-line force, F_H, is evaluated as the sum of a drag component, F_D, due to the resistance of a solid body to the incident flow motion, and an inertia component, F_{HI}, depending on the horizontal acceleration of the oscillatory flow. The total horizontal force, F_H, is calculated as follows [10]:

$$F_H = F_D + F_{HI} = \frac{1}{2}\rho D C_D u|u| + \frac{\pi}{4} D^2 \rho C_{MH} a_H \tag{9}$$

where C_D represents the drag coefficient and C_{MH} is the horizontal inertia coefficient. The values of u and a_H are the ambient horizontal velocity and acceleration at the transversal axis of the cylinder, respectively.

The total vertical load, F_V, is given as the superimposition of a lift component, F_L, generated by the increased flow velocity across the cylinder induced by the blocking of the flow, and an inertia component, F_{VI}, depending on the vertical acceleration of the external flow at the transversal axis of the body. The transverse force is then written as [13]:

$$F_V = F_L + F_{VI} = \frac{1}{2}\rho D C_L u^2 + \frac{\pi}{4} D^2 \rho C_{MV} a_V \tag{10}$$

where C_L is the lift coefficient and C_{MV} is the vertical inertia coefficient. The value of a_V is the free stream vertical acceleration. It can be observed that, for $e/D = 0$, the contribution of F_{VI} is usually considered negligible (e.g., [14,20]) even if this force contribution affects the magnitude and the shape of the total vertical force and it is here considered as in the case with $e/D = 1$ [22]. Indeed, for $e/D > 0$, the value of F_{VI} becomes relevant in modelling F_V, as highlighted by Aristodemo et al. [22].

The undisturbed kinematic field, i.e., u, a_H, and a_V, in the Morison and transverse schemes was determined by Equation (3) from the experimental values of the surface elevation, η, deduced from the wave gauge placed at the vertical section of the cylinder.

5. Results

5.1. Surface Elevation and Kinematic Field

The evaluation of the solitary wave loads at the horizontal cylinder placed on the bottom channel depends on the suitable values of surface elevation at the cylinder section and the related free stream kinematic field at the cylinder, in addition to the correct generation of the solitary wave by the experimental piston paddle. Figure 3 highlights the comparison between the analytical solution given by Equation (1) and the experimental values of the surface elevation at the vertical axis of the cylinder for test number 5 (A = 0.037 m and T = 3.91 s, i.e., solitary wave with low amplitude and broad shape) and test number 30 (A = 0.071 m and T = 2.86 s, i.e., solitary wave with high amplitude and narrow shape), respectively. For both cases, a general good agreement on the incident solitary waves can be noticed, particularly for the higher values of η. The reference time instant $t = 0$ refers to the passage of the solitary wave crest at the vertical section of the cylinder.

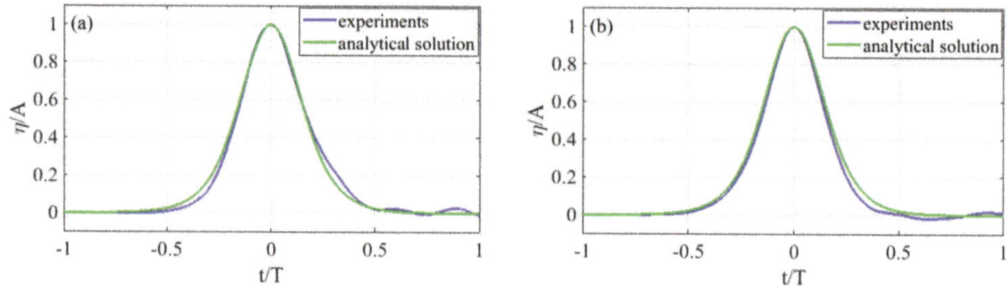

Figure 3. Time variation of surface elevation, η, in correspondence to the vertical axis of the cylinder: comparison between analytical solution and experiments. (**a**) Test number 5; (**b**) Test number 30.

With reference to test number 30, Figure 4 shows the comparison between analytical solutions and laboratory tests for the time variation of the ambient kinematic field in correspondence with the transversal axis of the cylinder, namely $z = D/2$, where the semi-empirical schemes will be applied. Starting from the surface elevations (see Figure 3b), u was directly determined by applying Equation (3), while a_H and a_V were respectively derived from u and v (see Equation (5)). Specifically, Figure 4a describes the time history of the horizontal velocity u where it is possible to notice the same shape of η (see Figure 3b). The theoretical horizontal acceleration, a_H, shows equal positive and negative peaks (Figure 4b), while the analytical vertical acceleration, a_V, presents a double positive peak and a greater negative one (Figure 4c). Small experimental deviations from the reference analytical solutions occur in the final part of the passage of the solitary wave across the cylinder, leading to slight non-symmetrical features of the values of u, a_H, and a_V. However, these discrepancies are essentially not relevant for stability purposes of the cylinder in which the force peaks play a fundamental role. As later highlighted, the relevance of the ambient kinematic field at the cylinder arises from the influence on the shape of the horizontal and vertical hydrodynamic loads as well as in the application of Morison and transverse semi-empirical schemes in which the various force components are directly proportional to u, a_H, and a_V.

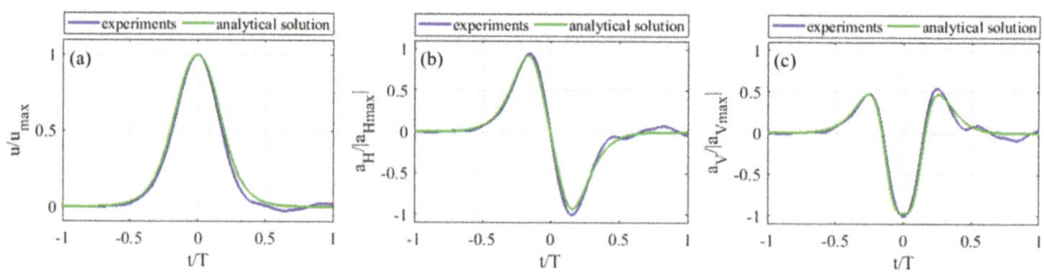

Figure 4. Time variation of undisturbed kinematic field at the transversal axis of the cylinder: comparison between analytical solution and experiments (test number 30). (**a**) Horizontal velocity, u; (**b**) Horizontal acceleration, a_H; (**c**) Vertical acceleration, a_V.

5.2. Hydrodynamic Forces

In this section, the time history of the solitary wave loads acting on the bottom-mounted cylinder deduced from the experimental tests are analyzed. As previously shown in Figure 3 for the surface elevation, two reference test cases characterized by a different wave amplitude and period are considered.

Figure 5a,b highlight the experimental values of the horizontal (F_H) and vertical (F_V) hydrodynamic forces induced by solitary waves for test numbers 5 and 30, respectively. The black dashed vertical line refers to the occurrence of the wave crest at the vertical section of the cylinder, i.e., the time instant $t = 0$ in which the maximum surface elevation and horizontal velocity appear. It is interesting to observe that, in terms of maximum peaks, F_H is greater than F_V for test number 5 (lower solitary wave), while $F_V > F_H$ for test number 30 (higher solitary wave). Moreover, a prevalence of positive values of the forces can be noticed, revealing that the cylinder is substantially subjected to the coupled action of a forward motion in the direction of solitary wave propagation and a lifting one towards the free surface. For test number 5, the maximum peak of F_H is more back-shifted than the other case if compared to the passage of the solitary wave crest. In both cases, there is a prevalence of the inertia component with respect to the contribution of the resistance offered by the presence of the cylindrical structure. The above findings are substantially in agreement with experimental observations related to the interaction between regular or random waves and cylinders placed on the bed when the parameter KC is considered (e.g., [15,36]). It can be observed that the shapes of F_H generally follow those related to a_H, with a less relevant contribution of the drag force related to the decreasing of the negative peak of F_H and the forward shift of the positive peak of F_H. Apart from a small contribution of the vertical inertia component for low values of F_V, the shape of the vertical load, for reference test numbers 5 and 30, follows that related to the horizontal velocity where the peak appears very close to the solitary wave crest. This situation arises when an external flow interacts with a bottom-mounted cylinder in which the lift component dominates the features of F_V (e.g., [14,19]). The occurrence of drag and lift forces will be better highlighted when Morison and transverse semi-empirical schemes are applied. However, it is important to notice that these contribution are linked to the formation of vortex patterns around the cylinder and the consequent deviation from a pure inertial field instead characterized by a potential flow (e.g., [37,38]).

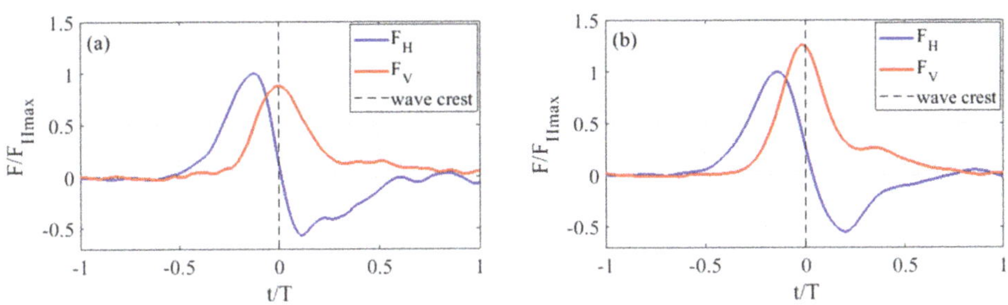

Figure 5. Time variation of experimental horizontal (F_H) and vertical (F_V) hydrodynamic forces. (**a**) Test number 5; (**b**) Test number 30.

Taking into account all experimental tests, the positive and negative maximum horizontal forces ($F_{Hmax,p}$ and $F_{Hmax,n}$) and the positive maximum vertical forces (F_{Vmax}) as a function of A/d are respectively shown in Figure 6. Note that these peaks are respectively normalized with respect to the positive maximum peak of the vertical force, F_{Vmax}*. As commonly carried out in the field of solitary waves interacting with offshore and coastal structures (e.g., [39]), the non-dimensional wave amplitude, A/d, will be considered as simple representative parameter to evaluate the features of the hydrodynamic forces and coefficients compared to Re and KC since these parameters are dependent on the indirect knowledge of the undisturbed horizontal velocity at the transversal axis of the cylinder. Moreover, a more stable trend of the involved quantities when A/d is adopted was observed. In general, the positive peaks increase almost linearly with A/d, while the negative ones highlight a higher variation for $A/d > 0.15$. The values of $F_{Hmax,n}$ are lower than the positive ones and those referring to F_{Vmax}. It is interesting to note that, for $A/d < 0.105$, $F_{Hmax,p}$ values are slightly greater than F_{Vmax}, while F_{Vmax} values are greater than $F_{Hmax,p}$ values for $A/d > 0.105$ and, particularly, for high A/d. The threshold corresponding to $A/d = 0.105$ is highlighted in Figure 6 with a grey dashed vertical line. Considering the whole experimental range of A/d, the values of $F_{Hmax,p}$ and $F_{Hmax,n}$ respectively exhibit an increase of 57% and 56%, with F_{Vmax} growth of about 69%.

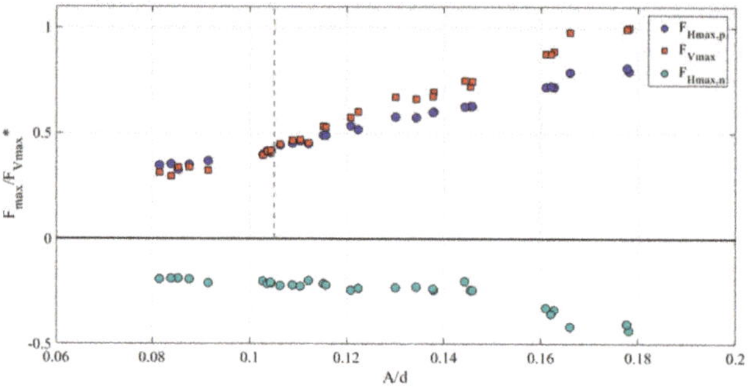

Figure 6. Maximum positive and negative peaks of experimental horizontal forces ($F_{Hmax,p}$ and $F_{Hmax,n}$, respectively), and maximum positive peaks of experimental vertical forces (F_{Vmax}), vs. A/d.

5.3. Calibration of Semi-Empirical Formulas

The calibration of Morison and transverse semi-empirical formulas to evaluate the solitary wave forces at bottom-mounted cylinders in an easy way is linked to the correct evaluation of the hydrodynamic coefficients. The above time-constant coefficients can be viewed as representative parameters of the complex flow field around the cylinder. In order to minimize the differences between experimental forces and those calculated by Morison and transverse schemes within the adopted apparent wave period, time-domain methods for evaluating the hydrodynamic coefficients were considered. The knowledge of the ambient kinematics field (i.e., horizontal velocity and horizontal and vertical acceleration) at the transversal axis of the cylinder and the hydrodynamic loads acting on it deduced through the experimental tests allowed for the calculation of in-line (C_D and C_{MH}) and transverse (C_L and C_{MV}) hydrodynamic coefficients. In this context, the ordinary and weighted least square methods were used (e.g., [40]). In the weighted least square method, the difference between the measured and the semi-empirical force is multiplied by F_H^k, with k a positive index. Within the adopted apparent wave period of the solitary wave at the cylinder, the hydrodynamic coefficients C_D and C_{MH} related to the Morison scheme are calculated as:

$$\begin{cases} C_D = \dfrac{\sum_{i=1}^{M} F_H^{2k+1} u|u| \sum_{i=1}^{M} F_H^{2k} a_H^2 - \sum_{i=1}^{M} F_H^{2k+1} a_H \sum_{i=1}^{M} F_H^{2k} u|u| a_H}{K_D \left[\sum_{i=1}^{M} F_H^{2k} u^4 \sum_{i=1}^{M} F_H^{2k} a_H - \left(\sum_{i=1}^{M} F_H^{2k} u|u| a_H \right)^2 \right]} \\ \\ C_{MH} = \dfrac{\sum_{i=1}^{M} F_H^{2k+1} a_H \sum_{i=1}^{M} F_H^{2k} u^4 - \sum_{i=1}^{M} F_H^{2k+1} u|u| \sum_{i=1}^{M} F_H^{2k} u|u| a_H}{K_{MH} \left[\sum_{i=1}^{M} F_H^{2k} a_H^2 \sum_{i=1}^{M} F_H^{2k} u^4 - \left(\sum_{i=1}^{M} F_H^{2k} u|u| a_H \right)^2 \right]} \end{cases} \quad (11)$$

where $K_D = \frac{1}{2}\rho D$ and $K_{MH} = \frac{1}{4}\pi D^2 \rho$. The value of M represents the number of force and kinematic values within the adopted wave period.

Similarly, the expressions to determine the hydrodynamic coefficients C_L and C_{MV} for the transverse formula read as:

$$\begin{cases} C_L = \frac{\sum_{i=1}^{M} F_V^{2k+1} u^2 \sum_{i=1}^{M} F_V^{2k} a_V^2 - \sum_{i=1}^{M} F_V^{2k+1} a_V \sum_{i=1}^{M} F_V^{2k} u^2 a_V}{K_L \left[\sum_{i=1}^{M} F_V^{2k} u^4 \sum_{i=1}^{M} F_V^{2k} a_V - \left(\sum_{i=1}^{M} F_V^{2k} u^2 a_V\right)^2\right]} \\ C_{MV} = \frac{\sum_{i=1}^{M} F_V^{2k+1} a_V \sum_{i=1}^{M} F_V^{2k} u^4 - \sum_{i=1}^{M} F_V^{2k+1} u^2 \sum_{i=1}^{M} F_V^{2k} u^2 a_V}{K_{MV} \left[\sum_{i=1}^{M} F_V^{2k} a_V^2 \sum_{i=1}^{M} F_V^{2k} u^4 - \left(\sum_{i=1}^{M} F_V^{2k} u^2 a_V\right)^2\right]} \end{cases} \tag{12}$$

where $K_L = K_D$ and $K_{MV} = K_{MH}$. Equations (11) and 12 recover the ordinary least square approach by setting $k = 0$.

The performances of the ordinary and weighted least square approaches for calculating C_D, C_{MH}, C_L, and C_{MV} are analysed on the basis of the mean square error percent (MSEP). The MSEP was obtained by comparing Morison and transverse forces and those calculated experimentally in the following way:

$$MSEP = \frac{1}{M} \sum_{i=1}^{M} \left(\frac{F_i^e - F_i^s}{F_i^e} \right)^2 \tag{13}$$

where F^s represents the generic semi-empirical force and F^e is the generic experimental one.

For practical aims, attention was paid to the maximum peaks of the wave forces, i.e., positive and negative for the horizontal force and only positive for the vertical one, and the related phase shifts, $\phi = 2\pi t_m / T$, where t_m is the occurrence time of the maximum positive or negative peak within the wave period. Figure 7 describes the mean values of MSEP for all 30 laboratory tests calculated by the ordinary least square (OLS), the weighted least square using $k = 1$ (WLS1), $k = 2$ (WLS2), $k = 3$ (WLS3), $k = 4$ (WLS4), $k = 5$ (WLS5), and $k = 6$ (WLS6). The choice to test the weighted least square method up to $k = 6$ is to capture the maximum positive peaks of the horizontal and vertical hydrodynamic loads without any overestimation of the above quantities. The values of MSEP linked to the positive peaks of both forces are lower than those related to the negative horizontal forces. When k increases, MSEP for $F_{Hmax,p}$ tends to decrease, ranging from about 4.5% for $k = 0$ to 0.7% for $k = 6$. The same feature refers to F_{Vmax} for which the MSEP ranges from about 3.6% for $k = 0$ to 0.2% for $k = 6$. Conversely, MSEP strongly increases proportionally to $F_{Hmax,p}$, moving from 24% for $k = 0$ up to 54% for $k = 6$. With regard to the phase shift associated with the force peaks, the resulting values of MSEP prove to be generally low and oscillate between 0.9% and 4.9%, with lowest values for $F_{Hmax,p}$ associated with $k = 6$ and lowest values for $F_{Hmax,n}$ using $k = 0$. Taking into account the mean values of MSEP related to all force peaks and associated phase shifts, it is possible to observe that the OLS method ($k = 0$) gives the lowest MSEP, equal to 6.7%. Although the use of $k > 1$ leads to good values of the maximum peaks of both forces, a relevant overestimation of the negative peak of F_H is noted. Then, the ordinary least square method was considered to determine the hydrodynamic coefficients in the Morison and transverse equations.

Figure 8 highlights the experimental values of C_D, C_{MH}, C_L, and C_{MV} in the Morison and transverse semi-empirical equations as a function of A/d ranging from 0.08 to 0.18. The 95% prediction intervals are also plotted in Figure 8 and based on second-order polynomial fitting curves for C_D, C_{MH}, and C_{MV} and on an exponential fitting equation for C_L. Similar to the case with $e/D = 1$ [22], the hydrodynamic coefficients C_{MH}, C_L and C_{MV} show a general decreasing trend when A/d increases. This feature is particularly evident for the stable trend of horizontal inertia coefficient, while for the corresponding vertical one the trend is quite scattered, i.e., with the highest uncertainty, even if this kind of force component has a small weight in calculating the vertical load as compared to the lifting load, as successively highlighted in the application of semi-empirical force models. The values of C_D generally tend to increase up to approximately $A/d = 0.15$ with a corresponding $C_D = 1.4$, followed by a tendential reduction. This particular feature can be heuristically explained through the formation of vortex patterns around the cylinder that allows the occurrence of drag and lift force components at the cylinder. Indeed, for low A/d and KC, the increasing trend of C_D is linked to a predominant transverse direction of the vortices compared to the incoming flow direction, while the decreasing trend of C_D is related to a resulting movement of the vortices in the direction of the incident flow, as observed experimentally for regular waves by Sumer et al. [41] and numerically for solitary waves by Aristodemo et al. [22]. Regarding the magnitude of the hydrodynamic coefficients, C_D values range from 0.7 to 1.4, C_{MH} from 2.4 to 3, C_L from 3.9 to 4.7, and C_{MV} from 3.7 to 6.9.

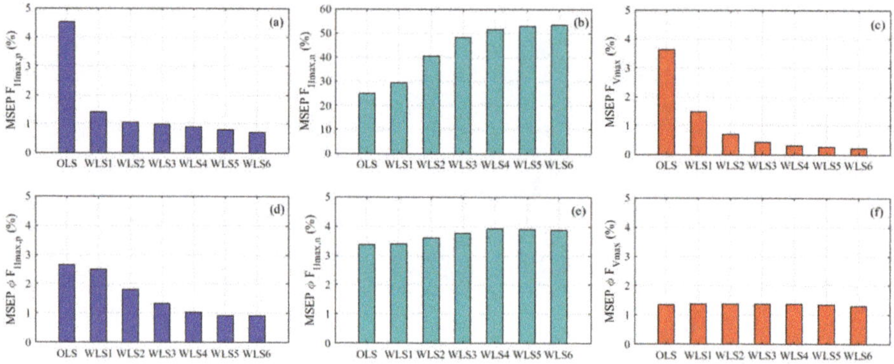

Figure 7. Experimental mean square error percent (MSEP) through OLS, WLS1, WLS2, WLS3, WLS4, WLS5, and WLS6 methods. (**a**) Maximum positive horizontal force, $F_{Hmax,p}$; (**b**) Maximum negative horizontal force, $F_{Hmax,n}$; (**c**) Maximum vertical force, F_{Vmax}; (**d**) Phase shift, ϕ, associated with $F_{Hmax,p}$; (**e**) Phase shift, ϕ, associated with $F_{Hmax,n}$; (**f**) Phase shift, ϕ, associated with F_{Vmax}. OLS: ordinary least square; WLS: weighted least square.

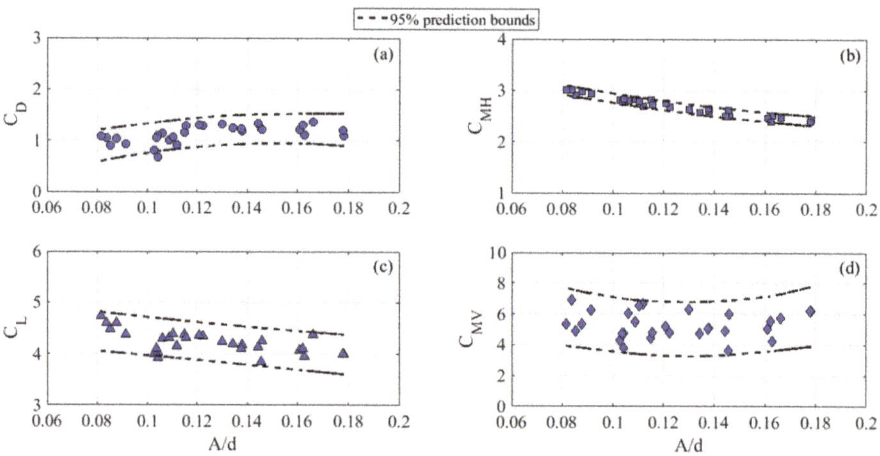

Figure 8. Experimental hydrodynamic coefficients vs. A/d. (**a**) C_D; (**b**) C_{MH}; (**c**) C_L; (**d**) C_{MV}.

In a slightly wider experimental range ($4 < KC < 8$), the hydrodynamic force coefficients C_D, C_{MH}, and C_L are also plotted vs. KC in Figure 9 and compared with literature experimental results referring to regular waves interacting with a bottom-mounted cylinder. Only the vertical inertia coefficients were not reported since no experiments were conducted in the present KC range. To the author's knowledge, the unique experiments by Cheong et al. [13] to determine C_{MV} were performed out of the investigated range, i.e., $0.05 < KC < 1.25$, even if a decreasing trend when KC increases was noted as in the present dataset. A more scattered range for KC compared to A/d can be observed since the Keulegan–Carpenter number depends on the apparent wave period which suffers from small oscillations near the free surface due to the appearance of spurious trailing waves. Figure 9a shows the present values of C_D as a function of KC plotted against the experiments conducted by Sarpkaya and Rajabi [12], and Bryndum et al. [14] through a non-linear fit of the laboratory data (Neill and Hinwood [15] and Chevalier et al. [16]). Comparing the values of C_D, the authors observed an overall good agreement with the non-linear fit carried out by Bryndum et al. [14] and a slight underestimation of the present experiments as compared to those performed by Sarpkaya and Rajabi [12], Neill and Hinwood [15], and Chevalier et al. [16], even if few values of C_D are present in the above works. With regard to C_{MH}, Figure 9b shows the comparison between the current dataset and the values obtained by Sarpkaya and Rajabi [12], Bryndum et al. [14], Neill and Hinwood [15], Chevalier et al. [16], and Aristodemo et al. [19]. The present values of C_{MH} are lower than those calculated by Sarpkaya and Rajabi [12], Bryndum et al. [14] and, in particular, by Aristodemo et al. [19], while they are slightly higher than those obtained by Neill and Hinwood [15] and Chevalier et al. [16]. However, the current values of C_{MH} tend to an asymptotic value equal to 3.29 occurring for the potential flow characterized by only inertia forces (very low KC numbers) deduced from the studies of Bryndum et al. [14] and Sumer and Fredsoe [9]. Paying attention to C_L (see Figure 9c), the present data are compared with those obtained by Sarpkaya and Rajabi [12], Bryndum et al. [14], Neill and Hinwood [15], and Aristodemo et al. [19]. The agreement with the fitting curve proposed by Bryndum et al. [14] is very good, while the current values of C_L underestimate those proposed by Sarpkaya and Rajabi [12] and strongly overestimate those obtained by Neill and Hinwood [15] and Aristodemo et al. [19]. As for C_{MH}, it can be noted that C_L tends to an asymptotic value equal to 4.49 for the fully inertial regime, i.e., when $KC \to 0$.

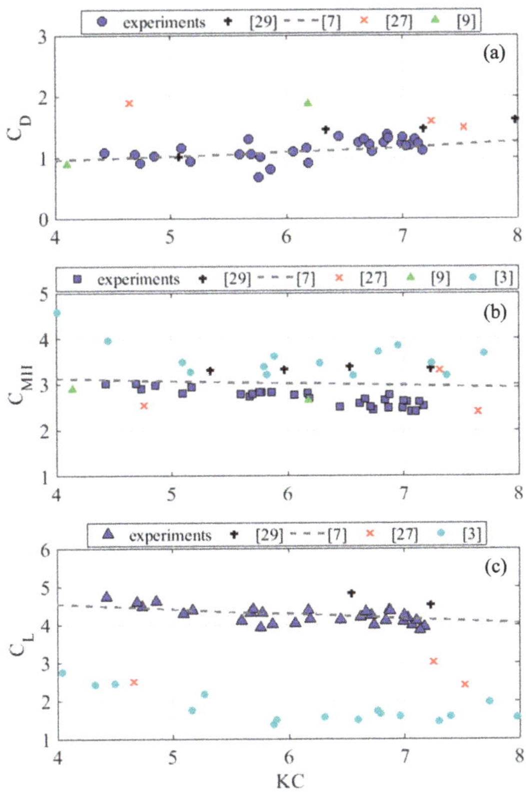

Figure 9. Experimental hydrodynamic coefficients vs. KC and comparison with literature coefficients for regular waves. (**a**) C_D; (**b**) C_{MH}; (**c**) C_L.

5.4. Use of Semi-Empirical Formulas

The Morison and transverse equations calibrated by experimental values of C_D, C_{MH}, C_L, and C_{MV} and deduced from OLS approach are here illustrated in the time domain to highlight the features of the different force components in horizontal (drag and inertia) and vertical (lift and inertia) directions. For the considered test (case numbers 5 and 30 shown in Figure 5), Figure 10a,b respectively describe the comparisons between the time variation of the drag ($F_D M$) and inertia ($F_{HI} M$) as force components, and the total force ($F_H M$) determined by the calibrated Morison equation (see Equation (9)) and the horizontal one (F_H) deduced through the laboratory experiments. For the same tests, Figure 10c,d respectively show the comparisons between the time history of the lift ($F_L T$) and inertia ($F_{VI} T$) force components, and the total ($F_V T$) determined by the calibrated transverse equation (see Equation (10)) and the vertical one (F_V) calculated by the laboratory experiments. It is worth noting that Morison and transverse component and total forces are respectively specified through the symbols M and T in Figure 10. An overall good agreement between semi-empirical methods and experiments is noted. This is evident particularly for the maximum positive peak of the horizontal and vertical force modelled by the Morison and transverse scheme for test number 5, respectively. A less accurate reproduction of the time variation of the experimental loads can be observed for the negative peak of the horizontal force, particularly for test number 30, related to the higher solitary wave. With regard to the phase shifts linked to the force peaks, the Morison and transverse forces are lower and forward-shifted as compared to the experimental ones, particularly for test number 30. This is linked to the complex patterns of vortical structures around the cylinder that are not included in the simple expressions of the adopted semi-empirical schemes. It can be noted that the horizontal force is dominated by the inertia contribution depending on the undisturbed horizontal acceleration if compared to the drag, which is related to the free stream horizontal velocity. The vertical force field is conversely dominated by the lift force and the effect of the vertical inertia is linked to a lowering of the former contribution to give the modelling of the vertical load. In general, the shape of the vertical force follows that related to the ambient horizontal velocity in which the peak is substantially in phase with the surface elevation. Owing to the presence of spurious trailing waves, it is also possible to observe a low contribution of a positive vertical load in its final part that is not modelled by the transverse scheme.

The four force components, i.e., F_D, F_{HI}, F_L, and F_{VI} used to calculate the total horizontal and vertical loads using the calibrated Morison and transverse formulas through the present laboratory experiments are also analysed in terms of positive and negative peaks. As highlighted in Figure 11 as a function of A/d, the force components are weighted with respect to the corresponding maximum peak of the semi-empirical horizontal and vertical force in order to show their contribution to model the total loads. For the peaks of horizontal forces (see Figure 11a), the inertia component ranges about from 90% for low A/d to 80% for high A/d, leading to a progressive reduction of an inertia-dominated regime and a growth of the weight of the drag up to 45%. Paying attention to the peaks of vertical loads (see Figure 11b), a major role is linked to the maximum lift force, F_{Lmax}, which shows a slight decrease when A/d increases. It can be observed that the ratio between the peaks of the lift component and the total vertical one is generally greater than 1. The weight of the positive ($F_{VImax,p}$) and negative ($F_{VImax,n}$) peaks compared to the maximum vertical force is quite low. Both contributions exhibit a very small increase when A/d increases. Specifically, the values of $F_{VImax,p}$ range about from 5 to 9%, while the values of $F_{VImax,n}$ range about from 10 to 20%. It is worth noting that the tendencies of the force peaks generally reflect the features of the hydrodynamic coefficients, as illustrated in Figure 8.

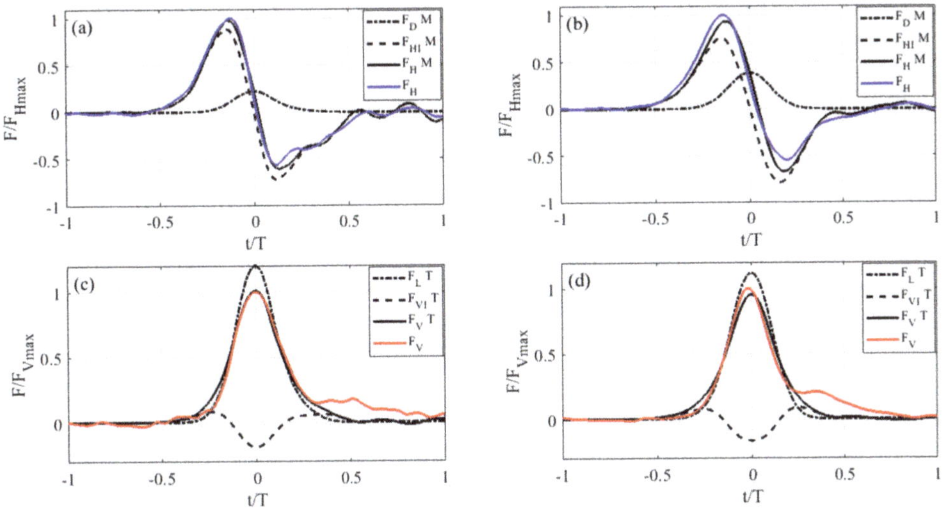

Figure 10. Time history of hydrodynamic loads determined by semi-empirical formulas and experiments. (**a**) Comparison between Morison F_DM, $F_{HI}M$, and F_HM calibrated by experiments and experimental F_H (test number 5); (**b**) Comparison between Morison F_DM, $F_{HI}M$, and F_HM calibrated by experiments and experimental F_H (test number 30); (**c**) Comparison between transverse F_LT, $F_{VI}T$, and F_VT calibrated by experiments and experimental F_V (test number 5); (**d**) Comparison between transverse F_LT, $F_{VI}T$, and F_VT calibrated by experiments and experimental F_V (test number 30).

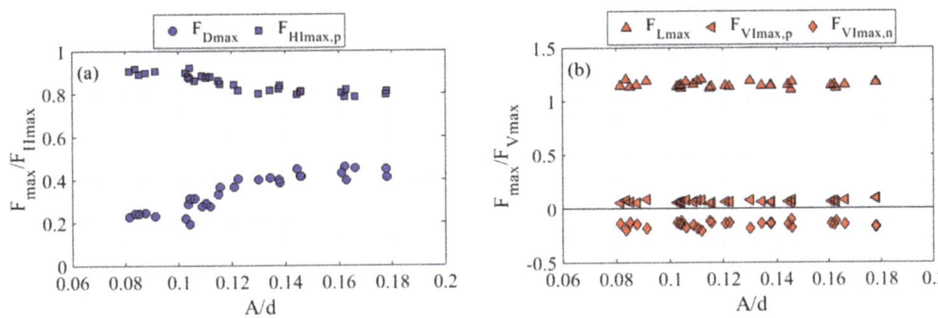

Figure 11. (**a**) Peaks of experimental Morison force components vs. A/d: positive drag F_{Dmax} and positive inertia $F_{HImax,p}$; (**b**) Peaks of experimental transverse force components vs. A/d: positive lift F_{Lmax}, positive inertia $F_{VImax,p}$, and negative inertia $F_{VImax,n}$.

6. Conclusions

The horizontal and vertical hydrodynamic forces induced by solitary waves on a horizontal cylinder placed on a horizontal sea bed have been investigated by means of 2D laboratory experiments. For this purpose, 30 laboratory tests were performed in a wave flume in which a battery of 12 pressure sensors allowed the hydrodynamic loads to be deduced.

In the present experimental flow regime (A/d ranging from about 0.08 to 0.18, $4 < KC < 7$ and Re of order of 10^4), both the peaks and the shapes of the total wave forces are strongly influenced by the inertia component in the horizontal direction and by the lift component in the vertical direction. Concerning the force peaks, it has been observed that, for $A/d < 0.105$, the positive horizontal peaks are higher than the positive vertical ones. For $A/d > 0.105$, the physical process is inverted and the positive vertical peaks are greater than the positive horizontal ones. The above feature is evident for high A/d.

To provide engineering indications, the overall good agreement between analytical solutions and laboratory tests in terms of surface elevation at the vertical section of the cylinder and free stream kinematic at the transversal axis of the considered structure has led, in conjunction with the experimental forces, to the calibration of the hydrodynamic coefficients in the Morison and transverse equations. The analysis of the hydrodynamic coefficients given by the experimental forces and the ambient kinematic field has highlighted that C_D initially increases and then slightly decreases when A/d increases, while C_{MH}, C_L, and C_{MH} show an overall decreasing trend. The application of Morison and transverse schemes has led to a satisfactory evaluation of the maximum peaks and the associated phase shifts of the hydrodynamic forces, particularly for the positive peaks.

Acknowledgments: The authors would like to thank the laboratory technicians of the University of Calabria Fabio De Napoli and Salvatore Straticó (GMI Lab. of the Department of Civil Engineering), and Renato Bentrovato and Ernesto Ramundo (Department of Mechanical, Energy and Management Engineering) for their help in developing the physical model for the experiments.

Author Contributions: F.A. and G.T. conceived and designed the experiments; G.T. performed the experiments; G.T. and F.A. analyzed the data; F.A. wrote the paper; and P.V. supervised the research.

Conflicts of Interest: The authors declare no conflict of interest.

References

1. Di Risio, M.; De Girolamo, P.; Bellotti, G.; Panizzo, A.; Aristodemo, F.; Molfetta, M.G.; Petrillo, A.F. Landslide-generated tsunamis runup at the coast of a conical island: New physical model experiments. *J. Geophys. Res. Ocean.* **2009**, *114*, doi:10.1029/2008JC004858.
2. Schimmels, S.; Sriram, V.; Didenkulova, I. Tsunami generation in a large scale experimental facility. *Coast. Eng.* **2016**, *110*, 32–41.
3. Filianoti, P.; Di Risio, M. Solitary wave loads on submerged breakwater: Laboratory tests. In Proceedings of the 22nd International Offshore and Polar Engineering Conference, Rhodes, Greece, 17–22 June 2012; pp. 1–6.
4. Wu, Y.-T.; Hsiao, S.-C. Propagation of solitary waves over double submerged barriers. *Water* **2017**, *9*, 917.
5. Sibley, P.; Coates, I.E.; Arumugam, K. Solitary wave forces on horizontal cylinders. *Appl. Ocean Res.* **1982**, *4*, 113–117.
6. Chian, C.; Ertekin, R.C. Diffraction of solitary waves by submerged horizontal cylinders. *Wave Motion* **1992**, *15*, 121–142.
7. Xiao, H.; Huang, W.; Tao, J.; Liu, C. Numerical modeling of wave-current forces acting on horizontal cylinder of marine structures by VOF method. *Ocean Eng.* **2013** *67*, 58–67.
8. Sarpkaya, T.; Isaacson, M. *Mechanics of Wave Forces on Offshore Structures*; Van Nostrand Reinhold Company: New York, NY, USA, 1981; pp. 1–650.
9. Sumer, B.M.; Fredsoe, J. *Hydrodynamics Around Cylindrical Structures*; Advanced Series on Coastal Engineering; World Scientific: Singapore, 2006; pp. 1–546.
10. Morison, J.R.; O'Brien, M.P.; Johnson, J.W.; Schaaf, S.A. The forces exerted by surface waves on piles. *Pet. Trans.* **1950**, *189*, 149–156.
11. Grace, R.A.; Zee, G.T.Y. Wave forces on rigid pipes using ocean test data. *J. Waterw. Port Coast. Ocean Eng.* **1981**, *107*, 71–92.
12. Sarpkaya, T.; Rajabi, F. Hydrodynamic drag on bottom-mounted smooth and rough cylinders in periodic flow. In Proceedings of the 11th Annual Offshore Technology Conference, Houston, TX, USA, 5–8 May 1980; pp. 219–226.
13. Cheong, H.F.; Jothi-Shankar, N.; Subbiah, K. Inertia dominated forces on submarine pipelines near seabed. *J. Hydraul. Res.* **1989**, *27*, 5–22.
14. Bryndum, M.B.; Jacobsen, V.; Tsahalis, D.T. Hydrodynamic forces on pipelines: model tests. *J. Offshore Mech. Arct. Eng.* **1992**, *114*, 231–241.

15. Neill, I.A.R.; Hinwood, J.B. Wave and wave-current load on a bottom-mounted circular cylinder. *Int. J. Offshore Polar* **1998**, *2*, 122–129.
16. Chevalier, C.; Lambert, E.; Bélorgey, M. Efforts sur une conduite sous-marine en zone côtière. *Revue Française de Génie Civil* **2001**, *5*, 995–1014. (In French)
17. Aristodemo, F.; Tomasicchio, G.R.; Veltri, P. Modelling of periodic and random wave forces on submarine pipelines. In Proceedings of the 25th International Conference on Mechanics and Arctic Engineering, Hamburg, Germany, 4–9 June 2006; pp. 1–10.
18. Det Norske Veritas. *DNV-RP-F109, On-Bottom Stability Design of Offshore Pipelines*; Det Norske Veritas: Hovik, Norway, 2010; pp. 1–41.
19. Aristodemo, F.; Tomasicchio, G.R.; Veltri, P. Wave and current forces at a bottom-mounted submarine pipeline. *J. Coast. Res.* **2013**, *65*, 153–158.
20. Aristodemo, F.; Tomasicchio, G.R.; Veltri, P. New model to determine forces at on-bottom slender pipelines. *Coast. Eng.* **2011**, *58*, 267–280.
21. Kuznetsov, K.I.; Harris, J.; Germain, N.; Aristodemo, F. *Modification of a Wake Model for Hydrodynamic Forces on Submarine Cables with a Rough Seabed*; EGU2018-19847; Geophysical Research Abstracts: Vienna, Austria, 2018; Volume 20.
22. Aristodemo, F.; Tripepi, G.; Meringolo, D.D., Veltri, P. Solitary wave-induced forces on horizontal circular cylinders: Laboratory experiments and SPH simulations. *Coast. Eng.* **2017**, *129*, 17–35.
23. Liu, X.; Lin, P.; Shao, S. An ISPH simulation of coupled structure interaction with free surface flows. *J. Fluids Struct.* **2014**, *48*, 46–61.
24. Aristodemo, F.; Meringolo, D.D.; Groenenboom, P.; Lo Schiavo, A.; Veltri, P.; Veltri, M. Assessment of dynamic pressures at vertical and perforated breakwaters through diffusive SPH schemes. *Math. Prob. Eng.* **2015**, *2015*, 305028.
25. Aristodemo, F.; Marrone, S.; Federico, I. SPH modeling of plane jets into water bodies through an inflow/outflow algorithm. *Ocean Eng.* **2015**, *105*, 160–175.
26. Gui, Q.; Dong, P.; Shao, S. Numerical study of PPE source term errors in the incompressible SPH models. *Int. J. Num. Methods Fluids* **2015**, *77*, 358–379.
27. Shadloo, M.S.; Weiss, R.; Yildiz, M.; Dalrymple, R.A. Numerical simulation of long wave runup for breaking and nonbreaking waves. *Int. J. Offshore Polar Eng.* **2015**, *25*, 1–7.
28. Shadloo, M.S.; Oger, G.; Le Touzé, D. Smoothed particle hydrodynamics method for fluid flows, towards industrial applications: Motivations, current state, and challenges. *Comput. Fluids* **2016**, *136*, 11–34.
29. Nasiri, H.; Jamalabadi, M.Y.A.; Sadeghi, R.; Safaei, M.R.; Nguyen, T.K.; Shadloo, M.S. A smoothed particle hydrodynamics approach for numerical simulation of nano-fluid flows: Application to forced convection heat transfer over a horizontal cylinder. *J. Therm. Anal. Calorim.* **2018**, 1–9, doi:10.1007/s10973-018-7022-4.
30. Rayleigh, L. On waves. *Phil. Mag.* **1876**, *1*, 257–279.
31. Guizien, K.; Barthélemy, E. Accuracy of solitary wave generation by a piston wave maker. *J. Hydraul. Res.* **2002**, *40*, 321–331.
32. Dingemans, M.W. *Water Wave Propagation Over Uneven Bottoms*; Advanced Series on Coastal Engineering; World Scientific: Singapore, 1997; pp. 1–703.
33. Madsen, P.A.; Fuhrman, D.R.; Schäffer, H.A. On the solitary wave paradigm for tsunamis. *J. Geophys. Res.* **2008**, *113*, C12012.
34. Lee, J.J.; Skjelbreia, J.E.; Raichlen, F. Measurements of velocities in solitary waves. *J. Waterw. Port Coast. Ocean Eng.* **1982**, *108*, 200–218.
35. Tripepi, G.; Aristodemo, F.; Veltri, P.; Pace, C.; Solano, A.; Giordano, C. Experimental and numerical investigation of tsunami-like waves on horizontal circular cylinders. In Proceedings of the 36th International Conference on Ocean, Offshore and Arctic Engineering, Trondheim, Norway, 25–30 June 2017; pp. 1–10.
36. Mattioli, M.; Mancinelli, A.; Brocchini, M. Experimental investigation of the wave-induced flow around a surface-touching cylinder. *J. Fluids Struct.* **2013**, *37*, 62–87.
37. Lin, M.Y.; Liao, G.Z. Vortex shedding around a near-wall circular cylinder induced by a solitary wave. *J. Fluids Struct.* **2015**, *58*, 127–151.
38. Filianoti, P.; Aristodemo, F.; Tripepi, G.; Gurnari, L. Wave flume test to check a semi-analytical method for calculating solitary wave loads on horizontal cylinders. In Proceedings of the 36th International Conference on Ocean, Offshore and Arctic Engineering, Trondheim, Norway, 25–30 June 2017; pp. 1–9.

39. Seiffert, B.; Hayatdavoodi, M.; Ertekin, R.C. Experiments and computations of solitary-wave forces on a coastal-bridge deck. Part I: Flat Plate. *Coast. Eng.* **2014**, *88*, 194–209.
40. Wolfram, J.; Naghipour, M. On the estimation of Morison force coefficients and their predictive accuracy for very rough circular cylinders. *Appl. Ocean Res.* **1999**, *21*, 311–328.
41. Sumer, B.M.; Jensen, B.L.; Fredsoe, J. Effect of a plane boundary on oscillatory flow around a circular cylinder. *J. Fluid Mech.* **1991**, *225*, 271–300.

water

MDPI

Article

Extreme Wave Analysis by Integrating Model and Wave Buoy Data

Fabio Dentale [1,2], Pierluigi Furcolo [2], Eugenio Pugliese Carratelli [1,2], Ferdinando Reale [2,*], Pasquale Contestabile [1,3] and Giuseppe Roberto Tomasicchio [4]

1 Inter-University National Consortium for Marine Sciences (CoNISMa), 00196 Roma, Italy; fdentale@unisa.it (F.D.); epc@unisa.it (E.P.C.)
2 Civil Engineering Department, University of Salerno, 84084 Fisciano, Italy; p.furcolo@unisa.it
3 Department of Engineering, University of Campania "Luigi Vanvitelli", 81100 Caserta, Italy; pasquale.contestabile@unicampania.it
4 Department of Innovation Engineering, Ecotekne Centre, University of Salento, 73100 Lecce, Italy; giuseppe.tomasicchio@unisalento.it
* Correspondence: freale@unisa.it; Tel.: +39-089-964-112

Received: 12 February 2018; Accepted: 21 March 2018; Published: 24 March 2018

Abstract: Estimating the extreme values of significant wave height (H_S), generally described by the H_S return period T_R function $H_S(T_R)$ and by its confidence intervals, is a necessity in many branches of coastal science and engineering. The availability of indirect wave data generated by global and regional wind and wave model chains have brought radical changes to the estimation procedures of such probability distribution—weather and wave modeling systems are routinely run all over the world, and H_S time series for each grid point are produced and published after assimilation (analysis) of the ground truth. However, while the sources of such indirect data are numerous, and generally of good quality, many aspects of their procedures are hidden to the users, who cannot evaluate the reliability and the limits of the $H_S(T_R)$ deriving from such data. In order to provide a simple engineering tool to evaluate the probability of extreme sea-states as well as the quality of such estimates, we propose here a procedure based on integrating H_S time series generated by model chains with those recorded by wave buoys in the same area.

Keywords: wave extreme events; Mediterranean Sea; North Atlantic Spanish coasts; Gulf of Mexico; wave modeling; small scale storm variations

1. Introduction

The analysis of extremes arises in many branches of science and engineering. Hurricane winds for suspension bridge design and storm surge heights for coastal and offshore works are well-known examples in civil engineering. Extreme value analysis (EVA) is a branch of statistics dealing with the extreme deviations from the median of probability distributions. Knowledge of the value of an extreme event for a given return period T_R is the result of the EVA.

In particular, in ocean and coastal engineering, extreme events are described in terms of the function $H_S(T_R)$, which links the significant wave height (in the following: SWH, or H_S) of a sea state with different return periods T_R [1,2].

The traditional—and probably also the best—sources of data for such analyses are the historical in-situ (in the following: "direct" or "experimental") wave measurements provided by wave buoy recorders. Buoys measure the motion of the sea surface, and modern buoys also measure slope and lateral motion [3]. Properly analysed [4,5], these data allow an estimate of sea wave properties in terms of wave spectrum or, more simply, in terms of the main sea state parameters such as significant wave height H_S, mean and peak period T_m, T_p and mean direction θ_m.

However, the number of wave buoys is necessarily limited. For example, the entire Italian Data Buoy Network (in the following Rete Ondametrica Nazionale, RON), operational from 1989 until 2014, consisted of only 15 stations positioned along the more than 7000 km of Italian coasts [6,7].

As a consequence, it has become a common practice [8,9] to use data (in the following indicated as "indirect data" or "model data") originated from global or regional wave models, which are in turn driven by meteorological wind models. Global and regional wave data are readily available and this favours their extensive use by ocean engineers and researchers. The wave part of the model chain implicitly takes into account the geographical and morphological aspects of the wave formation and propagation, such as the water depth (when applicable) and the variability of fetch, while the meteorological part includes all the information related both to large scale air circulation as well as to "regional" (in a meteorological sense) orography.

However, substantial differences can be found when performing an EVA that considers directly observed and model wave data [10–14]. This difference is primarily due to the obvious fact that direct data, even though affected by errors—like any experiment—are still a more reliable estimate of the true values, compared to a model chain that, despite all the efforts and the care taken, is the result of enormously complicated calculations. The numerical diffusion [13] present in all the models tends to smoothen the results and thus to decrease the peak values, particularly in areas with strong spatial horizontal gradients. Besides, most of the time the wave model results are only calibrated ("assimilated") using of data from altimeter satellites, whose timing and location is uncorrelated with the weather or the sea state, so the calibration is biased against extreme weather or sea states.

It must also be noted here [14] that the scatter of results for wind velocity is always larger than for waves: this implies that the atmospheric variability [15] is the basic reason for the differences between model and experimental extreme data.

Another important aspect is the bias in the evaluation of extremes present in all sources of data whenever the sampling of the relevant parameter (SWH in our case) is carried out with too long a time interval compared with the inherent time constant of the phenomenon. The results shown for instance in [10,16], prove that use of data with a low time resolution (such as a 3 or 6 h) causes a considerable undervaluation of the extreme SWH values for a given return time T_R. The following Figure 1 [16] illustrates this point—by degrading the original buoy data from a sampling rate of 30′ (full data set) to a sampling rate of 6 h, there is an important reduction of the estimated $H_S(T_R)$. This raises the problem of deciding what would be the "right" interval to choose in order to compute the $H_S(T_R)$ curves. Current engineering practice is oriented towards 30′ or 1 h intervals, mostly because that is the commonly available sampling rate for buoy data, but also because what is normally important is not the extreme single wave, but some kind of average wave height; after all, this is the reason for adopting the concept of "significant wave height". The opportunity of this choice seems to be confirmed by the apparent convergence of the curves (1 h curve in blue is very close to the 30′ curve in red).

Investigating the use of even shorter sampling rates would lead to a different problem, i.e., the determination of the probability of single extreme waves: an issue which also has relevance, but seems to be not yet clarified, see for instance [17].

As a consequence of all this, the main problem in ocean and coastal engineering is that while long series of model data are available practically everywhere, the quality of such data is inadequate for the purpose of evaluating extreme SWHs; on the other hand, measurements taken at buoy wave meters provide reliable data, but for a limited number of sites.

The objective of this work is to propose a procedure by which model data can be integrated with experimental data in order to provide a better estimate of extreme SWH values. This is done by considering model-derived data at a given location as estimators—in a statistical sense—of the true values; and, in order to improve the estimate as well as to evaluate the error, information on their statistical distribution is obtained by analyzing the wave buoy data series in the area.

In the following such a procedure is discussed and applied to three different sea areas:

1. South Mediterranean Italian coast;

2. North Atlantic Spanish coast;
3. Gulf of Mexico.

Global or regional wave datasets are provided by different organizations, i.e., ECMWF (European Centre for Medium-Range Weather Forecasts) and NOAA (National Oceanic and Atmospheric Administration).

The NOAA National Center for Environmental Prediction (NCEP) developed the Climate Forecast System (CFS), a fully coupled model representing the interaction between the Earth's atmosphere, oceans, land and sea ice. A reanalysis of the sea and atmosphere state for the period of 1979–2009 has been conducted, resulting in the CFS Reanalysis (CFSR) dataset. Using the CFSR dataset, the NOAA Marine Modelling and Analysis Branch (MMAB) has produced a wave hindcast for the same period. The wave hindcast dataset has been generated using the WAVEWATCH III (WW3) model (v3.14) (NOAA/National Weather Service, College Park, MD, U.S.A.) and is suitable for use in climate studies. The wave model resolves 50 wave frequencies (from 0.035 to 0.963 Hz) and 36 wave directions (directional resolution of 10°). Data are given at a three-hourly time resolution and are available both on a global grid (spatial resolution of 0.5° × 0.5°) and on 16 different regional nested grids with a variable spatial resolution.

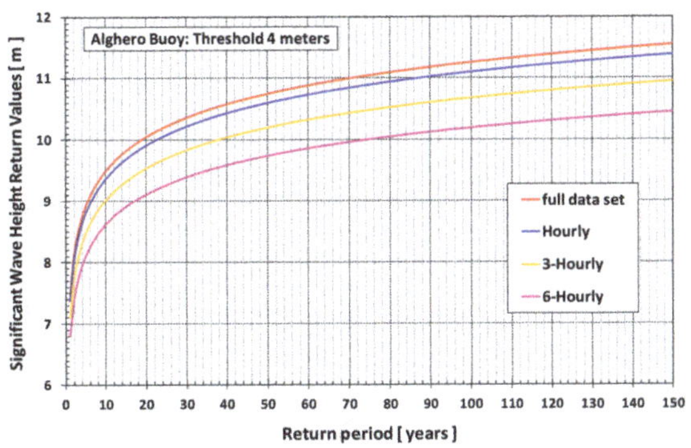

Figure 1. Effect of the varying sampling rate on the SWH extreme values.

The ECMWF produces the ERA-Interim dataset, another global atmospheric reanalysis dataset starting from 1979 and continuously updated. It also models oceanographic variables, including waves. The wave model used by ECMWF is based on the WAM (WAve Model) approach [18], resolving 30 wave frequencies and 24 wave directions. Furthermore, the wave model contains corrections for treating unresolved bathymetry effects and a reformulation of the dissipation source term. ERA-Interim produces four analysis data per day (at 00:00, 06:00, 12:00 and 18:00 UTC) and two 10-day forecast data per day, initialized from analysis at 00:00 and 12:00 UTC. Both wave products are distributed on a global 1.0° × 1.0° latitude/longitude grid.

Many commercial companies provide a model series with smaller sampling intervals, and often with a finer spatial resolution; however only NOAA and ECMWW provide well-tested and public reanalysis data, so in the following only those two sources have been considered.

2. Considered Datasets

Six-hour ECMWF ERA-Interim re-analysis data were used throughout the work. However, since in two locations (South Mediterranean Italian coasts and Gulf of Mexico) NOAA data with adequate resolution are also available, these have been added to provide additional elements.

2.1. South Mediterranean Italian Coasts

Direct data for this area are from the six Italian Data Buoy Network (RON) buoys reported in Figure 2.

Figure 2. Position of the six considered Italian buoys—(RON).

More specific details about RON can be found in [6]. Table 1 reports the exact location and dataset time extension for the six considered buoys.

Table 1. South Mediterranean Italian coast: position, data sampling and dataset time extension for the different data sources considered.

Data Source	Latitude	Longitude	From	To	Sample	Grid
Alghero buoy	40°32′55″ N	08°06′25″ E	1 July 1989	31 December 2007	30 min.	-
Catania buoy	37°26′24″ N	15°08′48″ E	1 July 1989	30 September 2006	30 min.	-
Cetraro buoy	39°27′12″ N	15°55′06″ E	1 February 1999	31 December 2007	30 min.	-
Crotone buoy	39°01′25″ N	17°13′12″ E	1 July 1989	31 July 2007	30 min.	-
Mazara buoy	37°31′05″ N	12°32′00″ E	1 July 1989	31 December 2007	30 min.	-
Ponza buoy	40°52′00″ N	12°57′00″ E	1 July 1989	31 December 2007	30 min.	-
ECMWF Global	36° N–42° N	7° E–20° E	1 January 1979	31 December 2007	6 h	1° × 1°
NOOA med_10m	30° S–48° N	7° W–43° E	1 January 1979	31 December 2007	3 h	10′ × 10′

For indirect data, both the global six-hour analysis ECMWF ERA-Interim and the Mediterranean (med_10m) nested grid three-hourly NOAA-MMAB products were used. The former, provided on a 1° × 1° latitude/longitude grid, were acquired for an area between 36° N–42° N of latitude and 7° E–20° E of longitude; the latter, on a 10′ × 10′ (about 0.167° × 0.167°) latitude/longitude grid, cover an area between 30° S–48° N of latitude and 7° W–43° E of longitude (see Table 1).

2.2. North Atlantic Spanish Coast

For this area, direct data were sampled from the five buoys reported in Figure 3. The exact buoy location and relative dataset time extension are reported in Table 2.

Table 2. North Atlantic Spanish coast: position, data sampling and dataset time extension for the different data sources considered.

Data Source	Latitude	Longitude	From	To	Sample	Grid
Bilbao buoy	43°38′24″ N	03°05′24″ W	1 January 2002	30 November 2010	1 h	-
C. De Penhas buoy	43°45′00″ N	06°09′36″ W	1 January 1998	30 November 2010	1 h	-
C. Silleiro buoy	42°07′12″ N	09°25′48″ W	1 July 1998	30 November 2010	1 h	-
Estaca Bares buoy	44°07′12″ N	07°40′12″ W	1 January 1998	31 August 2010	1 h	-
Villano Sisar. buoy	43°30′00″ N	09°12′36″ W	12 May 1998	30 November 2010	1 h	-
ECMWF Global	40° N–46° N	12° W–2° W	1 January 1998	31 December 2010	6 h	1° × 1°

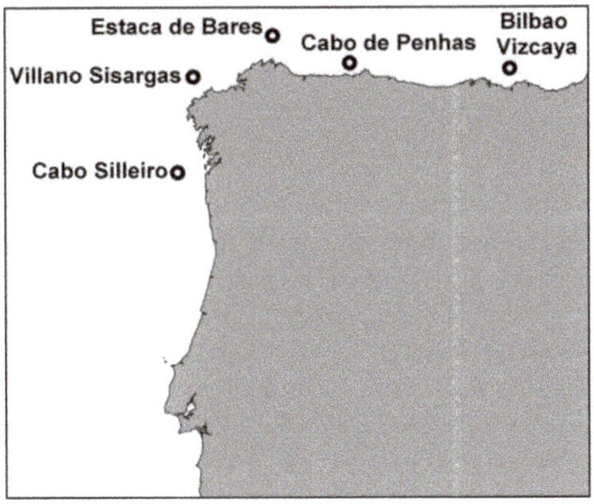

Figure 3. Buoy positions along the North Atlantic Spanish coast.

ECMWF ERA-Interim (global four analyses per day) indirect data were used. They are provided on a 1° × 1° latitude/longitude grid and were acquired for an area between 40° N–46° N and 12° W–2° W of longitude (see Table 2).

2.3. Gulf of Mexico

Direct data were collected at seven National Data Buoy Center (NDBC) buoys whose locations are shown in Figure 4 and Table 3. Table 3 also reports the dataset time span for each buoy.

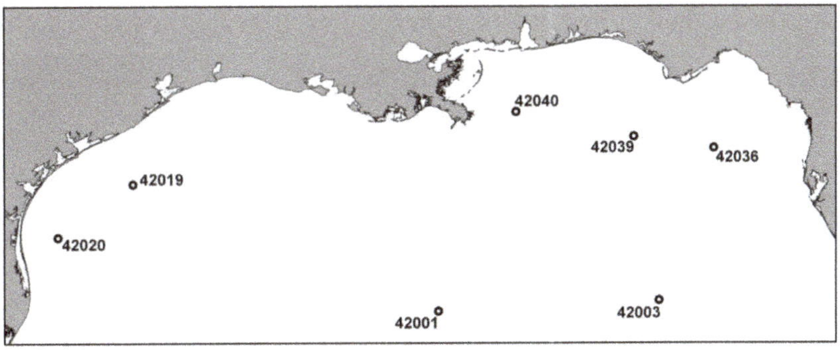

Figure 4. Position of seven NDBC buoys in the Gulf of Mexico.

Table 3. Gulf of Mexico: position, data sampling and dataset time extension for the different data sources considered.

Data Source	Latitude	Longitude	From	To	Sample	Grid
42001 buoy	25°53′48″ N	89°40′06″ W	1 March 1979	31 December 2007	1 h	-
42003 buoy	26°00′25″ N	85°38′54″ W	1 January 1979	31 December 2007	1 h	-
42019 buoy	27°54′24″ N	95°21′09″ W	1 May 1990	31 December 2007	1 h	-
42020 buoy	26°58′04″ N	96°41′39″ W	1 May 1990	31 December 2007	1 h	-
42036 buoy	28°30′03″ N	84°30′56″ W	1 January 1994	31 December 2007	1 h	-
42039 buoy	28°47′18″ N	86°00′31″ W	1 December 1995	31 December 2007	1 h	-
42040 buoy	29°12′30″ N	88°13′33″ W	1 December 1995	31 December 2007	1 h	-
ECMWF Global	24° N–32° N	100° W–80° W	1 January 1979	31 December 2007	6 h	1° × 1°
NOAA ecg_10m	0° N–55° N	100° W–50° W	1 January 1979	31 December 2007	3 h	10′ × 10′

ECMWF ERA-Interim were considered as well as NOAA-MMAB. The former were acquired for an area between 24° N–32° N of latitude and 100° W–80° W of longitude on a grid of 1° × 1°, the latter, in particular, were acquired from the Gulf of Mexico and NW Atlantic (ecg_10m) nested grid that covers an area between 0° N–55° N of latitude and 100° W–50° W of longitude with a 10′ × 10′ (about 0.167° × 0.167°) spatial resolution.

3. Methods

As stated above, substantial differences can be found when performing an EVA considering directly observed and indirect wave data. Therefore, for practical coastal engineering purposes a procedure is necessary to correct and to evaluate the reliability of $H_S(T_R)$ curves derived from model data for any location at sea. In the following, one such procedure is proposed and tested and shown to be reliable in areas where an adequate number of in situ wave buoys are available. Unlike most calibration or validation procedures currently being performed, all our elaborations were carried out on the $H_S(T_R)$ functions rather than on the single raw extreme recorded significant wave heights H_S. The basic concept is that the parameters of any $H_S(T_R)$ function are themselves randomly distributed, and that the distribution of such parameters can be estimated by analyzing in situ data for any given area, thus providing a regional assessment of the error associated with such an estimate.

It is worth noting that the wave part of the model chain takes care of physical marine aspects such as fetch and depth, so that the difference between the data from the buoy and the data from the relevant model grid point depends partly on the meteorological uncertainty (i.e., the error of the wind part of the chain) and partly from the inherent error of the wave part of the model.

This approach is similar to what is done in many fields where distributed data (from a model or from a remote sensor) need to be assimilated and corrected with field data. In hydrology, for example, regional models for hydro-meteorological variables, such as extreme or annual rainfall, are often obtained by coupling a deterministic indicator, based on models, with a spatial model of the residuals measured at gauged sites [19–22]. A similar approach is also used for extracting the estimated rainfall rate at ground through meteorological radars and rain gauge measurements [23,24]. Several methods have been introduced to this purpose [23–27]. In all such procedures, a regional assessment of statistical uncertainty is carried out by making use of spatial estimates of the error distributions.

For SWH extremes, the limited information coming from the historical data can be largely compensated by the available indirect model archive data. In order to do so, in this work the model data are used as indicators, and the buoy data are used for the correction of biases and the evaluation of uncertainties.

Extreme SWH values $HM(T_R)$ derived from model (indirect) time series are thus integrated with extreme SWH values $HD(T_R)$ derived from whatever wave buoy (direct) time series available in the same geographical area. This provides a tool to derive an estimated function $H_S(T_R)$ for any point in the area, as well as an assessment of the quality of the whole model chain.

There are many possible alternatives which can be selected for this purpose—a first important decision to be taken is whether the observation period upon which the parameters are estimated should be the same for both the model and the wave buoy, or should encompass the whole length of the available period of observed or modeled data. On the one hand, for the sake of consistency, the time spans of the observation should coincide; on the other hand, since the final scope of the work is to provide a reliable tool, it would instead make sense to compare all the available in situ data with all the available model data (normally, the simulated data series are much longer than the experimental ones). As stated above, in the present paper, since the objective is to illustrate a procedure in the simplest possible way, the first alternative has been pursued so the series considered for each location overlap as much as possible.

A further choice is the actual threshold value to adopt if a POT (peak over threshold) procedure is used to compute the $HD(T_R)$ and $HM(T_R)$ functions. In this work, among the various possible alternatives, the thresholds were chosen by making sure that the number N_T of extreme events considered would be roughly equal for all the samples. Obviously, N_T increases by decreasing the threshold because more events are taken into account; on the other hand, if the threshold is too low the quality of the estimate is compromised, since the events will no longer be statistically independent. A compromise has to be found, and a measure of such compromise is given by the value of the parameter $\lambda = N_T/n$, n being the number of available observation years. This procedure is normally adopted in coastal engineering activities [28] and is meant to provide a higher number of extremes, compared with annual maxima.

In any case, since the present paper is not aimed at evaluating, discussing or recommending one particular form of $H_S(T_R)$, or any particular procedure to estimate its parameters, the only requirement is that such a form and procedure should be uniform throughout the whole analysis.

In the following, the peak over threshold (POT) method in the form described for instance in [29–33] was followed to produce a set of extreme significant wave height values. There is here a choice to be made as to which extreme value distribution should be fitted to the data. For instance [34,35] suggest that a GPD (generalized Pareto distribution) should be employed for EVA. However, according to current ocean engineering practice [28,36], the Weibull distribution was adopted (Equation (1)):

$$F(H_S) = 1 - \exp\{-[(H_S - B)/A]^k\}, \quad (1)$$

where A, B and k are known respectively as scale, position and shape parameters. Once the distribution parameters are known, the H_S return value for a given return period T_R (in years) is computed by making use of Equation (2):

$$H_S(T_R) = B + A[\ln(\lambda T_R)]^{1/k}. \quad (2)$$

The λT_R term derives from the POT techniques, where λ extreme values are considered on average for each observation year.

The same operation is carried out with both historical experimental direct datasets $HD(T_R)$ and indirect data $HM(T_R)$ at the same locations—naturally most of the time the positions do not exactly coincide with model grid points, so a spatial bi-linear interpolation (co-location) as described in [37] (pp. 10–11) has to be carried out. As shown in Figure 5, for the four model grid points (black crosses) around each buoy location (filled red circle), a linear interpolation (red circles) between each pair of grid points has been computed. These two, in either the latitudinal or longitudinal direction, have then been used for linear interpolation to the requested boy location (filled red circle).

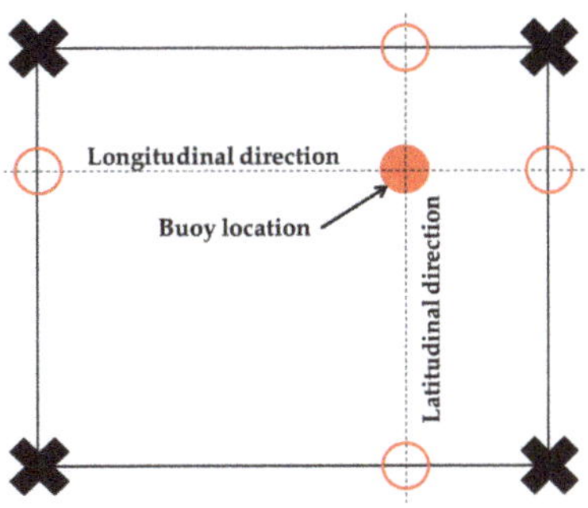

Figure 5. Scheme of bi-linear spatial interpolation of the four model grid points (black crosses) to the requested buoy location (filled red circle). First a linear interpolation (red circles) between each pair of grid points is computed; then a further linear interpolation between these points, in either the latitudinal or longitudinal direction, provides the value to the requested position.

Since wave buoy time series are often incomplete, the experimental and model data series cannot be made to coincide exactly; some care must thus be taken to make sure that the extent of time they refer to are not too far apart. The mean rate values λ for the indirect data has been kept close enough to the λ value of the in-situ data, which generally means adopting a lower threshold.

Once the experimental HD(T_R) and the model-derived HM(T_R) curves have been computed, no matter how good the data or how careful their elaboration is, they will certainly differ for the reasons stated above.

In the following Figure 6, the results are shown for a test carried out in the three sea regions.

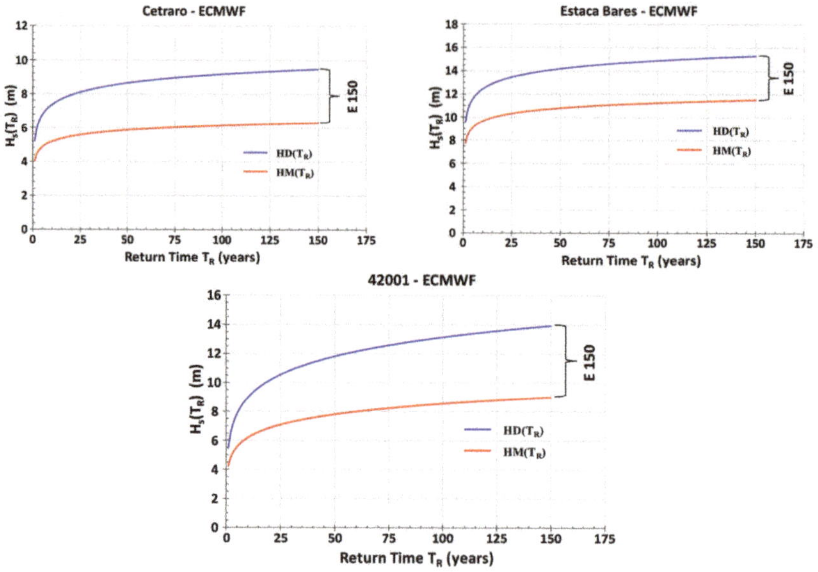

Figure 6. HD(T_R) values computed with direct (blue curves) and indirect (red curves) data for various buoy location.

Models generally underestimate extreme values compared to experimental data, partly due to the inevitable smoothing of the results due to the numerical interpolation, and partly due to inherent limitations of the model chain, as it was shown quantitatively in [13–16]. As a consequence, it is to be expected for a given return time T_R that the corresponding significant wave height return value computed with the indirect data HM(T_R) is lower than the value obtained by making use of direct data HD(T_R).

The HM(T_R) values are then assumed to be indicators of the unknown true values, and a statistical correlation must thus be found between HM(T_R) and HD(T_R) in order to provide a reliable estimator $H_S(T_R)$. We have then [22]:

$$H_S(T_R) = HM(T_R) + HM(T_R)\cdot e(\mu,\sigma), \quad (3)$$

$e(\mu,\sigma)$ being the relative error distribution.

Following Equation (3), the expected values H_S for each T_R are:

$$H_S(T_R) = HM(T_R) + \mu\cdot HM(T_R) \quad (4)$$

and its upper and lower bounds of the 95% confidence intervals, corresponding at the 97.5% an 2.5% percentiles, are respectively:

$$H_S(T_R) = HM(T_R) + \mu(T_R)\cdot HM(T_R) \pm 2\sigma(T_R)\cdot HM(T_R), \quad (5)$$

$e(\mu,\sigma)$ represents the error caused by many reasons: meteorological uncertainty, model inaccuracy, and, as often happens, the different sampling rate between the experimental and the model data. The value of the parameters μ and σ can be evaluated by making use of a spatial analysis, i.e., by comparing the model result $HM_i(T_R)$ with the experimental $HD_i(T_R)$ values at the buoy location i of the m buoys available in the area. The relative error $e_i(T_R)$ at location i is then given by Equation (6):

$$e_i(T_R) = [HD_i(T_R) - HM_i(T_R)]/HM_i(T_R). \quad (6)$$

Its expected value $\mu(T_R)$ can be estimated as:

$$\mu(T_R) = \frac{1}{m}\sum_{i=1}^{m} e_i \quad (7)$$

and its root mean square $\sigma(T_R)$ as

$$\sigma(T_R) = \sqrt{\frac{1}{m-1}\sum_{i=1}^{m}(e_i - \mu)^2} \quad (8)$$

the sums over the index i being extended to the m wave buoys in the region.

These estimates can be used to quantify the accuracy of the model chain, as well as to provide a way to compute $H_S(T_R)$ curves for geographical locations not coinciding with wave buoys by making use of Equation (4), and of the standard deviation $\sigma(T_R)$ given by Equation (8).

Figure 7 reports some examples. Curves $H_S(T_R)$ are shown together with the σ_{68} and σ_{96} confidence interval curves respectively equal to:

$$H_{68} = H_S(T_R) \pm \sigma(T_R) \cdot H_S(T_R) \quad (9)$$

$$H_{96} = H_S(T_R) \pm 2\sigma(T_R) \cdot H_S(T_R) \quad (10)$$

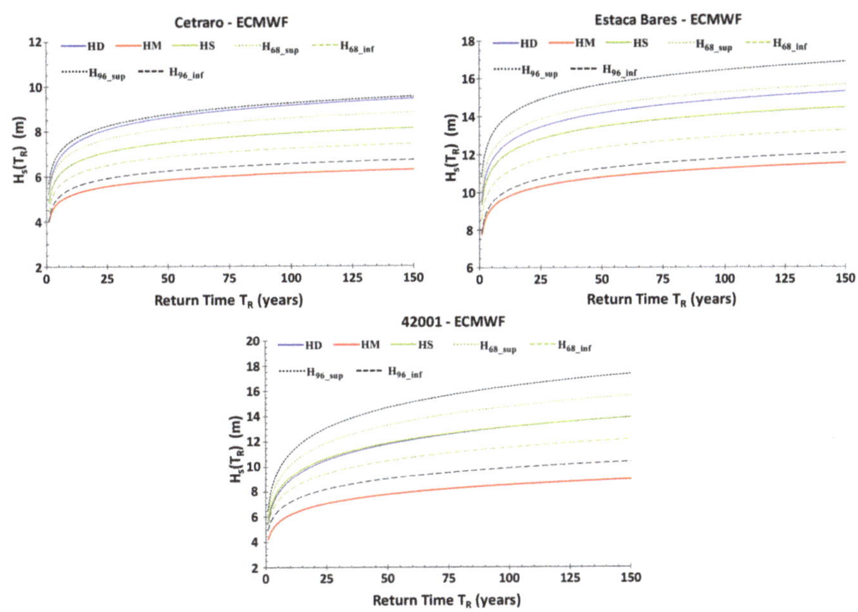

Figure 7. $H_S(T_R)$ curves for direct (HD) and indirect (HM) data for various locations with σ_{68} and σ_{96} confidence intervals.

Similar analyses have been carried out for the three regions, and the results are reported in the following section.

4. Results

The computations were separately carried out as described above for all the buoys in each of the three regions. The buoys in each region are close enough to be considered similar from a meteorological point of view; it is also worth noting that in one of the regions (Southern Italian coast) the position of each of the buoys with respect to the coast are very different from each other.

4.1. South Mediterranean Italian Coasts

The dataset extension, threshold and mean rate values both for direct and indirect data relative to each location are reported in Table 4. The NOAA model data are included only as a comparison and were not used in the estimation procedure.

Table 4. South Mediterranean Italian coast: data relevant to each location for direct (buoy) and the corresponding co-located indirect (model) datasets.

Source	Location	From	To	Sample	Threshold (m)	N_t	Years (n)	λ
Buoy	Alghero	1 July 1989	31 December 2007	30′	5.00	143	16.92	8.45
	Catania	1 July 1989	30 September 2006	30′	2.50	83	14.90	5.57
	Cetraro	1 February 1999	31 December 2007	30′	3.00	52	7.58	6.86
	Crotone	1 July 1989	31 July 2007	30′	3.00	105	16.34	6.43
	Mazara	1 July 1989	31 December 2007	30′	3.50	109	15.55	7.01
	Ponza	1 July 1989	31 December 2007	30′	3.50	94	15.95	5.89
ECMWF	Alghero	1 July 1989	31 December 2007	6 h	4.00	117	18.5	6.32
	Catania	1 July 1989	30 September 2006	6 h	3.00	81	17.25	4.70
	Cetraro	1 February 1999	31 December 2007	6 h	3.00	39	8.92	4.37
	Crotone	1 July 1989	31 July 2007	6 h	3.00	91	18.08	5.03
	Mazara	1 July 1989	31 December 2007	6 h	3.00	116	18.5	6.27
	Ponza	1 July 1989	31 December 2007	6 h	3.00	82	18.5	4.43
NOAA	Alghero	1 July 1989	31 December 2007	3 h	4.50	106	18.5	5.73
	Catania	1 July 1989	30 September 2006	3 h	2.00	68	17.25	3.94
	Cetraro	1 February 1999	31 December 2007	3 h	3.00	47	8.92	5.27
	Crotone	1 July 1989	31 July 2007	3 h	3.00	76	18.08	4.20
	Mazara	1 July 1989	31 December 2007	3 h	3.50	105	18.5	5.68
	Ponza	1 July 1989	31 December 2007	3 h	3.00	114	18.5	6.16

Results are shown in Tables 5 and 6 (respectively for the ECMWF and NOAA model data), which report the various parameters of the curves for all the buoy locations in the area.

Table 5. South Mediterranean Italian coast: comparison between buoy (HD), ECMWF model (HM) and estimated (H_S) significant wave height for various values of the return period T_R.

T_R and Related σ Values	T_R = 25 Years			T_R = 50 Years			T_R = 75 Years			T_R = 100 Years		
	σ = 0.1043			σ = 0.1070			σ = 0.1087			σ = 0.1100		
Location	HD	HM	H_S	HD	HM	H_S	HD	HM	H_S	HD	HM	H_S
Alghero	9.9	7.2	9.2	10.3	7.5	9.7	10.6	7.7	9.9	10.8	7.9	10.1
Catania	6.3	5.3	6.8	6.7	5.6	7.1	6.9	5.7	7.4	7.1	5.8	7.5
Cetraro	8.1	5.6	7.1	8.6	5.9	7.5	8.9	6.0	7.7	9.2	6.1	7.9
Crotone	6.5	5.3	6.8	6.8	5.6	7.1	7.0	5.7	7.3	7.1	5.8	7.5
Mazara	7.0	6.0	7.6	7.3	6.3	8.0	7.5	6.4	8.2	7.6	6.5	8.4
Ponza	6.9	5.5	7.0	7.3	5.8	7.5	7.5	6.0	7.7	7.7	6.1	7.9

Table 6. South Mediterranean Italian coast: comparison between buoy (HD), NOAA model (HM) and estimated (H_S) significant wave height for various values of the return period T_R.

T_R and Related σ Values	T_R = 25 Years			T_R = 50 Years			T_R = 75 Years			T_R = 100 Years		
	σ = 0.0867			σ = 0.0893			σ = 0.0908			σ = 0.0918		
Location	**HD**	**HM**	**H_S**	**HD**	**HM**	**H_S**	**HD**	**HM**	**H_S**	**HD**	**HM**	**H_S**
Alghero	9.9	7.4	8.9	10.3	7.7	9.3	10.6	7.9	9.5	10.8	8.0	9.6
Catania	6.3	5.1	6.1	6.7	5.4	6.5	6.9	5.6	6.8	7.1	5.8	6.9
Cetraro	8.1	6.6	7.9	8.6	7.0	8.4	8.9	7.2	8.7	9.2	7.4	8.9
Crotone	6.5	6.2	7.4	6.8	6.5	7.8	7.0	6.7	8.1	7.1	6.8	8.2
Mazara	7.0	6.1	7.3	7.3	6.3	7.6	7.5	6.5	7.8	7.6	6.6	7.9
Ponza	6.9	5.8	6.9	7.3	6.1	7.3	7.5	6.2	7.5	7.7	6.4	7.6

4.2. North Atlantic Spanish Coast

The dataset extension, threshold and mean rate values λ both for direct and indirect data relative to each location are reported in Table 7.

Table 7. North Atlantic Spanish coast: data relevant to each location for direct (buoy) and the corresponding co-located indirect (model) datasets.

Source	Location	From	To	Sample	Threshold (m)	N_t	Years (n)	λ
Buoy	Bilbao	1 January 2002	30 November 2010	1 h	5.50	60	7.30	8.22
	C. De Penhas	1 January 1998	30 November 2010	1 h	5.00	89	10.17	8.75
	C. Silleiro	1 July 1998	30 November 2010	1 h	6.00	82	10.38	7.90
	Estaca Bares	1 January 1998	31 August 2010	1 h	6.00	68	8.68	7.83
	Villano Sisar	12 May 1998	30 November 2010	1 h	6.00	70	9.34	7.49
ECMWF	Bilbao	1 January 2002	30 November 2010	6 h	4.50	61	8.92	6.84
	C. De Penhas	1 January 1998	30 November 2010	6 h	4.50	102	12.92	7.90
	C. Silleiro	1 July 1998	30 November 2010	6 h	5.00	84	12.42	6.77
	Estaca Bares	1 January 1998	31 August 2010	6 h	5.50	81	12.67	6.39
	Villano Sisar.	1 May 1998	30 November 2010	6 h	6.00	73	12.58	5.80

Results are shown in Table 8, which reports the various parameters of the curves for all the buoy locations in the area.

Table 8. North Atlantic Spanish coast: comparison between buoy (HD), ECMWF model (HM) and estimated (H_S) significant wave height for various values of the return period T_R.

T_R and Related σ Values	T_R = 25 Years			T_R = 50 Years			T_R = 75 Years			T_R = 100 Years		
	σ = 0.0741			σ = 0.0753			σ = 0.0760			σ = 0.0765		
Location	**HD**	**HM**	**H_S**	**HD**	**HM**	**H_S**	**HD**	**HM**	**H_S**	**HD**	**HM**	**H_S**
Bilbao	14.0	10.2	12.7	14.8	10.8	13.5	15.3	11.1	14.0	15.6	11.4	14.3
C. De Penhas	11.2	9.8	12.1	11.8	10.3	12.8	12.2	10.5	13.2	12.4	10.7	13.5
C. Silleiro	12.5	9.5	11.8	13.1	9.9	12.4	13.5	10.2	12.7	13.8	10.3	13.0
Estaca Bares	13.4	10.3	12.8	14.2	10.8	13.5	14.6	11.0	13.8	14.9	11.2	14.1
Villano Sisar.	13.7	11.0	13.6	14.4	11.5	14.3	14.8	11.8	14.7	15.1	12.0	15.0

4.3. Gulf of Mexico

Data relevant to each location for direct and indirect data are reported in the following Table 9. NOAA model data are included only as a comparison and have not been used in the estimation procedure.

Table 9. Gulf of Mexico: data relevant to each location for direct (buoy) and the corresponding co-located indirect (model) datasets.

Source	Location	From	To	Sample	Threshold (m)	N_t	Years (n)	λ
Buoy	42001	1 March 1979	31 December 2007	1 h	3.50	187	26.28	7.12
	42003	1 January 1979	31 December 2007	1 h	3.50	158	24.50	6.45
	42019	1 May 1990	31 December 2007	1 h	3.50	119	14.83	8.02
	42020	1 May 1990	31 December 2007	1 h	3.50	116	14.49	8.01
	42036	1 January 1994	31 December 2007	1 h	3.50	81	11.83	6.85
	42039	1 December 1995	31 December 2007	1 h	3.50	87	11.51	7.56
	42040	1 December 1995	31 December 2007	1 h	3.50	66	11.51	5.73
ECMWF	42001	1 March 1979	31 December 2007	6 h	3.00	200	28.83	6.94
	42003	1 January 1979	31 December 2007	6 h	3.00	179	29.00	6.17
	42019	1 May 1990	31 December 2007	6 h	3.00	60	17.67	3.40
	42020	1 May 1990	31 December 2007	6 h	3.00	100	17.67	5.66
	42036	1 January 1994	31 December 2007	6 h	3.00	77	14.00	5.50
	42039	1 December 1995	31 December 2007	6 h	3.00	53	12.08	4.39
	42040	1 December 1995	31 December 2007	6 h	3.00	49	12.08	4.06
NOAA	42001	1 March 1979	31 December 2007	3 h	3.50	215	27.92	7.70
	42003	1 January 1979	31 December 2007	3 h	3.50	171	28.08	6.09
	42019	1 May 1990	31 December 2007	3 h	3.00	132	17.33	7.62
	42020	1 May 1990	31 December 2007	3 h	3.00	92	17.33	5.31
	42036	1 January 1994	31 December 2007	3 h	3.00	81	13.67	5.93
	42039	1 December 1995	31 December 2007	3 h	3.00	95	12.00	7.92
	42040	1 December 1995	31 December 2007	3 h	3.00	80	12.00	6.67

Results are shown in Tables 10 and 11 (respectively for the ECMWF and NOAA model data), which report the various parameters of the curves for all the buoy locations in the area.

Table 10. Gulf of Mexico: comparison between buoy (HD), ECMWF model (HM) and estimated (H_S) significant wave height for various values of the return period T_R.

T_R and Related σ Values	T_R = 25 Years			T_R = 50 Years			T_R = 75 Years			T_R = 100 Years		
	σ =0.1722			σ = 0.1815			σ = 0.1862			σ = 0.1893		
Location	HD	HM	H_S	HD	HM	H_S	HD	HM	H_S	HD	HM	H_S
42001	10.5	7.1	10.7	11.8	7.8	11.9	12.6	8.2	12.6	13.1	8.5	13.1
42003	10.9	7.6	11.5	12.3	8.5	12.9	13.1	9.0	13.8	13.7	9.4	14.4
42019	6.9	4.8	7.2	7.5	5.1	7.8	7.8	5.4	8.2	8.0	5.5	8.5
42020	8.0	5.2	7.8	8.8	5.5	8.4	9.2	5.8	8.9	9.6	6.0	9.2
42036	9.2	7.3	11.1	10.1	8.1	12.4	10.7	8.6	13.2	11.2	9.0	13.8
42039	12.3	8.0	12.1	13.9	9.1	13.8	14.8	9.7	14.8	15.5	10.1	15.6
42040	16.9	9.0	13.6	19.5	10.3	15.7	21.2	11.1	17.0	22.4	11.6	17.9

Table 11. Gulf of Mexico: comparison between buoy (HD), NOAA model (HM) and estimated (H_S) significant wave height for various values of the return period T_R.

T_R and Related σ Values	T_R = 25 Years			T_R = 50 Years			T_R = 75 Years			T_R = 100 Years		
	σ = 0.1071			σ = 0.1145			σ = 0.1184			σ = 0.1210		
Location	HD	HM	H_S	HD	HM	H_S	HD	HM	H_S	HD	HM	H_S
42001	10.5	8.3	10.6	11.8	9.2	11.7	12.6	9.7	12.4	13.1	10.0	12.9
42003	10.9	9.1	11.5	12.3	10.1	12.9	13.1	10.7	13.8	13.7	11.2	14.4
42019	6.9	5.5	7.0	7.5	5.9	7.6	7.8	6.2	7.9	8.0	6.4	8.2
42020	8.0	5.9	7.5	8.8	6.5	8.3	9.2	6.8	8.7	9.6	7.0	9.1
42036	9.2	8.3	10.6	10.1	9.3	11.9	10.7	9.8	12.7	11.2	10.3	13.2
42039	12.3	9.8	12.5	13.9	11.1	14.2	14.8	11.8	15.2	15.5	12.3	15.9
42040	16.9	11.5	14.6	19.5	13.1	16.7	21.2	14.0	18.1	22.4	14.7	19.0

5. Discussion

The first—if perhaps expected—result of the work is that the difference between the model derived curve $HM(T_R)$ and the experimental curve is always negative, i.e., the model results present a consistent negative bias in the estimation of extreme events. There are of course various reasons why this is the case—in the first place the already mentioned limitations due to the model's inherent limits [38–40]; a further practical reason is the effect, also discussed above, caused by the longer sampling time of the model data in comparison with the high data sampling of modern wave buoys (30′), which also coincide with the standard engineering practice.

Once the bias curve for a given area is computed, $H_S(T_R)$ can easily be obtained by adding the bias to the model values $HM(T_R)$. $H_S(T_R)$ always presents a marked improvement in comparison with the simple model data $HM(T_R)$, so it always make sense to correct model derived data with an estimate of the error distribution in the local buoy wave meters.

Another important aspect is the information gained on the uncertainty of the model-derived extreme value curves by considering the variation of statistical parameters over a given area (spatial analysis). Confidence intervals have been provided for 68% ($\pm 1\sigma$) and 95% ($\pm 2\sigma$), and it was found that most of the $HD(T_R)$ curves derived from real data at the buoy locations fall well within the $\pm 1\sigma$ interval of the estimated $H_S(T_R)$, all of them however being actually within the $\pm 2\sigma$ curves. Given the relatively small number of buoys in each sea region, this behavior it is coherent with what could have been expected. More importantly, perhaps, $H_S(T_R)$ always showed a marked improvement in comparison with the simple model data $HM(T_R)$. This also holds for the particular case of the Southern Italian buoys, which are all located in the same area, but face different directions due to the complex coast morphology. The fact that the results are not much different from those in the other directions, seems to confirm that the wave models are accurate enough to take the wave generation into account, and that most of the uncertainty derives from the wind modelling part of the chain.

6. Conclusions

The paper shows that a probabilistic estimate $H_S(T_R)$ of the significant wave height H_S as a function of the return time T_R is possible by comparing wave buoy (direct) and model (indirect) data in a given area. $H_S(T_R)$ functions obtained from historical buoy data have been compared with similar curves derived from indirect data from both ECMWF and NOAA archives for three distinct areas: the Gulf of Mexico, the North Atlantic waters along the Spanish coasts and part of the Mediterranean Sea surrounding Italy. The comparison shows a systematic negative bias, thus proving that model curves always underrate experimental ones. The error distribution of the model of the data was studied on a geographical basis, so that a $H_S(T_R)$ curve and its confidence values can be evaluated for any model grid point in the area. Widely diffused, freely available and reliable model data archives such as ECMWF and NOAA analysis and reanalysis data can thus be used for engineering purposes.

While the objective of the work is not to suggest a particular extreme value distribution or estimation methodology, a general procedure has been provided to improve the reliability of model data for EVA; such a procedure, already in the present form, can also be used to evaluate the suitability of a given model data archive to the estimation of the probability of extreme sea states.

Further development should include testing the procedure with some of the new commercially available model data sets which present a higher spatial and temporal resolution. This would allow an independent assessment of their quality as well as—possibly—a better estimate of the extreme value SWH.

Acknowledgments: Most of the work described in the paper was carried out within CUGRI (University Joint Research Centre on Major Hazards). The authors are grateful to Giovanni Besio (University of Genoa), Gabriele Nardone (APAT-ISPRA, Italian Environmental Agency), and Lucio Torrisi (CNMCA-Italian Air Force Meteorological Office), for useful data and helpful discussions. Model data from ECMWF (European Centre for Medium-Range Weather Forecast) and from NOAA NCEP. Wave buoys data from ISPRA (Italian coasts), Puertos de l'Estado (Spanish North Atlantic) and from NOAA-National Data Buoy Center (Mexican Gulf).

Author Contributions: F.D. and F.R. carried out the elaborations on the wave data and developed the maritime engineering part; P.F. developed the statistical parts; P.C. revised the text and developed the marine engineering part; E.P.C. and G.R.T. supervised the work.

Conflicts of Interest: The authors declare no conflict of interest.

References

1. Salvadori, G.; Tomasicchio, G.R.; D'Alessandro, F. Multivariate approach to design coastal and off-shore structures. *J. Coast. Res.* **2013**, *1*, 386–391. [CrossRef]
2. Salvadori, G.; Tomasicchio, G.R.; D'Alessandro, F. Practical guidelines for multivariate analysis and design in coastal adn off-shore engineering. *Coast. Eng.* **2014**, *88*, 1–14. [CrossRef]
3. Cavaleri, L.; Sclavo, M. The calibration of wind and wave model data in the Mediterranean Sea. *Coast. Eng.* **2006**, *53*, 613–627. [CrossRef]
4. Longuet-Higgins, M.S.; Cartwright, D.E.; Smith, N.D. Observations of the directional spectrum of sea waves using the motion of a floating buoy. In *Ocean Wave Spectra*, 1st ed.; Prentice-Hall: Englewood Cliffs, NJ, USA, 1963; pp. 111–132.
5. Kuik, A.J.; Ph van Vledder, G.; Holthuijsen, L.H. A Method for the Routine Analysis of Pitch-and-Roll Buoy Wave Data. *J. Phys. Oceanogr.* **1988**, *18*, 1020–1034. [CrossRef]
6. Bencivenga, M.; Nardone, G.; Ruggiero, F.; Calore, D. The Italian Data Buoy Network. *WTI Trans. Eng. Sci.* **2012**, *74*, 321–332. [CrossRef]
7. Vicinanza, D.; Contestabile, P.; Ferrante, V. Wave energy potential in the north-west of Sardinia (Italy). *Renew. Energy* **2013**, *50*, 506–521. [CrossRef]
8. Iuppa, C.; Cavallaro, L.; Vicinanza, D.; Foti, E. Investigation of suitable sites for wave energy converters around Sicily (Italy). *Ocean Sci.* **2015**, *11*, 543–557. [CrossRef]
9. Iuppa, C.; Cavallaro, L.; Foti, E.; Vicinanza, D. Potential wave energy production by different wave energy converters around Sicily. *J. Renew. Sust. Energy* **2015**, *7*. [CrossRef]
10. Arena, F.; Laface, V.; Barbaro, G.; Romolo, A. Effects of Sampling between Data of Significant Wave Height for Intensity and Duration of Severe Sea Storms. *Intern J. Geosci.* **2013**, *4*, 240–248. [CrossRef]
11. Reale, F.; Dentale, F.; Pugliese Carratelli, E.; Torrisi, L. Remote Sensing of Small-Scale Storm Variations in Coastal Seas. *J. Coast. Res.* **2014**, *30*, 130–141. [CrossRef]
12. Mentaschi, L.; Besio, G.; Cassola, F.; Mazzino, A. Performance evaluation of Wavewatch III in the Mediterranean Sea. *Ocean Model.* **2015**, *90*, 82–94. [CrossRef]
13. Cavaleri, L.; Bertotti, L. The improvement of modelled wind and wave fields with increasing resolution. *Ocean Eng.* **2006**, *33*, 553–565. [CrossRef]
14. Bertotti, L.; Cavaleri, L.; Soret, A.; Tolosana-Delgado, R. Performance of global and regional nested meteorological models. *Cont. Shelf Res.* **2014**, *87*, 17–27. [CrossRef]
15. Abdalla, S.; Cavaleri, L. Effect of wind variability and variable air density on wave modelling. *J. Geophys. Res.* **2002**, *107(C7)*. [CrossRef]
16. Dentale, F.; Reale, F.; D'Alessandro, F.; Damiani, L.; Di Leo, A.; Pugliese Carratelli, E.; Tomasicchio, G.R. Sampling Bias in the Estimation of Significant Wave Height Extreme Values. In Proceedings of the 35th Conference on Coastal Engineering, Antalya, Turkey, 17–20 November 2016.
17. Benetazzo, A.; Barbariol, F.; Bergamasco, F.; Carniel, S.; Sclavo, M. Space-time extreme wind waves: Analysis and prediction of shape and height. *Ocean Model.* **2017**, *113*, 201–216. [CrossRef]
18. Komen, G.J.; Cavaleri, L.; Donelan, M.; Hasselmann, K.; Hasselmann, S.; Janssen, P.A.E.M. *Dynamics and Modelling of Ocean Waves*, 1st ed.; Cambridge University Press: Cambridge, UK, 1994; p. 532.
19. Chua, S.-H.; Bras, L.R. *Estimation of Stationary and Non-stationary Random Fields: Kriging in the Analysis of Orographic Precipitation*, 1st ed.; Massachusetts Institute of Technology, Department of Civil Engineering, Ralph M. Parsons Laboratory for Water Resources and Hydrodynamics: Cambridge, MA, USA, 1980; p. 171.
20. Slimani, M.; Obled, C. Regionalization of extreme rainfall parameters through kriging and correlation with topography descriptors. In Regional Flood Frequency Analysis, Proceedings of the International Symposium on Flood Frequency and Risk Analysis, Louisiana State University, Baton Rouge, LA, USA, 14–17 May 1986.
21. Prudhomme, C.; Reed, D.W. Mapping extreme rainfall in a mountainous region using geostatistical techniques: A case study in Scotland. *Int. J. Climatol.* **1999**, *90*, 1337–1356. [CrossRef]

22. Pelosi, A.; Furcolo, P. An Amplification Model for the Regional Estimation of Extreme Rainfall within Orographic Areas in Campania Region (Italy). *Water* **2015**, *7*, 6877–6891. [CrossRef]
23. Seo, D.-J.; Breidenbach, J.P.; Johnson, E.R. Real-time estimation of mean field bias in radar rainfall data. *J. Hydrol.* **1999**, *223*, 131–147. [CrossRef]
24. Harrison, D.L.; Scovell, R.W.; Kitchen, M. High-resolution precipitation estimates for hydrological uses. *Proc. Inst. Civ. Eng. Water Manag.* **2009**, *162*, 125–135. [CrossRef]
25. Krajewski, W.F. Cokriging radar-rainfall and rain gage data. *J. Geophys. Res. Atmos.* **1987**, *92*, 9571–9580. [CrossRef]
26. Todini, E. A Bayesian technique for conditioning radar precipitation estimates to rain-gauge measurements. Hydrol. *Earth Syst. Sci.* **2001**, *5*, 187–199. [CrossRef]
27. Chumchean, S.; Sharma, A.; Seed, A. An Integrated Approach to Error Correction for Real-Time Radar-Rainfall Estimation. *J. Atmos. Ocean. Tech.* **2006**, *23*, 67–79. [CrossRef]
28. Goda, Y. Statistical Analysis of Extreme Waves. In *Random Seas and Design of Maritime Structures*, 1st ed.; Liu, P.L.-F., Ed.; World Scientific Publishing Co. Pte. Ltd.: Singapore, 2000; Volume 15, pp. 377–425, ISBN 981-02-3256-X.
29. Thompson, P.; Cai, Y.; Reeve, D.; Stander, J. Automated threshold selection methods for extreme wave analysis. *Coast. Eng.* **2009**, *56*, 1013–1021. [CrossRef]
30. Feng, X.; Tsimplis, M.N.; Quartly, G.D.; Yelland, M.J. Wave height analysis from 10 years of observations in the Norwegian Sea. *Cont. Shelf Res.* **2014**, *72*, 47–56. [CrossRef]
31. Sartini, L.; Mentaschi, L.; Besio, G. Comparing different extreme wave analysis models for wave climate assessment along the Italian coast. *Coast. Eng.* **2015**, *100*, 37–47. [CrossRef]
32. Sartini, L.; Cassola, F.; Besio, G. Extreme waves seasonality analysis: An application in the Mediterranean Sea. *J. Geophys. Res. Oceans* **2015**, *120*, 6266–6288. [CrossRef]
33. Sartini, L.; Besio, G.; Dentale, F.; Reale, F. Wave Hindcast Resolution Reliability for Extreme Analysis. In Proceedings of the 26th International Ocean and Polar Engineering, Rhodes, Greece, 26 June–2 July 2016; International Society of Offshore and Polar Engineers (ISOPE): Cupertino, CA, USA, 2016.
34. Leadbetter, M.R. On a basis for 'Peaks over Threshold' modelling. *Stat. Probabil. Lett.* **1991**, *12*, 357–362. [CrossRef]
35. Abild, J.; Andersen, E.Y.; Rosbjerg, D. The climate of extreme winds at the great belt, Denmark. *J. Wind Eng. Ind. Aerod.* **1992**, *41*, 521–532. [CrossRef]
36. Mathiesen, M.; Goda, Y.; Hawkes, P.J.; Mansard, E.; Martín, M.J.; Peltier, E.; Thompson, E.F.; Van Vledder, G. Recommended practice for extreme wave analysis. *J. Hydraul. Res.* **1994**, *32*, 803–814. [CrossRef]
37. Anderson, E. User Guide to ECMWF Forecast Products. Available online: https://www.ecmwf.int/sites/default/files/User_Guide_V1.2_20151123.pdf (accessed on 28 February 2018).
38. Moeini, M.H.; Etemad-Shahidi, A.; Chegini, V.; Rahmani, I.; Moghaddam, M. Error distribution and correction of the predicted wave characteristics over the Persian Gulf. *Ocean Eng.* **2014**, *75*, 81–89. [CrossRef]
39. Sulis, A.; Cozza, R.; Annis, A. Extreme wave analysis methods in the gulf of Cagliari (South Sardinia, Italy). *Ocean Coast. Manag.* **2017**, *140*, 79–87. [CrossRef]
40. Contestabile, P.; Di Lauro, E.; Buccino, M.; Vicinanza, D. Economic Assessment of Overtopping BReakwater for Energy Conversion (OBREC): A Case Study in Western Australia. *Sustainability* **2017**, *9*, 51. [CrossRef]

water

MDPI

Article

Storm Energy Flux Characterization along the Mediterranean Coast of Andalusia (Spain)

Rosa Molina [1], Giorgio Manno [2], Carlo Lo Re [2,*], Giorgio Anfuso [1] and Giuseppe Ciraolo [2]

[1] Department of Earth Sciences, Faculty of Marine and Environmental Sciences, University of Cádiz, Polígono del Río San Pedro s/n, 11510 Puerto Real, Spain; r.molina.gil@gmail.com (R.M.); giorgio.anfuso@uca.es (G.A.)

[2] Department of Engineering, University of Palermo, Viale delle Scienze, Bd. 8, 90128 Palermo, Italy; giorgio.manno@unipa.it (G.M.); giuseppe.ciraolo@unipa.it (G.C.)

* Correspondence: carlo.lore@unipa.it; Tel.: +39-238-965-24

Received: 13 February 2019; Accepted: 5 March 2019; Published: 11 March 2019

Abstract: This paper investigates wave climate and storm characteristics along the Mediterranean coast of Andalusia, for the period 1979–2014, by means of the analysis of wave data on four prediction points obtained from the European Centre for Medium-Range Weather Forecasts (ECMWF). Normally, to characterize storms, researchers use the so-called "power index". In this paper, a different approach was adopted based on the assessment of the wave energy flux of each storm, using a robust definition of sea storm. During the investigated period, a total of 2961 storm events were recorded. They were classified by means of their associated energy flux into five classes, from low- (Class I) to high-energetic (Class V). Each point showed a different behavior in terms of energy, number, and duration of storms. Nine stormy years, i.e., years with a high cumulative energy, were recorded in 1980, 1983, 1990, 1992, 1995, 2001, 2008, 2010, and 2013.

Keywords: energy flux; storm classification; stormy year; coastal erosion; Andalusia coast

1. Introduction

Coastal areas are extremely important in Mediterranean countries since they host the majority of their population and economic activities [1]. Over the last few decades, one of the faster urban developments has occurred along the Spanish Mediterranean coast, especially at the Costa del Sol [2]. As a result of this expansion, human activities and buildings were placed extremely close to the shore [3]; therefore, they are now threatened by natural hazards influenced by climate change-related processes such as sea-level rise and increases in storm frequency and intensity [4,5]. To reduce storm impacts, it is necessary to understand specific coastal characteristics and sensibilities as well as to fully comprehend storm nature. In recent years, several researchers have studied these aspects from different viewpoints. In Spain, Rodrıguez-Ramırez et al. [6] studied storm records on the Huelva coast to obtain appropriate future development and management strategies; Mendoza and Jiménez [7] and Mendoza et al. [8] presented an intensity scale for wave storms on the Catalan coast to characterize their spatial and temporal variability; Guisado and Malvárez [9] and Pintado and García [10] used extreme wave conditions to complete the characterization of the morpho-dynamic environments of the Costa del Sol and Huelva areas; Anfuso et al. [11] and [12,13] characterized storms along the Atlantic side of Andalusia. In recent decades, coastal scientists have used several indexes to characterize storms, e.g., Halsey [14] ranked north-east Atlantic coastal storms (northeasters or nor'easters) based on a damage potential index and Dolan and Davis [15] proposed an intensity scale index to classify nor'easters into five classes, from weak to extreme, based on wave height and storm duration. Orford et al. [16] and Orford and Carter [17] used the role of storm surge to develop a new storm index. Kriebel et al. [18] proposed a nor'easter risk index by combining the effects of storm surge, wave, and duration and

Zhang et al. [19] developed a storm erosion potential index by combining the effect of storm tide, wave energy, and duration. This paper analyzes a 35-year wave climate dataset obtained from the European Centre for Medium-Range Weather Forecasts (ECMWF) for four available prediction points equally spaced along the Mediterranean coast of Andalusia (south of Spain). This allowed the definition and assessment of storm characteristics and their spatial and temporal distribution along the investigated area. To characterize the storms, a new approach was adopted, assessing the real wave energy flux of each storm, using a robust definition of the storm itself. During the investigated period, a total of 2961 storm events were recorded. These were classified according to five classes of storms, from low (Class I) to high-energetic (Class V). Results obtained are useful to understand potential impacts of both single and grouped storms, and hence put in place the appropriate prevention and mitigation strategies.

2. The Study Area

This study is focused on the wave climate of the Andalusia Mediterranean coast, a very populated area whose land cover has experienced important changes during recent decades [20]. Málaga is the province that has experienced the most important coastal occupation, in particular due to the construction of structures related to national and international tourism [20]. The coast is a micro-tidal environment (tidal range < 20 cm, [9]), about 546 km long, and including four provinces, i.e., Cádiz, Málaga, Granada, and Almería (Figure 1).

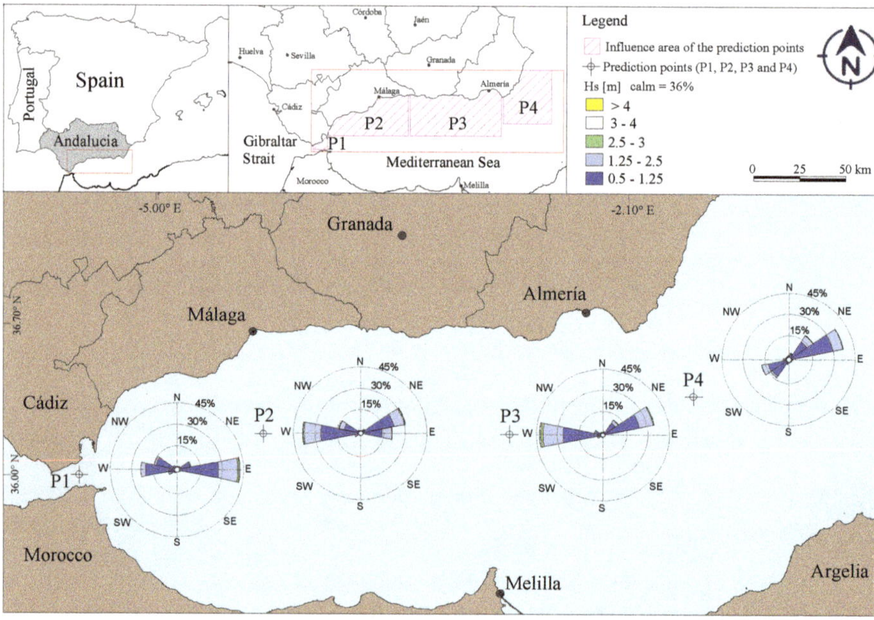

Figure 1. Geographic location of the study area and wave rise (1979–2014) for each prediction point.

There are many Andalusian coastal areas that suffer erosion problems [21], typically linked to very energetic storms producing severe damage to coastal structures. For example, Figure 2 shows the damage produced in winter 2015 by a western storm at Alumuñecar (Granada Province). This storm particularly affected the beaches of San Cristobal and La Herradura. At the former, the extreme wave run-up broke the facilities for summer tourism, and at the latter, storm waves reached the road at several points, depositing cobbles and sand that endangered people the circulation of vehicles.

Figure 2. Severe damage after a western storm in winter 2015 at the Reina Sofia Promenade (Almuñecar), photo by europapress.

On the Mediterranean Andalusian coast, the beaches are rectilinear and composed of medium to fine and dark to golden sands and, especially in Granada and Almería, cliffed sectors are observed. The near-shore area generally shows high slope values and intermediate and reflective morpho-dynamic states linked to a narrow continental shelf [22]. This sector is exposed to winds from E and SE with minimum and maximum wind speed values that range from 0.4 to 0.9 m/s [9].

3. Methods

With the aim of characterizing the wave climate of the studied area, four prediction points were identified along the Mediterranean Andalusian coast; at each point, wave rises and the monthly means of maximum wave height $H_{m0,max}$ were calculated. The wave rises give information about the direction and intensity of incoming waves, whereas the monthly wave height means give information about the seasonal characteristics of the Mediterranean Andalusian coast. To identify each single storm, the definition of [23] was adopted, which allowed calculation of the energy flux by using the linear deep-water wave theory and, finally, classification of energy flux, preferring this parameter to empirical ones.

3.1. Wave Climate Preliminary Analysis

Wave climate analysis was carried out using wave data modelled by the ECMWF by means of the WAve Model (WAM). This numerical model, which solves the energy balance equation, forecasts wave climate that is then subjected to quality controls ensuring its consistency [24] (https://www.ecmwf.int/en/elibrary/16951-wave-model accessed on January 2019). This paper used MATLAB scripts to analyze wave characteristics along the Mediterranean coast of Andalusia for the period 1979–2014, obtained by the ECMWF within the framework of the ERA-INTERIM project. The four predictions points, from W to E, used in this paper, are Point 1, close the Strait of Gibraltar, Point 2, east of Málaga, and Points 3 and 4 in front and east of Almería, respectively (Figure 1). Each temporal series is formed by 51,860 data points recorded over a period of 35 years (1 January 1979–30 January 2014) which is enough to analyze potential trends of increasing wave heights, the presence of climate-controlled cycles, or annual variations due to climate events [5].

3.2. Storm Classification Using the Energy Flux

To determine storm events, the definition used was the one given by [23], i.e., a Mediterranean Sea storm is a sequence of sea states in which the spectrally significant wave height exceeds the threshold h_t and does not fall below this threshold for a continuous time interval greater than 12 h. He also considered that the time interval between consecutive single storms must be greater than 12 h. The government agency "Puertos del Estado" [25] suggested, for the Mediterranean coast of Andalusia, adopting a threshold value $h_t = 1.5$ m. Boccotti [23] suggests adopting the same value for the threshold (h_t) because it is 1.5 times the mean yearly significant wave height. Following the aforementioned criteria, 2961 storms were selected for the period 1979–2014.

Many studies (e.g., [7,8,11–13]) based the classification of storms on the use of the Dolan and Davis [15] "Storm Power Index". In this paper, a physically based parameter was preferred, namely, wave energy flux [26,27]. Wave energy flux, or wave power per unit of wave-front length (P), was calculated using the following equation:

$$P = \frac{\rho g^2}{64\pi} T_e H_{m0}^2 \left[\frac{W}{m}\right] \tag{1}$$

where ρ is water density, g is the gravity acceleration, T_e is the energy period that represents the period of the sinusoidal wave with the same energy as a real sea state (for which a JONSWAP spectrum is about 90% of the peak period) and H_{m0} is the spectrally significant wave height. To obtain an accurate estimation of the total energy (E^i_{tot}) of each storm [26–29] the energy flux was time-integrated:

$$E^i_{tot} = \int_0^{d_i} P dt \left[\frac{Wh}{m}\right] \tag{2}$$

where the d^i is the duration of i-th storm. Using Equations (1) and (2) the total energy of each of the 2961 storms was calculated.

For example, two storms at the prediction Point P3 are shown in Figure 3. The first storm started on 25 February 2009 and the second on 4 March 2009. Figure 3a shows the H_{m0} values of the two storms identified by means of the threshold h_t = 1.5 m. The corresponding energy flux P and the total energy E^i_{tot} (Equations (1) and (2)) is shown in Figure 3b.

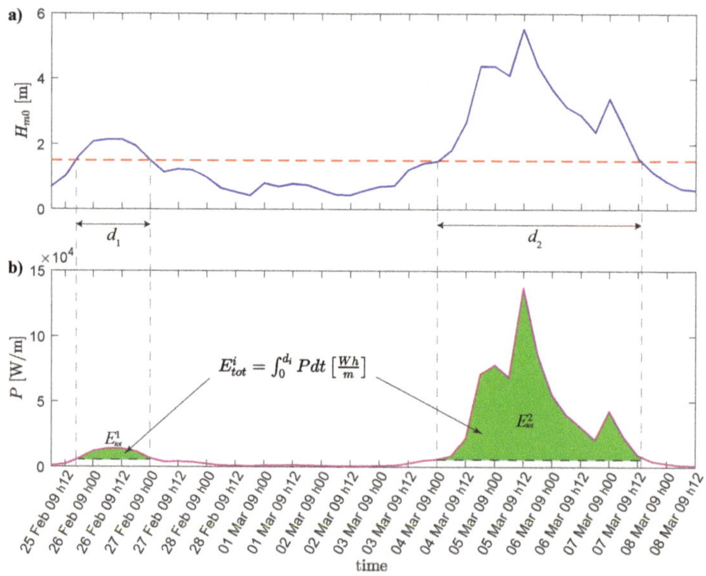

Figure 3. (**a**) Spectrally significant wave height H_{m0}, during the storms of 25 February and 4 March 2009 at the prediction Point P3. The dashed red line is the spectrally significant wave height threshold h_t = 1.5 m, d_1 and d_2 are storm duration. (**b**) Energy flux during time and total storm energy.

The method proposed by [30], i.e., "the natural breaks" function, was used to classify storm events into five classes, from Class I (low-energetic events) to Class V (high-energetic events).

The Return Period of the Energy Flux of Storms

For the estimation of the return period of the energy flux (P) of each storm class, the probability of exceedance was fitted with a modification of the Weibull Cumulative Distribution Function (CDF) described in [23,31]:

$$F(P) = e^{-\left(\frac{P}{w}\right)^{u}}\ [\%] \tag{3}$$

where w and u parameters that depend on the location under examination. Using the auxiliary variables $X = 100 \cdot ln(2.5 \cdot P)$ and $Y = 100 \cdot ln\,[ln\,(1/F(P))]$ [23,31], in the coordinate system X, Y, the data points should lie on a straight line. Thus, the parameters of Equation (3) can be easily estimated by fitting those data points by means of the least-squares method.

The above-mentioned Weibull CDF is related to the return period T_r by means of

$$T_r = \frac{1}{\lambda \cdot F(P)}\ [year] \tag{4}$$

where λ is the mean number of events per year. Consequently, for the lower and upper limit of each storm class, the probability of exceedance and the return period were calculated.

3.3. Stormy Year

For the sake of continuity with previous research [11,32], the definition of stormy year was adopted here. In particular, for each prediction point, stormy year empirical recurrence period (R_i) and annual frequency of occurrence (f_o) were assessed by using the equations:

$$R_i = \frac{(n+1)}{m}\ [year] \tag{5}$$

$$f_o = \frac{1}{R_i}\ [\%] \tag{6}$$

where R_i is the recurrence interval calculated for n number of years (35 in this paper), m is the number of events that occurred within the date-range of interest, and f_o is the yearly frequency of occurrence of the event. For each prediction point, two values of R_i and f_o were calculated using a minimum and a maximum threshold, respectively. The minimum threshold is the mean of total energy (μ) calculated within the whole period of 35 years, the maximum threshold is $\mu + \sigma$, where σ is the standard deviation of the total energy.

4. Results and Discussion

4.1. Wave Height Characterization

Wave height data did not present a general trend along the investigated period, but a clear seasonal behavior was recognized, as observed by Rangel-Buitrago and Anfuso [13] on the Atlantic sector of Andalusia. As a general trend, higher average values of monthly maximum spectrally significant wave heights ($H_{m0,max}$) were observed during the winter season, i.e., the December–March period and, specifically, Points 1, 2 and 4 recorded maximum values in March and Point 3 in February (Figure 4). High values recorded in March are linked to the great importance of eastern waves due to regnant winds during such months. Lower average $H_{m0,max}$ values were observed during the summer time, i.e., July, August, and September, ranging from 1.7 m at Point 1 to 2.1 m at Point 3. Dealing with the behavior of each analyzed point, Point 3 showed the highest values in all years, always followed by Point 2; but during June, July, and August, Point 4 recorded greater values than Point 2 (Figure 4).

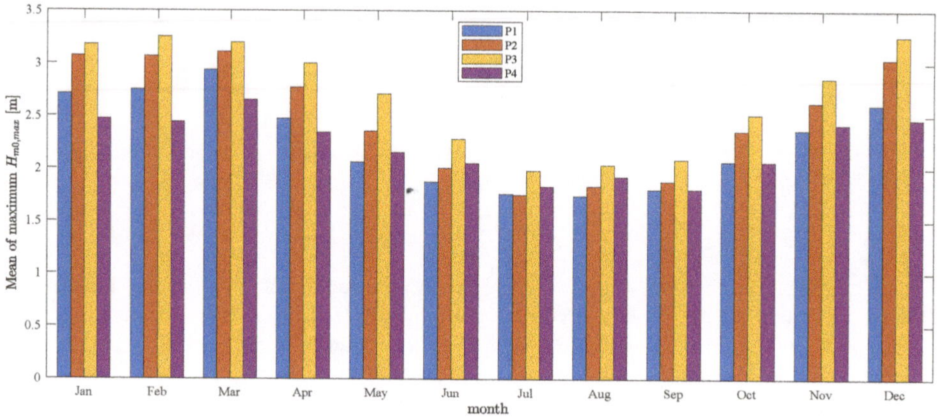

Figure 4. Average values of monthly maximum spectral significant wave heights $H_{m0,max}$.

Approaching directions observed at different points clearly reflected coastal orientation and prevailing marine climate (Figure 1). Point 1 is close to the Strait of Gibraltar but sheltered to the Atlantic swell waves. East approaching fronts, with prevailing wave height classes of 0.5–1.25 m (c. 30%) and 1.25–2.5 (c. 10%), clearly prevailed at this prediction point (Figure 1) and were linked to the strong easterly winds associated with a surface pressure gradient over the Gibraltar Strait when the Azores high pressure is located over the Iberian Peninsula, while there is pronounced low pressure over northern Africa [33]. Points 2 and 3 are situated at the central part of the investigated area, so they are exposed to winds and waves from W and E and E-NE directions; specifically, at Point 2, E and N approaching waves are equivalent to W approaching fronts. At Point 3, the western component prevailed on the north-east approaching direction because of the increase of the western geographic fetch; an increment of the NE component was also observed since this point is more exposed to this approaching direction with respect to Point 2 because of coastal orientation (Figure 1). At Point 4, despite the prevalence of the E-NE approaching direction, the NE component becomes even more evident than at Point 3. Furthermore, since this latter point is sheltered to the west because of coastal orientation (NNE-SSW oriented), waves approaching from the third quadrant essentially present SW and W-SW components.

4.2. Storm Characterization

During the investigated period, a total of 2961 storm events were categorized into five classes, i.e., Class I (weak), Class II (moderate), Class III (significant), Class IV (severe), and Class V (extreme). Points 3 and 2 recorded the highest number of storms (Figure 5). The distribution is similar at all points, with a clear dominance of events belonging to Classes I and II: 87.4% in Point 1, 86.6% in Point 2, 83.1% in Point 3, and 90.4% in Point 4 (Table 1, Figure 5). Mean wave height value of each class did not present great spatial variations, i.e., all points recorded similar values of wave height for the same class. Concerning maximum and minimum wave height values per class, a clear and general trend was not observed, e.g., Point 4 presented the lowest wave height value for Class III but the highest for Classes IV and V (see Table 1).

Table 1. Storm characteristics at each prediction point: class, energy (E_{tot}), frequency of storms, significant wave height (H_{m0}) peak period (T_p), and duration (D).

Point	Class	E_{tot} [Wh/m]			Frequency [%]	H_{m0} [m]		T_p [s]	D [days]
		Min	Max	Mean		Mean	Standard Deviation	Mean	Mean
	I	108	503	265	59.6	2.01	0.26	6.5	0.9
	II	503	1100	730	27.8	2.61	0.37	7.4	1.9
1	III	1100	2102	1509	8.3	3.27	0.48	8.1	2.9
	IV	2102	4179	2624	3.5	3.64	0.65	8.5	4.2
	V	4179	9165	5635	0.7	4.68	0.98	9.7	4.8
	I	108	503	272	58.9	2.05	0.29	6.5	0.9
	II	503	1100	749	27.6	2.65	0.38	7.4	2.0
2	III	1100	2102	1484	9.9	3.13	0.50	7.9	3.3
	IV	2102	4179	2752	2.3	3.76	0.62	8.6	4.8
	V	4179	9165	5551	0.4	4.73	0.76	9.2	6.9
	I	108	503	270	55.0	2.05	0.28	6.5	0.9
	II	503	1100	739	28.0	2.69	0.39	7.4	1.9
3	III	1100	2102	1460	12.4	3.22	0.47	8.0	3.1
	IV	2102	4179	2775	3.4	3.99	0.52	8.9	4.4
	V	4179	9165	5632	0.9	4.87	0.74	9.5	7.0
	I	108	503	280	67.1	2.03	0.27	6.8	0.9
	II	503	1100	712	23.2	2.52	0.35	7.6	2.0
4	III	1100	2102	1408	8.3	3.12	0.46	8.3	3.0
	IV	2102	4179	2802	0.9	4.34	0.60	9.6	3.3
	V	4179	9165	5248	0.3	5.18	0.49	10.6	3.7

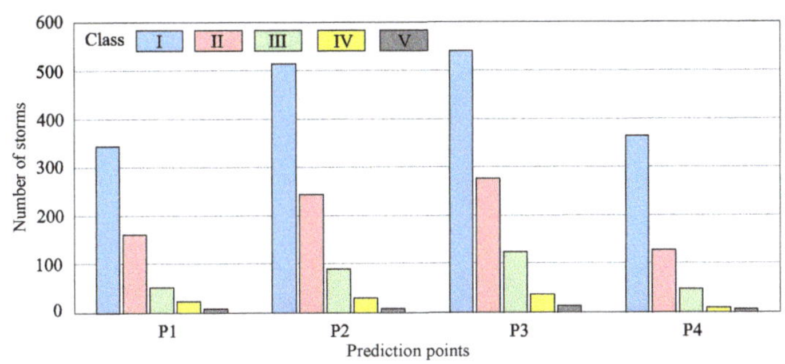

Figure 5. Distribution of storm events per class at each point.

Regarding wave period, the same storm classes of different points presented similar values. Higher wave period variations among classes were recorded at Points 2 and 3 (Table 1). The mean storm duration presented an increase from Class I to Class V at each point (especially Points 2 and 3) and important differences among points (Table 1). Figure 6 reports the monthly distribution of all storms (i.e., the sum of all events recorded at each points) per class. Results are very similar to those obtained by [7,8,11–13,15,34] in their respective studies. Concerning temporal distribution, Classes I and II storms were observed along the whole year. Class III storms were recorded in winter and spring seasons (from October to May), with a minimal occurrence during summer months, i.e., June (8 storms), August (2 storms), and September (2 storms). Class IV storms were observed from November to March, and Class V storms were only recorded from December to March (especially in February, with 7, and March, with 5 events). It is interesting to observe that Classes II and III are very frequent in March, because it relates to approaching waves generated by E and SE winds that are quite frequent in springtime.

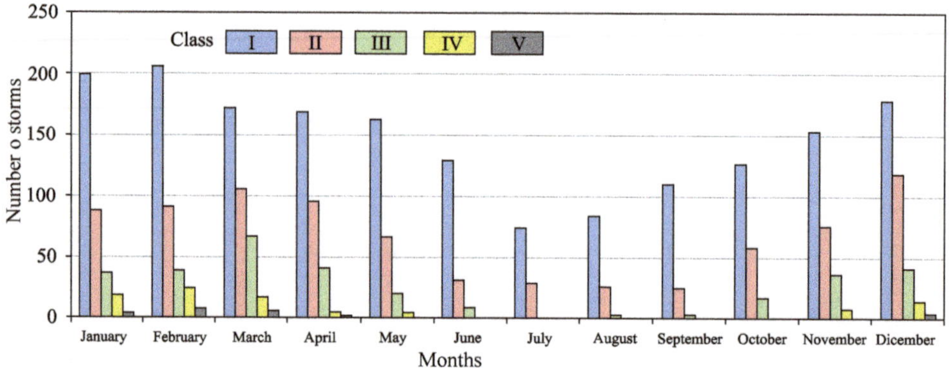

Figure 6. Monthly distribution of all storm events per class and month.

Concerning approaching directions, at Point 1, Classes I and II mainly approached from the east with a small western component (Figure 7) and Classes III, IV and V almost exclusively (>95% of records for each class) approached from E and were linked to the predominant eastern storm waves (Figure 1). At Point 2, the sum of the E and E-NE directions is broadly equivalent to the western component (Figure 7) and clearly reflects, with a slight increase of the eastern component, the wave rise shown in Figure 1. At Point 3, storm energy classes approaching directions (Figure 7) broadly reflect wave rise presented in Figure 1 with an increase of the importance of the western component for all storm classes. At Point 4 (Figure 7), the storm approaching directions are consistent with the wave direction pattern observed in Figure 1. At Point 4, more numerous storms approach from the E-NE direction and disappear in the SW direction. The return period (T_r) of Classes III, IV, and V events at each prediction point are shown in Table 2.

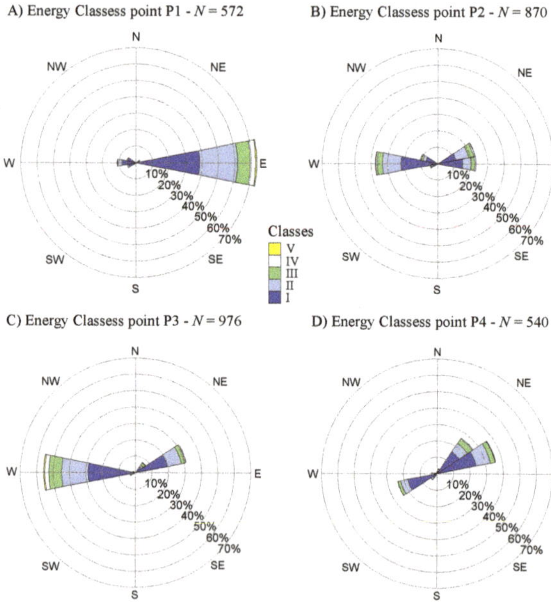

Figure 7. Energy flux roses at all prediction points. (**A**) Point P1, (**B**) Point P2, (**C**) Point P3, (**D**) Point P4. N is the total number of storm events.

At Points 2 and 3 the yearly the probability of energy flux exceedance for Class III storms was 100%, varying from 76.9% to 100% at Point 1 and from 43.5% to 100% at Point 4 (Table 2). For Class IV, the probability of occurrence ranged from 41.7% to 100% and from 58.8% to 100% at Points 2 and 3 respectively, and from 27.8% to 76.9% at Point 1 and from 12.7% to 43.5% at Point 4. Class V probability of exceedance have minimum percentages of 12.7% and 18.5% at Points 2 and 3, and 8.4% at Point 1 and 3.3% at Point 4. Mentioned values are broadly similar to observations carried out by Anfuso et al. [11] near the areas of Huelva and Cádiz, i.e., the less energetic zones of Cádiz Gulf. It is observed that the most energetic points (i.e., Points 2 and 3) have the highest percentages of probability of energy flux exceedance (Figure 8), so the facing littoral, i.e., the coast between Málaga and Almería, is very exposed to storms belonging to all classes, and especially to most energetic ones (III, IV, and V) that can have a great impact on both natural and urbanized sectors.

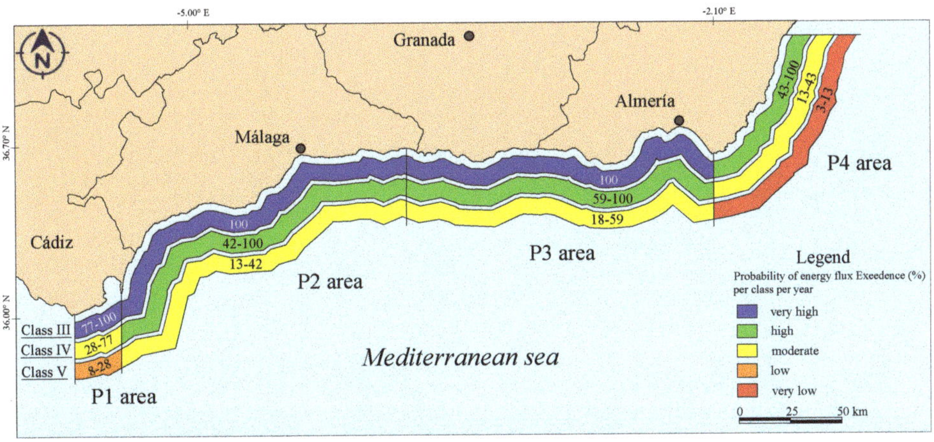

Figure 8. Energy flux probability of exceedance (%) of storms per class III, IV, and V at each point in the coastal area of influence. The percentage values per class as reported on Table 2 are superimposed in the colored stripes.

4.3. Characterization of Stormy Years

In this paper, a special attention was devoted to the yearly cumulative characteristics of storms, i.e., their number, duration, and energy. In fact, as observed by Ferreira [35], the relationship between storms and beach erosion (and/or damage to human structures) varies according to single storm characteristics, storm grouping, and coastal response/morpho-dynamic behavior. Concerning the characterization/quantification of stormy years, only Classes III, IV, and V were considered since these events are the ones that produce important coastal damage according to [12]. With respect to energy data, a similar trend was recorded at the four points, with a similarity observed among Points 1, 2, and 3 (Figure 9).

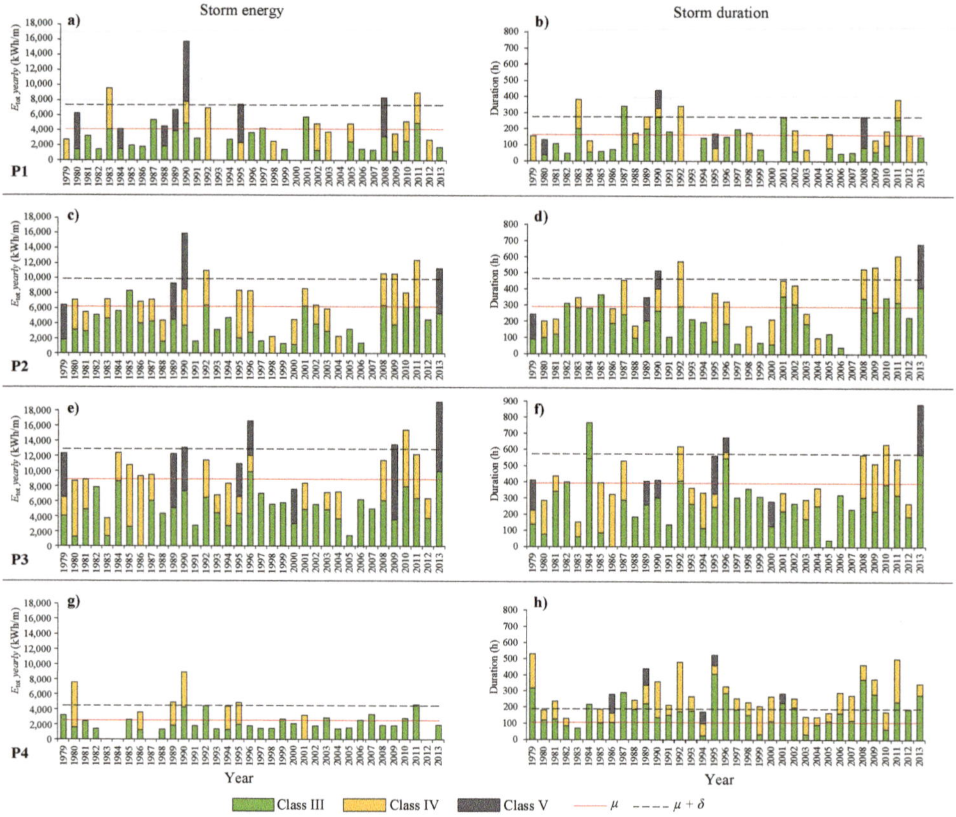

Figure 9. In the left column: the total energy of storms per year for Classes III, IV, and V. At (**a**) Point P1, (**c**) Point P2, (**e**) Point P3, and (**g**) Point P4. In the right column: the corresponding durations of storms (**b**) Point P1, (**d**) Point P2, (**f**) Point P3, and (**h**) Point P4. Red and dashed lines respectively represent the average value (μ) and the average plus one standard deviation ($\mu + \sigma$).

Dealing with yearly energy distribution, it was possible to observe nine energetic years, i.e., 1980, 1983, 1990, 1992, 1995, 2001, 2008, 2010, and 2013. Anomalies were recorded at Point 3, which differed in four of the nine aforementioned years; at Point 1, which recorded a low-energy year in 2013 (i.e., essentially Classes I, II, and III), and at Point 4, which recorded low energy values in 1983, 1984, and 2008 and high values in 1986 and 2007 (Figure 9). Analyzed data of stormy years did not present a clear trend; as an example, yearly distribution of cumulative energy presented a correlation factor (r^2) that ranged from 10^{-5} for Point P2 to 2×10^{-4} for Point P4, but showed a cyclical behavior as highlighted by the calculation of the return period. The recurrence interval and the yearly frequency of occurrence are presented in Table 3 according to the distribution of yearly cumulative energy values (Figure 9). Per each Point, two values of recurrence interval (and frequency of occurrence) are presented, and refer to the mean value and the mean plus one standard deviation of yearly cumulative energy. The frequency of occurrence of stormy year was higher at Point 2 with 17% and 53%, respectively, for the lower and higher energy value, i.e., a low-energetic storm year is observed every 1.9 years and a high-energetic year presents a recurrence interval of 6 years. Point 4 recorded longer recurrence intervals, which ranged from 2.6 to 9 years (39% and 11% values of frequency), respectively, for higher and lower energetic stormy years. Points 1 and 3 presented average values (Table 3).

Table 2. Return period (T_r) and probability of energy flux exceedance ($F(P)$) for the lower and upper limit of each Class (III, IV, V) calculated using Weibull CDF Equation (3). The minimum values are highlighted with blue-colored font.

Point		Class III		Class IV		Class V	
		min	max	min	max	min	max
P1	T_r [year]	<1	1.3	1.3	3.6	3.6	11.9
	$F(P)$ [%]	76.9	100	27.8	76.9	8.4	27.8
P2	T_r	-	<1	<1	2.4	2.4	7.9
	$F(P)$	-	100	41.7	100	12.7	41.7
P3	T_r	-	<1	<1	1.7	1.7	5.4
	$F(P)$	-	100	58.8	100	18.5	58.8
P4	T_r	<1	2.3	2.3	7.9	7.9	30.1
	$F(P)$	43.5	100	12.7	43.5	3.3	12.7

Table 3. Stormy year recurrence interval (R_i) and yearly frequency of occurrence (f_o) calculated from Equation (4), considering yearly cumulative energy values (Figure 9), for each prediction point (Table 1).

Point	Stormy Year			
	R_i [year]		f_o [%]	
	min	max	min	max
P1	2.2	7.2	14	44
P2	1.9	6.0	17	53
P3	2.2	7.2	14	44
P4	2.6	9.0	11	39

Regarding the cumulative storm duration (Figure 9), the trend was not clear. From one side, there is a good correspondence between the number of storms and storm duration within each point, that is, years with many storms also recorded a high cumulative yearly storm duration. Such correspondence was also observed with yearly energy values distribution. On the other hand, no correspondence was observed among different points. A comparison of stormy year distribution recorded in this paper with the one observed by [12] for the Atlantic region of Andalusia showed the opposite behavior linked to the fact that storms on the Atlantic side of Andalusia are essentially related to the predominance of western approaching fronts; meanwhile, on the Mediterranean side [36], they are essentially linked to eastern approaching fronts.

5. Conclusions

To prevent and reduce coastal erosion and damages to human structures, as well as the comprehension of past and future climate trends and beaches' annual and seasonal behavior and evolution, it is very important to characterize the local sea climate. This paper shows as wave climate considerably varies along the Mediterranean coast of Andalusia, from a relatively sheltered area (Point 1) close to the Gibraltar Strait, exclusively affected by easterly wave fronts, to a very energetic central area (Points 2 and 3) exposed to both western and eastern fronts, and to a low-energy area (Point 4) that, because of coastal orientation, is sheltered to western and very exposed to eastern fronts. This behavior is evident when analyzing wave approaching data: main approaching storm directions for high-energy events ranged from E at Point 1, from E to W at Point 2 (only from W for Class V), from W with a small E-NE component at Point 3, and from NE at Point 4 with secondary W-SW components for Classes III and IV (13.33% W for Class III and 20% SW for Class IV). Points 2 and 3 were the most energetic points due to high-energy events approaching from western directions. Stormy years were considered years with a great cumulative energy of Classes III to V events. Nine stormy years were selected during the investigated period, i.e., 1980, 1983, 1990, 1992, 1995, 2001, 2008, 2010, and 2013. Finally, this study highlighted that the yearly probability of energy flux exceedance of more energetic events is higher in the central sector of the studied area, i.e., at Points 2 and 3, Class III events have a 100% probability; Class IV have a 41.7 to 100% probability; and Class V have a 12.7 to 58.8% probability of energy flux exceedance, so this area has a high sensitivity to storm impacts. Future investigations should be devoted to the analysis of the effects on natural and urbanized coastal environments of single representative storms of each storm class to better understand how an increase in energy and storm duration affects coastal behavior, which is strictly linked to beach morpho-dynamic state, and damages to human structures. Secondly, the effects of storm groupings should be better investigated since erosion and damages to coastal structures greatly depend on cumulative storm energy, on separation between storm events, and on natural beach recovery rates.

Author Contributions: Data curation, R.M., G.M. and C.L.R.; Formal analysis, R.M., G.M. and C.L.R.; Methodology, R.M., G.M., C.L.R., G.A. and G.C.; Software, C.L.R. and G.M.; Supervision, G.A., G.C.; Writing—original draft, R.M., G.M., C.L.R., G.A. and G.C.; Writing—review & editing, C.L.R., G.M., G.A. and G.C.

Funding: This research received no external funding.

Acknowledgments: This work is a contribution to the Andalusia P.A.I. Research Group no. RNM-328 and has been partially developed at the Centro Andaluz de Ciencia y Tecnología Marinas (CACYTMAR), Puerto Real (Cádiz, Spain).

Conflicts of Interest: The authors declare no conflict of interest.

Abbreviations

The following abbreviations and symbols are used in this manuscript:

CDF	Cumulative Distribution Function
ECMWF	European Centre for Medium-Range Weather Forecasts
ERA-Interim	ECMWF Reanalysis (from January 1979 onward)
POT	Peak Over Threshold
WAM	Wave Model

Symbols

d^i	Duration of i-th storm
E^i_{tot}	Total energy i-th storm
$F(P)$	Weibull Cumulative Distribution Function of energy flux
f_o	Frequency of occurrence related to the empirical recurrence interval
g	Gravity acceleration
H_{m0}	Spectral significant wave height
$H_{m0,max}$	Monthly mean of maximum spectral wave height within the studied date-range (35 years)
h_t	Wave height threshold for storm definition
λ	Mean number of events per year
m	Number of events (upper threshold) within the studied date-range (35 years)
μ	Mean of total energy within the studied date-range (35 years)
N	Number of storm events
P	Energy flux or wave power per unit of wave-front length
R_i	The stormy year empirical recurrence interval
ρ	Water volumetric mass density
σ	Standard deviation of total energy within the studied date-range (35 years)
T_e	Energy period
T_r	Return period related to the probability of exceedance
u	Weibull function second parameter
w	Weibull function first parameter
X	Abscissa of auxiliary variable for linear fitting
Y	Ordinate of auxiliary variable for linear fitting

References

1. Ministerio de Medio Ambiente y Medio Rural y Marino. *Evolución de las Zonas Costeras en Europa*; Editorial Ministerio de Medio Ambiente: Madrid, Spain, 2008; Volume 1, p. 108.
2. Garcia, G.M.; Pollard, J.; Rodriguez, R.D. Origins, management, and measurement of stress on the coast of southern Spain. *Coast. Manag.* **2000**, *28*, 215–234.
3. Manno, G.; Lo Re, C.; Ciraolo, G. Uncertainties in shoreline position analysis: The role of run-up and tide in a gentle slope beach. *Ocean Sci.* **2017**, *13*, 661–671. [CrossRef]
4. Bacon, S.; Carter, D.T. Wave climate changes in the North Atlantic and North Sea. *Int. J. Climatol.* **1991**, *11*, 545–558. [CrossRef]
5. Komar, P.D.; Allan, J.C. Increasing hurricane-generated wave heights along the US east coast and their climate controls. *J. Coast. Res.* **2008**, *24*, 479–488. [CrossRef]
6. Rodrıguez-Ramırez, A.; Ruiz, F.; Cáceres, L.; Vidal, J.R.; Pino, R.; Muñoz, J. Analysis of the recent storm record in the southwestern Spanish coast: Implications for littoral management. *Sci. Total Environ.* **2003**, *303*, 189–201. [CrossRef]
7. Mendoza, E.T.; Jiménez, J.A. Clasificación de tormentas costeras para el litoral catalán (Mediterráneo NO). *Tecnología y Ciencias del Agua* **2008**, *23*, 21–32.
8. Mendoza, E.; Jimenez, J.; Mateo, J.; Salat, J. A coastal storms intensity scale for the Catalan sea (NW Mediterranean). *Nat. Hazards Earth Syst. Sci.* **2011**, *11*, 2453–2462. [CrossRef]
9. Guisado, E., Malvárez, G.C. Multiple scale morphodynamic mapping: methodological considerations and applications for the coastal atlas of Andalusia *J. Coast. Res.* **2009**, *56*, 1513–1517.
10. Pintado, E.G.; García, G.C.M. El estado morfodinámico de las playas a través de modelización numérica de propagación y asomeramiento del oleaje: El frente litoral de Doñana. *GeoFocus. Revista Internacional de Ciencia y Tecnología de la Información Geográfica* **2015**, *1*, 163–180.
11. Anfuso, G.; Rangel-Buitrago, N.; Cortés-Useche, C.; Castillo, B.I.; Gracia, F. Characterization of storm events along the Gulf of Cadiz (eastern central Atlantic Ocean). *Int. J. Climatol.* **2016**, *36*, 3690–3707. [CrossRef]
12. Rangel-Buitrago, N.; Anfuso, G. Coastal storm characterization and morphological impacts on sandy coasts. *Earth Surf. Processes Landf.* **2011**, *36*, 1997–2010. [CrossRef]
13. Rangel-Buitrago, N.; Anfuso, G. Winter wave climate, storms and regional cycles: The SW Spanish Atlantic coast. *Int. J. Climatol.* **2013**, *33*, 2142–2156. [CrossRef]
14. Halsey, S. Proposed classification scale for major northeast storms: East coast, USA based on extent of damage. In *Geological Society of America, Abstracts with Programs (Northeastern Section)*; Geological Society of America: Boulder, CO, USA, 1986; Volume 18, p. 21.
15. Dolan, R.; Davis, R.E. An intensity scale for Atlantic coast northeast storms. *J. Coast. Res.* **1992**, *8*, 840–853.
16. Orford, J.; Hinton, A.; Carter, R.; Jennings, S. A tidal link between sea level rise and coastal response of a gravel-dominated barrier in Nova Scotia. *Geophys. Monogr.* **1992**, *11*, 71–79.
17. Orford, J.; Carter, R. Examination of mesoscale forcing of a swash-aligned, gravel barrier from Nova Scotia. *Mar. Geol.* **1995**, *126*, 201–211. [CrossRef]
18. Kriebel, D.; Dalrymple, R.; Pratt, A.; Sakovich, V. Shoreline risk index for northeasters. In Proceedings of the 1996 Conference on Natural Disaster Reduction, Washington, DC, USA, 3–5 December 1996.
19. Zhang, K.; Douglas, B.; Leatherman, S. Beach erosion potential for severe nor'easters. *J. Coast. Res.* **2001**, *17*, 309–321.
20. Manno, G.; Anfuso, G.; Messina, E.; Williams, A.; Suffo, M.; Liguori, V. Decadal evolution of coastline armouring along the Mediterranean Andalusia littoral (South of Spain). *Ocean Coast. Manag.* **2016**, *124*, 84–99. [CrossRef]
21. Fernandez-Nunez, M.; Díaz-Cuevas, P.; Ojeda, J.; Prieto, A.; Sánchez-Carnero, N. Multipurpose line for mapping coastal information using a data model: The Andalusian coast (Spain). *J. Coast. Conserv.* **2015**, *19*, 461–474. [CrossRef]
22. Guisado, E.; Malvárez, G.C.; Navas, F. Morphodynamic environments of the Costa del Sol, Spain. *J. Coast. Res.* **2013**, *65*, 500–505. [CrossRef]
23. Boccotti, P. *Wave Mechanics for Ocean Engineering*; Elsevier: Amsterdam, The Netherlands, 2000; Volume 64.

24. Dee, D.; Uppala, S.; Simmons, A.; Berrisford, P.; Poli, P.; Kobayashi, S.; Andrae, U.; Balmaseda, M.; Balsamo, G.; Bauer, P.; et al. The ERA-Interim reanalysis: Configuration and performance of the data assimilation system. *Q. J. R. Meteorol. Soc.* **2011**, *137*, 553–597. [CrossRef]
25. Puerto del Estado *Recomendaciones para Obras Marítimas R.O.M. 0.3-91. Oleaje. Anejo I. Clima Marítimo en el Litoral Español*; Technical Report; Ministerio de Obras Públicas y Transportes: Madrid, Spain, 1992.
26. Monteforte, M.; Lo Re, C.; Ferreri, G.B. Wave energy assessment in Sicily (Italy). *Renew. Energy* **2015**, *78*, 276–287. [CrossRef]
27. Lo Re, C.; Manno, G.; Ciraolo, G.; Besio, G. Wave Energy Assessment around the Aegadian Islands (Sicily). *Energies* **2019**, *12*, 333.
28. Iglesias, G.; Carballo, R. Wave energy potential along the Death Coast (Spain). *Energy* **2009**, *34*, 1963–1975. [CrossRef]
29. Vicinanza, D.; Cappietti, L.; Ferrante, V.; Contestabile, P. Estimation of the wave energy in the Italian offshore. *J. Coast. Res.* **2011**, *64*, 613–617.
30. Jenks, G.; Caspall, F. Error on Choroplethic Maps: Definition, Measurement, Reduction. *Ann. Assoc. Am. Geogr.* **1971**, *61*, 217–244. [CrossRef]
31. Lo Re, C.; Cannarozzo, M.; Ferreri, G.B. Present-day use of an empirical wave prediction method. *Proc. Inst. Civ. Eng. Marit. Eng.* **2016**, *169*, 3–14. [CrossRef]
32. Almeida, L.; Vousdoukas, M.; Ferreira, Ó.; Rodrigues, B.; Matias, A. Thresholds for storm impacts on an exposed sandy coastal area in southern Portugal. *Geomorphology* **2012**, *143–144*, 3–12. [CrossRef]
33. Dorman, C.; Beardsley, R.; Limeburner, R. Winds in the strait of gibraltar. *Q. J. R. Meteorol. Soc.* **1995**, *121*, 1903–1921. [CrossRef]
34. Moritz, H.; Moritz, H. Evaluating extreme storm power and potential implications to coastal infrastructure damage, Oregon Coast USA. In Proceedings of the 9th International Workshop on Wave Hindcasting and Forecasting, Victoria, BC, Canada, 24–29 September 2006.
35. Ferreira, Ó. Storm groups versus extreme single storms: Predicted erosion and management consequences. *J. Coast. Res.* **2005**, *42*, 221–227.
36. Menéndez, M.; Mendez, F.J.; Losada, I.J.; Medina, R.; Abascal, A.J. Variaciones del régimen extremal del clima marítimo en el litoral español en el periodo 1958–2001. In *El Clima entre el Mar y la Montaña. Asociación Española de Climatología y Universidad de Cantabria, Santander, Serie A*; García Codron, J.C., Diego Liaño, C.F., De Arróyabe Hernáez, P., Garmendia Pedraja, C., Rasilla Álvarez, D., Eds.; Asociación Española de Climatología (AEC): Santander, Spain, 2004; Volume 4, pp. 73–84.

water

MDPI

Technical Note

Wave Climate at Shallow Waters along the Abu Dhabi Coast

Waleed Hamza [1], Letizia Lusito [2,*], Francesco Ligorio [2], Giuseppe Roberto Tomasicchio [2] and Felice D'Alessandro [2]

1 Biology Department, College of Science, United Arab Emirates University, 15551 Al Ain, UAE; w.hamza@uaeu.ac.ae
2 Department of Engineering for Innovation, University of Salento, Campus Ecotekne, 73100 Lecce, Italy; francesco.ligorio@unisalento.it (F.L.); roberto.tomasicchio@unisalento.it (G.R.T.); felice.dalessandro@unisalento.it (F.D.)
* Correspondence: letizia.lusito@unisalento.it

Received: 3 July 2018; Accepted: 18 July 2018; Published: 26 July 2018

Abstract: High-resolution, reliable global atmospheric and oceanic numerical models can represent a key factor in designing a coastal intervention. At the present, two main centers have the capabilities to produce them: the National Oceanic and Atmospheric Administration (NOAA) in the U.S.A. and the European Centre for Medium-Range Weather Forecasts (ECMWF). The NOAA and ECMWF wave models are developed, in particular, for different water regions: deep, intermediate, and shallow water regions using different types of spatial and temporal grids. Recently, in the Arabian Gulf (also named Persian Gulf), the Abu Dhabi Municipality (ADM) installed an ADCP (Acoustic Doppler Current Profiler) to observe the atmospheric and oceanographic conditions (water level, significant wave height, peak wave period, water temperature, and wind speed and direction) at 6 m water depth, in the vicinity of the shoreline of the Saadiyat beach. Courtesy of Abu Dhabi Municipality, this observations dataset is available; the recorded data span the period from June 2015 to January 2018 (included), with a time resolution of 10 min and 30 min for the atmospheric and oceanographic variables, respectively. At the ADCP deployment location (ADM*ins*), the wave climate has been determined using wave propagation of the NOAA offshore wave dataset by means of the Simulating WAves Nearshore (SWAN) numerical model, the NOAA and ECMWF wave datasets at the closest grid point in shallow water conditions, and the SPM '84 hindcasting method with the NOAA wind dataset used as input. It is shown that the best agreement with the observed wave climate is obtained using the SPM '84 hindcasting method for the shallow water conditions.

Keywords: wave hindcasting; Abu Dhabi; shallow waters; Shore Protection Manual; wave climate

1. Introduction

The United Arab Emirates (UAE) is an Arabian Gulf (also named Persian Gulf) nation with an extended coastal area, comprising more than 700 km of coastline and embracing many different shallow water wetland habitats. Some areas of the Abu Dhabi coastline are undergoing a large development with residential, business, cultural, and tourism infrastructure (http://government.ae/en/about-the-uae/uae-future). As a consequence, there is an increasing need for a detailed knowledge of the wave conditions in order to design the coastal interventions [1–4].

The present study focuses on Saadiyat (Figure 1), a large low-lying 27 km^2 island situated in the Gulf within the Emirate of Abu Dhabi. Saadiyat island hosts a vast national park of mangroves and a 9 km SW–NE oriented natural sandy beach, Saadiyat beach, populated by numerous flora and fauna species. Part of the island is undergoing a considerable development program. In March 2015, the "North Beach Phase 1 Development Plans" document was submitted to the Abu Dhabi Urban

Planning Council as a part of the Shoreline Protection Works Master Plan, an addendum to the Concept Master Plan of the Saadiyat Cultural District. The urban plan includes the construction of new luxury hotels, a private luxury residential area, and the Cultural District with the Louvre and Guggenheim museums (Figure 2).

Figure 1. Aerial view of Saadiyat beach.

Figure 2. Concept Master Plan of Saadiyat Cultural District (courtesy Tourism Development & Investment Company).

At present, professionals involved in the Abu Dhabi coastal development typically use two main sources of wind and wave data: the National Oceanic and Atmospheric Administration (NOAA) and the European Centre for Medium-Range Weather Forecasts (ECMWF), following the well consolidated scientific methods already developed for the analysis of shallow water coastal areas [5–9].

To overcome the scarcity of in situ observations in the area, the Abu Dhabi Municipality (ADM) has recently installed several instruments in the vicinity of the Abu Dhabi coastline at different water depths. Within a research project granted by the National Water Center at the United Arab Emirates University, the new in situ observed wave conditions dataset has been made available.

The in situ observed wave dataset allows for the comparison and verification of data from other sources, such as the NOAA and ECMWF global wave models, third-generation shallow water models such as the Simulating WAves Nearshore (SWAN) [10,11], and second-generation local wave models [12]. The comparison and verification method used is the Shore Protection Manual '84 (SPM '84) method [1], a second-generation wave hindcasting method widely used for local hindcasting. SPM '84 has been selected because of its simplicity in the algorithms and because it allowed for the consideration of different wave regions. SWAN has been already successfully used to model waves in coastal regions [13–19].

The SPM '84 method allows the estimation of the wave characteristics using wind data (intensity, direction and duration of the event) at a reference height of 10 m above sea water level, and the effective fetch value. It has been used to calculate the wave conditions at the location where the ADM instrument (ADM*ins*) is deployed from the NOAA wind data, which are provided at 10 m above sea water level.

The comparison and verification of data from NOAA, ECMWF, and SPM '84 has been conducted under the assumption that the ADM*ins* and the closest NOAA and ECMWF nodes are exposed to homogeneous atmospheric and oceanographic conditions [20,21].

The present paper is organized as follows: Section 2 presents a review of the state-of-the-art of atmospheric and oceanographic global numerical models and the description of the application of the SPM '84 parametric local wave growth method, Section 3 describes the results, and Section 4 discusses the results from the use of the SPM '84 wave hindcasting and the adopted verification procedure.

2. Materials and Methods

Recently, the ADM installed an instrument to observe atmospheric (wind speed and direction) and oceanographic (water level, significant wave height, peak wave period, water temperature) characteristics close to the shoreline of the Saadiyat beach. The instrument, an Argonaut-XR produced by the company "SonTek—A Xylem Brand" (San Diego, CA, USA), is a 3-D up-looking monostatic Acoustic Doppler Current Profiler (ADCP) that uses sonar for precise water velocity measurements. Monostatic refers to the fact that the same transducer is used as a transmitter and receiver, with a sampling frequency of 1.5 MHz. Recorded data from June 2015 to January 2018 (included) were available at a time resolution of 10 and 30 min for the atmospheric and oceanographic variables, respectively. The deployed Argonaut-XR presented a limitation: the profiler had not been equipped with the software/hardware necessary to also measure wave spectra; therefore, the instrument could not measure the wave directions directly.

Direct wave measurements are considered the most reliable source of information. In the Gulf area, this type of information is rare or even missing. The Abu Dhabi Municipality started to monitor the wave conditions only four years ago. Other Municipalities in the UAE do not provide directly observed data. As a consequence, at present, the only reliable source of wave data in Abu Dhabi is to consider the new in situ observed wave conditions dataset of sufficient length to represent a mean annual wave climate in relation to the weather conditions in the area, which are quite mild, characterized by the absence of extreme wind conditions (e.g., hurricanes) and extreme wave conditions, due to the fact that the Gulf is an enclosed basin. The wave observations, recorded by the ADCP, have been compared to wave numerical models, designed by some of the major oceanography research centers all over the world.

The first considered model is designed by NOAA. The NOAA National Centers for Environmental Prediction (NCEP) developed the Climate Forecast System (CFS), a fully coupled model representing the interaction between the Earth's atmosphere, oceans, land, and sea ice. A reanalysis of the sea and atmosphere state for the period of 1979–2009 has been conducted, resulting in the CFS Reanalysis (CFSR) dataset [22]. The purpose of the CFSR is to produce multiyear global, state-of-the-art gridded representations of atmospheric and oceanic states, generated by a consistent model and data assimilation system. The vertical discretization of the atmosphere consists of 64 layers. The temporal resolution for the atmospheric variables is 3 h. Using the CFSR dataset, the NOAA Marine Modeling and Analysis Branch (MMAB) has produced a wave hindcast for the same period. The wave hindcast dataset has been generated using the WAVEWATCH III (WW3) model (v3.14), and it is suitable for use in climate studies. The wave model suite consists of global and regional nested grids. The rectilinear grids were developed using ETOPO-1 bathymetry [23], together with v1.10 of the Global Self-consistent Hierarchical High-resolution Shoreline (GSHHS) database. The higher-resolution North West Indian Ocean bathymetry grid, adopted in the considered data, has a resolution of 10 arc-minutes (1/6°). The WW3 model is run using a 30 arc-minute computational resolution, but the results are interpolated on a 10 arc-minute numerical grid. The spatial resolution of the considered data is, therefore, 1/6°, which corresponds to roughly 20 km. The North West Indian Ocean computational grid, adopted in the considered data, extends in longitude from 30° E to 70° E (with 241 grid nodes) and in latitude from 20° S to 31° N (307 grid nodes). As wave characteristics are dominated by wind dynamics, it is possible to achieve an accurate wave hindcast by using statistically homogeneous wind fields from a long-term reanalysis such as the CFSR, without the need for wave data assimilation [24–26]. The NOAA datasets (both wind and waves) are freely available. The NOAA WAVEWATCH III/CFSR webpages [27,28] present additional details about the datasets.

The second of the considered wave hindcasting methods is designed by the European Centre for Medium-Range Weather Forecasts (ECMWF) and it is named ERA-Interim. The ERA-interim dataset is another global atmospheric reanalysis, starting from 1979 and is continuously updated. It is based on the Cy31r2 version of the Integrated Forecasting System (IFS). It also models oceanographic variables, including waves. The atmospheric configuration uses 60 vertical model levels and a 6-hourly temporal resolution [29,30]. The wave model is based on WAM [31,32], resolving 30 wave frequencies and 24 wave directions. The wave model contains corrections for treating unresolved bathymetry effects and a reformulation of the dissipation source term. The bathymetric information is based on the ETOPO-2 bathymetry, with a 2 arc-minute resolution (1/30°). The computational grid resolution for the wave model is 80 km [33], but the output grid has been interpolated onto a finer mesh, of 1/8° horizontal resolution. The ECMWF ERA-Interim dataset is available upon request on the ECMWF servers. Further information on how to access the data can be found at Reference [34].

The SPM '84 method [1] is a parametric wave hindcasting method; it allows for the determination of the significant wave height (H_s) and peak period (T_p) from (i) the geographical extent of the area where the constant wind is blowing, indicated as fetch (F), (ii) the wind duration (t), and (iii) the depth of water in the generation area (d) [35]. Fetches are related to the curvature of the isobars describing the weather system at the origin of the wave growth. The extent of a typical weather system results in an upper limit for fetches of roughly 500 km.

Given the relative values of F, T_p, and t, the sea state is classified as follows:

- Fully Arisen Sea (FAS), where any added wind energy is balanced by wave energy dissipation and the waves have the maximum possible height;
- Fetch-Limited Sea (FL), where the wave height is limited by the length of the fetch over which the wind has blown; or
- Duration-Limited Sea (DL), where the wave height is limited by the duration of the wind conditions.

The SPM '84 method determines the spectral significant wave height H_{m_0} from σ, the variance of the sea surface elevation, according to:

$$H_{m_0} = 4\sigma = 4\sqrt{m_0} \quad (1)$$

where m_0 is the zero moment of the wave spectrum, i.e., the area under the spectrum itself. In deep water, H_{m_0} is approximately equal to H_s; in shallow water, H_{m_0} becomes less than H_s.

The wave growth formulas at both deep and shallow waters are given in terms of the wind stress factor U_A (adjusted windspeed or wind stress), which is related to the normal windspeed at a height of 10 m, U, by the following equation:

$$U_A = 0.71\ U^{1.23} \quad (2)$$

where both U_A and U are measured in m/s. The SPM '84 method is based on the assumption that the wave growth is entirely caused by a wind blowing at a constant speed and direction for a given duration over the specific fetch area. The wind direction is considered constant if it varies from the wind direction mean by less than 15°, and the wind speed is considered constant if it varies from the wind speed mean by less than 2.5 m/s. With these assumptions and considerations, the equations for wave growth hindcasting/forecasting at shallow water conditions for fully arisen, fetch-limited, and duration-limited sea conditions have been determined. These equations are derived from the analogous deep-water equations [1] with the additional condition that the wave energy is further reduced due to additional effects like bottom friction.

In the present application of wave hindcasting in the Arabian Gulf area, the SPM '84 input data are the ADM*ins* effective fetches, F [36,37]; the water depth at the ADM*ins*, d (equal to 6 m); the wind stress and direction of the wind generating the waves in proximity of ADM*ins*, taken from the NOAA CFSR grid point which is the closest to the ADM*ins*; and the wind duration, t. In the case of Saadiyat beach, there are two NOAA CFSR grid points that are similarly distant from ADM*ins* (Figure 3), located at the north and at the south of the ADM*ins*; in the following, these two grid points will be indicated as NOAA North (NOAA N) and NOAA South (NOAA S). Figure 3 shows the location of the ADM*ins*, the NOAA N and NOAA S grid nodes and the location of the ECMWF grid point. Table 1 reports the coordinates of these grid points, with their corresponding water depths.

Table 1. ADM*ins*, NOAA N, NOAA S, ECMWF, NOAA Offshore coordinates, and water depths.

Name	Longitude (°E)	Latitude (°N)	Water Depth (m)
ADMins	54°24′29″	24°34′17″	6
NOAA North	54°30′00″	24°40′00″	6
NOAA South	54°19′59″	24°30′00″	3
ECMWF	54°22′30″	24°37′30″	11
NOAA Offshore	54°00′00″	25°00′00″	21

Figure 4 shows the NOAA N (a) and NOAA S (b) wind rose, indicating that the atmospheric conditions at the two locations can be considered uniform. In fact, the NOAA N and NOAA S nodes are exposed to the same atmospheric conditions, given the relatively small distance between the two nodes. Because of this similarity, for the SPM '84 application, only the data from the north node have been considered; in particular, the wind stress $U_{A,N}$, and wind direction, θ.

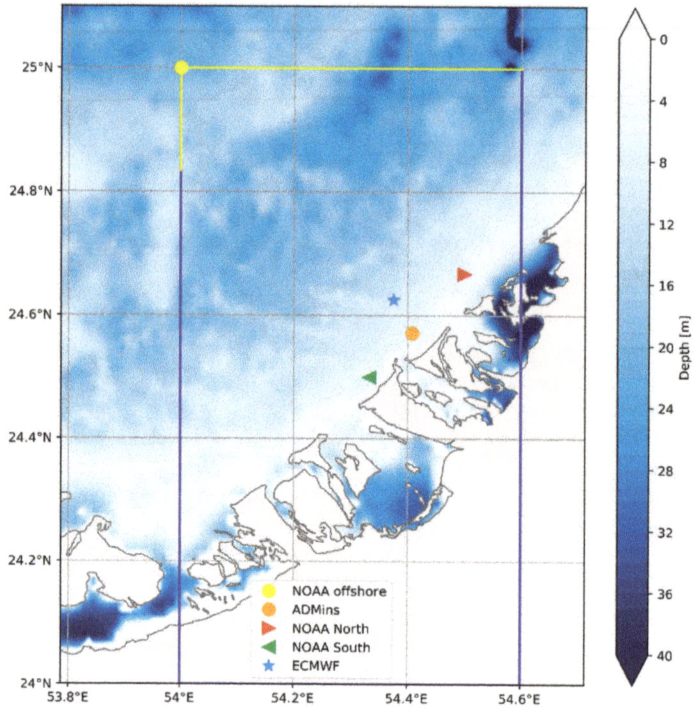

Figure 3. Bathymetry of the Saadiyat area, where markers show the position of ADM*ins*, the NOAA N, the NOAA S, the NOAA offshore and the ECMWF model nodes.

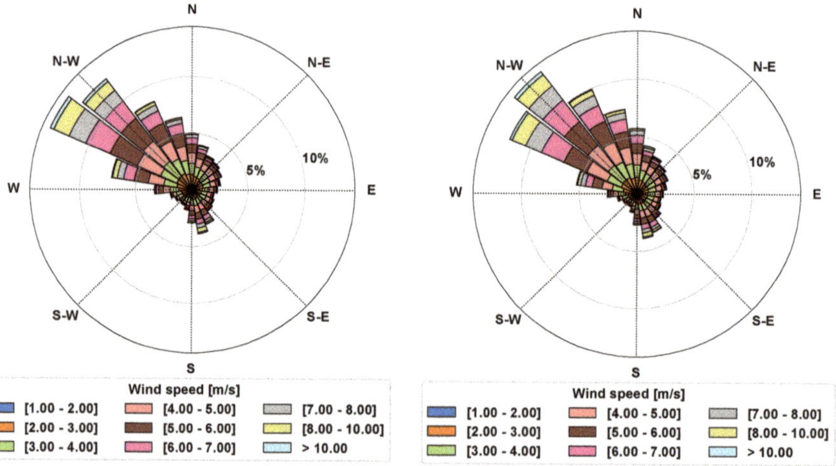

Figure 4. NOAA wind rose at the north node (**a**), and at the south node (**b**), in proximity of the ADM*ins*, indicating that the atmospheric conditions along this stretch of coast can be considered uniform.

The NOAA N wind time series, consisting of approximately 90,000 points, was used to determine an averaged time series of steady wind in both magnitude and direction. A wind measurement datum, *i*, was considered to belong to a steady weather phenomenon if the following is true:

- $| U_i - < U_j > | \leq 2.5$ m/s
- $| \theta_i - < \theta_j > | \leq 15°$

where the symbol $<U_j>$ indicated the average of the wind speed calculated over the *j* points before the *i*th, already previously "clustered" in a steady wind condition. The same procedure applied for $<\theta_j>$. If the *i*th point met the steady state condition, a new average was calculated, aggregating the *i*th point to the other *j*. The cluster then contained *j* + 1 points. Then, the successive point was examined. If a point did not satisfy these conditions, the time-cluster was closed and the location was considered to be subject to a new different wind condition afterwards [38]. The procedure described above is indicated in the following as a "dynamical running average" (DRA). Figure 5 shows the result of the DRA procedure for August 2009. The top panel shows the time series of the wind speed (red) and the dynamically averaged wind speed (blue). The bottom figure shows time series of the wind direction (red) and the dynamically averaged wind direction (blue).

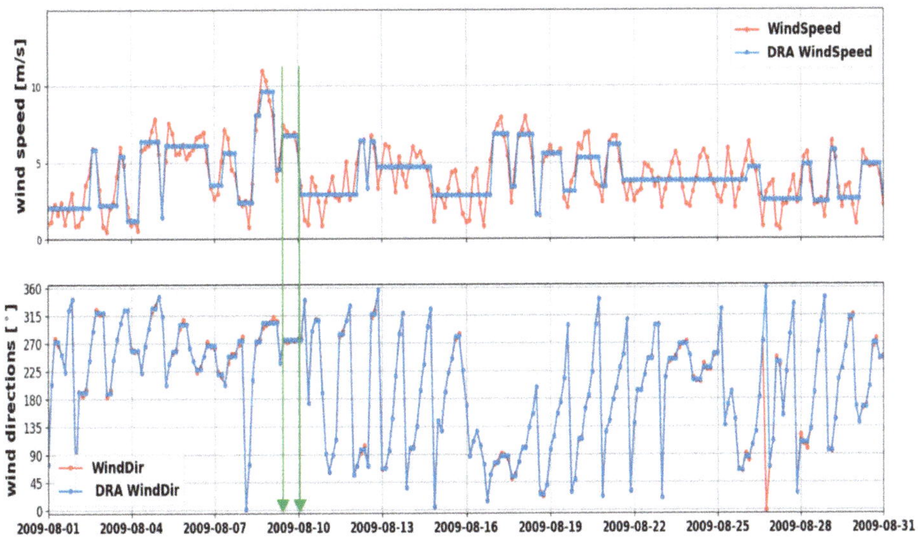

Figure 5. Top panel: time series of the wind speed (red) and the dynamical running averaged wind speed (blue). Bottom panel: time series of the wind direction (red) and the dynamically averaged wind direction (blue).

Comparing the DRA wind speed and direction time series (Figure 5, blue lines), only segments where DRA time series are flat in both the wind speed and direction time series were considered (Figure 5, green vertical line). These segments indicated periods of time where the wind conditions were steady (constant). Therefore, the time period between the beginning and the end of these segments has been assumed as the wind duration t_{ADMins} of the steady wind phenomenon generating the waves (Figure 6, orange horizontal line). In the example in Figure 6, the resulting duration of the particular wind phenomenon was 15 h (5 data points with 3 h time resolution). If no flat segment could be identified in both the DRA time series, the wind duration was assumed to be 3 h (equal to the time resolution of the original time series).

Having determined the duration of the wind generating the waves, it was possible to apply the SPM '84 method at the ADM*ins* location, as explained in the following. The wave direction at the ADM*ins* $\theta_{wave,ADMins}$ was calculated considering the wind–wave direction correlation [39]. Sea regimes (FAS, FL, DL) at the ADM*ins* location, where the water depth d was 6 m, were determined as in Reference [40]. If the factor:

$$\frac{U_{A,N}}{(g \cdot d)^{\frac{1}{2}}} < 0.83 \quad (3)$$

FAS exists at the ADM*ins*.

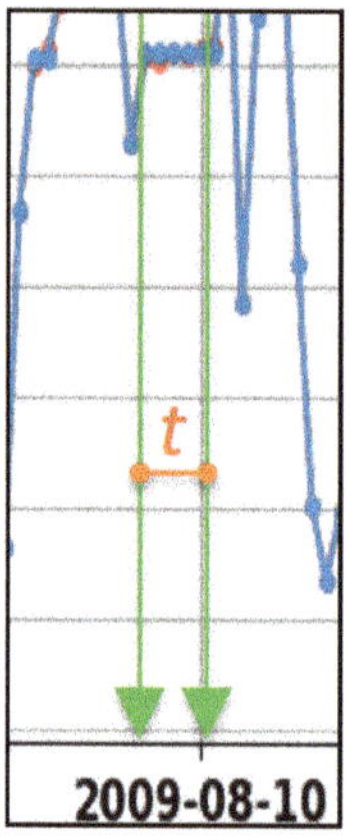

Figure 6. Determination of the duration of the steady wind generating the waves (zoom of Figure 5).

If the ADM*ins* is under the FAS condition, the significant wave height, $H_{s,ADMins}$, was calculated as:

$$H_{s,ADMins} = \frac{U_{A,N}^2}{g} 0.283\, tanh\left[0.530\left(\frac{g \cdot d}{U_{A,N}^2}\right)^{\frac{3}{4}}\right] \quad (4)$$

where all the SI units are adopted. If the ADM*ins* was not under the FAS condition, the duration t_{ADMins} was compared to the minimum duration, t_{min}, necessary for establishing the FL condition. The minimum duration was calculated as in Reference [38]:

$$t_{min} = 2.59 \cdot \frac{F_{eff,ADMins}}{\hat{d}} \cdot \left(\frac{g \cdot \hat{d}}{U_{A,N}^2}\right)^{\frac{2}{3}} \quad (5)$$

where $\hat{d}$ is an average water depth along the fetch extension. In this case, $\hat{d} = 30\ m$. If $t_{ADMins} < t_{min}$, the location was in the DL condition, otherwise it was in the FL condition.

In the FL condition, the significant wave height at the ADM*ins*, $H_{s,ADMins}$, was calculated as:

$$H_{s,ADMins} = \frac{U_{A,N}^2}{g} 0.283\, tanh\left[0.530\left(\frac{g \cdot d}{U_{A,N}^2}\right)^{\frac{3}{4}}\right] \cdot tanh\left[\frac{0.00565 \cdot \left(g \cdot F_{eff,T}/U_{A,N}^2\right)^{\frac{1}{2}}}{\tan\text{h}\left[0.530\left(\frac{g \cdot d}{U_{A,N}^2}\right)^{\frac{3}{4}}\right]}\right] \quad (6)$$

In the DL condition, from Equation (5), a new effective fetch $F\prime_{eff,ADMins}$ was calculated by considering $t_{min} = t_{ADMins}$ and used in Equation (6) to calculate $H_{s,ADMins}$ in DL conditions.

Another approach to calculate the wave climate at the ADM*ins* location consisted of propagating the wave dataset at the offshore NOAA grid point towards the coastline. Wave propagation from offshore to ADM*ins* was performed by means of the numerical model SWAN [10,11], which is a third generation fully-spectral wind-wave model developed at the Delft University of Technology (The Netherlands) that simulates random, short-crested wind generated waves in coastal regions.

The offshore wave time series was extracted from the NOAA CFSR model at coordinates 54°00′00″ E, 25°00′00″ N, with water depth equal to 21 m (Figure 3 and Table 1). SWAN was run in stationary mode, in which waves are assumed to propagate instantaneously throughout the model domain. This assumption was reasonable for the small domain and slowly varying forcing conditions for this case. Nonlinear triad interactions [41,42] were significantly larger than quartet interactions for the short spatial scales and shallow depths considered here, and thus quartet interactions were neglected. The model runs included the Madsen expression for bottom friction dissipation [43] using a coefficient of 0.001 m. Breaking wave dissipation was estimated with a bore-based model [44], with the depth-induced constant wave breaking parameter $\gamma = 0.73$, found as the mean value of the dataset of Battjes and Stive [45,46]. The frequency, f, and the directional resolution, δ, were defined using the SWAN default values, resulting in 24 logarithmically distributed frequency bins (e.g., $\Delta f \approx 0.14f$ [47] over $0.04 \leq f \leq 1.00$ Hz and 36 10°-wide directional bins $\Delta\delta$ evenly spaced over $0° \leq \delta < 360°$ [47]). Table 2 summarizes the main SWAN configuration parameters.

The boundary conditions on open boundaries in the north and north-north-west of the computational domain (Figure 3, yellow and blue lines) have been taken from the NOAA WW3 re-analyses, at the coordinates 54°00′00″ E, 25°00′00″ N (Figure 3, yellow dot). Past similar works, e.g., Gorrel et al. [13], suggest that negligible errors in the computing area grids result from the assumption of uniform waves on the N and NNW cross-shore open water boundaries of the grid. The bathymetry information, provided through the "General Bathymetric Chart of the Ocean" (GEBCO) consortium, consists of a gridded terrain model for ocean and land with a spatial resolution of 30 arc-seconds. It was generated by combining quality-controlled ship depth soundings with interpolation between sounding points guided by satellite-derived gravity data [48,49]. Figure 3 shows also the GEBCO bathymetry of the considered area.

Table 2. Main SWAN configuration parameters.

Model Version	**SWAN 41.20**
Time and Spatial mode	Stationary 2-dimensional
Bathymetry	GEBCO_2014 (30 arc-second)
Computational domain	LON: 54.00417° E, 54.59584° E
	LAT: 24.00417° N to 25.00417° E
	71 × 120 computational nodes
Wave frequency grid	0.04 to 1.00 Hz, 24 frequencies
Directional grid	0° to 360°, 36 directions
Physics	
Breaking	constant breaker index, $\gamma = 0.73$
Whitecapping	Komen et al., 1984 [50]
Bottom Friction	Madsen et al., 1988 [43]
	equiv. roughness length scale of the bottom 0.001 m
Triads	included
Diffraction	excluded
Quadruplets	excluded
Boundary conditions	N boundary: from 54.00417° E to 54.59584° E
provided by NOAA WW3	NNW boundary: from 24.834° N to 25.00417° N
Boundary wave spectra shape	Jonswap [51]

Table 3 summarizes the information about the bathymetry and the computational resolution used in SWAN, and in the NOAA and ECMWF wave models.

Table 3. Bathymetry and the computational resolution.

Name	Bathymetry		Computational Grid	Output Grid
	Name	Resolution	Resolution	Resolution
NOAA	ETOPO-1 (1 arc-minute resolution)	10 arc-minutes = 1/6° (interpolated)	10 arc-minutes = 1/6° (interpolated from 1/2°)	10 arc-minutes = 1/6° (interpolated from 1/2°)
ECMWF	ETOPO-2 (2 arc-minute resolution)	2 arc-minutes = 1/30°	80 km	1/8° (interpolated from T255)
SWAN	GEBCO	30 arc-seconds = 1/120°	30 arc-seconds = 1/120°	30 arc-seconds = 1/120°

3. Results

Field measurements showed that the Argonaut-XR appeared to be unable to measure wave peak periods smaller than 3 s, although the instrumental sensitivity range is 2–20 s, according to the system manual technical specifications [52]. Figure 7 shows the wave period normalized probability density function distribution (normalised p.d.f.) as measured by the ADM*ins*, as determined by the NOAA and the ECMWF wave models at the closest location to the ADM*ins* position, and as estimated by the SPM '84. Figure 7 shows a cut-off at 3 s in the peak period distribution measured by the ADM*ins*. This is not in agreement with the results from the NOAA and ECMWF wave numerical models indicating that waves with a 3 s peak period appear in the area.

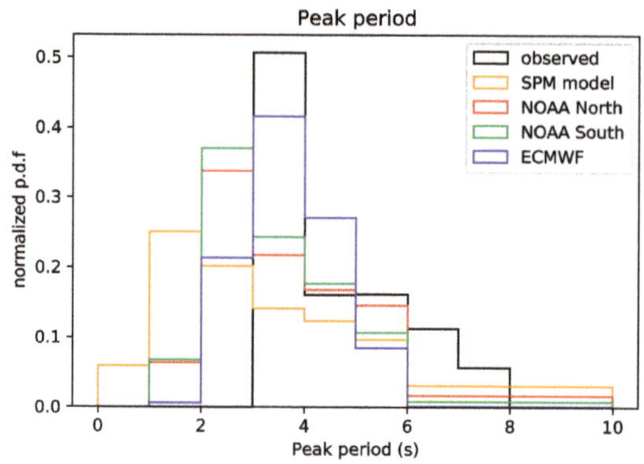

Figure 7. Wave period distribution (normalized p.d.f.) by ADM*ins*, by NOAA and ECMWF datasets and by SPM '84.

Figure 8 shows the wave roses from SPM '84 (1979–2009) and ADM*ins* (June 2015–January 2018). The maximum value of the significant wave height, H_s, recorded at a 6 m water depth in front of Saadiyat beach, was around 2 m. The SPM '84 underestimates the low (0.25–0.50 m) and high (1–2 m) waves, respectively. For each wave height class, Figure 9 depicts the difference between values of appearance frequency (*af*) in %, determined by SPM '84 and observed by ADM*ins*; the difference appears to be small, therefore the *af* determined by SPM '84 and observed by ADM*ins* are in fairly good agreement. The largest differences were found for values of wave height between 1 and 1.5 m.

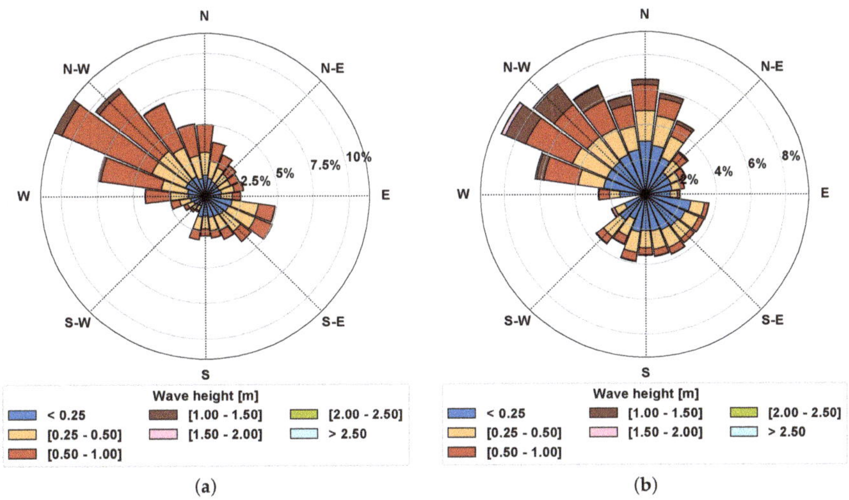

Figure 8. Wave roses from SPM '84 (**a**) and from ADM*ins* (**b**).

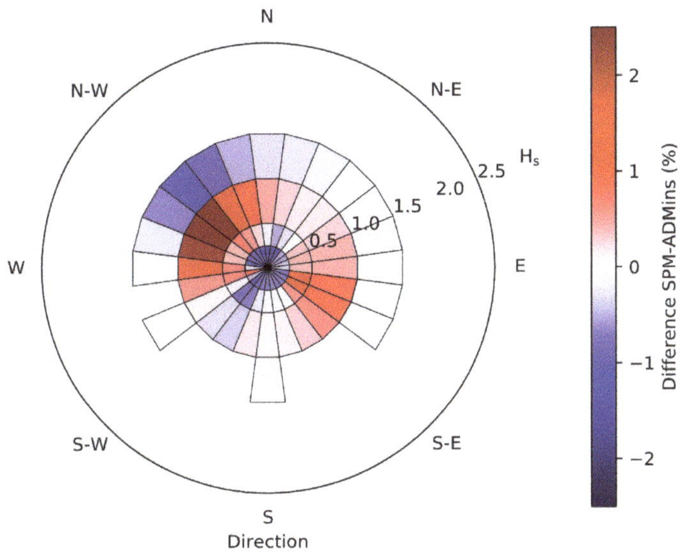

Figure 9. Difference in values of appearance frequency determined by SPM '84 and observed by ADM*ins*.

Figure 10 shows the wave roses from the NOAA and ECMWF wave models at the nodes, which were closest to ADM*ins*. Comparing the wave roses obtained from the NOAA and ECMWF nodes at shallow water conditions (Figure 10), with the wave roses estimated by SPM '84 and observed by ADM*ins* (Figure 8) shows that the NOAA and ECMWF models did not sufficiently capture the variability of the wave climate at the ADM*ins* location in shallow water conditions; in particular, for the directional distribution. The higher horizontal resolution (1/8°) of the ERA-Interim wind and waves hindcasted datasets allowed ECMWF to model the waves having H_s between 1.5 and 2 m with better precision compared to NOAA.

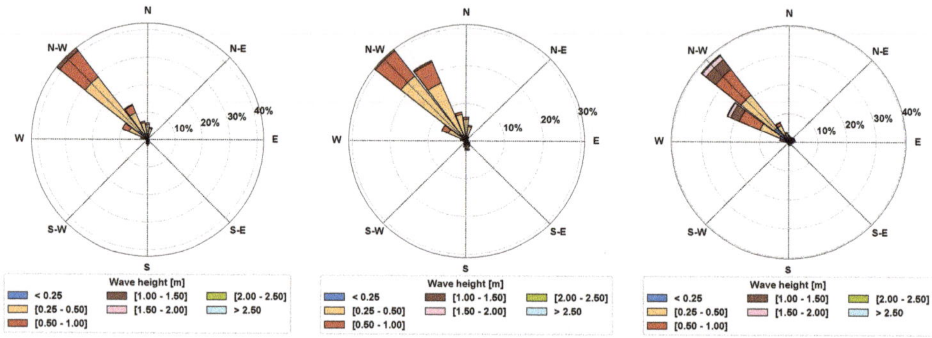

(**a**) (**b**) (**c**)

Figure 10. NOAA wave rose at the north node (**a**), NOAA wave rose at the south node (**b**), and ECMWF (**c**) wave rose in proximity of the ADM*ins*.

Figure 11 shows the SWAN-computed wave rose at the ADM*ins* location. As for the ADM*ins* and SPM '84, also in this case, the majority of the waves at Saadiyat beach came from the north-west direction, having a wave height of up to 2 m.

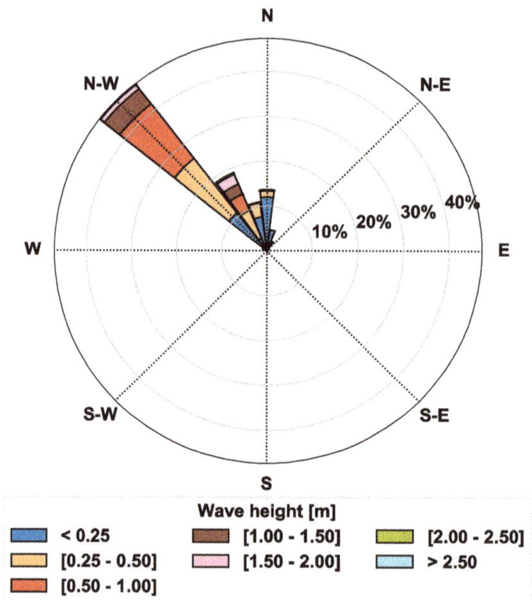

Figure 11. SWAN-computed wave rose at the ADM*ins* location.

Comparing the results from SWAN and SPM '84 with the observations at the ADM*ins* location, it followed that SWAN was able to reproduce waves with low and high H_s better than the SPM '84, but the SPM '84 wave directional distribution presented a better agreement with the observations.

To summarize, Table 4 shows the observed/modelled appearance frequency (*af*) (in %) in the main sector (NW or numerically 315 $\pm$ 22.5°), divided in seven classes of H_s.

Table 4. NW *af*, for classes of H_s.

Dataset	Classes of H_S (m)						
	[0, 0.25]	[0.25, 0.5]	[0.5, 1.0]	[1.0, 1.5]	[1.5, 2.0]	[2.0, 2.5]	>2.5
ADM*ins* (O)	7.23	5.59	7.01	3.53	0.38	0.05	0.01
SWAN	14.88	22.25	19.33	6.68	3.48	0.70	0.00
SPM '84	4.48	7.26	15.45	0.97	0.00	0.00	0.00
NOAA N	7.90	38.99	17.63	1.68	0.03	0.00	0.00
NOAA S	4.76	39.01	16.47	0.86	0.00	0.00	0.00
ECMWF	12.66	25.29	20.16	7.49	2.85	0.59	0.10

With reference to the observations at the ADMins (O), Table 5 shows the *af BIAS*, *af* Root Mean Square Error (*RMSE*), *af* Normalized BIAS (*NBIAS*), *af* Normalized RMSE (*NRMSE*), Symmetric Mean Absolute Percentage Error (*SMAPE*), and symmetric RMSE (*SRMSE*) for the modelled values (M) by: SPM' 84, SWAN, NOAA N, NOAA S, and ECMWF with respect to the observations, calculated according to:

$$BIAS = \frac{\sum_{i=1}^{7}(M_i - O_i)}{N} \quad (7)$$

$$RMSE = \sqrt{\frac{\sum_{i=1}^{7}(M_i - O_i)^2}{N}} \quad (8)$$

$$NBIAS = \frac{\sum_{i=1}^{7}[(M_i - O_i)/O_i]}{N} \quad (9)$$

$$NRMSE = \sqrt{\frac{\sum_{i=1}^{7}[(M_i - O_i)/O_i]^2}{N}} \quad (10)$$

$$SMAPE = \frac{1}{N}\sum_{1=1}^{7}\left|\frac{M_i - O_i}{\frac{(M_i+O_i)}{2}}\right| \quad (11)$$

$$SRMSE = \sqrt{\frac{\sum_{i=1}^{7}\left\{\frac{M_i-O_i}{\left[\frac{(M_i+O_i)}{2}\right]}\right\}^2}{N}} \quad (12)$$

where the index i indicates the H_s class (i = 1, 2, ... , 7) and N = 7, the number of classes of H_s.

SPM '84 presented the lowest *af RMSE* (3.55%), indicating that the standard deviation of the *af* residuals, i.e., the model error was equal to 3.55% for events in the NW sector. The SWAN *af RMSE* (8.52%) was lower compared to NOAA N (13.27%) and S (13.20%), and ECMWF (9.36%), confirming the good performance of SWAN under shallow water conditions with respect to the global scale wave models, WAM/WW3, which are more suitable for oceanic large-scale applications.

The main drawbacks of the use of RMSE solely in calculating model performance are the scale dependency (if the model includes variables with different scales or magnitudes, then absolute error measures could not be applied), the high influence of outliers in data on the model performance evaluation, and the low reliability (the results could be different depending on the different fraction of data) [53]. For these reasons, Table 5 shows not only absolute model error statistical indicators, such as *BIAS* and *RMSE*, but also indicators based on percentage errors, such as *NBIAS* and *NRMSE*. The advantage of such indicators is that they do not depend on the scale of the observations. The disadvantages are that (i) they include a division by 0 if observed data values are very small, (ii) the very high weight of outliers in the final result, (iii) the "asymmetry issue", and (iv) the error values differ whether the predicted value is bigger or smaller than the actual. For these reasons, a third group of statistical indicators is considered: the so-called "symmetric error" indicators, which are the *SMAPE* and the *SRMSE*.

Table 5. Wave model errors' statistical indicators.

Dataset	*af BIAS*	*af RMSE*	*af NBIAS*	*af NRMSE*	*af SMAPE*	*af SRMSE*
SWAN	6.22	8.52	3.84	5.98	1.25	1.35
SPM '84	0.62	3.55	-0.37	0.86	1.23	1.42
NOAA N	6.06	13.27	0.59	2.42	1.27	1.44
NOAA S	5.33	13.20	0.46	2.43	1.42	1.54
ECMWF	6.48	9.36	4.80	6.07	1.19	1.27

Table 5 shows that the performance of SPM '84 was the second best considering the SMAPE, after only ECMWF, and followed closely by SWAN. This was due to the fact that both ECMWF and SWAN modelled the contribution of waves with higher H_s better than SPM '84 in the NW sector, while SPM'84 modelled the directional distribution better.

Analyzing the correlation in each class of H_s between the observations and each model, Figure 12 shows scatter plots of each hindcasting model *af* versus the observed *af*. Each subplot shows the values of the coefficient of determination, R^2 calculated as:

$$R^2 = 1 - \frac{SS_{res}}{SS_{tot}} \tag{13}$$

where SS_{res} is the sum of squares of residuals while SS_{tot} is the total sum of squares (proportional to the variance of the data).

The best performance in terms of the correlation was shown by ECMWF (R^2 = 0.856) and SWAN (R^2 = 0.831); SPM '84 was third (R^2 = 0.618).

To facilitate the comparative assessment of the different hindcasting models, Figure 13 shows a Taylor diagram [54] representing the performance of each model. Taylor diagrams are used to quantify the degree of correspondence between the modelled and observed behavior in terms of three statistics: the Pearson correlation coefficient, related to the azimuthal angle (in blue), the *RMSE* (green), and the standard deviation (black). The Pearson correlation coefficient (gauging similarity in pattern between the simulated and observed *af*) is related to the azimuthal angle; the centered *RMSE* in the modelled *af* is proportional to the distance from the point on the x-axis identified as "reference"; and the standard deviation of the modelled *af* distribution is proportional to the radial distance from the origin. Therefore, Figure 13 shows that all the models had a standard deviation (grey dotted contours) much larger than the observed data, which have a standard deviation of 3.03. The Pearson correlation coefficient was high for SWAN and ECMWF (>90%) compared to SPM '84 (80%), but SWAN and ECMWF also presented a much higher RMSE (>5%) and a larger standard deviation (>7.5%) compared to the SPM '84. Therefore, the Taylor diagram showed that the SPM '84 exhibited the best performance among the considered models.

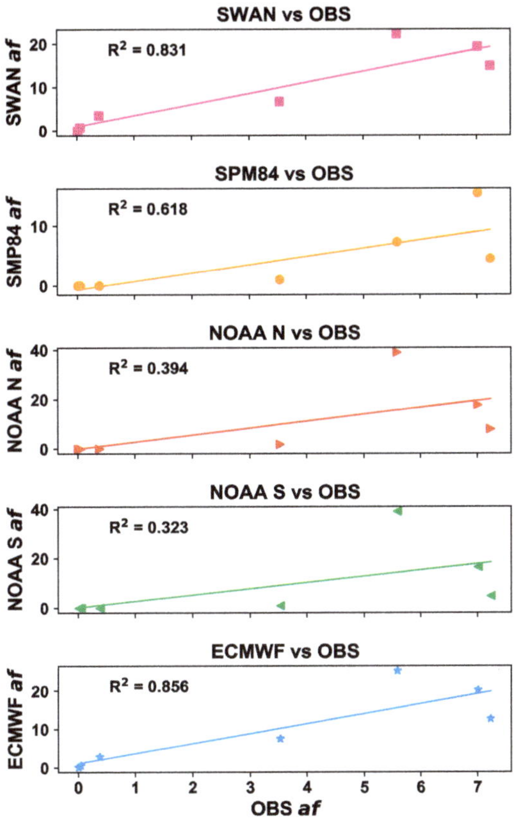

Figure 12. Scatter plots of each hindcasted model *af* versus the observed *af*.

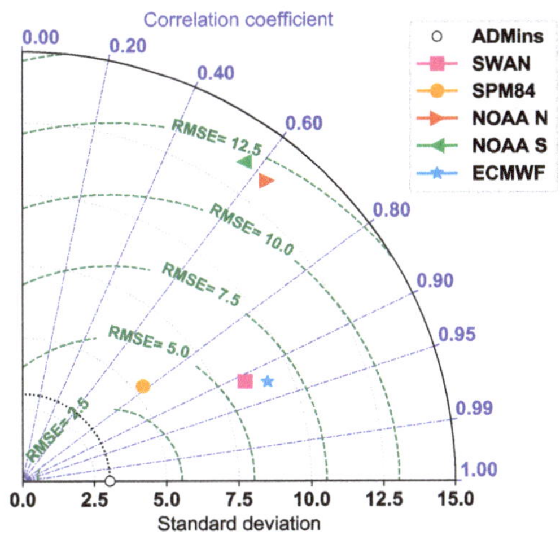

Figure 13. Taylor diagram of the performance of each of the considered models.

4. Discussion and Conclusions

Where the ADCP instrument was deployed, the observed wave climate had been obtained and compared with the wave climate determined by means of different models at the same location: wave propagation of the NOAA offshore wave dataset conducted by the SWAN numerical model, assuming wave conditions at the NOAA offshore grid point as boundary conditions; the NOAA and ECMWF wave dataset at the closest grid point; and the SPM '84 [1] hindcasting method with the closest NOAA wind dataset used as input. The analysis of the wave characteristics has been expressed in terms of distributions of individual wave heights and directions.

The predictive capability of the SPM '84 has been favorably verified against the observed and calculated wave climates. The estimated *af RMSE* showed that SPM '84 resulted in a better agreement with the observed data compared to the other investigated models, and SPM '84 exhibited the smallest *af RMSE* (3.55%), followed by SWAN (8.52%), ECMWF (9.36%), and NOAA N and S nodes (13.27% and 13.20%, respectively). Considering a symmetric statistical indicator, such as the SMAPE, SPM '84 showed a comparable performance to ECMWF and SWAN. The reason for this was that ECMWF and SWAN simulated waves well in the NW sector with high H_s. The principal limitation of the latter two models was their limited representation of the wave directional distribution; mainly due to the assumed NOAA offshore directional distribution and to the limited bathymetrical grid, respectively.

Although SWAN, similarly to WW3 and WAM, used by NOAA and ECMWF, respectively, was a third-generation wave model, it showed better performance in determining the wave characteristics at Saadiyat with respect to NOAA and ECMWF models. This was due to the fact that SWAN contains some additional parameterizations primarily for shallow water [46], different numerical techniques, and different formulations for the wind input and the white-capping with respect to WW3. In addition, it used the higher resolution computational grid and bathymetry in the model implementation, as was shown in Table 3.

A Taylor diagram (Figure 13) of the overall performance of each of the considered wave hindcasting methods showed that the best agreement with the observed wave climate in the vicinity of Saadiyat beach at Abu Dhabi was obtained using the SPM '84 hindcasting method for the shallow water condition. On the contrary, more sophisticated atmosphere–ocean numerical models, such as those used by NOAA and ECMWF, presented some limitation.

The SPM '84 best performance could also be due to the fact that the SPM '84 was based on exact equations (cfr. Equations (3)–(6)) and exact input terms, such as fetches, water depths, wind characteristics, and therefore, not influenced by the resolution of the computational grid or the bathymetric information. The SPM '84 could be improved, aiming at better modelling waves with high H_s by recalculating the numerical factors that appear in Equations (3)–(6), or calibrating the method before the application by using a subset of the wave observations. In the case of Saadiyat beach, it was not possible to perform this calibration for the time being due to the limited statistics of wave observations collected so far.

The reliability in wave modelling from the application of the SPM '84 can lead to a better planning of the ambitious coastal interventions that are foreseen in the Gulf area in the next future.

Author Contributions: G.R.T. and L.L. collected all the data, conceived and designed the application of the SPM '84 method; L.L. analyzed the data, and implemented SWAN and the application of the SPM '84 method; F.L., W.H., and F.D'A. contributed to the discussion regarding the results; G.R.T. and L.L. wrote the paper.

Funding: This work was funded by the United Arab Emirates University research grant, through the National Water Center, Grant # 31R115; "Impact of coastal (long-shore) currents on erosion/deposition and consequent water/sediments quality variations along the coastal area of Abu Dhabi City" and by the Apulia Region (Italy), through the Regional Cluster Projects "Start", Grant # 0POYPE3 and "Eco-Smart Breakwater", Grant # S6LU5I7.

Acknowledgments: The authors thank the Abu Dhabi Municipality for providing in situ data relative to wave conditions at Saadiyat beachfront.

Conflicts of Interest: The authors declare no conflict of interest and the funding sponsors had no role in the design of the study; in the collection, analyses, or interpretation of data; in the writing of the manuscript; or in the decision to publish the results.

References

1. United States Army, Corps of Engineers; Coastal Engineering Research Center (USACE). *Shore Protection Manual*; Department of the Army, Waterways Experiment Station, Corps of Engineers, Coastal Engineering Research Center: Washington, DC, USA, 1984.
2. Goda, Y. *Random Seas and Design of Maritime Structures*, 3rd ed.; Advanced Series on Ocean Engineering; World Scientific: Singapore, 2010.
3. Van der Meer, J.W. Rock Slopes and Gravel Beaches under Wave Attack. Ph.D. Thesis, Delft University of Technology, Delft, The Netherlands, 1988.
4. Tomasicchio, G.R.; D'Alessandro, F. Wave energy transmission through and over low crested breakwaters. *J. Coast. Res.* **2013**, *1*, 398–403. [CrossRef]
5. Cavaleri, L.; Bertotti, L.; Lionello, P. Shallow water application of the third-generation WAM wave model. *J. Geophys. Res.* **1989**, *94*, 8111–8124. [CrossRef]
6. Stopa, J.E.; Ardhuin, F.; Babanin, A.; Zieger, S. Comparison and validation of physical wave parameterizations in spectral wave models. *Ocean Model.* **2016**, *103*, 2–17. [CrossRef]
7. Rakha, K.; Al-Salem, K.; Neelamani, S. Hydrodynamic Atlas for the Arabian Gulf. *J. Coast. Res.* **2007**, *50*, 550–554.
8. Mazaheri, S.; Ghaderi, Z. Shallow Water Wave Characteristics in Persian Gulf. *J. Coast. Res.* **2011**, *64*, 572–575.
9. Anoop, T.R.; Sanil Kumar, V.; Shanas, P.R. Spatial and temporal variation of surface waves in shallow waters along the eastern Arabian Sea. *Ocean Eng.* **2014**, *81*, 150–157. [CrossRef]
10. Booij, N.; Ris, R.C.; Holthuijsen, L.H. A third-generation wave model for coastal regions, Part I, Model description and validation. *J. Geophys. Res.* **1999**, *104*, 7649–7666. [CrossRef]
11. Ris, R.C.; Booij, N.; Holthuijsen, L.H. A third-generation wave model for coastal regions, Part II, Verification. *J. Geophys. Res.* **1999**, *104*, 7667–7681. [CrossRef]
12. Holthuijsen, L.H. *Waves in Oceanic and Coastal Waters*; Cambridge University Press: Cambridge, MA, USA, 2007; p. 387. ISBN 978-0521860284.
13. Gorrell, L.; Raubenheimer, B.; Elgar, S.; Guza, R.T. SWAN predictions of waves observed in shallow water onshore of complex bathymetry. *Coast. Eng.* **2011**, *58*, 510–516. [CrossRef]
14. Brown, J.M. A case study of combined wave and water levels under storm conditions using WAM and SWAN in a shallow water application. *Ocean Model.* **2010**, *35*, 215–229. [CrossRef]
15. Dally, W.R. Comparison of a mid-shelf wave hindcast to ADCP-measured directional spectra and their transformation to shallow water. *Coast. Eng.* **2018**, *131*, 12–30. [CrossRef]
16. Muliati, Y.; Lukman Tawekal, R.; Wurjanto, A.; Kelvin, J.; Pranowo, W.S. Application of SWAN model for hindcasting wave height in Jepara coastal waters, North Java, Indonesia. *Int. J. GEOMATE* **2018**, *15*, 114–120. [CrossRef]
17. Xu, F.; Perrie, W.; Solomon, S. Shallow Water Dissipation Processes for Wind Waves off the Mackenzie Delta. *Atmos.-Ocean* **2013**, *51*, 296–308. [CrossRef]
18. Rogers, W.E.; Kaihatu, J.M.; Hsu, L.; Jensen, R.E.; Dykes, J.D.; Todd Holland, K. Forecasting and hindcasting waves with the SWAN model in the Southern California Bight. *Coast. Eng.* **2007**, *54*, 1–15. [CrossRef]
19. Bottema, M.; van Vledder, G.P. A ten-year data set for fetch- and depth-limited wave growth. *Coast. Eng.* **2009**, *56*, 703–725. [CrossRef]
20. Goda, Y. Spread Parameter of Extreme Wave Height Distribution for Performance-Based Design of Maritime Structures. *J. Waterw. Port C ASCE* **2004**, *130*. [CrossRef]
21. Gencarelli, R.; Tomasicchio, G.R.; Veltri, P. Wave height long term predictions based on the use of the spread parameter. In *Coastal Engineering Proceedings, Proceedings of the 30th Conference on Coastal Engineerings (ICCE), Dan Diego, CA, USA, 3–8 September 2006*; McKee Smith, J., Ed.; World Scientific: Stevens Point, WI, USA, 2007; pp. 701–713.
22. Saha, S.; Moorthi, S.; Pan, H.L.; Wu, X.; Wang, J.; Nadiga, S.; Tripp, P.; Kistler, R.; Woollen, J.; Behringer, D.; et al. The NCEP Climate Forecast System Reanalysis. *Bull. Am. Meteorol. Soc.* **2010**, *91*, 1015–1057. [CrossRef]
23. Amante, C.; Eakins, B.W. ETOPO1 1 Arc-Minute Global Relief Model: Procedures, Data Sources and Analysis. In *NOAA Technical Memorandum NESDIS*; NGDC-24; National Geophysical Data Center: Boulder, CO, USA, 2009; p. 19.

24. Tolman, H.L. Validation of a new global wave forecast system at NCEP. In *Ocean Wave Measurements and Analysis*; Edge, B.L., Helmsley, J.M., Eds.; ASCE: Reston, VA, USA, 1998; pp. 777–786.
25. Tolman, H.L. *User Manual and System Documentation of WAVEWATCH-III Version 2.22*; NOAA/NWS/NCEP/OMB Technical Note; US Department of Commerce: Washington, DC, USA, 2002; Volume 222, p. 133.
26. Tolman, H.L. *Testing of WAVEWATCH III Version 2.22 in NCEP's NWW3 Ocean Wave Model Suite*; NOAA/NWS/NCEP/OMB Technical Note; US Department of Commerce: Washington, DC, USA, 2002; Volume 214, p. 99.
27. WAVEWATCH III 30-Year Hindcast Wave Model Developed by NOAA (Phase 2). Available online: http://polar.ncep.noaa.gov/waves/hindcasts/nopp-phase2.php (accessed on 4 October 2017).
28. The Climate Forecast System Reanalysis [1979–2010]. Available online: http://cfs.ncep.noaa.gov/cfsr/ (accessed on 4 October 2018).
29. Berrisford, P.; Dee, D.; Poli, P.; Brugge, R.; Fielding, K.; Fuentes, M.; Kållberg, P.; Kobayashi, S.; Uppala, S.; Simmons, A. *The ERA-Interim Archive Version 2.0*; ERA Report Series; ECMWF: Shinfield Park, UK, 2011; Volume 1.
30. Dee, D.P.; Uppala, S.M.; Simmons, A.J.; Berrisford, P.; Poli, P.; Kobayashi, S.; Andrae, U.; Balmaseda, M.A.; Balsamo, G.; Bauer, D.P.; et al. The ERA-Interim reanalysis: Configuration and performance of the data assimilation system. *Q. J. R. Meteorol. Soc.* **2011**, *137*, 553–597. [CrossRef]
31. WAMDI Group. The WAM Model—A Third Generation Ocean Wave Prediction Model. *J. Phys. Oceanogr.* **1988**, *18*, 1775–1810. [CrossRef]
32. Komen, G.J.; Cavaleri, L.; Donelan, M.; Hasselmann, K.; Hasselmann, S.; Janssen, P.A.E.M. *Dynamics and Modelling of Ocean Waves*; Cambridge University Press: Cambridge, MA, USA, 1994; p. 532.
33. ECMWF. *Part VII: ECMWF Wave Model*; ECMWF IFS Documentation—Cy43r3 Operational Implementation 11 July 2017; Technical Report; ECMWF: Reading, UK, 2017.
34. ERA-Interim. Available online: https://www.ecmwf.int/en/forecasts/datasets/reanalysis-datasets/era-interim (accessed on 10 April 2018).
35. Kamphuis, J.W. *Introduction to Coastal Engineering and Management*; World Scientific: Stevens Point, WI, USA, 2000; p. 437. ISBN 978-981-283-484.
36. Seymour, R.J. Engineers Estimating Wave Generation on Restricted Fetches. *ASCE J. Waterw. Port. C* **1977**, *103*, 251–264.
37. Smith, J. *Wind-Wave Generation on Restricted Fetches*; Department of the Army, Waterways Experiment Station, Corps of Engineers, Coastal Engineering Research Center: Washington, DC, USA, 1991.
38. Karimpour, A.; Chen, Q.; Twilley, R.R. Wind Wave Behavior in Fetch and Depth Limited Estuaries. *Sci. Rep.* **2017**, *7*, 40654. [CrossRef] [PubMed]
39. Leenknecht, D.A.; Szuwalski, A.; Sherlock, A.R. *Automated Coastal Engineering System (ACES)*; Department of the Army, Waterways Experiment Station, Corps of Engineers, Coastal Engineering Research Center: Washington, DC, USA, 1992; p. 373.
40. Hanley, K.E.; Belcher, S.E.; Sullivan, P.P. A Global Climatology of Wind–Wave Interaction. *J. Phys. Oceanogr.* **2010**, *40*, 1263–1282. [CrossRef]
41. Freilich, M.; Guza, R. Nonlinear effects on shoaling surface gravity waves. *Philos. Trans. R. Soc. Lond. A* **1984**, *311*, 1–41. [CrossRef]
42. Elgar, S.; Guza, R. Observations of bispectra of shoaling surface gravity waves. *J. Fluid Mech.* **1985**, *161*, 425–448. [CrossRef]
43. Madsen, O.S.; Poon, Y.-K.; Graber, H.C. Spectral wave attenuation by bottom friction: Theory. In Proceedings of the 21th International Conference on Coastal Engineering, Costa del Sol-Malaga, Spain, 20–25 June 1988; ASCE: Reston, VA, USA, 1988; pp. 492–504.
44. Battjes, J.; Janssen, J. Energy loss and set-up due to breaking of random waves. In Proceedings of the 16th International Conference on Coastal Engineering, Hamburg, Germany, 27 August–3 September 1978; pp. 569–587.
45. Battjes, J.; Stive, M. Calibration and verification of a dissipation model for random breaking waves. *J. Geophys. Res.* **1985**, *90*, 9159–9167. [CrossRef]
46. The SWAN Team. *Scientific and Technical Documentation SWAN Cycle III Version 41.20A*; Delft University of Technology: Delft, The Netherlands, 2018.

47. Booij, N.; Haagsma, I.J.; Holthuijsen, L.; Keiftenburg, A.; Ris, R.; van der Westhuysen, A.; Zijlema, M. *SWAN Cycle-III Version 40.41 User Manual*; Delft University of Technology: Delft, The Netherlands, 2004; 39p.
48. GEBCO. The GEBCO_2014 Grid, Version 20150318. Available online: www.gebco.net (accessed on 19 July 2018).
49. Weatherall, P.; Marks, K.M.; Jakobsson, M.; Schmitt, T.; Tani, S.; Arndt, J.E.; Rovere, M.; Chayes, D.; Ferrini, V.; Wigley, R. A new digital bathymetric model of the world's oceans. *Earth Space Sci.* **2015**, *2*, 331–345. [CrossRef]
50. Komen, G.J.; Hasselmann, K.; Hasselmann, H. On the Existence of a Fully Developed Wind-Sea Spectrum. *J. Phys. Oceanogr.* **1984**, *14*, 1271–1285. [CrossRef]
51. Hasselmann, K.; Barnett, T.P.; Bouws, E.; Carlson, H.; Cartwright, D.E.; Enke, K.; Ewing, J.A.; Gienapp, H.; Hasselmann, D.E.; Kruseman, P.; et al. Measurements of wind-wave growth and swell decay during the Joint North Sea Wave Project (JONSWAP). *Ergnzungsheft zur Deutschen Hydrographischen Zeitschrift Reihe* **1973**, *8*, 95.
52. Sontek. *2000 SonTek/YSI ADP®Acoustic Doppler Profiler Technical Documentation*; SonTek/YSI: San Diego, CA, USA, 2000.
53. Shcherbakov, M.V.; Brebels, A.; Shcherbakova, N.L.; Tyukov, A.P.; Janovsky, T.A.; Kamaev, V.A. A Survey of Forecast Error Measures. *World Appl. Sci. J.* **2013**, *24*, 171–176. [CrossRef]
54. Taylor, K.E. Summarizing multiple aspects of model performance in a single diagram. *J. Geophys. Res.* **2001**, *106*, 7183–7192. [CrossRef]

water

Article

Multi-Collocation-Based Estimation of Wave Climate in a Non-Tidal Bay: The Case Study of Bagnoli-Coroglio Bay (Tyrrhenian Sea)

Pasquale Contestabile [1,2], Fabio Conversano [3], Luca Centurioni [4], Umberto Mario Golia [1], Luigi Musco [3], Roberto Danovaro [3,5] and Diego Vicinanza [1,3,*]

1 Department of Engineering, University of Campania Luigi Vanvitelli, 81031 Aversa, Italy; pasquale.contestabile@unicampania.it (P.C.); UmbertoMario.golia@unicampania.it (U.M.G.)
2 CoNISMa National Inter-University Consortium of Marine Sciences, Piazzale Flaminio 9, 00196 Roma, Italy
3 Stazione Zoologica Anton Dohrn, Villa Comunale, 80121 Napoli, Italy; fabio.conversano@szn.it (F.C.); luigi.musco@szn.it (L.M.); r.danovaro@univpm.it (R.D.)
4 Lagrangian Drifter Laboratory, Scripps Institution of Oceanography, University of California San Diego, 9500 Gilman Drive, MC 0213, La Jolla, CA 92093, USA; lcenturioni@ucsd.edu
5 Department of Life and Environmental Sciences (DiSVA), Polytechnic University of Marche, Via Brecce Bianche, 60100 Ancona, Italy
* Correspondence: diego.vicinanza@unicampania.it; Tel.: +39-328-482-0770

Received: 24 May 2020; Accepted: 2 July 2020; Published: 7 July 2020

Abstract: In this paper, the advantages of shaping a non-conventional triple collocation-based calibration of a wave propagation model is pointed out. Illustrated through a case study in the Bagnoli-Coroglio Bay (central Tyrrhenian Sea, Italy), a multi-comparison between numerical data and direct measurements have been carried out. The nearshore wave propagation model output has been compared with measurements from an acoustic Doppler current profiler (ADCP) and an innovative low-cost drifter-derived GPS-based wave buoy located outside the bay. The triple collocation—buoy, ADCP and virtual numerical point—make possible an implicit validation between instrumentations and between instrumentation and numerical model. The procedure presented here advocates for an alternative "two-step" strategy. Indeed, the triple collocation technique has been used solely to provide a first "rough" calibration of one numerical domain in which the input open boundary has been placed, so that the main wave direction is orthogonally aligned. The need for a fast and sufficiently accurate estimation of wave model parameters (first step) and then an ensemble of five different offshore boundary orientations have been considered, referencing for a more detailed calibration to a short time series of a GPS-buoy installed in the study area (second step). Such a stage involves the introduction of an enhancement factor for the European Centre for Medium-Range Weather Forecasts (ECMWF) dataset, used as input for the model. Finally, validation of the final model's predictions has been carried out by comparing ADCP measurements in the bay. Despite some limitations, the results reveal that the approach is promising and an excellent correlation can be found, especially in terms of significant wave height.

Keywords: wave numerical model; directional wave spectra drifter (DWSD); ADCP; GPS wave buoy; triple collocation; Bagnoli-Coroglio Bay

1. Introduction

1.1. Motivations and Perspective

Many coastal engineering applications require robust estimates of the "design sea state" with a certain return period, and incorrect estimates can have dramatic effects on the flood risk analysis or on the structural design of maritime structures. Therefore, trustworthy and robust wave datasets are required [1,2]. In the last few decades, satellite observations and meteorological reanalysis have resulted in considerable improvements in weather and wave climate forecasting. Their use is gradually increasing, a day at a time. Moreover, in Italy, where there is a long history of wave measurement [3], datasets such as those provided by the European Centre for Medium-Range Weather Forecasts (ECMWF) [4] have become widely used to improve/substitute the dataset provided by the Italian Wave Buoy Network. The reasons can be addressed as:

- the presence of some missing values (the percentage of missing data can severely reduce the representativeness of the sample and disturb the conclusions drawn from the dataset) [5];
- the spatial resolution, the Italian Wave Network (IWN) consisting of only 15 stations positioned along the more than 7000 km of Italian coasts [6];
- the temporal window, since the oldest buoy of the IWN in operation from 1989 until 2014 [1,7].

As result, nowadays, ECMWF, which covers the period from 1 January 1979 onward and continuously extends forward in near real time, is assumed as the only source for wave climate assessment [8–14]. Several papers have discussed how to validate hindcast data (e.g., [15–24]). For coastal engineers and marine scientists, it is important to take into account the tendency to underestimate significant wave height values during severe storm conditions performed by the ECMWF dataset, as evidenced in several studies [6,8,25–35]. Biased estimates of wave heights will affect [36,37] both long-term return level estimates for extreme wave analysis and the short/medium time wave climate in nearshore areas, resulting from the wave model being forced with a hindcast dataset.

The detailed validation of the ECMWF hindcast model and coastal propagation model are beyond the scope of the present paper. The goals of this study are to:

- perform a comparison of different calibration stages based on the triple collocation method [38–49] and on the use of multiple numerical geographical domains with different orientations;
- highlight the discrepancies and errors in the use of different sources of wave data for both offshore and nearshore wave climate analysis from the perspective of coastal engineering measures and especially in the assessment of non-extreme wave conditions.

The latter are of particular importance for the study site, the Bagnoli-Coroglio Bay, because it represents one of the most polluted areas in the world but is nestled between two marine protected areas (the Gaiola and the Baia marine protected areas). This large bay at the north-western end of the Gulf of Naples (Tyrrhenian Sea) is included within the contaminated Sites of National Interest (SIN) for the high levels of environmental contamination by heavy industrial activities by the Ilva, Eternit, Cementir and Federconsorzi industrial factories and plants [50,51]. Due to the limited exchange of water, the accumulation of pollutants poses major concerns for human and environmental health [52]. In 2015, the Italian central government took over the planning competences over the area. By the end of 2015, the remediation of soil surfaces and the marine area has not yet been completed. In the former industrial area, most of the buildings have been demolished, while the surface and subsoil have been remediated by only 50%. In the southern portion of the area, which hosted the asbestos industry, only 30% of the remediation has been completed [53]. In the period of 2016–2018, researchers collected updated information to develop the next phase of the restoration project. This research phase was granted by the ABBaCo project ("Sperimentazioni pilota finalizzate a restauro Ambientale e Balneabilità del SIN Bagnoli-Coroglio") [54,55], in which the present study takes form.

In addition to wave data for both littoral drift/shoreline modeling, future wave climate assessment should include other detailed eco-hydraulic analyses in order to respond to the nascent requests of marine biologists and ecologists (e.g., the coupled turbulence-dissolved oxygen dynamics modeling and forecasting, [56]; projected changes in wave climate [57,58], nearshore velocity field and related dynamics of deep chlorophyll [59–63], habitat mapping purposes [64–68] and ecosystem-based coastal defence [69–73]). Therefore, results from a high-resolution coastal propagation model have been compared with the in situ measurements of an innovative economical GPS-based wave buoy and with an acoustic Doppler current profiler (ADCP) in order to calibrate the numerical model itself. The measurements have been carried out by placing the pair of instruments very close to each other. In particular, a wave buoy called the directional wave spectra drifter, (DWSD) designed and fabricated by the Lagrangian Drifter Laboratory (LDL) of the Scripps Institution of Oceanography (SIO) [74], is examined, exploring its significant potential use in a low-cost drifter for measuring waves in coastal areas.

1.2. Approach and Challenges

The triple collocation—DWSD buoy, ADCP and virtual numerical point—makes possible an implicit validation between instrumentations and between instrumentation and numerical model. Considering the recent depletion of the IWN, as well as in all of Europe, mainly due to the high costs of maintenance of the traditional wave buoy systems, the opportunity to develop cost-effective and sustainable technologies to monitor waves is of strong interest to researchers and engineers. In the last decade, global positioning system (GPS) technology has been introduced in wave buoys as a cheaper alternative to traditional instruments which mainly utilized accelerometers to measure the pitch, heave and roll of the buoy [75–80].

Technological advancements of the GPS receivers have helped the development of reliable GPS-tracked wave buoys, which are currently gradually complementing conventional sensor-based wave buoys, offering the same high-quality data as traditional, well-established, accelerometer-based buoys such as the Datawell directional wave-rider buoys [81–93]. GPS technology has also been largely adopted in the development of surface drifters that track the world ocean surface circulation [86–88], while other authors [89–91] have recently proposed that the GPS drifter is particularly suited for nearshore or surf-zone applications. The use of a GPS receiver, as opposed to an autonomous instrument package, results not only in considerable cost saving but it enables also the development of smaller buoys, which can be easily transported, deployed and handled from a small boat. This wave buoy has been developed, moving on from the experience acquired from the Global Drifter Program (GDP) [86–89]. Its small size (40-cm diameter) also has the advantage of coping with a higher wave frequency, extending the range of measurement [93].

The idea to have multiple lines of evidence agree has always fascinated climate scientists and ocean modelers, and a cluster of wave buoys goes right in that direction. Therefore, this work describes an experience of a calibration procedure in which multiple numerical simulations, called ensembles, are calibrated by means of the DSWD buoy.

The method presented in this work allows an enclosing calibration procedure to be a building block in a single two-step approach. The triple collocation technique (applied in a point outside the study area) has been used solely to provide a first "rough" calibration. Having this fast calibration (first step), then (second step) the tuning of wave parameters in the numerical model, is refined by an ensemble of five numerical domains running in different wave sectors, in which time series are compared with another DWSD buoy located within the study site. Finally, we demonstrate the method via direct comparison with the wave time series measured by an ADCP installed in the bay. The final dataset obtained from the calibrated model has been used to describe the local wave climate. Quali-quantitative considerations from the whole historical dataset are drawn. The results suggest that the numerical model's calibrations, based on short-term wave buoy measurements, can be easily applied in different areas where detailed wave data are not available.

The paper is organized as follows: the next section provides detailed information on the hindcast model and the instruments used at the study site, as well as a description of the numerical model and the underpinning assumptions that were used to carried out the calibration. In Section 3, the validation results of the DSWD buoy against the ADCP are reported. Moreover, the final dataset obtained from the calibrated model has been used to describe the local annual wave climate. Sources of uncertainty, relevant shortcomings and contradictions between the calibrated and uncalibrated numerical model are also highlighted. Section 4 is devoted to an overall discussion, with remarks on the future perspective. Finally, some conclusions are drawn.

2. Wave Data and Methodology

2.1. Offshore Wave Dataset

The present work has been based on two sources of offshore wave data: wave buoy records and hindcast data. The first have been supplied by pitch-roll type directional buoys operating offshore in Ponza (central Tyrrhenian Sea). The records are available from 1 July 1989 [94,95], as a part of the IWN. From 1989 to about 2002, the wave buoys collected 30 min of wave measurements every 3 h, but when in the presence of wave heights greater than 1.5 m, the measurements were continuous. From 2002 to 31 December 2014, the wave measurements were always continuous and the wave characteristic parameters refer to 30-min time intervals. In any case, the dataset comprises the spectrum zero-moment wave height (H_{m0}), the mean wave period (T_m) and the mean wave direction (θ).

A gross stochastic error detection phase has been applied. The data processing has firstly regarded the missing data problem. Missing values reduce the representativeness of the sample and they can severely disturb the conclusions drawn from the data. For Ponza buoys, about 10% missing data, covering about 20 years of observation, have been detected. In order to get a conservative estimation in case of a lack in the time series, missing data or values of wave height of less than 0.2 m for several hours have been considered as errors and removed. However, to test the sensitivity of the results, H_{m0} = 1 m and 2 m have also been used. This analysis has shown that the estimated wave energy flux does not differ substantially (i.e., less than 12%) if wave heights of 1 m or 2 m are used to fill the missing data. Therefore, by considering missing data, unrealistic calm conditions and spikes, of the approximately 126 thousand available data of the whole dataset, only 96,879 values were considered useful.

In addition to these buoy records, the dataset was compared/complemented with the ECMWF dataset [4], in which historical observational data spanning an extended period are implemented through a single consistent analysis in forecast models. The ECMWF dataset is composed of a coupled ocean atmosphere and a general circulation model, i.e., an atmospheric reanalysis coupled with a wave model integration where no wave parameters are assimilated, making the wave part a hindcast run.

The dataset used is termed ERA-Interim, continuously updated in real time. Significant wave height (H_{m0}), mean period (T_m) and mean direction (θ), ranging from January 1979 to December 2018, were extracted from the ERA-Interim archive, available for download online [46].

The ECMWF internal WAve Model (WAM) covers the Mediterranean Sea by a base model grid with a resolution of 0.75° × 0.75°. ERA-Interim and WAM products are publicly available on the ECMWF Data Server. The WAM provides wave characteristics assimilated every 6 h. Here, 12 grid points (E1–E12) were considered. The geographical coordinates and distance from the seabed of all offshore points that are of interest to the present study are shown in Table 1. The position of point O, as representative of the "offshore" of Gulf of Naples, and of point W (offshore Pozzuoli's Gulf), are also reported. Geographical information is graphically represented in Figure 1.

Table 1. Geographical information of ECMWF grid points E1–E12, Ponza wave buoy and reference point O (offshore of the Gulf of Naples) and point W (offshore Pozzuoli's Gulf).

Point	Latitude	Longitude	Depth
E1	41.25 N	12.75 E	122
E2	40.50 N	12.75 E	3601
E3	40.50 N	13.50 E	1688
E4	40.50 N	14.25 E	1017
E5	39.75 N	12.75 E	3591
E6	39.75 N	13.50 E	3072
E7	39.75 N	14.25 E	2377
E8	39.75 N	15.00 E	1755
E9	39.00 N	12.75 E	3020
E10	39.00 N	13.50 E	3179
E11	39.00 N	14.25 E	3438
E12	39.00 N	15.00 E	2781
Ponza buoy	40°52′0.10″ N	12°57′0.00″ E	100
O	40°29′45.06″ N	13°47′46.70″ E	1037
W	40°45′56.49″ N	14° 7′41.22″ E	100

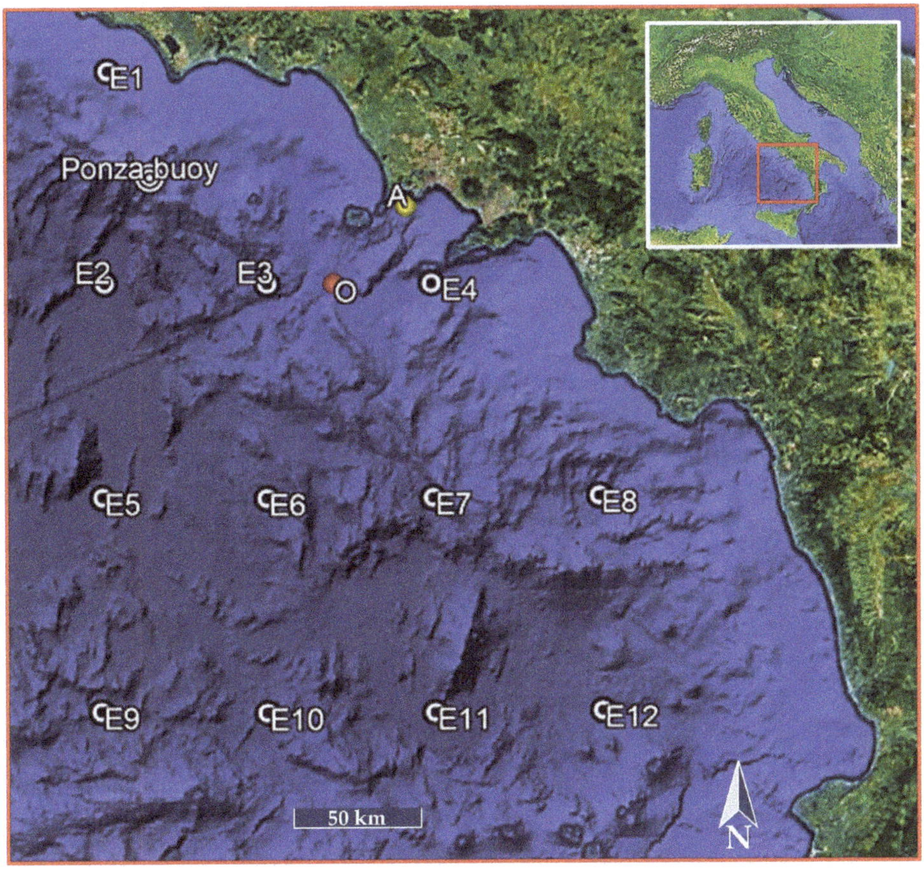

Figure 1. Map of mid-Tyrrhenian Sea, showing the location of the Ponza wave buoy, European Centre for Medium-Range Weather Forecasts (ECMWF) grid points E1–E12 and reference point O (offshore of the Gulf of Naples) and point W (offshore Pozzuoli's Gulf).

2.1.1. Comparison of Offshore Wave Data

The assessment of the whole dataset available for the Ponza wave buoy and for ECMWF time (point E3) is graphically represented with polar diagrams assembled in Figure 2. The wave dataset obtained from the Ponza wave buoy by means of the "geographic transposition of wave gauge data" to point O is also reported. The geographic transposition has been applied according to the method originally formulated by Contini and De Girolamo [96].

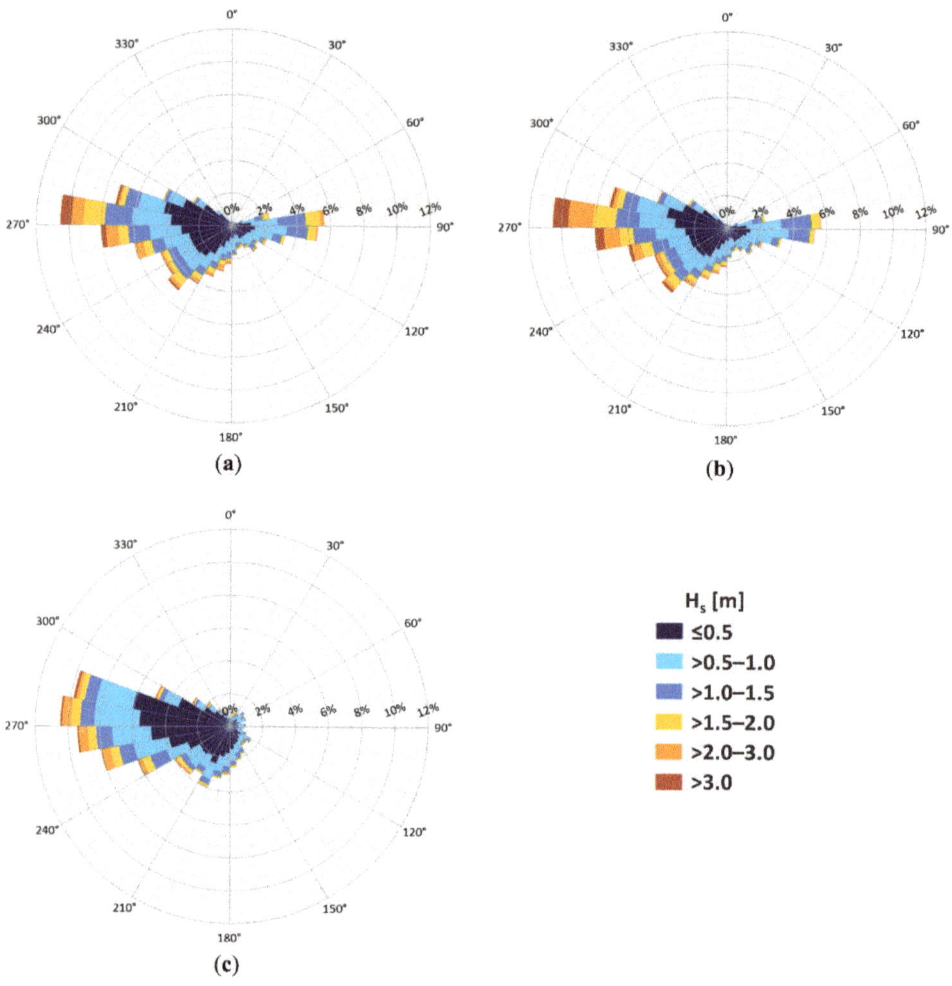

Figure 2. Wave climate referenced to different wave height classes (in legend): (**a**) at Ponza buoy; (**b**) at point O, as obtained by transposition of Ponza wave buoy data; (**c**) at ECMWF point E3.

The method is based on the following hypotheses:

(a) the wind speed and direction are the same at both real and "virtual" (transposed) stations;
(b) the extent of the wave generation region can be described by the effective fetches [97];
(c) the wind blows over the fetch long enough to assume that wave conditions are independent of the wind duration (fetch-limited conditions);
(d) both real and virtual stations are in deep water.

Under the above conditions, the spectral significant wave height H_{m0} and the peak period T_m can be estimated using the Sverdruv Munk Bretschneider (SMB) method [98]:

$$\frac{g \times H_{m0}}{U_A^2} = 1.6 \times 10^{-3} \times \left|\frac{g \times F}{U_A}\right|^{\frac{1}{2}} \quad (1)$$

$$\frac{g \times T_m}{U_A^2} = 2.857 \times 10^{-1} \times \left|\frac{g \times F}{U_A}\right|^{\frac{1}{3}} \quad (2)$$

where g is the gravity acceleration, U_A represents the wind-stress factor and F is the effective fetch.

Equations (1) and (2) can be written at real and virtual sites and, under the assumption that wind conditions are the same for both stations, the following equations can be derived:

$$\frac{H_{m0,V}}{H_{m0,R}} = \left|\frac{F_V}{F_R}\right|^{\frac{1}{2}} = K_H \quad (3)$$

$$\frac{T_{m,V}}{T_{m,R}} = \left|\frac{F_V}{F_R}\right|^{\frac{1}{2}} = K_T \quad (4)$$

where the subscripts R and V denote the variables referring to the real and virtual station, respectively.

The "transposition coefficients" K_H and K_T allow us to calculate from the real wave buoy records the transposed wave gauge dataset at the virtual station. For the sake of completeness, the virtual station O was selected between the ECMWF grid points E3 and E4 before the hindcast data analysis. In this way, once we selected the best ECMWF reference point from one of the two points, a sufficient comparison with the transposed dataset was ensured. More than 65% of the annual wave energy comes from the sector 220°–280°, in accordance with the long fetch facing the Gulf and with the mesoscale climate conditions, with swells approaching from distant storms coming from the NW sector of the Mediterranean Sea.

The comparison between the ECMWF dataset and buoy records (both real and transposed) shows some differences. In particular, the rate of waves coming from the east is significantly reduced. Then, the lowest values of energy in the ECMWF points are noticeable, particularly in the highest power class. It can be seen that the average wave power moves from 3.85 kW/m computed at the Ponza wave buoy (4.73 kW/m for point O) to 2.19 kW/m at point E3 (Table 2).

Table 2. Main wave climate parameters (based on the whole datasets) at Ponza wave buoy, point O and ECMWF grid point E3.

Dataset	$H_{s,mean}$ (m)	$H_{s,max}$ (m)	σ_H (m)	$T_{p,\ mean}$ (s)	$T_{p,\ mean}$ (s)	σ_T (s)	θ_m (°)	σ_θ (°)	$P_{,mean}$ (kW/m)
Ponza buoy	0.89	7.9	0.71	5.49	33.33	2.34	215.92	70.95	3.85
Point O	0.92	9.46	0.79	5.63	38.16	2.49	215.92	70.95	4.73
E3	0.65	6.133	0.53	5.07	11.54	1.44	222.62	76.51	2.19

Moreover, Table 3 shows the average differences between the measurements carried out by the buoy (at Ponza and after transposition at point O, respectively) and the hindcast data. Such values of $H_{s,PONZA\ BUOY}$ / $H_{s,ECMWF}$ and $H_{s,POINT\ O}$ / $H_{s,ECMWF}$ are organized by wave class (in terms of H_s ranges).

Table 3. Average differences between the buoy records (at Ponza and transposed at point O, respectively) and ECMWF hindcast data for point E3.

H_s (m)	$H_{s,PONZA\ BUOY}/H_{s,ECMWF}$ (m)	$H_{s,POINT\ O}/H_{s,ECMWF}$ (m)
<0.5	1.04	1.06
0.5–0.75	1.75	1.77
0.75–1	1.47	1.50
1–2	1.41	1.43
2–3	1.35	1.37
3–4	1.21	1.36
Mean	1.37	1.42

It is possible to note that for calm conditions (H_s < 0.5 m), the datasets are very similar. The highest discrepancy was found for 0.5 m < H_s < 0.75 m, where values of H_s recorded by the Ponza wave buoy (point O, respectively) were on average 1.75 times (1.77, respectively) higher than those reported for the E3 hindcast data. The overall mean discrepancy between the Ponza buoy and the ECMWF data was 1.37, while between point O and ECMWF, it was 1.42. A comparison of the wave height time series obtained from the different datasets is highlighted in Figure 3.

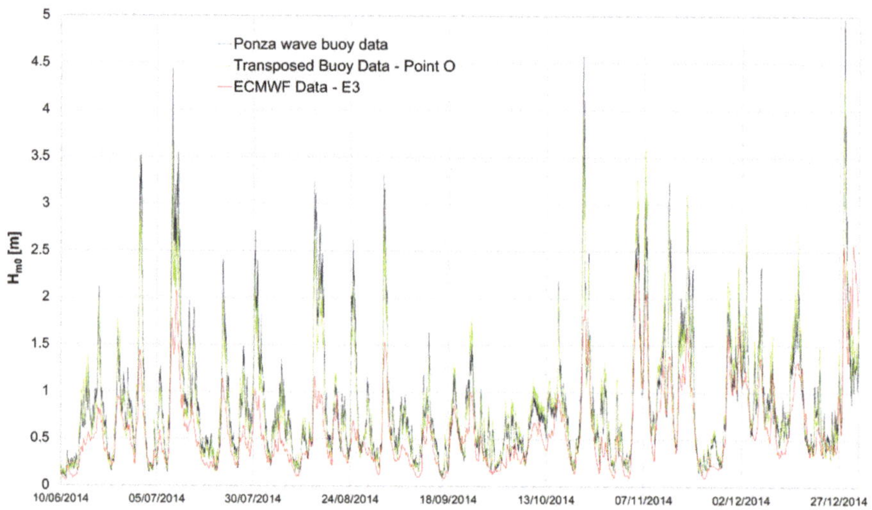

Figure 3. Comparison of the wave height obtained from buoy records, transposition of buoy records at point O and ECMWF data for the reference point E3.

The bulk of these differences can be attributed to the dissimilar measurement conditions. The smaller sampling frequency for the hindcast data involves peak attenuation, acting as a band-pass filter and smoothing the signal. The underestimate of the ECMWF data was previously highlighted within the WW-Medatlas projects [28]. Moreover, through intercomparison with NCEP (National Centers for Environmental Prediction) Climate Forecast System Reanalysis [26] and with wave buoy data, the lowest values of energy in the ECMWF points were detected, especially in the highest power class (e.g., [6,10,27,36]). Hence, the use of the ERA-Interim dataset could be considered adequate for slightly conservative wave power potential and studying long-term variations in wave height [10] but, at the same time, should be examined carefully during detailed resource assessments or for arriving at the design wave condition or to build a detailed nearshore wave model. The main parameters of the wave climate at each grid point are reported in Table 4.

Table 4. Main wave climate parameters (based on 39-year average) at ECMWF grid points.

Point	$H_{s,mean}$ (m)	$T_{m,\ mean}$ (s)	$\theta_{,\ mean}$ (°)	P_{mean} (kW/m)
E1	0.64	4.43	222.63	2.19
E2	0.77	4.71	229.82	2.94
E3	0.65	4.43	222.62	2.19
E4	0.67	4.55	237.02	2.49
E5	0.85	4.86	235.25	3.77
E6	0.79	4.83	237.98	3.69
E7	0.74	4.69	242.95	2.91
E8	0.69	4.68	240.89	2.56
E9	0.95	4.87	238.58	4.73
E10	0.95	4.87	238.59	4.11
E11	0.86	4.72	249.12	4.11
E12	0.75	4.70	254.91	3.13

A tentative contour map (based on interpolation of power rate at 12 grid points) has been provided in Figure 4, where wave power isolines are depicted, ranging from 2.5 to 5 kW/m.

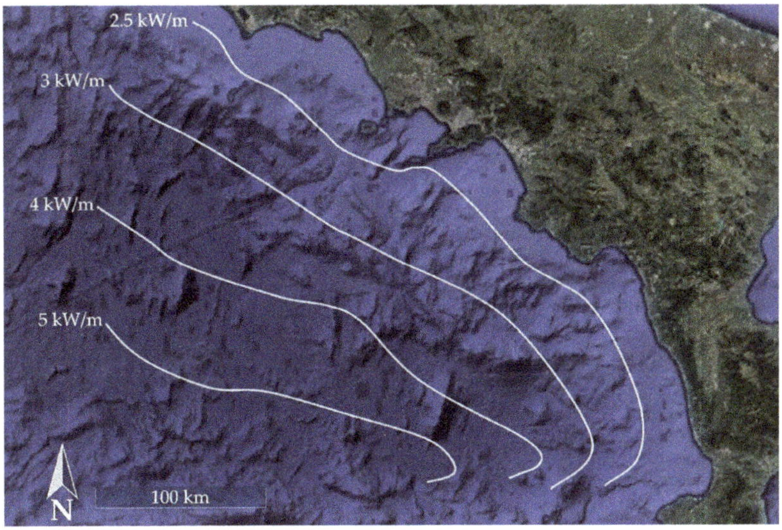

Figure 4. The 18-year averaged energy flux for the 12 ECMWF grid points and contour lines of the estimated mean wave power flux per unit crest on the central and southern Tyrrhenian Sea.

2.2. Study Site and Nearshore Wave Instrumentation

The nearshore study site is represented by the Bagnoli-Coroglio Bay, located within the Gulf of Naples, a natural semi-enclosed embayment within the Gulf of Pozzuoli (also known as the Gulf of Puteoli). Its mean depth is ca 60 m, with a maximum depth of 110 m and a surface area of 33 kmq.

Due to the proximity to the city of Naples, the whole area historically represents one of the best studied coastal areas of the Mediterranean Sea [99].

Thanks to the presence of the Stazione Zoologica "Anton Dohrn" (SZN) since 1872, marine investigations have been carried out for more than a century and half [100]. Recently, two Monitoring and Environmental Data Units (MEDA) of the SZN have been installed in the Gulf of Naples and in the Gulf of Pozzuoli (Figure 5). These MEDA units are mainly used for the chemical, biological and environmental monitoring of the marine ecosystem and both are equipped with an ADCP [61]. This shallow marine area is also famous, as it is the most highly active volcanic district in the coastal zone of SW Italy [101]. The geographical coordinates and water depth for MEDA A (Gulf of Pozzuoli) and MEDA B (Gulf of Naples) are indicated in Table 5. Close to MEDA B, a DWSD wave buoy (DWSD-B hereafter), provided by the University of Campania, was installed during a field campaign which took place from May to June 2016 [93]. Instead, in the period of February–March 2017, a DWSD buoy (DWSD-A) was placed close to MEDA A.

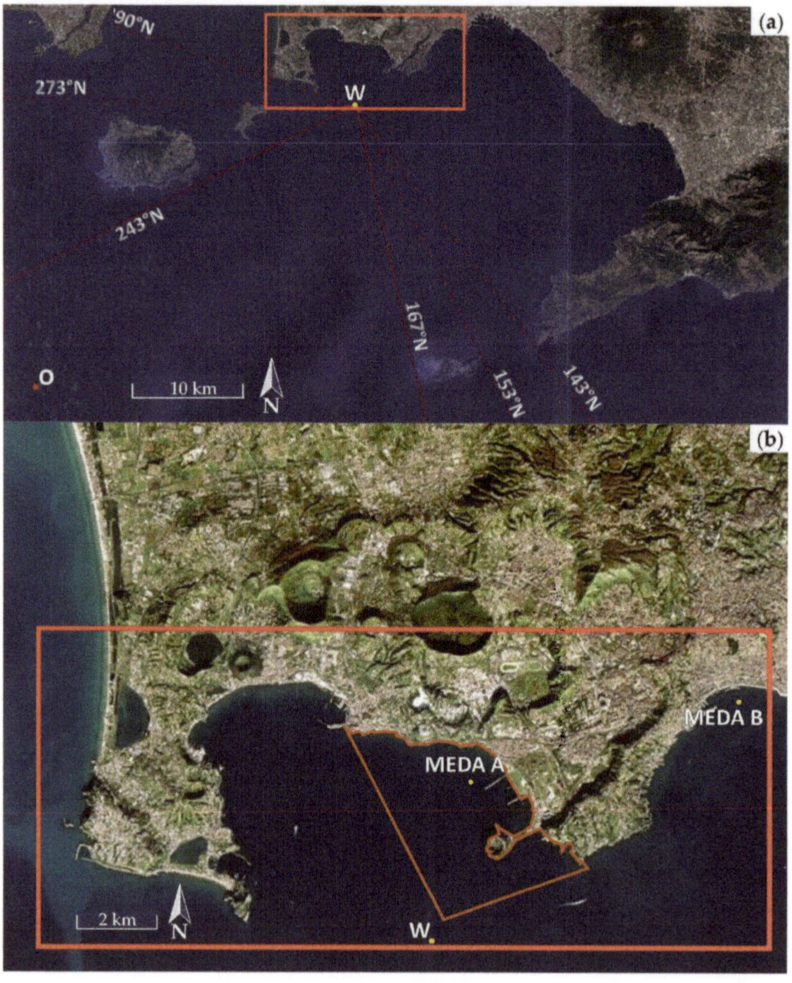

Figure 5. (**a**) Map of the Gulf of Naples and location of the study site. (**b**) Zoom on the study area. The positions of point W (offshore Pozzuoli's Gulf), Monitoring and Environmental Data Units, MEDA-A and MEDA-B, are also depicted. The brown contour line defines the remediation site boundaries.

Table 5. Geographical information of MEDA units.

Point	Latitude	Longitude	Depth
MEDA A	40°49.668′ N	14°13.984′ E	19.0
MEDA B	40°48.550′ N	14°09.300′ E	17.5

2.2.1. Acoustic Doppler Current Profiler

The ADCPs used for the test campaign are one of the most widely used instruments in oceanographic research for measuring the wave velocity profile. Such instruments are also able to provide wave information. ADCP-A and ADCP-B are part of the aforementioned MEDA A and MEDA B, respectively. The ADCP (Figure 6) is a bottom-mounted upward-looking instrument which takes the measurements of the waves basically using three independent techniques.

(a) (b) (c)

Figure 6. Stazione Zoologica "Anton Dohrn" (SZN) instruments: (**a**) MEDA A; (**b**) MEDA B; (**c**) acoustic Doppler current profilers (ADCPs).

The first method is wave measuring using the basic principle of Doppler shifting to evaluate the orbital velocities of waves, ensonifying the entire water column along four inclined beams. The orbital velocity measured by the ADCP along each distant beam provides information above the directional and non-directional wave spectrum. In addition to wave orbital velocity measurement, the ADCP also measures the non-directional spectra through echo ranging (surface track) and bottom pressure with a pressure transducer, providing redundant measurements of wave height and water depth.

In Tables 6 and 7, the ADCP specification and the parameters used for the spectral analysis are described. Details of the ADCP wave measurements are described in [102–106].

Table 6. ADCP configuration.

Time between Full Ensemble Records	(15 min)
Frequency	(600 kHz)
Size of the depth cell	(50 cm)
Number of bins in the current profile	(49 bins)
ADCP altitude above bottom	(50 cm)
Number of beams	(4 beams)

Table 7. The spectral analysis parameters of the ADCP.

Frequency band width	(0.0078 Hz)
Maximum upper cutoff frequency	(0.5 Hz)
Sea-swell transition frequency	(0.11 Hz)
Minimum lower cutoff	(0.039 Hz)
Number of direction frequency bands	(128 bands)
Number of frequency bands	(128 bins)

2.2.2. Directional Wave Spectra Drifter-Derived Wave Buoy

The DWSD buoy uses the GPS sensor package in order to measure $w(t)$, $u(t)$ and $v(t)$, which represent respectively the vertical, horizontal E-W and horizontal S-N buoy velocity components, from changes in the frequency of the GPS signal [86,93,107,108]. The measurements are made for a 17 min long sample of $u(t)$, $v(t)$ and $w(t)$ every hour, divided into overlapping 4-min segments with 1 Hz of sampling frequency. The power spectral density, co-spectra and quadrature-spectra parameters are derived from a Fourier transform of the correlation functions related to each pair of the three aforementioned signals, giving the first five independent Fourier coefficients (a_0, a_1, a_2, b_1, b_2) and thus the wave spectra for each hourly sea state. For each measured sea state, the three velocity components, the computed first five Fourier coefficients and the main wave data parameters are transmitted in real time through the Iridium satellite system. All these wave data, including data on battery voltage and the pressure, temperature and humidity of the hull, are accessible in real time from a dedicated website.

The DWSD buoy (Figure 7) has a simple spherical geometry with a diameter of 0.39 m and weight of 12 daN, reducing in this way the installation and maintenance costs, being very easy to handle and to install.

(**a**) (**b**)

Figure 7. The directional wave spectra drifter (DWSD) buoy: (**a**) ashore, prior to launching; (**b**) right after deployment at MEDA-B. The orange float required for the mooring system is also shown.

2.3. *Wave Propagation and Model Calibration*

In order to consider the intricate variations in wave energy density occurring from offshore of the Gulf of Naples to the Bagnoli-Coroglio Bay, the nearshore energetic patterns have been studied by means of the numerical suite MIKE 21 SW spectral wave model, developed by DHI Water and Environment [109]. The model takes into account the effects of refraction and shoaling due to varying depths and local wind generation and energy dissipation due to bottom friction and wave-breaking, and it has been validated by comparison with data from buoys and satellites by several authors [8,110–113]. Moreover, several scientific papers (e.g., [114–117]) discussed the overall satisfactory agreement between MIKE 21 SW and SWAN, TOMAWAC and STWAVE. In particular, Ilia and O'Donnell [118] found that both MIKE 21 SW and SWAN were largely consistent in their observations during storms, even if MIKE 21 SW predicted some of the storm peaks slightly better than SWAN. Therefore, the results from this study can be of interest for applications with other spectral wave models.

The basic equations in the model are derived from the conservation equation for the spectral wave action density Z, based on the approach proposed by the authors of [119]. In fact, in the presence of currents, wave action is conserved whilst the wave energy is not [120]. The source/sink term that represents all physical processes which generate, dissipate or redistribute energy, S_{tot}, can be written as:

$$S_{tot} = S_{in} + S_{surf} + S_{dw} + S_{bot} + S_{nl} \quad (5)$$

where S_{in} represents the energy transfer from wind to waves, S_{surf} is the dissipation of wave energy due to depth-induced breaking, S_{dw} is the dissipation of wave energy due to whitecapping, S_{bot} is the dissipation due to bottom friction and S_{nl} is the energy transfer due to non-linear triad (three-wave) interactions. The following approaches/models are used in the model:

- for the wave bathymetric breaking, the formulation proposed by Battjes and Janssen [121];
- the formulation of Kofoed-Hansen and Rasmussen [122] for bottom friction dissipation;
- the Komen et al. [123] dissipation model for whitecapping;
- the triad-wave interaction is modeled using the simplified approach proposed by Eldeberky and Battjes [124,125];
- for *Snl* parameterization, the discrete interaction approximation developed by Hasslmann et al. [126].

Operatively, the models compute the evolution of Z by solving the action balance equation [127], which in the Cartesian co-ordinates can be written as:

$$\frac{\partial Z}{\partial t} + \nabla_{x,y}\left[\left(C_g + U\right)Z\right] + \frac{\partial}{\partial \sigma}(C_\sigma Z) + \frac{\partial}{\partial \theta}(C_\theta Z) = \frac{S_{tot}}{\sigma} \tag{6}$$

where $Z = V/\sigma$, V being the variance density and σ the relative angular frequency, θ is the mean wave direction measured clockwise from true north, C_g is the group velocity, U is the current velocity vector and C_σ and C_θ are the propagation velocities in spectral space (σ,θ). The left-hand side of the above equation represents the local rate of change of the wave energy density, propagation in geographical space and shifting of frequency and refraction due to the spatial variation of the depth and current.

For wave propagation over slowly varying depths h, σ can be written by means of the linear dispersion relation

$$\sigma = \sqrt{g\,k\,tanh(kh)} \tag{7}$$

in which k is the wave number.

The magnitude of the group velocity C_g is given by

$$C_g = \frac{\partial \sigma}{\partial k} = \frac{1}{2}\left[1 + \frac{2kh}{\sin h(2kh)}\right]\sqrt{\frac{g}{h}\tan h(kh)} \tag{8}$$

The implicit assumption of these equations is that properties of the medium (water depth and current) as well as the wave field itself vary on time and space scales that are much larger than the variation scales of a single wave.

The model takes into account diffraction by using the approximation proposed by Holthuijsen et al., [128], based on the revised version of the mild slope equation model of Berkhoff [129] that was proposed by Porter [130].

It is worth noting that the source functions S_{in}, S_{nl} and S_{ds} in MIKE 21 SW are similar to the source functions implemented in the WAM Cycle 4 model [131,132]. The latter provides the basis for the ECMWF wave hindcast dataset [16,133,134]. One of the main restrictions of the model is that when propagation leads to waves moving nearly parallel and close to the coast, there is an unrealistic loss of energy caused by the large second-order diffusion error [135]. In this case, moreover, the main assumption that the source integration time step has to be shorter than or equal to the propagation time step is at fault. Hence, an intrinsic sensibility to direction can be detected, representing a warning if significant diffraction-reflection conditions can be found.

The model solves the governing equation by means of finite element-type methods to discretize geographical and spectral space. A parameterization of the conservation equation in the frequency domain is performed by introducing the zeroth and the first moment of the action spectrum as dependent variables.

The computational domain was discretized using an unstructured grid with meshes based on linear triangular elements (Figure 8) and performed using the cell-centered finite volume method.

The seabed was performed by interpolating at the grid nodes the information provided by the General Bathymetric Chart of the Oceans (GEBCO) database [136]. The grid resolution was assumed to be variable linearly between the maximum depth to 150 m for depths in the range of 500 m to 100 m. Constant values of 150 m and 1000 m of the grid resolution have been assumed for water depth shallower than 100 m and deeper than 500 m, respectively.

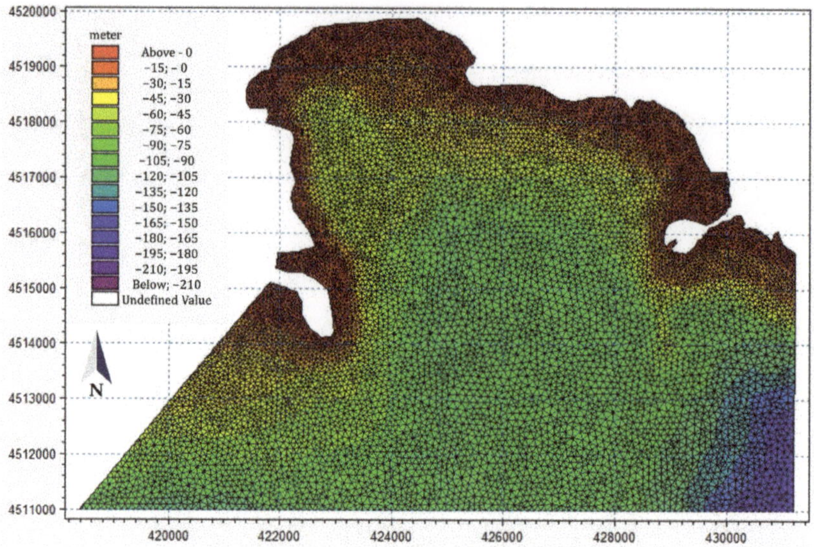

Figure 8. Zoom on the Bagnoli-Coroglio Bay with focus on the computational mesh implemented in MIKE 21 SW:

The wave model was run as forced wave-by-wave with data from the ECMWF internal WAve Model (WAM) with the ERA-Interim dataset related to source point E3. The basic data necessary to fulfill the offshore requirements are the significant wave height (H_{m0}), mean wave period (T_m) and mean wave direction (θ), provided by 6-h hindcast wave data. Wave power series was calculated from the resulting dataset provided by the transformation model. For natural sea states, where waves are random in height and period (and direction), the spectral parameters have to be used. The wave energy flux can be defined as:

$$P = \rho g \int_0^{2\pi} \int_0^{\infty} C_g(f,h) S(f,\theta) df d\theta \tag{9}$$

where ϱ is the sea water density, $S(f,\theta)$ denotes the 2D wave spectrum as a function of the spectral wave frequency f and mean wave direction θ and $C_g(f,h)$ denotes the wave group velocity, expressed by Equation (8).

2.3.1. First Step: Triple Collocation

Although a phase-decoupled approach is employed in order to reproduce the qualitative behavior of changes and spatial redistribution in the wave direction, when significant diffraction conditions are detected (e.g., in front of reflecting obstacles like rocky coasts), a Boussinesq-type model is required. However, due to the large extension of the spatial domain and the large wave dataset and computing effort normally required for the computation of diffraction in arbitrary geophysical conditions, a different approach was applied in the present study. Taking into account the way in which MIKE 21 SW solves the equations, better results are normally obtained with waves that are parallel to one of the coordinate axes or (if just one boundary is used to force the model) perpendicular to the offshore boundary from which the input waves are coming [137]. Therefore, the input open boundary should be placed so that the main wave direction is orthogonally aligned. Looking to a large scale, the exchanges between the Gulf of Naples and the Tyrrhenian Sea occur along the Bocca Grande, the main aperture of the gulf between Ischia and Capri. Considering that Bocca Grande opens to the west into the Mediterranean Sea and the bulk of the waves are provided by westerly waves [5], firstly, the open boundary of the numerical domain was orthogonally aligned along the 220° N direction. For convenience, we will call this domain M220°. Under this domain, the numerical model was forced with the hindcast data by ECMWF (grid reference point E3). Since two sources of direct meausurement of the "physical truth" (with certain systematic deviations and random errors) are available (i.e., ADCP-B and DWSD-B datasets), the search for the best set of model parameters (breaking parameter, bottom friction and white capping) was iteratively obtained by firstly applying a triple collocation procedure. This is the singularity of the present study, for which the third dataset is not univocally defined: the triple collocation is not used just for error estimation purposes but to help the search for the best calibration.

The term "triple collocation" indicates a methodology used to characterize systematic biases and random errors in satellite observations, model fields and in situ measurements. It attempts to segregate the measurement uncertainties, spatial and temporal representation and sampling differences in the different datasets by an objective method [137,138]. A frequent and often biased assumption is that all errors are due to the system that is being tested against a reference system, that is in turn assumed perfect. In this vein, Stoffelen [38] also refers to the biases associated with regression and with error distributions. These issues cannot be clearly resolved in dual comparisons, as scatter will be caused simultaneously by all the above issues for both observing systems, and there is no clear objective way to assign errors to one or the other [139]. In triple collocation, instead, three (ideally) independent datasets are brought together, so that three scatter plots can be made. Focusing on the specific application to coastal engineering, and referring to studies provided by Robertson et al. [45], Muraleedharan et al. [42] and McColl et al. [46], the technique needs the assumption of a linear relationship between the measured value and true value. The following equation can be considered:

$$X_i = \alpha_i + \beta_i T + e_i \tag{10}$$

where the Xi ($i \in \{1, 2, 3\}$) are collocated measurement systems linearly related to the true underlying value T with additive random errors e_i, respectively. The terms α_i and β_i are unknown calibration parameters representing bias and the linear calibration coefficient (i.e., the ordinary least squares intercepts and slopes, respectively).

Given that the true value T is unknown, this method requires that one of the datasets is defined as the reference. However, as noted by Janssen et al. [40], the choice of T does not affect the results. The ADCP-B dataset was defined as the reference dataset; hence, its α and β will be set to 0 and 1, respectively.

The first step removes α from the datasets by introducing the following new variables:

$$X_i' = X_i - \alpha_i \tag{11}$$

A new set of equations, without T, result from inserting (11) into (10). The uncertainty term e in (10) can be modified to:

$$X_i'' - T = \frac{X_i'}{\beta_i} - T = \frac{e_i}{\beta_i} = e_i'' \tag{12}$$

By calculating the difference between any two of above equations, the true value can be eliminated, obtaining the following:

$$X_1'' - X_2'' = e_1'' - e_2''$$

$$X_3'' - X_2'' = e_3'' - e_2'' \tag{13}$$
$$X_1'' - X_3'' = e_1'' - e_3''$$

Assuming the errors from the independent sources have zero mean and are uncorrelated with each other and with T, error terms can then be calculated by multiplying any of the two equations above, introducing the mean values:

$$\overline{(X_1'' - X_2'')(X_3'' - X_2'')} = \left(e_1'' - e_2''\right)\left(e_3'' - e_2''\right) = \left(e_2''\right)^2$$
$$\overline{(X_1'' - X_3'')(X_2'' - X_3'')} = \left(e_1'' - e_3''\right)\left(e_2'' - e_3''\right) = \left(e_3''\right)^2 \tag{14}$$
$$\overline{(X_1'' - X_3'')(X_1'' - X_2'')} = \left(e_1'' - e_3''\right)\left(e_1'' - e_2''\right) = \left(e_1''\right)^2$$

Then, according to Janssen et al., [40], it is possible calculate the linear calibration coefficient for the X_2 and X_3 datasets (the DWSD-B datasets and the numerical output, respectively, in the present study). It can be calculated as:

$$\beta_2 = \left(-B_2 + \frac{\sqrt{B_2^2 - 4A_2C_2}}{2A_2}\right) \tag{15}$$

$$\beta_3 = \left(-B_3 + \frac{\sqrt{B_3^2 - 4A_3C_3}}{2A_3}\right) \tag{16}$$

where $A_2 = r_2\overline{X_1'X_2'}$, $r_2 = \overline{e_1}/\overline{e_2}$, $B_2 = \overline{X_1'^2} - r_2\overline{X_2'^2}$, $C_2 = \overline{X_1'X_2'}$, and $A_3 = r_3\overline{X_1'X_3'}$, $r_3 = \overline{e_1}/\overline{e_3}$, $B_3 = \overline{X_1'^2} - r_3\overline{X_3'^2}$, $C_3 = \overline{X_1'X_3'}$.

Hence, the bias is calculated as:

$$\alpha_2 = \overline{X_2} - \beta_2\overline{X_1} \tag{17}$$

$$\alpha_3 = \overline{X_3} - \beta_3\overline{X_1} \tag{18}$$

assuming as initial values for the iterative method that $\alpha_2 = \alpha_3 = 0$ and $\beta_2 = \beta_3 = 1$. The iterative process ends when either of the bias, beta or error variance converge [45]. For this study, convergence was based on the error variance.

Operatively, the triple collocation procedure has been repeated several times with different numerical outputs (X_3) obtained each time after arbitrary modification of model parameters. It was assumed, in particular, that the calibration of the numerical output was achieved when the error variance estimated by means of the triple collocation between X_3 and X_2 and X_3 and X_1 were both smaller than the error variance between X_1 and X_2.

For the qualitative evaluation of the comparison results, statistical indicators such as bias and root mean square error (*RMSE*) were used. These parameters are defined as:

$$Bias\left(X_i, X_j\right) = \frac{1}{N}\sum_{n=1}^{N}\left(x_{j,n} - x_{i,n}\right) \tag{19}$$

$$RMSE\left(X_i, X_j\right) = \sqrt{\frac{1}{N-1}\sum_{n=1}^{N}\left(x_{j,n} - x_{i,n}\right)^2} \tag{20}$$

where x_j and x_i $(i,j \in \{1, 2, 3\}, i \neq j)$ indicate the wave parameters at the n-th hourly sea state, respectively, measured by the DWSD buoy, the ADCP or provided by numerical runs, and N is the total number of hourly sea states considered for the field test campaign. The notation is such that capital letters represent random variables, and lower-case letters represent realizations of random variables. We also may derive the bivariate correlations between the measurement sources as:

$$\Gamma = \frac{\mathrm{COV}\left(X_i, X_j\right)}{\sigma_{X_i}\sigma_{x_j}} \tag{21}$$

obtained by defining the second-order quantities that are estimable directly from sample measurements, i.e., the covariance, $\mathrm{COV}(X_i,X_j)$, and the standard deviation, σ, of two datasets.

The key benefits of the first step can be summarized as follows:

- firstly, qualitative evaluations of the input time series amount to cheap and almost instantaneous forward runs of a pretrained network of direct measurements;
- secondly, fast and accurate approximations of numerical output with respect to the model parameters returns, since short recordings by DWSD-B and ADCP-B are taken into account.

These benefits, used together under the triple collocation technique, allow for an efficient rough calibration.

2.3.2. Second Step: Ensemble of Multiple Runs

The small gulf of Bagnoli-Coroglio Bay represents a sub-basin in the northwestern end of the Gulf of Naples. Water exchange occurs between Pozzuoli Bay and the Gulf of Naples through a section that is 100 m deep and 2 km wide. Considering its wave sector, in addition to M220°, the other four different offshore boundary orientations have been applied (M180°, M240°; M260°; M280°), as shown in Figure 9.

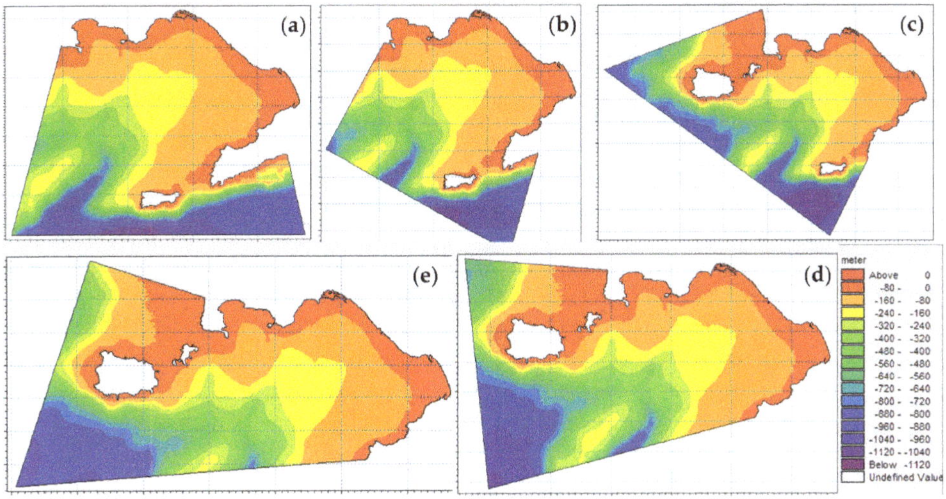

Figure 9. Bathymetry implemented for each boundary orientation: (**a**) model M180°; (**b**) model M220°; (**c**) model M240°; (**d**) model M260°; (**e**) model M280°.

The main idea was that the model results coming from M220° can be improved by weighing the contributions of each sub-model. Hence, equal numbers of wave sectors from the whole offshore dataset have been propagated within the respective oriented model, as explained in Table 8. In other words, the 1-year wave time series was divided into five sub-series, collecting waves by the five directional ranges. The angular width of such wave sectors is not equal but is chosen concurrently, considering that the morphological characteristics of the bay and offshore directional wave rose.

Table 8. Geographical information of nearshore study sites.

Name of the Model	Wave Sector
M180°	From 0° N to 190° N
M220°	From 190° N to 220° N
M240°	From 220° N to 250° N
M260°	From 250° N to 270° N
M280°	From 270° N to 360° N

At their first run, the set of model parameters found for M220° was considered. The procedure is synthesized in the scheme reported in Figure 10.

The whole dataset of resulting wave patterns obtained by means of the multi-domain procedure can be seen as not significantly affected by neglecting diffraction (i.e., not significantly different in comparison to adopting a diffraction model using just five scenarios along the mean five directions).

The calibration of each domain has been carried out, comparing the numerical results of a 1-year wave time series at the control point corresponding to the DWSD-A location, focusing only on waves coming from the reference wave sector. Then, the search for the optimal tuning parameters was carried out iteratively.

Concerning this iterative operation, it is of significative importance that an enhancement factor is introduced into the input series. In fact, after a first run for all the five boundary orientations, a sort of large bias was detected. Even if a calibration in phase one was carried out, the gross error remained too high, especially for larger wave heights. This discrepancy was not found when the geographically transposed dataset (at point O) was used to force the numerical model. Therefore, in order to use the ECMWF dataset (with the advantage of 40 years of continuous wave data), a rough correction parameter had to be applied (i.e., the enhancement factor) and then the finer calibration procedure could be applied. As previously highlighted, the use of a reanalysis product (the ERA-Interim dataset) as input for the numerical model leads to a general underestimation of the wave height. Considering that these differences can be attributed mainly to the dissimilar measurement conditions, an estimation of the discrepancies between ECMWF and IWN buoy records (the one at Ponza buoy and at point O, by transposition) is available in Section 2.1.1. In this work, the value of 1.42 (see Table 3) has been applied to amplify the wave height time series at point E3. The amplification factor was not necessary in the first step. This can be attributed to the small range of wave heights collected at MEDA B. As highlighted in Section 2.1.1, no relevant differences between reanalysis and direct measurements were detected for $H_{m0} < 0.75$ m.

Finally, the resulting five numerical series were assembled in order to reconstruct the 1-year nearshore dataset. The main information for the assemblage was represented by the offshore wave direction provided at point E3. The logic description of the assemblage algorithm is reported in Appendix A.

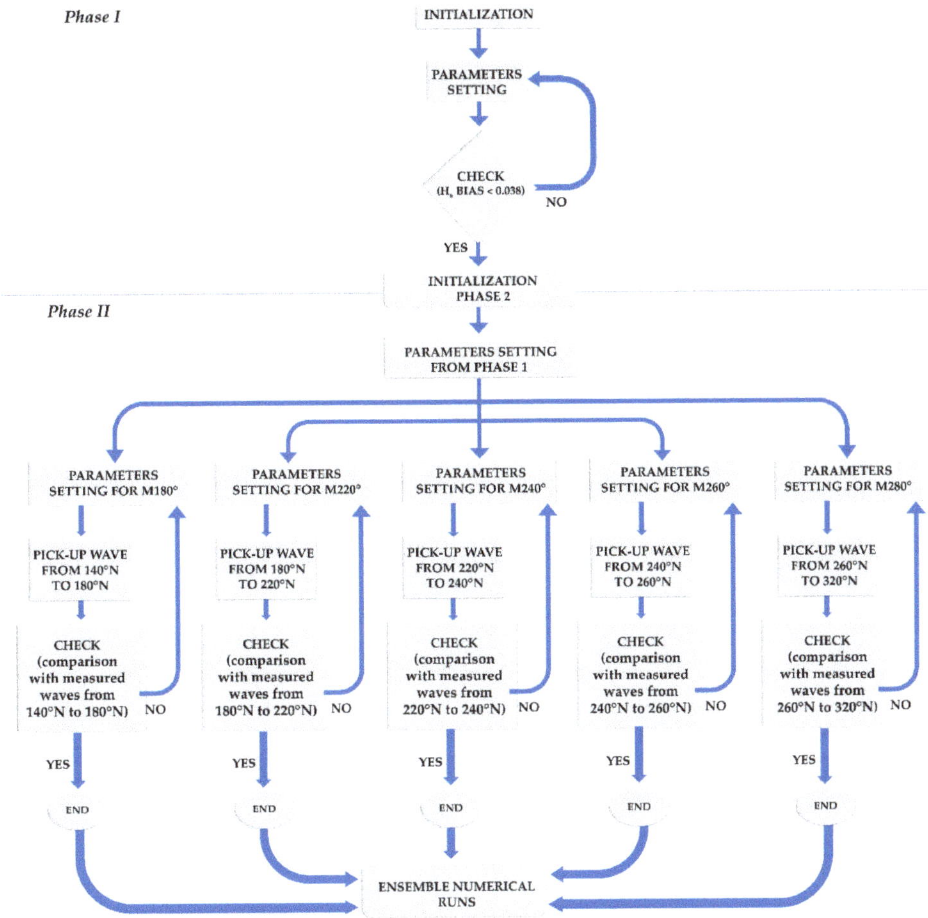

Figure 10. Flow chart of the two-phase calibration procedure.

3. Results

3.1. DWSD Buoy Compared to the ADCP

Figure 11a shows the time series of the significant wave height H_{m0} measured by the two wave instruments, presenting very good agreement considering the very different instrumental techniques that were used. It can be noted that the time series plot between the DWSD-B and ADPC-B looks very similar, without any significant deviation, especially when the significant wave height exceeded 0.5 m. In Figure 11b, the comparative analysis of the significant wave height clearly shows a good correlation between the simultaneous buoy data and the ADCP, with a bias of 0.038 m, *RMSE* of 0.07 m and a correlation coefficient *R* of 0.96. Such a strong agreement between two wave sensors of totally different natures confirms the high quality of this cheap DWSD. For practical coastal engineering applications, sea states with values of H_{m0} lower than 0.50 m are in many cases considered as calm conditions. If only the sea states greater than this threshold are considered in the comparison between DWSD and ADCP, then the results show excellent correlation, with a bias of of 0.011 m and *RMSE* of 0.05 m. The wave height correlation coefficient between sources is $\Gamma = 0.87$.

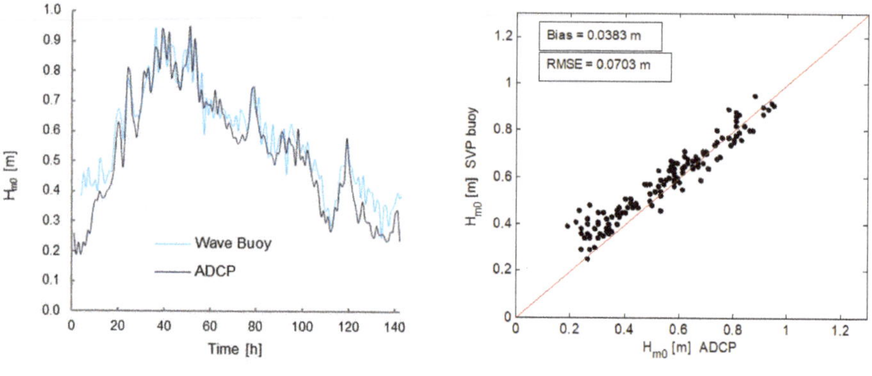

(a) (b)

Figure 11. (**a**) Data series of the significant wave height H_{m0} between DWSD-B buoy and ADCP-B (from 12 May to 18 May 2016); (**b**) comparison of the significant wave height H_{m0} between the two instruments.

The correlation of the peak period T_p between the ADCP-B and the DWSD-B is shown in Figure 12a. The peak period considered in the analysis refers to the peak related to the wind–sea spectra (0.11 Hz < f < 0.49 Hz). The results of peak periods show some small differences between the two instruments, mainly when the calculated wave spectra have multiple peaks in the wind sea frequency range (0.11 Hz < f < 0.49 Hz), of approximately equal magnitude, leading to some difficulties in the correct evaluation of the T_p. Small stochastic effects may easily modify the spectral peak, yielding a slightly different peak period. Figure 12b shows the comparison of the peak period for each sea state, showing a bias of −0.1 sec and an *RMSE* for the 1.1 section.

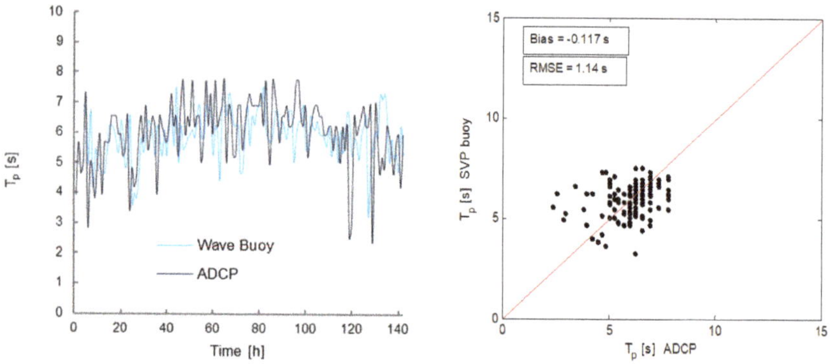

(a) (b)

Figure 12. (**a**) Data series of the peak period T_p measured with the DWSD-B buoy and the ADCP-B (from 12 May to 18 May 2016); (**b**) comparison of the peak period T_p between the two instruments.

3.2. Triple Comparison

After the implicit validation between ADCP and the DWSD buoy was obtained, the M220° MIKE 21 model was calibrated. The breaking parameter, bottom friction and white capping were tuned in order to provide better wave predictions. In the present work, the iterative procedure ends when the bias related to H_s between numerical output and DWSD buoy is less than the one between ADCP-B and DWSD-B (i.e., <0.038). In particular, a bias = 0.03 m was reached.

Figure 13 shows a brief time series of the different datasets. The original ECMWF dataset (without the enhancement factor) is depicted. It is worth noting that the available data for triple comparison are few and, in particular, the range of measured wave height is between 0.4 m and about 1 m. This reinforces the need for a second stage of calibration.

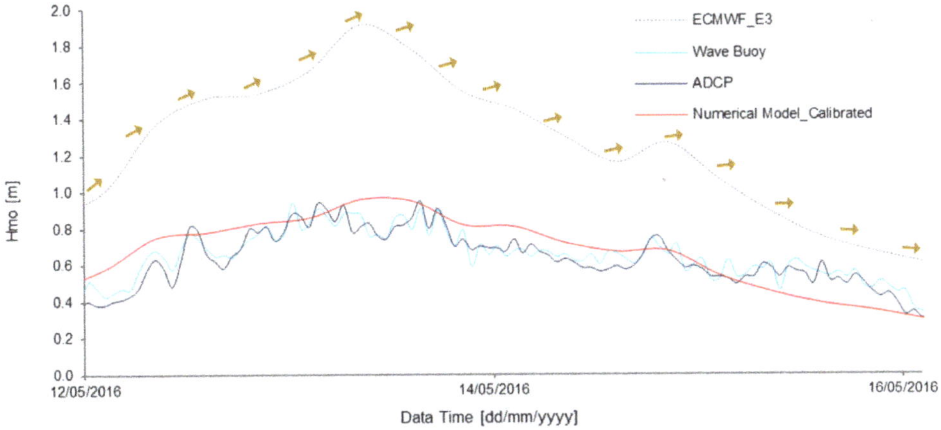

Figure 13. Observed and modeled time series of hourly averaged wave height. The observed time series of 6-hourly averaged wave height at the ECMWF grid point E3 (offshore of Gulf of Naples) is also reported.

It is also important to note the smoothed signal for the hindcast data and, hence, the numerical model, due to the smaller sampling frequency. By means of the small arrows in Figure 13, we also reported the offshore wave direction at point E3, which ranged between 232° N to 274° N. It can be noted that for waves coming from 230°–250° N, the numerical model fits very well with the direct measurements (in this vein, it could be seen as the envelope curve for those time series). On the other hand, when the direction became higher than 250°–255° N, the error significantly increased. This evidence corroborates the use of a multi-domain approach in order to overcome the intrinsic limits of the numerical model in relation to diffraction issues.

3.3. Final Calibration

According to the main wave direction, the five numerical domains were run. The second step of the calibration procedure was obtained by comparing the numerical output with the records at the DWSD buoy located at MEDA A. Unfortunately, during the working period of the Buoy-A, the ADCP-A did not record any data due to malfunction. A unique time series from the five datasets has been re-constituted.

Figure 14 shows the comparison between the ensemble of numerical runs as reconstituted by the five numerical domains and the Buoy-A and related to 1 year of data. This period, however, is not a consecutive time, but it was reconstituted considering the available period of measurements at ADCP-A. In other words, the fictitious 1-year ADCP-A time series was built by linking the following timeslots: 1 May 2016–30 November 2016, 1 December 2017–8 March 2018, 22 March 2017–27 April 2017. Obviously, for the numerical runs, the same temporal windows from the hindcast data were used.

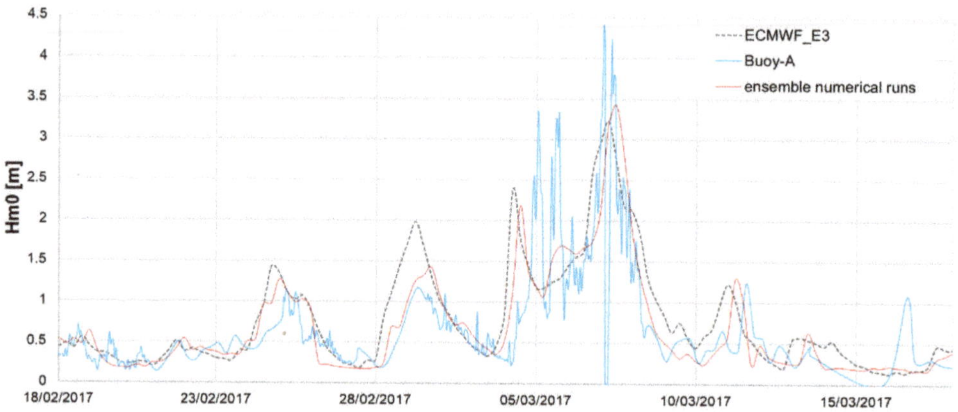

Figure 14. Comparison of time series obtained with the ensemble numerical runs, buoy records at MEDA A and ECMWF data for the reference point E3.

The results related to 1 year are graphically represented with polar diagrams, assembled in Figure 15. The wave rose of Figure 15a shows that the 1-year predominant offshore waves are the ones coming from WNW to WSW. Depending on the system morphology, the wave rose measured at the ADCP-A (Figure 15b) undergoes a radical transformation, both in predominant direction and wave energy. The comparison between ADCP-A and wave climate obtained by using a single numerical domain (i.e., M220°, in Figure 15c) shows some differences.

The loss of directional information is noticeable for all wave height classes, resulting in an unrealistic predominant wave direction (210°–225° N): the occurrence frequency of these waves is about 48%.

The ensemble numerical runs (Figure 15d) and the energy flux from each wave class are consistent with the ADCP-A measurements and the relationship between one wave sector and another is well evidenced. As highlighted in Figure 14, the smaller sampling frequency (6-h) for the numerical model leads to lower peak values and to a reinforcement of the lowest wave height class (i.e., <0.25 m), which also occurs if the enhancement factor for the input dataset is applied. If a moving average filter (spanning 6 h) is applied to the 1-year ADCP-A measurements in order to have a time series with the same sampling frequency of the numerical ensemble, an excellent correlation can be found, with a bias of 0.009 m and an *RMSE* of 0.52 m. The main wave climate parameters and main statistics obtained for the various sources are reported in Table 9.

The results shown as the mean energy flux at MEDA A are not negligible, in contrast with the intricate morphology of the Bagnoli-Coroglio Bay.

A tentative wave energy flux density contour map is shown along the analyzed coastline in Figure 16. It is stressed that the mean wave power is able to provide its effect up to the outer surf zone. The explanation is straightforward: when the wave front is parallel to the bathymetry and, in particular, if a favourable funnel shape of the coast is recognizable, the main phenomenon governing the wave transformation is energetic refraction, and shoaling could be easily recognized as an energy-conserving, non-dissipative mechanism. This effect could in part explain the validity of the results even though a Boussinesq-type model was not used.

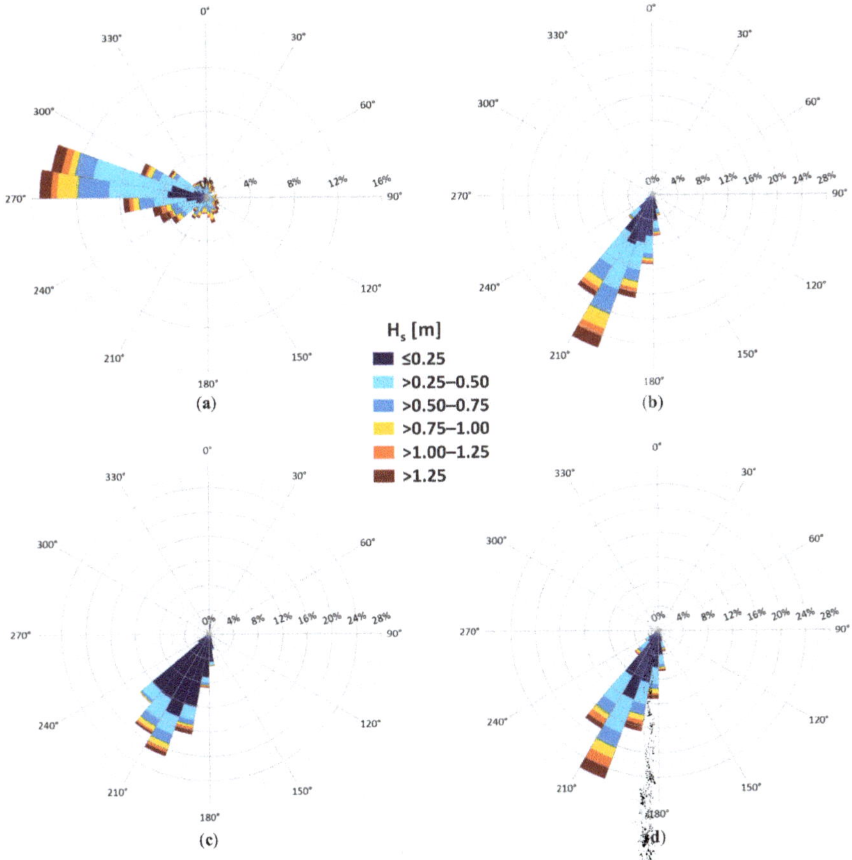

Figure 15. Wave rose referenced to different wave height classes (in legend) and related to the fictitious 1-year period for: (**a**) ECMWF grid point E3; (**b**) ADCP at MEDA A; (**c**) numerical model by using M220° only; (**d**) reconstituted time series by the five numerical domains.

Table 9. Wave statistics obtained for the various sources.

	ECMWF	**Ensemble**	**ADCP**
	H_s (m)	H_s (m)	H_s (m)
Mean	0.62	0.38	0.39
Median	0.45	0.24	0.26
Max	3.95	2.96	3.27
Min	0.07	0.00	0.00
Standard deviation	0.5	0.41	0.38
	T_p (s)	T_p (s)	T_p (s)
Mean	5.14	4.54	5.47
Median	4.94	4.55	5.4
Max	10.33	10.16	28.1
Min	1.96	0.23	0.00
Standard deviation	1.71	2.61	2.27

Table 9. *Cont.*

	ECMWF	Ensemble	ADCP
	θ (°)	θ (°)	θ (°)
Mean	234.39	201.7	198.65
Median	262.39	202.21	203.00
Max	359.98	359.02	359.00
Min	0.03	0.00	0.00
Standard deviation	77.07	35.99	35.72
	P (kW/m)	P (kW/m)	P (kW/m)
Mean	1.73	0.93	0.96
Median	0.4	0.09	0.13
Max	67.88	33.95	43.19
Min	0	0	0
Standard deviation	4.18	2.70	2.80

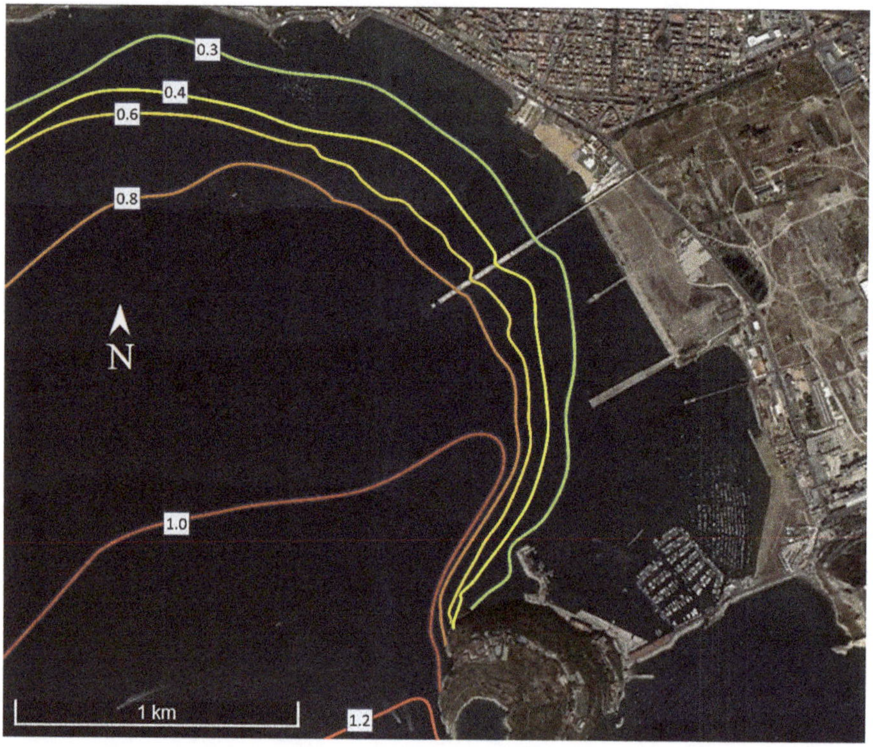

Figure 16. Mean wave power flux per unit crest (expressed in kW/m) in the Bagnoli-Coroglio Bay.

4. Additional Considerations and Future Perspectives

Waves are a concentrated form of solar energy. This energy flows through the Earth's climate system, and its components respond. The response (change in energy flow) usually has impacts on other parts of the climate system. This is known as feedback. Such feedback can be positive (it leads to reinforce a small change) or negative (it acts as a stabilizing force, pushing the system back to its original state). In the context of climate assessment (also in the perspective of climate change detection), this has the opposite connotation: positive feedback destabilizes the system (which is usually bad), while negative feedback acts against the perturbation [140]. Whereas several climate factors have been classified as positive or negative feedback, storm waves and sea level rises are not univocally defined. In the study of environmental restoration projects, like the one in Bagnoli-Coroglio Bay, the identification of feedback is important in order to better understand the sensitivity of local environmental parameters to changes. The authors of [141], in analyzing the impact of sea level rises and storms along the U.S East Coast, stumbled across a variation in sea surface height associated with the Gulf Stream. As is typical of many applications, wave climate changes sit in context with internal variability and other local causes. The non-tidal condition for the Bagnoli-Coroglio area makes it possible to avoid the influence of internal modes of variability in the atmosphere. Therefore, the results of the present study may be assumed as the basis for building other models for "climate" purposes, like those for sea level rises, water circulation and heat exchange at the water surface.

A second point concerns the effectiveness of the numerical model calibration by means of a short (thus affordable and feasible) period of in situ buoy measurement. A heuristic explanation could be that the available dataset respects the hypothesis of the representativeness of the sample, i.e., it is adequate to the distribution of wave parameters. In this vein, it appears crucial that measurements cover the majority of the wave height range. The fact that measurements at MEDA A were carried out during an intense winter storm, in fact, has proven to be essential.

It is worth noting that, in the present study, a "true" triple collocation method (e.g., [39,41]) is not possible due to the short overlap of time series at MEDA-B and lack of simultaneous measurements from ADCP-A and DWSD-A.

However, the reliability of the DWSD buoy, its versatility and cost-effectiveness allows the implementation of a sustainable global array of wave sensors that will support the validation of satellite products and enhanced climatological studies, as well as providing an indispensable tool for the calibration and validation of numerical models in coastal areas.

5. Conclusions

Any eco-restoration actions require a high level of accuracy in the assessment of the nearshore wave patterns and in the definition of wave climate scenarios for the following decades. Due to this delicate issue, the accuracy and reliability of the techniques and instrumentation used to define the waves are crucial. The present study provides a non-conventional application of the multi-collocation method. In fact, essentially due to a brief overlap of sea state observations carried out by the ADCP and the DWDS buoy, a full triple collocation method is impossible to apply. However, the direct measurements available at two different locations allowed, by means of a two-step strategy, the calibration of a numerical model. In particular, during the second calibration phase, a nonparametric wave height enhancement factor was required in order to achieve the best optimization of the numerical model. The enhancement factor consists of an amplification of each value of the WAM dataset provided by ECMWF. To estimate such an amplification, it was proposed as the assumption of the average discrepancy observed between the WAM hindcast dataset at a point located offshore of the study area and the time series, obtained by the transposition of the available offshore wave buoy records. Then, the ECMWF time series was used as input for the numerical model. It was remarked that the second step used a highly representative sample consisting of measurements collected in a period experiencing a large range of significant wave height. Hence, the "need for speed" is by no means limited to rough calibration, and the path of the two-step method exposed here moves in this direction.

To summarize, two main outcomes can be considered from this study:

1. the procedure here proposed, in which every sea state is subject to five (one for each grid model orientation) numerical propagations by a simpler spectral wave model, allows researchers to reach a good level of accuracy, similarly to a more time-consuming Boussinesq-type wave model which, nowadays, represents a state-of-the-art modeling technique if a very detailed wave disturbance in an enclosed coastal area needs to be explored;

2. the effectiveness of numerical model calibration by means of a short period of direct measurement, opening up opportunities to use low-cost GPS buoys. To bridge the gap of abundant direct measurements in the sea from traditional wave buoy networks, the capability of GPS-buoy clusters to provide data for assimilation, calibration and validation of both climate and weather models could be optimally leveraged.

Specifically for the study area, the results shown were not negligible values of wave energy flux at the study site, also if a significant variability of punctual wave power could be envisaged. The evaluation of wave climate here presented would provide the opportunity for careful eco-engineering solutions against storm control and for restoration purposes. In this vein, it is worth noting a first application of the method to provide wave data for a source apportionment assessment of marine sediment contamination in the study area [142]. Reliable nearshore wave assessment, in fact, makes it possible to assess restoration practices from the perspective of projected medium/long term changes in sea state characteristics; for instance, those stemming from climate change (both at global and local levels). Future field campaigns will help to increase confidence in the technologies and in the approach presented in this work.

Author Contributions: D.V. and R.D. carried out the conceptualization; P.C., L.C. and L.M. conceived the investigation; P.C., U.M.G. and F.C. made the formal analysis; L.C. and P.C. performed the methodology for buoy data analysis; F.C. and P.C. performed the methodology for ADCP data analysis; P.C. and U.M.G. wrote the original draft; D.V. L.M., and R.D. performed the review; P.C. and F.C. performed the editing stages. All authors have read and agreed to the published version of the manuscript.

Funding: This research was funded by the Italian Ministry for Education, University and Research (MIUR) through the ABBaCo project, grant number C62F16000170001.

Acknowledgments: The support of the University of Campania "Luigi Vanvitelli" through the VALERE program (VAnviteLli pEr la RicErca) is gratefully acknowledged. A special thanks is addressed to Augusto Passerelli (SZN), Enrico Di Lauro and Vincenzo Ferrante (University of Campania) for actively supporting the field campaign, the pre-treatment of data and the set-up of the numerical model.

Conflicts of Interest: The authors declare no conflict of interest.

Appendix A

This appendix contains logic details about the algorithm used to assemble the five numerical runs.

Input

1. Sea states, S, datasets (in terms of triple significant wave height H_{mo}, peak period T_p, wave direction θ) from the five numerical models (differing for the open boundary orientation) M180°, M220°, M240°, M260°, M280°, i.e., $S_{M180°}$, $S_{M220°}$, $S_{M240°}$, $S_{M260°}$, $S_{M280°}$, respectively;
2. Original hindcast wave direction dataset from ECMWF grid point E3, θ_{ECMWF} (expressed in degrees measured clockwise from true north).

Result

- Ensemble dataset, S_{EN}, of 1-year numerical runs with $N = 1460$ data (8760 h /6 h = number of ERA-Interim data spanning 365 days at six-hour time slots).

Algorithm

For each n-th hourly sea state (n=1, . . . ,N)
if $\theta_{ECMWF,n} \in]0,190]$ then $S_{EN,n} = S_{M180°,n}$
if $\theta_{ECMWF,n} \in]190,220]$ then $S_{EN,n} = S_{M220°,n}$
if $\theta_{ECMWF,n} \in]220,250]$ then $S_{EN,n} = S_{M240°,n}$
if $\theta_{ECMWF,n} \in]250,270]$ then $S_{EN,n} = S_{M260°,n}$
if $\theta_{ECMWF,n} \in]270,360]$ then $S_{EN,n} = S_{M280°,n}$
end

References

1. Dentale, F.; Furcolo, P.; Pugliese Carratelli, E.; Reale, F.; Contestabile, P.; Tomasicchio, G.R. Extreme wave analysis by integrating model and wave buoy data. *Water* **2018**, *10*, 373. [CrossRef]
2. Fortunato, A.B.; Li, K.; Bertin, X.; Rodrigues, M.; Miguez, B.M. Determination of extreme sea levels along the Iberian Atlantic coast. *Ocean Eng.* **2016**, *111*, 471–482. [CrossRef]
3. Franco, L. History of coastal engineering in Italy. In *History and Heritage of Coastal Engineering*; ASCE: Reston, VA, USA, 1996; pp. 275–335.
4. European Centre for Medium-Range Weather Forecasts. Available online: http://www.ecmwf.int/ (accessed on 10 May 2020).
5. Vicinanza, D.; Cappietti, L.; Ferrante, V.; Contestabile, P. Estimation of the wave energy in the Italian offshore. *J. Coast. Res.* **2011**, *64*, 613.
6. Vicinanza, D.; Contestabile, P.; Ferrante, V. Wave energy potential in the north-west of Sardinia (Italy). *Renew. Energy* **2013**, *50*, 506–521. [CrossRef]
7. Bencivenga, M.; Nardone, G.; Ruggiero, F.; Calore, D. The Italian Data Buoy Network. *WTI Trans. Eng. Sci.* **2012**, *74*, 321–332.
8. Contestabile, P.; Ferrante, V.; Vicinanza, D. Wave energy resource along the coast of Santa Catarina (Brazil). *Energies* **2015**, *8*, 14219–14243. [CrossRef]
9. Contestabile, P.; Lauro, E.D.; Galli, P.; Corselli, C.; Vicinanza, D. Offshore wind and wave energy assessment around Malè and Magoodhoo Island (Maldives). *Sustainability* **2017**, *9*, 613. [CrossRef]
10. Shanas, P.R.; Kumar, V.S. Temporal variations in the wind and wave climate at a location in the eastern Arabian Sea based on ERA-Interim reanalysis data. *Nat. Hazards Earth Syst. Sci.* **2014**, *14*, 1371–1381. [CrossRef]
11. Iuppa, C.; Cavallaro, L.; Vicinanza, D.; Foti, E. Investigation of suitable sites for wave energy converters around Sicily (Italy). *Ocean Sci.* **2015**, *11*, 543–557. [CrossRef]
12. Contestabile, P.; Vicinanza, D. Coastal defence integrating wave-energy-based desalination: A case study in Madagascar. *J. Mar. Sci. Eng.* **2018**, *6*, 64. [CrossRef]
13. Iuppa, C.; Cavallaro, L.; Foti, E.; Vicinanza, D. Potential wave energy production by different wave energy converters around Sicily. *J. Renew. Sustain. Energy* **2015**, *7*, 061701. [CrossRef]
14. Caloiero, T.; Aristodemo, F.; Ferraro, D.A. Trend analysis of significant wave height and energy period in southern Italy. *Theor. Appl. Climatol.* **2019**, *138*, 917–930. [CrossRef]
15. Sterl, A.; Komen, G.K.; Cotton, P.D. Fifteen years of global wave hindcasts using winds from the European Centre for Medium-Range Weather Forecasts reanalysis: Validating the reanalyzed winds and assessing the wave climate. *J. Geophys. Res. Ocean.* **1998**, *103*, 5477–5492. [CrossRef]
16. Janssen, P.A.E.M.; Bidlot, J.R.; Abdalla, S.; Hersbach, H. Progress in ocean wave forecasting at ECMWF. In *ECMWF Technical Memory*; ECMWF: Reading, UK, 2005; p. 27.
17. Palm, S.P.; Benedetti, A.; Spinhirne, J. Validation of ECMWF global forecast model parameters using GLAS atmospheric channel measurements. *Geophys. Res. Lett.* **2005**, *32*. [CrossRef]
18. Klein, S.A.; Jakob, C. Validation and sensitivities of frontal clouds simulated by the ECMWF model. *Mon. Weather Rev.* **1999**, *127*, 2514–2531. [CrossRef]
19. Molteni, F.; Buizza, R.; Palmer, T.N.; Petroliagis, T. The ECMWF ensemble prediction system: Methodology and validation. *Q. J. R. Meteorol. Soc.* **1996**, *122*, 73–119. [CrossRef]
20. Hanna, E.; Valdes, P. Validation of ECMWF (re) analysis surface climate data, 1979–1998, for Greenland and implications for mass balance modelling of the ice sheet. *Int. J. Climatol. A J. R. Meteorol. Soc.* **2001**, *21*, 171–195. [CrossRef]
21. Mace, G.G.; Jakob, C.; Moran, K.P. Validation of hydrometeor occurrence predicted by the ECMWF model using millimeter wave radar data. *Geophys. Res. Lett.* **1998**, *25*, 1645–1648. [CrossRef]
22. King, J.C. Validation of ECMWF sea level pressure analyses over the Bellingshausen Sea, Antarctica. *Weather Forecast.* **2003**, *18*, 536–540. [CrossRef]
23. Hagemann, S.; Gates, L.D. Validation of the hydrological cycle of ECMWF and NCEP reanalyses using the MPI hydrological discharge model. *J. Geophys. Res. Atmos.* **2001**, *106*, 1503–1510. [CrossRef]

24. Janssen, P. ECMWF wave modeling and satellite altimeter wave data. In *Satellites, Oceanography and Society*; Halpern, D., Ed.; Elseviers Oceanography Series; Elsevier: Amsterdam, The Netherlands, 2000; Volume 63, pp. 35–56.
25. Tokmakian, R.; Challenor, P.G. On the joint estimation of model and satellite sea surface height anomaly errors. *Ocean Model.* **1999**, *1*, 39–52. [CrossRef]
26. Stopa, J.E.; Cheung, K.F. Intercomparison of wind and wave data from the ECMWF Reanalysis Interim and the NCEP Climate Forecast System Reanalysis. *Ocean Model.* **2014**, *75*, 65–83. [CrossRef]
27. Sanil Kumar, V.; Naseef, T.M. Performance of ERA-Interim wave data in the nearshore waters around India. *J. Atmos. Ocean. Technol.* **2015**, *32*, 1257–1269. [CrossRef]
28. Cavaleri, L.; Bertotti, L. Accuracy of the modelled wind and waves in enclosed seas. *Tellus* **2004**, *56*, 167–175. [CrossRef]
29. Bertotti, L.; Cavaleri, L.; Soret, A.; Tolosana-Delgado, R. Performance of global and regional nested meteorological models. *Cont. Shelf Res.* **2014**, *87*, 17–27. [CrossRef]
30. Mentaschi, L.; Besio, G.; Cassola, F.; Mazzino, A. Performance evaluation of Wavewatch III in the Mediterranean Sea. *Ocean Model.* **2015**, *90*, 82–94. [CrossRef]
31. Cavaleri, L. Wave modeling missing the peaks. *J. Phys. Oceanogr.* **2009**, *39*, 2757–2778. [CrossRef]
32. Arena, F.; Laface, V.; Barbaro, G.; Romolo, A. Effects of Sampling between Data of SignificantWave Height for Intensity and Duration of Severe Sea Storms. *Int. J. Geosci.* **2013**, *4*, 240–248. [CrossRef]
33. Reale, F.; Dentale, F.; Pugliese Carratelli, E.; Torrisi, L. Remote Sensing of Small-Scale Storm Variations in Coastal Seas. *J. Coast. Res.* **2014**, *30*, 130–141. [CrossRef]
34. Wang., J.; Li., B.; Gao., Z.; Wang., J. Comparison of ECMWF significant wave height forecasts in the China sea with buoy data. *Weather Forecast.* **2019**, *34*, 1693–1704. [CrossRef]
35. Cavaleri, L.; Bertotti, L. The improvement of modelled wind and wave fields with increasing resolution. *Ocean Eng.* **2006**, *33*, 553–565. [CrossRef]
36. Sartini, L.; Besio, G.; Dentale, F.; Reale, F. Wave Hindcast Resolution Reliability for Extreme Analysis. In Proceedings of the 26th International Ocean and Polar Engineering Conference, Rhodes, Greece, 26 June–2 July 2016.
37. Contestabile, P.; Di Lauro, E.; Buccino, M.; Vicinanza, D. Economic assessment of Overtopping BReakwater for Energy Conversion (OBREC): A case study in Western Australia. *Sustainability* **2017**, *9*, 51. [CrossRef]
38. Stoffelen, A. Toward the true near-surface wind speed: Error modeling and calibration using triple collocation. *J. Geophys. Res. Ocean.* **1998**, *103*, 7755–7766. [CrossRef]
39. Nearing, G.S.; Yatheendradas, S.; Crow, W.T.; Bosch, D.D.; Cosh, M.H.; Goodrich, D.C.; Seyfried, M.S.; Starks, P.J. Nonparametric triple collocation. *Water Resour. Res.* **2017**, *53*, 5516–5530. [CrossRef]
40. Janssen, P.A.; Abdalla, S.; Hersbach, H.; Bidlot, J.R. Error estimation of buoy, satellite, and model wave height data. *J. Atmos. Ocean. Technol.* **2007**, *24*, 1665–1677. [CrossRef]
41. Caires, S.; Sterl, A. Validation of the ERA-40 ocean wave dataset using triple collocation. *J. Geophys. Res.* **2003**, In Press. [CrossRef]
42. Muraleedharan, G.; Rao, A.D.; Sinha, M.; Mahapatra, D.K. Analysis of Triple Collocation Method for validation of model predicted significant wave height data. *J. Ind. Geophys. Union* **2006**, *10*, 79–84.
43. Schulz-Stellenfleth, J. A multi-collocation method for coastal zone observations with applications to Sentinel-3A altimeter wave height data. *Ocean Sci.* **2019**, *15*, 249–268. [CrossRef]
44. Caires, S.; Sterl, A. A new nonparametric method to correct model data: Application to significant wave height from the ERA-40 re-analysis. *J. Atmos. Ocean. Technol.* **2005**, *22*, 443–459. [CrossRef]
45. Robertson, B.; Lin, Y.; Buckham, B. Application of triple collocation technique to wave resource assessments and wave energy converter energy production. In Proceedings of the 14th Workshop on Wave Hindcasting and Forecasting, Key West, FL, USA, 8–13 November 2015; p. 23.
46. McColl, K.A.; Vogelzang, J.; Konings, A.G.; Entekhabi, D.; Piles, M.; Stoffelen, A. Extended triple collocation: Estimating errors and correlation coefficients with respect to an unknown target. *Geophys. Res. Lett.* **2014**, *41*, 6229–6236. [CrossRef]
47. Scipal, K.; Doubkova, M.; Hegyova, A.; Dorigo, W.; Wagner, W. An empirical understanding of triple collocation evaluation measure. In Proceedings of the EGU General Assembly Conference Abstracts, Vienna, Austria, 7–12 April 2013; Volume 15.

48. Wang, H.; Zhu, J.; Yang, J. Error analysis on ESA's Envisat ASAR wave mode significant wave height retrievals using triple collocation model. *Remote Sens.* **2014**, *6*, 12217–12233. [CrossRef]
49. Wang, H.; Shi, C.Y.; Zhu, J.H.; Huang, X.Q.; Chen, C.T. Validation of significant wave height product from Envisat ASAR using triple collocation. In Proceedings of the IOP Conference Series: Earth and Environmental Science, Beijing, China, 22–26 April 2013; IOP Publishing: Bristol, UK, 2014; Volume 17, p. 012279.
50. Bertocci, I.; Dell'Anno, A.; Musco, L.; Gambi, C.; Saggiomo, V.; Cannavacciuolo, M.; Lo Martire, M.; Passarelli, A.; Zazo, G.; Danovaro, R. Multiple human pressures in coastal habitats: Variation of meiofaunal assemblages associated with sewage discharge in a post-industrial area. *Sci. Total Environ.* **2019**, *655*, 1218–1231. [CrossRef] [PubMed]
51. Morroni, L.; d'Errico, G.; Sacchi, M.; Molisso, F.; Armiento, G.; Chiavarini, S.; Rimauro, J.; Guida, M.; Siciliano, A.; Ceparano, M.; et al. Integrated characterization and risk management of marine sediments: The case study of the industrialized Bagnoli area (Naples, Italy). *Mar. Environ. Res.* **2020**, *160*, 104984. [CrossRef]
52. Ruocco, N.; Bertocci, I.; Munari, M.; Musco, L.; Caramiello, D.; Danovaro, R.; Zupo, V.; Costantini, M. Morphological and molecular responses of the sea urchin Paracentrotus lividus to highly contaminated marine sediments: The case study of Bagnoli-Coroglio brownfield (Mediterranean Sea). *Mar. Environ. Res.* **2020**, *154*, 104865. [CrossRef] [PubMed]
53. Ragozino, S.; Varriale, A.; De Vita, G.E. Self-organized practices for complex urban transformation. The case of Bagnoli in Naples, Italy. Tracce Urbane. *Riv. Ital. Transdiscipl. Studi Urbani* **2018**, *2*. [CrossRef]
54. ABBaCo Project. Sperimentazioni pilota finalizzate al restauro ambientale e balneabilità del SIN Bagnoli-Coroglio. Italian Ministry for Education, University and Research Grant number C62F16000170001. 2018. Available online: http://www.szn.it/index.php/en/research/integrative-marine-ecology/research-projects-emi/abbaco (accessed on 6 July 2020).
55. Musco, L.; Bertocci, I.; Buia, M.C.; Cannavacciuolo, M.; Conversano, F.; Gallo, A.; Gambi, M.C.; Ianora, A.; Iudicone, D.; Margiotta, F.; et al. Restauro ambientale e balneabilità a del SIN di Bagnoli Coroglio-Progetto ABBaCo Workshop SiCon2017. Siti contaminati. In Proceedings of the Esperienze Negli Interventi di Risanamento. Roma, Facoltà Ingegneria Civile ed Industriale. La Sapienza, Roma, Italy, 8–10 February 2017.
56. Chatelain, M.; Guizien, K. Modelling coupled turbulence—Dissolved oxygen dynamics near the sediment–water interface under wind waves and sea swell. *Water Res.* **2010**, *44*, 1361–1372. [CrossRef]
57. Semedo, A.; Weisse, R.; Behrens, A.; Sterl, A.; Bengtsson, L.; Günther, H. Projection of global wave climate change toward the end of the twenty-first century. *J. Clim.* **2012**, *26*, 8269–8288. [CrossRef]
58. Casas-Prat, M.; Sierra, J.P. Projected future wave climate in the NW Mediterranean Sea. *J. Geophys. Res. Ocean.* **2013**, *118*, 3548–3568. [CrossRef]
59. Valenti, D.; Denaro, G.; Spagnolo, B.; Mazzola, S.; Basilone, G.; Conversano, F.; Brunet, C.; Bonanno, A. Stochastic models for phytoplankton dynamics in Mediterranean Sea. *Ecol. Complex.* **2016**, *27*, 84–103. [CrossRef]
60. Valenti, D.; Denaro, G.; Spagnolo, B.; Conversano, F.; Brunet, C. How diffusivity, thermocline and incident light intensity modulate the dynamics of deep chlorophyll maximum in Tyrrhenian sea. *PLoS ONE* **2015**, *10*, 1–31. [CrossRef]
61. Van Dijk, M.A.; Passarelli, A.; Conversano, F.; Casotti, R. Phytoplankton dynamics by autonomous high-frequency flow cytometry from a floating buoy in the Gulf of Naples. In Proceedings of the IMEKO TC19 Workshop on Metrology for the Sea (MetroSea 2017), Naples, Italy, 11–13 October 2017.
62. Fanelli, E.; Aguzzi, J.; Casotti, R.; Conversano, F.; D'Aiello, D.; Iudicone, D.; Marini, S.; Stefanni, S. NEREA, the Naples Ecological REsearch for Augmented observatories: Towards an end-to-end transdisciplinary approach for the study of marine ecosystems. In Proceedings of the 2019 IMEKO TC-19 International Workshop on Metrology for the Sea, Genova, Italy, 3–5 October 2019.
63. Margiotta, F.; Balestra, C.; Buondonno, A.; Casotti, R.; D'Ambra, I.; Di Capua, I.; Gallia, R.; Mazzocchi, M.G.; Merquiol, L.; Pepi, M. Do plankton reflect the environmental quality status? The case of a post-industrial Mediterranean Bay. *Mar. Environ. Res.* **2020**, *160*, 104980. [CrossRef]
64. Gerovasileiou, V.; Smith, C.W.J.; Sevastou, K.; Papadopoulou, N.; Dailianis, T.; Bekkby, T.; Fiorentino, D.; Mcowen, C.; Amaro, T.; Bengil, E.G.T. Habitat mapping in the European Seas-is it fit for purpose in the marine restoration agenda? *Mar. Policy* **2019**, *106*, 103521. [CrossRef]

65. Azzellino, A.; Lanfredi, C.; D'amico, A.; Pavan, G.; Podestà, M.; Haun, J. Risk mapping for sensitive species to underwater anthropogenic sound emissions: Model development and validation in two Mediterranean areas. *Mar. Pollut. Bull.* **2011**, *63*, 56–70. [CrossRef] [PubMed]
66. Azzellino, A.; Lanfredi, C.; Contestabile, P.; Ferrante, V.; Vicinanza, D. Strategic environmental assessment to evaluate WEC projects in the perspective of the environmental cost-benefit analysis. In Proceedings of the The Twenty-first International Offshore and Polar Engineering Conference, Maui, HI, USA, 19–24 June 2011; International Society of Offshore and Polar Engineers: Mountain View, CA, USA, 2011.
67. Azzellino, A.; Gaspari, S.A.; Airoldi, S.; Lanfredi, C. Biological consequences of global warming: Does sea surface temperature affect cetacean distribution in the western Ligurian Sea. *J. Mar. Biol. Assoc. UK* **2008**, *88*, 1145–1152. [CrossRef]
68. Danovaro, R.; Fanelli, E.; Canals, M.; Ciuffardi, T.; Fabri, M.C.; Taviani, M.; Argyrouh, M.; Azzurrobi, E.; Bianchellia, S.; Cantafaroj, A.; et al. Towards a marine strategy for the deep Mediterranean Sea: Analysis of current ecological status. *Mar. Policy* **2020**, *112*, 103781. [CrossRef]
69. Borsje, B.W.; Van Wesenbeeck, B.K.; Dekker, F.; Paalvast, P.; Bouma, T.J.; Van Katwijk, M.M.; de Vries, M.B. How ecological engineering can serve in coastal protection. *Ecol. Eng.* **2011**, *37*, 113–122. [CrossRef]
70. Chung, C.H. Forty years of ecological engineering with Spartina plantations in China. *Ecol. Eng.* **2006**, *27*, 49–57. [CrossRef]
71. Contestabile, P.; Aristodemo, F.; Vicinanza, D.; Ciavola, P. Laboratory study on a beach drainage system. *Coast. Eng.* **2012**, *66*, 50–64. [CrossRef]
72. Cheong, S.M.; Silliman, B.; Wong, P.P.; Van Wesenbeeck, B.; Kim, C.K.; Guannel, G. Coastal adaptation with ecological engineering. *Nat. Clim. Chang.* **2013**, *3*, 787–791. [CrossRef]
73. Temmerman, S.; Meire, P.; Bouma, T.J.; Herman, P.M.; Ysebaert, T.; De Vriend, H.J. Ecosystem-based coastal defence in the face of global change. *Nature* **2013**, *504*, 79–83. [CrossRef]
74. Lagrangian Drifer Laboratory, SCRIPPS Institution of Oceanography. Available online: https://gdp.ucsd.edu/ldl/ (accessed on 15 May 2020).
75. Longuet-Higgins, M.S.; Cartwright, D.E.; Smith, N.D. Observations of the directional spectrum of sea waves using the motion of a floating buoy. In *Ocean Wave Spectra*, 1st ed.; Prentice-Hall: Englewood Cliffs, NJ, USA, 1963; pp. 111–132.
76. Kuik, A.J.; Ph van Vledder, G.; Holthuijsen, L.H. A Method for the Routine Analysis of Pitch-and-Roll Buoy Wave Data. *J. Phys. Oceanogr.* **1988**, *18*, 1020–1034. [CrossRef]
77. Krogstad, H.E.; Barstow, S.F.; Haug, O.; Markussen, P.Ø.; Ueland, G.; Rodriguez, I. SMART-800: A GPS based directional wave buoy. In *Ocean Wave Measurement and Analysis*; ASCE: Reston, VA, USA, 1997; pp. 1182–1195.
78. Herbers, T.H.C.; Jessen, P.F.; Janssen, T.T.; Colbert, D.B.; MacMahan, J.H. Observing ocean surface waves with GPS-tracked buoys. *J. Atmos. Ocean. Technol.* **2012**, *29*, 944–959. [CrossRef]
79. Kato, T.; Terada, Y.; Itoh, T.; Nagata, S.; Fujita, T.; Abe, T.; Miuake, T.; Nagai, T.; Koshimura, S.; Miyazaki, S. A new tsunami detection system using RTK-GPS. In Proceedings of the International Tsunami Symposium, Seattle, WA, USA, 7–10 August 2001; pp. 645–651.
80. Nagai, T.; Satomi, S.; Terada, Y.; Kato, T.; Nukada, K.; Kudaka, M. GPS buoy and seabed installed wave gauge application to offshore tsunami observation. In Proceedings of the fifteenth International Offshore and Polar Engineering Conference, Seoul, Korea, 19–24 June 2005; pp. 292–299.
81. De Vries, J.J.; Waldron, J.; Cunningham, V. Field tests of the new datawell dwr-g gpa wave buoy. *Sea Technol.* **2003**, *44*, 50–55.
82. Colbert, D. *Field Evaluation of Ocean Wave Measurements with GPS Buoy*; NAVAL Postgraduate School: Monterey, CA, USA, 2010.
83. Jeans, G.; Bellamy, I.; de Vries, J.J.; van Weert, P. Sea trial of the new Datawell GPS directional Waverider. In Proceedings of the IEEE/OES Seventh Working Conference on Current Measurement Technology, San Diego, CA, USA, 13–15 March 2003; pp. 145–147.
84. Patra, S.K.; Jena, B.K. Inter-comparison of wave measurement by accelerometer and GPS wave buoy in shallow water off Cuddalore, East Coast of India. *Indian J. Geo-Mar. Sci.* **2014**, *43*, 45–49.
85. Datawell, B.V. Mini Directional Waverider GPS Specifications. Haarlem, The Netherlands. Available online: http://www.datawell.nl (accessed on 15 May 2020).

86. Centurioni, L.; Turton, J.; Lumpkin, R.; Braasch, L.J.; Brassington, G.B.; Chao, Y.; Charpentier, E.; Chen, Z.; Corlett, G.K.; Dohan, K.; et al. Global in situ Observations of Essential Climate and Ocean Variables at the Air–Sea Interface. *Front. Mar. Sci.* **2019**, *6*, 419. [CrossRef]
87. Niiler, P.P. The world ocean surface circulation. In *Ocean Circulation and Climate*; Siedler, G., Church, J., Gould, J., Eds.; Academic Press: Cambridge, MA, USA, 2001; Volume 77, pp. 193–204.
88. Maximenko, N.; Lumpking, R.; Centurioni, L. Chapter 12—Ocean Surface Circulation". In *International Geophysics*; Griffies, S.M., Siedler, J.G.G., Church, A.J., Eds.; Academic Press: Cambridge, MA, USA, 2013; pp. 283–304.
89. Centurioni, L.R. Drifter technology and impacts for sea surface temperature, sea-level pressure, and ocean circulation studies. In *Observing the Oceans in Real Time*; Venkatesan, R., Tandon, A., D'Asaro, E., Atmanand, M.A., Eds.; Springer International Publishing: Cham, Switzerland, 2018; pp. 37–57.
90. MacMahan, J.; Brown, J.; Thornton, E. Low-cost handheld Global Positioning System for measuring surf-zone currents. *J. Coast. Res.* **2009**, *25*, 744–754. [CrossRef]
91. Postacchini, M.; Centurioni, L.R.; Braasch, L.; Brocchini, M.; Vicinanza, D. Lagrangian Observations of Waves and Currents From the River Drifter. *IEEE J. Ocean. Eng.* **2016**, *41*, 94–104.
92. Johnson, D.; Stocker, R.; Head, R.; Imberger, J.; Pattiaratchi, C. A compact, low-cost GPS drifter for use in the oceanic nearshore zone, lakes, and estuaries. *J. Atmos. Ocean. Technol.* **2003**, *20*, 1880–1884. [CrossRef]
93. Centurioni, L.; Braasch, L.; Di Lauro, E.; Contestabile, P.; De Leo, F.; Casotti, R.; Franco, L.; Vicinanza, D. A new strategic wave measurement station off Naples port main breakwater. *Coast. Eng.* **2017**, *1*, 36. [CrossRef]
94. Piscopia, R.; Inghilesi, R.; Panizzo, A.; Corsini, S.; Franco, L. Analysis of 12-year wave measurements by the Italian Wave Network. In *Coastal Engineering 2002, Cardiff, Wales, 7–12 July 2002*; Smith, J.M., Ed.; World Scientific Publishing Company: Singapore, 2003; pp. 121–133.
95. Corsini, S.; Franco, L.; Inghilesi, R.; Piscopia, R. *Atlante delle onde nei mari Italiani–Italian Waves Atlas*; Agenzia per la Protezione dell'Ambiente e per i Servizi Tecnici (APAT) and University of Rome: Rome, Italy, 2004; p. 3.
96. Contini, P.; De Girolamo, P. Impatto morfologico di opere a mare: Casi di studio. In Proceedings of the Atti VIII Convegno AIOM, Lerici, Italy, 28–29 May 1998. (In Italian).
97. Seymour, R.J. Estimating Wave Generation on Restricted Fetches. *J. Waterw. Port Coast. Ocean Div.* **1977**, *103*, 251–264.
98. SPM. *Shore Protection Manual*; Waterways Experiment Station, Coastal Engineering Research Center, U.S. Army Corps of Engineers: Vicksburg, MS, USA, 1984.
99. Gaglioti, M.; Vega Fernandez, T.; Musco, L.; Gambi, M.C. Habitat and benthic diversity in the bay of Bagnoli and surrounding areas (Gulf of Naples, Italy): A historical baseline for environmental restoration. *Mar. Environ. Res.* **2020**, *157*, 104925. [CrossRef] [PubMed]
100. Groeben, C. The Stazione Zoologica: A clearing house for marine organisms. In *Oceanographic History. The Pacific and Beyond*; Benson, K.R., Rehbock, P.F., Eds.; University of Washington Press: Seattle, WA, USA, 2002; pp. 537–547.
101. Somma, R.; Iuliano, S.; Matano, F.; Molisso, F.; Passaro, S.; Sacchi, M.; Troise, C.; de Natale, G. High-resolution morpho-bathymetry of Pozzuoli Bay, southern Italy. *J. Maps* **2016**, *12*, 222–230. [CrossRef]
102. Pinkel, R.; Smith, J.A. Open ocean surface wave measurements using Doppler sonar. *J. Geophys. Res.* **1987**, *92*, 12967–12973. [CrossRef]
103. Smith, J.A. Doppler sonar and surface waves: Range and resolution. *J. Atmos. Ocean. Technol.* **1989**, *6*, 680–696. [CrossRef]
104. Terray, E.A.; Brumley, B.H.; Strong, B. Measuring waves and currents with an upward-looking ADCP. In Proceedings of the IEEE 6th Working Conference on Current Measurement, San Diego, CA, USA, 13 March 1999; pp. 66–71.
105. Strong, B.; Brumley, B.H.; Terray, E.A.; Stone, G.W. The performance of ADCP derived wave directional spectra and comparison with other independent measurements. In Proceedings of the OCEANS 2000 MTS/IEEE Conference and Exhibition, Providence, RI, USA, 11–14 September 2000; pp. 1195–1203.
106. Teledyne Marine. Available online: http://www.teledynemarine.com/adcps/marine-measurements (accessed on 6 July 2020).

107. Poulain, P.M.; Centurioni, L. Direct measurements of world ocean tidal currents with surface drifters. *J. Geophys. Res. Ocean.* **2015**, *120*, 6986–7003. [CrossRef]
108. Centurioni, L.; Horányi, A.; Cardinali, C.; Charpentier, E.; Lumpkin, R. A global ocean observing system for measuring sea level atmospheric pressure: Effects and impacts on numerical weather prediction. *Bull. Am. Meteorol. Soc.* **2017**, *98*, 231–238. [CrossRef]
109. DHI Water and Environment. Available online: https://www.dhigroup.com/ (accessed on 15 May 2020).
110. Appendini, C.M.; Urano-Latorre, C.P.; Figueroa, B.; Dagua-Paz, C.J.; Torres-Freyermuth, A.; Salles, P. Wave energy potential assessment in the Caribbean Low Level Jet using wave hindcast information. *Appl. Energy* **2015**, *137*, 375–384. [CrossRef]
111. Henfridsson, U.; Neimane, V.; Strand, K.; Kapper, R.; Bernhoff, H.; Danielsson, O. Wave energy potential in the Baltic sea and the Danish part of the North Sea, with reflections on the skagerrak. *Renew. Energy* **2007**, *32*, 2069–2084. [CrossRef]
112. Venugopal, V.; Nemalidinne, R. Wave resource assessment for Scottish waters using a large scale North Atlantic spectral wave model. *Renew. Energy* **2015**, *76*, 503–525. [CrossRef]
113. Aydogan, B.; Ayat, B.; Yüksel, Y. Black Sea wave energy atlas from 13 years hindcasted wave data. *Renew. Energy* **2013**, *57*, 436–447. [CrossRef]
114. Strauss, D.; Mirferendesk, H.; Tomlinson, R. Comparison of two wave models for Gold Coast, Australia. *J. Coast. Res.* **2007**, *50*, 312–316.
115. Samaras, A.G. High-resolution wave and hydrodynamics modelling in coastal areas: Operational applications for coastal planning, decision support and assessment. *Nat. Hazards Earth Syst. Sci.* **2016**, *16*, 1499. [CrossRef]
116. Fonseca, R.B.; Gonçalves, M.; Guedes Soares, C. Comparing the performance of spectral wave models for coastal areas. *J. Coast. Res.* **2017**, *33*, 331–346. [CrossRef]
117. Samaras, A.G.; Vacchi, M.; Archetti, R.; Lamberti, A. Wave and hydrodynamics modelling in coastal areas with TELEMAC and MIKE21. In Proceedings of the XXth TELEMAC-MASCARET User Conference, Karlsruhe, Germany, 16–18 October 2013; pp. 59–64.
118. Ilia, A.; O'Donnell, J. An Assessment of Two Models of Wave Propagation in an Estuary Protected by Breakwaters. *J. Mar. Sci. Eng.* **2018**, *6*, 145. [CrossRef]
119. Holthuijsen, L.H.; Booij, N.; Herbers, T.H.C. A prediction model for stationary, short-crested waves in shallow water with ambient currents. *Coast. Eng.* **1989**, *13*, 23–54. [CrossRef]
120. Ozer, J.; Padillahernandez, R.; Monbaliu, J.; Fanjul, E.A.; Albiach, J.C.C.; Osuna, P.; Yu, J.C.S.; Wolf, J. A coupling module for tides, surges and waves. *Coast. Eng.* **2000**, *41*, 95–124. [CrossRef]
121. Battjes, J.A.; Janssen, J.P.F.M. Energy loss and set-up due to breaking of random waves. In Proceedings of the 16th International Conference on Coastal Engineering, Hamburg, Germany, 27 August–3 September 1978; pp. 569–587.
122. Kofoed-Hansen, H.; Rasmussen, J.H. Modelling of nonlinear shoaling based on stochastic evolution equations. *Coast. Eng.* **1998**, *33*, 203–232. [CrossRef]
123. Komen, G.J.; Cavaleri, L.; Donelan, M.; Hasselmann, K.; Hasselmann, S.; Janssen, P.A.E.M. *Dynamics and Modelling of Ocean Waves*; Cambridge University Press: Cambridge, UK, 1994.
124. Eldeberky, Y. Parameterization of triad interaction in wave energy model. In Proceedings of the International Conference on Coastal Dynamics, Gdansk, Poland, 4–8 September 1995.
125. Eldeberky, Y.; Battjes, J.A. Spectral modeling of wave breaking: Application to Boussinesq equations. *J. Geophys. Res. Ocean.* **1996**, *101*, 1253–1264. [CrossRef]
126. Hasselmann, S.; Hasselmann, K.; Allender, J.H.; Barnett, T.P. Computations and parameterizations of the nonlinear energy transfer in a gravity-wave specturm. Part II: Parameterizations of the nonlinear energy transfer for application in wave models. *J. Phys. Oceanogr.* **1985**, *15*, 1378–1391. [CrossRef]
127. Booij, N.; Ris, R.C.; Holthuijsen, L.H. A third-generation wave model for coastal regions: 1. Model description and validation. *J. Geophys. Res.* **1999**, *104*, 7649–7666. [CrossRef]
128. Holthuijsen, L.H.; Herman, A.; Booij, N. Phase-decoupled refraction–diffraction for spectral wave models. *Coast. Eng.* **2003**, *49*, 291–305. [CrossRef]
129. Berkhoff, J.C.W. Computation of combined refraction—Diffraction. In Proceedings of the 13th International Conference on Coastal Engineering, Vancouver, BC, Canada, 10–14 July 1972; pp. 471–490.
130. Porter, D. The mild-slope equations. *J. Fluid Mech.* **2003**, *494*, 51–63. [CrossRef]

131. MIKE 21 SM Scientific Documentation. Available online: https://manuals.mikepoweredbydhi.help/2017/Coast_and_Sea/M21SW_Scientific_Doc.pdf (accessed on 22 June 2020).
132. Group, T.W. The WAM model—A third generation ocean wave prediction model. *J. Phys. Oceanogr.* **1988**, *18*, 1775–1810. [CrossRef]
133. Gunther, H.; Hasselmann, S.; Janssen, P.A.E.M. *The WAM Model Cycle 4*; Report No. 4; Modellberatungsgruppe: Hamburg, Germany, 1992.
134. Janssen, P.; Janssen, P.A. *The Interaction of Ocean Waves and Wind*; Cambridge University Press: Cambridge, UK, 2004.
135. Monbaliu, J.; Padilla-Hernandez, R.; Hargreaves, J.C.; Albiach, J.C.C.; Luo, W.; Sclavo, M.; Guenther, H. The spectral wave model, WAM, adapted for applications with high spatial resolution. *Coast. Eng.* **2000**, *41*, 41–62. [CrossRef]
136. General Bathymetric Chart of the Oceans. Available online: http://www.gebco.net/ (accessed on 15 May 2020).
137. Klein, R. *Hydrodynamic Simulation with Mike 21 of Mele Bay and Port Vila, Vanuatu*; Technical Report 263; SOPAC: Suva, Fiji, 1998.
138. Vogelzang, J.; Stoffelen, A. *Triple Collocation*; KNMI: De Bilt, The Netherlands, 2012. [CrossRef]
139. Chakraborty, A.; Kumar, R.; Stoffelen, A. Validation of ocean surface winds from the OCEANSAT-2 scatterometer using triple collocation. *Remote Sens. Lett.* **2013**, *4*, 84–93. [CrossRef]
140. Gettelman, A.; Rood, R.B. Climate change and global warming. In *Demystifying Climate Models*; Springer: Berlin/Heidelberg, Germany, 2016.
141. Ezer, T.; Atkinson, L.P. Accelerated flooding along the US East Coast: On the impact of sea-level rise, tides, storms, the Gulf Stream, and the North Atlantic Oscillations. *Earths Future.* **2014**, *2*, 362–382. [CrossRef]
142. Giglioli, S.; Colombo, L.; Vicinanza, D.; Contestabile, P.; Musco, L.; Somma, R.; Azzellino, A. Source apportionment assessment of marine sediment contamination in a post-industrial area (Bagnoli, Naples). *Water* **2020**, Submitted.

water

MDPI

Article

Threshold Values of Extreme Hydrometeorological Events on the Polish Baltic Coast

Jacek Tylkowski [1,*] and Marcin Hojan [2]

1 Institute of Geoecology and Geoinformation, Faculty of Geographical and Geological Sciences, Adam Mickiewicz University, Krygowski 10, 61-680 Poznań, Poland

2 Institute of Geography, Department of Landscape History Research, Kazimierz Wielki University, Kościeleckich Square 8, 85-033 Bydgoszcz, Poland; homar@ukw.edu.pl

* Correspondence: jatyl@amu.edu.pl

Received: 30 August 2018; Accepted: 25 September 2018; Published: 27 September 2018

Abstract: The main aim of this study is to determine the threshold values for extreme sea and weather events on the Polish Baltic coast. The study is based on daily hydrometeorological data on the sea level; air temperature and atmospheric precipitation collected between 1965–2014 from six coastal sites (Świnoujście; Kołobrzeg, Ustka, Łeba, Hel, and Gdynia/Gdańsk). Threshold values for the occurrence of extreme events (with a probability of 10% and 95%, and a return rate of once every 10 years) and exceptionally extreme events (with a probability of 1% and 99%, and a return rate of once every 100 years) were determined using probability distribution and quantile analysis. Hydrometeorological absolute extremes were also determined. The methodology used to determine these extreme events and the time-space analysis of hydrometeorological extremes reveal significant geohazards for the functioning of the Baltic coastal zone, including the erosion of coastal dunes and cliffs and the destruction of technical infrastructure.

Keywords: extreme events; threshold values; probability; hydrometeorological conditions; Baltic coast

1. Introduction

It is expected that global warming will continue in the 21st century. In countries around the Baltic Sea, climate change and sea level rise have been studied since at least the 1990s [1,2]. In the Baltic Sea region, global warming is likely to be higher than the global average. Global warming will be accompanied by an increase in precipitation in the winter and uncertainty of weather changes in the summer, with a high probability of a high frequency of droughts in the southern zone of the region. The forecasted atmospheric changes will also be accompanied by an increase in the sea level and its temperature. Hydrometeorological changes will affect the natural environment, for example, marine biogeochemistry and coastal erosion [3]. Climate changes and physical properties of the sea favor the occurrence of extreme hydrometeorological events. The functioning and transformation of the natural environment in the coastal zone are especially determined by extreme hydrometeorological events. Extremely high storm surges and atmospheric precipitation intensify hydrological and geomorphological processes (including storm floods and mass movements). Extreme weather events therefore pose a significant threat to the natural environment and human activity. On the Polish Baltic coast, the greatest geohazards are storm floods. The frequency thereof is rising, causing significant economic losses across long stretches of land [4].

Literature contains neither a uniform definition of extreme events nor a methodology for their designation [5–10]. Extreme weather events are most often determined using probability characteristics [11–13]. The extreme characteristics of meteorological, hydrological, and geomorphological events include threshold values whose probability of being exceeded is lower than 10%, thus with a return period of once per decade. Exceptionally extreme weather events are considered those with a probability of less than 1%, thus with a return period of once per century. Probability characteristics are often used to estimate threats to hydrotechnical infrastructure in the coastal zone. Also used to determine extreme events is the quantile method, which most often accounts for threshold values using percentiles 5% and 95% [14–16]. Regardless of the methodology employed to determine extreme weather events (whether probabilistic or quantile), the events must be assigned absolute maximums and minimums.

The main aim of this study is to determine the threshold values for extreme sea and weather events (in terms of sea level, air temperature, and atmospheric precipitation) on the Polish Baltic coast. This paper presents absolute values of hydrological and meteorological conditions, as well as threshold values for extreme weather events calculated using the probabilistic method (for return period) and quantile analysis (for frequency of occurrence).

Aside from their scientific significance, extreme hydrometeorological events have a utilitarian significance. Extreme thermal and precipitation conditions, as well as high sea level, are the main determinants of geomorphological changes in the sea coastal zone (sometimes with disastrous and irreversible consequences). Exceeding the threshold values defining extreme hydrometeorological events may pose a significant threat to the functioning of the coastal zone. Therefore, the determination of threshold values and return periods for extreme events is important for various areas of human activity, especially those related to the protection, management, and development of the coastal region.

2. Materials and Methods

The research area was the Polish Baltic coast (about 500 km long), where there are two basic types of shoreline: dunes and cliffs. The dune coastal zone, which covers around 85% of the coast, has been forming since the wane of the last glacial period, i.e., the beginning of the Holocene. It consists of sand barriers (spits) that separate coastal lakes from the sea, and proglacial wetlands made up of dunes of varying origin and height (2–30 m). The cliff coastal zone, which covers around 15% of the coast, was formed during the Pleistocene Epoch and is a type of high coastline (up to 95 m). It is made up of glacial clays accumulated by the continental glacier in the form of head moraines, as well as fluvioglacial loams and sands accumulated mainly in the form of bottom moraines. Poland's Baltic coastal zone is home to coastal islands (e.g., Wolin and Uznam), peninsulas (e.g., Hel), bays (e.g., the Bay of Pomerania and Bay of Gdańsk), and lagoons (e.g., the Szczecin and Vistula lagoons). Sections of the coast near the Szczecin and Vistula lagoons are floodplains and areas of organic accumulation. According to the Köppen classification system, the South Baltic coast is in a temperate zone with a humid, continental climate; warm summers; and an even distribution of rainfall in all seasons. In the last half-century, in the research area, the average annual air temperature was 8.3 °C and the average annual rainfall was around 605 mm. A statistically significant increase trend was observed in the mean annual air temperature of 0.3 °C/10 years. There was no statistically significant tendency for atmospheric precipitation [17]. The Atlantic climate and the sea have an important influence on the climate of the region, which is manifested in frequent cloud cover, increased humidity, and increased wind speed. Most notable are winds from the N-W sector, which most frequently cause a rise in sea level and wind waves. Storm surges most frequently occur during cyclonic circulations from the direction: Northwest 20.6%, West 18.3%, and North 12.7% [18]. In winter, the Polish coast is usually on the warm side of the jet stream which, during positive NAO (North Atlantic Oscillation), causes warming, increased precipitation, and strong winds. In summer, the influence of the subtropical high-pressure system is important, which causes more extreme thermal conditions. The Polish Baltic coast can therefore be classified as a modern marine climate [19].

Daily data from a 50-year homogeneous measurement series (1965–2014) were used to determine the threshold values for extreme hydrometeorological events. The data included average, maximum, and minimum air temperature; total atmospheric precipitation; and average and maximum sea level. Meteorological data (from meteorological stations in Świnoujście, Kołobrzeg, Ustka, Łeba, Hel, and Gdynia) and sea level data (from mareographical stations in Świnoujście, Kołobrzeg, Ustka, Hel, and Gdańsk) were obtained from the Institute of Meteorology and Water Management in Warsaw

(Figure 1). In Poland, the Baltic High System (BHS) applies, which accounts for long-term observations of average sea level using the Kronstadt sea gauge. The estimated difference between the system based on the Normaal Amsterdams Peil (NAP) and the BHS is around 15 cm (the Kronstadt system is higher). Although Poland uses a high system based on the Kronstadt sea gauge (zero level = −508 cm), the registration and recording of sea levels is based on the NAP (zero level = −500 cm). This is why the sea gauges do not need to be adjusted [20]. In order to obtain a reference to the 0.00 NN ordinate, it would be necessary to reduce the sea-level values presented in this paper by 500 cm. An analysis of the average sea levels in the Polish Baltic coast in 1965–2014 showed a rate of increase of about 2 mm per year. This dynamic is similar to that found for the global average increase of the sea level, which was estimated to be 1–2 mm per year [21].

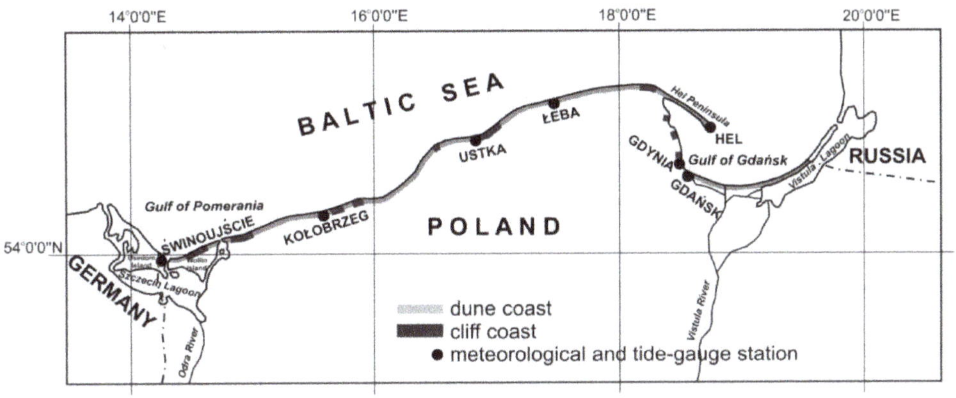

Figure 1. Research area and locations of meteorological and mareographical stations in the coastal zone of the Polish Baltic.

The definitions and criteria for determining extreme weather events are based on global research [22–24]. The following criteria were used in this study:

- Absolute extreme. The highest or lowest value that has been empirically evidenced for a given meteorological and hydrological event.
- Extreme event. Values close to the absolute extremes for a given meteorological and hydrological event, whose probability of being exceeded is lower than 10%, i.e., the chances of their occurrence (or "return period") is at most once every ten years. In quantile terms, extreme threshold values were designated using percentiles 5/100 and 95/100.
- Exceptionally extreme event. Values close to the absolute extremes for a given meteorological and hydrological event, whose probability of being exceeded is lower than 1%, i.e., the chances of their occurrence (or "return period") is at most once every hundred years. In quantile terms, exceptionally extreme threshold values were designated using percentiles 1/100 and 99/100.

For the probabilistic approach, the following procedure was used to determine extreme events: verification of the completeness and homogeneity of data (using Alexandersson's homogeneity test); adjustment of the theoretical distribution to empirical distribution (using the Kolmogorov − Smirnov test); and determination of the probability density distribution from the cumulative probability density function. This study specifies the threshold values for hydrometeorological events with a return period of 1 year P(X) 100%, 5 years P(X) 20%, 10 years P(X) 10%, 50 years P(X) 2%, 100 years P(X) 1%, 500 years P(X) 0.2%, and 1000 years P(X) 0.1%. Return periods were determined by the following formula [12].

The probability of the threshold values being exceeded was estimated from the inverse probability density function. Calculations were made in Easy Fit Professional (Version 5.6, MathWave Technologies, Washington, DC, USA).

3. Results

In the coastal zone of the Polish Baltic, the average annual air temperature from 1965–2014 ranged from 7.9 °C in Łeba to 8.6 °C in Świnoujście. Annually, on average, the most precipitation occurred in Ustka (704.2 mm) and the least in Świnoujście (535.7 mm). The average sea level ranged from 501.0 cm in Świnoujście to 509.6 cm in Gdańsk. The eastern coastal region is therefore cooler and more humid, and has a higher sea level than the western coastal region.

3.1. Absolute Extreme

The absolute values of thermal and precipitation extremes in the Baltic coastal zone are low compared to those recorded further away from the coast. The highest absolute maximum air temperature (38.0 °C) was recorded in Kołobrzeg (10 August 1992), and the lowest (33.7 °C) in Hel. The lowest absolute minimum air temperature ranged from −25.0 °C in Łeba (6 February 2012) to −18.7 °C in Hel. The highest absolute amplitude of air temperature (≥60 °C) occurred in the central part of the coastal zone (Ustka and Łeba), and the lowest (<55 °C) in the eastern part (Hel and Gdynia). The absolute values for average daily air temperature followed a similar pattern. The daily precipitation efficiency in the coastal zone is fairly low. The highest daily efficiency was recorded in the middle of the coastal zone, especially near Ustka and Łeba (141.0 mm, 24 July 1988). In the western (Świnoujście and Kołobrzeg) and eastern (Hel and Gdynia) coastal zone, the maximum daily precipitation was less than 90 mm. The absolute extreme of the maximum sea level was 661 cm in Świnoujście (4 November 1995), which was higher than the average sea level by 1.6 m. The highest average daily sea level was recorded at around 620 cm at bay stations in Świnoujście (Zatoka Pomorska) and Gdańsk (Zatoka Gdańska). The maximum average daily sea level was about 620 cm, which was higher than the average sea level by around 1.1 m. The absolute minimum of the average sea level on the entire Baltic coast was lower than the average level by 1.0–0.7 m (Table 1).

The absolute extremes presented in Table 1 occurred once in the 50-year period studied. The occurrence of such events in the future (especially maximum sea level and atmospheric precipitation) could have catastrophic consequences for human activity and the natural environment.

Table 1. Absolute extremes of air temperature (°C), atmospheric precipitation (mm), and sea level (cm) in the Polish coastal zone of the South Baltic (1965–2014). AGL: About ground level.

Place	Maximum Air Temperaturę at 2 m AGL	Mean Air Temperaturę at 2 m AGL		Minimum Air Temperaturę at 2 m AGL	Precipitation	Maximum Sea Level	Mean Sea Level	
	Max	Max	Min	Min	Max	Max	Max	Min
Świnoujście	37.4	27.7	−16.5	−22.2	76.6	661	624	401
Kołobrzeg	38.0	28.1	−16.5	−21.9	85.2	644	618	401
Ustka	37.8	29.8	−16.1	−22.2	94.2	636	611	428
Łeba	37.2	28.4	−16.5	−25.0	141.0	-	-	-
Hel	33.7	26.0	−14.6	−18.2	77.1	620	604	431
Gdynia	35.0	27.2	−16.0	−18.7	82.1	-	-	-
Gdańsk	-	-	-	-	-	644	620	436

3.2. Extreme Events—Probabilistic Analysis

When using the probabilistic method to determine extreme threshold values, it was very important to match the theoretical distribution to empirical distribution (by selecting the best theoretical distribution from among the 40 available in Easy Fit 5.6 Professional). Matches were indicated by the statistically significant (significance level $p < 0.001$) lowest value of the Kolmogorov − Smirnov (K − S) test (Table 2), which was additionally confirmed by the lowest Corrected Akaike Information Criterion (AIC_C).

For data on daily air temperature, the best match was the Error (Exponential Power) Distribution, which has the following parameters and functions of cumulative distribution and probability density:

k—shape parameter; σ—scale parameter (standard deviation); μ—location parameter (mean); z—standardized value; Γ—scale parameter.

$$F(x) = \begin{cases} 0.5\left(1 + \frac{\Gamma_{|c_0 z|^k}(1|k)}{\Gamma(1|k)}\right) & x \geq \mu \\ 0.5\left(1 - \frac{\Gamma_{|c_0 z|^k}(1|k)}{\Gamma(1|k)}\right) & x < \mu \end{cases} \quad (1)$$

$$f(x) = c_1 \sigma^{-1} \exp\left(-|c_0 z|^k\right) \quad (2)$$

$$c_0 = \left(\frac{\Gamma(3|k)}{\Gamma((1|k))}\right)^{1/2} \quad c_1 = \frac{k c_0}{2\Gamma(1|k)} \quad z \equiv \frac{x-\mu}{\sigma} \quad (3)$$

For data on daily atmospheric precipitation, the matching of distributions and the subsequent determination of threshold values for a specific return period were heavily flawed. This is why data for maximum daily precipitation in all months from 1965–2014 were taken into account. The best match was the Log-Gamma Distribution, which had values of $p = 0.59$ and $K - S = 0.108$. The probability of daily precipitation for a given return period estimated later for this distribution indicated values that could possibly occur (Table 3). The Log-Gamma Distribution has the following parameters and functions of cumulative distribution and probability density:

α—continuous parameter ($\alpha > 0$); β—continuous parameter ($\beta > 0$).

$$0 < x < \infty \quad (4)$$

$$F(x) = \frac{\Gamma_{\ln(x)/\beta}(\alpha)}{\Gamma(\alpha)} \quad (5)$$

$$f(x) = \frac{(\ln(x))^{\alpha-1}}{x\beta^{\alpha}\Gamma(\alpha)} \exp(-\ln(x)|\beta) \quad (6)$$

For daily data on the sea level, the best match was the Generalized Logistic Distribution. This distribution has the following parameters and functions of cumulative distribution and probability density:

k—shape parameter; σ—scale parameter (standard deviation > 0); μ—location parameter (mean); z—standardized value.

$$1 + k\frac{(x-\mu)}{\sigma} > 0 \quad \text{for } k \neq 0 \quad (7)$$

$$-\infty < x < +\infty \quad \text{for } k = 0 \quad (8)$$

$$F(x) = \begin{cases} \frac{1}{1+(1+k\,z)^{-1/k}} & k \neq 0 \\ \frac{1}{1+\exp(-z)} & k = 0 \end{cases} \quad (9)$$

$$f(x) = \begin{cases} \frac{(1+k\,z)^{-1-1/k}}{\sigma\left(1+(1+k\,z)^{-1/k}\right)^2} & k \neq 0 \\ \frac{\exp(-z)}{\sigma(1+\exp(-z))^2} & k = 0 \end{cases} \quad (10)$$

$$z \equiv \frac{x-\mu}{\sigma} \quad (11)$$

Table 2. Matching theoretical distributions to empirical data (using the K − S test) on thermal conditions, precipitation conditions, and sea level in the Polish coastal zone of the South Baltic (1965–2014).

Place	Hydrometeorological Conditions					
	Maximum Air Temperature	Mean Air Temperature	Minimum Air Temperature	Precipitation	Maximum Sea Level	Mean Sea Level
	Statistical Distribution					
	Error (Exponential Power) Distribution			Log-Gamma Distribution	Generalized Logistic Distribution	
Świnoujście	k = 3.91 σ= 8.19 μ = 11.95 K − S = 0.021 p < 0.001	k = 3.39 σ= 7.18 μ = 8.62 K − S = 0.033 p < 0.001	k = 2.42 σ = 6.72 μ = 5.70 K − S = 0.042 p < 0.001	α = 70.59 β = 0.048 K − S = 0.108 p = 0.590	k = 0.10 σ = 11.55 μ = 506.46 K − S = 0.015 p < 0.001	k = 0.05 σ = 10.78 μ = 499.80 K − S = 0.015 p < 0.001
Kołobrzeg	k = 3.50 σ = 7.98 μ = 11.67 K − S = 0.025 p < 0.001	k = 3.17 σ = 7.12 μ = 8.34 K − S = 0.039 p < 0.001	k = 2.30 σ = 6.68 μ = 5.34 K − S = 0.044 p < 0.001	α = 127.11 β = 0.028 K − S = 0.123 p = 0.429	k = 0.10 σ = 12.01 μ = 507.11 K − S = 0.014 p < 0.001	k = 0.06 σ = 11.06 μ = 500.73 K − S = 0.013 p < 0.001
Ustka	k = 3.35 σ = 7.97 μ = 11.36 K − S = 0.031 p < 0.001	k = 3.14 σ = 7.08 μ = 8.20 K − S = 0.036 p < 0.001	k = 2.30 σ = 6.75 μ = 5.41 K − S = 0.041 p < 0.001	α = 89.93 β = 0.040 K − S = 0.090 p = 0.797	k = 0.09 σ = 11.98 μ = 508.11 K − S = 0.017 p < 0.001	k = 0.05 σ = 11.13 μ = 502.18 K − S = 0.018 p < 0.001
Łeba	k = 3.74 σ = 8.01 μ = 11.22 K − S = 0.031 p < 0.001	k = 3.31 σ = 7.12 μ = 7.92 K − S = 0.038 p < 0.001	k = 2.16 σ = 6.82 μ = 4.61 K − S = 0.041 p < 0.001	α = 87.05 β = 0.040 K − S = 0.101 p = 0.672	-	-
Hel	k = 5.58 σ = 8.08 μ = 11.30 K − S = 0.027 p < 0.001	k = 4.47 σ = 7.07 μ = 8.24 K − S = 0.031 p < 0.001	k = 3.12 σ = 6.64 μ = 5.52 K − S = 0.037 p < 0.001	α = 72.66 β = 0.049 K − S = 0.066 p = 0.975	k = 0.09 σ = 12.15 μ = 508.28 K − S = 0.015 p < 0.001	k = 0.042 σ = 11.15 μ = 503.29 K − S = 0.020 p < 0.001
Gdynia	k = 4.30 σ = 7.81 μ = 11.28 K − S = 0.023 p < 0.001	k = 3.75 σ = 7.23 μ = 8.51 K − S = 0.031 p < 0.001	k = 2.87 σ = 6.97 μ = 6.00 K − S = 0.039 p < 0.001	α = 83.83 β = 0.041 K − S = 0.094 p = 0.758	-	-
Gdańsk	-	-	-	-	k = 0.07 σ = 12.17 μ = 513.45 K − S = 0.018 p < 0.001	k = 0.02 σ = 11.64 μ = 508.29 K − S = 0.019 p < 0.001

Analysis of the probability density function revealed potential threshold values for thermal-precipitation conditions and sea level for the expected return period (Table 3, Figure 2). Threshold values for extreme events are determined by a 10% probability (once every ten years), and for exceptionally extreme events, by a 1% probability (once every hundred years). For maximum air temperature (T_{max}), the highest values were recorded in Ustka in the central coastal area (10% = 33.2 °C, 1% = 35.5 °C), and the lowest in Hel in the eastern coastal area (10% = 30.0 °C, 1% = 31.3 °C). For minimum air temperature (T_{min}), the highest values were recorded in Łeba (10% = −18.1 °C, 1% = −21.6 °C), and the lowest in Hel (10% = −13.1 °C, 1% = −15.3 °C). For average daily air temperature (T), the highest threshold values were recorded in Kołobrzeg (10% = 28.2 °C, 1% = 30.5 °C), and the lowest in Hel (10% = 25.7 °C, 1% = 27.2 °C). The range of air temperature threshold values between stations was <50%. For atmospheric precipitation (P), the highest threshold values were recorded in Hel (10% = 58.9 mm, 1% = 96.7 mm), and the lowest in Świnoujście (10% = 49.8 mm,

1% = 80.5 mm). The spatial variation of extreme threshold values for atmospheric precipitation in the Polish coastal zone is <20%. For the sea level, the highest threshold values were recorded in Kołobrzeg (maximum sea level H_{max} 10% = 660.1 cm, 1% = 731.0 cm; and average daily sea level H 10% = 617.0 cm, 1% = 661.0 cm), and the lowest in Gdańsk (maximum sea level H_{max} 10% = 648.9 cm, 1% = 703.5 cm; and average daily sea level H 10% = 612.3 mm, 1% = 644.7 cm). The spatial differences between extreme threshold values for sea level are very small, and do not exceed 5%.

In the Polish Baltic coastal zone, spatial variation between threshold values for extreme hydrometeorological events were recorded. The highest probability values for thermal extremes and sea level were recorded in the western part of the zone (Świnoujście, Kołobrzeg), and for atmospheric precipitation, in the central part of the zone (Ustka, Łeba), as shown in Table 2.

Table 3. Probability of extreme hydrometeorological events in the Polish coastal zone of the South Baltic (daily data from 1965–2014).

Place	Parameter	P(X) [%]									
		0.1	0.2	0.5	1	2	5	10	20	500	100
		T [Years]									
		1000	500	200	100	50	20	10	5	2	1
Świnoujście	T_{max}	36.7	36.2	35.6	35.0	34.5	33.7	33.1	32.4	31.3	30.5
	T	31.9	31.5	30.8	30.2	29.6	28.8	28.2	27.4	26.4	25.5
	T_{min}	−21.0	−20.2	−19.2	−18.4	−17.5	−16.3	−15.4	−14.4	−12.9	−11.8
	P	117.2	105.4	90.8	80.5	70.8	58.6	49.8	41.1	29.0	1.0
	H_{max}	817.0	787.6	752.0	727.0	703.9	675.8	656.3	638.1	616.0	600.6
	H	686.4	673.3	656.7	644.5	632.7	617.7	606.7	596.1	582.5	572.6
Kołobrzeg	T_{max}	37.1	36.6	35.9	35.3	34.7	33.8	33.1	32.3	31.2	30.2
	T	32.3	31.8	31.1	30.5	29.8	29.0	28.2	27.5	26.3	25.4
	T_{min}	−22.1	−21.3	−20.2	−19.4	−18.5	−17.2	−16.2	−15.1	−13.6	−12.3
	P	101.1	93.4	83.6	76.4	69.3	60.0	53.0	45.7	34.8	1.0
	H_{max}	820.5	791.3	755.9	731.0	707.9	679.7	660.1	641.7	619.4	603.7
	H	711.3	695.4	675.5	661.0	647.2	629.7	617.0	604.9	589.6	578.6
Ustka	T_{max}	37.4	36.9	36.1	35.5	34.8	33.9	33.2	32.3	31.2	30.2
	T	32.2	31.6	30.9	30.3	29.7	28.8	28.0	27.3	26.1	25.2
	T_{min}	−22.4	−21.6	−20.5	−19.6	−18.7	−17.4	−16.3	−15.3	−13.7	−12.5
	P	129.3	117.3	102.3	91.6	81.3	68.3	58.7	49.1	35.4	1.0
	H_{max}	786.0	762.0	732.5	711.5	691.8	667.4	650.2	634.0	614.0	599.8
	H	707.6	692.5	673.4	659.5	646.1	629.2	617.0	605.2	590.3	579.5
Łeba	T_{max}	35.9	35.5	34.8	34.3	33.7	32.9	32.2	31.5	30.4	29.5
	T	31.3	30.8	30.1	29.6	29.0	28.1	27.5	26.7	25.6	24.7
	T_{min}	−24.7	−23.8	−22.6	−21.6	−20.6	−19.2	−18.1	−16.9	−15.2	−13.8
	P	120.3	109.2	95.2	85.3	75.7	63.6	54.7	45.8	33.1	1.0
Hel	T_{max}	32.3	32.0	31.6	31.3	30.9	30.4	30.0	29.5	28.8	28.2
	T	28.3	28.0	27.5	27.2	26.7	26.2	25.7	25.2	24.4	23.7
	T_{min}	−17.1	−16.5	−15.8	−15.3	−14.7	−13.8	−13.1	−12.4	−11.3	−10.5
	P	142.2	127.5	109.4	96.7	84.7	69.7	58.9	48.4	33.8	1.0
	H_{max}	801.6	775.7	743.8	721.3	700.2	674.2	656.0	638.8	617.7	602.9
	H	692.3	679.3	662.7	650.5	638.6	623.5	612.5	601.7	588.0	577.9
Gdynia	T_{max}	33.9	33.5	32.9	32.5	32.0	31.4	30.8	30.2	29.3	28.6
	T	30.8	30.4	29.8	29.3	28.8	28.0	27.4	26.8	25.8	25.0
	T_{min}	−18.8	−18.2	−17.4	−16.8	−16.1	−15.1	−14.3	−13.4	−12.2	−11.2
	P	117.1	106.1	92.4	82.6	73.3	61.5	52.8	44.1	31.8	1.0
Gdańsk	H_{max}	767.7	747.2	721.8	703.5	686.1	664.4	648.9	634.2	615.8	602.6
	H	678.8	668.3	654.8	644.7	634.8	621.9	612.3	602.8	590.4	581.3

These theoretical values for the probability of hydrometeorological events (Figure 2, Table 3) are highly varied in comparison to the empirical absolute extremes (Table 2). For maximum and minimum air temperature (T_{max} and T_{min}), the absolute extremes had a theoretical return period of once every 500 years (Kołobrzeg, Ustka, Gdynia) and once every 1000 years (Świnoujście, Łeba, Hel).

For average temperature (T), the absolute extremes had a theoretical return period of once every 5 years (Świnoujście, Kołobrzeg, Gdynia), once every 10 years (Hel), once every 20 years (Łeba), and once every 50 years (Ustka). For atmospheric precipitation, the absolute extremes had a theoretical return period of once every 20 years (Hel), once every 50 (Świnoujście, Gdynia), once every 100 years (Ustka), once every 200 years (Kołobrzeg), and once every 1000 years (Łeba). For maximum sea level (H_{max}), the absolute extremes had a theoretical return period of once every 2 years (Hel), once every 5 years (Kołobrzeg, Ustka, Gdańsk), and once every 10 years (Świnoujście). For maximum average sea level (H), the absolute extremes had a theoretical return period of once every 5 years (Ustka, Hel), once every 10 years (Kołobrzeg, Gdańsk), and once every 20 years (Świnoujście).

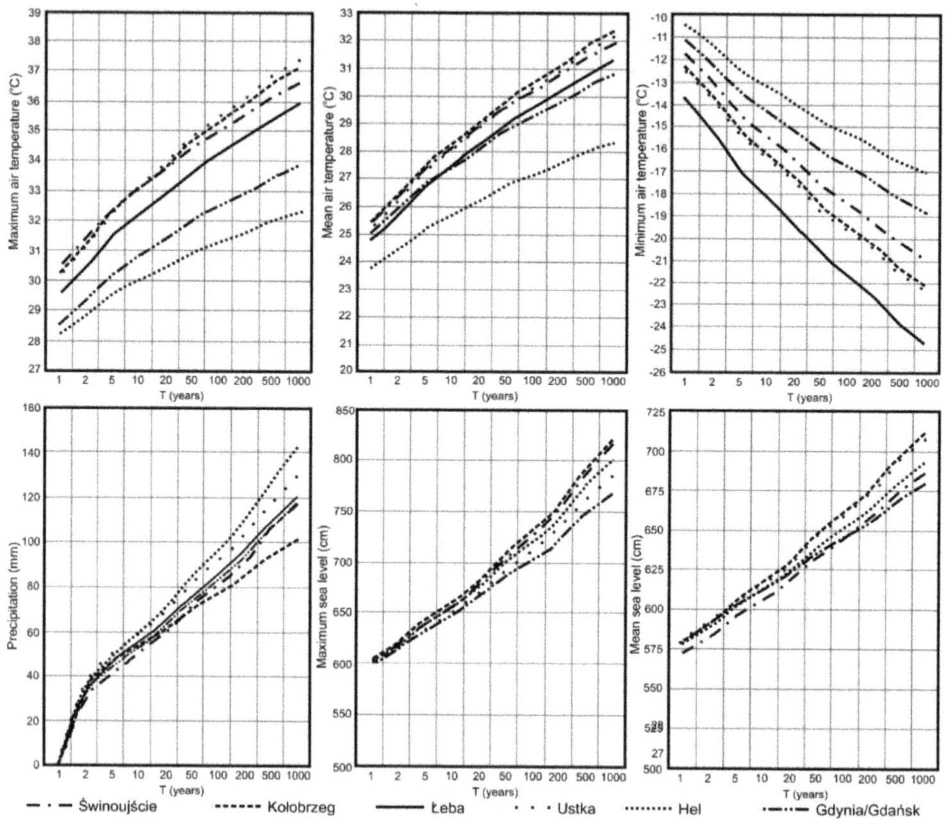

Figure 2. Threshold values for the probability (return period) of extreme hydrometeorological events in the Polish Baltic coastal zone.

The occurrence of extreme hydrometeorological events with a 10% probability in the Polish Baltic coastal zone is marked by high episodicity and randomness (Figure 3). For the period from 1965–2014, no statistically significant tendency for the increase or decrease in the occurrence of extreme events can be determined. The least common were days on which the sea level was extremely high—there were only three, 12 years apart (in 1983, 1995, and 2012). More common was extremely high atmospheric precipitation, which occurred on 22 days, the greatest concentration of which was in 1991–1993, and 2010–2012. Most common were extreme thermal events, especially maximum and minimum air temperature (49 and 67 days, respectively). Exceptional was 1987, in which there were 11 extremely cold days (T_{min}). Then, in 2010, there was an exceptional number of extremely hot days (T_{max}), i.e., five.

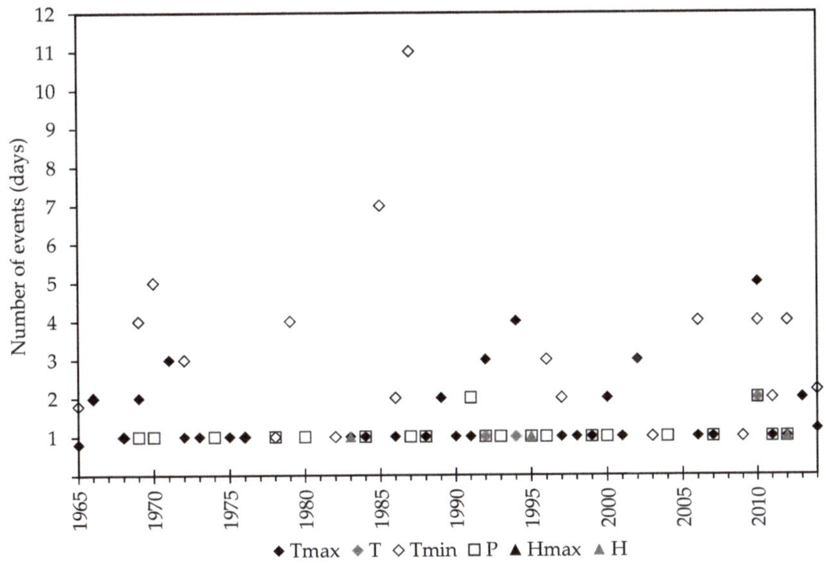

Figure 3. Occurrence of extreme hydrometeorological events with a 10% probability in the Polish Baltic coastal zone (collectively).

3.3. Extreme Events—Quantile Analysis

Quantile analysis (Table 4) determines the threshold values for the occurrence of extreme events in reference to empirical data using percentages 1%, 5%, 95%, and 99%. In probabilistic analysis, theoretical threshold values for extreme hydrometeorological events are much higher than in quantile analysis, resulting in the episodic occurrence of extreme events with long return periods. In quantile analysis, threshold values are lower, and extreme events occur much more frequently than in probabilistic analysis. For example, for the 95th percentile, they concern 5% of the data set (900–1000 cases in the 50-year collection of hydrometeorological data). Quantile analysis is therefore more helpful than probabilistic analysis for the study of time trends and spatial variability. Due to its much higher frequency of extreme events, quantile analysis enables the determination of a statistically significant linear trend in the 50-year period (Figure 4).

For the Polish Baltic coastal zone, analysis revealed a statistically significant (significance level $p < 0.05$, correlation coefficient $r > 0.3$) positive trend in the occurrence of extreme events (95%), air temperature, and sea level (Figure 4). For average air temperature, as well as maximum and average sea level, analysis revealed an increase in the number of events to three every 10 years. For maximum air temperature, the increase was lower, i.e., to two days every 10 years. For the quantile analysis, the annual numbers of extreme thermal events ranged as follows: from 10 in 1977 and 1980 to 45 in 2006 for maximum temperature; from 70 in 1977 to 48 in 2006 for average temperature; and from 1 in 1977 and 1990 to 52 in 1969 for minimum temperature. Extreme precipitation occurrences ranged from seven in 1975, 1982, and 2005, to 23 in 2002. Extreme maximum sea level occurrences ranged widely from three in 1966 to 67 in 1983.

Table 4. Extreme threshold values (percentiles 5/100 and 95/100) and exceptionally extreme threshold values (percentiles 1/100 and 99/100) of hydrometeorological events in the Polish coastal zone of the South Baltic (1965–2014).

Place	Percentile	Maximum Air Temperature at 2 m AGL [°C]	Mean Air Temperature at 2 m AGL [°C]	Minimum Air Temperature at 2 m AGL [°C]	Precipitation [mm]	Maximum Sea Level [cm]	Mean Sea Level [cm]
Świnoujście	1/100	−4.6	−7.7	−11.2	0.1	464	455
	5/100	−0.5	−2.9	−5.4	0.1	478	471
	95/100	24.7	19.4	15.8	12.0	548	535
	99/100	29.0	22.2	18.1	22.3	576	557
Kołobrzeg	1/100	−4.6	−8.2	−12.0	0.1	464	458
	5/100	−0.4	−3.2	−6.0	0.1	478	472
	95/100	24.2	18.9	15.3	13.0	550	538
	99/100	28.8	21.7	17.4	26.1	577	558
Ustka	1/100	−4.7	−8.2	−11.8	0.1	465	459
	5/100	−0.6	−3.1	−6.0	0.1	478	473
	95/100	24.3	18.8	15.5	13.8	550	539
	99/100	29.0	21.7	17.8	25.7	573	558
Łeba	1/100	−5.0	−8.4	−12.9	0.1	–	–
	5/100	−0.8	−3.5	−6.8	0.1	–	–
	95/100	24.1	18.6	14.9	13.5	–	–
	99/100	28.5	21.5	17.3	26.2	–	–
Hel	1/100	−4.8	−7.0	−9.9	0.1	464	460
	5/100	−1.0	−2.8	−5.0	0.1	478	474
	95/100	23.8	18.9	15.6	12.2	548	539
	99/100	26.9	21.3	17.9	23.6	570	556
Gdynia	1/100	−4.8	−7.5	−10.5	0.1	–	–
	5/100	−0.9	−3.0	−5.4	0.1	–	–
	95/100	23.2	19.4	16.4	12.4	–	–
	99/100	26.7	21.9	18.9	23.4	–	–
Gdańsk	1/100	–	–	–	–	468	463
	5/100	–	–	–	–	483	478
	95/100	–	–	–	–	555	545
	99/100	–	–	–	–	578	563

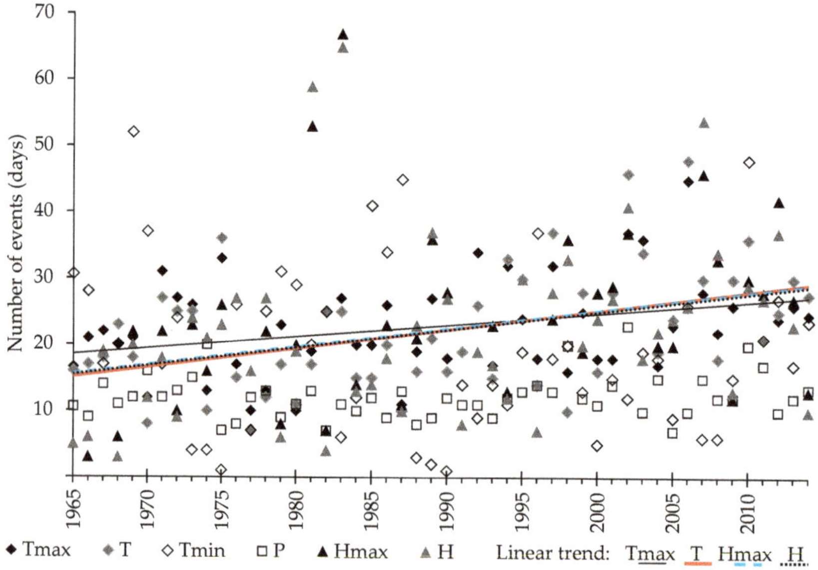

Figure 4. Occurrence of extreme hydrometeorological events in the Polish Baltic coastal zone.

Analysis of occurrence of extreme hydrometeorological events showed a high variability in time and space. In the whole coastal zone, the most extreme thermal events (over 100 cases in the five-year period) took place in the last three pentads (at the beginning of the 21st century) of the analyzed long-term period 1965–2104. This situation confirms the regularity of climate warming on the South Baltic coast. However, the occurrence of extreme precipitation and sea level was characterized by a large time-space heterogeneity. For example, a high frequency (over 50 days in the five-year period) of extreme rainfall events was recorded in Świnoujście in the last pentad (2010–2014) of the long-term period (1965–2014). On the other hand, in Ustka, the most extreme precipitation events were found in the periods 1969–1974 and 1995–1999. In the case of high sea level, the most extreme events (over 100 in the five-year period) occurred in the pentad 1980–1984 and the beginning of the twenty-first century (2000–2014) (Figure 5).

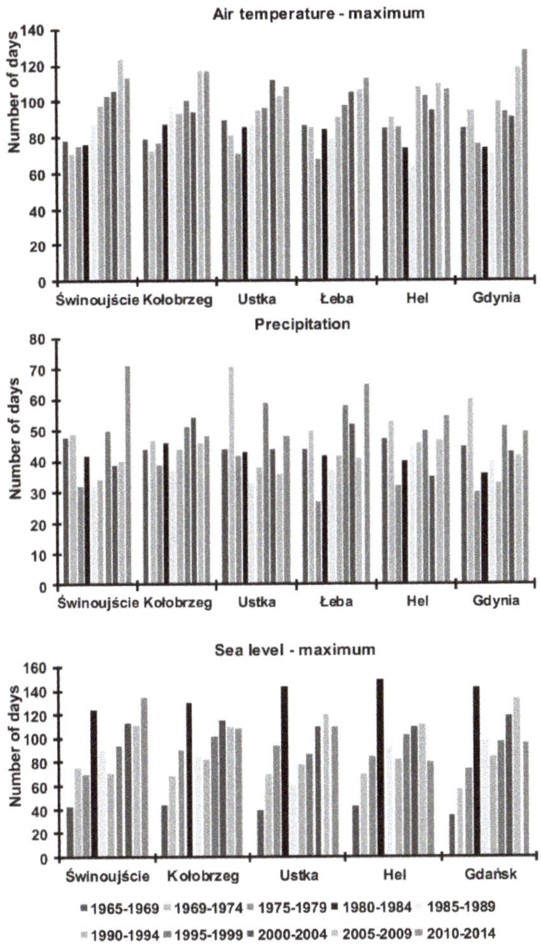

Figure 5. Temporal and spatial variability of the occurrence of extreme hydrometeorological events in the Polish coastal zone of the Baltic Sea.

4. Discussion

The methodology of separation, the threshold values for extreme events, and the temporal and spatial regularities of hydrometeorological conditions in this work are consistent with those in other works, especially those from the Baltic countries [9,13,25–31].

Certain quantitative discrepancies between threshold values may be a result of the theoretical distribution in the probability distribution analysis and the varied extent of input data. The Log-Gamma distribution for maximum monthly precipitation was similar to the gamma distribution for maximum monthly precipitation (i.e., $\geq$0.1 mm) for 1961–2010 [16]. For sea level, the best-suited theoretical distribution was the Generalized Logistic Distribution. In other works, the Gumbel Distribution is most often used [12,19]. The higher values of probability (excess) in this work were due to varying periods of input data and measurement intervals. This study analysed data from a uniform period (50 years, 1965–2014) measured at uniform daily intervals for each station (the number of entries for each hydrometeorological phenomenon at each station was ~18,250). In other works (e.g., [10]), the time period for sea level data varied (e.g., Świnoujście 1901–2006, Ustka 1948–2006) and the quantity of input data was significantly lower (mainly monthly data, e.g., Świnoujście < 1400 and Ustka ~800). The probability values for extreme sea levels obtained in this work were around 25 cm higher for a probability of 10% and 40 cm higher for a probability of 1% than those in other works on the probability of maximum seal levels in the South Baltic being exceeded [12,32]. The threshold values obtained for maximum sea level are realistic. For example, those with a probability of 0.1 (once per 1000 years) ranged from 2.68 (Gdańsk) to 3.21 (Świnoujście) meters above the average sea level (MAMSL). Similar sea levels with a 1000-year return period have been recorded in the Polish Baltic coastal zone, including 2.2 MAMSL on 13 November 1972 in Kołobrzeg [33], as well as in estuary sections of rivers during ice blockages, such as 2.67 MAMSL on 16 March 1956 in Wisła Świbno and 3.10 MAMSL on 21 March 1888 in Nogat [12].

Due to their probability and return period, extreme events rarely occur. This is why quantile analysis is more appropriate when using statistical methods, time decomposition, and extreme hydrometeorological event forecasting. It revealed an increase in the frequency of extreme sea level and air temperature events within the last 50 years. However, it did not reveal the same for atmospheric precipitation, which confirms that changes in precipitation, as well as in the number of days with extreme precipitation, are insignificant and spatially inconsistent [31,34]. Extreme hydrometeorological events occur at an especially high frequency during positive North Atlantic Oscillation (NAO). Deep, low-pressure centres from the Atlantic moving rapidly over the Polish Baltic coast are conducive to storm surges, effective atmospheric precipitation, and significant warming, especially in winter [35,36].

The extreme threshold values obtained using quantile analysis (e.g., of percentiles 5% and 95%) are similar to those obtained using the 3σ-rule, a common method for determining extreme events. The 3σ-rule can be used as a warning system for danger and abnormal hydro-meteorological conditions, preferably for normally distributed data. In this study, threshold values of the 95% and 99% percentiles (using the Generalized Logistic Distribution) for the maximum average sea level are very similar (+1% error) to values $+2\sigma$ from μ and $+3\sigma$ from μ (σ—standard deviation, μ—average). To determine the range of a specific data percentile for deviations of $\pm2\sigma$ and $\pm3\sigma$ from μ for thermal and precipitation data, specific distribution tables can be used (for temperature—Error (Exponential Power) Distribution, for precipitation—Log-Gamma Distribution). Nevertheless, the quantile classification used in this study to determine threshold values for extreme thermal and precipitation events is more appropriate than typical classifications based on the 3σ-rule.

Southern Baltic coasts are eroded during an extremely high sea level, especially during high storm waves and intense precipitation. Exceeding the set threshold values for medium and maximum sea level can be treated as a threat to extreme erosion of coastal dunes. An analysis of the correlation between the sea level and loss of coastal dunes in Poland has made it possible to define threshold values for potential erosion of the dune coast. Extreme values of sea level determined with the probabilistic method coincide with empirical studies favoring the erosion of coastal dune. Conditions during an average sea level of 602 cm and maximum of 636 cm are potentially beneficial to intensive erosion of the dune coast (>100,000 m^3) [18]. Thus, threshold values of the mean sea level > 596 cm and maximum sea level > 638 cm (Table 3) determined in this report based on the probabilistic method

(10% probability, once every ten years) constitute an important geoindicator for extreme erosion of dunes on the Polish coast of the Baltic Sea.

The threshold values for extreme mean sea level determined in the study (e.g., for Świnoujście 10% probability = 596 cm) are also a good indicator of cliff coast erosion (e.g., for the island of Wolin). Empirical cliff top recession rates studies in 1984–2014 showed that the statistically defined extreme sea level is responsible for significant transformation of the foot of the cliff and initiates mass movements [37]. This value is consistent with empirical research on coastal cliffs abrasion [38]. What is more, the thresholds for extreme precipitation, as defined in this paper, are a good indicator of the initiation of mass movements on the cliff coast. For example, the precipitation threshold value for Świnoujście is 49.8 mm with a probability of 10% (Table 3), which coincides with research on the impact of weather conditions on the dynamics of landslide processes on the shores of the Wolin Island, conducted in 2006–2009 by Winowski, et al. [38]. This author stated that the initiation of mass movements occurs when the daily sum of rainfall is higher than 40 mm, while for regular landslide processes, when daily rainfall is at least 60 mm.

Analyzing the 21st century Regional climate models (RCM) dealing with, among other things, changes in air temperature, precipitation, sea level, and storm surges, an increased frequency or crossing of thresholds for extreme hydrometeorological events in the Baltic Sea zone—set out in Table 3, may be expected in the nearest time. In the South Baltic region, the average annual air temperature is expected to increase in the 21st century by 2–3 °C, with an additional increase in atmospheric precipitation by 0–10% in the summer and 10–20% in the winter season [39]. It is also expected that the periods of heat will become more frequent and will last longer [40]. For the southern Baltic coast, most RCMs indicate an increase in frequent daily rainfall, both in winter and summer [41]. On the other hand, model simulations suggest a decrease in the number of days with atmospheric precipitation [42], which may increase the risk of long periods of drought [43]. It is predicted that in the 21st century, the sea level of the ocean will increase due to the loss of land ice masses and the thermal expansion of ocean water from 28 to 61 cm [44]. An absolute increase in sea level in the Baltic Sea is estimated at 80% of the global average [3]. For the south and south-west coast of the Baltic Sea, the estimated relative increase in the level would be particularly high, around 50–60 cm [45]. The hydrodynamic modeling performed assumes an increase of storm surges for the entire Baltic Sea in all seasons [46]. The size of the predicted increase in the case of long-term extremes by 5% or 10 cm is relatively small. Therefore, the absolute change of extremely high sea levels during storm surges will probably depend on the size of the average sea level change. This is particularly important for the Southern Baltic, where the largest increase in the average sea level is forecasted [47]. This will result in the increased erosion of dunes and cliffs and may contribute to damage to the coast's technical infrastructure (descent to the beach, technological conurbations, or existing bank fortifications).

5. Conclusions

Delimitation of threshold values for extreme hydrometeorological events in the Polish coastal zone of the Southern Baltic performed in the study allows for the following conclusions:

Threshold values were determined using the probabilistic method and quantile approach. The probabilistic method that captures the probability and return period of extreme events (10%, 1 time/100 years) is particularly useful for determining threshold values of extreme hydrometeorological events in the context of their impact on the dynamics of real geomorphological changes on the seacoast. Statistical threshold values of extreme sea level and atmospheric precipitation determined with this method are reflected in the empirical intensity of the erosion of coastal dunes and cliffs, e.g., an average daily sea level > 596 cm (10% probability) will generate the potential erosion of coastal dunes in the entire Polish coastal zone of the Baltic >100,000 m^3. Therefore, threshold values of hydrometeorological extremes determined by the probabilistic method are a good indicator of the risk of occurrence of erosion processes. On the other hand, the quantile approach due to a much higher frequency of events is very helpful for the time decomposition of the occurrence of extreme hydrometeorological events. It is then possible to determine the trend, cyclical, and seasonal variability of extreme hydrometeorological events. A positive trend (+3 days/10 years) of the occurrence of extreme events (95% percentile) as regards to air temperature and sea level was found.

Temporal and spatial diversity of threshold values of extreme hydrometeorological events was found in the Polish coastal zone of the Baltic Sea. The highest values of thermal extremes and sea level were found for the western part of the coast, in the Szczecin Pobrzeża region (Świnoujście and Kołobrzeg), and in the case of atmospheric precipitation in the central part of the coast, in the Pobrzeże Koszalińskie region (Ustka, Łeba). The eastern region, Gdańsk Coast, demonstrated the lowest threshold values of extreme hydrometeorological events. The most extreme hydrometeorological events occurred in the western region, on the Szczecinskie Coast, and the least in the central part of the coast, on the Koszalin Coast. Coasts of the Pomeranian Bay (Świnoujście and Kołobrzeg) and, to a lesser extent, Gdańsk Bay (Hel, Gdynia/Gdańsk), are the most threatened by the occurrence of extreme hydrometeorological events. The smallest threat occurs in the coastal zone of the open sea (Ustka, Łeba).

The beginning of the 21st century (2000–2014) had the highest frequency of extreme hydrometeorological events—average and maximum air temperature, as well as mean and maximum sea level. Such a situation is a manifestation of climate warming and sea level rise in the coastal zone of the South Baltic Sea.

Taking into account global and especially regional climate models prepared for the 21st century, one can expect an increase in the frequency of extreme hydrometeorological events in the near future (exceeding threshold values). This applies in particular to the maximum and average daily air temperature, daily sum of atmospheric precipitation, and maximum and average sea level.

Delimitation of extreme hydrometeorological events is significantly practical. Threshold values for seal level and thermal and precipitation events determine the extreme ranges of certain hydrometeorological conditions to which the economy should adapt. The threshold values presented in this study can be used to forecast changes in climatic and hydrological conditions in the Baltic coastal zone for the 21st century.

Author Contributions: Conceptualization, J.T. and M.H.; Data curation, J.T. and M.H.; Formal analysis, J.T. and M.H.; Funding acquisition, J.T. and M.H.; Investigation, J.T. and M.H.; Methodology, J.T.; Project administration, J.T. and M.H.; Resources, J.T. and M.H.; Validation, J.T.; Visualization, J.T. and M.H.; Writing—original draft, J.T. and M.H.; Writing—review & editing, M.H.

Funding: This research was funded by the Polish Ministry of Science (Project Supporting Maintenance of Research Potential of the Department of Physical Edu., Health and Tourism at Kazimierz Wielki University No. BS/2016/N1).

Acknowledgments: Hydrometeorological data were obtained from the Institute of Meteorology and Water Management, National Research Institute in Warsaw.

Conflicts of Interest: The authors declare no conflict of interest.

References

1. Rotnicki, K.; Borzyszkowska, W. Accelerated sea level rise and its components at the Polish Baltic Coast in the years 1951–1990. In *Ewolucja Geosystemów Nadmorskich Południowego Bałtyku*; Borówka, R.K., Młynarczyk, Z., Wojciechowski, A., Eds.; Bogucki Wydawnictwo Naukowe: Poznań, Poland; Szczecin, Poland, 1999; pp. 141–160, ISBN 83-88163-02-7.
2. Masria, A.; Negm, A.; Iskander, M.M.; Saavedra, O.C. Coastal protection measures, case study (Mediterranean zone, Egypt). *J. Coast. Conserv.* **2015**, *19*, 281–294. [CrossRef]
3. Räisänen, J. Future Climate Change in the Baltic Sea Region and Environmental Impacts. *Oxf. Res. Encycl. Clim. Sci.* **2017**. [CrossRef]
4. Paprotny, D.; Terefenko, P. New estimates of potential impacts of sea level rise and coastal floods in Poland. *Nat. Hazards* **2017**, *85*, 1249–1277. [CrossRef]

5. Heino, R.; Brázdil, R.; Førland, E.; Tuomenvirta, H.; Alexandersson, H.; Beniston, M.; Pfister, C.; Rebetez, M.; Rosenhagen, G.; Rösner, S.; et al. Progress in the study of climate extremes in northern and central Europe. *Clim. Chang.* **1999**, *42*, 151–181. [CrossRef]
6. Frich, P.; Alexander, L.V.; Della-Marta, P.; Gleason, B.; Haylock, M.; Klein Tank, A.M.G.; Peterson, T. Observed coherent changes in climatic extremes during the second half of the twentieth century. *Clim. Res.* **2002**, *19*, 193–212. [CrossRef]
7. Alexander, L.V.; Zhang, X.; Peterson, T.C.; Caesar, J.; Gleason, B.; Klein Tank, A.M.G.; Haylock, M.; Collins, D.; Trewin, B.; Rahimzadeh, F.; et al. Global observed changes in daily climate extremes of temperature and precipitation. *J. Geophys. Res.* **2006**, *111*. [CrossRef]
8. Przybylak, J.; Vízi, Z.; Araźny, A.; Kejna, M.; Maszewski, R.; Uscka-Kowalkowska, J. Poland's Climate Extremes Index, 1951–2005. *Geogr. Pol.* **2007**, *80*, 47–58.
9. Friederichs, P. Statistical downscaling of extreme precipitation events using extreme value theory. *Extremes* **2010**, *13*, 109–132. [CrossRef]
10. Ustrnul, Z.; Czekierda, D.; Wypych, A. Extreme values of air temperature in Poland according to different atmospheric circulation classifications. *Phys. Chem. Earth* **2010**, *35*, 429–436. [CrossRef]
11. Kysely, J. Probability estimates of extreme temperature events: Stochastic modelling approach vs. extreme value distributions. *Studia Geophysica et Geodaetica* **2002**, *46*, 93–112. [CrossRef]
12. Wiśniewski, B.; Wolski, T. Occurrence probability of maximum sea levels in Polish ports of Baltic Sea coast. *Pol. Marit. Res.* **2009**, *3*, 62–69. [CrossRef]
13. Haigh, I.D.; Nicholls, R.; Wells, N. A comparison of the main methods for estimating probabilities of extreme still water levels. *Coast. Eng.* **2010**, *57*, 838–849. [CrossRef]
14. IPCC. *Climate Change. The Scientific Basis. Contribution of the Working Group I to the Third Assessment Report of the Intergovernmental Panel on Climate Change*; Houghton, D., Griggs, N., Van der Linden, D., Maskell, J., Eds.; Cambridge University Press: Cambridge, UK; New York, NY, USA, 2001; pp. 1–881, ISBN 0-521-01495-6.
15. Czernecki, B.; Miętus, M. Comparison of thermal classification on the example in selected regions of Poland. *Rev. Geophys.* **2011**, *LVI*, 201–233.
16. Wójcik, R.; Pilarski, M.; Miętus, M. Statistical downscaling of probability density function of daily precipitation on the Polish coast. *Meteorol. Hydrol. Water Manag.* **2014**, *2*, 27–36. [CrossRef]
17. Tylkowski, J. Temporal and spatial variability of air temperature and precipitation at the Polish coastal zone of the southern Baltic Sea. *Baltica* **2013**, *26*, 83–94.
18. Tylkowski, J. The temporal and spatial variability of coastal dune erosion in the Polish Baltic coastal zone. *Baltica* **2017**, *30*, 97–106. [CrossRef]
19. Wolski, T. *Temporal and Spatial Characterization of Extreme Baltic Sea Levels*; Scientific Publisher of the University of Szczecin: Szczecin, Poland, 2017; pp. 1–265, ISBN 978-83-7972-091-0.
20. Wiśniewski, B.; Wolski, T.; Giza, A. Adjustment of the European Vertical Reference System for the representation of the Baltic Sea water surface topography. *Sci. J.* **2014**, *38*, 106–117.
21. Russell, G.L.; Gornitz, V.; Miller, J.R. Regional sea level changes projected by the NASA/GISS atmosphere-ocean model. *Clim. Dyn.* **2000**, *16*, 789–797. [CrossRef]
22. Horton, E.B.; Folland, C.K.; Parker, D.E. The changing incidence of extremes in worldwide and Central England temperatures to the end of the twentieth century. *Clim. Chang.* **2001**, *50*, 267–295. [CrossRef]
23. Milly, P.C.D.; Wetherald, R.T.; Dunne, K.A.; Delworth, T.L. Increasing risk of great floods in a changing climate. *Nature* **2002**, *415*, 514–517. [CrossRef] [PubMed]
24. Zwoliński, Z. Selected extreme phenomena of Polish lakelands. *Landf. Anal.* **2008**, *8*, 98–106.
25. Miętus, M. Temperature and Precipitation Variability in the Polish Baltic Coast and Its Expected Course until 2030. *Res. Mater. IMGW Meteorol. Ser.* **1996**, *26*, 1–72.
26. Hupfer, P.; Harff, J.; Sterr, H.; Stigge, H.J. *Die Küste. Die Wasserstände an der Ostseeküste. Entwicklung-Sturmfluten-Klimawandel*; Kuratorium für Forschung im Küsteningenieurwesen: Cuxhaven, Germany, 2003; pp. 1–331, ISBN 3-8042-1057-0.
27. Suursaar, U.; Kullas, T.; Otsmann, M.; Kouts, T. Extreme sea level events in the coastal waters of western Estonia. *J. Sea Res.* **2003**, *49*, 295–303. [CrossRef]
28. Linderson, M.L.; Achberger, C.; Chen, D. Statistical downscaling and scenario construction of precipitation in Scania, southern Sweden. *Nord. Hydrol.* **2004**, *35*, 261–278. [CrossRef]

29. Jensen, J.; Müller-Navara, S.H. *Die Küste. Storm Surges on the German Coast*; Kuratorium für Forschung im Küsteningenieurwesen: Cuxhaven, Germany, 2008; Volume 74, pp. 92–124, ISBN 3-8042-1057-0.
30. Kriaučiūnienė, J.; Meilutytė-Barauskienė, D.; Reihan, A.; Koltsova, T.; Lizuma, L.; Šarauskienė, D. Variability of regional series of temperature, precipitation and river discharge in the Baltic States. *Boreal Environ. Res.* **2012**, *2*, 150–162.
31. Marosz, M.; Wójcik, R.; Pilarski, M.; Miętus, M. Extreme daily precipitation totals in Poland during summer: The role of regional atmospheric circulation. *Clim. Res.* **2013**, *56*, 245–259. [CrossRef]
32. Wróblewski, A. Analysis and forecast of long term sea level changes along the polish Baltic Sea coast, Part I Annual sea level maxima. *Oceanology* **1992**, *33*, 65–85.
33. Wolski, T.; Wiśniewski, B.; Giza, A.; Kowalewska-Kalkowska, H.; Boman, H.; Grabbi-Kaiv, S.; Hammarklint, T.; Holfort, J.; Lydeikaite, Z. Extreme sea levels at selected stations on the Baltic Sea coast. *Oceanology* **2014**, *56*, 259–290. [CrossRef]
34. Łupikasza, E. Spatial and temporal variability of extreme precipitation in Poland in the period 1951–2006. *Int. J. Climatol.* **2010**, *30*, 991–1007. [CrossRef]
35. Degirmendžić, J.; Kożuchowski, K.; Żmudzka, E. Changes of air temperature and precipitation in Poland in the period 1951–2000 and their relationship to atmospheric circulation. *Int. J. Climatol.* **2004**, *24*, 291–310. [CrossRef]
36. Marsz, A.A. Influence of the North Atlantic Oscillation (NAO) on the increase in air temperature over Poland under conditions of variable heat resources in the North Atlantic. *Rev. Geophys.* **2013**, *LVIII*, 127–143.
37. Kostrzewski, A.; Zwoliński, Z.; Winowski, M.; Tylkowski, J.; Samołyk, M. Cliff top recession rate and cliff hazards for the sea coast of Wolin Island (Southern Baltic). *Baltica* **2015**, *28*, 109–120. [CrossRef]
38. Winowski, M. Cliff landslides activity under the influence of extreme meteorological and hydrological conditions, Wolin Island—Southern Balic. *Landf. Anal.* **2015**, *28*, 87–102. [CrossRef]
39. Collins, M.; Knutti, R.; Arblaster, J.; Dufresne, J.-L.; Fichefet, T.; Friedlingstein, P.; Wehner, M. Long-term climate change: Projections, commitments and irreversibility. In *IPCC, Climate Change: The Physical Science Basis. Contribution of Working Group I to the Fifth Assessment Report of the Intergovernmental Panel on Climate Change*; Cambridge University Press: Cambridge, UK, 2013; pp. 1029–1136.
40. Nikulin, G.; Kjellström, E.; Hansson, U.; Jones, C.; Strandberg, G.; Ullerstig, A. Evaluation and future projections of temperature, precipitation and wind extremes over Europe in an ensemble of regional climate simulations. *Tellus* **2011**, *63A*, 41–55. [CrossRef]
41. Christensen, O.B.; Kjellström, E.; Zorita, E. Projected Change—Atmosphere. In *Second Assessment of Climate Change for the Baltic Sea Basin*; The BACC II Author Team, Ed.; Regional Climate Studies; Springer: Cham, Switzerland, 2015; ISBN 978-3-319-16006-1.
42. Lehtonen, I.; Ruosteenoja, K.; Jylhä, K. Projected changes in European extreme precipitation indices on the basis of global and regional climate model ensembles. *Int. J. Climatol.* **2014**, *34*, 1208–1222. [CrossRef]
43. Orlowsky, B.; Seneviratne, S.I. Global changes in extreme events: Regional and seasonal dimension. *Clim. Chang.* **2012**, *116*, 669–696. [CrossRef]
44. Church, J.A.; Clark, P.U.; Cazenave, A.; Gregory, J.M.; Jevrejeva, S.; Levermann, A.; Merrifield, M.A.; Milne, G.A.; Nerem, R.S.; Nunn, P.D.; et al. Sea level change. In *Climate Change: The Physical Science Basis. Contribution of Working Group I to the Fifth Assessment Report of the Intergovernmental Panel on Climate Change*; Stocker, T.F., Qin, D., Plattner, G.K., Tignor, M., Allen, S.K., Boschung, J., Nauels, A., Xia, Y., Bex, V., Midgley, P.M., Eds.; Cambridge University Press: Cambridge, UK, 2013; pp. 1137–1216.
45. Grinsted, A. Projected Change—Sea Level. In *Second Assessment of Climate Change for the Baltic Sea Basin*; BACC II Author Team, Ed.; Springer: London, UK, 2015; pp. 253–263, ISBN 978-3-319-16006-1.
46. Vousdoukas, M.I.; Voukouvalas, E.; Annunziato, A.; Giardino, A.; Feyen, L. Projections of extreme storm surge levels along Europe. *Clim. Dyn.* **2016**, *47*, 3171–3190. [CrossRef]
47. Gräwe, U.; Burchard, H. Storm surges in the western Baltic Sea: The present and a possible future. *Clim. Dyn.* **2012**, *39*, 165–183. [CrossRef]

water

MDPI

Article

Hydrometeorological Conditions for the Occurrence of Aeolian Processes on the Southern Baltic Coast in Poland

Marcin Hojan [1,*], Jacek Tylkowski [2] and Mirosław Rurek [1]

[1] Institute of Geography, Department of Landscape History Research, Kazimierz Wielki University in Bydgoszcz, 85-033 Bydgoszcz, Poland; mirur@ukw.edu.pl

[2] Institute of Geoecology and Geoinformation, Faculty of Geographical and Geological Sciences, Adam Mickiewicz University, Krygowski 10, 61-680 Poznań, Poland; jatyl@amu.edu.pl

* Correspondence: homar@ukw.edu.pl

Received: 15 September 2018; Accepted: 26 November 2018; Published: 27 November 2018

Abstract: This article presents the temporal and spatial variability of hydrometeorological conditions conducive aeolian processes on the Southern Baltic coastal zone in Poland. The analysis made use of daily meteorological (wind, temperature, and rainfall) and hydrological (sea level) data from 1961 to 2010. Data for four stations (Świnoujście, Kołobrzeg, Ustka, Hel) were provided by the Institute of Meteorology and Water Management in Warsaw. A time decomposition of hydrometeorological conditions conducive to the initiation and intensification of aeolian processes in the coastal zone was also performed. In addition to their scientific significance, the temporal and spatial conditions for aeolian processes on the Baltic coast of Poland have an essential utilitarian significance. Modern aeolian processes on the Baltic coast limit the development potential of the coastal zone. Aeolian processes have a positive and negative impact on geomorphological transformation of the sea coast. They take part in the reconstruction of the beach and foredunes after storms. In periods between storms, coastal wind is seen to decrease the balance of beach sediments and lowers the beach area. On the other hand, onshore wind favors, among other things, filling of tourist infrastructure and development located at the hinterland of the beach and dunes. Hydrometeorological conditions especially favorable to the intensification of aeolian processes are the main determinants of geomorphological changes in the coastal zone (some of which can be extreme). Temporal and spatial analysis of hydrometeorological conditions conducive to aeolian processes is important for many areas of human activity, especially those concerning protection, management, and development of the coast.

Keywords: aeolian processes; hydrometeorological conditions; onshore and offshore winds; Southern Baltic coast

1. Introduction

To determine how geo-ecosystems of the coastal zone function, it is necessary to identify in detail the hydrometeorological conditions that most influence the trend, cyclicality, and seasonality of geomorphological processes. An example of such a process is aeolian transport. It constitutes an important factor for shaping beaches, coastal dunes and sandy cliff slopes. Research on aeolian processes was carried out irregularly and in various parts of the Polish Baltic coast, including: Świna Gate Sandbar [1], Wolin island [2,3], Kołobrzeg [4], Mierzeja Łebska [4–11], Lubiatowo [12], and Hel [13]. The dynamics of aeolian transport depend on hydrometeorological factors and the type of substrate surface, especially its roughness and moisture [4,9,10,14,15]. Meteorological conditions (including the direction and speed of wind, atmospheric precipitation, and air and ground temperature) determine the initiation, intensity, and duration of aeolian processes. Sea level, on the other hand, determines the availability of beach sediments, which may be subject to deflation, transport, and aeolian accumulation.

Since the internal climate of the Earth varies constantly, climate models are devoted to analyze the non-linear interactions and different time responses of the components in the climate system. With regard to the atmosphere, changes in temperature, rainfall, wind speed and direction are mainly subject to particular analysis [16–18]. Atmospheric parameters may vary over the year. Changes in wind direction affect regional changes in atmospheric factors, such as rainfall and cloudiness [19]. What is more, changes in wind direction and other meteorological factors affect the morphological changes of the coast [20]. Statistical modeling is also used; it takes into account temperature and atmospheric precipitation in the Baltic Sea basin [21]. Research has shown that there are changes in wind directions and speeds in the Estonia region [22]. Sandy beaches, separated from the land by foredunes, favor transport of sand along the coast. Prolonged, above-average winds have a greater impact on sand transport than relatively short-lived strong winds during storms [23]. The analysis of available meteorological data allows to observe changes on the coast occurring as a result of intensifying aeolian processes.

The following take part in development of the beach: Coastal currents in the littoral zone, waves in the inflow zone causing beach erosion during the development of storms and accumulation of sand on the beach during the weakening of storms and aeolian processes in interstorm periods [9]. Material provided by marine alongshore currents comes from eroded cliffs [4,24].

Wind speed, air temperature, as well as air and substrate humidity affect the fraction of exhausted mineral material [10]. Granulation of dunes and sands of the Southern Baltic coast is mainly made up of sands. Research on sediments of sandy beaches in the Baltic Sea coastal zone in Lithuania [25] and in Poland [12] has shown that beaches in these countries are made of fine-grained sands. Other studies on beach sediments on the Curonian Spit, Vistula and Hel have shown that the beaches are dominated by medium- and coarse-grained sand, whereas foredunes are built of fine-grained sands [13,26]. This is in line with the results of a study by Reference [12], which indicated that the graining of beach sediments in Świnoujście, Kołobrzeg, Ustka and Hel is dominated by medium- and fine-grained sands. Coarse sands are found on beaches along which currents and sea waves blur clay thresholds [12].

So far, research carried out in the Baltic Sea basin has been related to various hydrometeorological factors. The rise of sea level [27,28], storm frequencies [29–32] and climatic conditions (also those in other parts of the world) [33] may affect the functioning of aeolian processes in the geosystem of the South Baltic coasts. This work aims to analyze meteorological data, which will allow to establish to answer to the following question: Do hydrometeorological conditions changing in time and space translate into variability of conditions for potential initiation and intensity of aeolian processes in the Polish coastal zone of the Southern Baltic? For this purpose, hydrometeorological threshold values for the potential favorable conditions of aeolian processes were defined, time dynamics and spatial variability of hydrometeorological conditions were indicated, and frequency of intensification of aeolian processes in relation to particularly conducive hydrometeorological conditions was presented. In the Polish coastal zone of the Baltic Sea, this type of research has not yet been done.

2. Materials and Methods

2.1. Aim and Methods

The main aim of this work was to conduct a temporal and spatial analysis of hydrometeorological conditions potentially favorable (initiating) and particularly conducive (intensifying) to aeolian processes in the Southern Baltic's Polish coastal zone. The occurrence of these conditions was accompanied by fulfillment of all threshold values. The conditions were classified into two groups (threshold values):

- Potentially favorable to initiation of aeolian processes (maximum wind speed $\geq$4 m·s^{-1}, no precipitation in the last two days, average daily air temperature >0 °C, maximum sea level (storm level) <570 cm).
- Particularly conducive to intensification of aeolian processes (maximum wind speed $\geq$10 m·s^{-1}, precipitation <5 mm/2 days, minimum air temperature >0 °C, maximum sea level (average level over many years) <~502 cm).

The above meteorological threshold values were derived from direct field studies of aeolian transport dynamics on the Polish Baltic coast. To determine the impact of weather conditions on the occurrence of aeolian processes in the South Baltic coastal zone, field research was conducted in both the dune coastal zone [1,3,4,17,18,34,35] and the cliff coastal zone [2,36,37]. It is assumed that the mobilization of sandy sediments on the beach begins at wind speeds of 4–5 m·s^{-1}. This threshold estimated to be 4.4 m·s^{-1} for the finest dry sands, and 10 m·s^{-1} for moist material [38]. Marked intensification of aeolian transport occurs during winds blowing from the sea, as well as parallel to the coast, as they buffet the entire beach zone and the slopes of dunes and cliffs. Aeolian processes are limited by atmospheric precipitation, which increases the moisture of surface sediments and limits their ability to move. Field studies [2] and regression analysis (atmospheric precipitation—aeolian transport) revealed that aeolian transport occurs when the sum of atmospheric precipitation from the last two days is <5 mm. In the case of pluviometric conditions, the mobilization of sand as a result of splashing was not accounted for. Apart from wind speed higher than 8 m·s^{-1}, movement of sand grains due to splashing is responsible for many factors related to, among others, kinetic energy of rain, morphology and forms of land cover during rainfall [10]. Many factors are responsible for the movement of sand as a result of splashing during rainfall. These include the kinetic energy of the rain, and the morphology and types of surface cover. Splashing of sand on the beach takes place even in situations when wind shear speeds are too low to initiate aeolian transport [39]. Aeolian transport is also determined by thermal conditions [7,10]. When surface-level sediments are frozen, the movement of sand is impossible. The initiation of aeolian transport therefore begins when the average daily air temperature is >0 °C. Yet, there are no limitations on the movement of sand when the temperature is above freezing. The dynamics of aeolian processes are also related to the availability of beach sediments. During storm surges ($\geq$570 cm in the Polish Baltic coastal zone), the sea level is high, making the availability of beach sediments for aeolian processes low (beach sediment may be available in the upper beach area), and sometimes cancelling it altogether (e.g., a narrow beach in the cliff coastal zone). For this reason, it was assumed that the conditions for initiation of aeolian processes occurred on days when sea level was below 570 cm. Particularly conducive to intensification of aeolian processes were wide beaches and a maximum sea level lower than the average sea level over the analyzed period, i.e., ~502 cm. This temporal and spatial analysis of how sea conditions affect aeolian processes was based on daily sea level data from mareographic stations in Świnoujście, Kołobrzeg, Ustka, and Hel. Only the Baltic High System (BHS) (based on the Kronstadt sea-gauge) was used. The estimated difference between the Normal-Null (NN)- and BHS-based systems is about 15 cm (the Kronstadt system is higher). Although Poland uses a high system based on the Kronstadt sea gauge, the registration and recording of sea levels is based on the Normal-Null reference system [29].

2.2. Hydrometeorological Data

This study is based on daily hydrometeorological data from 1961 to 2010 concerning: Average, maximum, and minimum air temperature; average and maximum wind speed; the sum of atmospheric precipitation; and average, maximum, and minimum sea level. The data was provided by the Institute of Meteorology and Water Management in Warsaw (https://danepubliczne.imgw.pl). Data was collected from four coastal stations in Świnoujście, Kołobrzeg, Ustka, and Hel (Figure 1). One drawback of these stations is that they are located behind the coastal dunes, which means that recorded wind speeds may be lower than those actually occurring on the beach.

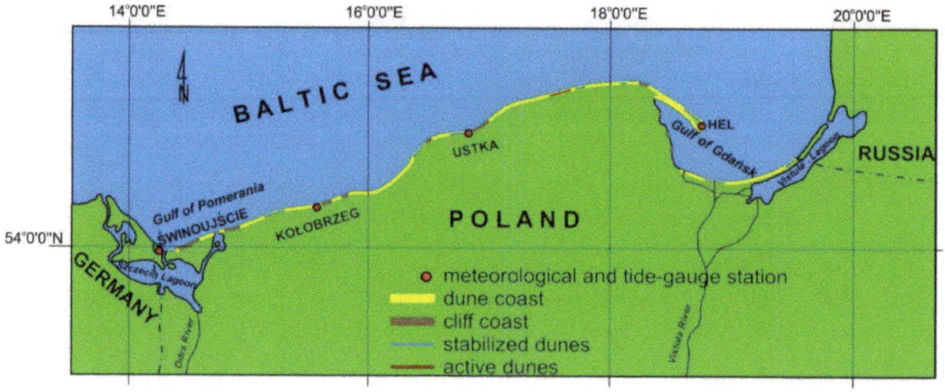

Figure 1. Area of research—location of measurement stations in the Polish Baltic coastal zone.

Wind directions were not taken into account in the study. It was assumed that aeolian processes operating on the beach occur with all directions of wind: Onshore, offshore, alongshore.

2.3. Study Area

The Polish coast has a length of 500 km [40]. It represents two basic types of coastline: Dune and cliff (Figure 1) [41]. Out of this, 80% is constituted by dune, developing in the Holocene. These are usually sandy barriers with a height of 2 to 35 m. When it comes to dune shores, about 15% are constituted by accumulative stretches and around 35% by erosional stretches [1]. Cliffs occupy about 15% of the sea shore [40]. They were created during the Holocene and contemporary erosion of frontal moraines. Moraines are built of glacial clays, usually 10–95 m high, as well as fluvioglacial sand and clays accumulated in the form of ground moraines, most often 6–15 m high. A small part (a few %) in estuarine sections of rivers is constituted by a low coast of flood plains or organic accumulation. The height of this type of coastline is 0–5 m above sea level. Aeolian processes occur on all coastal types, especially in the beach zone. Aeolian processes are also observed on the slopes of dunes and cliffs.

Four areas were selected for the analysis: Świnoujście, Kołobrzeg, Ustka, Hel. Distances between these areas are around 90–130 km. The Świnoujście area covers the section located east of Świnoujście. It is a dune section of the coast, which currently comprises an accumulative section. The Kołobrzeg zone west of the Parsęta River is a dune section. On the other hand, there is a clay shore east of the Parsęta River, with a height of 2–6 m above sea level. The western part of the Ustka area is a dune section. There are sand and clay cliffs up to 41 m above sea level in the eastern part of this area near Dębina. The Hel area, on the other hand, constitutes a spit section of the coast, which is 36 km long. There are also inland dunes in the final part of the Hel Peninsula.

3. Results

3.1. Hydrometeorological Conditions

Analysis of meteorological data revealed that from 1961 to 2010, the maximum daily wind speed was lower than 10 m·s^{-1} only in Kołobrzeg in 1989, 1991, 2005, 2009, and 2010. In these years, the proper anemometric conditions for aeolian processes in beach sediments wet from precipitation did not occur. The highest maximum wind speed occurred on the eastern coast (Ustka 26 m·s^{-1} in 2004, Hel 23 m·s^{-1} in 1999). The average annual wind speed was highest in Hel 4.4 m·s^{-1} and Ustka 4.3 m·s^{-1} (exceeded the threshold value of 4 m·s^{-1} for initiation of aeolian processes in dry sandy sediments), and significantly lower in Świnoujście 3.7 m·s^{-1} and Kołobrzeg 3.2 m·s^{-1}. The anemometric conditions in the eastern zone of the Polish Baltic coast were more favorable to aeolian processes (Figure 2).

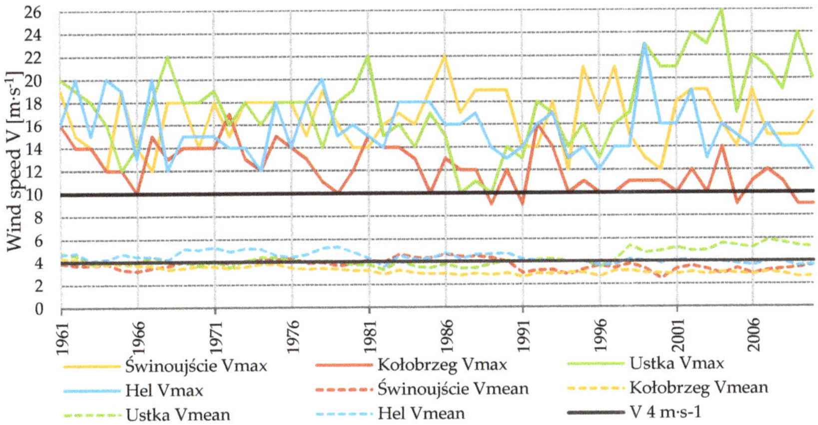

Figure 2. Annual dynamics of wind speed in the Polish Baltic coastal zone.

Pluviometric conditions in the open coastal zone of the Baltic (Ustka and Kołobrzeg) were less favorable to aeolian processes than those in bay areas (Świnoujście and Hel). In Ustka and Kołobrzeg, the following were confirmed to be significantly higher: Average annual sum of precipitation (702.2 mm and 636.6 mm); average annual number of days with precipitation (181 and 180 days); and maximum daily sum of precipitation (94.2 and 85.2 mm). In Świnoujście and Hel (near the Pomeranian and Gdańsk Bay), the following were confirmed to be significantly lower: Average annual sum of precipitation (561.5 and 586.1 mm); average annual number of days with precipitation (169 and 177 days); and maximum daily sum of precipitation (58.7 and 77.1 mm). Pluviometric conditions in the central zone of the Baltic coast (Kołobrzeg and Ustka) were thus less favorable to aeolian processes than those in the western (Świnoujście) and eastern (Hel) zones (Figure 3).

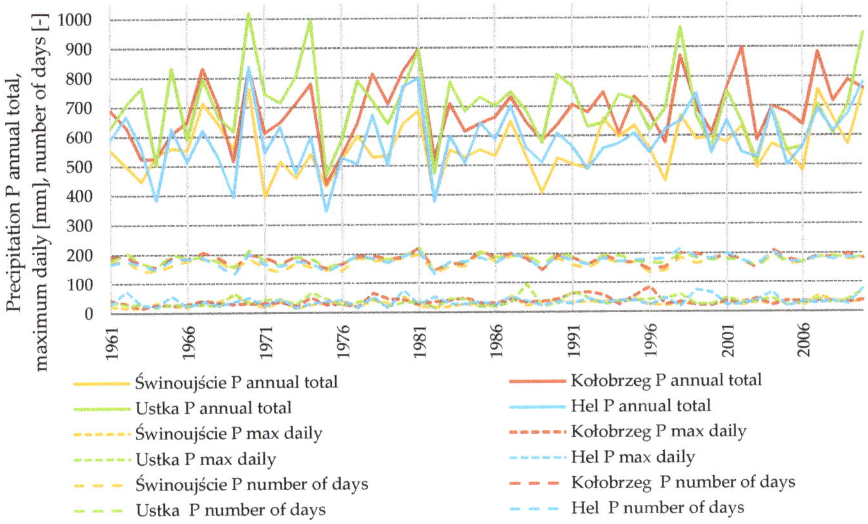

Figure 3. Annual dynamics of precipitation in the Polish Baltic coastal zone.

Each year, the average annual air temperature was similar across the entire Baltic coast zone. This confirms the rule that thermal conditions are more spatially homogenous than pluviometric and anemometric conditions. The average annual air temperature for the analyzed period decreased from

8.5 °C in Świnoujście, to 8.2 °C in Kołobrzeg, to 8.1 °C in Ustka and Hel. The occurrence of aeolian processes is limited by the freezing of surface settlements ($t_{max} \leq 0$ °C). The most days with freezing temperatures were recorded in the eastern coastal zone in Hel, with an annual average of 28 days. Slightly fewer days with freezing temperatures were recorded in the remaining coastal zone, with an annual average of 23 to 25 days. The conditions for aeolian processes were particularly unfavorable in 1963, 1969, 1997, and 2010, when there were over 50 days with freezing temperatures recorded on the entire coast (Figure 4).

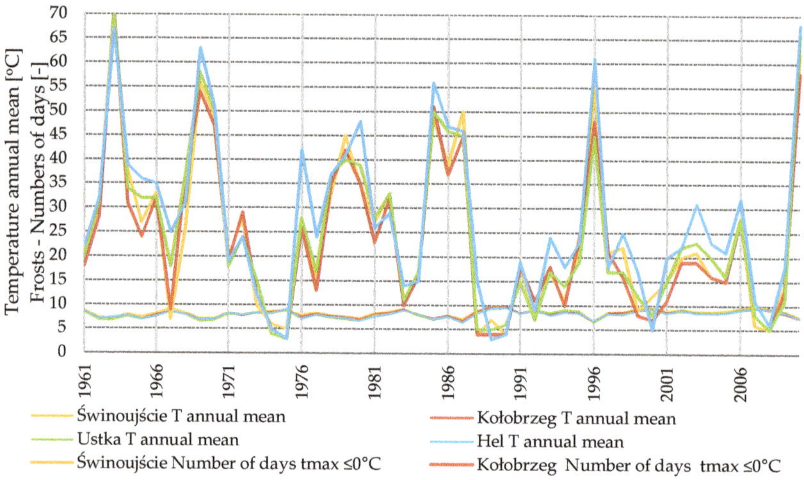

Figure 4. Annual dynamics of air temperature in the Polish Baltic coastal zone.

The annual dynamics of the maximum, average, and minimum sea level from 1961 to 2010 are very temporally and spatially varied (Figure 5). The average sea level ranged from 500 cm in Świnoujście to 504 cm in Hel. The absolute maximum sea level was 661 cm (Świnoujście—4 November 1995), and the absolute minimum sea level was 375 cm (Świnoujście—4 November 1979). The amplitude of sea-level fluctuations in the Polish Baltic coastal zone was, thus, 286 cm. It was only in 1966 that no excess of storm level $H_{max} > 570$ cm was recorded at any of the stations. There were no limitations on the availability of beach sediments for aeolian processes on any day of that year.

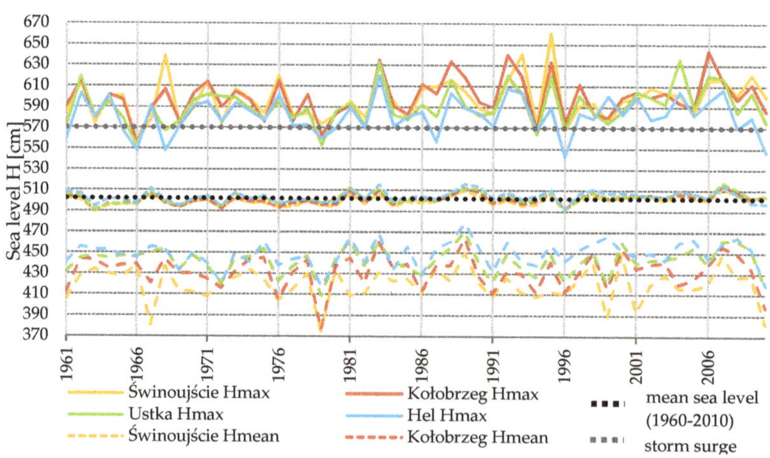

Figure 5. Annual dynamics of sea level in the Polish Baltic coastal zone.

Particularly favorable conditions for intensification of aeolian processes occurred in the entire coastal zone when the maximum sea level was lower than the average sea level from the analyzed period (~502 cm), and especially when the sea level was exceptionally low (H_{min} < 430 cm). It was then that surface-level beach sediments were most available for aeolian processes. Such events (H_{min} < 430 cm) occurred in the entire coastal zone in 1972, 1979, and 2010. The greatest differences in sea level were recorded on the west coast (Świnoujście and Kołobrzeg), where storm surges and post-storm level drops were particularly significant. Relatively small differences in maximum and minimum sea level occurred on the east coast in Hel.

3.2. *Potentially and Particularly Favorable Hydrometeorological Conditions for the Occurrence of Aeolian Processes*

Temporal and spatial analysis of hydrometeorological conditions for initiation of aeolian processes revealed a greater annual average of events on the coast of the Pomeranian Bay (Świnoujście, 92 days) and the Gdańsk Bay (Hel, 96 days) than in the open coastal zone (Kołobrzeg, 64 days; Ustka, 87 days) (Table 1). Along the entire coast, the greatest number of days potentially beneficial for initiation of aeolian processes was found in the decade of 1971–1980 (annual average of 93 days) and the least days in the decade of 2001–2010 (annual average of 79 days). This happened on the most days in 1975 (average of 125 days) and the least days in 2004 (average 65 days). For the occurrence of hydrometeorological conditions potentially favorable to initiation of aeolian processes, all thermal, precipitation, wind, and sea criteria must be fulfilled simultaneously.

Table 1. Average number of days per year with conditions potentially favorable to initiation of aeolian processes in the Polish Baltic coastal zone.

Hydrometeorological Criteria for Aeolian Processes Initiation	Place	Time Period					
		1961–1970	1971–1980	1981–1990	1991–2000	2001–2010	1961–2010
		Mean Number of Days in Year					
Wind speed $\geq$4 m·s^{-1}	Świnoujście	293	315	344	264	262	295
	Kołobrzeg	305	288	253	214	192	250
	Ustka	312	306	291	302	351	312
	Hel	325	337	336	319	319	327
Mean air temperature >0 °C	Świnoujście	303	322	323	327	322	319
	Kołobrzeg	301	322	324	327	321	319
	Ustka	300	320	321	329	323	318
	Hel	301	318	322	326	321	318
Sea level <502 cm	Świnoujście	362	361	358	360	358	360
	Kołobrzeg	362	360	358	360	358	360
	Ustka	363	361	358	361	359	360
	Hel	364	362	360	361	361	361
Precipitation 0 mm/2 days	Świnoujście	137	141	131	141	127	135
	Kołobrzeg	122	127	123	128	122	124
	Ustka	123	128	119	126	125	124
	Hel	131	133	126	123	122	127
Potential favorable days for aeolian processes initiation	Świnoujście	85	105	107	86	79	92
	Kołobrzeg	75	81	66	56	44	64
	Ustka	80	85	76	88	104	87
	Hel	94	103	100	92	88	96

This is why the occurrence of all these conditions at once in the entire Baltic coastal zone is significantly rarer (annual average of 85 days) than their occurrence individually (sea level, average 360 days; air temperature, average 319 days; wind speed, average 296 days; and precipitation, average 128 days). During the analyzed period, there was a statistically significant ($p < 0.05$) increase in the number of days per year with an average air temperature of >0 °C (r ~ +0.3), and a decrease in the number of days per year with a maximum wind speed of $\geq$4 m·s^{-1} (r ~ −0.5). Only in Ustka was there an increase in the number of days per year with a wind speed of $\geq$4 m·s^{-1} (r ~ +0.9). The remaining hydrometeorological conditions, i.e., precipitation of 0 mm/2 days and sea level <570 cm, exhibited a statistically insignificant ($p > 0.05$) downward trend in the number of days per year, with a low correlation (r < −0.3).

Analysis of the number of days per year with favorable conditions for initiation of aeolian processes revealed a statistically insignificant downward trend for Świnoujście and Hel (r ~ −0.2), and a statistically significant downward trend for Kołobrzeg (r ~ −0.7). Only for the central coast in the region of Ustka was a statistically significant upward trend recorded (r ~ +0.4) (Figure 6).

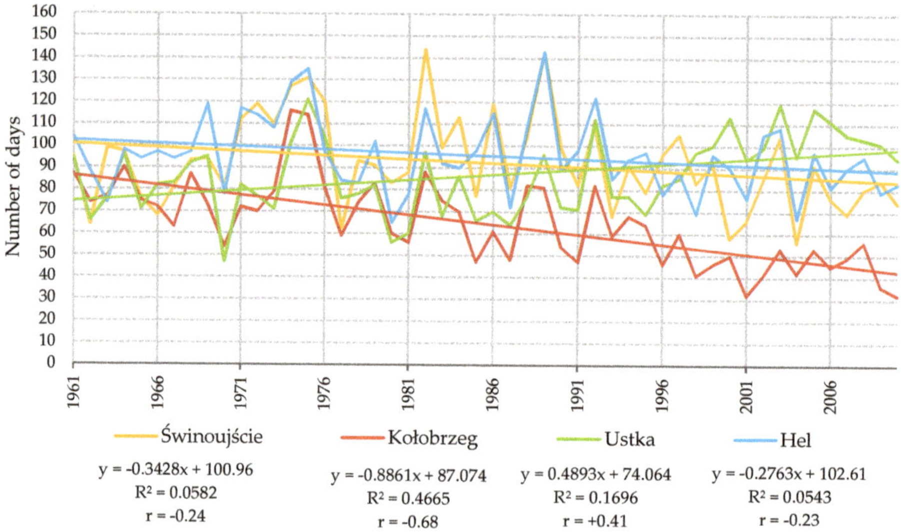

Figure 6. Annual trend of days with potentially favorable conditions for initiation of aeolian processes in the Polish Baltic coastal zone.

Temporal and spatial analysis of hydrometeorological conditions particularly conducive to intensification of aeolian processes revealed a greater average annual number of events in Świnoujście and Hel (4 days) than in Ustka (3 days) and Kołobrzeg (1 day) (Table 2). Conditions particularly conducive to intensification of aeolian processes occurred on the most days in the decade of 1971–1980 (annual average of 5 days), and on the least days in the decade of 1991–2000 (annual average of only 1 day). This happened on the most days in 1976 (average of 9 days) and the least in 1990, 1991, and 1997 (no days). Intensification of aeolian processes in the coastal zone may only occur when all hydrometeorological criteria are fulfilled simultaneously. Such events in the entire Baltic coastal zone are thus very rare (annual average of 3 days). The individual hydrometeorological conditions of these events occur more frequently (air temperature—average of 290 days, precipitation—average of 281 days, sea level—average of 140 days, and wind speed—average of just 10 days). During the analyzed period, there was a statistically significant ($p < 0.05$) increase in the number of days per year with a minimum daily air temperature of >0 °C (r ~ +0.3), and a decrease in the number of days per year with a maximum wind speed of $\geq$10 m·s^{-1} (r ~ −0.6). Only in Ustka was there an increase in the number of days per year with a wind speed of $\geq$10 m s^{-1} (r ~ +0.5). The remaining hydrometeorological conditions, i.e., precipitation of <5 mm/2 days and sea level <502 cm, exhibited a statistically insignificant ($p > 0.05$) downward trend in the number of days per year, with a low correlation (r < −0.3).

Table 2. Average number of days per year with conditions particularly conducive to intensification of aeolian processes in the Polish Baltic coastal zone.

Hydrometeorological Criteria for Aeolian Processes Intensification	Place	Time Period					
		1961–1970	1971–1980	1981–1990	1991–2000	2001–2010	1961–2010
		Mean Number of Days in Year					
Maximum wind speed ≥10 m s^{-1}	Świnoujście	24	41	47	22	28	32
	Kołobrzeg	14	10	4	3	2	6
	Ustka	34	24	15	34	95	40
	Hel	64	85	39	18	11	43
Minimum air temperature >0 °C	Świnoujście	275	287	290	292	289	286
	Kołobrzeg	268	285	287	291	286	283
	Ustka	308	308	308	308	308	308
	Hel	267	282	287	293	292	284
Mean sea level <502 cm	Świnoujście	169	174	134	149	112	148
	Kołobrzeg	159	177	126	139	110	142
	Ustka	158	156	121	137	109	136
	Hel	144	158	114	123	122	132
Precipitation <5 mm/2 days	Świnoujście	286	293	289	286	285	288
	Kołobrzeg	272	279	279	275	267	274
	Ustka	271	272	273	278	278	275
	Hel	291	293	286	290	282	288
Particularly favorable days for aeolian processes intensification	Świnoujście	4	6	5	2	2	4
	Kołobrzeg	1	1	0	0	0	0
	Ustka	3	1	0	3	9	3
	Hel	6	11	1	1	0	4

Analysis of the number of days per year with conditions conducive to intensification of aeolian processes revealed a statistically significant downward trend for Kołobrzeg and Hel (r ~ −0.6), and a statistically significant upward trend for Ustka (r ~ +0.5) (Figure 7). For Świnoujście, this trend was statistically insignificant.

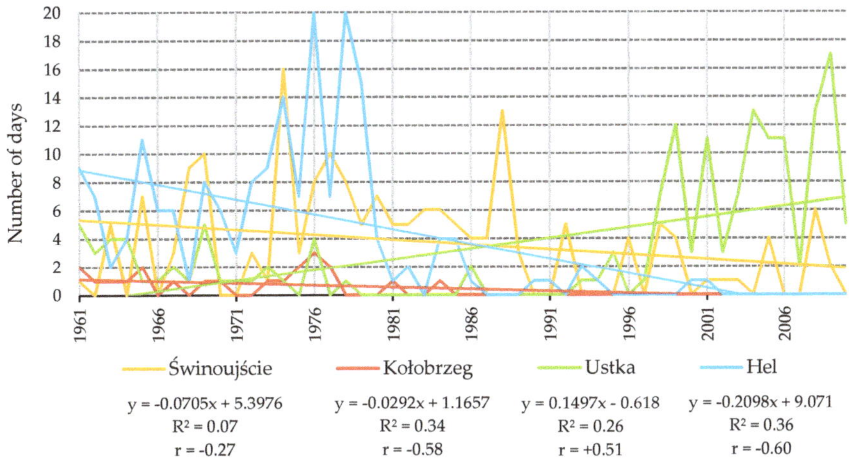

Figure 7. Annual trend of days with conditions particularly conducive to intensification of aeolian processes in the Polish Baltic coastal zone.

The occurrence of hydrometeorological conditions potentially favorable to initiation and particularly conducive to intensification of aeolian processes in the Polish Baltic coastal zone clearly depended on the seasons (Figure 8). During the cold half-year, especially from November to March, the number of days with conditions for the occurrence of aeolian processes was lower than from April to October. The cold half-year is the storm period, during which, despite higher wind speeds, storm surges are not conducive to aeolian processes, as they limit the availability of sandy beach sediments. Furthermore, during this period, there is a higher frequency of days with precipitation, as well as low temperatures causing ground frost. In the warm half-year, there are significantly fewer storm surges and days with precipitation, and no days with freezing temperatures. Conditions particularly favorable to the initiation and intensification of aeolian processes occurred especially in the spring, from April to June. At the beginning of spring, hydrometeorological conditions for the occurrence of aeolian processes were exceptionally favorable, as vegetation was not fully developed at the beginning of the season, and therefore did not limit the availability of sandy sediments for aeolian processes. Additionally, the inflow of dry masses of continental air from the Northeast were frequently recorded during this time.

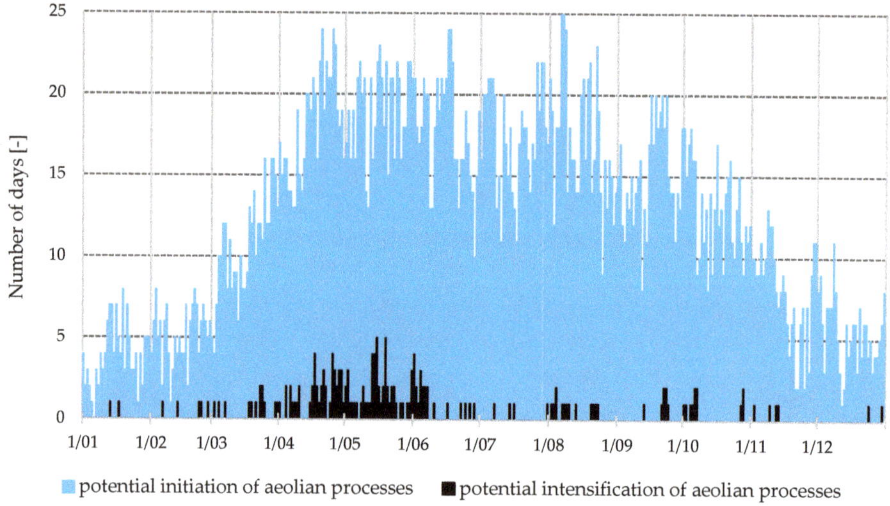

Figure 8. Seasonal dynamics of the number of days with conditions potentially favorable to initiation and particularly conducive to intensification of aeolian processes in the Polish Baltic coastal zone—a case study on Świnoujście 1961–2010.

For the Polish Baltic coastal zone, analysis of the general trend in occurrence of hydrometeorological conditions potentially favorable to initiation and particularly conducive to intensification of aeolian processes (based on the maximum number of days per year with conditions for aeolian processes recorded by all four measurement stations) revealed statistically significant patterns (Figure 9). For hydrometeorological conditions potentially favorable to initiation of aeolian processes, an upward trend was observed. However, for hydrometeorological conditions particularly conducive to initiation of aeolian processes, a downward trend was observed. The hydrometeorological conditions from 1961 to 2010 were thus characterized by a certain increase in the number of events (days) with conditions for the occurrence of aeolian processes, albeit of relatively low intensity.

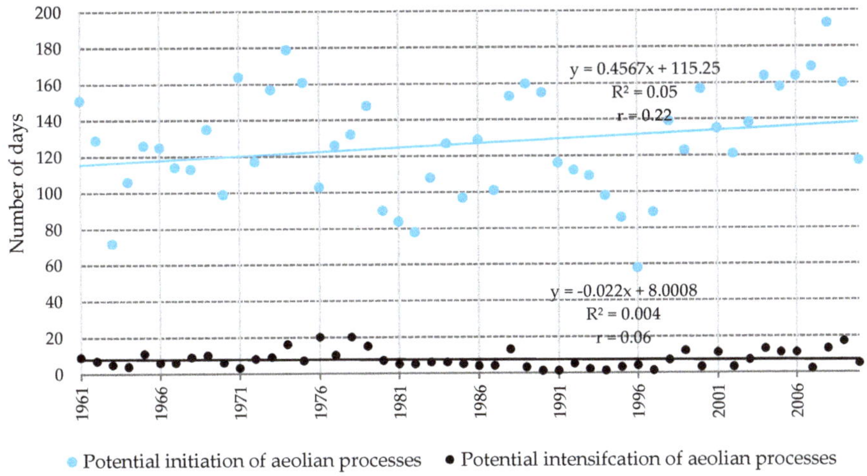

Figure 9. Trend in the number of days with hydrometeorological conditions potentially favorable to initiation and particularly conducive to intensification of aeolian processes in the Polish Baltic coastal zone.

4. Discussion

Research on the mechanism of aeolian processes in the coastal zone (especially quantitative values and the laws governing the movement of particles by wind) conducted in natural conditions [11,42,43], as field experiments [5,10,15,34,44,45], and especially as laboratory experiments [46,47] has often yielded divergent results. The number of variables determining the mechanism of aeolian processes in the coastal zone is much higher than in desert areas, especially in laboratory conditions (wind tunnels).

Field observations have shown that aeolian processes have positive and negative effects in geomorphological transformations of the coastal zone. The first group contains reconstruction of the beach after storms and during interstorm periods [9]. At this time, sand can be transported by winds from the beach to the dunes [1,9,10,34,45,48]. Ripple marks form on the surface of the beach, as well as aeolian shadows behind various obstacles, transverse sand patches, or even small barchans [10]. The second group contains lowering of the beach surface by seaward and alongcoast winds [4,9,49]. The visible effect of lowering of the beach is aeolian pavement and microshadows behind small obstacles [10].

During a storm build-up, the transverse profile of the beach (width and height) is of great importance; the beach can be flooded completely. During weakening of the storm, the aeolian processes act the most quickly on the upper beach fragment at the base of the dunes or cliff. During large storms, at sea level of 570 cm, sea waves cause erosion of sand dunes [1,41]. Such a situation occurs in many places on the Polish coast, even on beaches with a width of up to 70 m, which have low altitudes.

The threshold hydrometeorological conditions potentially favorable to initiation and particularly conducive to intensification of aeolian processes determined in the study are a kind of generalization. The threshold values were determined based on a review of literature concerning field research on the dune coastal zone [5,9,10,38], and original research on the cliff coastal zone [2,3]. We can assume that the hydrometeorological criteria used in this study are very general, but appropriate for the South Baltic coastal zone.

Analysis of the variability of hydrometeorological conditions over time on the Polish coast revealed patterns similar to those found in other studies on the South Baltic coast [50–52]. Current trends and forecasts of climatic conditions in the South Baltic coastal zone indicate that for every 10-year interval, the average annual air temperature will increase by 1.9 °C, and the annual sum of precipitation will decrease by 4% (meaning around 30 mm) [53]. The trend of increasing temperatures, especially during winter (1.7–2.3 °C), will limit the number of days on which freezing temperatures freeze the ground, and as a consequence, improvement of conditions unfavorable to the functioning of aeolian processes. Furthermore, the drop in precipitation will most likely reduce the number of days with precipitation, and thus increase the occurrence of conditions favorable to aeolian processes. Additionally, the observed increase in sea level by 3–4 cm every 10 years [41] may reduce the width of beaches and the availability of sandy sediments for aeolian processes. The increase in the level of the South Baltic Sea is insignificantly higher than the forecasted rise in global sea level, which is estimated at 1–2 mm·a^{-1} [54].

The high number of days with conditions potentially favorable to aeolian processes from 1971 to 1980 was accompanied by high annual intensity of potential aeolian transport. In this decade, the value of aeolian transport in Świnoujście was estimated at around 1075 t·m^{-1}—from ~80 t·m^{-1} in 1972 to ~150 t·m^{-1} in 1977 [4]. During the same period in Kołobrzeg, there were significantly fewer hydrometeorological events favorable to initiation of aeolian processes. The average annual value of aeolian transport was around 395 t·m^{-1}—from ~30 t·m^{-1} in 1980 to ~60 t·m^{-1} in 1975 [4]. According to Reference [4], 80% of the total aeolian transport takes place during 8% of the year (for about 1 month). The period of potential increase in aeolian transport (1 month) therefore constituted 30% to 50% of the average annual number of days with conditions favorable to aeolian processes (92 days in Świnoujście and 64 days in Kołobrzeg).

5. Conclusions

The relationship between hydrometeorological conditions and aeolian processes in the coastal zone is not linear. Extremely high wind speed does not always generate extreme aeolian erosion, transport, and accumulation on beaches and the slopes of dunes and cliffs, especially when strong winds are accompanied by storm surges and significant precipitation. The geomorphological effects of wind in the coastal zone are determined by many other factors that disturb the relationship between hydrometeorological conditions and aeolian processes. Among the most important of these factors are: Morpholytic conditions on beaches and the slopes of dunes and cliffs; surface exposure to wind; land cover from vegetation in different seasons; the dynamics and frequency of previous extreme hydrometeorological events; and human activity (e.g., hydraulic engineering).

Temporal and spatial analysis of hydrometeorological conditions determining initiation and intensification of aeolian processes in the South Baltic coastal zone in Poland yielded the following conclusions:

- The hydrometeorological and morpholithodynamic conditions of the coastal zone are conducive to aeolian processes. The average annual number of days with conditions favorable to initiation of aeolian processes is particularly high in terms of sea level (360 days) and air temperature (319 days). This number is slightly lower for wind speed (296 days), and lowest for precipitation (128 days). All of these factors must occur simultaneously for the initiation of aeolian processes. For this reason, the average annual number of potentially favorable events was 85. The number of days with conditions particularly conducive to intensification of aeolian processes was significantly lower (only three). While the average annual number of days with conducive conditions in terms of air temperature and precipitation was high (290 and 281 days), this number was significantly lower for sea level (140 days), and negligible for wind speed (only 10 days). Therefore, the average annual number of days with conditions potentially favorable to initiation of aeolian processes is nearly 30 times greater than the number of days with conditions particularly conducive to their intensification.
- The Pomeranian Bay (Świnoujście) and Gdańsk Bay (Hel) are particularly predisposed to the occurrence of hydrometeorological conditions potentially favorable to aeolian processes. In these areas, aeolian processes can occur for over 3 months per year on average. In the open coastal zone, aeolian processes can occur for around 2 months per year on average, e.g., in Kołobrzeg.

- The upwards trend of conditions particularly favorable to the intensification of aeolian processes for Ustka is probably related to the coastline's exposure to seaward winds. An additional effect may be the increase in share of winds associated with storms.
- Clustering of high frequencies of potential aeolian processes occurred in the 1970s. The lowest frequency of hydrometeorological conditions for the occurrence of aeolian processes was recorded in the first decade of the 21st century. Aeolian processes may occur on as few as 32 days per year (Kołobrzeg, 2010), and as many as 143 days per year (Świnoujście and Hel, 1989). In seasonal terms, the hydrometeorological conditions most conducive to initiation and intensification of aeolian processes occurred in the spring, especially from April to June.
- Conditions conducive to intensification of aeolian processes indicate that there is positive and negative wind activity within the coastal zone. Seaward wind directions cause formation of forms on the beach and build up foredunes. In addition, they cause flooding of promenades, pavements and streets in seaside resorts. Inland wind directions cause dissipation of dunes and clear off sandy material to the sea. Seaward and alongshore directions contribute to lowering of the beach area and transporting sand to other sections of the beach.

Regional climate models (RCM) predict climate change in the upcoming few decades involving the increase of temperature and precipitation [55]. The efficiency of daily rainfall is also to increase [56], and their frequency will be lower [57]. More frequent and longer periods of heat waves are predicted [16] as well as the possibility of long periods of drought [58], which will be conducive to the functioning of aeolian processes. The number of days potentially beneficial and particularly favorable to aeolian processes is likely to increase. An increase in their intensity is also expected.

Considering the current trend of rising sea levels, climatic changes, and the increasing frequency of extreme hydrometeorological events in the South Baltic coastal zone, it is not possible to reliably forecast the frequency of aeolian processes. However, the results of this study may be useful for determining how the coast functions, especially its beaches, coastal dunes, and moraine cliffs, whose low resistance to aeolian erosion is characteristic for the Baltic coast from Germany to Estonia.

Author Contributions: Conceptualization, M.H. and J.T.; Data curation, M.H. and J.T.; Formal analysis, M.H. and J.T.; Funding acquisition, M.H. and J.T.; Investigation, M.H., J.T. and M.R.; Methodology, M.H.; Project administration, M.H. and J.T.; Resources, M.H. and J.T.; Validation, J.T.; Visualization, M.H., J.T. and M.R.; Writing—original draft, M.H. and J.T.; Writing—review and editing, M.H and M.R.

Funding: The APC was funded by the Polish Ministry of Science (Project Supporting Maintenance of Research Potential of the Department of Physical Edu., Health and Tourism at Kazimierz Wielki University no. BS/2016/N1).

Acknowledgments: Hydrometeorological data were obtained from the Institute of Meteorology and Water Management, National Research Institute in Warsaw. We would like to thank the four reviewers for very helpful comments that improved our paper substantially.

Conflicts of Interest: The authors declare no conflict of interest.

References

1. Łabuz, T.A. Polish coastal dunes—Affecting factors and morphology. *Landf. Anal.* **2013**, *22*, 33–59. [CrossRef]
2. Hojan, M. Aeolian processes on the cliffs of Wolin Island. In *Quaestiones Geographicae*; Adam Mickiewicz University Press: Poznań, Poland, 2009; Volume 28, pp. 39–46. ISBN 978-83-232-2133-3.
3. Hojan, M.; Więcław, M. Influence of meteorological conditions on aeolian processes along the Polish cliff coast. *Baltica* **2014**, *27*, 61–72. [CrossRef]
4. Borówka, R.K. Multi-year trends of change in the intensity of potential aeolian transport on the West Pomeranian Coast of the Baltic in context of the morphology and present-day development of coastal dunes. In *Quaternary Studies in Poland SI*; Polish Scientific Publishers PWN: Warszawa, Poland, 1999; pp. 67–75.
5. Hildebrandt-Radke, I. The effect of meteorological factors on aeolian transport on the beach of the Gardno-Łeba Barrier. In *Quaternary Studies in Poland SI*; Polish Scientific Publishers PWN: Warszawa, Poland, 1999; pp. 109–114.

6. Hildebrandt-Radke, I. Rola szerokości plaży w nasyceniu strumienia wiatrowo-piaszczystego na plaży Mierzei Gardnieńsko-Łebskiej. *Badania Fizjograficzne nad Polską Zachodnią* **2002**, *53*, 43–56.
7. Borówka, R.K. Present day dune processes and dune morphology on the Łeba Barrier, polish coast of the Baltic. *Geografiska Annaler* **1980**, *62A*, 75–82. [CrossRef]
8. Borówka, M.; Rotnicki, K. Balance of the aeolian sand transport on the beach and the problem of sand nourishment of the active dune field on the Łeba Barrier. *J. Coast. Res.* **1995**, *22*, 257–265.
9. Borówka, M.; Rotnicki, K. Main directions of aeolian sand transport and its budget on barier sandy beach (Łeba Barrier case study). In *Ewolucja Geosystemów Nadmorskich Południowego*; Bałtyku, R.K., Borówka, Z., Młynarczyk, A., Eds.; Bogucki Wyd. Nauk.: Szczecin, Poland, 1999; pp. 17–24. ISBN 83-88163-02-7.
10. Rotnicka, J. *Aeolian Sand Transport on a Tideless Beach: Rate, Controlling Factors and Influence on Foredune Formation (Łeba Barrier Case, Poland)*; Bogucki Wyd. Nauk.: Poznań, Poland, 2013; ISBN 978-83-63400-71-2.
11. Rotnicka, J. Aeolian vertical mass flux profiles above dry and moist sandy beach surfaces. *Geomorphology* **2013**, *187*, 27–37. [CrossRef]
12. Zawadzka-Kahlau, E. *Morphodynamics of Southern Baltic Dune Coasts*; Wydawnictwo Uniwersytetu Gdańskiego: Gdańsk, Poland, 2012; ISBN 978-83-7865-016-4.
13. Žilinskas, G.; Jarmalavièius, D.; Pupienis, D. The influence of natural and anthropogenic factors on grain size distribution along the south eastern Baltic spits. *Geol. Q.* **2018**, *62*, 375–384.
14. Bauer, B.O.; Davidson-Arnott, R.G.D.; Hesp, P.A.; Namikas, S.L.; Ollerhead, J.; Walker, I.J. Aeolian sediment transport on a beach: Surface moisture, wind fetch, and mean transport. *Geomorphology* **2009**, *105*, 106–116. [CrossRef]
15. Rotnicka, J. Impact of beach surface type on the rate of sand transport by wind. In Proceedings of the 11th International Coastal Symposium, Szczecin, Poland, 9–13 May 2011; pp. 2058–2062.
16. Nikulin, G.; Kjellström, E.; Hansson, U.; Jones, C.; Strandberg, G.; Ullerstig, A. Evaluation and future projections of temperature, precipitation and wind extremes over Europe in an ensemble of regional climate simulations. *Tellus* **2011**, *63A*, 41–55. [CrossRef]
17. Bierstedt, S.E.; Hünicke, B.; Zorita, E. Variability of wind direction statistics of mean and extreme wind events over the Baltic Sea region. *Tellus* **2015**, *67*, 29073. [CrossRef]
18. Bierstedt, S.E.; Hünicke, B.; Zorita, E.; Ludwig, J. A wind proxy based on migrating dunes at the Baltic coast: Statistical analysis of the link between wind conditions and sand movement. *Earth Syst. Dyn.* **2017**, *8*, 639–652. [CrossRef]
19. Rutgersson, A.; Jaagus, J.; Schenk, F.; Stendel, M.; Bärring, L.; Briede, A.; Claremar, B.; Hanssen-Bauer, I.; Holopainen, J.; Moberg, A.; et al. Recent Change—Atmosphere. In *Second Assessment of Climate Change for the Baltic Sea Basin*; Springer International Publishing: Berlin, Germany, 2015; pp. 69–97.
20. Clemmensen, L.B.; Hansen, K.W.T.; Kroon, A. Storminess variation at Skagen, northern Denmark since ad 1860: Relations to climate change and implications for coastal dunes. *Aeolian Res.* **2014**, *15*, 101–112. [CrossRef]
21. Hünicke, B.; Zorita, E. Influence of temperature and precipitation on decadal Baltic Sea level variations in the 20th century. *Tellus* **2006**, *58A*, 141–153. [CrossRef]
22. Jaagus, J.; Kull, A. Changes in surface wind directions in Estonia during 1966–2008 and their relationships with large-scale atmospheric circulation. *Estonian J. Earth Sci.* **2011**, *60*, 220–231. [CrossRef]
23. Reimann, T.; Tsukamoto, S.; Harff, J.; Osadczuk, K.; Frechen, M. Reconstruction of holocene coastal foredune progradation using luminescence dating—An example from the Swina Barrier (Southern Baltic sea, NW Poland). *Geomorphology* **2011**, *132*, 1–16. [CrossRef]
24. Subotowicz, W. Transformation of the cliff coast in Poland. *J. Coast. Res.* **1995**, *22*, 57–62.
25. Molodkov, A.; Bitinas, A. Sedimentary record and luminescence chronology of the Lateglacial and Holocene aeolian sediments in Lithuania. *Boreas* **2006**, *35*, 244–254. [CrossRef]
26. Kovaleva, A.; Chubarenko, B.; Pupienis, D. Grain size variability as an indicator of sediment transport alongshore the Curonian Spit (south-eastern Baltic Sea). *Baltica* **2016**, *29*, 145–155. [CrossRef]
27. Rotnicki, K.; Borzyszkowska, W. Accelerated sea level rise and its components at the Polish Baltic Coast in the years 1951–1990. In *Ewolucja Geosystemów Nadmorskich Południowego Bałtyku*; Borówka, R.K., Młynarczyk, Z., Wojciechowski, A., Eds.; Bogucki Wydawnictwo Naukowe: Poznań, Poland; Szczecin, Poland, 1999; pp. 141–160. ISBN 83-88163-02-7.

28. Rosentau, A.; Bennike, O.; Uścinowicz, S.; Miotk-Szpiganowicz, G. The Baltic Sea Basin. In *Submerged Landscapes of the European Continental Shelf: Quaternary Paleoenvironments*, 1st ed.; Flemming, N.C., Harff, J., Moura, D., Burgess, A., Bailey, G.N., Eds.; John Wiley & Sons Ltd.: Hoboken, NJ, USA, 2017; pp. 103–133. [CrossRef]
29. Wolski, T. *Spatial and Temporal Characteristics of the Extreme Sea Levels of the Baltic Sea*; Wydawnictwo Naukowe Uniwersytetu Szczecińskiego: Szczecin, Poland, 2017; ISBN 978-83-7972-091-0.
30. Sztobryn, M.; Stigge, H.J. *Storm Surges on the Southern Baltic Sea*; IMGW Press: Warszawa, Poland, 2005; ISBN 83-88897-61-6.
31. Wiśniewski, B.; Wolski, T. Occurrence probability of maximum sea levels in Polish ports of Baltic Sea coast. *Pol. Mar. Res.* **2009**, *3*, 62–69. [CrossRef]
32. Łabuz, T.A.; Grunewald, R.; Bobykina, V.; Chubarenko, B.; Česnulevičius, A.; Bautrenas, A.; Morkunaite, R.; Tõnisson, H. Coastal dunes of the Baltic Sea shores: A review. *Quaest. Geogr.* **2018**, *37*, 47–71. [CrossRef]
33. Koraim, A.S.; Heikal, E.M.; AboZaid, A. Different methods used for protecting coast from sea level rise caused by climate change. *Curr. Dev. Oceanogr.* **2011**, *3*, 33–66.
34. Rotnicka, J. Factors controlling the development of foredunes along the Łeba Barrier on the south Baltic coast of Poland. In Proceedings of the 11th International Coastal Symposium, Szczecin, Poland, 9–13 May 2011.
35. Zhang, W.; Schneider, R.; Kolb, J.; Teichmann, T.; Dudzinska-Nowak, J.; Harff, J.; Hanebuth, T.J.J. Land–sea interaction and morphogenesis of coastal foredunes—A modeling case study from the Southern Baltic Sea coast. *Coast. Eng.* **2015**, *99*, 148–166. [CrossRef]
36. Kostrzewski, A.; Zwoliński, Z.; Winowski, M.; Tylkowski, J.; Samołyk, M. Cliff top recession rate and cliff hazards for the sea coast of Wolin Island (Southern Baltic). *Baltica* **2015**, *28*, 109–120. [CrossRef]
37. Tylkowski, J. Hydro-meteorological conditions underpinning cliff-coast erosion on Wolin Island, Poland. *Przegląd Geograficzny* **2018**, *90*, 111–135. [CrossRef]
38. Ludwig, J.; Lindhorst, S.; Betzler, C.; Bierstedt, S.E.; Borówka, R.K. Sedimentary rhythms in coastal dunes as a record of intra-annual changes in wind climate (Łeba, Poland). *Aeolian Res.* **2017**, *27*, 67–77. [CrossRef]
39. Riksen, M.J.P.; Goossens, D. The role of wind and splash erosion in inland drift-sand areas in the Netherlands. *Geomorphology* **2007**, *88*, 179–192. [CrossRef]
40. Subotowicz, W. *Litodynamika Brzegów Klifowych Wybrzeża Polski*; Gdańskie Towarzystwo Naukowe, Ossolineum: Wrocław, Poland, 1982; ISBN 83-04-01301-0.
41. Tylkowski, J. The temporal and spatial variability of coastal dune erosion in the Polish Baltic coastal zone. *Baltica* **2017**, *30*, 97–106. [CrossRef]
42. Kadib, A.A. Mechanism of sand movement on coastal dunes. Proceedings of the American Society of Civil Engineers. *Waterw. Div.* **1966**, *2*, 27–44.
43. Svasek, J.; Terwindt, J. Measurements of sand transport by wind on a natural beach. *Sedimentology* **1974**, *21*, 311–322. [CrossRef]
44. Horikawa, K.; Hotta, S.; Kubota, S.; Katori, S. On the sand transport rat by wind on beach. *Coast. Eng. Jpn.* **1983**, *26*, 101–120. [CrossRef]
45. Psuty, N.P. Sediment budget and dune/beach interaction. *J. Coast. Res. SI* **1988**, *3*, 1–4.
46. McEwan, I.K.; Wilets, B.B. On the prediction of bed-load sand transport rate in air. *Sedimentology* **1994**, *41*, 1241–1251. [CrossRef]
47. Sherman, D.J.; Li, B.; Ellis, J.T.; Farrell, E.J.; Maia, L.P.; Granja, H. Recalibrating Aeolian sand transport models. *Earth Surf. Processes Landf.* **2013**, *38*, 169–178. [CrossRef]
48. Hesp, P.A. Foredune formation in Southeast Australia. In *Coastal Geomorphology in Australia*; Thom, B.G., Ed.; Academic Press: Sydney, Australia, 1984; pp. 69–97.
49. Łabuz, T.A. Morphodynamics and rate of Cliff erosion in Trzęsacz (1997–2017). *Landf. Anal.* **2017**, *34*, 29–50. [CrossRef]
50. Świątek, M. Precipitation changes on the Polish coast of the Baltic Sea (1954–2003) due to changes in intensity of westerlies over Europe. *Clim. Res.* **2011**, *48*, 23–29. [CrossRef]
51. Tylkowski, J. Temporal and spatial variability of air temperature and precipitation at the Polish coastal zone of the Southern Baltic Sea. *Baltica* **2013**, *26*, 83–94. [CrossRef]
52. Wiśniewski, B.; Wolski, T.; Musielak, S. A long-term trend and temporal fluctuations of the sea level at the Polish Baltic coast. *Oceanol. Hydrobiol. Stud.* **2011**, *40*, 96–107. [CrossRef]

53. Miętus, M. *Variability of Air Temperature and Precipitation in the Polish Baltic Coast and its Expected Course until 2030*; Materiały Badawcze: Meteorologia, Poland, 1996.
54. Russell, G.L.; Gornitz, V.; Miller, J.R. Regional sea level changes projected by the NASA/GISS atmosphere-ocean model. *Clim. Dyn.* **2000**, *16*, 789–797. [CrossRef]
55. Collins, M.; Knutti, R.; Arblaster, J.; Dufresne, J.L.; Fichefet, T.; Friedlingstein, P.; Wehner, M. Long-term climate change: Projections, commitments and irreversibility. In *Climate Change the Physical Science Basis (1029–1136), Contribution of Working Group I to the Fifth Assessment Report of the Intergovernmental Panel on Climate Change*; Cambridge University Press: Cambridge, UK, 2013.
56. Christensen, O.B.; Kjellström, E.; Zorita, E. Projected Change—Atmosphere. In *The BACC II Author Team Second Assessment of Climate Change for the Baltic Sea Basin*; Regional Climate Studies; Springer: Cham, Switzerland, 2015; ISBN 978-3-319-16006-1.
57. Lehtonen, I.; Ruosteenoja, K.; Jylhä, K. Projected changes in European extreme precipitation indices on the basis of global and regional climate model ensembles. *Int. J. Climatol.* **2014**, *34*, 1208–1222. [CrossRef]
58. Orlowsky, B.; Seneviratne, S.I. Global changes in extreme events: Regional and seasonal dimension. *Clim. Chang.* **2012**, *110*, 669–696. [CrossRef]

water

MDPI

Article

Low Frequency Waves Detected in a Large Wave Flume under Irregular Waves with Different Grouping Factor and Combination of Regular Waves

Luigia Riefolo [1], Pasquale Contestabile [2,*], Fabio Dentale [3] and Guido Benassai [4]

1 Department of Civil and Environmental Engineering DICA, Politecnico di Milano, 20133 Milan, Italy; luigia.riefolo@polimi.it
2 Department of Engineering, University of Campania "Luigi Vanvitelli", 81031 Aversa, Italy
3 MEDUS (Maritime Engineering Division University of Salerno) Department of Civil Engineering, University of Salerno, 84084 Fisciano (SA), Italy; fdentale@unisa.it
4 Department of Engineering, Parthenope University of Naples, 80133 Naples, Italy; guido.benassai@uniparthenope.it
* Correspondence: pasquale.contestabile@unicampania.it; Tel.: +39-081-501-0245

Received: 26 January 2018; Accepted: 19 February 2018; Published: 23 February 2018

Abstract: This paper describes a set of experiments undertaken at Universitat Politècnica de Catalunya in the large wave flume of the Maritime Engineering Laboratory. The purpose of this study is to highlight the effects of wave grouping and long-wave short-wave combinations regimes on low frequency generations. An eigen-value decomposition has been performed to discriminate low frequencies. In particular, measured eigen modes, determined through the spectral analysis, have been compared with calculated modes by means of eigen analysis. The low frequencies detection appears to confirm the dependence on groupiness of the modal amplitudes generated in the wave flume. Some evidence of the influence of low frequency waves on runup and transport patterns are shown. In particular, the generation and evolution of secondary bedforms are consistent with energy transferred between the standing wave modes.

Keywords: spectral analysis; low frequency; wave grouping; eigen analysis; eigenmode; random waves; combination waves

1. Introduction

Extreme storms may significantly affect the coastal environment, especially in terms of erosion and sediment transport. They can provoke disastrous consequences such as sediment transport beyond the surf zone to unusual depths [1]. Waves reaching a coastline release the majority of their energy and momentum within the surf zone as intense turbulence generated at the front face of the breaker. However, a portion of that energy is transferred to low frequency modes [2], like Low Frequency Waves, longshore currents, rip-currents and shear waves, that are oscillations generated by a shear instability of the mean longshore current profile, especially on barred beaches [3,4]. Low-frequency waves can be generated from intense interaction between short waves and between short waves and long waves at the surf-swash boundary [5,6].

Cross-shore standing long wave swash oscillations are usually forced by infragravity frequency ($f < 0.05$ Hz) waves [7–14] as are swash oscillations due to waves traveling or oscillating along the water edge, parallel to the mean-water line (edge waves) [15–19]. Although wind-waves or short-waves (typical frequency of about 0.1 Hz) are the major force behind the swash zone (SZ) dynamics, the importance of the SZ for the generation/transformation of low frequency motions has been recognized [6,20–23]. In the SZ, in fact, while the final dissipation of short-wave (wind and swell) energy occurs, the Low Frequency Wave (LFW hereinafter) energy (typical wave frequencies between 0.03 and 0.003 Hz) is, generally, reflected back seaward. In addition, intense interaction between short waves and between short waves and long waves at the surf-swash boundary can lead to the generation and reflection of further LFW [5,6]. Superficial SZ hydrodynamics, subsurface SZ hydrodynamics, sediment dynamics and co-related beachface morphodynamics determine and are strongly determined by the frequency of swash motion in no-tidal seas [24–31]. Nonlinearity is a major mechanism responsible for LFW. However, amplitude modulation of waves surface elevation (also termed wave grouping or groupiness) represents another predominant feature in short wave processes affecting long wave generation.

The Groupiness Factor (GF) is a measure of the degree of grouping or rather than of the amplitude modulation of the incident wave field. It is defined as the standard deviation of Smoothed Instantaneous Wave Energy History (SIWEH) normalized with respect to its mean value. A technique which involves no low-pass filtering to compute the low frequency part of the square of the water surface elevation is the Hilbert transform technique. Groupiness Factor is a global measure of the variance contained, in the entire low frequency, part of the square of the water surface elevation [32,33].

Some questions still remain as to which mechanisms dominate under different surf zone conditions. For example, while it is well known that more regular waves (swell waves) tend to promote recovery but irregular waves (sea waves) tend to promote erosion, it is not clear if this is related to changes in nonlinearity or groupiness or to changes in energy [34–36].

In this context, the project SUSCO (Swash zone response under grouping Storm Conditions) used the Hydralab III facility at the large wave flume of the Maritime Engineering Laboratory (LIM), Catalonia University of Technology (UPC) [37]. The objectives of the project were to compare the shoreline response and the SZ hydrodynamics of various wave regimes. In particular, the beach response between monochromatic conditions and wave groups were investigated as a direct result of the wave groupiness. The effects of forced and free long waves induced by the groupiness were also examined.

The data provided by the SUSCO campaign represent a comprehensive and controlled series of tests for evaluating in detail many complex phenomena affecting the hydro-morphodynamic in the surf and SZ. Starting from that dataset, the main aim of this study is to compare the effect of various wave regimes on very low frequencies generations. In particular, wave flume seiching is investigated as a potential effect of low-frequency energy during the experiments. As known, wave generation in an enclosed flume could cause seiching owing to wave reflections or wave grouping effects that can transfer wave energy to low-frequencies [38].

Often wave flume experiments involve wave reflection generated by structure/beaches. Nowadays this problem is solved by controlling the paddle through an active reflection system, which is not applicable to long wave absorption. When active absorption is applied, the reflected waves approaching the generator are predicted in real time and paddle control signals are modified to absorb the waves approaching the generator. The result is that the control of the incident waves is maintained throughout the test. The wave resonance in the wave flume is generated by the reflection phenomena occurring when the wave frequency begins to be equal to the fundamental or harmonic resonant frequency of the flume. Lengthwise oscillations of long waves in a flume can be troublesome as they take a long time to dissipate, owing to their high reflectivity [39].

The wave generator does not have the capability to absorb long wave energy, it is first necessary to see if any of the long wave activity is due to resonance of certain frequencies with the wave flume (seiche). The natural frequencies of the wave flume have to be determined by eigenvalue analysis.

Bellotti et al. [40], on the basis of data from tests performed at the same large wave flume of LIM, found a non-negligible low-frequency component that can be addressed to seiches of the flume. Resonant seiche response in a wave flume (or wave tank) represents an unfortunate drawback, especially for mobile bed experiments. In a resonant condition, the excitation of low-frequency (resonant) modes of the wave flume results in components whose amplitudes can increase to be of the same order as the primary waves [41]. Some authors are concerned about resonant cross-tank seiching that would occur in a wave flume under different wave conditions [42,43]. The variation in seiching has been found to modify the final equilibrium profile and the bar-trough morphology [35,36].

For these reasons, it is important to quantify any influence of the low-frequency for erosive and accretive wave conditions especially concerning the interpretation of different hydro-morphodynamic effects. Furthermore, among the generated waves of the experimental tests described in the Section 2, the results for random and combination tests are presented here.

Low frequencies are studied through the Spectral and Eigen analysis as presented in Section 3, estimating the frequency of the dominant seiches in the wave flume. Furthermore, the influence of the low frequencies on the wave energy spectra is investigated using a numerical solution, which adopts as input the measured data of the beach profiles of the case studies. Finally, combined analysis between modes for surface displacement and net sediment transport has been used to clarify the influence of LFW on morphodynamics. The results of these analyses are shown in Sections 4 and 5; then the main findings are summarized in Section 6.

2. Experimental Setup

In this paper, wave interaction with a rouble mound breakwater has been modeled using numerical simulations. The CIEM (Canal d'Investigació i Experimentació Marítima) large-scale wave flume is 100 m long, 3 m wide and up to 4.5 m deep. The beach consisted of commercial well-sorted sand with a medium sediment size (d_{50}) of 0.25 mm, with a narrow grain size distribution (d_{10} = 0.154 mm and d_{90} = 0.372 mm) and a measured settling velocity (w_s) of 0.034 m/s. The experimental profile and equipment distribution is presented in Figure 1. The x-coordinate origin is at the wave paddle at rest condition before starting the waves and positive toward the shoreline. The movable bed profile started after 31 m of concrete with a section 1:20 slope from x = 31 to 37 m prior to a plane bed, from x = 37 to 42 m, followed by a 1:15 slope plane beach (Figure 1). Prior to running each wave condition, the water depth was performed by manual reshaping and then compacted by running 10 minutes of 'smoothing' wave conditions, in order to return almost the same initial profile. The water depth at the toe of the wedge-type wave paddle was 2.5 m [44].

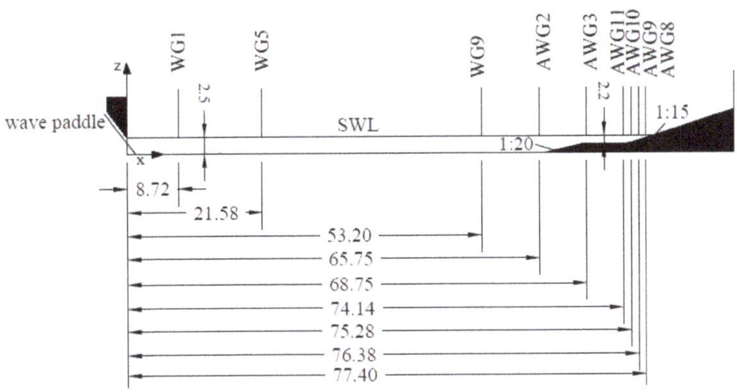

Figure 1. Longitudinal cross-section of the Catalonia University of Technology (UPC) flume and detail of acoustic wave gauges and wave gauges location (length in m).

Among all the instruments installed and used in the controlled area [45], 6 Acoustic Wave Gauges (AWGs), 10 Resistant Wave Gauges (WGs), and 1 beach profiler were acquired. These instruments were sampled at 20 Hz according to the characteristics of the acquisition system (Figure 1).

The bottom profile information was acquired by means of a mechanical bed profiler that measures the emerged and submerged profile along a central line of the flume. The mechanical profiler consists of a wheel at the end of a pivoting arm which is mounted on a moving platform along the flume.

A computer records the velocity of the platform and the angle rotation of the arm. Those data are used to extract the X and Z flume bathymetry. The overall vertical profile accuracy is therefore estimated to be ±10 mm. Such accuracy is comparable with more recent experimental study on beach profile evolution [46], where the standard deviation of the vertical profile was found to be equal to 0.1 cm and 0.2 cm for the emerged and submerged beach profile, respectively. The vertical datum for the profile data is the sea water level (SWL in Figure 1). The horizontal datum is that of the reference point in the flume, which is 7.4 m from the wavemaker and approximately 36 m seaward of the beach toe.

As previously mentioned, the tests were carried out for two different levels of energy flux, defined as "erosive" and "accretive", covering a broad range of wave amplitudes, short and long wave frequencies, modulation rates and group frequencies. The behavior of the two wave regimes was forecasted by mean of the dimensionless sediment fall velocity number (Dean number) [47] and on the basis of previous experiments in the CIEM flume using similar wave conditions. The final profiles and net sediment transport are consistent with these initial estimates. Therefore, hereinafter "erosive test" and "accretive test" refer to wave conditions able to produce morphological patterns (over the whole beach profile) in which erosive or accretive conditions dominate, respectively. The following waves were appointed:

- random waves with different Grouping Factors (GF);
- combination of free partial standing long waves plus monochromatic short waves (hereinafter, combined waves);
- regular monochromatic;
- bichromatic waves (including bound long waves).

Among all the tests generated during the test campaign, random and combination waves in both cases erosive and accretive are here examined. In the Tables 1 and 2 the wave characteristics such as wave height H (m), related to the different wave components, and wave period T (s) for those tests in erosive and accretive conditions are listed. Considering the measured sediment fall velocity of 0.034 m/s, the beach was morphologically characterized by intermediate and reflective conditions for erosive and accretive waves regimes respectively [48]. According to Wright and Short [49], the beach state is a function of breaker height, period and of the sediment size.

Table 1. Wave characteristics for accretive conditions.

Test	H (m)	T (s)	Wave Type
CA_1	0.226 0.038	6 30	Combination
CA_2	0.226 0.038	6 15	Combination
RA_1	0.319	6.7	Random GF = 0.96
RA_2	0.319	6.7	Random GF = 1.08

Four random wave trains, RE_1, RE_2, RA_1, RA_2, were generated with the same variance based on wave height and the same peak frequency as their corresponding monochromatic pair. On the other hand, to generate long-wave short-wave combinations, the monochromatic conditions have been perturbed by small amplitude long waves, added to the control signal. Hence, cases CE_1, CE_2, CA_1, and CA_2 represent the addition of free long waves to otherwise monochromatic wave conditions (Tables 1 and 2). Controlled wave generation was achieved by a wedge-type wave paddle, particularly suited for intermediate-depth waves. The wave generation software used for controlling the wave paddles was AWASYS5 [50].

Table 2. Wave characteristics for erosive conditions.

Test	H (m)	T (s)	Wave Type
CE_1	0.370 0.038	3.7 30	Combination
CE_2	0.370 0.038	3.7 15	Combination
RE_1	0.530	4.1	Random GF = 1
RE_2	0.530	4.1	Random GF = 1.1

The groupiness was slightly varied as well as the phases, obtaining random waves of identical energy spectrum (conforming to JONSWAP spectrum with γ = 3.3) [48]. The Grouping Factor, which measures the degree of grouping, or rather than of the amplitude modulation of the incident wave field, is computed according with Hald [51] (Equation (1)) as the standard deviation σ of half the squared wave surface elevation envelope curve $E(t)$ relative to the squared variance σ^2 of the surface elevation $\eta(t)$ as follows:

$$\mathrm{GF} = \frac{\sigma[E(t)]}{\sigma^2[\eta(t)]} \quad (1)$$

In particular, the half squared envelope signal is computed by the Hilbert transformation. The standard approach of wave generation is to use random uncorrelated phases which in the average leads to GF = 1.0 along with $\sigma^2 \approx 0.13$ for 500 waves. [48–51]. Greater values of GF suggest that small waves tend to be succeeded by small waves, and large waves by other large waves.

All tests started from a similar initial 1/15 handmade slope. For the erosive conditions the reshaping took place along the active profile, while the reshaping occurred from the landward edge of the bar trough to the run-up limit on the accretive conditions. The tests were composed of four steps which included five bottom profiles. Each test lasted 24 min and was repeated 6 times. Consequently, the final profile (P4) was generated after a total active wave time equal to 144 min.

Different bottom profiles where acquired during each wave testing condition. One at the beginning of the experiments (P0), and consecutively at the end of the 1st (P1), 2nd (P2), 4th (P3) and 6th (P4) runs for each of the 13 wave conditions.

Tests were a compromise between the desire to reach an equilibrium profile and the available experimental time.

3. Methods

Data treated in this work concern the first step of 24 min duration for each test. The wave conditions started with a similar initial beach profile for both wave conditions, accretive and erosive, respectively, as shown in Figures 2 and 3. For this reason, the results discussed below can be considered without significant influence from morphodynamic feedback during the measurements. Such conditions afford to obtain reliable observations of hydrodynamic behaviors of LFW also in movable bed experiments.

Furthermore, before carrying out the computation of net sediment volume variation ΔV_{SZ}, bed elevation data have been corrected. In fact, it is worth mentioning that the total beach volume of each profile was not the same. Due to profiler measurement errors, in particular the inability to accurately measure ripple volumes and some non-uniformity of the profile, calculation of ΔV along the whole profile never returns identically zero. Therefore, errors in the calculated ΔV were corrected by distributing the mismatch in sediment volume along the whole profile, leading to a zero value of ΔV. Generally, the error distributed is of a few millimeters, hence the correction does not significantly affect the volume computation. This approach derives from the method proposed by and Baldock et al. [18] for calculation of the net time-averaged sediment transport. The analysis assumed a depth of closure for the sediment transport calculations at x = 60 m (at a water depth of approximately 1 m), and applied the sediment continuity correction over the active profile, that is, landward of x = 60 m or shallower than 1 m. This enables greater resolution and improved accuracy in the ΔV calculations. Closure errors corresponded to a mean error in vertical elevation across the profile that ranged from 3 mm to 15 mm, with an average of 9 mm over all tests, which is consistent with the estimated accuracy of the bed profiler. However, other methodologies could be applied to reconstruct transverse profiles and to study coastal evolution, in particular performing a comparison of aerial photographs according to Muñoz-Pérez et al. [52].

In the present paper, two kinds of analysis have been carried out, as described in the following two sub-sections.

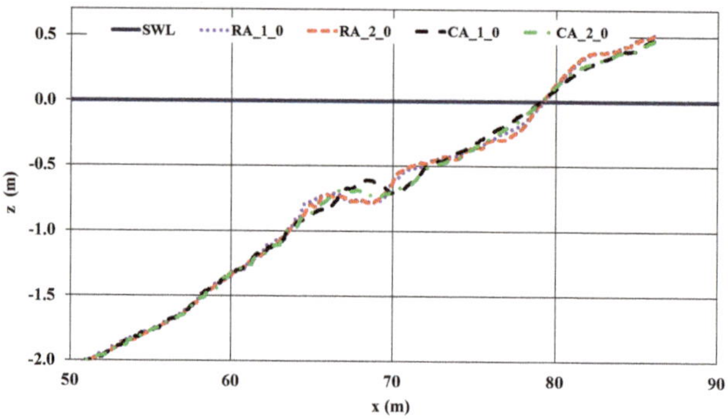

Figure 2. Initial beach profiles of the accretive tests, random and combination, respectively.

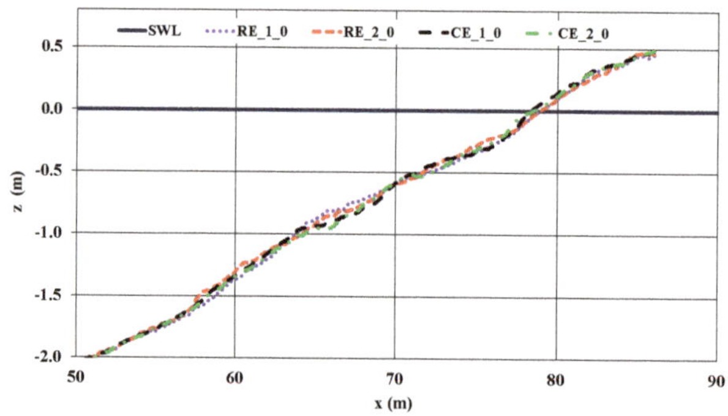

Figure 3. Initial beach profiles of the erosive tests, random and combination, respectively.

3.1. Spectral Analysis

On the basis of the time series of surface elevations the spectral analysis has been performed to evaluate the energy characteristics induced by wave motion. The response of low frequency variations in the large wave flume for random and combination tests has been analyzed. In addition, the spectra response is observed taking into account the time series of the water surface elevation in accretive and erosive wave conditions.

According to Molloy [53], the spectral density represents the momentum corresponding to the specific frequencies. A higher momentum or spectral density corresponds to a larger amplitude oscillation.

Operatively, a Fast Fourier Transform (FFT) is applied to execute spectral analysis for each test. Once the peaks in the spectra corresponding to low-frequencies and harmonics are identified, they are later compared with wave modes determined by the Eigen analysis, as described in the following.

3.2. Eigen Analysis

It is first necessary to study if any of the long wave activity is due to resonance of specific frequencies with the wave flume (hereinafter "seiche"). According to Rabinovich [54] the resonance occurs when the dominant frequencies of the external forcing match the Eigen frequencies of the flume.

In particular, seiche is long-period standing oscillation in an enclosed basin. The resonant (eigen) frequency of seiche is determined by basin geometry and water depth. The set of eigen frequencies and associated modes are a fundamental property of a specific basin [54]. The mode of a seiche is the number of nodes it has within the system. The period of a seiche with "n" nodes is given by Merian's formula [55]. This assumes that the basin is rectangular, with a uniform depth. The related period can be computed as:

$$Tn = \frac{2L}{n\sqrt{gh}} \tag{2}$$

where: Tn is the period of an nth mode seiche; L is the wavelength of the seiche (length of the basin); n is the number of nodes/modes of the seiche; g is the acceleration due to gravity 9.81 m/s^2, and h is the average water depth.

The period derived from Equation (2) is the time it takes for the waveform to oscillate from one end of the basin to the other and back. That is, to travel a distance of twice the basin length. Obviously, the different seiche modes are not mutually exclusive. Seiches with various different modes can occur together in a system. However, the fundamental oscillation is usually dominant, as first shown by Wilson [56].

In the next section the results of the calculated Eigenvalues and Eigenmodes through a numerical approach are shown. This method was presented by Kirby et al. [57] to determine the family of Eigenmodes for measured wave flume geometry. The resulting matrix eigenvalue problem, derived from the next equation (Equation (3)), is solved using the EIG routine in MATLAB™.

$$q_{xx} - \lambda z^{-1} q = 0 \tag{3}$$

where $q = zu$ is the volume flux, u the horizontal velocity, z the water depth, $\lambda = \omega^2/g$ represents the Eigenvalue for the problem and ω the angular frequency. Equation (3) is finite differenced using centered second-order derivatives. The corresponding expansion, orthogonality condition, and dispersion relation are given by:

$$q(x) = \sum_{n=1}^{\infty} q_n F_n(x) \tag{4}$$

$$\int_0^L z^{-1} F_n F_m dx = 0; \quad n \neq m \tag{5}$$

$$\omega_n^2 = -g \frac{\int_0^L (F_n')^2 dx}{\int_0^L z^{-1} F_n^2 dx} \tag{6}$$

in which F_n and F_m represent the family of eigenmodes of order n and m respectively.

In this study, the four families of eigenmodes, F_1, F_2, F_3 and F_4, have been determined by solving numerically Equations (3)–(6).

4. Results and Discussion

Spectral power density (expressed in m^2/Hz) and water surface elevation signal plots (related to the wage gauge WG5 at 21.58 m in the wave flume) are presented in Figure 4.

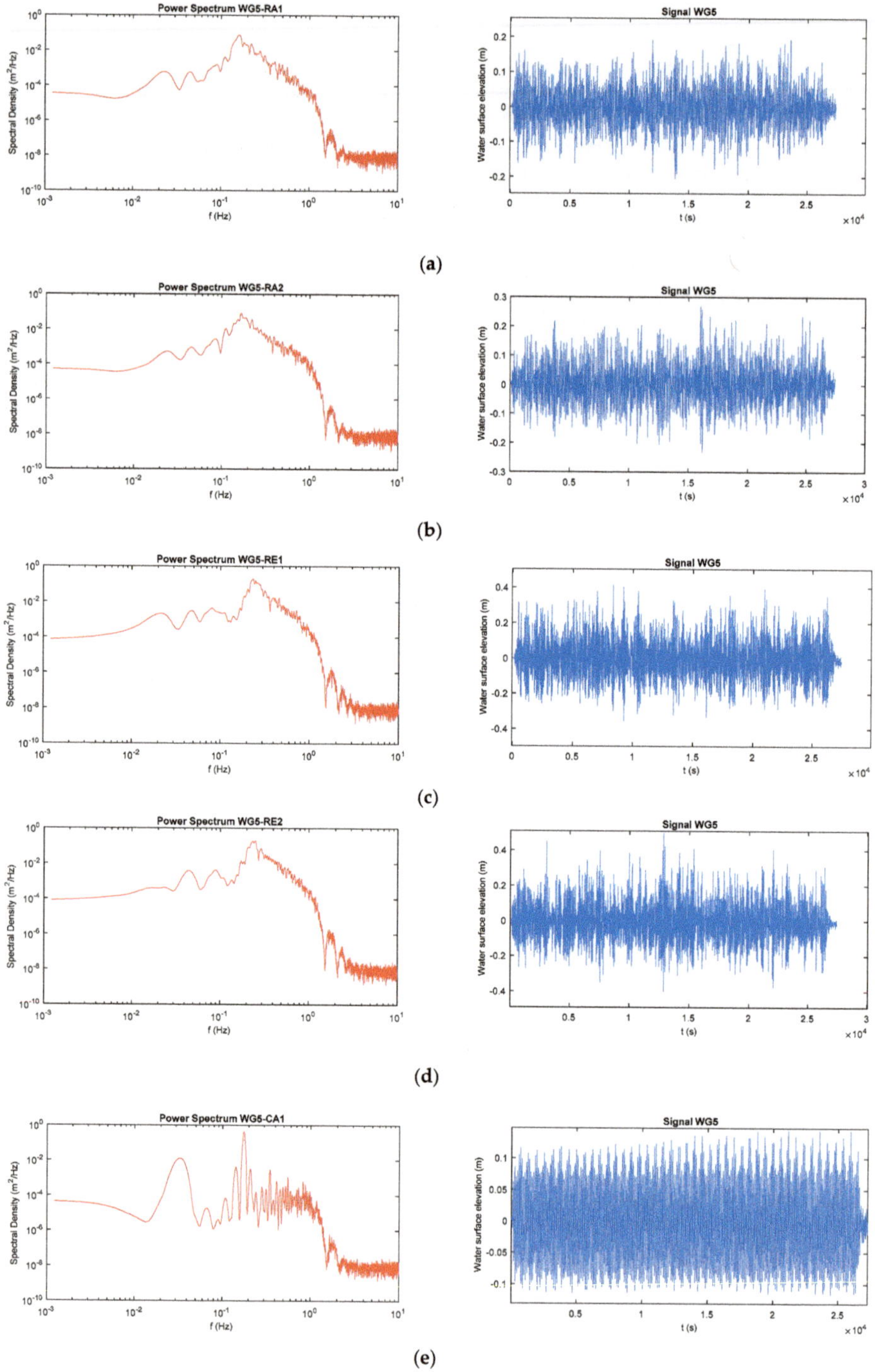

Figure 4. *Cont.*

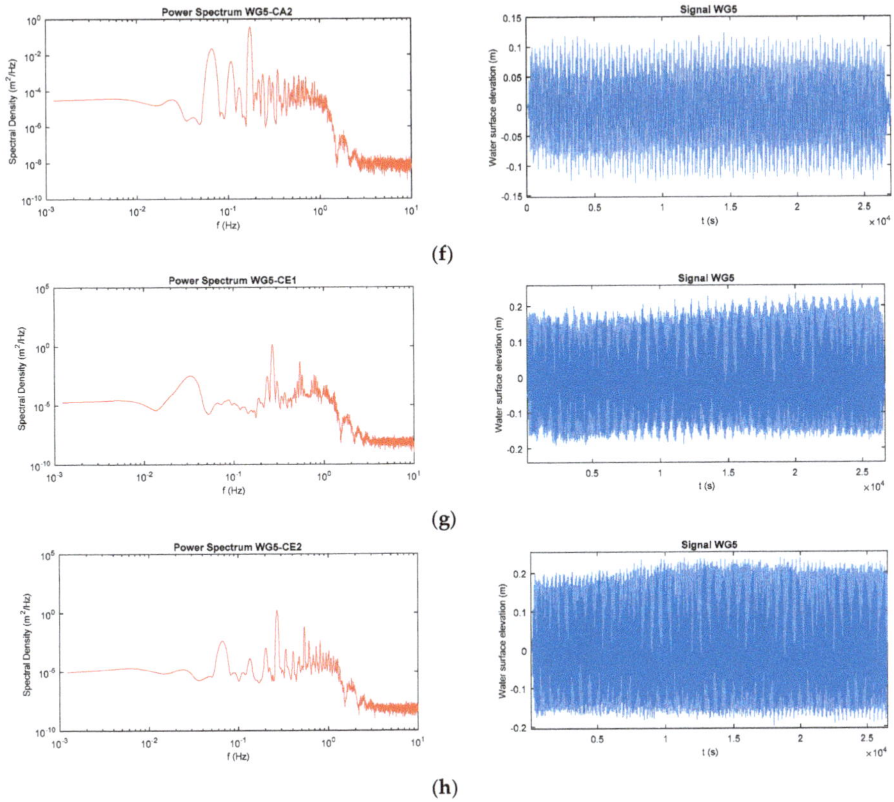

Figure 4. Power spectral density and Time series water surface elevation for test: (**a**) RA_1 (GF = 0.96); (**b**) RA_2 (GF = 1.08); (**c**) RE_1 (GF = 1); (**d**) RE_2 (GF = 1.1); (**e**) CA_1; (**f**) CA_2; (**g**) CE_1 and (**h**) CE_2.

It is possible to note the great variability of the oscillation modes even in the region of low frequencies, according to Cáceres and Alsina [58]. In fact, the authors found that random waves need longer time periods to reach a stationary position. These are wave conditions with wider breaking areas. Furthermore, they found that the wave group and short wave period ratio plays a significant role in the suspended sediment fluxes through the generation of harmonics with longer periods than the wave group. This is due to the fact that the time evolution of morphological features is affected by the length of the breaking area.

Focusing on frequencies under 0.1 Hz, the power spectral density peak values and the corresponding frequencies for random tests in both the erosive and accretive case are reported in Tables 3 and 4. Those values are related to the first, second and third harmonic. Water surface elevation time series are between −0.2 and 0.2 m meanly for all the studied tests.

Furthermore, it is noted that increasing the grouping factor the spectral density associated to the 1st harmonic strongly decrease in random waves. In fact, for cases RA_2 and RE_2 the lowest frequency peak is relatively less pronounced (Tables 3 and 4). That energy density is moved in the higher part of the spectra, as further demonstrated by Table 5, where power levels associated with various sections of the spectra are reported. In particular, for erosive tests, no relevant difference in the 2nd and 3rd harmonic energy density is observed. Moreover, beyond the first harmonic, a monotone trend in wave spectra is found, while for accretive conditions the spectral density slightly increases moving toward lower frequencies. It is notable as for tests RE_1 and RE_2 the third harmonic represents the highest peak before the carrier wave (frequency of about 0.25 Hz). For accretive conditions a slight peak appears at about 0.11 Hz.

Table 3. Peak frequencies, periods and power spectral density for RA_1 and RA_2 tests.

	Harmonics	f (Hz)	T(s)	E (m^2/Hz)
Test RA_1	1st harmonic	0.02319336	43.1157895	0.00059542
	2nd harmonic	0.04516602	22.1405405	0.00058067
	3rd harmonic	0.09033203	11.0702703	0.00155648
Test RA_2	1st harmonic	0.02441406	40.9600042	0.00052997
	2nd harmonic	0.04516602	22.1405405	0.00089923
	3rd harmonic	0.08544922	11.7028571	0.00267999

Table 4. Peak frequencies, periods and power spectral density for RE_1 and RE_2 tests.

	Harmonics	f (Hz)	T(s)	E (m^2/Hz)
Test RE_1	1st harmonic	0.02197266	45.5111111	0.00211399
	2nd harmonic	0.04760742	21.0051282	0.00310199
	3rd harmonic	0.08056641	12.4121212	0.00424879
Test RE_2	1st harmonic	0.02319336	43.1157895	0.00041757
	2nd harmonic	0.04516602	22.1405405	0.0038171
	3rd harmonic	0.08911133	11.2219178	0.0040941

Table 5. Power spectral density for various sections of the spectra for the random tests.

E (m^2/Hz)	RE_1	RE_2	RA_1	RA_2
total	1.22×10^{1}	1.20×10^{1}	3.37×10^{0}	4.00×10^{0}
<0.1 Hz	1.45×10^{-1}	1.19×10^{-1}	3.97×10^{-2}	6.34×10^{-2}
<0.03 Hz	2.18×10^{-2}	6.68×10^{-3}	5.26×10^{-3}	5.07×10^{-3}

In the case of combination waves the 1st harmonic does not decrease strongly, as for the combination cases, but at about 0.8 Hz. Moreover, a power spectral response corresponding to twice the wave frequency is noted, especially for the erosive wave regimes (Figure 4e–h). In Tables 6–8 the peak frequencies, periods and power spectral density are summarized, also related to various sections of the spectra. In particular, the difference for the combination waves in accretive and erosive conditions is higher when is considered the power spectral density at frequencies less than 0.03 Hz.

Table 6. Peak frequencies, periods and power spectral density for CA_1 and CA_2 tests.

	Harmonics	f (Hz)	T(s)	E (m^2/Hz)
Test CA_1	1st harmonic	0.032958984	30.34074074	0.01233396
	2nd harmonic	0.065917969	15.17037037	0.00001770
	3rd harmonic	0.108642578	9.20449438	0.00004718
Test CA_2	1st harmonic	0.025634766	39.00952381	0.00002758
	2nd harmonic	0.041503906	24.09411765	0.00000737
	3rd harmonic	0.067138672	14.89454545	0.02348628

For tests RA_1 and RA_2 a sort of spreading around the carrier wave should be noted (frequency of about 0.158 Hz and 0.167 Hz respectively). In particular, this spreading takes the form of a second downshifted peak, more evident for RA_2. The shift from the main wave is of about ±0.045 Hz, that is, an amount approximately equal to the second harmonic. It is worth noting that the presence of 2nd harmonic disturbance on the carrier wave is also perceptible for RE_2. The latter turns out to be important looking at Figure 5, where spectral density ratio between third and second mode and between second and first harmonic are summarized. A heuristic explanation for the singular behavior of RE_2 and the influence of the second harmonic could be addressed to the enhanced non-linear interactions due to wave groupiness forcing a long wave at the frequency of second harmonic. In this vein, it is worth considering that the largest wave height (both maximum and significant) compared to others wave conditions with the same energy flux is formed precisely during RE_2. And just like the undertow largely follows the instantaneous wave height [59], the larger waves in the groups could dominate the whole hydrodynamics.

Table 7. Peak frequencies and power spectral density for CE_1 and CE_2 tests.

	Harmonics	f (Hz)	T(s)	E (m^2/Hz)
Test CE_1	1st harmonic	0.03295898	30.34074074	0.00314475
	2nd harmonic	0.06713867	14.89454545	0.00000949
	3rd harmonic	0.10805664	12.41212121	0.00002413
Test CE_2	1st harmonic	0.023193359	43.11578947	0.00000319
	2nd harmonic	0.040283203	24.82424242	0.00000234
	3rd harmonic	0.067138672	14.89454545	0.00393993

Table 8. Power spectral density for various sections of the spectra for the combination tests.

E (m^2/Hz)	CE_1	CE_2	CA_1	CA_2
total	1.11×10^{1}	1.32×10^{1}	3.43×10^{0}	3.17×10^{0}
<0.1 Hz	2.53×10^{-2}	3.14×10^{-2}	9.80×10^{-2}	1.86×10^{-1}
<0.03 Hz	4.50×10^{-3}	1.35×10^{-4}	1.81×10^{-2}	3.83×10^{-4}

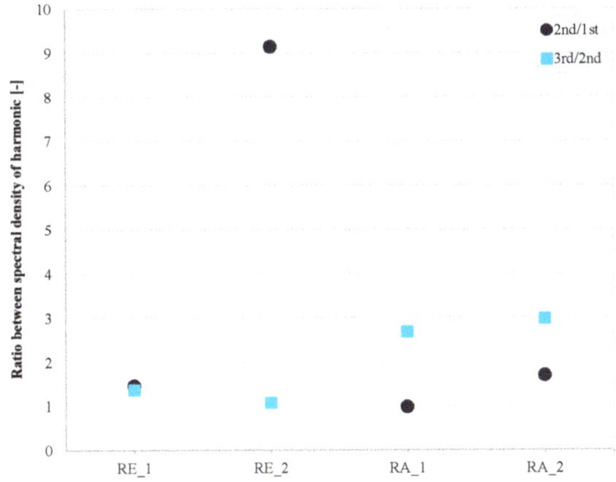

Figure 5. Ratio between spectral density level at different harmonics of the random tests.

The ratios between spectral density level at different harmonics of the combination tests show a similar behavior in the cases of erosive and accretive regime for the monochromatic waves perturbed with larger (CE_1 and CA_1) and smaller (CE_2 and CA_2) long waves, respectively (Figure 6).

In order to clarify the origin of these waves, the frequencies of the longest standing waves in the flume have been evaluated and compared with the measured lowest frequencies. The natural frequencies of the wave flume are determined by Eigenvalue analysis. The calculated periods and frequencies using formulas found in the literature were published in recent research conducted by Riefolo et al. [60]. Measured and numerically predicted mode periods are compared in Figures 7 and 8, where a linear deviation from the bisector line is highlighted (see black dashed line). The frequencies calculated with the formula proposed by Merian [55] are shown in Table 9. Appreciably, correlation between the measured and calculated frequencies has been found. These tests, in fact, show a good correspondence between the calculated and measured frequencies especially for the first and second harmonic, for which the influence of wave flume-generated seiching can be definitively highlighted in the case of random waves (Figure 7). On the other hand, the monochromatic wave perturbed with the larger long waves for the erosive condition (CE_1) gives different variation, in the measured and calculated Eigenmodes, of the case in the accretive conditions (CA_1). Instead, a similar variation of the measured and calculated Eigenmodes for the tests CA_2 and CE_2 is highlighted where the monochromatic wave was perturbed with smaller long waves, for both analyzed conditions, accretive and erosive, respectively (Figure 8). This similarity has been confirmed by the ratios between spectral density level at different harmonics of the combination tests, as previously described in Figure 6.

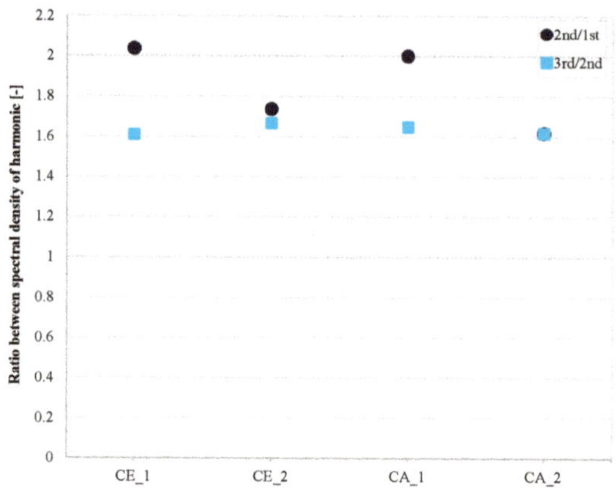

Figure 6. Ratio between spectral density level at different harmonics of the combination tests.

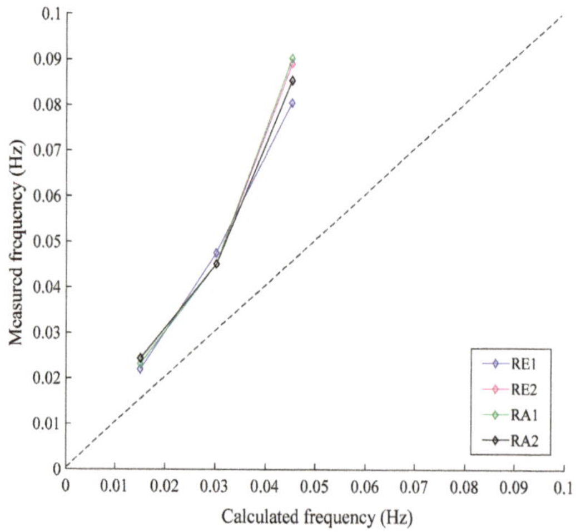

Figure 7. Frequencies plot of measured versus calculated Eigenmodes with formula for rectangular basin for random tests in the case of accretive and erosive condition.

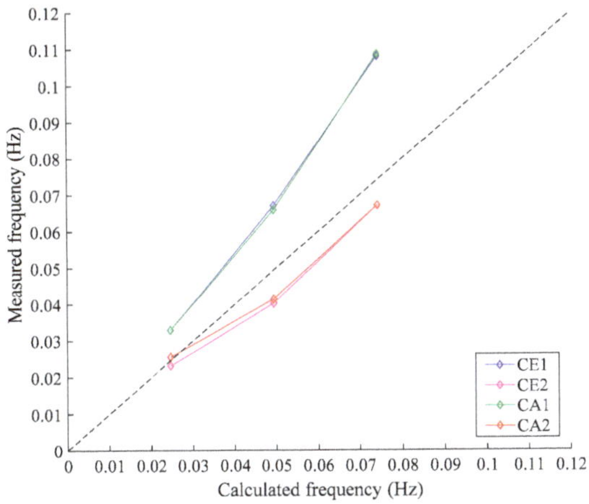

Figure 8. Frequencies plot of measured versus calculated Eigenmodes with formula for rectangular basin for combination tests in the case of accretive and erosive condition.

Table 9. Eigen values calculated through the formula proposed by Merian [55].

Mode	f (Hz)
1st	0.02474874
2nd	0.04949747
3rd	0.07424621

5. Additional Considerations

The effects of the LFWs have been assessed by a set of the four lowest computed modes (F_1–F_4) for volume flux in the wave flume, shown in Figures 9 and 10. Mode periods are computed by the eigen analysis based on water depth *z*. Very interesting results can be identified:

- A strong non-linear pattern of F_1–F_4 is identified for all the tests in proximity of breaking zone;
- Clear opposite behaviours of volume flux eigenmodes are shown for accretive and erosive wave conditions in the case of random and combination waves, except for the test CE_2;
- A different variation of the Eigenmodes for the combination tests in the erosive, clearly, due to the non-linearity effects;
- The monochromatic wave perturbed with the larger long waves for the erosive condition (CE_1) has an opposite variation of the Eigenmodes, than the monochromatic wave perturbed with smaller long waves (CE_2).

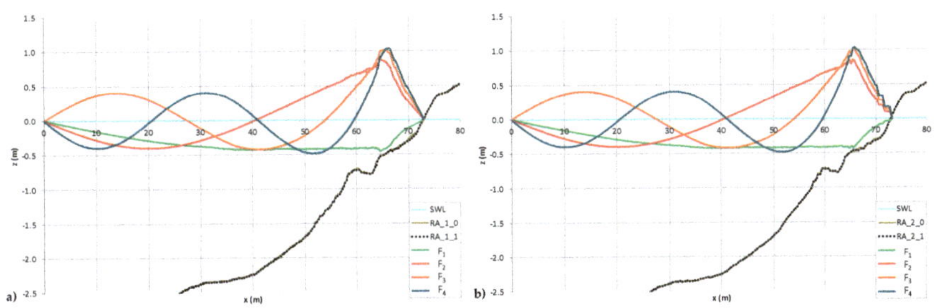

Figure 9. *Cont.*

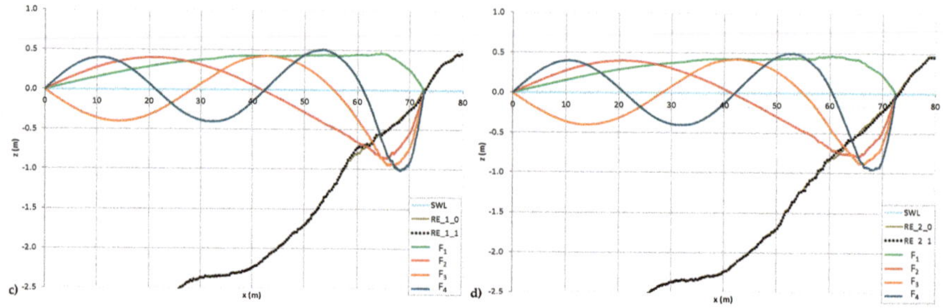

Figure 9. Eigenmodes for volume flux q related to the profile measured in random test: (**a**) RA_1; (**b**) RA_2; (**c**) RE_1 and (**d**) RE_2.

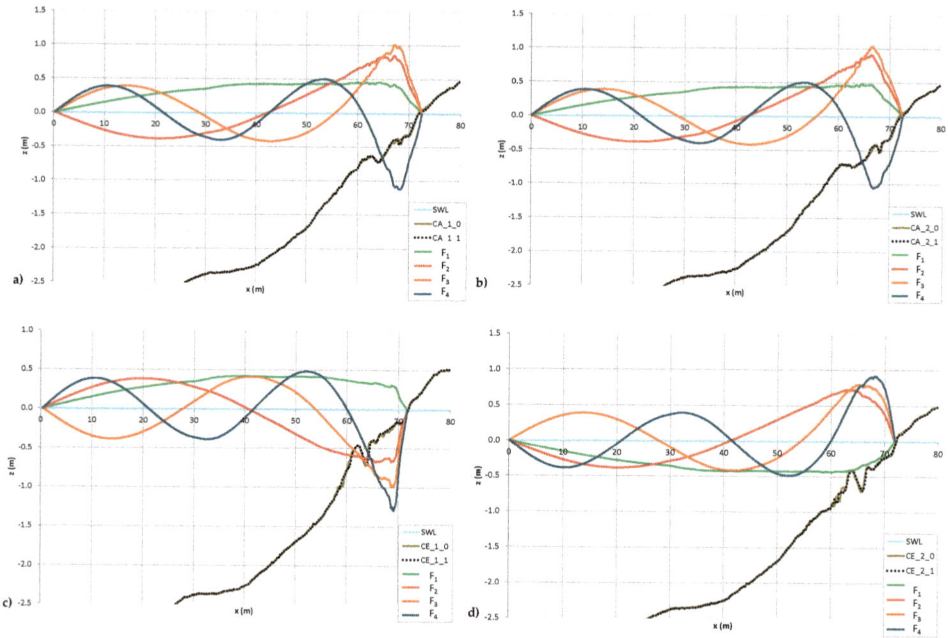

Figure 10. Eigenmodes for volume flux q related to the profile measured in random test: (**a**) CA_1, (**b**) CA_2, (**c**) CE_1 and (**d**) CE_2.

Moreover, the spreading around the carrier wave seen for RE_2, and probably due to the 2nd harmonic disturbance, leads to more homogeneous erosion phenomena compared to RE_1, where the formation of a well visible bar is recognizable. Final profiles and net sediment transport were consistent with these initial estimates [48]. Despite the fact that RE_1, RE_2, CE_1 and CE_2 have identical spectral energy, their local effect on beach profile can be significantly different due the presence of LFWs effectively influencing the spectra. The greater concentration of power at certain frequencies due to larger wave grouping could promote a stronger influence of LF motions in the SZ, in particular for two reasons:

1. specific eigenmode of the wave flume (generated seiches) induces spreading or downshift of carrier wave frequency, as foreseen;
2. grouping of short waves in the inner surf zone could directly induce low-frequency oscillations of the shoreline.

As termed by Russell [61] and Smith and Mocke [62], LFWs are powerful agents of sediment transport as they remove large amounts of the sediment which is put into suspensions by the short (wind) waves.

5.1. *Influence on Morphodynamic*

Figures 11 and 12 summarize the net sediment volume variation, ΔV_{SZ}, in the SZ (here approximated as the emerged beach) between the test start and end. These were obtained using the changes in bed elevation between profiles, above z = 0. Focus on the emerged SZ, during the analyzed first step of experiments, all wave conditions lead to a landward net sediment transport, except cases RE_2 and CE_1. For these wave conditions, the energy density associated with the second harmonic has been found about ten times the power level of the first mode, increasing the disturbance effect of the 2nd harmonic itself. Although erosive random waves have the same wave height and mean period, their morphological effect is quite different. It is noted that the grouping factor could promote the presence of multiple low-frequency motions responsible of nonlinear interactions. For this reason, there is a significant contribution that determines the evolution of surface changes. Unfortunately, the case CE_1 formed part of the tests that developed lateral cross-flume asymmetry; this does not significantly alter the wave height on WG 5, but the profile correction may not be sufficient for SZ sediment volume here computed. Hence, results for that case are not reported in Figure 12.

It is important to note as case CE_2 shows a positive net sand volume variation in the emerged SZ greater than accretive cases. Such conditions, however, are not in contrast with the typical erosion pattern along the whole profile (see [20]) of case CE_2. The reason for such strong "apparent" accretive behaviour in the SZ should be addressed to the initial beach profiles that, in such experiments, were usually the same or very similar for all wave conditions. Hence, in erosive conditions the beach is moving more rapidly toward an approximate equilibrium profile, and the mean beachface slope change induced by erosion of sediment provide some local deposition/positive slope change.

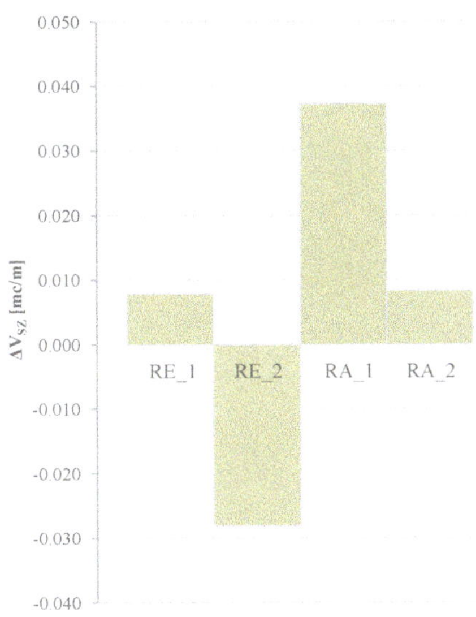

Figure 11. Net sand volume variation in the emerged swash zone (water depth > 0) after 24 min of wave generation: positive values represent accretion or landward transport; negative values represent erosion or seaward transport, for random tests.

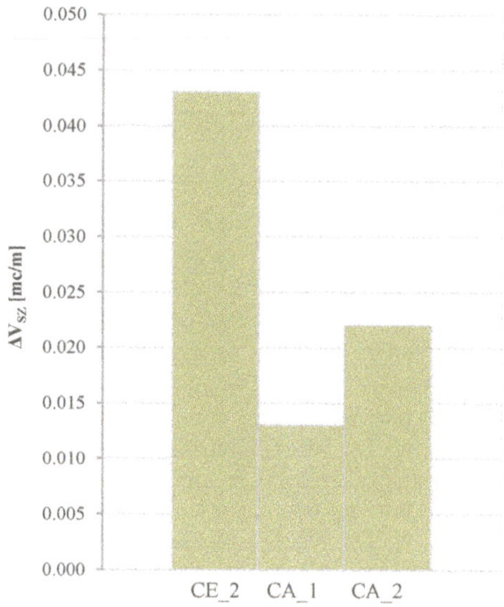

Figure 12. Net sand volume variation in the emerged swash zone (water depth > 0) after 24 min of wave generation: positive values represent accretion or landward transport for combination tests.

5.2. *Influence on Swash Hydrodynamics*

A tentative correlation between detected LFWs and runup is reported in the following. The measurements of runup have been conducted by a PC-based acquisition of digital camera records. Data processing has been carried out by using Image Processing Toolbox™ in MATLAB™ (version 2017, Natickk, MA, USA). The camera was mounted on the profiler's carriage at rest, close to the board of the beach and focused on the wave flume. The camera was calibrated every time that it was removed from the waterproof housing (i.e., when profiler was operating). The calibration procedure consists of determining the relative position of a minimum of four points ("targets") in order to find the initial location and orientation of the camera relative to the calibration frame, as described by [63–67]. Since the position of instruments (captured in the video records) was known, eight "target" points were used, enhancing the quality of analysis. Then, the process was automatically completed by the definition of all remaining pixel position on the frame. In particular, the relative position of runup (as separation between dry and wet zones) was identified.

Runup measurements took place within the first three minutes of each step in order to identify the role of spectral wave components and wave grouping on runup, starting from the same underlying beach conditions.

The good agreements between runup values form video analysis and the ones derived by the extrapolation of measurements from micro-acoustic wave gauges (which only provide the height of the swash lens) on some uprush/backwach cycle give confidence in the technique.

The maximum runup measured are reported in Table 10. Looking at the table, some preliminary considerations can be drawn:

- "accretive" conditions do not necessarily involve smaller runup;
- despite comparable energy levels, random waves give a runup twice higher than combination cases;
- the higher the grouping factor the higher the maximum runup.

Hence, the data demonstrate that increasing differences in the spectral wave components through perturbations of monochromatic waves with long waves or increasing the grouping factor in random waves promotes the enhancement of wave run up, except for case R_E2.

Table 10. Maximum runup measured within the first three min of each step.

Test	Measured (m)	Test	Measured (m)
R_E1	0.36	R_A1	0.30
R_E2	0.41	R_A2	0.37
C_E1	0.16	C_A1	0.18
C_E2	0.12	C_A2	0.23

6. Conclusions

Experimental data from CIEM large-scale wave flume were specifically used to investigate low frequencies generated by random and combination waves in accretive and erosive conditions.

By examining the spectral distribution of wave power, a secondary downshifted peak around the carrier wave has been noted for conditions (both erosive and accretive) with higher grouping factor. This spreading takes the form of a second downshifted peak of about ± 0.045 Hz, an amount approximately equal to the second harmonic. Through the Eigen analysis it has been possible to estimate the periods of the dominant seiches in the wave flume and detect the low frequencies for combination and random waves. Then, comparing measured and calculated Eigenvalues for a rectangular wave flume, a good correspondence for the first and second harmonic has been found. Consequently, the wave flume-generated nature of first and second harmonic has been definitively clarified.

During the analyzed first step of experiment, all wave conditions lead to a landward net sediment transport in the swash zone, except cases RE_2 and CE_1 (Figures 11 and 12). For these wave conditions, the energy density associated to the second harmonic has been found to be about ten times the power level of the first mode, increasing the disturbance effect of the 2nd harmonic itself. However, case CE_1 contained errors due to cross-flume asymmetry of the evolving beach profile. Direct comparison of this test with the other erosive conditions is, therefore, not possible.

In the case of the erosive random waves, the presence of multiple low-frequency motions is responsible for nonlinear interactions, promoted by the grouping factor. Therefore, LFWs contribute significantly to the evolution of surface displacements.

The hydro-morphodynamic effects of the LFWs have been assessed by a volume flux Eigenmode analysis.

A clear opposite pattern of volume flux Eigenmode depending on accretive or erosive behaviour of wave conditions has been detected, except for the test CE_2. Despite the fact that the final profile and net sediment transport are consistent with the erosive trend, this test has shown an unexpected response on swash zone hydro-morphodynamics; the behavior is closer to a typical accretion wave condition.

A pattern of energy exchange between modes has been identified for all the tests in proximity of the breaking zone.

Results observed for swash zone volume variations measured during the first step (i.e., in comparable morphodynamic initial condition between tests) could be directly addressed to resonance phenomena in the wave flume. They may be also representative of real resonant conditions which could be generated landward of a submerged breakwater or in a natural enclosed basin (e.g. pocket beach with large secondary bar).

Clearly, the magnitude of net sediment volume in the swash zone suggests that low frequency motions could have a significant influence only on generation/development of secondary bedforms. In fact, morphodynamic effects of the seiches seem recognizable and significant only at a certain scale of observation. On the other hand, hydrodynamic influence on runup is more evident, but further analysis is required.

Future works will consider the last steps of each test, in order to investigate the low frequencies when affected by morphodynamic effects and vice versa (e.g. bedforms migration).

Acknowledgments: This work was supported by the European Community's Sixth Framework Programme through the grant to the budget of the Integrated Infrastructure Initiative HYDRALAB III within the Transnational Access Activities, Contract No. 022441. Special thanks go to Prof. James Kirby (University of Delaware) and Iván Cáceres (Universitat Politècnica de Catalunya) for their help to analyze data.

Author Contributions: Pasquale Contestabile performed the laboratory experiments; Luigia Riefolo conceived the analysis tools; Luigia Riefolo, Fabio Dentale and Guido Benassai contributed to analysis of the data; Luigia Riefolo and Pasquale Contestabile wrote the paper; Fabio Dentale and Guido Benassai revised the paper draft; Luigia Riefolo updated the paper according to their review.

Conflicts of Interest: The authors declare no conflict of interest.

References

1. Budillon, F.; Vicinanza, D.; Ferrante, V.; Iorio, M. Sediment transport and deposition during extreme sea storm events at the Salerno Bay (Tyrrhenian Sea): Comparison of field data with numerical model results. *Nat. Hazards Earth Syst. Sci.* **2006**, *6*, 839–852. [CrossRef]
2. Brocchini, M.; Bellotti, G. Integral flow properties of the swash zone and averaging. Part 2. Shoreline boundary conditions for wave-averaged models. *Int. J. Fluid Mech. Res.* **2002**, *458*, 269–281. [CrossRef]
3. Oltman-Shay, J.; Howd, P.A.; Birkemeier, W.A. Shear instabilities of the mean longshore current: 2. Field observations. *J. Geophys. Res. Oceans* **1989**, *94*, 18031–18042. [CrossRef]
4. Miles, J.R.; Russell, P.E.; Ruessink, B.G.; Huntley, D.A. Field observations of the effect of shear waves on sediment suspension and transport. *Cont. Shelf Res.* **2002**, *22*, 657–681. [CrossRef]
5. Watson, G.; Barnes, T.C.D.; Peregrine, D.H. The generation of low-frequency waves by a single wave group incident on a beach. *Coast. Eng.* **1994**, *1995*, 776–790.
6. Mase, H. Frequency down-shift on swash oscillations compared to incident waves. *J. Hydraul. Res.* **1995**, *33*, 397–411. [CrossRef]
7. Suhayda, J.N. Standing waves on beaches. *J. Geophys. Res.* **1974**, *79*, 3065–3071. [CrossRef]
8. Huntley, D.A.; Guza, R.T.; Bowen, A.J. A universal form for shoreline run-up spectra. *J. Geophys. Res.* **1977**, *82*, 2577–2581. [CrossRef]
9. Guza, R.T.; Thornton, E.B. Swash oscillations on a natural beach. *J. Geophys. Res.* **1982**, *87*, 483–491. [CrossRef]
10. Holland, K.T.; Raubenheimer, B.; Guza, R.T.; Holman, R.A. Runup kinematics on a natural beach. *J. Geophys. Res.* **1995**, *100*, 4985–4993. [CrossRef]
11. Raubenheimer, B.; Guza, R.T. Observations and predictions of run-up. *J. Geophys. Res.* **1996**, *101*, 25575–25587. [CrossRef]
12. Van Dongeren, A.; Lowe, R.; Pomeroy, A.; Trang, D.; Roelvink, D.; Symonds, G.; Ranasinghe, R. Numerical modeling of dynamics over a fringing coral reef. *Coast. Eng.* **2013**, *73*, 178–190. [CrossRef]
13. Ruju, A.; Lara, J.L.; Losada, I.J. Numerical analysis of run-up oscillations under dissipative conditions. *Coast. Eng.* **2014**, *86*, 45–56. [CrossRef]
14. De Bakker, A.T.M.; Tissier, M.F.S.; Ruessink, B.G. Beach steepness effects on nonlinear infragravity-wave interactions: A numerical study. *J. Geophys. Res. Oceans* **2015**, *121*. [CrossRef]
15. Bowen, A.J.; Inman, D.L. Edge waves and crescentic bars. *J. Geophys. Res.* **1971**, *76*, 8662–8671. [CrossRef]
16. Huntley, D.A.; Guza, R.T.; Thornton, E.B. Field observations of surf beat: 1. Progressive edge waves. *J. Geophys. Res.* **1981**, *86*, 6451–6466. [CrossRef]
17. Guza, R.T.; Thornton, E. Observations of surf beat. *J. Geophys. Res.* **1985**, *90*, 3161–3172. [CrossRef]
18. Oltman-Shay, J.; Guza, R.T. Infragravity edge wave observations on two California beaches. *J. Phys. Oceanogr.* **1987**, *17*, 644–663. [CrossRef]
19. Holland, K.T.; Holman, R.A. Wavenumber-frequency structure of infragravity swash motions. *J. Geophys. Res.* **1999**, *104*, 13479–13488. [CrossRef]
20. Baldock, T.E.; Alsina, J.; Caceres, I.; Vicinanza, D.; Contestabile, P.; Power, H.; Sanchez-Arcilla, A. Large-scale experiments on beach profile evolution and surf and swash zone sediment transport induced by long waves, wave groups and random waves. *Coast. Eng.* **2011**, *58*, 214–227. [CrossRef]

21. Watson, G.; Peregrine, D.H.; Toro, E.F. Numerical solution of the shallow-water equations on a beach using the weighted average flux method. In Proceedings of the First European Computational Fluid Dynamics Conference, Brussels, Belgium, 7–11 September 1992; Volume 1, pp. 495–502.
22. Baldock, T.E.; Holmes, P.; Horn, D.P. Low frequency swash motion induced by wave grouping. *Coast. Eng.* **1997**, *32*, 197–222. [CrossRef]
23. Brocchini, M. Eulerian and Lagrangian aspects of the longshore drift in the surf and swash zones. *J. Geophys. Res. Oceans* **1997**, *102*, 23155–23168. [CrossRef]
24. Masselink, G.; Puleo, J.A. Swash-zone morphodynamics. *Cont. Shelf Res.* **2006**, *26*, 661–680. [CrossRef]
25. Brocchini, M.; Baldock, T.E. Recent advances in modeling swash zone dynamics: Influence of surf-swash interaction on nearshore hydrodynamics and morphodynamics. *Rev. Geophys.* **2008**, *46*. [CrossRef]
26. Horn, D.P.; Baldock, T.E.; Li, L. The influence of groundwater on profile evolution of fine and coarse sand beaches. In Proceedings of the Sixth International Symposium on Coastal Engineering and Science of Coastal Sediment Process, New Orleans, LA, USA, 13–17 May 2007.
27. Ciavola, P.; Vicinanza, D.; Fontana, E. Beach drainage as a form of shoreline stabilization: case studies in Italy. In Proceedings of the 31st International Conference on Coastal Engineering, Hamburg, Germany, 31 August–5 September 2008.
28. Ciavola, P.; Vicinanza, D.; Aristodemo, F.; Contestabile, P. Large-scale morphodynamic experiments on a Beach Drainage System. *J. Hydraul. Res.* **2011**, *49*, 523–528. [CrossRef]
29. Damiani, L.; Aristodemo, F.; Saponieri, A.; Verbeni, B.; Veltri, P.; Vicinanza, D. Full-scale experiments on a beach drainage system: Hydrodynamic effects inside beach. *J. Hydraul. Res.* **2011**, *49*, 44–54. [CrossRef]
30. Contestabile, P.; Aristodemo, F.; Vicinanza, D.; Ciavola, P. Laboratory study on a beach drainage system. *Coast. Eng.* **2012**, *66*, 50–64. [CrossRef]
31. Ciavola, P.; Contestabile, P.; Aristodemo, F.; Vicinanza, D. Beach sediment mixing under drained and undrained conditions. *J. Coast. Res.* **2013**, *65*, 1503–1508. [CrossRef]
32. Mansard, E.P.D.; Sand, S.E. A Comparative Evaluation of Wave Grouping Measures. In Proceedings of the 24th International Conference on Coastal Engineering, Kobe, Japan, 23–28 October 1994.
33. Merrick, C.; Haller, C.; Dalrymple, R.A. Looking for Wave Groups in the Surf Zone. In Proceedings of the International Conference on Coastal Research in Terms of Large Scale Experiments, Gdansk, Poland, 4–8 September 1995; pp. 81–92.
34. Baldock, T.E.; Manoonvoravong, P.; Pham, K.S. Sediment transport and beach morphodynamics induced by free long waves, bound long waves and wave groups. *Coast. Eng.* **2010**, *57*, 898–916. [CrossRef]
35. Dally, W.R. Long Wave Effects in Laboratory Studies of Cross-Shore Transport. 1991. Available online: http://cedb.asce.org/CEDBsearch/record.jsp?dockey=0071660 (accessed on 21 February 2018).
36. Baldock, T.E.; Huntley, D.A. Long-Wave Forcing by the Breaking of Random Gravity Waves on a Beach. *Philos. Trans. R. Soc. Lond. Ser. A* **2002**, *458*, 2177–2201. [CrossRef]
37. Hydrolab III. EU Integrated Infrastructure Initiative, SUSCO Swash Zone Response under Grouping Storm Conditions. Available online: http://hydralab.eu/research--results/ta-projects/project/97/ (accessed on 21 February 2018).
38. Haller, M.C.; Dalrymple, R.A. Rip current instabilities. *J. Fluid Mech.* **2001**, *433*, 161–192. [CrossRef]
39. Sorensen, R.M. *Basic Wave Mechanics: For Coastal and Ocean Engineers*; John Wiley & Sons, Inc.: Hoboken, NJ, USA, 1993.
40. Bellotti, G.; Archetti, R.; Brocchini, M. Experimental validation of mean swash zone boundary conditions. *J. Geophys. Res.* **2003**, *108*, 3250. [CrossRef]
41. Young, I. Observations of triad coupling of finite depth wind waves. *Coast. Eng.* **1998**, *33*, 137–154. [CrossRef]
42. Losada, M.A.; Roldan, A.J.; Dalrymple, R.A. Eigenfunction analysis of water wave propagation down a wave flume. *J. Hydraul. Res.* **1994**, *32*, 371–385. [CrossRef]
43. Van Dongeren, A.; Battjes, J.; Janssen, T.; van Noorloos, J.; Steenhauer, K.; Steenbergen, G.; Reniers, A. Shoaling and shoreline dissipation of low-frequency waves. *J. Geophys. Res.* **2007**, *112*. [CrossRef]
44. Patel, M.H.; Ionnaou, P.A. Comparative Performance Study of Paddle- and Wedge-Type Wave Generators. *J. Hydronaut.* **1980**, *14*, 5–9. [CrossRef]

45. Riefolo, L.; Contestabile, P.; Ferrante, V.; Azzellino, A.; Centurioni, L.; Vicinanza, D. The role of spectral wave components, bandwidth and wave grouping on runup: Preliminary results from large-scale experiments. In Proceedings of the 5th International Conference on the Application of Physical Modelling to Port and Coastal Protection Coastlab14, Varna, Bulgaria, 29 September–2 October 2014.
46. Moreno, L.J. Perfil de Playas Sustentadas: Estudio Hidrodinámico y Experimental. Directrices de Diseño. Ph.D. Thesis, Technical University of Madrid, Madrid, Spain, 2016.
47. Dean, R.G. Heuristic models of sand transport in the surf zone. In Proceedings of the First Australian Conference on Coastal Engineering, Sydney, Australia, 14–17 May 1973.
48. Vicinanza, D.; Baldock, T.E.; Contestabile, P.; Alsina, J.; Cáceres, I.; Brocchini, M.; Conley, D.; Andersen, T.L.; Frigaard, P.; Ciavola, P. Swash Zone Response under Grouping Storm Conditions. *J. Hydraul. Res.* **2011**, *4*, 55–63. [CrossRef]
49. Wright, L.D.; Short, A.D. Morphodynamic variability of surf zones and beaches: A synthesis. *Mar. Geol.* **1984**, *56*, 93–118. [CrossRef]
50. Aalborg University. Software for Wave Laboratories. Available online: http://hydrosoft.civil.aau.dk/AwaSys/ (accessed on 17 February 2018).
51. Hald, T. *Wave Group Analysis by Means of the Hilbert Transform Technique*; Aalborg University: Aalborg, Denmark, 1995.
52. Munoz-Perez, J.J.; Caballero, I.; Tejedor, B.; Gomez-Pina, G. Reversal in longshore sediment transport without variations in wave power direction. *J. Coast. Res.* **2010**, 780–786. [CrossRef]
53. Molloy, E. Seiching in Cockburn Sound. Master's Thesis, University of Western Australia, Crawley, WA, Australia, November 2001.
54. Rabinovich, A.B. Seiches and Harbor Oscillations. In *Handbook of Coastal and Ocean Engineering*; Kim, Y.C., Ed.; World Scientific: Singapore, 2009.
55. Merian, J.R. Ueber die Bewegung tropfbarer Flüssigkeiten in Gefässen [On the Motion of Drippable Liquids in Containers]. Master's Thesis, Göttingen University, Göttingen, Germany, 1828. (In German)
56. Wilson, B. Seiches. In *Advances in Hydrosciences*; Academic Press: Urbana, IL, USA, 1972; Volume 8, pp. 1–94.
57. Kirby, J.T.; Özkan-Haller, H.T.; Haller, M.C. Seiching in a large wave flume. In Proceedings of the 30th International Conference on Coastal Engineering, San Diego, CA, USA, 3–8 September 2006.
58. Cáceres, I.; Alsina, J.M. Suspended sediment transport and beach dynamics induced by monochromatic conditions, long waves and wave groups. *Coast. Eng.* **2016**, *108*, 36–55. [CrossRef]
59. Svendsen, I.A. Mass flux and undertow in a surf zone. *Coast. Eng.* **1984**, *8*, 347–365. [CrossRef]
60. Riefolo, L.; Azzellino, A.; Ferrante, V.; Contestabile, P.; Vicinanza, D. Wave flume-generated seiching analysis. In Proceedings of the Twenty-fifth International Ocean and Polar Engineering Conference, Kona, Big Island, HI, USA, 21–26 June 2015.
61. Russell, P.E. Mechanisms for beach erosion during storms. *Cont. Shelf Res.* **1993**, *13*, 1243–1265. [CrossRef]
62. Smith, G.G.; Mocke, G.P. Interaction between breaking/broken waves and infragravity-scale phenomena to control sediment suspension transport in the surf zone. *Mar. Geol.* **2002**, *187*, 329–345. [CrossRef]
63. Benassai, G.; Aucelli, P.; Budillon, G.; De Stefano, M.; Di Luccio, D.; Di Paola, G.; Montella, R.; Mucerino, L.; Sica, M.; Pennetta, M. Rip current evidence by hydrodynamic simulations, bathymetric surveys and UAV observation. *Nat. Hazards Earth Syst. Sci.* **2017**, *17*, 1493. [CrossRef]
64. Nunziata, F.; Buono, A.; Migliaccio, M.; Benassai, G. Dual-polarimetric C-and X-band SAR data for coastline extraction. *IEEE J. Sel. Top. Appl. Earth Obs. Remote Sens.* **2016**, *9*, 4921–4928. [CrossRef]
65. Nunziata, F.; Migliaccio, M.; Li, X.; Ding, X. Coastline extraction using dual-polarimetric COSMO-SkyMed PingPong mode SAR data. *IEEE Geosci. Remote Sens. Lett.* **2014**, *11*, 104–108. [CrossRef]
66. Nunziata, F.; Buono, A.; Migliaccio, M.; Benassai, G. An Effective Method to Extract Coastline from Dual-Polarimetric C- and X-Band SAR Measurements. Available online: http://lps16.esa.int/page_paper0891.php (accessed on 21 February 2018).
67. Benassai, G.; Migliaccio, M.; Montuori, A. Sea wave numerical simulations with COSMO-SkyMed© SAR data. *J. Coast. Res.* **2013**, *65*, 660–665. [CrossRef]

water

Article

Laboratory Investigation on the Evolution of a Sandy Beach Nourishment Protected by a Mixed Soft–Hard System

Alessandra Saponieri [1,*,†], Nico Valentini [1,†], Marcello Di Risio [2,†], Davide Pasquali [2,†] and Leonardo Damiani [1,†]

1 Department of Civil, Environmental, Building Engineering and Chemistry (DICATECh), Politecnico di Bari, Bari 70126, Italy; nico.valentini@poliba.it (N.V.); leonardo.damiani@poliba.it (L.D.)
2 Department of Civil, Construction-Architectural and Environmental Engineering (DICEAA)—Environmental and Maritime Hydraulic Laboratory (LIam), University of L'Aquila, L'Aquila 67100, Italy; marcello.dirisio@univaq.it (M.D.R.); davide.pasquali@univaq.it (D.P.)
* Correspondence: alessandra.saponieri@poliba.it; Tel.: +39-080-596-3229
† These authors contributed equally to this work.

Received: 30 July 2018; Accepted: 27 August 2018; Published: 31 August 2018

Abstract: A new experimental campaign on a 2D movable-bed physical model, reproducing a typical nourishment sandy beach profile, is being carried out in the wave flume of the Laboratory of Coastal Engineering at Politecnico di Bari (Bari, Italy). The main aim is to assess the short-term evolution of a sandy beach nourishment, relying on a mixed solution built on the deployment of a Beach Drainage System (BDS) and a rubble-mound detached submerged breakwater. This paper aims at illustrating the experimental findings. Tests presented herein deal with both unprotected and protected configurations, focusing on the hydrodynamic and morphodynamic processes under erosive conditions. Results show that, with respect to the unprotected conditions, BDS reduces the shoreline retreat and the beach steepen within swash and surf zone as well. Moreover, a reduction of net sediment transport rate is observed. When BDS is coupled with the submerged sill, a reversal of the prevalent direction of the net sediment transport seaward occurs offshore the sheltered region. Less considerable positive effects on shoreline retreat are induced by the submerged structure, whereas the mean beach slope remains quite stable. Secondary effects of drain on the submerged sill performance are also highlighted. BDS reduces wave-induced setup on beach, by mitigating the mean water level raising, typically experienced by such structures.

Keywords: beach nourishment; beach drainage system; groundwater; submerged breakwater; cross-shore sediment transport; climate change

1. Introduction

Coasts are naturally affected by erosion processes enhanced by climate changes, raising urbanisation and exploitation of coastal zones. Beach size is decreasing with unfavourable consequences for the environment, society and economy. Coastal defence and mitigation interventions appear to be essential and the timing for decisions dealing with coastal protection against climate change should be taken at the early stage [1–3]. In such a context, it was observed that upgrading coastal defences and nourishing beaches would reduce the impacts related to climate change roughly by three orders of magnitude [4].

Several methods, ranging from standard rubble mound breakwaters (e.g., [5,6]) up to low-environmental impact solutions (e.g., beach nourishments, artificial reefs and by-pass systems (e.g., [7–10])), are typically deployed to recover and protect beaches from erosion. However, effects of defence work are not always sustainable under both environmental and economical points of view. Hard structures often cause a shift of erosion process along neighbouring areas, whereas soft solutions demonstrate to be long-term unsustainable. Among the latter, beach nourishments constitute the most used method. They can be considered as examples of environment-friendly solution for beach restoration and coastal preservation, basically consisting of filling the near-shore region with new sediment to restore or maintain an adequate beach width and ensure an appropriate dune protection, without compromising coastal environment and tourism.

To limit sediment losses during intervention lifetime, nourishment works are frequently combined with hard coastal defence structures, since they reduce sediment spreading and lengthen the time span between periodic re-nourishments for coastal restoration. On the other hand, defence structures influence wave propagation and reduce the environmental sustainability of the intervention. To reduce such impacts on environment, submerged breakwaters are widely used as shore protection system aimed at preventing shoreline retreat. They dissipate the incident wave energy, hence reducing the wave hydrodynamic action on coasts [11,12] and extending the residence time of sediments in the sheltered region [13,14], depending on configuration parameters (i.e., freeboard, distance from the shoreline, and transmission coefficient), wave climate and nearshore seabed [15–18]. Moreover, they have a lower impact on both hydrodynamic processes and nearshore zone morphodynamics with respect to the emerged detached breakwaters and groins, since they are able to enhance the water circulation, its renovations rates [19,20] and biological biodiversity [21].

Past studies show that sandy beach stability could also be increased by the deployment of a Beach Drainage System (BDS), which is counted among soft-engineering systems aimed at contrasting erosion. Previous works [22–24] demonstrate the importance of the close link between the swash zone sediment transport and groundwater. In particular, they demonstrate that a lower position of groundwater with respect to mean sea level can affect morphodynamics, by inducing a sediment stabilisation when infiltration inside the beach occurs. The BDS is able to increase the apparent sand permeability, by inducing within the beach an artificial lowering of the saturation line and an increase in the thickness of the unsaturated area. In this way, up-rush flux is more easily absorbed by the beach, whereas sea-ward flow is significantly reduced.

Currently, drainage efficacy in restoring eroded beaches is not well defined. BDS can be considered as an auxiliary system in coasts management, such as combined with a beach nourishment to increase sediment stability and, hence, the nourishment lifetime [25]. Field installations deployed around the world (e.g., Denmark, USA, UK, Japan, Spain, Sweden, Holland, France, Italy, and Malaysia) were not supported by an adequate long-term monitoring to highlight a full scientific evidence of beach stabilisation, mainly due to both the erroneous management and the non-existent maintenance of the systems. However, in some cases, an overall reasonable performance in short-medium term (1–5 years), was reported [26]. Moreover, even though the BDS concept was initiated about fifty years ago and many BDSs have been installed worldwide since 1981 [27], the first experimental studies on BDS morphodynamic and hydrodynamic performances were carried recently [28,29]. In 2010, full-scale laboratory experiments were carried out to overcome the limitations of previous studies [30–33]. The already tested BDS configurations were shown to stabilise the beach for medium and low energy conditions, while for high energy wave attacks the drainage system seemed to be inadequate in giving any stabilisation effect [34,35].

Another important issue related to beach nourishment design is related to the retrieval of nourished sediments with specific characteristics, fully compatible with the existing grain size and composition (i.e., mineralogy). Both the choice of sediments and the sampling area influence not only the further evolution of coastline and the beach response [36], but also the impact on environment. In general, the material necessary for nourishments comes from dredging operations (of navigation channels, harbour entrances or basins) or from mining sites (land or submarine). In all cases, the compatibility between added and native sediments is fundamental to assess the suitability of mining sites and the sediment volumes required to ensure nourishment stability, manage subsequent periodic recharging operations and prevent the intervention area, which could suffer dramatic consequences due to high turbidity during the works [37–39].

Due to both the high costs of sediment recovery and dredging operations and the significant time required for large nourishment projects, beach scraping may represent an attractive option [40–42], being widely undertaken and often privately supported by beach managers. Such a kind of small periodical interventions could constitute a valid alternative with respect to traditional large nourishment projects for recovering beaches [43], since small sediment volumes are required, with consequent reduced costs, working time and minimal impacts on coastal natural cycles. The sediment could be easily mined from the active littoral zone, ensuring the compatibility of the sediment source [44].

Beach nourishments, as well as beach scraping, can indeed be recognised as able to counteract beach erosion and, hence, limit the effects of climate changes on coastal flood risk due to beach erosion [45]. Of course, the approach should be environmentally sustainable.

Our research aims at investigating a mixed approach, relying on both standard and innovative beach defence systems, by gaining insight on the cross-shore, short term morphodynamic response of a beach nourishment protected by a rubble-mound detached submerged breakwater and a BDS deployed together. Unique and definitive design guidelines on BDS are still missing, since previous field installation did not provide an adequate long-term monitoring and laboratory experiments were mainly focused on the study of the hydrodynamics, by neglecting the role of longshore gradients in the morphodynamics evolution. The novelty of our research lies in assessing the reliability of the mixed approach in beach stabilisation. The basic idea is to switch high energy sea states propagating from offshore to medium/low energy waves in the surf zone, by means of the submerged breakwater, to enhance the efficiency of BDS and improve sandy nourishment performance by increasing the intervention lifetime, without affecting its low-environmental impact.

A modelling campaign on a 2D physical small scale model of a typical nourished sandy beach profile is being carried out. The tests presented herein dealt with the investigation of the profile evolution forced by erosive waves, by focusing on its morphological development up to equilibrium and its hydrodynamics. The configurations comprised the unprotected beach, BDS protected and the coupled BDS-breakwater beach cases. Protection systems efficiency is here reported in terms of cross-shore profile evolution, shoreline displacement, submerged bar migration and sediment transport rate. Furthermore, wave parameters and groundwater behaviour inside the beach in the different tested configurations are reported and discussed.

2. Materials and Methods

2.1. Experimental Setup

The experimental tests were performed in the wave flume at the Laboratory of Coastal Engineering of the Department of Civil, Environmental, Building Engineering and Chemistry of the Politecnico di Bari (Bari, Italy). The 2D wave flume used for the experiments is about 50 m long, 2.5 m wide and 1.2 m deep. It is equipped with a piston-type wave maker provided by Wallingford (UK) able to generate regular and irregular wave trains and extreme wave heights of 0.3 m with a maximum water level of 0.8 m. The nourishment profile was reproduced inside the flume as a 2D physical bed movable model. It may be viewed as a Froude scaled typical intervention with a prototype-to-model ratio equal to 1:10. However, as the results are intended to be general, model dimensions are illustrated and discussed as well hereinafter.

In Figure 1, a sketch of the model geometry adopted for the experiments is shown. The initial sandy profile began about 20 m from the wave paddle with a mean slope of 1/30 for 9 m, followed by a 1/8 sloped foreshore for 5.2 m and a horizontal emerged berm 2.5 m long, +0.15 m above the mean water level. The total sand volume was of about 18 m^3.

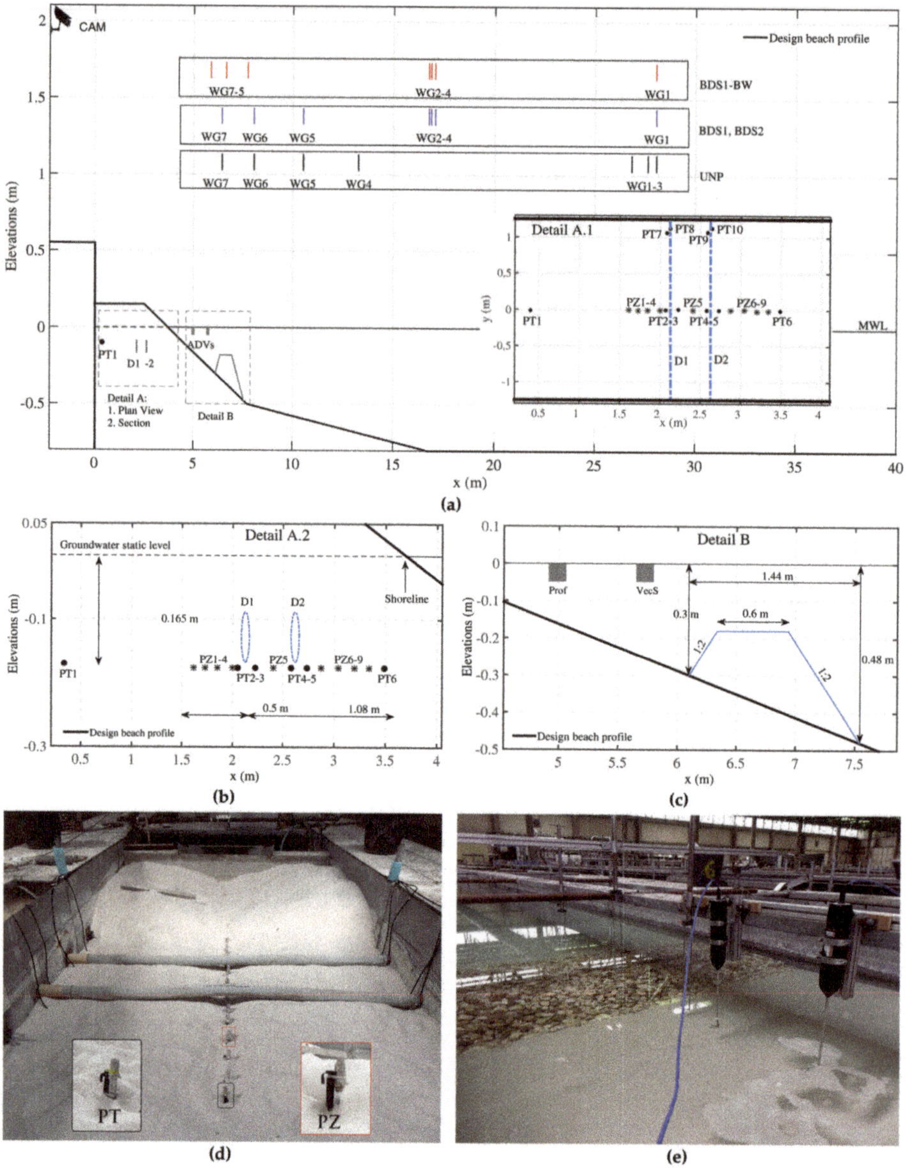

Figure 1. (**a**) Cross-section of the physical model built in LIC 2D flume with overlapped the plan view of the instruments and drains locations on the shoreface (Detail A.1); (**b**) detail of drain pipes, piezometers (*PZ*), pore pressure transducers (*PT*) placed inside the sand (Detail A.2); (**c**) detail of the submerged sill (Detail B); (**d**) perspective picture of drain pipes and instruments, during installation, with magnification of *PZ* and *PT*; and (**e**) lateral view of submerged sill, ADVs and wave gauges.

The adoption of a unique coordinate system was appropriate to analyse measurements derived from different instruments. The x-axis was aligned with the cross-shore direction, pointing to the wave paddle with the origin ($x = 0$) set at the onshore limit of the sandy beach profile at the flume centreline.

The z direction was aligned with the vertical direction, pointing upward with $z = 0$ at the mean water level. The y-axis was aligned with the long-shore direction.

Sand grain size curve was obtained at LIC, according to A.S.T.M. standard [46], whereas both permeability and water content were measured on reconstituted sand samples at the Geotechnical Engineering Laboratory of Politecnico di Bari. According to Wentworth grain size classes [47], the sand used was an almost uniform medium-fine sand with a D_{50} equal to 0.227 mm and D_{10} and D_{90} equal to 0.177 mm and 0.322 mm, respectively. The sand was characterised by a permeability of 4.1×10^{-5} m/s and a mean water content of about 27%.

The drainage system was constituted of a drain pipe placed inside the beach, at 0.165 m below the static groundwater level, parallel to the shoreline, covering the total model width (about 2.5 m). Two drain pipes (D_1 and D_2 in Figure 1) were installed to preliminarily evaluate the drainage performances at different distances from the initial shoreline (1 m and 1.5 m, respectively). Drains were 0.07 m diameter PVC pipes, with rectangular holes uniformly distributed along the side surface, allowing the water drainage. Moreover, to avoid the obstruction of the aforesaid holes by the sand, the pipes were covered by a geotextile membrane. The drains were connected through a blind pipe to a manifold well from which the water was removed by means of a pumping system and then fed back into the water circulation system. The on/off drains switch was made possible by means of two valves installed at the end of each pipe, outside the flume, before the connection with the blind pipes.

The dimensions of the rubble-mound submerged breakwater were chosen to switch high energy sea states propagating from offshore to medium/low within the surf zone, for which drainage seems to show a better efficiency [35]. This was achieved by testing a submerged breakwater characterised by a freeboard higher than conventional submerged defence structures. The breakwater was constructed of one layer of natural stones with a median diameter equal to 5 cm in model scale. The landward side of the structure was 2.4 m from the shoreline with a 0.6 m wide crest and a freeboard equal to −0.18 m below the mean water level. Both landward and seaward slopes were 1:2. Other significant dimensions (i.e., sea/landward water depths) are reported in Figure 1.

A constant JONSWAP wave spectrum with peak enhancement factor 3.3 and characterised by nominal significant wave height H_s and peak period T_p of 0.2 m and 1.5 s, respectively, was reproduced in the flume. The cross-shore nourishment evolution was investigated in both unprotected and protected conditions to compare the effects of the mixed configuration on both hydrodynamics and morphodynamics, forced by the same wave attack, reproducing erosive conditions. The empirical criteria proposed by [48–51] were used for the scope.

After testing beach profile evolution in unprotected conditions, each drain was tested without the submerged breakwater. Then, the mixed solution constituted by the coupled system of the drain D_1 and the rubble-mound detached submerged breakwater was tested. Each test was subdivided in steps, in accordance with the bottom measurements time-intervals, chosen to follow the quick bed variations occurring in the early stage. Accordingly, surveys were performed every 15 min in the first hour (Steps 1–4) and every 30 min for the second hour (Steps 5 and 6). Then, the profile was surveyed every hour until the fourth hour of test (Steps 7 and 8), every 2 h until the tenth and every 3 h up to equilibrium. In Table 1, tests are reported as performed in chronological order, with the relative reference name (*Test ID*). Moreover, the wave bulk parameters (significant wave height H_s, zero-order moment wave height H_{mo}, peak period T_p and zero-order moment m_0) estimated from the offshore wave gauge ($WG1$) for each test are reported as mean values of those calculated for each step. Tests were run until beach equilibrium condition was reached, approximately when variations in profile measurements were almost negligible.

Table 1. Test ID and main bulk parameters of wave attacks.

Test	Test ID	H_s (m)	H_{m0} (m)	T_p (s)	m_0
Unprotected	UNP	0.187	0.206	1.47	0.0027
Drain 1	BDS1	0.183	0.206	1.47	0.0027
Drain 2	BDS2	0.18	0.203	1.47	0.0026
Drain 1 + Submerged Breakwater	BDS1-BW	0.19	0.212	1.47	0.0028

2.2. Measurements

Wave transformations along the flume were investigated by means of seven resistive wave gauges placed in the flume central section ($WG1$–$WG7$ in Figure 1) by measuring water surface elevations with an acquisition frequency of 20 Hz. Wave parameters were calculated for each test step at wave gauge locations by means of standard zero-crossing and spectral analyses. The offshore wave gauge placed near the wave paddle ($WG1$) and gauges $WG2$–$WG3$ in the unprotected conditions (Figure 1) were used to check the pseudo-random wave trains generated in the flume, since no active absorption system was deployed. In the protected tests $BDS1$, $BDS2$, $BDS1$-BW (Table 1), wave gauges $WG2$–$WG4$ were moved and located at the foreshore toe in order to estimate the beach reflection coefficients by separating reflected from incident components by means of the method proposed by [52].

Water surface elevations measured from the gauges $WG4$–$WG7$ and $WG5$–$WG7$ in unprotected and protected conditions, respectively, were used to evaluate waves propagation along the nearshore. In particular, during tests performed with the submerged sill and the drainage system deployed together ($BDS1$-BW), three gauges were placed just seaward ($WG5$), landward ($WG7$) and over ($WG6$) the submerged structure to investigate the influence of the structure on waves energy exchange and propagation. Offshore wave spectra generated for all tests were compared, demonstrating that wave boundary conditions were almost the same in all configurations, since no differences in wave generation were observed. Moreover, wave reflection analysis was performed from the free surface elevation measured by $WG1$, $WG2$ and $WG3$, placed near the paddle in the unprotected configuration. A mean reflection coefficient overall the UNP test was estimated about 0.08, confirming that despite no active absorption system was deployed, the wave generation was only slightly affected by re-reflected waves.

Instantaneous local velocities were measured by means of two Acoustic Doppler Velocimeters ($ADVs$) located in the surf zone with an acquisition frequency of 20 Hz. The locations of $ADVs$, fixed for all tests, are reported in Figure 1, where $VecS$ refers to the Vectrino side-looking, whereas $Prof$ indicates the location of the Vectrino Profiler. The former measured the velocity components in x, y and z directions in a point. Velocity profiles were then obtained by moving the instruments along the vertical direction at 1 cm-spaced intervals for a total water column investigated of 3 cm, at about 6 cm above the bottom level, since during the first unprotected test, sediment accretion was observed in the area where both $ADVs$ were placed. A lower distance from the bottom would foreclose the velocity measurements during submerged bar formation. Every measurement was performed for 1 min at each vertical location (three points) and the time-averaged values were considered for deriving velocity profiles, for a total duration of measurement equal to $4 \div 5$ min. This allowed obtaining comparable results at each vertical location along the measurement section and neglect any bottom variation during each measurement, which could lead to erroneous analyses. $VecS$ measurements were performed at the beginning, at the end and at $t/2$ of each test step (where t indicates the duration of the test step), for steps not exceeding 1 h and every 30 min for test steps longer than 1 h. The $Prof$, located as depicted in Figure 1, continuously measured the x, y and z velocity components within a 3.4 cm high water column, with a vertical spatial resolution of 1 mm, from around 3.5 cm above the (varying) bed level at the beginning of each step.

To investigate the infiltration processes inside the porous medium and the effects of drainage on beach saturation degree (groundwater level) as well as on the swash zone hydrodynamics, an array of piezo-resistive pore pressure transducers (PT) and piezometers (PZ) were placed inside the beach (Figure 1). Six pore pressure transducers ($PT1$–$PT6$) were placed below the drain pipe along the flume centreline, up to the initial shoreline position. A transducer was also placed inside each drain ($PT8$–$PT10$) to investigate the hydraulic behaviour inside the drain pipes. Two more PTs were placed near each drains on the same side (positive y), as in Figure 1, to verify groundwater regime around the pipe, by comparing the water head inside and outside drains. Moreover, the $PT11$ was located in the pumping well in order to measure the average drained flow during drained tests, according to the procedure described in [30]. The transducers acquired at a frequency of 20 Hz. The static oscillations of water table were measured by means of nine piezometers ($P1$–$P9$ in Figure 1). Each of the piezometers was made of a brass filter covered by a geotextile layer placed inside the sand at the same elevation of the pore pressure transducers at the flume centreline, connected through high pressure PVC pipe (nominal diameter equal to 11 mm) to a cylindrical glass pipe placed outside the flume. The water level oscillations inside the glass pipes were measured by using water level gauges and manually recorded every 5 min for the entire duration of each test. Measurements from both instruments PTs and PZs were used to investigate the saturation line lowering in drained conditions and its raising in undrained configuration.

Considering both the different response times of PT and PZ instruments and the objective of analysing the slow oscillations of the water tables under the influence of wave groups, an averaging procedure of the PT values was conducted. The time window used for the procedure was 5 min around the acquisition datetime of PZ. Figure 2 highlights an example (without limitations) of the hydraulic pressure heads oscillations (Δh) induced by the external wave motion during swash cycles for configuration $BDS1$ with both initial (red circles) and final (gray circles) points used for averaging the time series, for the first 15 min long time step. A very fast decrease of signal at the $D1$ opening, in the range of 1–2 s, was noticeable. Specifically, $PT2$, $PT3$, $PT7$, and $PT8$ then exhibited slight fluctuations following wave groups motions. Meanwhile, $PT4$–$PT6$, $PT9$ and $PT10$ showed a lower decrease, followed by higher oscillations, with remarkable distinction of single wave influence, due to their closer position to the swash zone.

In Figure 3, an example of the spatial variability of saturation lines (Δh) with respect to the initial groundwater level (equal to the mean water level in the flume) is reported for all configurations, referring to the first temporal step (15 min). Circles refer to static (mean) groundwater level measured by the PZ, whereas inverted triangles correspond to the mean values derived from the dynamic pore pressure variations, measured by PT. For the UNP tests PZs derived measurements are not reported since no measurements were available due to problems which occurred with PZs.

Wave-induced run up on the beach (R) was derived by means of a high-resolution visible camera Sony Lens G (3D EYE camera, 18.2 MPX, lens-style DSC-Qx30). Timestack images for wave run up measurements [53] were generated in correspondence of each time step and configuration, from video with a duration of 30 min and a frame-rate frequency of 30 Hz.

The routines applied for projecting, as well as pre- and post-processing the recorded images and videos, were derived from [54,55] and suitably adapted for the experiments. Firstly, a geometrical correction for the lens distortion was applied by using the parameters derived from the intrinsic calibration procedure, carried out at the beginning of the extensive laboratory set-up. A perspective transformation matrix, 3×4 using homogeneous coordinates, for geo-referencing the camera [56,57] was calculated employing a set of Ground Control Point ($GCPs$), distributed in the FoV of the camera. The support of the $GCPs$, recorded by both the laser beam of the Total Station used for beach surveys and the camera, was built by means of a plastic spherical target placed on top of a steel pole. The perspective matrix was used to retrieve the image coordinates to be sampled from the frame, given the beach real-world coordinates of the transect. Then, pixel intensities were extracted along a selected central cross-shore transect from each frame during video progressions by means of Python scripting and OpenCV libraries.

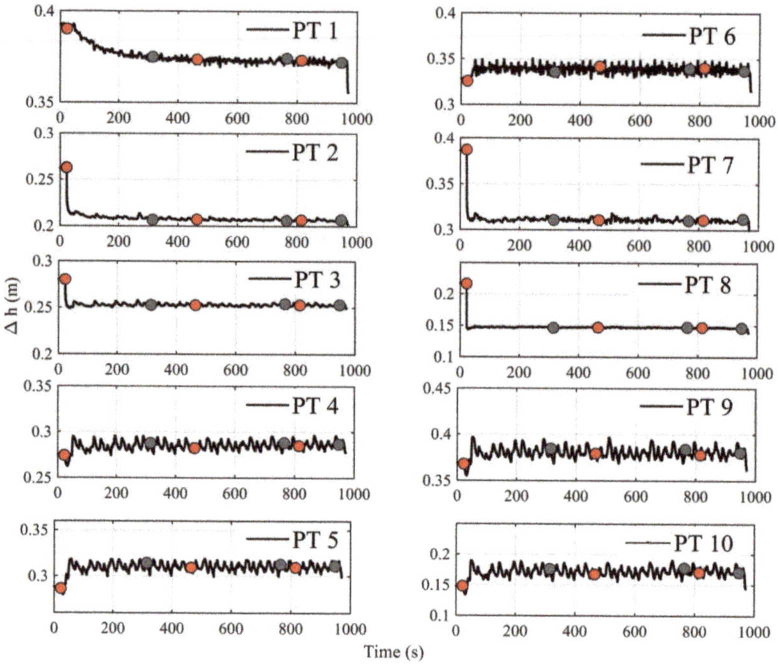

Figure 2. Time variation of *PT* during first time step (15 min) for configuration *BDS*1. Circle markers defines the initial (red-filled) and final (grey-filled) points used for signals averaging. Different ordinates limits are used to catch the time evolution of the signals.

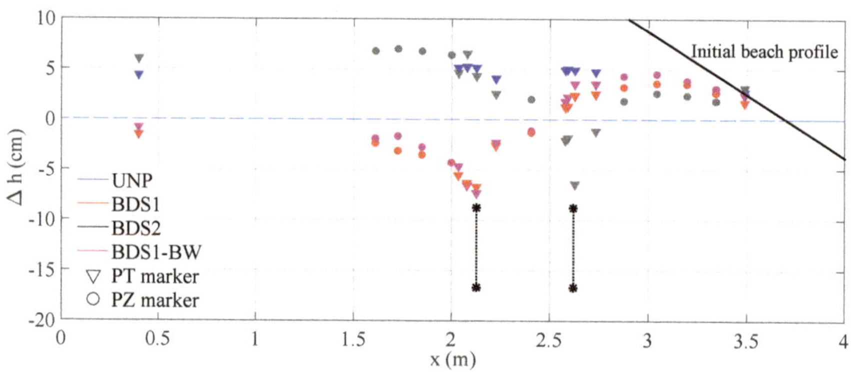

Figure 3. Spatial variation of water tables at the first step (15 min).

The timestack processing was basically performed by a procedure coded in Matlab and Fiji (Java). The step-by-step procedure was the following: (i) contrast enhancement with histogram equalisation; (ii) filter made of a Bi-Exponential Edge-Preserving Smoother (BEEPS, [58]); (iii) filter to reduce the effect on a non-uniform illumination by fast recursive Gaussian filters; Gabor filters on the gray-channel using five scales and eight orientations to enhance and localise major edges (output based on maximum intensity over the 40 images); (iv) despeckle and outliers filters; (v) an edge detector based on structured random decision forest [59]; and (vi) a final smoothing median filter on the edges detected, useful since the backwash is typically less distinguishable than the up-rush.

The evolution of beach profile was investigated by measuring the bottom elevations at the flume centreline, by means of a Leica FlexLine TS06plus Total Station with a uniform spatial resolution of 0.05 m for both emerged and submerged beach for a total number of measurement points equal to 295. According to time discretisation of tests previously described, at the end of each test step, bottom survey was performed, from which sediment transport rates, shoreline location, mean foreshore slope and bar position were derived.

3. Results and Discussion

3.1. Hydrodynamics

Free surface elevations time series collected by wave gauges were analysed in order to investigate wave transformations along the flume and compare the significant wave heights cross-shore distribution. Figure 4 reports the spatial variation of the mean significant wave height H_s in both protected and unprotected conditions calculated by means of the standard zero crossing method. At each wave gauge location the temporal variabilities of H_s are also reported for each configuration as error bars.

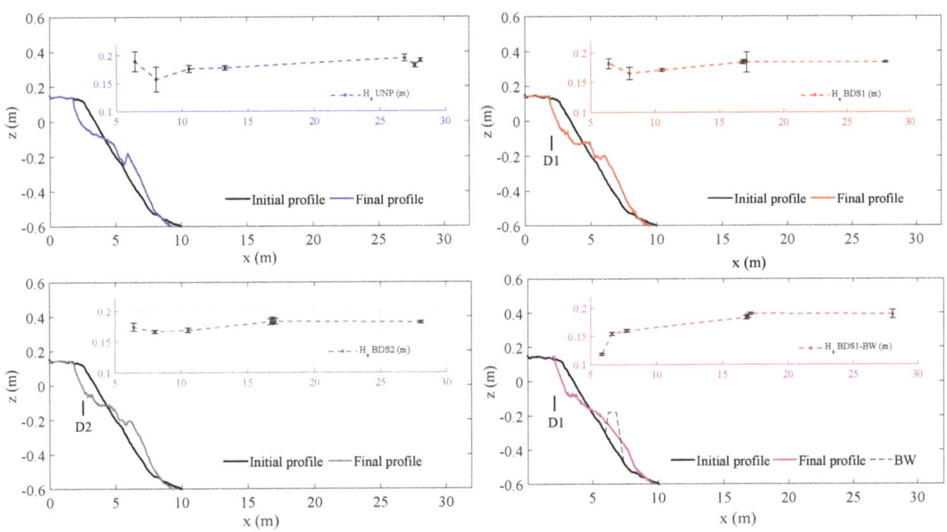

Figure 4. Significant wave height H_s spatial distribution.

In the unprotected configuration, H_s slightly decreased landward due to the interaction with the seabed up to the sand bar location and a raise of the significant wave height is recorded by the last gauge (*WG7*) in all steps, closer to the breakpoint. In fact, as observed in [60], an estimation of the breakpoint location can be determined at the maximum relative wave height, $\gamma = H_s/d$, where d is the local water depth, expected in the inner surf-zone and confirmed by the increasing γ spatial trend with the maximum at the *WG7* (time-averaged overall test steps $\gamma_{max} = 0.653$). Moreover, an increasing of the maximum relative wave height at the last wave gauge is observed over time in the range 0.50–0.87, demonstrating the coherent offshore movement of the breakpoint as the sand bar migrates offshore. Indeed, at the beginning of the tests (i.e., $\gamma \simeq 0.5$) the fraction of breaking waves could be argued to be very small (hence, the breakpoint is far from the location of *WG7*), whereas at the end of the tests (i.e., $\gamma \simeq 0.8$) the fraction of breaking waves highly increased, hence the breakpoint moved offshore to the location of WG7. The same behaviour was observed in drained conditions (*BDS*1 and *BDS*2), whereas differences could be highlighted in the presence of the submerged sill. The sill induced the breaking of the highest waves, leading to a substantial reduction of the significant wave height, with maximum values of γ evaluated at $WG6$ location, over the submerged sill.

In Figure 5, power amplitude spectra evolution along the profile is reported for all configurations at significant wave gauge locations (as reported in Figure 1), after 10 h of wave action, at sandbar equilibrium condition.

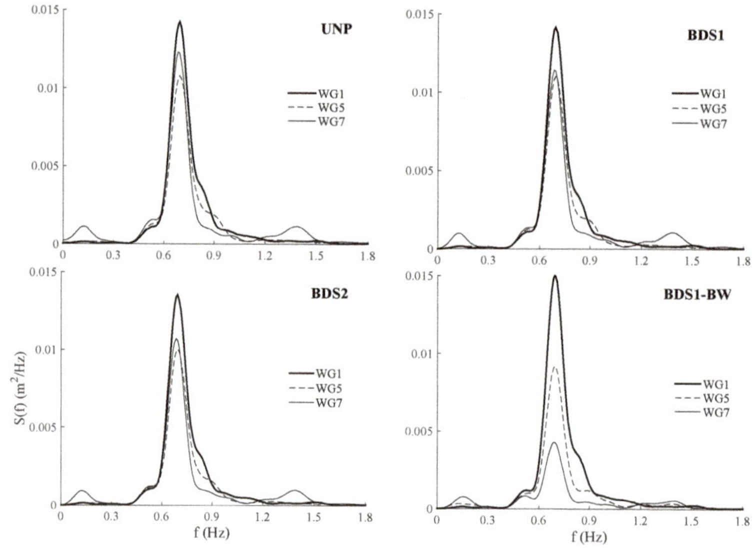

Figure 5. Power amplitude spectra variation for UNP, $BDS1$, $BDS2$ and $BDS1$-BW configurations.

Consistent with the H_s spatial variation observed in unprotected condition (UNP) (Figure 4), wave spectra showed an energy decay around the peak frequency (f_p), without any substantial modification of the spectral bandwidth until the location of the bar toe, except for higher harmonics components arising at about $2f_p$ due to the sandbar-induced breaking. Wave energy dissipation in drained conditions is slightly more pronounced, whereas no variation in peak frequency or other energy components is observed.

The submerged structure induces further noticeable effects on wave spectra spatial evolution, since well-known wave damping occurs when waves propagate over the sill. In particular, the Probability Density Functions derived from the short-term statistics of wave heights identified by zero-crossing analysis at both seaward and landward wave gauges location near the submerged sill, showed for all steps a decreasing of the wave heights (H_i) higher than 1.4 H_m (where H_m indicates the mean wave height at the toe of the structure). Being the mean ratio between the significant wave height at $WG5$ and $WG1$ calculated equal to 0.85 and the mean ratio between the significant and the mean wave height at the toe of the structure ($WG5$) of 0.65 (0.64–0.68), the breaking conditions occurred for wave heights $H_i > 0.76H_{so}$, with H_{so} the offshore significant wave height.

To analyse the swash zone behaviour, herein the wave setup (η) and run up (R) variations over time in all configurations are shown in Figure 6, derived from time-stacks processing. As highlighted in [31], the drainage system influenced the groundwater, leading to a reduction in η elevation on beach. A decreasing trend in time was clearly evident, and highlighted by the linear trend shown. The differences between the configurations were small. The influence of the $BDS1$ on the reduction of η could be observed, $BDS1$ exhibited smaller values at the first time steps, the linear trend then remained the lowest among the configurations, as well. The performance of $BDS2$ was poor, with results very similar to those observed in UNP, whereas the $BDS1$-BW highlighted, despite the submerged sill effect on wave momentum losses occurring during wave breaking and, hence, on onshore mass transport [61], a good efficiency in reduction of the η with respect to the unprotected conditions.

The vertical oscillations of R are analysed in terms of bulk parameters $R_{2\%}$ and R_{max}. It should be noted the maximum run up (either $R_{2\%}$ or R_{max}) was limited by the elevation of the horizontal emerged berm (i.e., +0.15 m). All the sill-unprotected cases (drained and undrained) showed time steps characterised by low values of both quantities, due also to a scarp formation at the shoreface. This condition was also due to the very steep profile (typical of artificial nourishment post-damping) and high-energy waves, which led to small beach scarp formation [62]. In these cases, the formation of the scarp did not allow the waves to reach the horizontal emerged berm. Swash waves steepened the beachface, subsequently they focused on the foot inducing its undercutting or removal. On the contrary, the $BDS1$-$BW1$ behaviour was characterised by high and almost constant values over the test duration. This was due to the morphodynamic differences highlighted, and the absence of the scarp formation at the shoreface, which highly influences the swash dynamics.

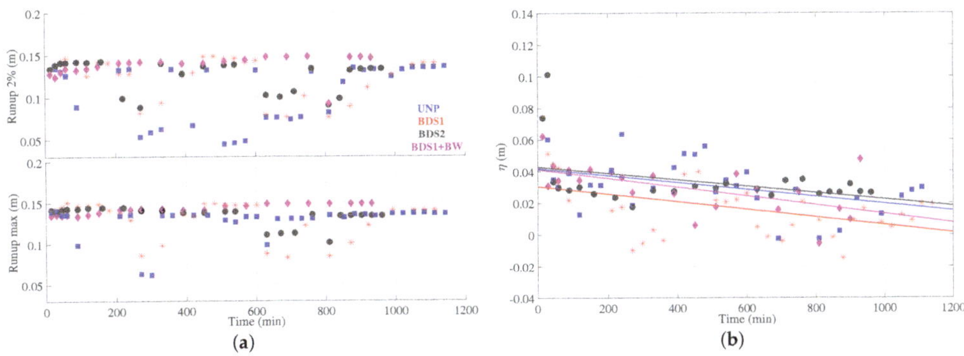

Figure 6. Bulk statistics of swash oscillations over time. (**a**) Wave runup $R_{2\%}$ (top), R_{max} (bottom) ; (**b**) Wave setup (η).

In view of investigating the potential contribution of the wave-induced setup in the cross-shore return currents, the temporal evolution of cross-shore component (v) velocity profiles are here reported as measured by the Vectrino Profiler, $Prof$ (Figure 1) close to the breaker zone and compared in different configurations. During waves breaking at the $ADVs$ sections, air bubbles penetrated into the water column, generating the large part of spike noises in recorded signals, by causing sometimes unreliable estimates of velocities. Few signal drop-outs were also found in correspondence to measuring points above the water level. The quality of $Prof$ velocity data, estimated in terms of Signal-to-Noise Ratio (SNR), were first used to discard values less than 15 dB. Then, the quasi 3D phase-space threshold method was used for the filtered data, according to the procedure described in [63,64]. Specifically, if any component of velocity u, v, or w was identified as a spike noise, all three components were eliminated from the dataset. The percentage of removed data was typically less than 14% out of the whole dataset. Beam velocities were recorded into the $ADV's$ orthogonal coordinate system, divided in long-shore u (y-axis), cross-shore v (x-axis) and vertical w velocities (z-axis), according to the reference system introduced in Section 2.1.

The location of the $ADVs$ was suitable for assessing the magnitude of the undertow currents, flowing in the lower section of the water column under breaking waves. The influence of the investigated defence systems on velocities field in the surf zone is here investigated, focusing the analyses on comparing time evolution of v components of velocity vectors, acquired by the $Prof$, whose results are in line with those of $VecS$, not reported here.

In Figure 7, light to dark profiles refer to the time evolution of the undertow currents, positive seaward, averaged in temporal intervals of 15 or 30 min, for all the configurations studied. The vertical axes z_b of each plot refers to distances from the bottom, positive upward, which varied over time due to morphological changes of the bed elevation. Few profiles were recorded at higher distances from the bottom, due to some inaccuracies in manual positioning of the instrument.

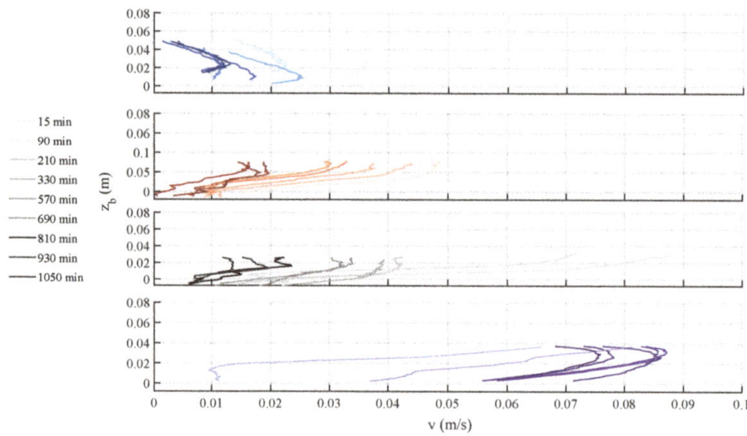

Figure 7. Time evolution of v-component of the velocity vectors for *UNP* (——); *BDS*1 (——); *BDS*2 (——); and *BDS*1-*BW* (——) tests. z_b refers to distances from the bottom, positive upward.

The undertow currents generally tended to decrease significantly, and after around 6 h their range of variability over the depth was less then 0.02 m/s. More specifically, drained configurations *BDS*1 were characterised by a very slight tendency in reduction with respect to *UNP* and *BDS*2. The phenomenon could be considered consistent with the observations in [31], where, for high-energy conditions, velocity profiles did not show any substantial variations. On the contrary, the configuration *BDS*1-*BW* exhibited a different behaviour. An increase of seaward currents was observed over time within the investigated depth, with a variability range at the later time steps of up to 0.04 m/s, followed by a lowering trend after 8 h, without reaching values less than 0.05 m/s. The authors related these results to differences of morphological response in presence of the sill, at its shoreward side, to be attributed to the feedback between breaking induced undertow and the location of the bar, its formation and migration (see Section 3.2).

The drains efficacy in increasing sea-water infiltration inside the beach during swash cycles was investigated by means of pore-pressure heads measured by both *PZ* and *PT*. As partly expected, all drained configurations showed a maximum water table lowering close to the drain. The effects on saturation degree decreases as the distance from the drain increases, so that the water table tended to the undisturbed groundwater level landward and to the mean water level seaward.

Figure 8 reports the groundwater behaviour at selected test steps in both unprotected and protected configurations. Moreover, beach profiles are reported as measured at the end of the same selected tests-steps. The groundwater dynamics, influenced by beach morphodynamics, for all configurations tested varied particularly in the first 120 min of the processes. The decrease was less than about 2 cm overall in this window. The process then proceeded slowly. Results highlighted that the drain with higher efficiency in water table lowering was the *D*1 due to its relative distance with respect to the shoreline, within the active infiltration zone [30,65]. The drain *D*2, closest to the shoreline, is able to intercept both the vertical infiltration flux through the porous sand and partly the water waves directly from the sea. Such effect is more evident as the beach profile evolves, since shoreline moves back and the drain *D*2 position is closer to the shoreline, so ineffective in reducing beach saturation degree. In such a condition, no reduction in backwash flow occurs.

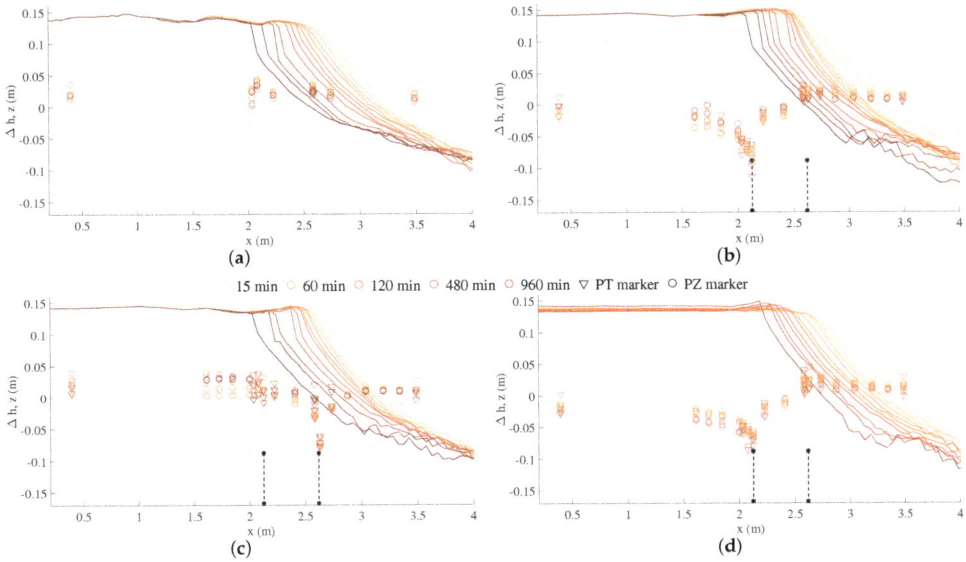

Figure 8. Spatial and temporal variation of groundwater table with the relative profiles measured at the end of the selected test step: (**a**) *UNP*; (**b**) *BDS*1; (**c**) *BDS*2; and (**d**) *BDS*1-*BW*.

3.2. Morphodynamics

High spatial as well as temporal resolution profile data were derived from the measurement of bed elevations in all tested configurations. In Figure 9, an example of the cross-shore time variation of the nourished beach, measured at the beginning of the test and at the end of each test step up to equilibrium condition is reported for test *UNP*. As expected, the morphological evolution of beach profile was faster in the first hours, whereas it tended to become slower up to the equilibrium condition, in correspondence of which any substantial variation in sediment transport was observed.

In Figure 10, the final profiles (Figure 10a) and the bottom changes with respect to the initial bed elevations (Figure 10b) are reported. Since the initial profiles were slightly different at the beginning of each test, the origin of the horizontal axes coincides with the initial shoreline location at the beginning of each test, in order to make final profiles comparable. In unprotected conditions temporal profile evolution showed a shoreline retreat since seaward sediment transport occurred along both swash and surf zones (Figure 10). Sediments moved offshore within the active zone with the formation of two submerged bars which evolved during tests until both equilibrium position and shape were reached. Such a behaviour was observed in unprotected and only drained conditions. The presence of the sill together with the drain *D*1 induced the submerged bar formation in the first stage of profile evolution, until the bar migrates near the structure landward toe and sediments accumulation occurred with a final S-shaped beach profile.

The high resolution bed elevation data in both spatial and temporal domains allowed the sediment transport rate ($m^3/s{\cdot}m$) to be analysed by applying Exner's equation (e.g., [12,66]). The sediment transport rate spatial variation ($q_s(x_i)$) is then expressed as a function of temporal bed elevation evolution ($\partial z/\partial t$) and the material porosity (p). Assuming the porosity constant along the profile since the sediment grain size was almost uniform without any variation of the mean diameter (D_{50}), Exner's equation can be written at each location x_i as follows (Equation (1)):

$$q_s(x_i) = q_s(x_{i-1}) - \frac{\Delta z}{\Delta t}\Delta x \tag{1}$$

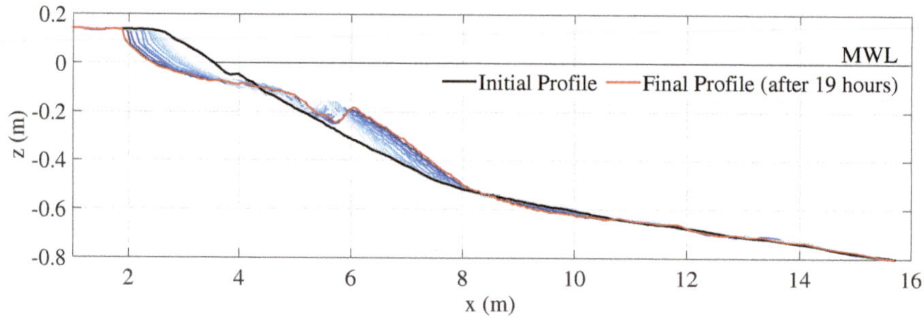

(a)

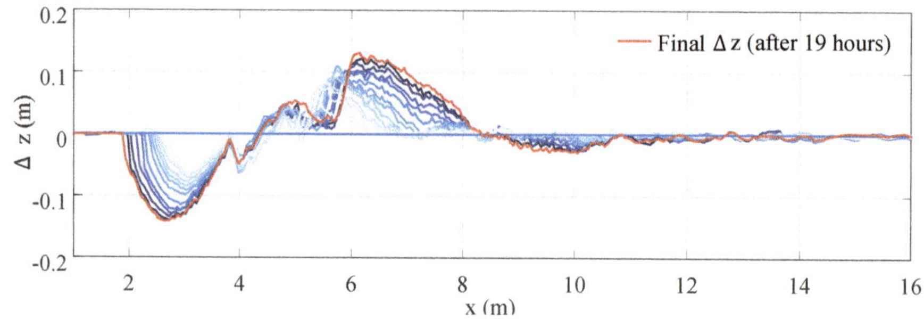

(b)

Figure 9. (**a**) Time-evolution of beach profile measured at the centreline (z, blue gradient lines). (**b**) Spatial and temporal evolution of bottom elevation changes (Δz) with respect to the initial profile.

Mass balance equation was solved on x_i locations of a regular grid with a spatial Δx equal to 0.01 m, interpolated from the bed elevation measurements acquired every 0.05 m, and a temporal Δt equal to the duration of each test step. During bed elevation measurements, some small errors in volumes calculation were found, with consequent unreliable values of q_s at boundaries. As discussed in [67], indeed, boundary conditions at both sides of the flume imposes that sediment flux has to be equal to zero at the beach toe and close to the run-up limit as well. Due to errors in bed elevation measurements (i.e., over the ripples), the intrinsic accuracy of the instruments ($2 \div 3$ mm), a not-perfect uniformity of profiles in long-shore direction and small sand losses in the flume, a correction on volumes calculation was needed. Accordingly, since it is not possible to determine where the mismatch occurs [67], a uniform redistribution of the closure errors calculated at the beach toe was applied across the profile where sediment transport rates were not zero.

In Figure 11, the corrected net sediment transport rates for unit length $q_s(x_i)$ ($m^3/s{\cdot}m$) are reported for all configurations at selected time steps to highlight the cross-shore variation of $q_s(x)$ over time and the differences induced by the defence system deployed. According to Equation (1) and local coordinate system with the x-axis positive seaward, negative values refer to onshore sediment transport, whereas positive values stand for offshore prevalent transport. Information about the net sediment transport fluxes at each location can be retrieved by considering the derivative of $q_s(x)$ with respect to x, which measures the change in the transport rate per unit increase in x along the curves. Therefore, an increasing of $q_s(x)$ in both positive or negative quadrants of the graphs indicates bottom erosion, mainly concentrated in the swash and surf zones, whereas negative derivatives can be observed at the sandbar location, where sediments settle.

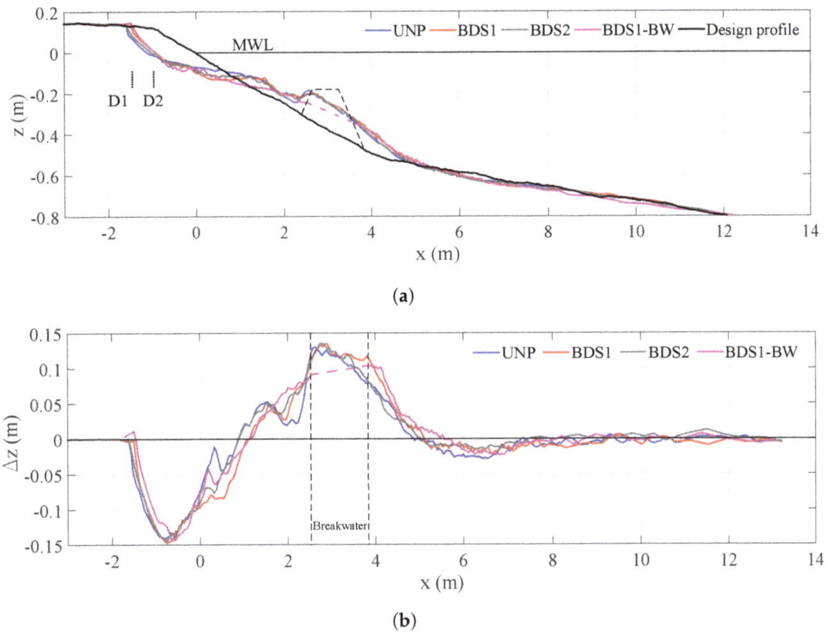

Figure 10. Comparison between final profiles (**a**) and bottom changes (**b**) for unprotected (*UNP*) and protected (*BDS*1, *BDS*2, and *BDS*1-*BW*) cases. The origin of the horizontal axis coincides with the location of the initial shoreline.

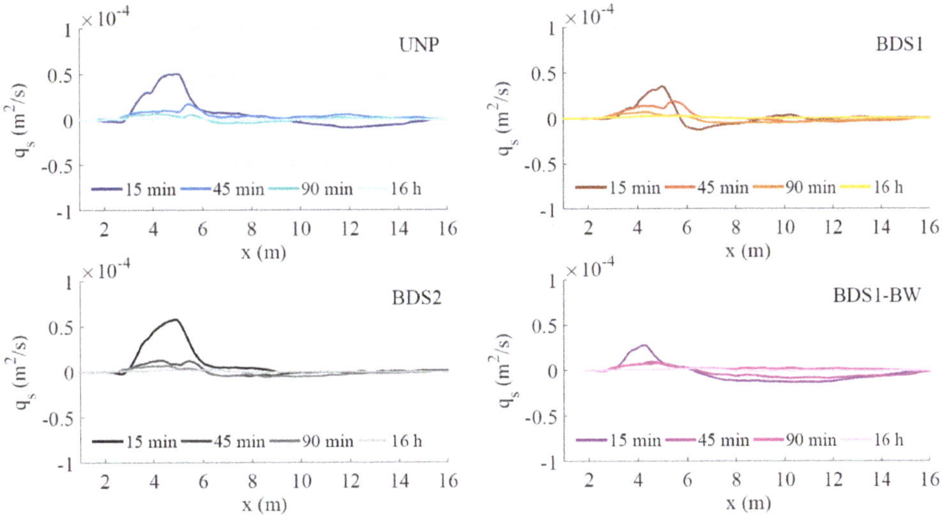

Figure 11. Sediment transport rates over time for the different configurations.

Figure 11 shows that in unprotected conditions the sediment transport was mainly seaward. In *BDS*1 and *BDS*1-*BW* configurations, both systems affect this trend, by reversing the direction of the net sediment transport onshore, in the area close to the sand bar and the sill, respectively. Such effect is more pronounced when the submerged sill is coupled with the drain. Sediment transport tended to an equilibrium condition in the first hour for all configurations.

The presence of the $BDS1$ mainly affected the morphodynamics in the first stage of these changes, whereas any substantial differences are highlighted with $BDS2$. During the first 15 min, a reduction of eroded sediments along the emerged beach in presence of $BDS1$ was observed, whereas any substantial difference occurred within the surf zone. The sediments moved offshore with the formation of the submerged bar at about the same position reached in the unprotected case. Moreover, positive effects were highlighted since sediment transport rate became negative in the seaward zone close to the sand bar. In presence of the $BDS2$, the erosion of the foreshore was comparable with that in unprotected conditions, since as beach eroded and shoreline moved back, the drain was completely submerged and, indeed, not suitable for reducing the beach saturation degree. After 60 min of wave exposure, transport rate was low in all configurations and sediments tended to be quite stable. Profile global shape remained quite constant and modifications were mainly due to the gradual redistribution of sediments from the emerged beach to shallow waters (swash and surf zones) and in correspondence of the offshore submerged bar.

As an example, in Figure 12, the comparison between q_s and the measured relative spatial variation of z_b within the active zone, at 15 min (Figure 12a), 90 min (Figure 12b), and 16 h (Figure 12c) are reported as representative of beach profile evolution for UNP test. Once the quasi-equilibrium condition after 1 h of test (Figure 12b) was reached, small q_s variations were observed along the emerged beach, within the swash zone and at bar location, dominated by slow seaward sediment transport. The redistribution of sand along the profile, indeed, induced a decrease of the mean foreshore slope with a consequent shoreline retreat, whereas the submerged bar migrated seaward, until it reached its local equilibrium under waves action (Figure 12c).

Figure 13 show the temporal variation of shoreline (Δx at 0 m depth isoline, solid line) and mean foreshore slope (β), respectively, for each configuration. In the Figure 13a the temporal variation of depth-lines -0.03 m and $+0.03$ m, with respect to the mean water level ($z = 0$ m) are reported (dotted lines). Results are presented with reference to the initial shoreline position up to the end, according to the time discretisation of profile measurements, evaluated as the intersection of each measured beach profile with the static water level in the flume. Meanwhile, the mean foreshore slope was calculated as the mean beach slope from the beach berm to the section where submerged bar formed.

As stated for profiles evolution, shoreline retreat and beach steepening were both faster at the beginning, for all configurations, since after 180 min the shoreline retreated about 50% of its final location. UNP and $BDS2$ configurations showed the maximum Δ_x at $z = 0$ m, demonstrating that the location of $D2$ was not useful and that beach behaviour was comparable with natural one. Corresponding to the sediment transport rate decrease, both shoreline onshore displacement and beach foreshore slope velocities decreased, even if both processes persisted. Even though a lower sediment transport rate was observed (Figure 11), shoreline did not stabilise, since slow sediments redistribution along the swash zone led to a decrease of beach slope (Figure 13a). When the drain $D1$ was activated, the shoreline recession was lower with and without the submerged sill, slightly lower in $BDS1$ configuration.

The temporal variation of the depth-lines around the shoreline reported in Figure 13a shows that the redistribution of sediment occurring after the early stage, induced a higher change of beach slope within the swash zone in unprotected conditions, with respect to the other configurations, as confirmed by the higher differences in depth-lines -0.03 m and 0.03 m retreats. In $BDS1$ and $BDS1$-BW configurations, temporal evolution of $+0.03$ m depth-lines shows that positive effects of both defence system could be also observed along the emerged beach, close to the shoreline, where higher sediment volumes with respect to unprotected conditions accumulated.

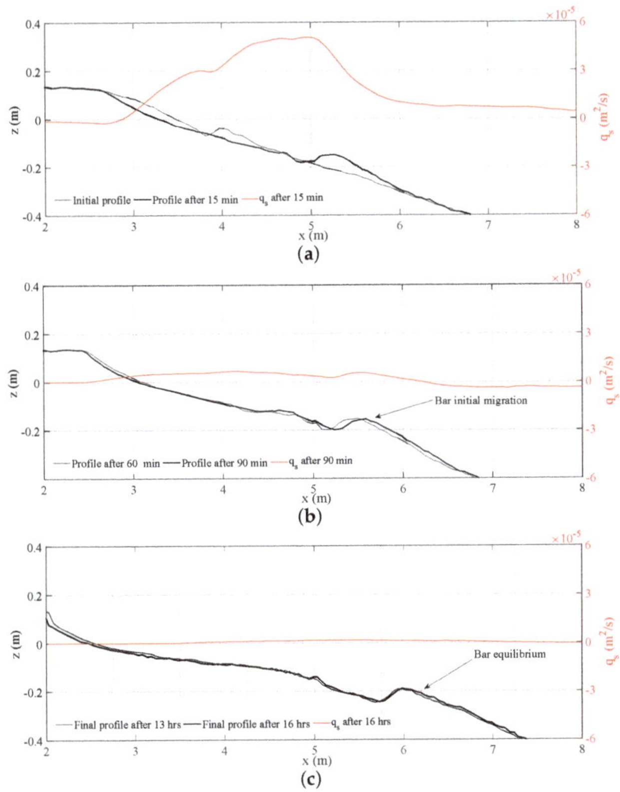

Figure 12. Sediment transport rates (q_s) and measured profiles (z_b) at 15 min (**a**), 90 min (**b**) and at the submerged bar equilibrium (**c**) in unprotected conditions.

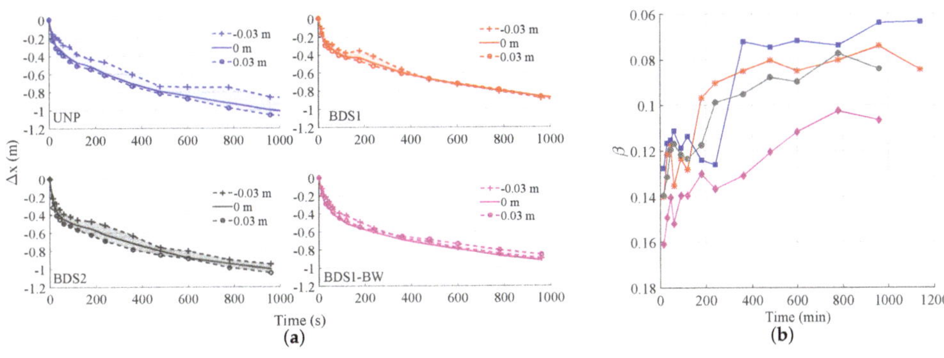

Figure 13. Temporal variation of depth-lines −0.03 m, 0 m, +0.03 m locations (**a**) and mean foreshore slope (**b**) for each configuration from the beginning of each test.

In Figure 14, the defence systems effects on profile evolution are analysed in terms of submerged bar behaviour. The bar is here sketched by means of its representative parameters reported in Figure 14a, where x_{bar} indicates the cross-shore distance of the bar crest elevation from the initial shoreline position, h_{bar} is the water column height over the crest bar and z_{bar} represents the maximum bar height with respect to the initial profile at x_{bar}. In Figure 14b, the bar migration (x_{bar}) over time is reported for *UNP*, *BDS*1, *BDS*2 and *BDS*1-*BW* configurations. To analyse bar evolution, a dimensionless bar crest height is introduced, ranging from 0 to 1, defined as follows (Equation (2)):

$$\zeta_{bar} = -\frac{z_{bar}}{z_{bar} + h_{bar}} \tag{2}$$

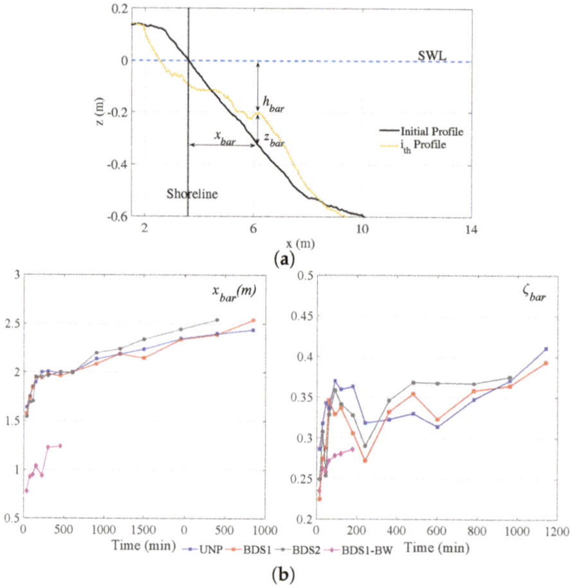

Figure 14. (**a**) Sketch of the main parameters useful for bar description. (**b**) Temporal variation of crest bar location, x_b (left panel) and dimensionless bar crest height, ζ_b (right panel).

Figure 14b provides evidence of quite similar behaviours of the breaker bars evolution over time under the wave condition tested for all cases, although some differences should be discussed. Results demonstrate that offshore bar migration velocity was higher in the first 3 h for all configurations. With respect to *UNP*, *BDS*1 and *BDS*2 conditions where sediments began to settle quite at the same water depth, the submerged sill induced the initial bar formation in the section closest to the shoreline. During such a phase the sand offshore movement from the swash zone also induced a timely increase of the bar height, more pronounced in unprotected conditions, as the temporal variation of ζ_{bar} shows. After the third hour, the bar migration showed a slow-down phase for around 40%, comparable for both natural and drained conditions. This was observed for the h_{bar} behaviour as well.

In presence of the submerged breakwater, bar formation only occurred in the early stage of the profile evolution. After about 3 h of test, as the bar moved offshore, the sediment began accumulating near the submerged sill and the bar parameters are not suitable for the analysis since the bar cannot be properly defined. For this reason, in Figure 14, temporal evolution of bar parameters for *BDS*1-*BW* is reported until 180 min of waves exposure. As the bar moved offshore, the bar height continuously increased until near a quasi-equilibrium condition. After 10 h, the bar position was almost stable.

The slight increase in x_{bar} was mainly due to the bar crest height arise together with a redistribution of sediments. The final location of the bar was almost the same for *UNP*, *BDS*1 and *BDS*2 tests. In *UNP*, with respect to the other cases, the sandbar height is slightly higher and the sharpening of its shape was observed. Notably, in unprotected conditions, an evident increase of ζ_{bar} was observed at the end of *UNP* test, against a lower increase of x_{bar}. This demonstrates that the increase of h_{bar} is not due to the offshore migration at greater depths, but to properly raise the bar crest height (z_{bar}).

4. Concluding Remarks

This paper presents a new experimental campaign aimed at assessing the cross-shore short-term evolution of a sandy beach nourishment in presence of an alternative mixed defence system, constituted by a Beach Drainage System and a submerged detached rubble-mound breakwater deployed together. The submerged sill is characterised by high freeboard and its purpose is twofold. It aims at improving BDS efficacy, by switching high energy sea states to medium low energy waves, since past studies [30] demonstrated BDS efficacy in stabilising sediments on beach for medium and low energy conditions, whereas any positive effects were observed under erosive wave conditions. Moreover, the submerged structure intercepts offshore sediment transport, resulting in a sediment accretion in the shallow waters and favouring the possibility of periodic interventions of beach scraping, with a consequent reduction of sand volumes, costs and working time, with a minimal impact on littoral natural cycles.

The experiments were performed at the Laboratory of Coastal Engineering of the Politecnico di Bari (Bari, Italy) on a 2D movable-bed physical model. A constant JONSWAP wave spectrum representative of erosive condition was used as forcing boundary wave conditions. To assess the performance of the proposed protection system, hydrodynamics and beach profile evolution up to equilibrium were firstly analysed in unprotected and only drained (without the submerged sill) conditions. Two different distances of drain pipe with respect to the initial shoreline position were tested. Wave transformation along the beach profile and over the submerged structure, surf zone velocities, beach groundwater as well as beach profile evolution, spatial and temporal sediment transport rates, shoreline, bar and mean foreshore slope evolution are presented and discussed.

Results show that, under tested conditions, both BDS and submerged sill affected hydrodynamics and morphodynamics along the surf and swash zones, with respect to the unprotected configuration. The general efficiency of the drain system mainly depends on its hydraulic regime, pipes characteristics, porous medium behaviour which influences flow resistance, and groundwater head. Beyond the full-scale experiments of [30], the comparison between the *PTs* placed inside both drains (*PT*8 and *PT*10) and those very close to their contour inside the sand (*PT*7 and *PT*9) allowed to state that a seamless hydraulic regime develops in the system sand-drain. No gaps between local pressure head inside both the pipe and the low transmissivity porous medium were observed, thus leading to a steep cone of depression. In such conditions, drains mainly worked under pressure. Such an outcome suggests the possibility of improving BDS efficacy in collecting sea-water by enhancing both design characteristics and porous medium infiltration capability and inserting, for example, a gravel layer around the pipe acting as a filter which can guarantee a stepwise increasing of permeability from sand to pipe.

The higher capability of the beach in absorbing run up flows in presence of the drain *D*1 induced lower shoreline retreat, a decrease of beach slope within the swash as well as surf zone and a reduction of net sediment transport rates. Moreover, the different form of sandbar visible in both *BDS*1 and *BDS*2 with respect to *UNP* tests justifies differences in wave energy dissipation over the sand bar.

Besides BDS effects on both hydrodynamics and morphodynamics observed when the system was tested alone, interesting outcomes were found in the jointly configuration with the submerged sill. The structure induced an evident wave energy reduction within the surf zone due to the wave breaking and a reversal of the prevalent direction of the net sediment transport seaward, offshore the sheltered region. However, the role of the structure in inducing further improvements in drainage efficacy is still doubtful, whereas the secondary effects of drain on the submerged sill performance were more clear. Results point out that BDS influenced swash zone hydrodynamics also in presence of the submerged breakwater, as the reduction of wave-induced setup with respect to unprotected tests demonstrated. It is widely accepted, indeed, that such submerged structures experiences mean water level raising on the shoreward side, which increases as structure freeboard (R_c) reduces [61,68]. Experiments showed that the drainage is able to mitigate such raising of the mean water level, even if further analyses are needed, since any influence on seaward undertow currents in the sheltered region was observed.

Different submerged sill and drains configurations (i.e., distance from the shoreline, berm width, freeboard, and presence of a gravel layer around the drain pipe) are planned for new experiments to clarify the mutual influence and dependency. Accordingly, these tests also represent a highly detailed database in space and time for further goals of developing numerical solutions useful for parameterisation of the main drainage parameters for a beach protection scheme which could potentially include a nourished beach profile protected by a breakwater.

Author Contributions: Conceptualization, A.S., N.V., M.D.R.; Methodology, A.S., N.V., M.D.R., D.P., L.D.; Software, A.S., N.V., M.D.R., D.P.; Validation, A.S., N.V., M.D.R., D.P.; Formal Analysis, A.S., N.V., M.D.R., D.P.; Investigation, A.S., N.V., M.D.R., D.P.; Resources, A.S., N.V., M.D.R., D.P.; Data Curation, A.S., N.V., M.D.R., D.P.; Writing—Original Draft Preparation, A.S., N.V.; Writing—Review & Editing, A.S., N.V., M.D.R., D.P., L.D.; Visualization, A.S., N.V.; Supervision, L.D.; Project Administration, A.S.; Funding Acquisition, A.S., L.D.

Funding: This research and the costs to publish in open access were funded by the Development Fund and Cohesion 2007–2013—APQ search Apulia Region *Regional program in support of smart specialization and social and environmental sustainability—Future In Research* and by Fondazione Puglia.

Conflicts of Interest: The authors declare no conflict of interest and that the founding sponsors had no role in the design of the study; in the collection, analyses, or interpretation of data; in the writing of the manuscript, and in the decision to publish the results.

References

1. Tsvetanov, T.G.; Shah, F.A. The economic value of delaying adaptation to sea-level rise: An application to coastal properties in Connecticut. *Climat. Chang.* **2013**, *121*, 177–193. [CrossRef]
2. Ruol, P.; Martinelli, L.; Favaretto, C. Vulnerability analysis of the Venetian littoral and adopted mitigation strategy. *Water* **2018**, *10*, 984. doi:10.3390/w10080984. [CrossRef]
3. Pasquali, D.; Di Risio, M.; De Girolamo, P. A simplified real time method to forecast semi-enclosed basins storm surge. *Estuar. Coast. Shelf Sci.* **2015**, *165*, 61–69. [CrossRef]
4. Intergovernmental Panel on Climate Change. *Climate Change 2014–Impacts, Adaptation and Vulnerability: Regional Aspects*; Cambridge University Press: New York, NY, USA, 2014.
5. Lamberti, A.; Archetti, R.; Kramer, M.; Paphitis, D.; Mosso, C.; Di Risio, M. European experience of low crested structures for coastal management. *Coast. Eng. J.* **2005**, *52*, 841–866. [CrossRef]
6. Celli, D.; Pasquali, D.; De Girolamo, P.; Di Risio, M. Effects of submerged berms on the stability of conventional rubble mound breakwaters. *Coast. Eng. J.* **2018**, *136*, 16–25. [CrossRef]
7. Bohnsack, J.A.; Sutherland, D.L. Artificial reef research: A review with recommendations for future priorities. *Bull. Mar. Sci.* **1985**, *37*, 11–39.
8. Capobianco, M.; Hanson, H.; Larson, M.; Steetzel, H.; Stive, M.; Chatelus, Y.; Aarninkhof, S.; Karambas, T. Nourishment design and evaluation: Applicability of model concepts. *Coast. Eng. J.* **2002**, *47*, 113–135. [CrossRef]
9. Vieira da Silva, G.; Toldo, E.; Klein, A.; Short, A.; Tomlinson, R.; Strauss, D. A comparison between natural and artificial headland sand bypassing in Santa Catarina and the Gold Coast. In Proceedings of the Australasian Coasts & Ports 2017 Conference, Cairns, QLD, Australia, 21–23 June 2017; pp. 1111–1117.
10. Becker, A.; Taylor, M.D.; Folpp, H.; Lowry, M.B. Managing the development of artificial reef systems: The need for quantitative goals. *Fish Fish.* **2018**, *19*, 740–752. [CrossRef]
11. Pilarczyk, K.W. Design of low-crested (submerged) structures: An overview. In Proceedings of the 6th COPEDEC (International Conference on Coastal and Port Engineering in Developing Countries), Colombo, Sri Lanka, Citeseer, 19 September 2003.
12. Di Risio, M.; Lisi, I.; Beltrami, G.; De Girolamo, P. Physical modeling of the cross-shore short-term evolution of protected and unprotected beach nourishments. *Ocean Eng.* **2010**, *37*, 777–789. [CrossRef]
13. Chiaia, G.; Damiani, L.; Petrillo, A. Evolution of a Beach with and without a Submerged Breakwater: Experimental Investigation. In Proceedings of the 23rd International Conference on Coastal Engineering, Venice, Italy, 4–9 October 1992; pp. 1959–1972.
14. Creter, R.E.; Garaffa, T.; Schmidt, C.J. Enhancement of beach fill performance by combination with an artificial submerged reef system. In Proceedings of the 7th National Conference on Beach Preservation Technology, Tallahassee, FL, USA, 9–11 February 1994; pp. 69–89.

15. Calabrese, M.; Vicinanza, D.; Buccino, M. Large-scale experiments on the behaviour of low crested and submerged breakwaters in presence of broken waves. In *Coastal Engineering 2002: Solving Coastal Conundrums*; World Scientific: Sinapore, 2003; pp. 1900–1912.
16. Armono, H.; Hall, K. Laboratory study of wave transmission on artifical reefs. In Proceedings of the Canadian Coastal Engineering Conference, Kingston, ON, Canada, 15–17 October 2003; Canadian Society for Civil Engineering: Kingston, Canada, 2003.
17. Stauble, D.K.; Tabar, J.R. The use of submerged narrow-crested breakwaters for shoreline erosion control. *J. Coast. Res.* **2003**, *19*, 684–722.
18. Wamsley, T.; Hanson, H.; Kraus, N.C. *Wave Transmission at Detached Breakwaters for Shoreline Response Modeling;* Technical report No. ERDC/CHL-CHETN-II-45; Engineer Research and Development Center Vicksburg MS Coastal and Hydraulics Lab: Vicksburg, MS, USA, 2002.
19. Malcangio, D.; Melena, A.; Damiani, L.; Mali, M.; Saponieri, A. Numerical study of water quality improvement in a port through a forced mixing system. In *Water Resources Management IX*; WIT Press: Southampton, UK, 2017; pp. 69–80.
20. Mali, M.; Malcangio, D.; Dell'Anna, M.M.; Damiani, L.; Mastrorilli, P. Influence of hydrodynamic features in the transport and fate of hazard contaminants within touristic ports. Case study: Torre a Mare (Italy). *Heliyon* **2018**, *4*, e00494. [CrossRef] [PubMed]
21. Sánchez-Arcilla, A.; Alsina, J.; Cáceres, I.; González-Marco, D.; Sierra, J.; Pena, C. Morphodynamics on a beach with a submerged detached breakwater. In *Coastal Engineering 2004*; World Scientific: Sinapore, 2005; pp. 2836–2848.
22. Bagnold, R. Beach formation by waves: Some model experiments in a wave tank. *J. Inst. Civ. Eng.* **1940**, *15*, 27–52. [CrossRef]
23. Grant, U. Influence of the water table on beach aggradation and degradation. *J. Mar. Res.* **1948**, *7*, 655–660.
24. Cartwright, N.; Baldock, T.E.; Nielsen, P.; Jeng, D.S.; Tao, L. Swash-aquifer interaction in the vicinity of the water table exit point on a sandy beach. *J. Geophys. Res. Ocean.* **2006**, *111*. [CrossRef]
25. Saponieri, A. Beach Drainage. In *Encyclopedia of Coastal Science*, 2nd ed.; Finkl, C.W., Makowski, C., Eds.; Springer International Publishing: New York, NY, USA, 2018; pp. 1–4, doi:10.1007/978-3-319-48657-4_370-2. [CrossRef]
26. Ciavola, P.; Vicinanza, D.; Fontana, E. Beach drainage as a form of shoreline stabilization: Case studies in Italy. In *Coastal Engineering 2008*; World Scientific: Sinapore, 2009; Volume 5, pp. 2646–2658.
27. Masria, A.; Iskander, M.; Negm, A. Coastal protection measures, case study (Mediterranean zone, Egypt). *J. Coast. Conserv.* **2015**, *19*, 281–294. [CrossRef]
28. Horn, D.P.; Baldock, T.E.; Li, L. The influence of groundwater on profile evolution of fine and coarse sand beaches. In Proceedings of the Sixth International Symposium on Coastal Engineering and Science of Coastal Sediment Process, New Orleans, LA, USA, 13–17 May 2007; pp. 506–519.
29. Damiani, L.; Petrillo, A.F.; Saponieri, A. Near shore Morphodynamic of drained beaches. *Coast. Eng. Proc.* **2011**, *1*, 8. [CrossRef]
30. Damiani, L.; Aristodemo, F.; Saponieri, A.; Verbeni, B.; Veltri, P.; Vicinanza, D. Full-scale experiments on a beach drainage system: Hydrodynamic effects inside beach. *J. Hydraul. Res.* **2011**, *49*, 44–54. [CrossRef]
31. Damiani, L.; Vicinanza, D.; Aristodemo, F.; Saponieri, A.; Corvaro, S. Experimental investigation on wave set-up and nearshore velocity field in presence of a BDS. *J Coast. Res.* **2011**, *SI 64*, 55–59.
32. Aristodemo, F.; Ciavola, P.; Veltri, P.; Saponieri, A. The influence of a Beach Drainage System on wave reflection and surf beat processes. *J Coast. Res.* **2011**, *SI 64*, 455–459.
33. Saponieri, A.; Damiani, L. Numerical analysis of infiltration in a drained beach. *Int. J. Sustain. Dev. Plan.* **2015**, *10*, 467–486. [CrossRef]
34. Ciavola, P.; Vicinanza, D.; Aristodemo, F.; Contestabile, P. Large-scale morphodynamic experiments on a beach drainage system. *J. Hydraul. Res.* **2011**, *49*, 523–528. [CrossRef]
35. Contestabile, P.; Aristodemo, F.; Vicinanza, D.; Ciavola, P. Laboratory study on a beach drainage system. *Coast. Eng.* **2012**, *66*, 50–64. [CrossRef]
36. Lisi, I.; Molfetta, M.; Bruno, M.; Di Risio, M.; Damiani, L. Morphodynamic classification of sandy beaches in enclosed basins: The case study of Alimini (Italy). *J. Coast. Res.* **2011**, *SI 64*, 180–184.
37. Krumbein, W.C.; James, W.R. *A Lognormal Size Distribution Model for Estimating Stability of Beach Fill Material;* Army Coastal Engineering Research Center: Washington, DC, USA, 1965.

38. Lisi, I.; Di Risio, M.; De Girolamo, P.; Gabellini, M. Engineering Tools for the Estimation of Dredging-Induced Sediment Resuspension and Coastal Environmental Management. In *Applied Studies of Coastal and Marine Environments*; IntechOpen: London, UK, 2016; pp. 55–83.
39. Di Risio, M.; Pasquali, D.; Lisi, I.; Romano, A.; Gabellini, M.; De Girolamo, P. An analytical model for preliminary assessment of dredging-induced sediment plume of far-field evolution for spatial non homogeneous and time varying resuspension sources. *Coast. Eng.* **2017**, *127*, 106–118. [CrossRef]
40. Bruun, P. Beach scraping—Is it damaging to beach stability? *Coast. Eng.* **1983**, *7*, 167–173. [CrossRef]
41. McNinch, J.; Wells, J. Effectiveness of beach scraping as a method of erosion control. *Shore Beach* **1992**, *60*, 13–20.
42. Henry, R. Preliminary Investigations into the Possible Effects of Beach Scraping on Wooli Beach With Specific Reference to Intertidal Macro-Infauna Diversity and Abundance. Bachelor's Thesis, Deakin University, Victoria, Australia, 1999.
43. Baldock, T.E.; Alsina, J. Impact of beach scraping on near shore sediment transport and bar migration. In Proceedings of the Coasts and Ports 2013: 21st Australasian Coastal and Ocean Engineering Conference and the 14th Australasian Port and Harbour Conference, Sydney, Australia, 11–13 September 2013; pp. 1–6.
44. Dare, J.L. Alternative Shore Protection Strategies: Innovative Options and Management Issues. Master's Thesis, Oregon State University, Corvallis, OR, USA, September 2003.
45. Di Risio, M.; Bruschi, A.; Lisi, I.; Pesarino, V.; Pasquali, D. Comparative analysis of coastal flooding vulnerability and hazard assessment at national scale. *J. Mar. Sci. Eng.* **2017**, *5*, 51, doi:10.3390/jmse5040051. [CrossRef]
46. Allen, R. Standard test methods for determining average grain size (F112). In *Annual Book of ASTM Standards, Metal-Mechanical Testing; Elevated and Low Temperature Tests; Metallography*; ASTM International: West Conshohocken, PA, USA, 1999; Volume 3.
47. Wentworth, C.K. A scale of grade and class terms for clastic sediments. *J. Geol.* **1922**, *30*, 377–392. [CrossRef]
48. Dean, R.G. Heuristic models of sand transport in the surf zone. In *First Australian Conference on Coastal Engineering, 1973: Engineering Dynamics of the Coastal Zone*; Institution of Engineers: Australia, 1973; p. 215. Avaliable online: https://search.informit.com.au/documentSummary;dn=971703171672500;res=IELENG (accessed on 28 August 2018).
49. Kriebel, D.; Dally, W.; Dean, R. Undistorted Froude model for surf zone sediment transport. In Proceedings of the 20th International Conference on Coastal Engineering, Taipei, Taiwan, 9–14 November 1986; pp. 1296–1310.
50. Kraus, N.C.; Larson, M. *Beach Profile Change Measured in the Tank for Large Waves 1956–1957 and 1962*; Technical report; Coastal Engineering Research Center: Vicksburg, MS, USA, 1988.
51. Dalrymple, R.A. Prediction of storm/normal beach profiles. *J. Water. Port Coast. Ocean Eng.* **1993**, *119*, 473–474. [CrossRef]
52. Mansard, E.P.; Funke, E. The measurement of incident and reflected spectra using a least squares method. In Proceedings of the 17th International Conference on Coastal Engineering, Sydney, Australia, 23–28 March 1980; pp. 154–172.
53. Salmon, S.A.; Bryan, K.R.; Coco, G. The use of video systems to measure run-up on beaches. *J. Coast. Res.* **2007**, *50*, 211–215.
54. Valentini, N.; Saponieri, A.; Damiani, L. A new video monitoring system in support of Coastal Zone Management at Apulia Region, Italy. *Ocean Coast. Manag.* **2017**, *142*, 122–135. doi:10.1016/j.ocecoaman.2017.03.032. [CrossRef]
55. Valentini, N.; Damiani, L.; Molfetta, M.G.; Saponieri, A. New coastal video-monitoring system achievement and development. *Coast. Eng. Proc.* **2017**, *1*, 11. [CrossRef]
56. Vousdoukas, M.I.; Velegrakis, A.F.; Dimou, K.; Zervakis, V.; Conley, D.C. Wave run-up observations in microtidal, sediment-starved pocket beaches of the Eastern Mediterranean. *J. Mar. Syst.* **2009**, *78*, S37–S47. [CrossRef]
57. Valentini, N.; Saponieri, A.; Molfetta, M.G.; Damiani, L. New algorithms for shoreline monitoring from coastal video systems. *Earth Sci. Inform.* **2017**, *10*, 495–506, doi:10.1007/s12145-017-0302-x. [CrossRef]
58. Thévenaz, P.; Sage, D.; Unser, M. Bi-exponential edge-preserving smoother. *IEEE Trans. Image Proc.* **2012**, *21*, 3924–3936, doi:10.1109/TIP.2012.2200903. [CrossRef] [PubMed]

59. Dollár, P.; Zitnick, C.L. Structured forests for fast edge detection. In Proceedings of the IEEE International Conference on Computer Vision, Sydney, NSW, Australia, 1–8 December 2013; pp. 1841–1848.
60. Eichentopf, S.; Cáceres, I.; Alsina, J.M. Breaker bar morphodynamics under erosive and accretive wave conditions in large-scale experiments. *Coast. Eng.* **2018**, *138*, 36–48, doi:10.1016/j.coastaleng.2018.04.010. [CrossRef]
61. Calabrese, M.; Vicinanza, D.; Buccino, M. 2D wave setup behind submerged breakwaters. *Ocean Eng.* **2008**, *35*, 1015–1028. [CrossRef]
62. Larson, M.; Hanson, H.; Kraus, N.C.; Newe, J. Short-and long-term responses of beach fills determined by EOF analysis. *J. Water. Port Coastal Ocean Eng.* **1999**, *125*, 285–293. [CrossRef]
63. Goring, D.G.; Nikora, V.I. Despiking acoustic Doppler velocimeter data. *J. Hydraul. Eng.* **2002**, *128*, 117–126. [CrossRef]
64. Meftah, M.B.; Malcangio, D.; De Serio, F.; Mossa, M. Vertical dense jet in flowing current. *Environ. Fluid Mech.* **2018**, *18*, 75–96. [CrossRef]
65. Cartwright, N.; Nielsen, P.; Jessen, O.Z. Swash zone and near-shore watertable dynamics. In *Coastal Engineering 2002: Solving Coastal Conundrums*; World Scientific: Singapore, 2003; pp. 1006–1015.
66. Baldock, T.; Birrien, F.; Atkinson, A.; Shimamoto, T.; Wu, S.; Callaghan, D.; Nielsen, P. Morphological hysteresis in the evolution of beach profiles under sequences of wave climates-Part 1; observations. *Coast. Eng.* **2017**, *128*, 92–105. [CrossRef]
67. Baldock, T.; Manoonvoravong, P.; Pham, K.S. Sediment transport and beach morphodynamics induced by free long waves, bound long waves and wave groups. *Coast. Eng.* **2010**, *57*, 898–916. [CrossRef]
68. Loveless, J.; Debski, D.; MacLeod, A. Sea level set-up behind detached breakwaters. In Proceedings of the 26th International Conference on Coastal Engineering, Copenhagen, Denmark, 22–26 June 1998; pp. 1665–1678.

water

MDPI

Article

Field and Numerical Study of Resistance and Resilience on a Sea Breeze Dominated Beach in Yucatan (Mexico)

Gabriela Medellín [1,2,*], Alec Torres-Freyermuth [1,2], Giuseppe Roberto Tomasicchio [3], Antonio Francone [4], Peter A. Tereszkiewicz [5], Letizia Lusito [3], Leonardo Palemón-Arcos [6] and José López [1,2]

1. Laboratorio de Ingeniería y Procesos Costeros, Instituto de Ingeniería, Universidad Nacional Autónoma de México, Sisal, Yucatán 97835, Mexico; atorresf@iingen.unam.mx (A.T.-F.); jlopezgo@iingen.unam.mx (J.L.)
2. Laboratorio Nacional de Resiliencia Costera, Laboratorios Nacionales CONACYT, Sisal, Yucatán 97835, Mexico
3. Department of Engineering, University of Salento, 73100 Lecce, Italy; roberto.tomasicchio@unisalento.it (G.R.T.); letizia.lusito@unisalento.it (L.L.)
4. Department of Engineering, University of Calabria, 87036 Arcavacata di Rende, Italy; antonio.francone@unical.it
5. Department of Earth and Environmental Sciences, University of West Florida, Pensacola, FL 32514, USA; petert@email.sc.edu
6. Departamento de Ingeniería Civil, Universidad Autónoma del Carmen, Cd. del Carmen, Campeche 24180, Mexico; leopalemon@hotmail.com

* Correspondence: gmedellinm@iingen.unam.mx; Tel.: +52-988-931-1000

Received: 3 November 2018; Accepted: 1 December 2018; Published: 8 December 2018

Abstract: The understanding of the beach capability to resist and recover from a disturbance is of paramount importance in coastal engineering. However, few efforts have been devoted to quantifying beach resilience. The present work aims to investigate the shoreline resistance and resilience, associated to a transient disturbance, on a sandy beach. A temporary groin was deployed for 24 h on a micro-tidal sea-breeze dominated beach to induce a shoreline perturbation. Morphological changes were measured by means of beach surveys to estimate the beach perturbation and the further beach recovery after structure removal. An Empirical Orthogonal Function (EOF) analysis of the shoreline position suggests that the first EOF mode describes the spatial-temporal evolution of the shoreline owing to the groin deployment/removal. A new one-line numerical model of beach evolution is calibrated with the field surveys, reproducing both the sediment impoundment and subsequent beach recovery after the structure removal. Thus, a parametric numerical study is conducted to quantify resistance and resilience. Numerical results suggest that beach resistance associated to the presence of a structure decreases with increasing alongshore sediment transport potential, whereas resilience after structure removal is positively correlated with the alongshore diffusivity.

Keywords: beach resilience; beach resistance; temporary groin; sea breezes; resilience index; GSb model; Yucatan peninsula

1. Introduction

The stability of an ecosystem depends on both resistance and resilience capability to withstand a given perturbation associated to either natural or anthropogenic disturbances [1–3]. The resilience concept has been widely employed in ecological [4] and social [5] sciences and disaster risk reduction [6,7]. However, studies incorporating resilience for coastal engineering applications are scarce [8–10] and hence further research is needed [11].

The beach resistance in the coastal vulnerability context can be associated to the amount of change produced by wave events and/or due to the presence of coastal infrastructure which alter the mean pattern of sediment transport in the coastal environment; thus, it is a measure of the beach capability to resist deviations with respect to an equilibrium morphological condition. On the contrary, the beach resilience determines the speed with which the beach morphology features (e.g., dune elevation and shoreline position) return to the pre-disturbed condition [2]. Thus, the knowledge of the beach stability (i.e., resistance and resilience) is fundamental for decision-making regarding mitigation measures against beach erosion. Beach erosion in the northern Yucatan coast is critical at many locations owing to the presence of coastal structures [12,13]. Meyer-Arendt [14] reported that construction of traditional groins, made of timber and rocks, began in the 1950s and increased significantly in the late 1960s with the construction of the port infrastructure. Furthermore, beach erosion has been exacerbated during the past decade owing to the use of impermeable groins and breakwaters. Therefore, structure removal has been considered as a mitigation measure against beach erosion in this region [13]. However, no information regarding the mechanisms controlling the recovery of the shoreline position after structure removal is available.

Temporary groins have been applied in previous studies [15–17] to measure alongshore sediment transport and to calibrate sediment transport formulations. More specifically, the beach morphology changes measured in such studies have been considered for the estimation of the K parameter in the CERC equation [18]. However, less efforts have been devoted to investigating beach evolution after coastal structures removal. Recent studies have focused on investigating the morphodynamic responses to seafloor artificial perturbations (e.g., excavated holes and channels) in the nearshore [19–21]. Moulton et al. [20] investigated the mechanisms controlling the infill of large excavated holes in the surf zone, finding that downslope gravity-driven bedload transport was important in morphological evolution for bathymetric features with large slopes.

The present work aims to investigate the shoreline resistance and resilience on a sea-breeze dominated beach by means of field observations and numerical modelling. The main findings are that, on sea breeze dominated environments, the: (i) beach resistance to the presence of a groin is negatively correlated with the alongshore sediment transport potential; and (ii) beach resilience after the structure removal is positively correlated with alongshore diffusivity.

The outline of the paper is as follows. Firstly, the study area is presented in Section 2. The experimental setup, numerical model and data analysis are described in the Materials and Methods section (Section 3). Section 4 presents the field observations during the experiment and the numerical model calibration and verification. Then, a discussion on the mechanisms controlling the shoreline resistance and resilience is presented (Section 5). Finally, concluding remarks are given in Section 6.

2. Study Area

The study area is located on a barrier island in the northern Yucatan Peninsula (Figure 1a), at the fishing village of Sisal (see Figure 1b). This coastal region is characterized by a micro-tidal range, intense sea breeze conditions, a mild continental shelf and low energy waves [22]. The field experiment was conducted between the Port and the Sisal Pier (see Figure 1c). Winds, offshore waves and mean sea level have been measured over the past years in order to characterize the main forcing mechanisms affecting the coastal region in this area (Figure 1b,c).

Wind conditions in the study area are dominated by synoptic scale patterns (i.e., Bermuda-Azores, easterly winds, cold-fronts and tropical waves) and local sea breezes [23]. The NE sea breeze winds are present throughout the year but are more frequent and intense in May. On the other hand, Central America Cold Surge events, associated with cold-front passages, are more frequent during winter months. Cold-fronts, usually originated in the Rocky Mountains [24], are characterized by sustained winds (W > 15 m s^{-1}) from the NNW and a high-pressure system (Figure 2a). Therefore, the mean wave climate in the study area is associated to locally generated NE waves owing to sea breeze events, with significant wave height $H_s < 1$ m (at $h = 10$ m water depth) (Figure 2b). More energetic NNW swell waves ($H_s > 2$ m and $T_p > 7$ s), associated to cold-fronts, occur during winter months (see Figure 2b). Furthermore, the presence of tropical storms is ubiquitous in the area [25,26]. Tidal regime is mixed, predominantly diurnal, with a spring and neap tidal range of 0.8 m and 0.1 m, respectively [27].

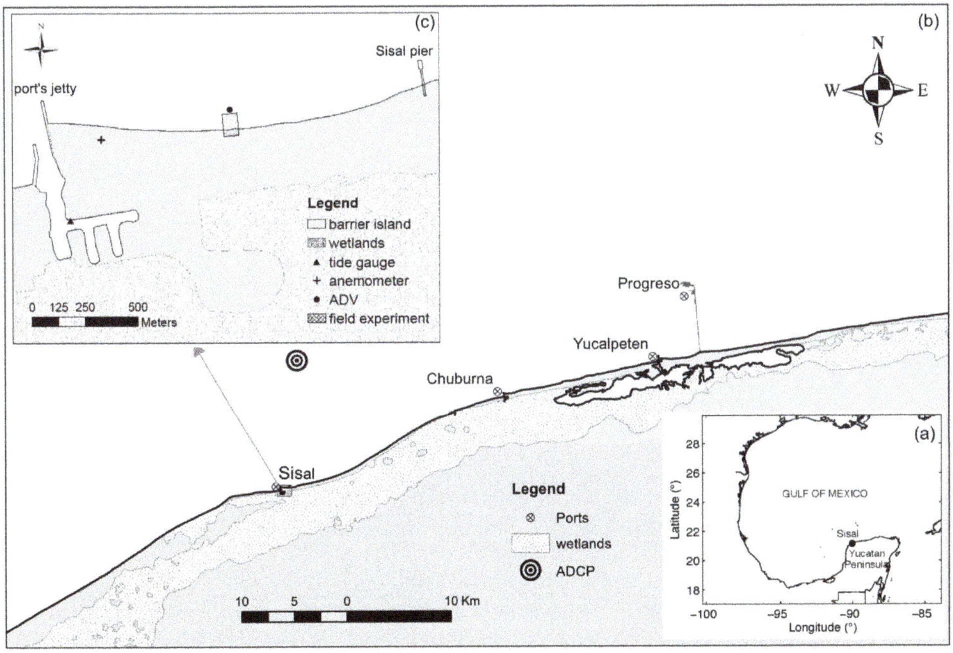

Figure 1. Location map showing (**a**) the Yucatan peninsula at the SE of the Gulf of Mexico, (**b**) a section of the Yucatan north coast showing the barrier island and wetlands and (**c**) the study area location, existing monitoring systems (ADCP: Acoustic Doppler Velocimeter Profile; ADV: Acoustic Doppler Velocimeter) and coastal structures.

Torres-Freyermuth et al. [22] conducted a field experiment to characterize the nearshore circulation during both intense local sea breeze events and synoptic Norte events. They found that during intense sea breeze events the alongshore currents significantly increase inside the surf and swash zones. Therefore, in the present case of a coast subjected to sea breeze conditions, the highly oblique winds (see Figure 2a) play a major role in driving longshore currents and consequently longshore transport [28–30].

The beach in the study area is composed of sand with median grain size, D_{50}, equal to 0.3 mm [31]. Furthermore, it presents a shoreline orientation 14° south of the E-W orientation [22], which is altered near the port's jetty and the Sisal Pier. The nearshore bathymetry is characterized by the presence of a sand bar system (Figure 3), where the outer bar is relatively alongshore uniform and inner bars present a high seasonal variability. Analysis of the shoreline variability, during the intense sea breeze season (May to September, 2015), at two transects bounding the study area suggests a small (<3 m) cross-shore variation at these locations (not shown). Therefore, the temporal groin experiment was conducted between these two transects located at the middle section between the two structures (Figure 3).

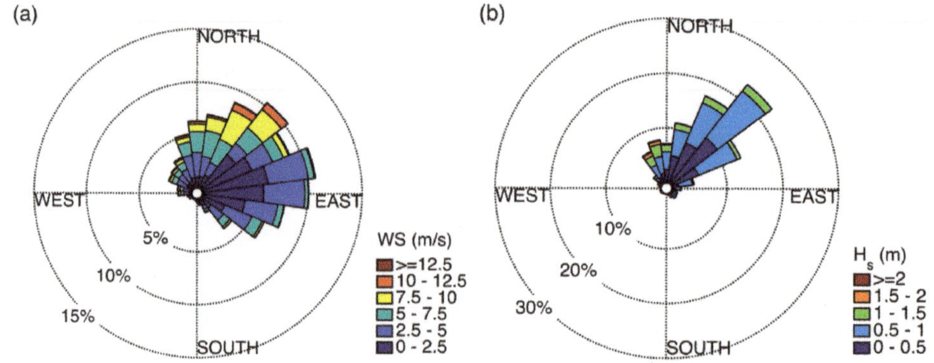

Figure 2. (**a**) Wind and (**b**) wave roses in the study area. Wind data are taken from the Sisal weather station MeteoSisal (www.weatherunderground.com 1 January 2009–17 June 2016) and the wave data were collected from an ADCP located at 10 m water depth in front of Sisal (10 December 2013–20 April 2016). *WS* and H_s stand for wind speed and significant wave height, respectively.

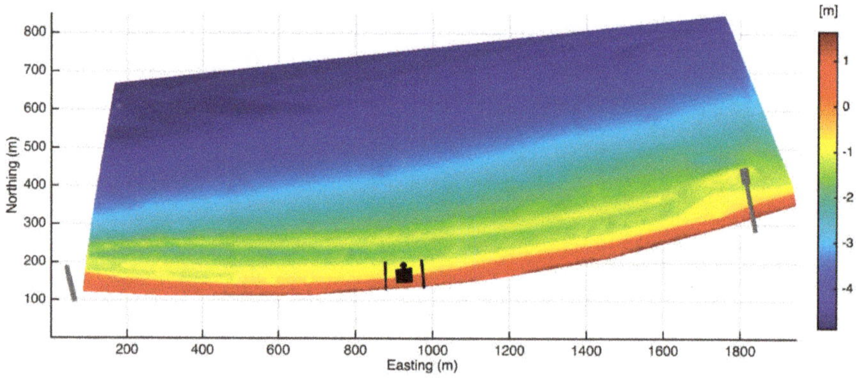

Figure 3. Bathymetry measured two days before the experiment (25 May 2015), showing the study area (■), the ADV (•) and the updrift/downdrift control lines (black solid lines) locations. The study site is located between the jetty of the Sisal port entrance channel (left-hand side, gray line) and the Sisal pier (right-hand side gray feature). The color bar indicates the elevation with respect to the mean sea level.

3. Materials and Methods

A description of field observations and data analysis is presented in this section. Furthermore, the numerical model employed in this work is also described.

3.1. Field Experiment

The field experiment was conducted in Spring 2015 to investigate the beach stability owing to the presence/removal of an artificial perturbation in the swash zone. Beach surveys before, during and after the structure deployment allow us to investigate shoreline resistance and resilience. We focused on a short-term sea breeze event due to: (i) the important role that sea-breeze events play in the sediment transport in the study area; (ii) the difficulties for conducting the beach surveys during more energetic wave conditions (storm conditions); and (iii) the labor-intensiveness required for obtaining high- spatial and temporal resolution morphology data for a longer period.

3.1.1. Temporary Groin

The temporary groin built for this study was based on the design proposed by [16]. The groin dimensions are consistent with the typical structures found along the northern Yucatan coast. The structure was made of 0.19 m thick wood-sections of dimensions 2.40 m by 1.20 m, with holes (0.08 m diameter) drilled and covered with a 63 μm sieve cloth in order to avoid a returning offshore flow near the structure [16]. The groin consisted of seven wood sections, installed over a frame made of iron pipes and clamps; 10 m are inside the surf/swash zone within the region of active transport and the remaining 4.4 m lie on the dry beach resulting on a total length of 14.4 m. Furthermore, sandbags were spread out along the base of the structure to avoid bed scouring that could lead to sediment bypassing underneath the structure. The sand bags consisted of polypropylene woven raffia bags (60 by 100 cm) filled at approximately 2/3 of their capacity with sand.

The deployment of each section started from the land toward the sea, allowing a 0.20 m overlap between sections. Each section consisted of three vertical pipes pounded 1.5 m into the sand bed using a hammer and one horizontal pipe, holding the three pipes with scaffold clamps (Figure 4a). The original design from [16] was improved by including two horizontal members, at the down-drift side of the structure (see Figure 4a), perpendicular to the groin and attached with clamps to an additional scaffold frame. This design provided additional resistance to alongshore forces induced by wind, waves and currents. The groin deployment took approximately three hours for a team of 12 people.

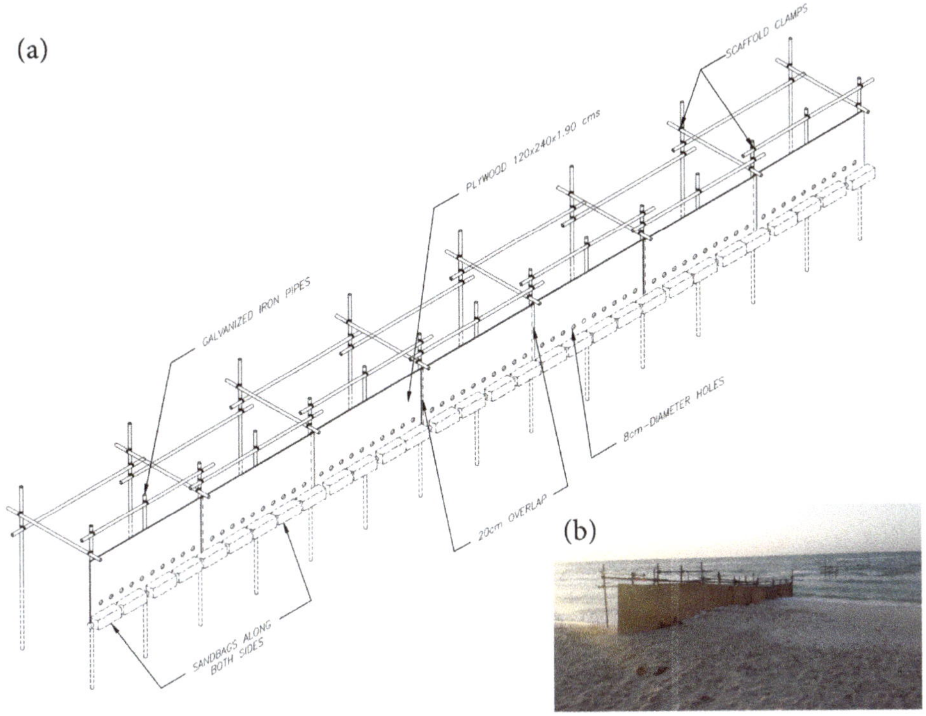

Figure 4. (**a**) Temporary groin design made of wood-sheets, lined with sand bags and a scaffold frame of iron pipes. (**b**) Picture of the shoreline perturbation 12-h after the groin deployment.

3.1.2. Data Collection

Different sensors were deployed to characterize the environmental conditions occurring during the field experiment. Wind data was measured every minute using a weather station located in a tower installed near the Sisal Port (Figure 1c). Offshore wave conditions were recorded at 10 m water depth using an RDI Acoustic Doppler Current Profiler (ADCP) located 11 km offshore (see Figure 1b for instrument location). During the experiment, three breaker lines were observed at the outer and inner bars and the inner-surf/swash zone transition. Moreover, an Acoustic Doppler Velocimeter (ADV) Nortek Vector, located onshore the inner bar (Figure 5a,b) at 0.2 m above the seabed, acquired high-frequency (16 Hz) velocity measurements during 48 h (27 May to 29 May). The instantaneous velocities were measured in the *XYZ* coordinate system, where velocities are defined such that u, v and w velocities correspond to the x (cross-shore), y (alongshore) and z (vertical) directions, respectively.

A temporary groin was deployed in the inner surf/swash zone on the morning (0800 local time) of 27 May and was removed 24-h later on the morning of 28 May 2015. Beach morphology was surveyed along 15 survey lines, covering the up- and down- drift sides of the temporary structure (Figure 5a). A Leica Differential Global Positioning System (DGPS) was employed using Real Time Kinematics (RTK) for conducting high-resolution topographic surveys. The equipment in RTK mode and Kinematic (phase) moving mode has a horizontal and vertical accuracy of 10 mm and 20 mm, respectively. The alongshore distance between transects varies from 2 to 6 m, with the highest resolution corresponding to those transects located close to the structure (Figure 5a). The DGPS beach surveys were conducted every two hours for 24-h to evaluate the beach resistance owing to the structure presence. Furthermore, measurements continued after the structure removal, with the same two-hour temporal resolution for 10 h and then were resumed with a lower temporal resolution (i.e., 29 May, 3 June), continuing until the beach was fully recovered. A total of 20 beach surveys were conducted. Control lines, located 50 m updrift and downdrift from the structure location (Figure 5a), were surveyed weekly to assess the natural beach variability in this area. It is worth to notice that the survey lines only cover the swash and inner surf zone and hence do not cover the entire surf zone which extends offshore. The field data is available via author's request following the instruction in the supplementary material section.

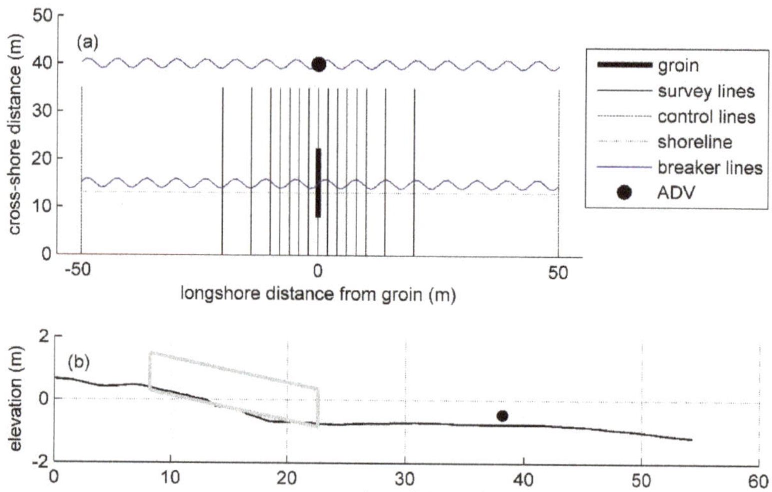

Figure 5. (**a**) Plan view of the survey lines and the structure location, showing the breaker lines and ADV location. (**b**) Beach profile of the middle transect showing the ADV location with respect to the structure.

The mean sea level was acquired every minute by a tidal gauge located inside the Sisal Port (Figure 1c). Mean sea level in the study area presents a cyclic annual variation [32], showing a minimum in July and a maximum in October. Therefore, when a short-term experiment is carried out (days to weeks), the seasonality of the mean sea level (msl) elevation should be considered. The vertical datum of the surveys corresponds to the MEX97 geoid [33] and the difference between the vertical datum and the mean water elevation during the experiment is approximately 0.2 m. Beach surveys in the present paper are referenced to the msl during the experiment period (May–June 2015). A summary of the field data collected is presented in Table 1.

Table 1. Measured data, sensor employed, sampling frequency and measured period.

Variable	Sensor	Manufacturer	Sampling Frequency	Start Date	End Date
Morphology	GPS	Leica Geosystems (Gallen, Switzerland)	2–24 h	27 May 2015	3 June 2015
Waves	ADCP	RD Instruments (Poway, CA, USA)	30 min	27 May 2015	3 June 2015
Surf zone currents	ADV Vector	Nortek (Rud, Norway)	16 Hz	27 May 2015	29 May 2015
Sea level	Radar level sensor	OTT HydroMet (Kempten Germany)	1 min	27 May 2015	4 June 2015
Winds	Weather station	Davis Vantage Pro 2 (Hayward, CA, USA)	1 min	27 May 2015	4 June 2015

3.1.3. Field Data Analysis

The shoreline position ($z = 0$) was extracted from each of the cross-shore DGPS survey lines. An Empirical Orthogonal Function (EOF) analysis of the shoreline position [34–36], with respect to the initial shoreline, was performed in order to investigate the dominant modes of variability during the experiment. The EOF analysis allows to represent the coastline data, $x(y, t)$, as a summation of n spatial ($e_n(y)$) and temporal functions ($c_n(t)$), where the variance explained decreases with the mode number, as follows:

$$x(y,t) = \sum_{n=1}^{N} c_n(t)e_n(y)$$

The shoreline variability associated to the presence of the temporal groin can be evaluated by analyzing the resulting spatial and temporal functions.

Wind magnitude and direction, from the meteorological tower, were averaged every 5 min. Wave statistics measured by the ADCP were computed every 60 min based on pressure and velocity measured at 2 Hz. On the other hand, the ADVs velocity information was removed when the correlation value was less than 80% with the aim of identifying potentially inaccurate measurements. The measured velocities were averaged over 512-s intervals to ensure stationary conditions and further estimate the (cross- and along- shore) currents inside the surf zone.

3.2. Numerical Model

3.2.1. Model Description

Numerical simulations have been conducted by means of a newly proposed morphodynamic model, named General Shoreline beach 1.0 (GSb), belonging to the one-line model typology [37]. This typology assumes that the beach cross-shore profile remains unchanged [38,39], thereby allowing beach change to be described uniquely in terms of the shoreline position. The peculiarity of the GSb model consists of simulating shoreline evolution based on a longshore transport formula/procedure suitable at any coastal mound: sand, gravel, cobbles, shingle and rock beaches [40–44]. The GSb model presents one calibration coefficient solely, K_{GSb}, which does not depend on the grain size diameter and depends on the alongshore gradient in breaking wave height [45]. The proposed general formula/procedure considers an energy flux approach combined with an empirical/statistical relationship between the wave-induced forcing and the number of moving units. GSb model allows to determine short-term (daily base) or long-term (years base) shoreline change for arbitrary combinations and configurations of structures (groins, jetties, detached breakwaters and seawalls) and beach fills that can be represented on a modelled reach of coast. A demo version of the numerical model can be downloaded by following the instructions in the supplementary material section.

3.2.2. Data Analysis

The results from the numerical simulations have been adopted to investigate the shoreline beach resistance and resilience at the considered stretch of coast. The resistance index, *RS* and resilience index, *RL*, proposed by [2], to investigate ecological stability, were adapted for the present study. *RS* is then defined as,

$$RS(t_0) = 1 - \frac{2|\Delta S_0|}{(l + |\Delta S_0|)} \tag{1}$$

where ΔS_0 represents the cross-shore distance, at t_0 = 720 h, between the perturbed shoreline and the unperturbed shoreline, in vicinity of the groin and l is the perturbation length which equals to the groin length. *RS* ranges between 0 and 1; minimal resistance (largest effect) to the beach perturbation corresponds to smaller values of *RS*. Similarly, the resilience index *RL*(*t*) is defined as,

$$RL(t) = \frac{2|\Delta S_0|}{(|\Delta S| + |\Delta S_0|)} - 1 \tag{2}$$

where ΔS represents the cross-shore distance, at time t, between the perturbed shoreline and the unperturbed shoreline with RL = 1 corresponding to a fully recovered shoreline (i.e., return to the pre-disturbed condition). Shoreline beach resistance and resilience are illustrated in Figure 6.

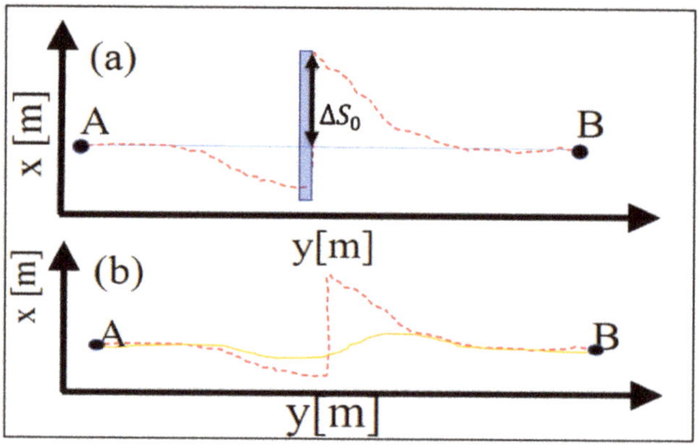

Figure 6. Definition sketch of shoreline evolution between points A and B for the cases of: (**a**) beach resistance owing to the groin (original shoreline: blue line; disturbed shoreline: red dashed line) and (**b**) beach resilience after the structure removal (perturbed shoreline: red-dashed line; shoreline position at time t_i after structure removal: yellow line).

4. Results

4.1. Field Observations

4.1.1. Forcing Conditions

Wind velocity time series shows a diurnal variability associated to sea-breeze events (Figure 7a). A maximum wind speed of 16 m s^{-1}, corresponding to the sea breeze event peak, was recorded during the experiment in the afternoon (1630 local time) of 27 May (Figure 7a). Offshore wave conditions, measured at 10 m water depth (Figure 7b), are highly correlated with local winds (Figure 7a). The wave height increased from H_s = 0.3 m measured at 10:00 to H_s = 1.0 m measured at 1800, with mean wave direction approaching from the NE (Figure 7b). The temporary structure was deployed during neap tides (light gray shade in Figure 7c) in order to decrease the influence of the tide on the effective length of the groin and hence restricting the swash zone width [16].

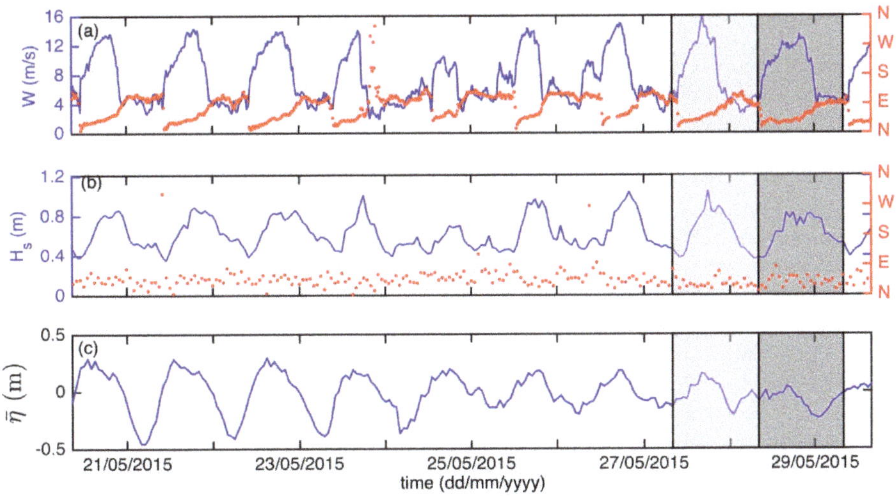

Figure 7. Measured (**a**) wind speed and wind direction (**b**) offshore significant wave height and wave direction at 10 m water depth and (**c**) mean sea level variation $\overline{\eta}$ from the Sisal gauge before, during (light gray shade) and after (dark gray shade) the groin deployment.

Sea breeze events increased the wave energy inside the surf zone (Figure 8a). The alongshore current velocity V_y becomes negligible during the land breeze (0300 h), whereas it reaches $V_y > 0.3$ m s^{-1} near the sea breeze peak (1800 h) (Figure 8b). Thus, westward currents dominated the surf zone hydrodynamics during the measured period (red solid-line in Figure 8b). On the other hand, the mean cross-shore current V_x (blue solid-line in Figure 8b) shows negligible current intensity at this location/elevation during the 48 h period.

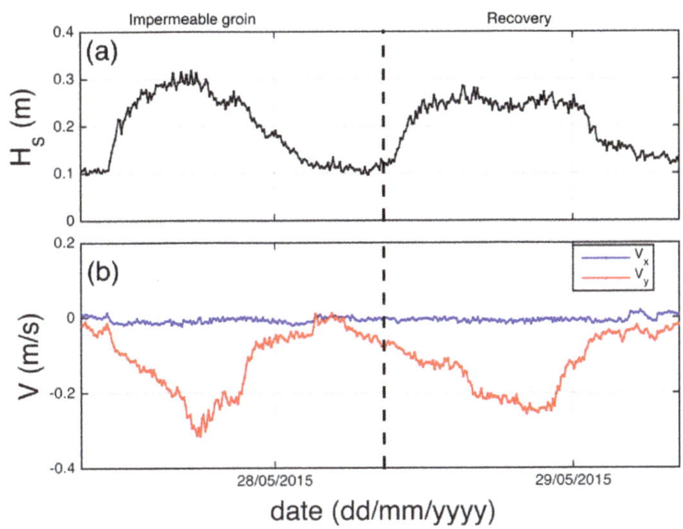

Figure 8. Surf zone conditions during and 24 h after the structure disturbance. (**a**) Significant wave height inside the surf zone and (**b**) cross- (blue line) and along- (red line) shore currents. Negative values in panel (**b**) indicate offshore/west-ward currents, while positive values indicate onshore/east-ward currents.

4.1.2. Observed Morphology Evolution

Beach morphology changes are evaluated by analyzing the high spatial and temporal resolution RTK DGPS survey data (Table 1 and Figure 9). The measured beach survey before the structure deployment (0730 local time) shows the alongshore uniformity of the beach contour lines (Figure 9a). However, the deployment of a temporary impermeable structure (at $x = 0$ m) induces a shoreline perturbation owing to the sediment impoundment at the east side of the structure and erosion at the downdrift (west) side (a time-lapse video of the experiment is included as supplementary material). The beach survey conducted 24 h after the structure deployment (0842 of the 28 May 2015) shows significant changes for $h > -0.6$ m (Figure 9b). These observations are consistent with the westward alongshore current associated to sea breeze events (Figure 8b). The shoreline contour reaches a maximum advance of 6 m and maximum retreat of less than -4 m at the updrift and downdrift side, respectively. The calculated sediment volume impoundment, at the updrift side of the structure, reached 70 m^3 in 24 h (2 to 6 m^3/m), whereas the volume loss at the downdrift for the same period is less than 40 m^3. Differences between up- and down- drift volume changes might be ascribed to the limited alongshore spatial coverage of the topographic measurements.

Alongshore uniformity for the submerged area ($h < 0$ m) was observed 24 h after the structure removal (0825 of the 29 May 2015). However, a clear perturbation in the subaerial beach profile was still present (Figure 9c). The disturbance smooths out during the following days, returning to the pre-disturbed condition (i.e., straight and parallel contours) 144 h after the structure removal (3 June 2015) (Figure 9d).

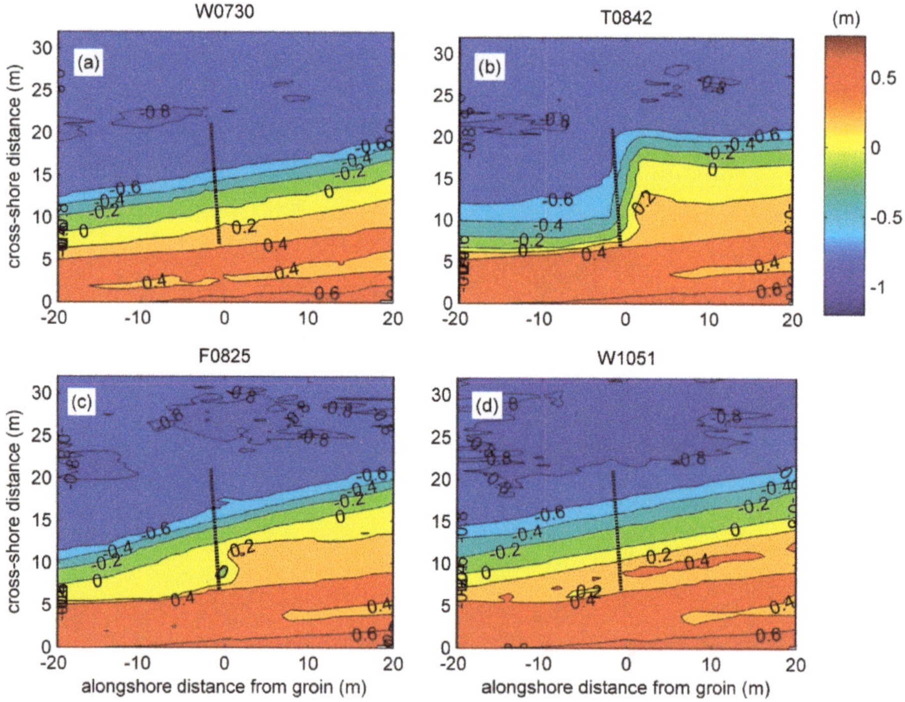

Figure 9. Beach survey (**a**) right before groin deployment, (**b**) right before groin removal (24 h later), (**c**) 24 h after groin removal, (**d**) 144 h after groin removal. The title of each subplot indicates the weekday and local time of each survey (W = Wednesday, T = Thursday, F = Friday).

The EOF analysis of the shoreline position allowed us to identify the spatial and temporal shoreline evolution before and after the structure removal. The first mode of variability, which represents more than 91% of the total variance, has been associated solely to the groin influence. The spatial function of the first mode ($e_1(y)$ in Figure 10a) describes the shoreline perturbation with positive/negative values associated to accretion/erosion at the updrift/downdrift side of the temporal groin. Furthermore, the temporal evolution ($c_1(t)$ in Figure 10b) shows an increase from zero to a maximum value in the first 24 h (before structure removal). After the structure removal, $c_1(t)$ shows slower decrease reaching a value close to zero at the end of the measured period. The second and third modes (not shown) represent 7.5% of the total variance and show spatial and temporal functions not associated to the groin presence. The beach recovery time can be calculated from the decay time of the perturbation described by the first temporal function ($c_1(t)$) which is nearly 144 h.

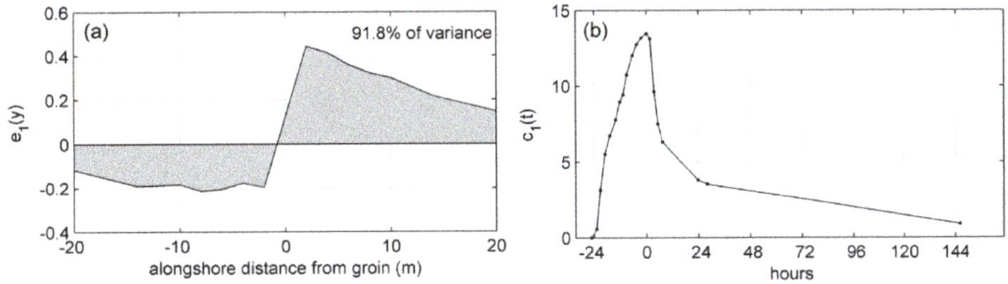

Figure 10. Empirical orthogonal function analysis of the shoreline data during the experiment (27 May–3 June, 2015). Only the (**a**) spatial function and (**b**) the temporal evolution of the first mode are shown.

4.2. Numerical Modelling

The field data were employed to calibrate the GSb model. Thus, a 14.4 m long groin (10 m wet plus 4.4 m dry) was positioned at the center of the domain. The alongshore model computational domain was assumed equal to 41 m. Model grid cell resolution, DX, has been set equal to 1 m with a total number of cells, NX, equal to 41, whereas the model experiment has been simulated adopting a calculation time step, DT, equal to 0.005 h. For a direct comparison with field measurements, the recording time step of the output files has been set to 1 h. The median grain size, D_{50}, has been set to 0.3 mm and the closure depth, h^*, 0.8 m. Hourly wave conditions recorded at the ADCP located 11 km offshore the beach have been adopted as input to the GSb model (Table 2).

Table 2. Observed wave conditions from 27 May 2015 to 3 June 2015; M = month, D = day, H = hour, H_s = significant wave height, T_p = peak wave period, θ = wave direction.

M	D	H	H_s (m)	T_p (s)	θ (°)	M	D	H	H_s (m)	T_p (s)	θ (°)	M	D	H	H_s (m)	T_p (s)	θ (°)
5	27	7	0.54	3.7	41	5	29	17	0.47	2.5	71	6	1	3	0.57	3.3	36
5	27	8	0.51	3.6	51	5	29	18	0.49	2.9	67	6	1	4	0.53	3.4	37
5	27	9	0.5	3.8	26	5	29	19	0.55	3	30	6	1	5	0.5	3.1	56
5	27	10	0.48	3.6	38	5	29	20	0.61	3.3	45	6	1	6	0.4	4.5	29
5	27	11	0.5	3.4	36	5	29	21	0.66	3.7	16	6	1	7	0.4	4.1	34
5	27	12	0.47	3.2	26	5	29	22	0.71	3.4	41	6	1	8	0.33	4.3	36
5	27	13	0.47	2.7	77	5	29	23	0.7	3.6	65	6	1	9	0.32	4.3	13
5	27	14	0.42	3	51	5	30	0	0.6	3.3	62	6	1	10	0.28	4.3	12
5	27	15	0.38	3.8	19	5	30	1	0.56	3.5	36	6	1	11	0.27	3.7	358
5	27	16	0.39	4.1	43	5	30	2	0.59	3.9	43	6	1	12	0.28	3.8	62
5	27	17	0.46	2.7	24	5	30	3	0.71	3.3	52	6	1	13	0.31	3.5	41
5	27	18	0.62	1.4	32	5	30	4	0.75	3.4	45	6	1	14	0.31	3.3	48
5	27	19	0.68	3.5	28	5	30	5	0.76	3.5	83	6	1	15	0.24	3.8	14
5	27	20	0.74	3.6	22	5	30	6	0.72	3.4	8	6	1	16	0.24	4.1	22
5	27	21	0.8	3.7	26	5	30	7	0.69	3.3	63	6	1	17	0.22	4.2	358
5	27	22	0.85	3.8	24	5	30	8	0.64	1.4	13	6	1	18	0.19	4.2	6
5	27	23	1.05	4.2	32	5	30	9	0.6	3.6	71	6	1	19	0.3	2.5	260
5	28	0	0.84	4.5	19	5	30	10	0.53	4.2	27	6	1	20	0.2	4.1	11
5	28	1	0.83	3.8	55	5	30	11	0.49	4.2	14	6	1	21	0.18	4.2	10
5	28	2	0.9	3.4	70	5	30	12	0.52	2.9	45	6	1	22	0.26	2.3	64
5	28	3	0.85	3.3	30	5	30	13	0.45	3.7	33	6	1	23	0.42	2.7	22
5	28	4	0.78	3.8	32	5	30	14	0.46	3.8	15	6	2	0	0.46	3.2	9
5	28	5	0.74	3.9	47	5	30	15	0.45	1.6	53	6	2	1	0.41	3.2	23
5	28	6	0.66	4.3	28	5	30	16	0.41	3.2	9	6	2	2	0.47	5.2	357
5	28	7	0.57	1.4	32	5	30	17	0.44	4.3	19	6	2	3	0.41	3.3	352
5	28	8	0.48	4.6	20	5	30	18	0.46	2.7	17	6	2	4	0.35	3	37
5	28	9	0.47	3.4	46	5	30	19	0.52	3.2	56	6	2	5	0.31	3.8	2
5	28	10	0.43	4.5	25	5	30	20	0.58	3	14	6	2	6	0.27	3.4	32
5	28	11	0.39	4.2	31	5	30	21	0.61	3.5	44	6	2	7	0.39	1.4	11
5	28	12	0.36	3.8	25	5	30	22	0.67	3.7	39	6	2	8	0.23	2.7	22
5	28	13	0.38	3.6	12	5	30	23	0.75	3.8	32	6	2	9	0.21	3.2	46
5	28	14	0.37	3.9	25	5	31	0	0.76	3.9	16	6	2	10	0.16	3.3	28
5	28	15	0.38	4.1	18	5	31	1	0.71	3.7	21	6	2	11	0.12	3	66
5	28	16	0.45	2.9	43	5	31	2	0.78	3.8	39	6	2	12	0.13	2.6	146
5	28	17	0.53	3.1	55	5	31	3	0.72	3.5	32	6	2	13	0.2	2	47

Table 2. *Cont.*

M	D	H	H_s (m)	T_p (s)	θ (°)	M	D	H	H_s (m)	T_p (s)	θ (°)	M	D	H	H_s (m)	T_p (s)	θ (°)
5	28	18	0.63	3.2	84	5	31	4	0.67	3.7	28	6	2	14	0.21	2.1	58
5	28	19	0.63	3.5	51	5	31	5	0.6	3.7	48	6	2	15	0.19	2.5	239
5	28	20	0.63	3.4	20	5	31	6	0.54	3.4	45	6	2	16	0.15	2.1	76
5	28	21	0.83	3.7	37	5	31	7	0.53	3.9	360	6	2	17	0.27	2.1	80
5	28	22	0.77	4.1	55	5	31	8	0.56	2.8	70	6	2	18	0.29	1.9	41
5	28	23	0.83	3.4	54	5	31	9	0.51	4.2	32	6	2	19	0.29	2.3	52
5	29	0	0.69	3.4	53	5	31	10	0.49	3.6	32	6	2	20	0.29	2.4	64
5	29	1	0.78	3.6	34	5	31	11	0.52	1.5	61	6	2	21	0.34	2.4	62
5	29	2	0.7	3.9	19	5	31	12	0.43	4.5	28	6	2	22	0.36	2.3	54
5	29	3	0.81	3.8	56	5	31	13	0.41	4.1	31	6	2	23	0.36	2.6	40
5	29	4	0.82	3.9	38	5	31	14	0.39	3.1	51	6	3	0	0.32	2.7	69
5	29	5	0.77	4.1	30	5	31	15	0.35	4.8	37	6	3	1	0.38	2.6	55
5	29	6	0.77	4.1	54	5	31	16	0.31	3.9	41	6	3	2	0.32	2.7	82
5	29	7	0.76	4.1	29	5	31	17	0.33	1.9	352	6	3	3	0.27	2.7	353
5	29	8	0.6	3.9	43	5	31	18	0.43	2.7	55	6	3	4	0.29	2.6	51
5	29	9	0.61	3.7	38	5	31	19	0.54	3	17	6	3	5	0.26	2.7	62
5	29	10	0.57	3.6	45	5	31	20	0.58	3.4	49	6	3	6	0.27	2	52
5	29	11	0.54	3.7	17	5	31	21	0.57	3.4	53	6	3	7	0.26	9.4	282
5	29	12	0.52	3.2	28	5	31	22	0.66	3.4	26	6	3	8	0.24	2.9	53
5	29	13	0.53	3.5	57	5	31	23	0.65	3.5	25	6	3	9	0.27	1.8	21
5	29	14	0.45	3.6	30	6	1	0	0.7	3.8	25	6	3	10	0.24	3.4	222
5	29	15	0.39	3.3	49	6	1	1	0.69	3.7	53						
5	29	16	0.44	1.5	25	6	1	2	0.67	3.7	49						

Table 3 shows the values of the *RMSE* (Root Mean Square Error) from the comparison of the observed and calculated (for different K_{GSb} values) shoreline positions, $y_{i,obs}$ and $y_{i,GSb}$, respectively, at 23 h after the groin deployment; the *RMSE* is defined as,

$$RMSE = \sqrt{\frac{\sum_{i=1}^{N}\left(y_{i,GSb} - y_{i,obs}\right)^2}{N}} \tag{3}$$

where N is the number of transects along the considered shoreline. A $K_{GSb} = 0.01$ value determined the lowest *RMSE*, showing a better agreement between the calculated and the observed shoreline positions.

Table 3. Calculated and observed shoreline positions at 14 transects along the shoreline and relative values of *RMSE* (Root Mean Square Error).

	Longshore Distance from Groin (m)	−20	−14	−10	−8	−6	−4	−2	2	4	6	8	10	14	20	*RMSE*
Cross-shore Distance from Baseline (m)	Initial Shoreline	11.3	11.6	11.5	11.8	11.9	12.0	12.8	12.7	13.1	13.5	13.6	13.5	14.2	14.9	
	Field data at 23 h	9.9	9.3	9.1	9.1	9.2	9.3	9.5	18.1	18.5	18.3	17.8	17.5	17.1	16.9	
	GSb, $K_{GSb} = 0.005$	10.1	9.8	9.5	9.4	9.5	10.0	11.2	15.7	16.2	16.4	16.5	16.4	16.4	16.6	3.825
	GSb, $K_{GSb} = 0.01$	10.1	9.8	9.4	9.4	9.4	10.2	11.4	15.8	16.3	16.4	16.4	16.4	16.4	16.6	3.817
	GSb, $K_{GSb} = 0.05$	10.1	9.4	9.3	9.8	9.8	11.6	12.5	17.0	16.7	16.5	16.4	16.4	16.4	16.6	4.127
	GSb, $K_{GSb} = 0.1$	10.1	8.9	10.2	11.2	11.2	12.8	13.3	18.1	16.8	16.8	16.6	16.5	16.5	16.6	5.269

Numerical simulations were conducted for two different lateral boundary conditions: pinned or moving lateral boundary. Figure 11 shows the comparison of the simulated shoreline positions obtained after 23 h for the two different lateral boundaries; as expected, the shoreline shapes differ in vicinity of the lateral boundaries but overlap in vicinity of the groin. If a moving lateral boundary condition is selected, the boundary will move a specified distance over a certain time period. GSb lateral boundaries have been selected as moving boundaries.

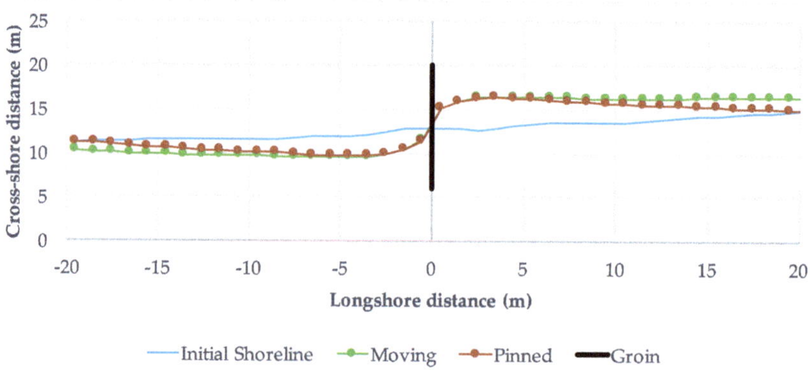

Figure 11. Comparison of the simulated shoreline positions obtained after 23 h for the two different lateral boundaries.

The first beach survey was assumed as the initial shoreline position in the numerical model. In particular, at the down-drift (x = 0 m) and up-drift (x = 41 m) boundaries of the computational domain, the observed specific distances from the first beach survey, equal to −1.4 m and 1.8 m, respectively, over a period of 24 h, have been assumed. On the other hand, for the post structure removal condition the computation duration has been extended to 168 h taking as initial condition the survey at t = 24 h.

Figure 12c–g show that the calibrated numerical model satisfactorily predicts the downdrift shoreline evolution, whereas the model is not capable of fully reproducing the shoreline advance, at t = 21 h, in the updrift side (Figure 12h).

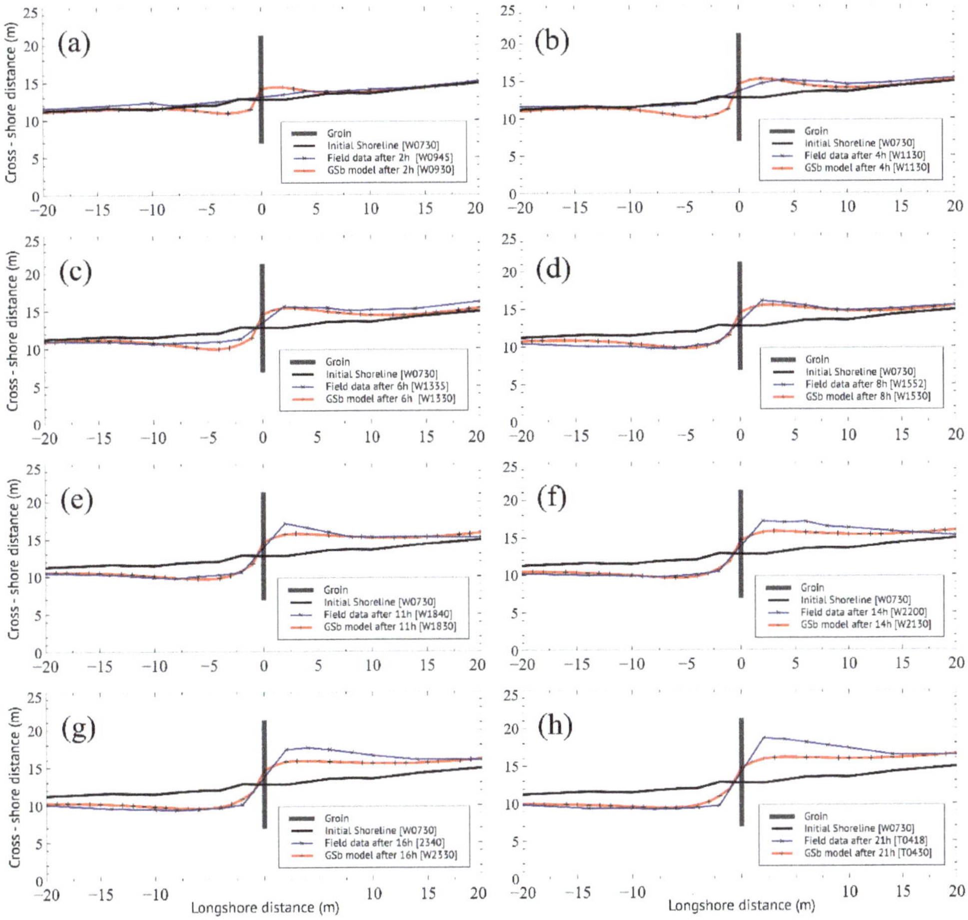

Figure 12. Measured and calculated shoreline position (**a**) 2 h, (**b**) 4 h, (**c**) 6 h, (**d**) 8 h, (**e**) 11 h, (**f**) 14 h, (**g**) 16 h and (**h**) 21 h after the structure deployment. Information on the brackets indicates the corresponding weekday (W = Wednesday, T = Thursday) and local time of field data and GSb simulation.

The model capability to predict the shoreline resilience after the structure removal was investigated (verification). The assumed initial shoreline position corresponds to the beach survey performed immediately before the structure removal and the numerical model is run without the structure using the daily mean conditions as measured for the following seven days. The numerical model calculated the drastic change occurring during the first 24 h after the groin removal (Figure 13a–e). Furthermore, it calculated the beach recovery occurring after approximately 7 days (Figure 13f). Therefore, within the framework of the field data gained in the investigation (sea-breeze conditions), the model can be considered as a reliable tool to conduct a numerical study on beach resistance and resilience for the adopted study area.

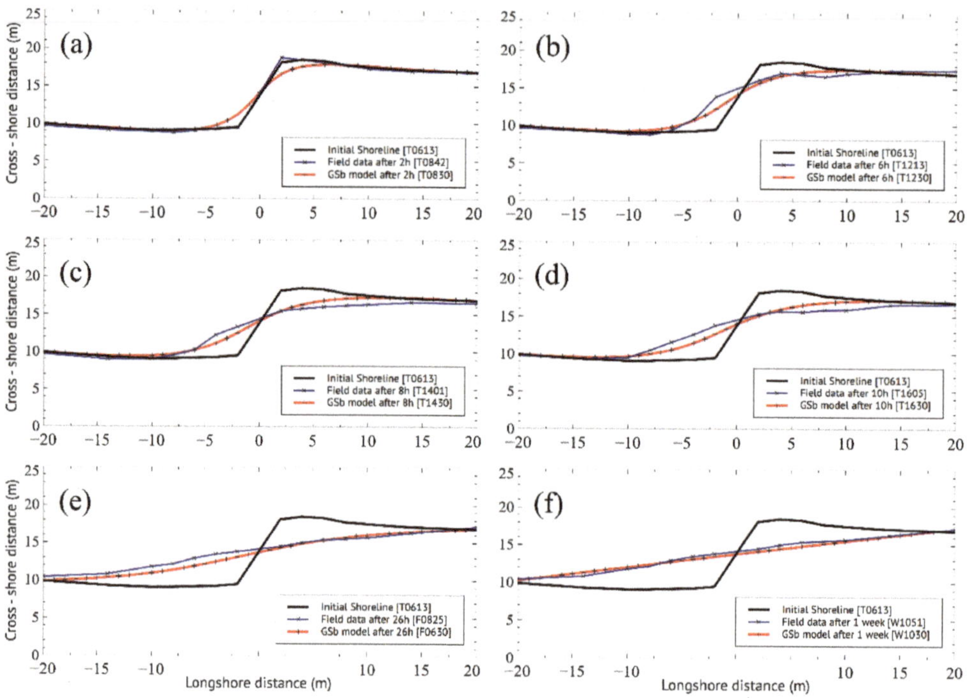

Figure 13. Measured and calculated shoreline position (**a**) 2 h, (**b**) 6 h, (**c**) 8 h, (**d**) 10 h, (**e**) 26 h and (**f**) 1 week after the structure removal. Information on the brackets indicates the corresponding weekday (T = Thursday, F = Friday) and local time of field data and GSb simulation.

5. Discussion: Shoreline Stability

Once the GSb model has been calibrated and verified, it has been used to investigate the beach resistance and resilience phenomena. The main limitation in the field data is related to the limited alongshore spatial coverage (40 m) of the topographic measurements. Therefore, for the numerical setup, we considered a 10 m long groin located in the middle of a 200 m long shoreline computational domain. For the parametric study we employed different cases (Table 4) encompassing constant (low-energy) wave conditions during a 720 h period. The latter allows us to assess the resistance and resilience sensitivity to different forcing conditions on a longer time-scale basis. For all cases, the shoreline position at the end of the simulation is shown in Figure 14a. A positive correlation between the shoreline distance change and the sediment transport was observed (Figure 14b) and hence $RS(t_0)$ decreases as the sediment transport rate Q [40,41] increases (Figure 14c). Values of $\Delta S(t_0)$ were estimated for wave conditions in Table 4.

Table 4. GSb simulated cases to investigate beach resistance to a 10 m groin, for waves of T_p = 3.5 s, after 720 h (t_0) of simulation.

Case	$H_{1/50}$ (m)	θ (°)	$\Delta S(t_0)$ (m)	$RS(t_0)$	$Q \times 10^{-5}$ (m^3/s)
Test 1A	0.232	15	1.33	0.765	1.52
Test 2A	0.232	30	2.28	0.629	2.45
Test 3A	0.232	45	2.62	0.585	2.46
Test 4A	0.310	15	2.47	0.604	4.03
Test 5A	0.310	30	3.75	0.454	6.47
Test 6A	0.310	45	5.11	0.324	6.50

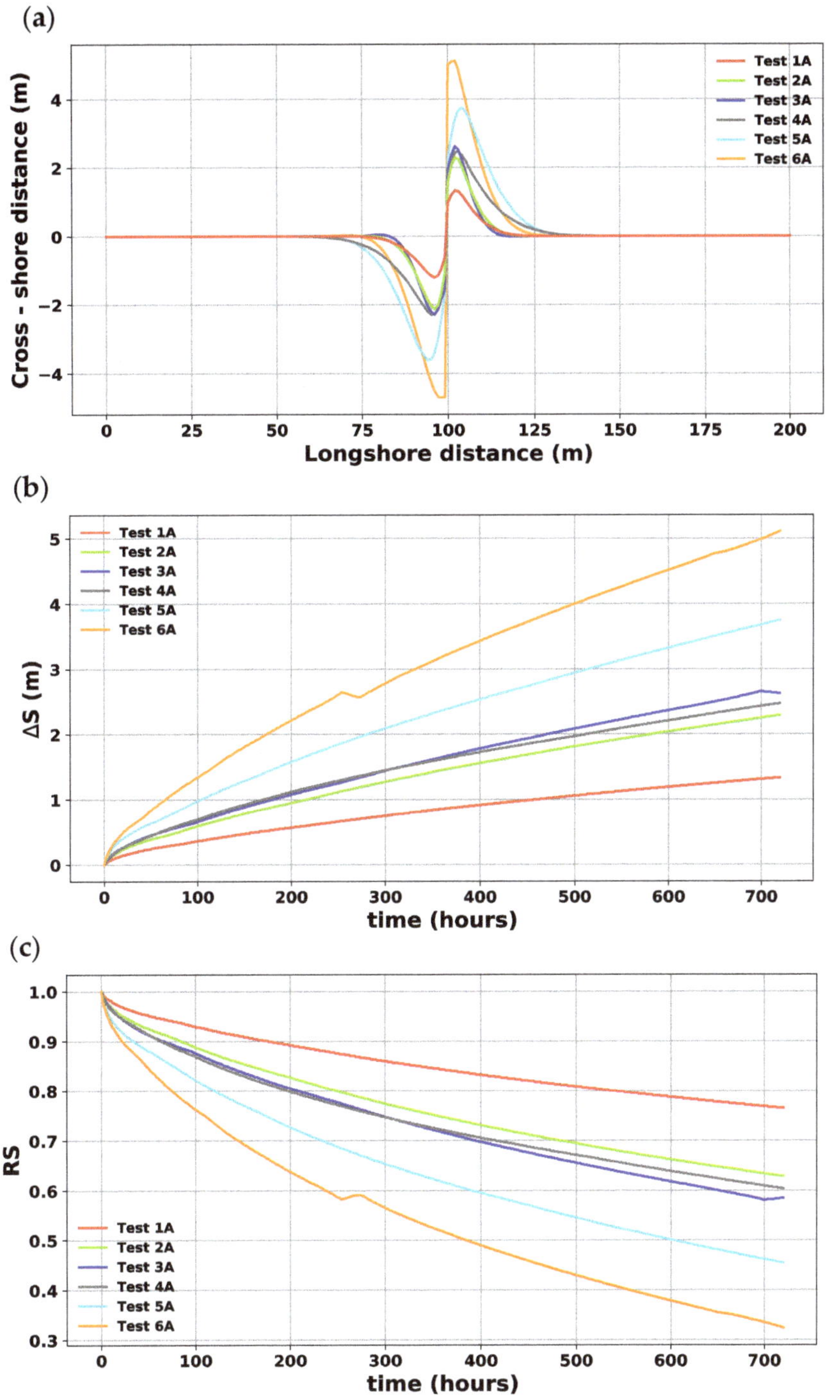

Figure 14. (**a**) Shoreline position, owing to the 10 m long groin presence, after 720 h (**b**) the corresponding shoreline distance increases for different wave conditions listed in Table 4 and (**c**) resistance index temporal evolution for different wave conditions listed in Table 4.

The numerical model was also used to simulate the beach recovery phenomena (i.e., resilience) after the structure removal. Thus, the numerical model is initialized with the shoreline from Test 6A at t_0 = 720 h (e.g., yellow line in Figure 14a), subjected to different wave forcing conditions (Table 5), without the presence of the structure. The shoreline recovery is significant during the first 24-h for all cases (Figure 15a), consistent with the field observations (Figure 9); afterward, it continues at a lower rate. The numerical results were used to compute the temporal evolution of RL using Equation (2) for each case (Figure 14b). Contrary to beach resistance, numerical results suggest that the beach resilience is not controlled by the longshore transport potential and it depends on the alongshore diffusivity G given by [40,41],

$$G = \frac{2\mu}{(h^* + B)} H_{1/50,b}^{5/2} \cos 2\theta_b \quad (4)$$

where μ is assumed equal to 0.15 $m^{1/2}$ s^{-1}, h^* is the closure depth, B is the berm elevation, $H_{1/50,b}$ is the value of $H_{1/50}$ at breaking [40,41], θ_b is the wave angle breaking with respect to the mean rectilinear trend of the shoreline. The diffusivity is associated to the longshore spreading of a shoreline perturbation owing to its departure from equilibrium for the existing forcing. The numerical simulations show that, for a given value of $H_{1/50}$, the beach resilience increases as the value of G increases (Figure 15b and Table 5). Therefore, alongshore diffusivity plays an important role on beach resilience in the study area.

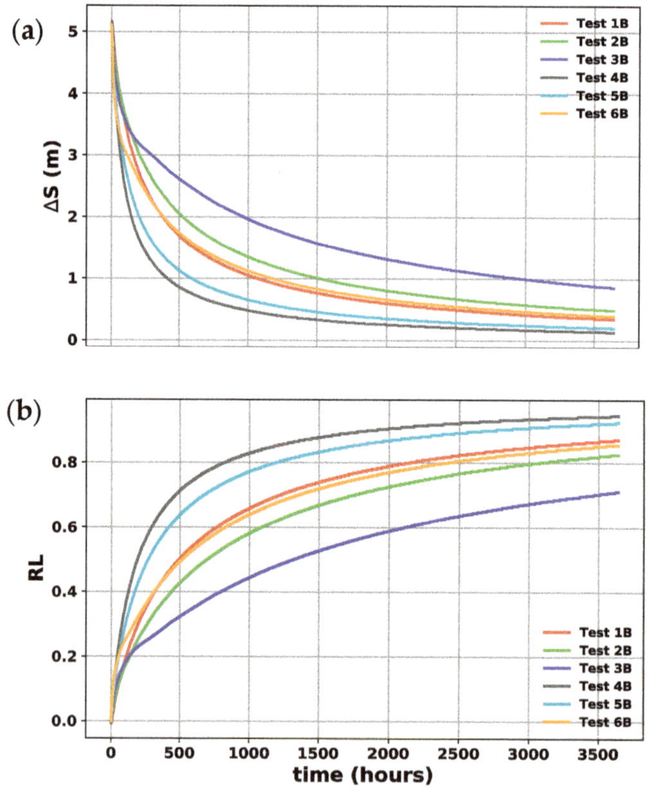

Figure 15. Beach resilience after structure removal: (**a**) shoreline change with respect to time after the structure removal and (**b**) resilience index temporal evolution.

Table 5. GSb simulated cases to investigate beach resilience associated to the groin removal. The initial condition for all simulations corresponds to the shoreline position for Test 6A at t = 720 h. θ_o is the deep-water wave angle.

Case	$H_{1/3}$ (m)	$H_{1/50}$ (m)	$H_{1/50,b}$ (m)	θ_o (°)	θ_b (°)	$G \times 10^{-2}$ (m²/s)	$RL(t_i)$
Test 1B	0.150	0.232	0.34	15	3.74	0.88	0.865
Test 2B	0.150	0.232	0.33	30	6.76	0.77	0.815
Test 3B	0.150	0.232	0.31	45	8.71	0.67	0.689
Test 4B	0.200	0.310	0.43	15	3.89	1.60	0.945
Test 5B	0.200	0.310	0.42	30	7.40	1.43	0.920
Test 6B	0.200	0.310	0.39	45	10.29	1.14	0.844

6. Conclusions

A field and numerical study of shoreline resistance and resilience was conducted on a sea-breeze dominated sandy beach. The following conclusions were found:

(1) Analysis of high- spatial and temporal resolution field observations showed a high growth rate of the perturbation due to a groin disturbance and a lower decaying rate after the disturbance removal.
(2) A new shoreline evolution model was calibrated with field observation and was able to reproduce both the perturbation growth and decay observed in the field.
(3) A parametric numerical study suggests that shoreline resistance decreases with wave conditions enhancing alongshore sediment transport; whereas the resilience increases as a function of the alongshore diffusivity.

Supplementary Materials: A demo version of the GSb numerical model, for Mac and Windows systems, has been made available for the scientific community and can be downloaded at: www.scacr.eu and field measurements are available at http://ocse.mx/en/experimento/beach-resilience-to-coastal-structures-brics upon request. A time-lapse video of the beach perturbation due to the temporary groin disturbance is included as supplementary material.

Author Contributions: Conceptualization, G.M., A.T.-F., P.A.T.; Writing-Original Draft Preparation, G.M., A.T.-F., G.R.T., A.F.; Writing-Review & Editing, G.M., A.T.-F., G.R.T., A.F., L.L.; Field work, G.M., A.T.-F., J.L., L.P.-A.; Field data Analysis, G.M., Structure design, L.P.-A., J.L., P.A.T., Numerical modeling, G.R.T., A.F., L.L.; Numerical data Analysis, G.R.T., A.F., L.L.; Funding Acquisition, G.M., A.T.-F.

Funding: This research was funded by CONACYT through the Cátedras-CONACyT (Project 1146), Investigación Científica Básica (Project 284819) and the Laboratorio Nacional de Resiliencia Costera (Project LN 293354). Additional financial support was provided by PAPIIT DGAPA UNAM (IN101218) and Instituto de Ingeniería UNAM.

Acknowledgments: We acknowledge the field support provided by students and researchers at the Laboratorio de Ingeniería y Procesos Costeros at UNAM, especially from Gonzalo Uriel Martín Ruiz, David A. Gracia, Elena Ojeda, Tonatiuh Mendoza, José Carlos Pintado-Patiño, José Alberto Zamora, Daniel Toxtega, Martín Ezquivelzeta, Miguel Ángel Valencia, Paola Espadas, Luis Ángel Gallegos, Daniel Pastrana, Jesús Aragón, Pedro Cabañas, Alejandro Astorga, Enna López, Rafael Meza, Wilmer Rey and Marcos García. Special thanks to Elena Ojeda and David Gracia for providing the time-lapse video of the experiment.

Conflicts of Interest: The authors declare no conflict of interest.

References

1. Pimm, S.L. The complexity and the stability of ecosystems. *Nature* **1984**, *307*, 321–326. [CrossRef]
2. Orwin, K.H.; Wardle, D.A. New indices for quantifying the resistance and resilience of soil biota to exogenous disturbances. *Soil Biol. Biochem.* **2004**, *36*, 1907–1912. [CrossRef]
3. Folke, C. Resilience: The emergence of a perspective for social-ecological systems analyses. *Glob. Environ. Chang.* **2006**, *16*, 253–267. [CrossRef]
4. Holling, C.S. Resilience and stability of ecological systems. *Ann. Rev. Ecol. Syst.* **1973**, *4*, 1–23. [CrossRef]
5. Garmezy, N.; Masten, A.S.; Tellegen, A. The study of stress and competence in children: A building block for developmental psychopathology. *Child Dev.* **1984**, *55*, 97–111. [CrossRef] [PubMed]

6. Alexander, D.E. Resilience against earthquakes: Some practical suggestions for planners and managers. *J. Seism. Earthq. Eng.* **2012**, *13*, 109–115.
7. Alexander, D.E. Resilience and disaster risk reduction: An etymological journey. *Nat. Hazards Earth Syst. Sci.* **2013**, *13*, 2707–2716. [CrossRef]
8. Tomasicchio, G.R.; Sanchez Arcilla, A.; D'Alessandro, F.; Ilic, S.; James, M.; Fortes, C.J.E.M.; Sancho, F.; Schüttrumpf, H. Large-scale flume experiments on dune erosion processes. *J. Hydraul. Res.* **2011**, *49*, 20–30. [CrossRef]
9. D'Alessandro, F.; Tomasicchio, G.R. Wave-dune interaction and beach resilience in large-scale physical model tests. *Coast. Eng.* **2016**, *116*, 15–25. [CrossRef]
10. Smith, E.R.; D'Alessandro, F.; Tomasicchio, G.R.; Gailani, J.Z. Nearshore placement of a sand dredged mound. *Coast. Eng.* **2017**, *126*, 1–10. [CrossRef]
11. Martinez, L.; Taramelli, A.; Silva, R. Resistance and Resilience: Facing the Multidimensional Challenges in Coastal Areas. *J. Coast. Res.* **2017**, *77* (Suppl. 1), 1–6. [CrossRef]
12. Lira-Pantoja, M.; Torres-Freyermuth, A.; Appendini, C.; Fernández, D.; Salles, P.; Mendoza, E.; López, J.; Pedrozo-Acuña, A. Chronic beach erosion induced by coastal structures in Chelem, Yucatan. *Coast. Eng. Proc.* **2012**, *1*, 125. [CrossRef]
13. Medellín, G.; Mariño-Tapia, I.; Euán-Ávila, J. The influence of a seawall on postnourishment evolution in a sea-breeze-dominated microtidal beach. *J. Coast. Res.* **2015**, *31*, 1449–1458. [CrossRef]
14. Meyer-Arendt, K.J. Recreational development and shoreline modification along the north coast of Yucatán, Mexico. *Tour. Geogr.* **2001**, *3*, 87–104. [CrossRef]
15. Bodge, K.R.; Dean, R.G. *Short-Term Impoundment of Longshore Sediment Transport*; TR CERC-87-7; WES: Vicksburg, MS, USA, 1987.
16. Wang, P.; Kraus, N.C. Longshore sediment transport rate measured by short-term impoundment. *J. Waterw. Port Coast. Ocean Eng.* **1999**, *125*, 118–126. [CrossRef]
17. Martin-Grandes, I.; Hughes, J.; Simmonds, D.J.; Chadwick, A.J.; Reeve, D.E. Novel methodology for one line model calibration using impoundment on mixed beach. *Coast. Dyn.* **2009**. [CrossRef]
18. Rosati, J.D.; Walton, T.L.; Bodge, K. *Longshore Sediment Transport, in Coastal Engineering Manual, Part III*; King, D.B., Ed.; Technical Report EM 1110-2-1100; U.S. Army Corps of Engineers: Washington, DC, USA, 2002.
19. Elgar, S.; Raubenheimer, B.; Thomson, J.; Moulton, M. Resonance in an evolving hole in the swash zone. *J. Waterw. Port Coast. Ocean Eng.* **2012**, *138*, 299–302. [CrossRef]
20. Moulton, M.; Elgar, S.; Raubenheimer, B. A surf zone morphological diffusivity estimated from the evolution of excavated holes. *Geophys. Res. Lett.* **2014**, *41*, 4628–4636. [CrossRef]
21. Moulton, M.; Elgar, S.; Raubenheimer, B.; Warner, J.C.; Kumar, N. Rip currents and alongshore flows in single channels dredged in the surf zone. *J. Geophys. Res. Oceans* **2017**, *122*, 3799–3816. [CrossRef]
22. Torres-Freyermuth, A.; Puleo, J.A.; DiCosmo, N.; Allende-Arandia, M.E.; Chardón-Maldonado, P.; López-González, J.; Figueroa, B.; Ruiz de Alegría-Arzaburú, A.; Figlus, J.; Roberts, T.; et al. Nearshore hydrodynamics on a sea breeze dominated beach during intense wind events. *Cont. Shelf Res.* **2017**, *151*, 40–52. [CrossRef]
23. Figueroa-Espinoza, B.; Salles, P.; Zavala, J. On the Wind Power Potential in the northwest of the Yucatan Peninsula in Mexico. *Atmosfera* **2014**, *27*, 77–89. [CrossRef]
24. Pérez, E.P.; Magaña, V.; Caetano, E.; Kusunoki, S. Cold surge activity over the Gulf of Mexico in a warmer climate. *Front. Earth Sci.* **2014**, *2*, 19. [CrossRef]
25. Appendini, C.M.; Pedrozo-Acuña, A.; Meza-Padilla, R.; Torres-Freyermuth, A.; Cerezo-Mota, R.; López-González, J.; Ruiz-Salcines, P. On the role of climate change on wind waves generated by tropical cyclones in the Gulf of Mexico. *Coast. Eng. J.* **2017**, *59*. [CrossRef]
26. Martell-Dubois, R.; Silva-Casarin, R.; Mendoza-Baldwin, E.G.; Muñoz-Pérez, J.J.; Cerdeira-Estrada, S.; Escalante-Mancera, E.; Laiz, I. Spectral bimodality of waves produced by hurricanes in the Caribbean coastal zone off Mexico. *Cienc. Mar.* **2018**, *44*, 33–48. [CrossRef]
27. Cuevas-Jiménez, A.; Euán-Ávila, J. Morphodynamics of carbonate beaches in the Yucatán Peninsula. *Cienc. Mar.* **2009**, *35*, 307–320. [CrossRef]
28. Hubertz, J.M. Observations of local wind effects on longshore currents. *Coast. Eng.* **1986**, *10*, 275–288. [CrossRef]

29. King, D.B. *Wave and Beach Processes Modeling for Sabine Pass to Galveston Bay, Texas, Shoreline Erosion Feasibility Study*; Technical Report ERDC/CHL TR-07-6; USACE Engineer Research and Development Center: Vicksburg, MS, USA, 2007; 150p.
30. Burnette, C.; Dally, W.R. The Longshore Transport Enigma and Analysis of a 10-Year Record of Wind-Driven Nearshore Currents. *J. Coast. Res.* **2018**, *34*, 26–41. [CrossRef]
31. Wellmann, N. Analysis of Near-Shore Sediment Samples from Sisal Beach (Mexico) Comparing Effects of Sea Breeze and el Norte Events. Ph.D. Thesis, Hochschule für Technik, Wirtschaft und Kultur Leipzig, Leipzig, Germany, 2014.
32. Zavala-Hidalgo, J.; Morey, S.L.; O'Brien, J.O. Seasonal circulation on the western shelf of the Gulf of Mexico using a high-resolution numerical model. *J. Geophys. Res.* **2003**, *108*, 3389. [CrossRef]
33. MEXICO97. Available online: https://www.ngs.noaa.gov/GEOID/MEXICO97/ (accessed on 20 August 2018).
34. Medellín, G.; Medina, R.; Falqués, A.; González, M. Coastline sand waves on a low-energy beach at "El Puntal" spit, Spain. *Mar. Geol.* **2008**, *250*, 143–156. [CrossRef]
35. Miller, J.K.; Dean, R.G. Shoreline Variability via Empirical Orthogonal Function Analysis: Part I Temporal and Spatial Characteristics. *Coast. Eng.* **2006**, *54*, 111–131. [CrossRef]
36. Muñoz-Pérez, J.J.; Medina, R.; Tejedor, B. Evolution of longshore beach contour lines determined by the E.O.F. method. *Sci. Mar.* **2001**, *65*, 393–402. [CrossRef]
37. Frey, A.E.; Connell, K.J.; Hanson, H.; Larson, M.; Thomas, R.C.; Munger, S.; Zundel, A. *GenCade Version 1 Model Theory and User's Guide*; Technical Report ERDC/CHL TR-12-25; U.S. Army Engineer Research and Development Center: Vicksburg, MS, USA, 2012.
38. Bruun, P. *Coast Erosion and the Development of Beach Profiles*; Technical Memorandum N. 44; Department of the Army, Beach Erosion Board, Corps of Engineers: Washington, DC, USA, 1954.
39. Dean, R.G. Equilibrium beach profiles: Characteristics and applications. *J. Coast. Res.* **1990**, *71*, 53–84.
40. Tomasicchio, G.R.; Lamberti, A.; Guiducci, F. Stone movement on a reshaped profile. In Proceedings of the 24th International Conference on Coastal Engineering, Kobe, Japan, 23–28 October 1994; Volume 2, pp. 1625–1640.
41. Lamberti, A.; Tomasicchio, G.R. Stone mobility and longshore transport at reshaping breakwaters. *Coast. Eng.* **1997**, *29*, 263–289. [CrossRef]
42. Tomasicchio, G.R.; Archetti, R.; D'Alessandro, F.; Sloth, P. Long-shore transport at berm breakwaters and gravel beaches. *Coast. Struct.* **2007**, 65–76. [CrossRef]
43. Tomasicchio, G.R.; D'Alessandro, F.; Barbaro, G.; Malara, G. General longshore transport model. *Coast. Eng.* **2013**, *71*, 28–36. [CrossRef]
44. Tomasicchio, G.R.; D'Alessandro, F.; Barbaro, G.; Musci, E.; De Giosa, T.M. Longshore transport at shingle beaches: An independent verification of the general model. *Coast. Eng.* **2015**, *104*, 69–75. [CrossRef]
45. Ozasa, H.; Brampton, A.H. Mathematical modeling of beaches backed by seawalls. *Coast. Eng.* **1980**, *4*, 47–64. [CrossRef]

water

MDPI

Article

Morphodynamic Response to Low-Crested Detached Breakwaters on a Sea Breeze-Dominated Coast

Alec Torres-Freyermuth [1,2,*], Gabriela Medellín [1,2], Ernesto Tonatiuh Mendoza [1,2], Elena Ojeda [1,2] and Paulo Salles [1,2]

1 Laboratorio de Ingeniería y Procesos Costeros, Instituto de Ingeniería, Unidad Académica Sisal, Universidad Nacional Autónoma de México, Sisal 97356, Mexico; GMedellinM@iingen.unam.mx (G.M.); EMendozaP@iingen.unam.mx (E.T.M.); EOjedaC@iingen.unam.mx (E.O.); PSallesA@iingen.unam.mx (P.S.)

2 Laboratorio Nacional de Resiliencia Costera, Laboratorios Nacionales CONACTY, Sisal 97356, Mexico

* Correspondence: ATorresF@iingen.unam.mx; Tel.: +52-988-931-1000

Received: 2 March 2019; Accepted: 21 March 2019; Published: 27 March 2019

Abstract: Low-crested detached breakwaters (LCDBs) have been widely employed as a mitigation measure against beach erosion. However, only a few studies have assessed their performance in sea-breeze-dominated environments. This work investigates the beach morphodynamics behind LCDBs deployed on a micro-tidal sea-breeze-dominated beach. The study area, located in the northern Yucatán peninsula, is characterized by low-energy, high-angle waves, which drive a persistent (westward) alongshore sediment transport ($O(10^4)$ m^3/year). High-resolution real-time kinematics global positioning system (GPS) beach surveys were conducted over a one-year period (2017–2018) to investigate the performance of LCDBs at three sites. Moreover, unmanned aerial vehicle flights were employed to evaluate far-field shoreline stability. Field observations revealed a distinct behavior in the three study sites, dependent on the breakwaters' transmission characteristics, geometry, stability, and shoreline orientation. Impermeable LCDBs, made of sand-filled geosystems, induced significant beach accretion (erosion) in up-(down-)drift areas. On the other hand, permeable LCDBs, made of Reef Ball™ modules, induced moderate beach changes and small erosion in down-drift areas owing to higher transmission coefficients. Measurements of LCDBs' freeboard height show that sand-filled geosystems' breakwaters presented a significant loss of sand during the study period, which explains the unexpected beach morphodynamic response on the lee side of the structure. Observations suggest that the study area is highly sensitive to the presence of LCDBs with low transmissivity.

Keywords: beach morphodynamics; UAV flights; beach surveys; Reef Balls™; sand-filled geosystems

---

1. Introduction

Sandy beaches and their associated front dunes provide both natural coastal protection against storm events and a habitat for different marine and terrestrial species. Therefore, beach erosion due to either natural or anthropogenic processes may cause environmental and economic impacts in the coast. This is particularly relevant in low-lying coastal areas, which are prone to climate change impacts, such as sea level rise and increasing storm activity [1].

Despite the increasing popularity of soft engineering [2,3], mixed soft–hard systems [4], and eco-engineering [5] solutions, so-called low-crested detached breakwaters (LCDBs), such as submerged rubble mounds, are popular in many locations around the world (e.g., [6]). The main parameters controlling shoreline response in the presence of LCDBs are distance offshore, length and orientation of the structure, transmission characteristics of the structure, depth at the structure, freeboard height, and wave characteristics [7,8]. Previous works have extensively investigated the stability, performance, and ecological impact of rubber-mound (conventional) LCDBs [8–11], but less effort has been devoted to understanding the performance of unconventional LCDBs (e.g., sand-filled geosystems and artificial reefs) until more recently (e.g., [12–15]).

Shoreline response to LCDBs has been investigated by means of numerical models (e.g., [16]), physical model tests (e.g., [11]), and field observations (e.g., [9]). However, monitoring studies involving measurements of subsequent detailed bathymetry surveys and far-field effects are scarce, owing to both economic and technical constraints [9]. Predicting the morphodynamic response on the lee side of LCDBs is challenging [7]. The wave transmission, controlled by the freeboard level, permeability, and wave conditions, determines the size and location of the shoreline salient [16], and hence, a transmission coefficient must be incorporated in the design formulas [7]. Black and Andrews [17] investigated the salient amplitudes for islands and reefs separately and developed a power curve relationship similar to the one for emergent breakwaters presented by [18].

In recent years, the construction of breakwaters with the use of synthetic materials, such as geotextile tubes filled with sand, has become widespread in some countries owing to their lower permanent impact on natural coastal processes [15]. Other approaches have considered the use of modular structures, such as artificial reefs (e.g., Reef Ball™), that mimic the wave dissipation effects of natural coral reefs. Artificial reefs have become popular in tropical regions, because they provide other ecosystem services, such as a habitat for marine species, while enhancing shoreline stability [19]. The use of sand-filled geosystems increased in popularity in the state of Yucatán (México) in the early 2000s, partially because of governmental environmental agencies that consider this alternative to be a soft engineering solution [20].

This work aims to identify the role of changing wave conditions, LCDB type, geometry, freeboard elevation, and distance and orientation of the structure on the morphological response on the northern Yucatan coast by means of a monitoring program conducted from March 2017 to May 2018 at three different sites. The outline of this paper is the following. First, the study area and LCDB characteristics are described in Section 2. The beach-monitoring program conducted at the study sites is described in Section 3. The results of the beach morphodynamics are presented in Section 4. A discussion on the factors affecting the shoreline salient prediction in the study sites is presented in Section 5. Finally, concluding remarks are given in Section 6.

2. Study Area

2.1. Description

The study area is characterized by a mild-slope continental shelf and a micro-tidal regime [21]. Intense sea breeze events generate short-period, high-incidence angle NE waves that are present all year. Thus, sea breezes play an important role in nearshore hydrodynamics [22] and sediment transport [23]. Furthermore, Central American cold surge (CACS) events, associated with cold-front passages, generate NNW swell waves that occur during winter months [24]. Appendini et al. [25] found that the net potential (westward) of sediment transport in the area is approximately 35,000 m^3/year and that such value is highly sensitive to shoreline orientation.

The northern Yucatán peninsula has been experiencing beach erosion over the past few decades, mainly associated with coastal development [26]. The construction of eleven shelter ports and a major offshore port from the 1960s to the 1980s solved many socioeconomic problems, but, at the same time, motivated the deployment of hundreds of unauthorized groins by beach homeowners at locations affected by the impoundment of the ports' jetties [27]. The presence of coastal structures increased during the following decades until the authorities enforced such structure removal, mainly consisting of permeable groins, in the early 2000s. Beach nourishment combined with the groin removal provided satisfactory results along a 5-km stretch of coast [28]. However, coastal erosion has been exacerbated in certain areas due to the lack of a long-term beach nourishment program, the removal of the primary dune due to the growth of the coastal urban area, and the continuous deployment of unauthorized structures [29]. Meyer-Arendt [27] estimated erosion rates to be between 0.3 and 1.0 m/year west of Progreso. A government-funded project considered the deployment of two breakwaters, made of sand-filled geosystems, as mitigation measures against beach erosion in two critical erosion spots located east of Progreso. At the same time, a privately funded project deployed breakwaters made of Reef Balls™ in the same area (Figure 1).

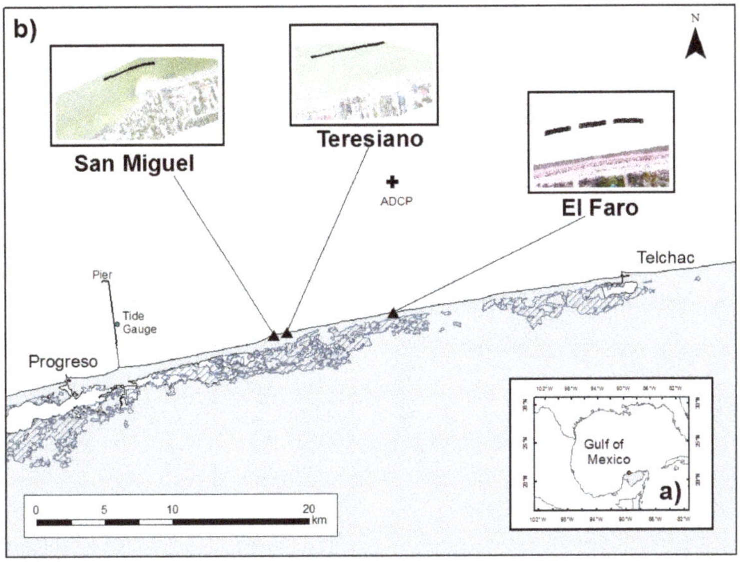

Figure 1. Study area location: (**a**) the breakwaters are located in the northern Yucatán Peninsula in the Gulf of Mexico; (**b**) the three sites (San Miguel, Teresiano, and El Faro) are between the ports of Progreso and Telchac. Waves and tides were measured with an Acoustic Doppler Current Profiler (ADCP) at a water depth of 10 m and a tidal gauge (www.mareagrafico.mx) installed near the 6 km-long Progreso Pier.

Low-crested structures were deployed between spring and summer 2017 in the northern Yucatán coast in Punta San Miguel, El Teresiano, and El Faro, located 10.5 km, 12 km, and 20 km from the Port of Progreso, respectively (Figure 1). The structures are all located within 9.5 km of the coast, and hence, environmental conditions (wind, waves, and water levels) are homogenous at all sites. Furthermore, the beach is characterized by the presence of beach houses constructed close to the shore (<15 m) over the primary dune.

San Miguel is located in a natural rocky headland where the shoreline orientation changes from 54 to 85° north (Figure 1). A 60-m rock revetment extends westward from the headland as a coastal protection measure to mitigate beach erosion. El Teresiano beach is located 1.5 km eastward from San Miguel and presents a uniform shoreline orientation of 80°. The beach width is less than 5 m at the most critical location, and waves reach the beach houses during high tides. Impermeable detached submerged breakwaters, made of geotextile tubes filled of sand, were installed at these two sites to increase the beach width (Figure 1).

El Faro has a shoreline orientation of 84° north (Figure 1), and the beach width at this site is <10 m. The coast is relatively pristine at the up-drift and down-drift side of the structure, but two beach houses are located behind the structure. The LCDBs deployed in El Faro are permeable.

2.2. Breakwater Characteristics

The low-crested detached breakwaters evaluated here are classified as impermeable and permeable based on the materials used for their construction. Differences in nearshore wave transformation and sediment transport patterns were expected on the lee side of the structures owing to the distance to the undisturbed shoreline, freeboard height, breakwater length and orientation, and shoreline orientation (Figure 2). The impermeable breakwaters are made of sand-filled geosystems, and the transmissivity depends only on the wave propagation above the freeboard crest. On the other hand, the artificial reef modules known as Reef Balls™ [30] are considered to be permeable structures, because they allow flow through the modules and the two 10-m gaps separating the three breakwater segments. An overview of the breakwaters' characteristics for each site is given in Table 1 and described in more detail below.

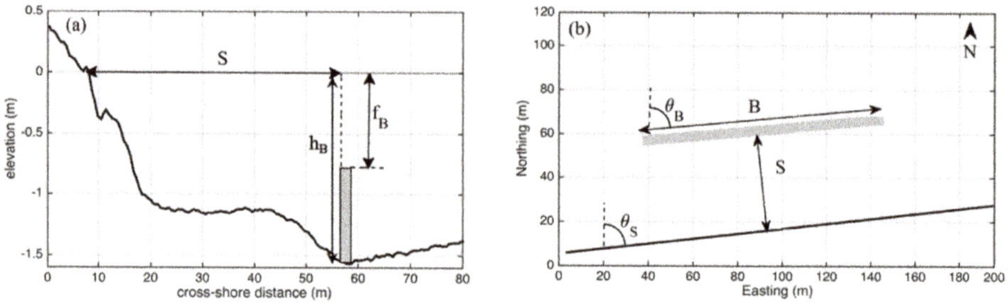

Figure 2. Description of structural parameters: (**a**) distance to the undisturbed shoreline (S), water depth at the breakwater (h_B), and freeboard elevation (f_B); (**b**) breakwater length (B), breakwater orientation (θ_B), and undisturbed shoreline orientation (θ_S).

Table 1. Breakwater characteristics at the three sites: breakwater length (B), distance from the shoreline (S), breakwater orientation (θ_B), shoreline orientation (θ_S), freeboard (f_B) of each section, and water depth at the breakwater (h_B).

Location	B (m)	S (m)	θ_B (°)	θ_S (°)	f_B (m)	h_B (m)
San Miguel	120	60	70	54 and 84	−0.58, −0.55, −0.64, −0.39, −0.37, −0.36	1.3
El Teresiano	140	90	80	80	−0.68, −0.68, −0.94, −1.02, −0.96, −0.69, −0.48, −0.62	1.6
El Faro	107	48	82	84	−0.89, −0.88, −0.78	1.5

2.2.1. Impermeable Breakwaters

The impermeable structures were installed at San Miguel and El Teresiano beaches (Figure 1). The structures were constructed with 10- and 20-m sections of propylene geotextile mesh filled with sand with a minimum resistance of 50 × 105 kN/m, 415 gr/m^2, and 1.3 mm thickness. According to the design, each section has a width of 1.83 m and a height of 0.90 m when filled to 70% of its capacity.

The breakwater at San Miguel was composed of 6 sections and had a total length of 120 m with a minimum distance of 60 m to the original shoreline position (Figure 3a). The structure was oriented 70° north. The three easternmost 20-m sections of the breakwater at San Miguel beach were constructed from 18 June to 25 June 2017. The fourth section was installed on 14 July 2017, and in the following days, the remaining two sections were deployed. The structure required 620 m^3 of sand, which was procured on site. The freeboard elevation (vertical distance between mean sea level and breakwater crest) in San Miguel varied between $z = -0.36$ and $z = -0.64$ m.

The LCDB at El Teresiano, composed of 8 sections, had a total length of 140 m with a distance to shore of 90 m and was oriented parallel to the shoreline (see Table 1). The structure required 720 m^3 of sand, which was taken from the submerged beach between the structure and the shoreline, at a water depth of 1.5 m. The structure construction took place from 8 May to 16 May 2017. The freeboard presented significant differences between each section of the breakwater. At the time of deployment, the breakwater at El Teresiano had a freeboard that varied between $z = -0.48$ and $z = -1.02$ m along the structure. This heterogeneity was not consistent with the design. Figure 3 shows the layout of the two impermeable structures.

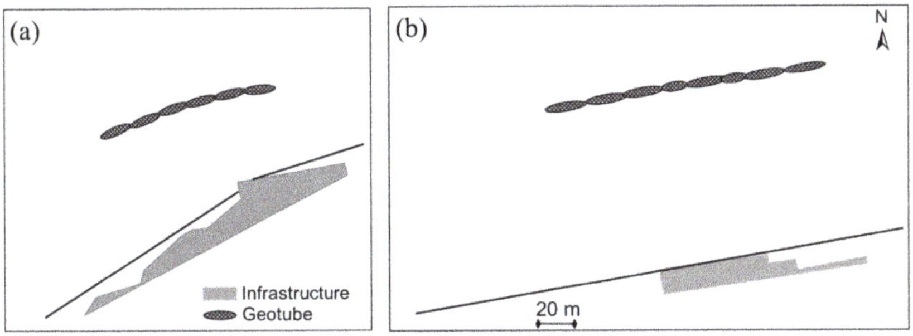

Figure 3. Impermeable detached breakwaters' layouts with respect to the mean shoreline orientation (black solid line) at (**a**) San Miguel and (**b**) El Teresiano.

2.2.2. Permeable Breakwaters

The permeable LCDB installed at El Faro beach was built using Reef Balls™. These are perforated hemispherical-shaped modules, available in different sizes, made of pH-neutralized concrete designed by Reef Ball Development Group, Ltd (Sarasota, FL, USA). The breakwater at El Faro has a total length of 107 m and consists of three 29-m sections separated by 10-m gaps (Figure 4). The structure was built with 135 elements of pallet balls arranged in two rows. The pallet balls are 800-kg elements, 0.90 m high, and with a 1.22-m diameter. The structure is oriented parallel to the shoreline, 48 m offshore (Figure 4). This field of detached breakwaters is considered, in this study, to be a single (permeable) breakwater.

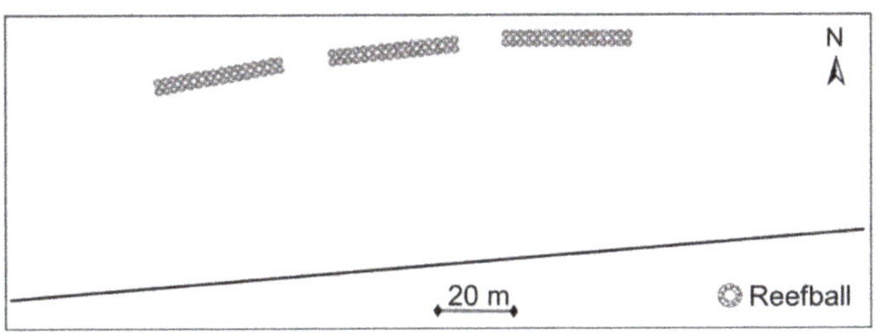

Figure 4. Reef Balls layout with respect to the mean shoreline orientation (black solid line) at El Faro.

3. Materials and Methods

Beach surveys were undertaken before the structure deployment in the three sites, and hence, beach morphology evolution could be evaluated with respect to the undisturbed condition. The freeboard elevation and breakwater length were periodically measured to assess structure variability. Furthermore, beach surveys and unmanned aerial vehicles (UAVs) flights were conducted to assess the functionality and far-field impact on adjacent beaches, respectively. Water levels and wave conditions were also measured to correlate the observed beach changes with the forcing conditions.

3.1. Beach Surveys

Beach surveys were conducted by means of a Leica™ real-time kinematics differential global positioning system (RTK-DGPS). Cross-shore transects were surveyed in the down-drift and up-drift areas and on the lee side of each structure with high spatial resolution. The GPS base and radio were installed inland on a fixed structure, specially built for this study, at each site, with known x, y, and z coordinate locations. The rover DGPS was installed in a backpack to facilitate walking along each transect, starting from the dry beach and extending offshore until reaching a water depth of approximately 1.6 m. The beach surveys covered an alongshore stretch ranging from 300 to 400 m, decreasing the separation between transects to 10 m on the lee side of the structures and increasing the separation to 20 m near the up-/down-drift boundaries of the surveyed area. A control point was measured at the beginning and at the end of each survey to correct the rover height on the backpack due to probable small vertical variations in its position during each survey. Typically, 20 cross-shore transects were surveyed at each site. However, the number of transects was increased in San Miguel from 20 to 27 due to the significant down-drift erosion observed beyond the original surveyed area (Figure 5).

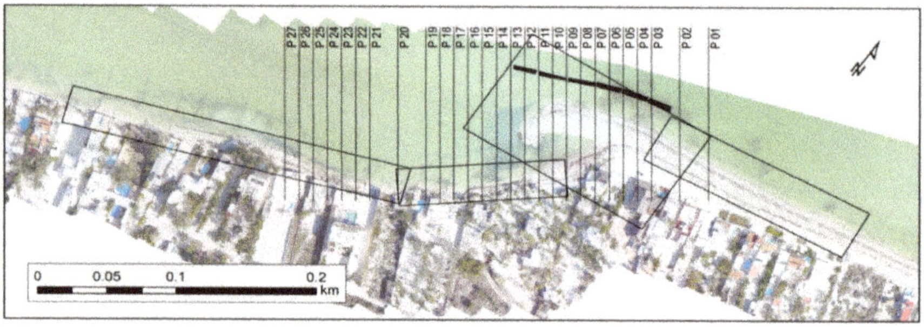

Figure 5. Aerial picture of San Miguel showing the survey transects (lines P01–P27) and the unmanned aerial vehicle (UAV) flight missions (rectangular shapes) employed for the monitoring. The aerial picture was taken on 11 April 2018.

The first survey at each site was taken as a reference to estimate relative sand volume changes. The shoreline position was estimated to be the cross-shore location with an elevation $z = 0$. Additionally, three-dimensional (3D) measurements allowed us to estimate the seabed changes in both the aerial and submerged beach profiles. A total of 43 beach surveys were conducted with approximately 14 surveys at each site (see symbols in Figure 6).

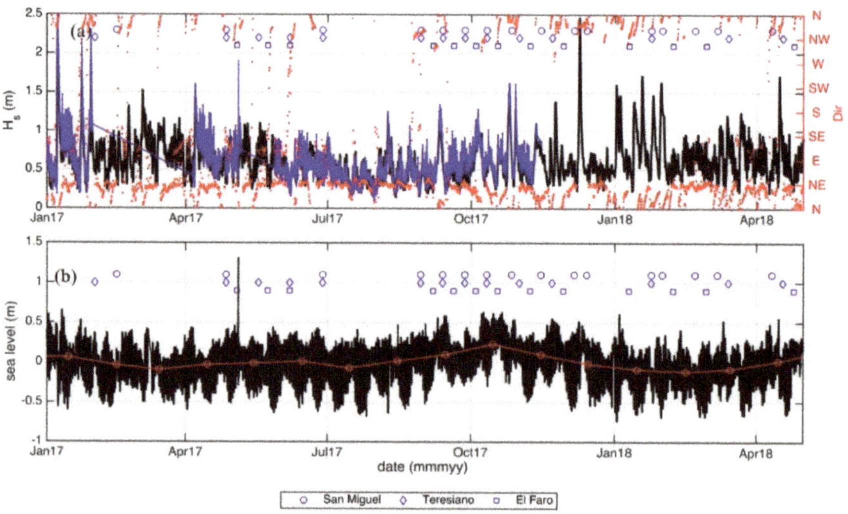

Figure 6. Time series of (**a**) significant wave height (black solid line: Wave Watch III hindcast; blue solid line: measured) and wave direction; and (**b**) mean sea level (black solid line: 1-min average; red circles: monthly average) during the study period. The blue symbols represent the surveys conducted at each site.

UAV flights have become a useful instrument in coastal studies [31]; the predominant advantages, including larger surveyed areas than GPS, graphic documentation through ortho-mosaics of the surveyed area, repeatability, and low cost, and, with the use of ground control points, centimeter-level precision digital surface models (DSMs) can be achieved. In this study, a Phantom 4 DJI quadcopter was used in combination with Pix4D Capture for flight planning and Pix4D Mapper to obtain digital surface models and subsequently extract the shoreline ($z = 0$).

Flights were conducted on a monthly basis for each site and when a scheduled engineering intervention or a storm event occurred. The mapped areas were rectangular with designed fixed parameters, such as a double grid to correctly compute the 3D models, 30-m flight altitude, 75% image overlap, 90-degree camera position, and medium-speed flight. Depending on the site and the complexity of the area, a series of missions was used (e.g., San Miguel was covered by 4 separate missions in each survey; see rectangles in Figure 5).

The accuracy of the ortho-mosaics and the DSMs depends largely on the quality and the coverage of the ground control points (GCPs) given that the surveyed images did not include a precise UAV location. Therefore, an average of 25 GCPs, precisely measured using RTK-DGPS, was used in each flight. Horizontal coordinates were referenced to the World Geodetic System of 1984 (WGS 1984) using the Universal Transverse Mercator coordinate system (UTM)Zone 16 N, and the vertical values were referenced to mean sea level using Gravimetric Mexican Geoid version 2010 (GGM10). The accuracy of the DSMs was validated using the RTK-DGPS data.

The obtained DSMs were integrated into the ArcGIS geographic information system, from which the coastline was obtained ($z = 0$) and put into the Digital Shoreline Analysis System [32], from which the shoreline change rate was obtained over the study period.

3.2. Waves and Water Level

Waves were measured with an RDI acoustic doppler velocimeter installed at 10-m water depth and located approximately 10 km offshore (see Figure 1). Wave measurements are depicted by the blue line in Figure 6a. Gaps in the record, due to instrument malfunction, were filled out using National Oceanic and Atmospheric Administration (NOAA) Wave Watch III hindcast data (black line in Figure 6a), corresponding with the closest node from the instrument's location. Maximum wave heights often occur during late fall and early spring and are associated with cold-front events, which drive NNW swells (Figure 6a). On the other hand, predominant, short-period NE waves associated with sea breeze events are present all year but are more frequent and intense during the spring–summer months, when cold fronts are absent (Figure 6a and see Figure 2 in [23]). Mean sea level was recorded with a tidal gauge from the National Mareographic Service (www.mareografico.unam.mx) located in the Progreso Pier (see Figure 1). A local storm called "turbonada" increased the mean sea level significantly on 4 May 2017 (Figure 6b).

4. Results

Environmental conditions, structure variability, and anthropogenic actions play an important role in beach morphodynamics. Thus, high spatial and temporal resolution monitoring allows us to understand a sea breeze-dominated beach's morphological responses to LCDBs. The analysis of the structure variability and the beach's morphological responses, based on field observations, is presented in this section.

4.1. Structures' Freeboard Variability

The impermeable LCDBs presented high spatial and temporal variability with respect to freeboard elevation. In the case of San Miguel, the deployment of the different sections was not uniform with the differences in elevation of 0.4 m between the sections (Figure 7a). Furthermore, the eastward sections' height changed up to −0.2 m in the two months following the installation due to loss of sand. The westmost section was torn apart in November 2017 to reduce down-drift erosion. By 11 April

2018, only the two middle sections remained, and by June, the breakwater was completely deflated (Figure 7a).

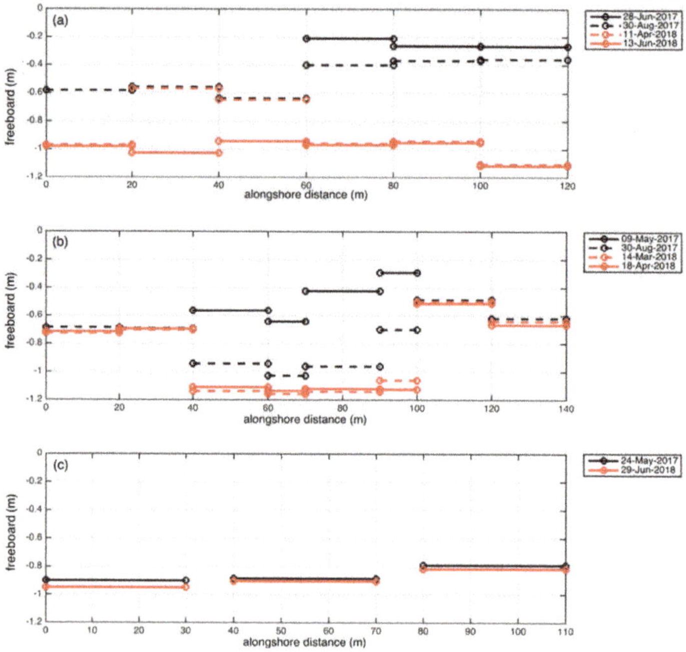

Figure 7. Freeboard elevation change at each of the sections of the low-crested breakwaters at (**a**) San Miguel, (**b**) El Teresiano, and (**c**) El Faro.

The breakwater at El Teresiano also presented significant differences with respect the freeboard height along the structure. The middle sections continuously lost elevation due to sand losses and were significantly reduced two months after deployment and almost completely deflated 10 months later. By 18 April 2018, only two 40-m sections remained with a gap between the sections of 60 m (Figure 7b). On the other hand, the breakwaters made of Reef Ball™ modules remined stable with freeboard changes of less than 0.10 m at the end of the study period (Figure 7c).

4.2. Morphodynamic Response

The importance of conducting high-resolution DGPS monitoring to explain beach changes is illustrated in San Miguel. The structure at this location was deployed with the aim of increasing the beach width at this natural headland to protect beach properties.

The beach morphology and structure history in San Miguel are shown in Figure 8. No significant changes were observed near the headland prior to the structure deployment (Figure 8a,b). The first half of the structure was deployed by 28 June 2017, and hence, some accumulation occurred at the eastward transects due to the net westward sediment transport (Figure 8c). The structure was fully deployed in July, and hence, the beach width increased behind the eastern half of the structure with low elevation, owing to the prevailing sea breeze conditions (Figure 8d). These conditions persisted until the beginning of the cold-front season, when swell waves decreased the salient length, increasing the beach elevation on the lee side of the structure and allowing sediment accretion in the western side (Figure 8e–i). However, significant beach erosion occurred at the down-drift (westward) beaches (Figure 8d–h). To mitigate down-drift beach erosion associated with sediment transport gradients due to wave diffraction, the westmost 20-m section of the geotextile was removed, inducing a positive recovery effect (Figure 8i–k). However, in January 2018, the authorities decided to remove sand from the salient to nourish the down-drift beaches and prevent the formation of a tombolo. The estimated volume extracted was 600 m^3, which was taken from the subaerial beach (decreasing beach elevation) and was placed in the down-drift transects with low elevation (Figure 8l). The effect of this action was negligible one week later (Figure 8m), and the salient size increased during the following month (Figure 8n). During the following months, the eastward sections of the breakwater were torn apart (Figure 7a), gradually emptying and decreasing their freeboard height. Only two 20-m sections remained by April 2018. The resulting configuration of the LCDB induced the formation of a sand spit (Figure 8o), reducing the sediment supply and eroding the beach as far as 500 m in the down-drift areas. Table 2 summarizes the chronology of the main environmental and anthropogenic events controlling the morphological response at this site.

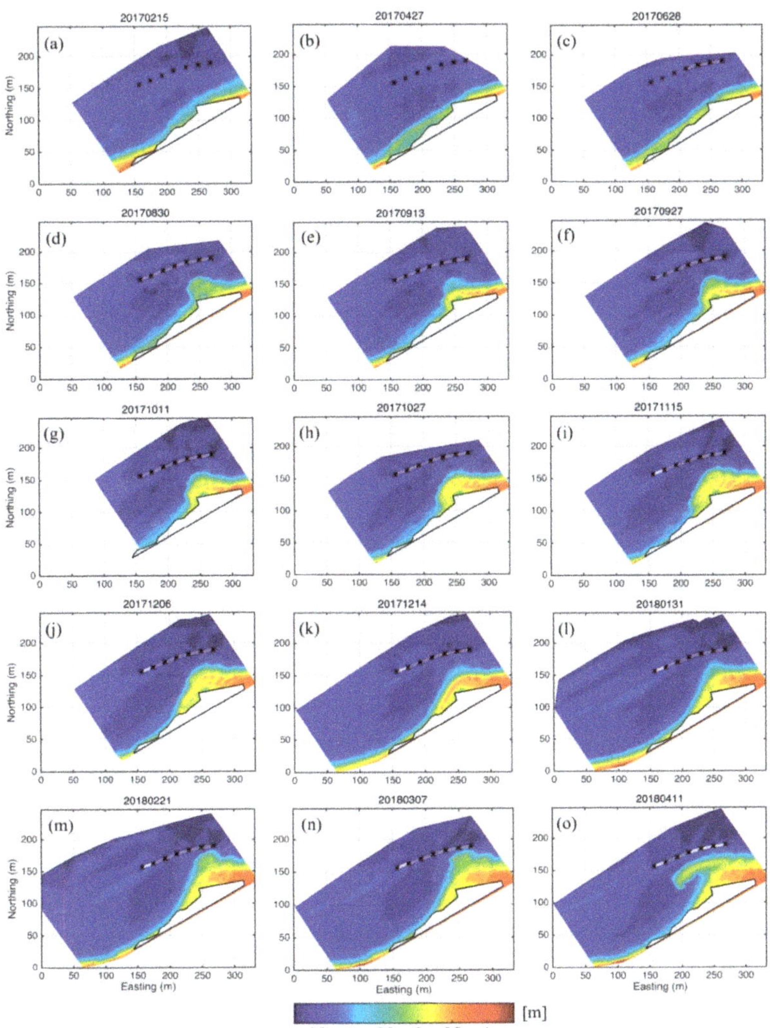

Figure 8. Topo-bathymetry and structure evolution at San Miguel from the (**a**) undisturbed condition (without structure) to (**o**) the last survey a year after. The title in each panel represents the survey date (*yyyymmdd*). The lighter color in breakwater sections indicates that the geotube section was deflated.

Table 2. Environmental and anthropogenic events affecting the beach evolution in San Miguel.

Event	Dates	Morphology Effect
Low-Crested Detached Breakwater construction	June–July 2017	Updrift accretion
Sea breeze events	July–August 2017 February–April 2018	Salient growth in the eastern half
Central American Cold Surge events	September–December 2017	Subaerial beach accretion and down-drift transport
Removal of westmost 20-m section	November 2017	Recovery of down-drift beaches
Mechanical sand extraction from the salient and placement in down-drift area	January 2018	No significant changes
Removal of eastward sections	March–April 2018	Sand spit formation

4.3. Spatial and Temporal Variability of Sand Volume

The volumetric changes were estimated using the high-resolution surveys from each site. The volume changes were computed for both the up-drift and down-drift areas taking as a reference the beach transect located at the center of the structure. Thus, the volume was integrated in the cross-shore (from the landward limit of each transect until $h > -1.5$ m) and alongshore direction (from the middle transect toward the down-drift and up-drift boundaries). The relative volume changes were estimated with respect to the reference volume obtained from the initial survey (without structures).

The up-drift (east, blue bars in Figure 9) area showed a volume increase in all sites, with a more significant increase with respect to the impermeable structures (Figure 9a,b). San Miguel beach showed a consistent accretion in the up-drift side, reaching more than 5000 m^3 in one year. Similar behavior was observed at El Teresiano but with a lower volume change (2500 m^3) over the same period. El Faro showed more sensitivity to seasonal changes, with a maximum accretion of 1500 m^3 but decreasing to less than 1000 m^3 by the end of the study period. On the other hand, the down-drift volume (red bars in Figure 9) showed a significant volume decrease with respect to the impermeable structures (Teresiano: 1000 m^3; San Miguel: 2000 m^3). The beach retreat in San Miguel was limited by a revetment that constrained the volume decrease. The permeable LCDB in El Faro, made of Reef Balls™, showed erosion at the down-drift side at about the same rate as the beach accretion during the first months. However, with the start of the cold-front season, the alongshore transport reversed, and the down-drift erosion decreased, turning into down-drift accretion at about the same volume as the up-drift accretion by the end of the study period (red bars in Figure 9c).

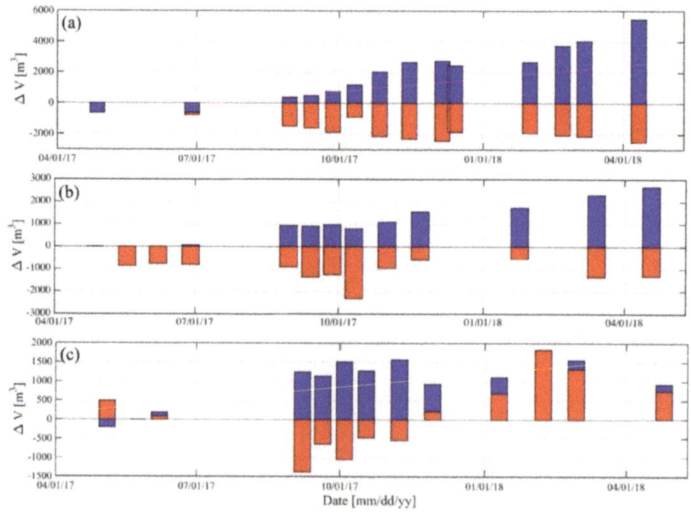

Figure 9. Relative volume change with respect to the LCDB central axis (east: blue; west: red) at (**a**) San Miguel, (**b**) El Teresiano, and (**c**) El Faro. Notice the difference in the vertical scales between panels.

The subaerial beach volume (z > −0.5 m) was also computed and evaluated at each beach profile (Figure 10). San Miguel showed a clear accretion/erosion trend in the eastern/western beach profiles (Figure 10a). This highlights the large impoundment produced by this LCDB. The negative impact on the down-drift side might have been enhanced by the drastic change in shoreline orientation at this site, large *B*/*S* ratio, and high freeboard elevation. El Teresiano showed a moderate down-drift effect, which was attenuated by the end of the study period due to changes in the structure's elevation (Figure 10b). However, with the start of the sea breeze season, the accretion increased in the up-drift area. The LCDB made of Reef Balls™ modules showed a negative effect in the down-drift side, followed by a sustained volume increase at the beginning of the cold-front season. The accretion induced by the permeable structure was moderate as compared with that of the impermeable ones.

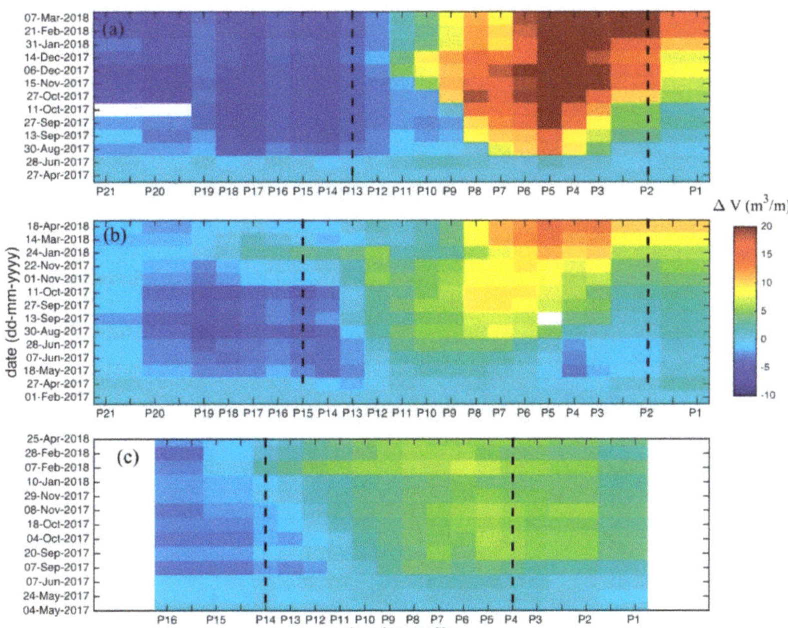

Figure 10. Subaerial volume change with respect to the reference beach survey (initial date, without LCDB) at (**a**) San Miguel, (**b**) El Teresiano, and (**c**) El Faro. The dashed vertical lines denote the limits of the LCDBs at each site. The date of the reference beach surveys corresponds to the initial date in the y-axis.

4.4. *Trends: Shoreline, Volume, and Seabed Elevation*

The trends of the shoreline position, emerged beach volume, and seabed elevation were estimated for the three sites during the study period. The rates of shoreline change were calculated using the shoreline (*z* = 0) extracted from the UAV flight data (San Miguel and El Teresiano) and DGPS surveys (El Faro). The rate of volume change and seabed elevation were calculated for the emerged section of each DGPS surveyed beach profile, considering *z* = −0.5 m to be the depth of the end of the emerged beach. The three beaches showed similar behavior with respect to trends in the shoreline position and in the emerged beach volume, with positive values in the eastern sections of the beach and less positive or negative values in the western sections. However, the magnitude and extension of the region with positive trends varied widely between sites.

San Miguel beach (Figure 11) showed the highest positive and negative shoreline trends of the three beaches. In the shadow area of the eastern section of the structure, the change rate was as high as 30 m/year. On the other hand, in the western section, the shoreline trends were negative—even in the shadow area of the structure—with values as low as −20 m/year in the down-drift section (Figure 11b). Although the original flight path considered approximately 350 m on the western side of the structure, there was clear evidence that the negative trend extended as far as 500 m in the down-drift side. The change in the emerged beach volume showed similar trends with values ranging between +39.1 and −8.3 m^3/m/year (Figure 11c). The beach presented a large accumulation of sediment in the eastern section of the study area. Figure 11d presents the rate of change in the elevation between 27 April 2017 (before the deployment of the structure) and 7 March 2018. In certain regions, the increase in beach elevation over the course of one year was larger than +1.5 m near the salient, whereas the decrease in the down-drift (western) locations reached values larger than −1.0 m.

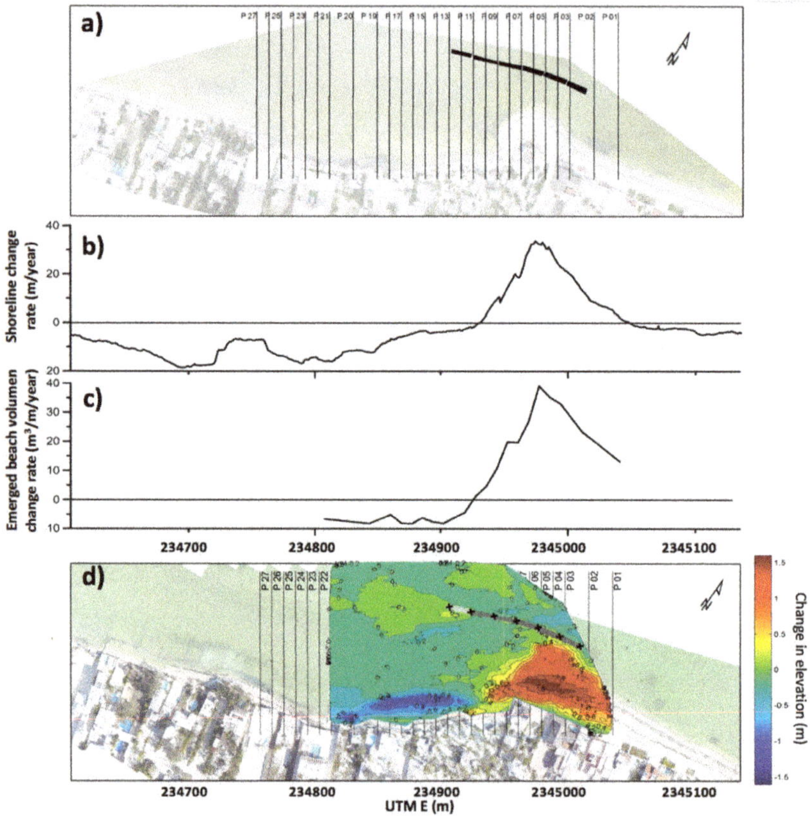

Figure 11. San Miguel beach (**a**) mosaic of the section of beach covered by the UAV flights with the location of the impermeable structure highlighted in black and parallel lines marking the high-resolution differential global positioning system (DGPS) profiles; (**b**) rate of change of the shoreline position during the study period obtained from UAV surveys; (**c**) rate of change of the emerged beach volume during the study period obtained from DGPS surveys; and (**d**) change in elevation between the last (7 March 2018) and first (27 April 2017, no structure) DGPS surveys.

El Teresiano (Figure 12) displayed a similar pattern, although (i) positive trends showing an advance in the shoreline position were below 20 m/year, (ii) the region where the shoreline trends were positive covered a wider section along the beach, and (iii) the negative trends also showed smaller values (>−5 m/year, Figure 12b). Regarding the volume, the trends varied between +10.8 and −3.5 m^3/m/year (Figure 12c). As presented in Figure 12d, the increase in elevation between 27 April and 18 April 2018 reached values similar to that in San Miguel (Figure 11d). However, the maximum decrease in elevation near the shoreline and in the emerged beach was found to be −0.4 m, less than half that in San Miguel.

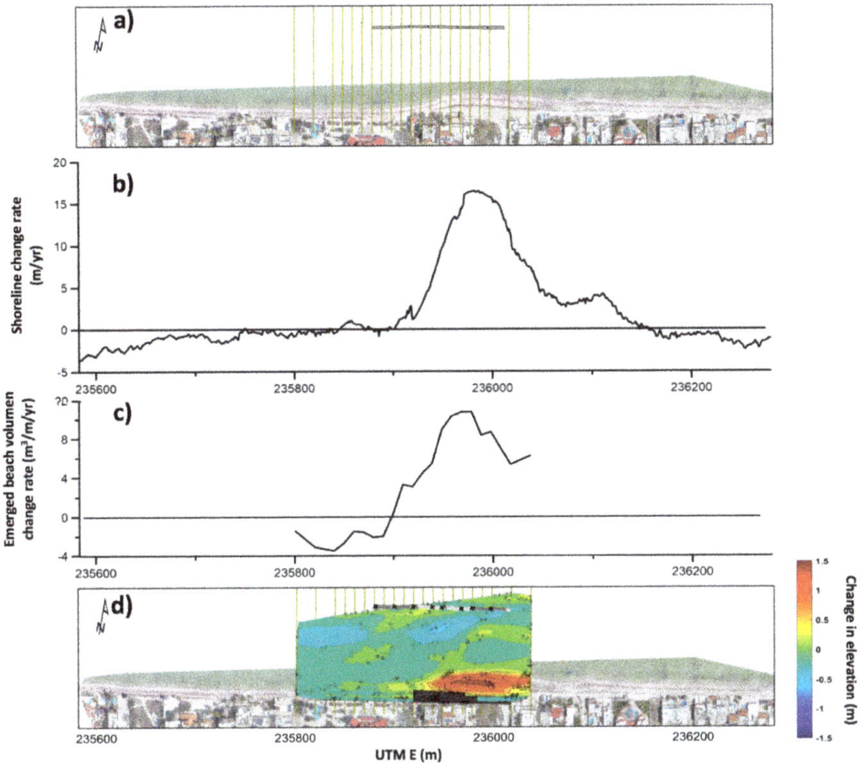

Figure 12. El Teresiano beach (**a**) mosaic of the section of beach covered by the UAV flights with the location of the impermeable structure highlighted in gray and parallel lines marking the high-resolution DGPS profiles; (**b**) rate of change of the shoreline position during the study period obtained from UAV surveys; (**c**) rate of change of the emerged beach volume during the study period obtained from DGPS surveys; and (**d**) change in elevation between the last (18 April 2018) and first (27 April 2017, no structure) DGPS surveys.

As for El Faro (Figure 13), the shoreline advance trends were positive on the lee side of the structure, with values below 6 m/year (Figure 13b). The trends in the emerged beach volume between 4 May 2017 and 25 April 2018 varied between +8.0 and −4.3 m^3/m/year. Changes in elevation after the deployment of the structure were less important at this site, with maximum increases of +0.6 m and decreases of −0.2 m.

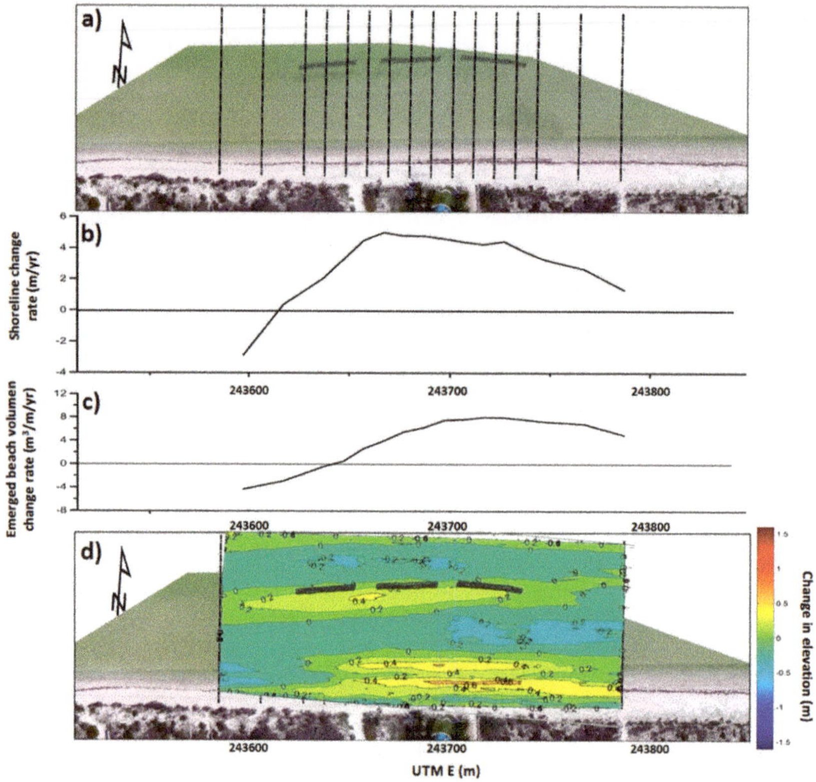

Figure 13. El Faro beach (**a**) aerial picture taken by the UAV flights with the location of the permeable structure highlighted in gray and parallel lines marking the high-resolution DGPS profiles; (**b**) rate of change of the shoreline position during the study period obtained from DGPS surveys; (**c**) rate of change of the emerged beach volume during the study period obtained from DGPS surveys; and (**d**) change in elevation between the last (25 April 2018) and the first survey (4 May 2017, no structure) DGPS surveys.

5. Discussion

A major problem with LCDB design is the difficulty of predicting the morphodynamic response on the lee side of the structure [7]. Empirical formulations, relating the distance to the tip of the salient X_{off}, the length of the structure B, and the distance to the undisturbed shoreline position S, predict a power curve relationship given by the following:

$$X_{off} = aB\left(\frac{B}{S}\right)^{b}$$

where the size of the salient $Y_s = S - X_{off}$ and the parameters a and b are those proposed by [18] and [17] for a single emergent breakwater (a = 0.68 and b = −1.22), reefs (a = 0.50 and b = −1.27), and islands (a = 0.40 and b = −1.52), respectively.

A distinct morphological response of the beach salient was observed in the three sites. The maximum shoreline salient size was measured in San Miguel, followed by El Teresiano and El Faro, respectively (Table 3 and Figures 11b, 12b and 13b). Empirical formulations (e.g., [17,18]) were employed, finding a satisfactory agreement for San Miguel (Table 3). This suggests that the beach response associated with a sand-filled geosystem can be predicted by the formulation developed by [17] for reefs. However, large differences between the observations and model predictions for El

Teresiano were found (Table 3). The latter can be ascribed to the continuous loss of sand in the middle sections of the geosystem (Figure 7b), which transformed a 140-m long breakwater into two 40-m long breakwaters separated by a 60 m gap. Applying the model developed by [17] to the case of El Teresiano for a 40-m breakwater predicted the size of the salient accurately (i.e., Y_s = 20 m). On the other hand, the shoreline salient on the lee side of the breakwater at El Faro was not predicted by any model, owing to the high transmissivity through the modules and the gaps between the sections. The formulations by [18] consistently underpredicted the salient size for impermeable LCDBs.

Table 3. Measurements and predictions of shoreline salient size Y_s at the three sites.

Location	Ys (m)	Hsu and Evans (1990)	Islands in Black and Andrews (2001)	Reefs in Black and Andrews (2001)
San Miguel	33	25	43	35
El Teresiano	16	34	61	50
El Faro	5	20	35	28

Sand-filled geosystems are highly vulnerable to vandalism in this area, and hence, the useful life of the structure can be drastically reduced. Furthermore, the tearing apart of the geotube sections plays an important role in beach evolution, leaving behind emptied geotextiles that are difficult to remove; moreover, the degradation of such geotextiles might have a negative ecological impact. It is not clear how the effect LCDBs with sand-filled geosystems can be correctly predicted if they are so prone to tearing and subsequent deflation.

6. Conclusions

A micro-tidal, sea-breeze-dominated beach's morphological responses to both impermeable and permeable low-crested detached breakwaters was analyzed based on high-resolution field observations. The following conclusions were found:

(1). Impermeable structures induced a salient located on the up-drift (east) side behind the structure, as well as down-drift erosion. The position and size of the salient is highly sensitive to the breakwater orientation with respect to the shoreline. Predicting the performance of sand-filled geosystems LCDBs is challenging owing to the high spatial and temporal variability of the structure.

(2). The permeable LCDB induced an order of magnitude less accretion and erosion than the impermeable structures, with a more symmetric salient on the lee side of the structure. The structures were found to remain stable during the study period. Based on observations, permeable (high transmissivity) structures are more suitable for the study area due to the persistent alongshore transport.

(3). The subaerial beach volume increase on the lee side of the LCDBs was strongly correlated with the beginning of the Central America cold surge season (i.e., October), owing to the combination of the maximum mean sea level and swell conditions. On the other hand, the freeboard elevation changes in the geotube sections showed no correlation with high energy conditions and hence can be ascribed to external factors.

(4). High spatial and temporal resolution measurements, combining DGPS and UAVs flights, were found to be important to explain far-field morphological changes.

(5). The design of sand-filled geosystems can be approximated with the formulation of [17] developed for reefs. On the other hand, the development of new formulations for high transmissivity structures, such as LCDB made of Reef Balls™ modules, is warranted.

Author Contributions: Conceptualization, A.T.-F., G.M., E.T.M., and E.O.; Original Draft Preparation, A.T.-F., G.M., E.T.M., and E.O.; Review and Editing of Manuscript, A.T.-F., G.M., E.T.M., E.O., and P.S.; Field Work, A.T.-F., G.M., E.T.M., and E.O.; Field Data Analysis, G.M., E.T.M., E.O., and A.T.-F.; and Funding Acquisition, P.S. and A.T.-F.

Funding: This research was funded by the Yucatán State Environmental Ministry of Yucatán and the Laboratorio Nacional de Resiliencia Costera (Project LN 293354). Additional financial support was provided by PAPIIT DGAPA UNAM (IN101218) and Investigación Científica Básica CONACYT (CB-2016-01-284819).

Acknowledgments: We are thankful for the use of Pix4D mapper Edu. NOAA is acknowledged for making Wave Watch III hindcast data available. The authors would like to thank Gonzalo Martín for IT support and José López Gonzalez and Juan Alberto Gómez for field support. Furthermore, a special thanks to the homeowners of the houses in San Miguel, El Teresiano, and El Faro for allowing the installation of the GPS base station. Anonymous reviewers provided fruitful comments that improved the manuscript.

Conflicts of Interest: The authors declare no conflict of interest.

References

1. Wong, P.P.; Losada, I.J.; Gattuso, K.-P.; Hinkel, J.; Kjattabi, A.; McInnes, K.L.; Saito, Y.; Sallenger, A. Coastal systems and low-liying areas. In *Climate Change 2014: Impactas, Adaptations, and Vulnerability, Part A: Global and Sectorial Aspects*; Contributions of Working Group II to the Fifth Assessment Report of the Intergovernmental Panel on Climate Change; Field, C.B., Barros, V., Dokken, D., Mach, K., Mastrandrea, M., Bilir, T., Chatterjee, M., Ebi, K., Estrada, Y., Genova, R., et al., Eds.; Cambridge University Press: Cambridge, UK; New York, NY, USA, 2014; pp. 361–409.
2. Dean, R.G. Beach Nourishment Theory and Practice. In *Advance Series on Ocean Engineering: Volume 18*; World Scientific: Singapore, 2003; p. 420.
3. Roberts, T.M.; Wang, P. Four-year performance and associated controlling factors of several beach nourishment projects along three adjacent barrier islands, West-Central Florida, USA. *Coast. Eng.* **2012**, *70*, 21–39.
4. Saponieri, A.; Valentini, N.; Di Risio, M.; Pasquali, D.; Damiani, L. Laboratory investigation on the evolution of a sandy beach nourishment protected by a mixed soft-hard system. *Water* **2018**, *10*, 1171. [CrossRef]
5. Borsje, B.W.; van Wesenbeeck, B.K.; Dekker, F.; Paalvast, P.; Bouma, T.J.; van Katwijk, M.M.; de Vries, M.B. How ecological engineering can serve in coastal protection. *Ecol. Eng.* **2011**, *37*, 113–122.
6. Lamberti, A.; Archetti, R.; Kramer, M.; Paphitis, D.; Mosso, C.; Di Risio, M. European experience of low crested structures for coastal management. *Coast. Eng.* **2005**, *52*, 841–866.
7. Pilarczyk, K.W. Design of low-crested (submerged) structures—An overview. In Proceedings of the 6th International Conference on Coastal and Port Engineering in Developing Countries, Colombo, Sri Lanka, 15–19 September 2003.
8. Van der Meer, J.W.; Briganti, R.; Zanuttigh, B.; Wang, B. Wave transmission and reflection at low-crested structures: Design formulae, oblique wave attack and spectral change. *Coast. Eng.* **2005**, *52*, 915–929.
9. Zyserman, J.A.; Johnson, H.K.; Zanuttigh, B.; Martinelli, L. Analysis of far-field erosion induced by low-crested rubble-mound structures. *Coast. Eng.* **2005**, *52*, 977–994.
10. Moschella, P.S.; Abbiati, M.; Aberg, P.; Airoldi, L.; Anderson, J.M.; Bacchiocchi, F.; Bulleri, F.; Dinesen, G.E.; Frost, M.; Garcia, E.; et al. Low-crested coastal defence structures as artificial habitats for marine life: Using ecological criteria in design. *Coast. Eng.* **2005**, *52*, 1053–1071.
11. Martinelli, L.; Zanuttigh, B.; Lamberti, A. Hydrodynamic and morphodynamic response of isolated and multiple low crested structrues: Experiments and simulations. *Coast. Eng.* **2006**, *53*, 363–379.
12. Isebe, D.; Azerad, P.; Bouchette, F.; Ivorra, B.; Mohammadi, B. Shape optimization of geotextile tubes for sandy beach protection. *Int. J. Numer. Methods Eng.* **2008**, *74*, 1262–1277.
13. Recio, J.; Oumeraci, H. Process based stability formulae for coastal structures made of geotextile sand containers. *Coast. Eng.* **2009**, *56*, 632–658.
14. Buccino, M.; Del Vita, I.; Calabrese, M. Engineering modeling of wave transmission of Reef Balls. *J. Waterw. Port Coast. Ocean Eng.* **2014**, *140*, 014010.
15. Das Neves, L.; Moreira, A.; Taveira-Pinto, F.; Lopez, M.I.; Veloso-Gomes, F. Performance of submerged nearshore sand-filled geosystems for coastal protection. *Coast. Eng.* **2015**, *95*, 147–159.
16. Hanson, H.; Kraus, N.C. Shoreline response to a single transmissive detached breakwater. In Proceedings of the Coastal Engineering Proceedings, Delft, The Netherlands, 2–6 July 1990.
17. Black, K.P.; Andrews, C.J. Sandy shoreline response to offshore obstacles part 1: Salient and tombolo geometry and shape. *J. Coast. Res.* **2001**, 82–93. [CrossRef]

18. Hsu, J.R.C.; Silvester, R. Accretion behind single offshore breakwater. *J. Waterw. Port Coast. Ocean Eng.* **1990**, *116*, 362–381.
19. Harris, L. Artificial Reef Structures for Shoreline Stabilization and Habitat Enhancement. In Proceedings of the 3rd Internatinal Surfing Reef Symposium, Raglan, New Zealand, 22–25 June 2003; pp. 176–178.
20. Alvarez, E.; Rubio, R.; Ricalde, H. Beach restoration with geotextile tubes as submerged breakwaters in Yucatan, Mexico. *Geotext. Geomembr.* **2007**, *25*, 233–241.
21. Enriquez, C.; Mariño-Tapia, I.J.; Herrera-Silveira, J.A. Dispersion in the Yucatan coastal zone: Implications for the red tide events. *Cont. Shelf Res.* **2010**, *30*, 127–137.
22. Torres-Freyermuth, A.; Puleo, J.A.; DiCosmo, N.; Allende-Arandia, M.E.; Chardón-Maldonado, P.; López-González, J.; Figueroa, B.; Ruiz de Alegría-Arzaburú, A.; Figlus, J.; Roberts, T.; et al. Nearshore hydrodynamics on a sea breeze dominated beach during intense wind events. *Cont. Shelf Res.* **2017**, *151*, 40–52.
23. Medellin, G.; Torres-Freyermuth, A.; Tomasicchio, G.R.; Francone, A.; Tereszkiewicz, P.A.; Luisito, L.; Palemón-Arcos, L.; López, J. Field and numerical study of resistance and resilience on a sea breeze dominated beach in Yucatan (Mexico). *Water* **2018**, *10*, 1806. [CrossRef]
24. Ojeda, E.; Appendini, C.M.; Mendoza, E.T. Storm-wave trends in Mexican waters of the Gulf of Mexico and Caribbean Sea. *Nat. Hazards Earth Syst. Sci.* **2017**, *17*, 1305–1317.
25. Appendini, C.M.; Salles, P.; Mendoza, E.T.; López, J.; Torres-Freyermuth, A. Longshore sediment transport on the northern coast of the Yucatan coast. *J. Coast. Res.* **2012**, *28*, 1404–1417.
26. Meyer-Arendt, K.J. Recreational landscape evolution along the north Yucatan coast. *Conf. Lat. Am. Geogr.* **1987**, *13*, 45–50.
27. Meyer-Arendt, K.J. Recreational development and shoreline modification along the north coast of Yucatán, Mexico. *Tour. Geogr.* **2001**, *3*, 87–104.
28. Medellín, G.; Mariño-Tapia, I.; Euán-Ávila, J. The influence of a seawall on postnourishment evolution in a sea-breeze-dominated microtidal beach. *J. Coast. Res.* **2015**, *31*, 1449–1458.
29. Tereszkiewicz, P.; McKinney, N.; Meyer-Arendt, K.J. Groins along the Northern Yucatán Coast. *J. Coast. Res.* **2018**, *34*, 911–919.
30. Barber, R.T. Reef Balls™: An Advanced Technique to Mimic Natural Reef System using Designed Artificial Reefs. In Proceedings of the International Conference on the Role of Divers in Protection of the Marine Environment, Kuwait City, Kuwait, 20–25 May 2000.
31. Turner, I.L.; Harley, M.D.; Drummond, C.D. UAVs for coastal surveying. *Coast. Eng.* **2016**, *114*, 19–24.
32. Thieler, E.R.; Himmelstoss, E.A.; Zichichi, J.L.; Ergul, A. *Digital Shoreline Analysis System (DSAS) version 4.0—An ArcGIS Extension for Calculating Shoreline Change*; US Geological Survey Open-File Report 2008-1278; US Geological Survey: Reston, VA, USA, 2009; p. 1278.

water

MDPI

Article

Application of a Coastal Vulnerability Index. A Case Study along the Apulian Coastline, Italy

Daniela Pantusa [1,*], Felice D'Alessandro [1], Luigia Riefolo [2], Francesca Principato [3] and Giuseppe Roberto Tomasicchio [1]

1 Innovation Engineering Department, University of Salento, I-73100 Lecce, Italy; felice.dalessandro@unisalento.it (F.D.); giuseppe.tomasicchio@unisalento.it (G.R.T.)
2 Department of Engineering, University of Campania "Luigi Vanvitelli", I-81031 Aversa, Italy; luigia.riefolo@unicampania.it
3 Department of Civil Engineering, University of Calabria, I-87036 Arcavacata di Rende, Italy; francesca.principato@unical.it
* Correspondence: daniela.pantusa@unisalento.it; Tel.: +39-0832-29-7795

Received: 4 July 2018; Accepted: 5 September 2018; Published: 10 September 2018

Abstract: The coastal vulnerability index (CVI) is a popular index in literature to assess the coastal vulnerability of climate change. The present paper proposes a CVI formulation to make it suitable for the Mediterranean coasts; the formulation considers ten variables divided into three typological groups: geological; physical process and vegetation. In particular, the geological variables are: geomorphology; shoreline erosion/accretion rates; coastal slope; emerged beach width and dune width. The physical process variables are relative sea-level change; mean significant wave height and mean tide range. The vegetation variables are width of vegetation behind the beach and posidonia oceanica. The first application of the proposed index was carried out for a stretch of the Apulia region coast, in the south of Italy; this application allowed to (i) identify the transects most vulnerable to sea level rise, storm surges and waves action and (ii) consider the usefulness of the index as a tool for orientation in planning strategies. For the case study presented in this work, the most influential variables in determining CVI are dune width and geomorphology. The transects that present a very high vulnerability are characterized by sandy and narrow beaches (without dunes and vegetation) and by the absence of Posidonia oceanica.

Keywords: coastal vulnerability index; climate change; sea level rise; storm surges; waves action; Mediterranean coasts

1. Introduction

The potentially massive impact of climate change on the world's coastal zones is globally recognized. The projections given by the Intergovernmental Panel on Climate Change (IPCC) indicate a globally averaged sea level rise (SLR) [1–3]; future storms are expected to become more intense with larger peak wind speeds; average wave conditions (wave height and direction) are also expected to be modified by climate change with frequent flooding events induced by severe overtopping and overwash. These climate change-driven variations in environmental forcing are likely to result in significant physical impact along the coasts [4]. The assessment of the vulnerability of coastal areas to climate change is therefore a topic of growing interest worldwide. There is an increasing need for a detailed knowledge of the wave conditions in order to design the coastal interventions [5–8]. In literature, there are different approaches and methodologies for the assessment of vulnerability and risk due to different types of hazard such as those related to climate change. A review of a multi-risk assessment for climate change impacts is discussed by [9] while in [10] are described the most commonly used methods to assess coastal vulnerability. According to [10] the methods to assess coastal vulnerability can be grouped into four main categories: index-based methods, indicator-based approach, GIS-based decision support systems, methods based on dynamic computer models.

Among the index-based methods, the coastal vulnerability index (CVI), originally presented by [11,12] is the first synthetic index to assess coastal vulnerability to climate change, in particular to SLR. The method uses a number of variables that affect coastal vulnerability and allows assessment of the relative coastal vulnerability of the different stretches of an investigated coastal area.

The CVI formulation proposed by [13], that modified the initial index proposed by [11,12], has been widely used for other applications and studies at different territorial scales [14–18]. In literature there are various applications of the CVI with modifications and integrations of physical parameters to adapt the index to the particular coastal area [19–26].

In this context, the present paper proposes a methodology and presents a case study for assessment of the physical vulnerability to coastal hazards; in particular, the paper proposes a CVI suitable for Mediterranean areas which considers 10 variables. Six variables replicate those proposed by [13], while the others 4 variables have been chosen to better characterize the Mediterranean coasts, especially the low-lying coastal areas.

Regarding the Mediterranean Sea, in literature there are several studies in relation to climate change. A review of climate change projections over the Mediterranean region based on global and regional climate change simulations is described in [27]. Storm surges and wind-waves constitute a further element of vulnerability and hazard for coastal areas in relation to erosion and dune breaching. Various studies have been carried out on this topic [28–32]. Projections of extreme storm surge levels along Europe have been investigated by [33]; the results obtained for the Mediterranean Sea predict changes mostly in the $\pm 5\%$ band, either positive or negative. As described by [33] these results are in line with the historical trends and there is consensus among different studies (e.g., [28,30,31]) for no changes, or even a decrease in the frequency and intensity of extreme events. Furthermore, as reported by [31] the increase of mean sea level and land subsidence, might significantly increase the hazard posed by coastal floods. Due to the concentration of economic activities in coastal areas, the European Environmental Agency [34] also consider the Mediterranean Sea region as one of the main climate change hotspots (i.e., one of the areas most responsive to climate change).

Regarding the application of CVI in the Mediterranean area, Doukakis [35] carried out a study to map the relative vulnerability of the western Pelleponese in Greece for a coastal length of about 50 km, while a recent application of the CVI index utilizing GIS technology is due to [36]. Another study carried out in Greece is that described by [37]; in this study, the classification of the southern coast of the Gulf of Corinth according to the sensitivity to the future sea level rise is attempted by applying the Coastal Sensitivity Index (CSI), with variable ranges specifically modified for the coastal environment of Greece, utilizing GIS technology. The results of the CVI application with an adaptation to the coast of Andalusia, Spain, are described in [38], a modified version of the CVI approach with an application to peninsular coastline of Spain is described in [4], while the Egyptian Mediterranean coast was examined for vulnerability to sea-level rise using the CVI by [39].

The study described in this paper, as mentioned above, presents an application of the CVI with the integration of four physical variables. The choice of these variables is due to the consideration that for low-lying coastal areas of the Mediterranean, which represent 46% of the Mediterranean coastline [40], coastal flooding generated by storm surge and wave-breaking represents one of the main destructive natural disasters in the Mediterranean [41].

In this direction, the four integrated variables, emerged beach width, dune width, width of vegetation behind the beach and Posidonia oceanica, are representative of the Mediterranean areas, and allow an evaluation of the ability of "natural systems" to dissipate the wave energy.

According to the CVI formulation proposed by [13], a relative vulnerability score is assigned to each variable based on the potential magnitude of its contribution to physical changes on the coast. Variables are ranked on a linear scale from 1–5 in order of increasing vulnerability and CVI values are classified in four different groups using percentiles as limits.

2. Methods and Data

2.1. Methods

The proposed CVI considers the following 10 variables:

1. Geomorphology
2. Coastal slope
3. Shoreline erosion/accretion rates
4. Emerged beach width
5. Dune width
6. Relative sea-level change
7. Mean significant wave height
8. Mean tide range
9. Width of vegetation behind the beach
10. Posidonia oceanica

In addition to the six variables described by [13], four new variables have been proposed: Emerged beach width; dune width; width of vegetation behind the beach and Posidonia oceanica. The new proposed variables, representative of the Mediterranean coast, allow us to evaluate the ability of "natural systems" to dissipate the wave energy. In particular, sandy beach-dune systems constitute the natural barrier protecting coastal areas against flooding due to storm surge and wave impacts. Furthermore, the effects of a well vegetated beach and seagrass Posidonia oceanica on wave energy have significant implications for coastal protection.

All variables have been divided into three typological groups: Geological, Physical process and Vegetation.

The Geological variables are:

- Geomorphology that expresses the relative erodibility of different landform types (e.g., rocky cliffs, sandy beaches) along the coast and requires information on the spatial distribution of landform types and their stability;
- Coastal slope that is an indicator of the relative vulnerability to inundation and of the potential rapidity of shoreline retreat;
- Shoreline erosion/accretion rates that allows to make assessments on the state of erosion or accretion;
- Emerged beach width is a variable related to the ability to dissipate wave energy; a wider beach has greater ability to dissipate the wave energy and therefore to reduce the impact of extreme events (e.g., storm surges).
- Dune width that represents an important variable for the conservation of the coastal zone, increasing its resilience [42–46]; in fact, dune can reduce the risk of erosion, as they constitute a reserve of sediment, and can counteract the risk of flooding of the hinterland.

The Physical process variables are:

- Relative sea level change that is derived from the time series of sea level records at each tide gauge stations along the coast; this variable includes both eustatic sea-level rise as well as regional sea-level rise due to isostatic and tectonic adjustments of the land surface;
- Mean significant wave height represents the potential for storm erosion. It is well known that storm erosion is directly related to the energy contained in storm waves and that the wave height has to be above a certain threshold (which depends on local conditions) to cause beach/dune erosion.
- Mean tide range that is linked to both permanent and episodic inundation hazards.

The Vegetation variables are:

- Width of vegetation behind the beach that is a variable related to vulnerability to storm events. In fact, the presence of vegetation is useful to dissipate wave energy and to reduce erosion in case of extreme events;
- Posidonia oceanica that is a marine phanerogam endemic to the Mediterranean basin which forms extended meadows along its coasts in a bathymetric surface to 0–40 m depth in clear waters [47]; several studies have shown the influence of these marine phanerogam on the nature and dynamics of coastal sediments [48–51]. The Posidonia oceanica colonize sandy beaches, [52] rocky substrates [53,54] and is generally absent in the depositional area of fine sediments such as the mouth of coastal rivers [55,56] in relation to the high rate of turbidity which causes a reduction of light penetration [57]. Recent papers [58,59] have confirmed that Posidonia oceanica forms a key coastal habitat, which plays a crucial role in the physical equilibrium of a large portion of the Mediterranean coasts. Therefore, the Posidonia oceanica is considered the resistance/resilience slime of the extreme weather events and SLR.

A stretch of coast is divided into a number of transects (or cross-sectional profiles of the beach) in order to assess its vulnerability. Each transect is characterized by a control area 0.5 km wide. Variables are ranked on a linear scale from 1–5 in order of increasing vulnerability.

The CVI is obtained by the square root of the product of the vulnerability scores assigned to each variable divided by the total number of variables:

$$\mathrm{CVI} = \sqrt{(\mathrm{a \cdot b \cdot c \cdot d \cdot e \cdot f \cdot g \cdot h \cdot i \cdot l})/10} \tag{1}$$

where a = Geomorphology, b = Coastal slope, c = Shoreline erosion/accretion rates, d = Emerged beach width, e = Dune width, f = Relative sea-level change, g = Mean significant wave height, h = Mean tide range, i = Width of vegetation behind the beach, l = Posidonia oceanica.

CVI values are classified in four different categories (low vulnerability, moderate vulnerability, high vulnerability and very-high vulnerability) using percentiles as limits.

2.2. Data

In the following, the data sources used to define the 10 variables are listed.

- Geomorphology has been derived from the map data (DTM) combined with the lithological map available on the Territorial Information System of Apulia Region, Sit-Apulia [60], for a 0.5 km grid cell;
- Coastal slope (%), has been estimated in accordance with [61]; it has been determined from a topographic and bathymetric grid extending 5 km landward and seaward of the shoreline. Elevation data have been obtained from the digital model available on the Sit-Apulia as gridded topographic and bathymetric elevation at 1 m vertical resolution for 8 m grid cells.
- Shoreline erosion/accretion rates (m/year), have been estimated as average values at the considered transect. The shorelines used were derived from the orthophotos available for years from 1992 to 2012, available on the Sit-Apulia [60];
- Emerged beach width (m), has been measured from the point where evidence of usual wave/tide impact ends to the point where vegetation or infrastructures begin. It has been evaluated considering the regional orthophotos, available on the Sit-Apulia [60];
- Dune width (m), has been evaluated considering the regional orthophotos, available on the Sit-Apulia [60]. Relative sea-level change (mm/year), has been derived considering the data reported for Mediterranean by National Oceanic and Atmospheric Administration, NOAA [62].
- Mean significant wave height (m), has been obtained with reference to the data of the Monopoli wave buoy belonging to the National Wave Metric Network (ISPRA—Institute for Environmental Protection and Research);

- Mean tide range (m), has been obtained from the European Environmental Agency- EEA data-base [63];
- Width of vegetation behind the beach (m), has been evaluated by considering the regional orthophotos, available on the Sit-Apulia [60]. The width of vegetation has been determined by clear and obvious signs of flora, indicated by the green area behind the beach; the measure was interrupted in the case of intersection with infrastructures such as roads, houses, etc.
- Posidonia oceanica (Boolean: presence/absence), has been evaluated on the basis of a research study carried out by Apulia Region [64].

Table 1 shows the range of vulnerability for the 10 variables. Regarding the variable geomorphology, the ranges of vulnerability considered are those proposed by [65]. The ranges of vulnerability for the variables of relative sea-level change, mean significant wave height and shoreline accretion/erosion rate have been chosen in agreement with those proposed by [37]. Regarding the variable coastal slope, the values have been chosen considering previous studies carried out for the Mediterranean coast [4,36,37]. In particular, the range chosen are those proposed by [4]. Regarding the variables emerged beach width, dune width and width of vegetation behind the beach, the ranges of vulnerability have been defined in consideration of the characteristics of the Italian and Mediterranean area [66]. Furthermore, the available data (regional orthophotos) made it possible to verify the similarity of these considered values with those typical of the Mediterranean environment. Finally, for the variable mean tide range, the ranges of vulnerability have been chosen in accordance with those proposed by [37] but the scores (linear scale from 1–5) are different. This assumption, in agreement with [13], is based on the concept that, in general, microtidal (tide range < 2.0 m) and macrotidal (tide range > 4.0 m) are characterized by high and low risk, respectively. The reasoning is based primarily on the potential influence of storms on coastal evolution, and their impact relative to the tide range. For example, on a tidal coast-line, there is only a 50 percent chance of a storm occurring at high tide. Thus, for a region with a 4.0 m tide range, a storm having a 3 m surge height is still up to 1 m below the elevation of high tide for half a tidal cycle. A microtidal coastline, on the other hand, is essentially always "near" high tide and therefore always at the greatest risk of inundation from storms [13]. Mediterranean area is a microtidal environment and the coast of Apulia has a tide range < 1 m. As such, the range of vulnerability, as mentioned above, are those proposed by [37] but the assigned scores are the inverse.

Other researchers (e.g., [12,67]) claimed the opposite; the large tidal range coast-lines were assigned a high-risk classification, and microtidal coasts received a low risk rating. The reasoning for this is that although a large tidal range dissipates wave energy, limiting beach or cliff erosion to a brief period of high tide, it also delineates a broad zone of intertidal area that will be most susceptible to inundation following long-term sea-level rise. Furthermore, the velocity of tidal currents depends partially on the tidal range. High tidal range is associated with stronger tidal currents that are capable of eroding and transporting sediment [67].

Table 1. Ranges of vulnerability for the considered variables.

Type Variables	Variables	Score				
		Very Low 1	**Low 2**	**Moderate 3**	**High 4**	**Very High 5**
Geologic	Geomorphology	Rocky, cliffed coasts	Medium cliffs, indented coasts	Low cliffs, alluvial plains	Cobble beaches, estuary, lagoon	Barrier beaches, sand beaches, salt marsh, mud flats, deltas, coral reefs
	Coastal slope (%)	>12	8–12	4–8	2–4	<2
	Shoreline Erosion/accretion (m/year)	>(+ 1.5)	(+1.5)–(+0.5)	(−0.5)–(+0.5)	(−0.5)–(−1.5)	<(−1.5)
	Emerged beach width (m)	>100	50–100	25–50	10–25	<10
	Dune width (m)	>100	75–100	50–75	25–50	<25
Physical process	Relative sea-level change (mm/year)	<1.8	1.8–2.5	2.5–3.0	3.0–3.4	>3.4
	Mean significant wave height (m)	<0.3	0.3–0.6	0.6–0.9	0.9–1.2	>1.2
	Mean tide range (m)	>0.8	0.6–0.8	0.4–0.6	0.2–0.4	<0.2
Vegetation	Width of vegetation behind the beach (m)	>400	200–400	100–200	50–100	<50
	Posidonia oceanica (Boolean: presence/absence)	Present				Absent

3. The Study Area

The proposed CVI index has been applied to a stretch of the coast of the Apulia Region, Southern Italy, between the marinas of Torre Canne and Villanova (Figure 1).

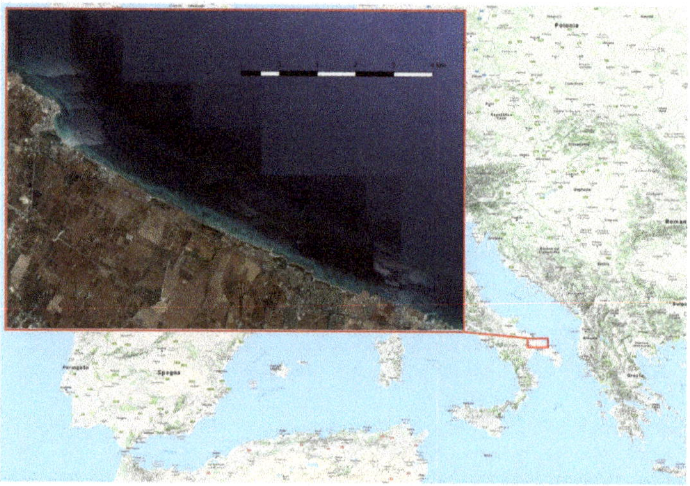

Figure 1. Case study area—Marinas of Torre Canne and Villanova, Apulia Region, Southern Italy.

Starting from the north, Torre Canne (Figure 2), the first stretch of about 7 km (up to Torre San Leonardo), which corresponds to the first 15 to 24 investigated transects, is characterized by beautiful beaches, which the Apulia Region has intended to protect by establishing with Regional Law No 31/2006, the Coastal Dunes Park. The park covers about 1.000 hectares. In the protected area there are many priority habitats, strongly threatened for their intrinsic fragility and for being located in areas at risk, but also habitats of the Community interest representative of the biogeographical reality of the Community territory. In the area there are beaches, consolidated dunes, retrodunal ponds and fossil dunes. The remaining 5 km of coastline, ranging from Torre San Leonardo to the port of Ostuni marina (Villanova) which make up the remaining nine transects of the study area, are jagged cliffs and consist of a series of coves with small beaches surrounded by Mediterranean vegetation.

Figure 2. Case study area and related transects.

The study area is mainly devoted to seaside tourism, and there are indeed many tourist accommodation facilities (hotels, resorts, etc.).

4. Results

A Geography Information System (GIS) platform has been used to better process the data. The stretch of coast has been divided into 24 transects; geomorphology includes very-high vulnerability and moderate vulnerability;coastal slope values are <2% varying between 0.75% (min) and 1.52% (max), while shoreline erosion/accretion rates is classified as very-low to very-high vulnerability. The value of the relative sea-level change is constant at low vulnerability and mean tide range is constant at very-high vulnerability. Mean significant wave height is constant at high vulnerability. Emerged beach width includes high vulnerability and moderate vulnerability, while dune width includes very-high vulnerability, high vulnerability and moderate vulnerability. Width of vegetation is classified as very-low to very-high vulnerability. Finally, Posidonia oceanica predominantly shows a low vulnerability since it is present in many transects.

The estimated minimum CVI value calculated for the case study is 30, while the maximum value is 300. The CVI mean is 123.40, the median is 84.85. The classes of CVI values have been divided into "low vulnerability" (green), "moderate vulnerability" (yellow), "high vulnerability" (orange) and "very-high vulnerability" (red) categories, respectively, on the basis of 25th, 50th, and 75th percentiles [13] as summarized in Table 2.

Table 2. Vulnerability categories.

Category	CVI Values
Low	<72.43
Moderate	72.43–84.85
High	84.85–163.62
Very-high	>163.62

Table 2 shows the vulnerability categories, while Table 3 shows the vulnerability value associated to each variable and the estimated CVI values for each transect (a = Geomorphology, b = Coastal slope, c = Shoreline erosion/accretion rates, d = Emerged beach width, e = Dune width, f = Relative sea-level change, g = Mean significant wave height, h = Mean tide range, i = Width of vegetation behind the beach, l = Posidonia oceanica).

Table 3. Vulnerability value associated to each variable and CVI values for each transect.

Transect	a	b	c	d	e	f	g	h	i	l	CVI Value	CVI Category
1	5	5	1	3	5	2	4	5	5	1	86.60	High
2	5	5	1	4	5	2	4	5	5	5	223.60	Very High
3	5	5	2	4	5	2	4	5	5	1	141.42	High
4	5	5	1	4	5	2	4	5	3	1	77.46	Moderate
5	5	5	1	3	4	2	4	5	5	1	77.46	Moderate
6	5	5	5	4	3	2	4	5	3	5	300.00	Very High
7	5	5	1	4	4	2	4	5	3	5	154.91	High
8	5	5	1	4	4	2	4	5	3	1	69.28	Moderate
9	5	5	4	4	5	2	4	5	2	1	126.49	High
10	5	5	2	4	4	2	4	5	3	1	97.98	High
11	5	5	2	4	4	2	4	5	2	1	80.00	Moderate
12	5	5	2	3	4	2	4	5	2	1	69.28	Moderate
13	5	5	3	3	3	2	4	5	2	1	73.48	Moderate
14	5	5	2	3	3	2	4	5	4	5	189.73	Very High
15	5	5	2	3	5	2	4	5	5	5	273.86	Very High
16	3	5	2	4	5	2	4	5	5	5	244.94	Very High
17	3	5	2	4	5	2	4	5	4	5	219.08	Very High
18	3	5	1	4	5	2	4	5	4	1	69.28	Moderate
19	3	5	2	3	5	2	4	5	4	1	84.85	Moderate
20	3	5	1	3	5	2	4	5	1	1	30.00	Low
21	3	5	2	3	5	2	4	5	4	1	84.85	Moderate
22	3	5	2	3	5	2	4	5	1	1	42.42	Low
23	3	5	1	3	5	2	4	5	5	1	67.08	Low
24	3	5	2	2	5	2	4	5	5	1	77.46	Moderate

Figure 3 shows a screenshot of the GIS page with the CVI values for the case study area.

Figure 3. CVI value for each transect.

5. Discussion

For the case study area, the most important variables are geomorphology, shoreline erosion and accretion rates, beach width, dune width, width of vegetation behind the beach and Posidonia oceanica, since the other variables are constant. As described above, the variable geomorphology mainly includes sandy beaches (very-high vulnerability) and low cliffs (moderate vulnerability), while shoreline erosion and accretion rates attain values between low vulnerability and moderate vulnerability.

The variable emerged beach width attains values between moderate vulnerability and very high vulnerability, as the beaches are not very large but rather narrow, while for the variable dune width in the area, the dune is present only in some transects characterized by no significant widths. Width of vegetation behind the beach is classified as very-low to very-high vulnerability while Posidonia oceanica is present in many transects.

In particular, transects from 1 to 3 are characterized by a sandy beach, with a low coastal slope and moderate emerged beach width; it should be noted that there is the absence of dune and vegetation, with constructions built close to the shoreline; for transect 1 and 3 the vulnerability is partly mitigated by the presence of Posidonia oceanica.

Transects 4 and 5 present a moderate vulnerability due to the presence of a modest dune, vegetation and Posidonia oceanica.

In transect 6 and 7, the value of vulnerability increases in relation to the absence of Posidonia oceanica, and for transect 6 a greater erosion is observed.

The transects from 8 to 13 present vulnerability that is predominantly moderate in relation, especially due to the presence of vegetation and Posidonia oceanica.

The transects from 14 to 17 (Torre San Leonardo), are characterized by the transition from sandy beach with dunes to low cliffs; in this stretch of coast there is an increase in vulnerability due to the absence of Posidonia oceanica and to the considerable reduction of the dune and the vegetation; this stretch is characterized also by an intensive land use with important population centers.

The transects from 18 to 24 are characterized by low and moderate vulnerability for the presence of Posidonia oceanica and vegetation.

It is important to highlight that in index-based methodologies, such as CVI, the availability of reliable and up-to-date databases is crucial. Variables like geomorphology and coastal slope can be considered stable since present negligible changes in time, while for the relative sea level change, mean tide range, and mean significant wave height, consolidated, international databases exist. For the variable shoreline erosion/accretion rates, reliable and up-to-date databases may not always be available [68]. For this variable, as well as for emerged beach width, dune width and width of vegetation behind the beach, it may be necessary to consider field measurements or the use of aerofotogrammetry. The field measurements are more precise but require significant investment and are limited in time and space. While the use of aerofotogrammetry is less accurate, it can be extended to larger areas. A more recent and useful tool for creating databases on these variables is the multispectral processing of images from satellites; these images present higher resolution with pixel processing and gradation based on photographic interpretation procedures (multispectral processing). This allows activating procedures for semi-automatic and/or automatic recognition of spatial elements. In fact, the increasing availability, resolution and spatial coverage of satellite imagery in recent years now provides a powerful alternative to derive reliable, global scale shoreline data. In this direction, in many recent studies the satellite images coupled with image processing techniques have been used [69–73].

6. Comparison between Two Index-Base Methods

In the present paper, the proposed CVI has been compared with the index-based method proposed by [37], postponing its verification to a later study by more complex process-based models (e.g., [74,75]).

The objective of the comparison is to evaluate, compared to an index similar in structure and range of vulnerability for each variable, the effects of further variables not yet taken into account. Specifically, it should be noted that the use of additional variables such as emerged beach width, dune width, width of vegetation behind the beach and posidonia oceanica, is based on the consideration that these variables can be useful to better characterize the Mediterranean coasts, especially the low-lying coastal areas.

Regarding the Coastal Sensitivity Index (CSI) proposed by [37], it uses the following physical variables: geomorphology, coastal slope, relative sea-level rise rate, shoreline erosion or accretion rate, mean tidal range and mean wave. This index was applied to the southern coast of the Gulf of Corinth, Greece. The obtained results are summarized in Table 4 and shown in Figure 4.

Table 4. Comparison between the proposed CVI and the CSI.

Transect	Proposed CVI	CSI Karymbalis et al. [37]
1	High	Moderate
2	Very high	Moderate
3	High	Very high
4	Moderate	Moderate
5	Moderate	Moderate
6	Very high	Very high
7	High	Moderate
8	Moderate	Moderate
9	High	Very high
10	High	Very high
11	Moderate	Very high
12	Moderate	Very high
13	Moderate	Very high
14	Very high	Very high
15	Very high	Very high
16	Very high	High
17	Very high	High
18	Moderate	Low
19	Moderate	High
20	Low	Low
21	Moderate	High
22	Low	High
23	Low	Low
24	Moderate	High

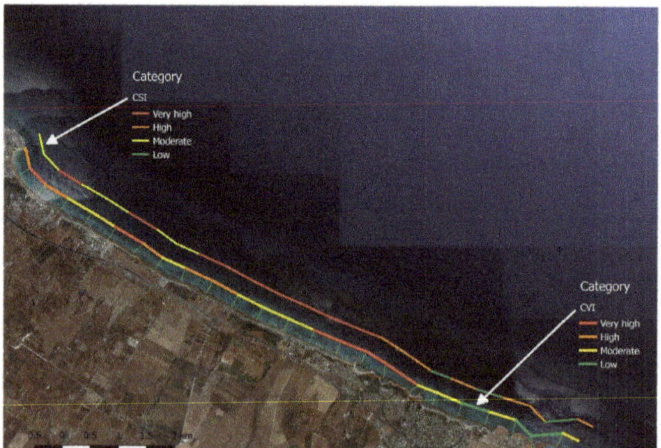

Figure 4. Proposed CVI and CSI values.

Generally, the two investigated methods show similar results. Some differences have been found likely due to the four variables proposed in the CVI to evaluate the ability of "natural systems" to dissipate the wave energy: emerged beach width; dune width; width of vegetation behind the beach and Posidonia oceanica.

In particular, for the cases of transects 9–13, the presence of Posidonia oceanica and the width of the vegetation determine a reduction of vulnerability in relation to the values obtained with the CSI. However, in the case of transects 16 and 17, the absence of Posidonia oceanica and dunes, the narrow beaches and the small width of the vegetation, determine an increase in vulnerability.

7. Conclusions

The CVI is a useful method for the assessment of the relative physical vulnerability of a stretch of coastline to the effects of climate change. The present paper proposes a CVI formulation, suitable for the Mediterranean coasts, that considers 10 variables and allows us to evaluate the vulnerability with respect to SLR, storm surges and waves action. In the following, the main conclusions of the study are as follows.

The tailored index CVI indicates that the dune width and the geomorphology are the most important drivers in building a regional index in terms of increasing the risk of flooding in this region. Regarding dune width, different transects are characterized by the absence of dunes or small-width dunes. Therefore, the relative vulnerability scores are mostly 5 (very-high vulnerability) and 4 (high vulnerability). For the geomorphology variable, most of the case study area consists of sandy beaches with a relative vulnerability score of 5 (very-high vulnerability).

On the contrary, width of vegetation behind the beach, shoreline erosion/accretion rates and Posidonia oceanica variables show a negligible influence. Width of vegetation behind the beach and shoreline erosion/accretion rates are classified as very-low to very-high vulnerability, while Posidonia oceanica is present in many transects.

The application of the proposed index shows the feasibility of the index and the possibility of using the CVI to make assessments on coastal vulnerability with respect to climate change.

The aim of the future research is to validate the proposed index by comparing it with the more complex numerical models in order to make the index a useful tool for coastal planning and management.

Author Contributions: D.P. Data curation, Formal analysis, Methodology, Validation, Writing-original draft, Writing-review & editing; F.D. Writing-review & editing; L.R. Validation (collaboration in the CVI application), Writing-original draft (collaboration in paper writing); F.P. Validation (collaboration in the CVI application); G.R.T. Conceptualization, Supervision, Writing-review & editing.

Funding: This work was funded by the Apulia Region (Italy), through the Regional Cluster Project "Eco-Smart Breakwater", Grant #S6LU5I7.

Conflicts of Interest: The authors declare no conflict of interest.

References

1. National Oceanic and Atmospheric Administration, NOAA. National Coastal Population Report, Population Trends from 1970 to 2020. 2013. Available online: http://stateofthecoast.noaa.gov/features/coastal-population-report.pdf (accessed on 12 February 2018).
2. IPCC. *Climate Change 2014: Impacts, Adaptation, and Vulnerability*; Field, C.B., Barros, V.R., Dokken, D.J., Mach, K.J., Mastrandrea, M.D., Bilir, T.E., Chatterjee, M., Ebi, K.L., Estrada, Y.O., Genova, R.C., et al., Eds.; Summaries, Frequently Asked Questions, and Cross-Chapter Boxes; A Contribution of Working Group II to the Fifth Assessment Report of the Intergovernmental Panel on Climate Change; World Meteorological Organization: Geneva, Switzerland, 2014; 190p, Available online: http://www.ipcc.ch/report/ar5/wg2/ (accessed on 12 February 2018)(In Arabic, Chinese, English, French, Russian, and Spanish).

3. Vousdoukas, M.I.; Mentaschi, L.; Voukouvalas, E.; Verlaan, M.; Jevrejeva, S.; Jackson, L.P.; Feyen, L. Global probabilistic projections of extreme sea levels show intensification of coastal flood hazard. *Nat. Commun.* **2018**, 1–12. [CrossRef] [PubMed]
4. López Royo, M.; Ranasinghe, R.; Jiménez, J.A. A Rapid, Low-Cost Approach to Coastal Vulnerability Assessment at a National Level. *J. Coast. Res.* **2016**, *32*, 932–945. [CrossRef]
5. United States Army, Corps of Engineers, Coastal Engineering Research Center (USACE). *Shore Protection Manual*; Department of the Army, Waterways Experiment Station, Corps of Engineers, Coastal Engineering Research Center: Washington, DC, USA, 1984.
6. Goda, Y. *Random Seas and Design of Maritime Structures*, 3rd ed.; Advanced Series on Ocean Engineering; World Scientific: Singapore, 2010.
7. Van der Meer, J.W. Rock Slopes and Gravel Beaches under Wave Attack. Ph.D. Thesis, Delft University of Technology, Delft, The Netherlands, 1988.
8. Tomasicchio, G.R.; D'Alessandro, F. Wave energy transmission through and over low crested breakwaters. *J. Coast. Res.* **2013**, *1*, 398–403. [CrossRef]
9. Gallina, V.; Torresan, S.; Critto, A.; Sperotto, A.; Glade, T.; Marcomini, A. A review of multi-risk methodologies for natural hazards: Consequences and challenges for a climate change impact assessment. *J. Environ. Manag.* **2016**, *168*, 123–132. [CrossRef] [PubMed]
10. ETC CCA. Methods for Assessing Coastal Vulnerability to Climate Change Technical Paper 1/2011. Available online: http://cca.eionet.europa.eu/docs/TP_1-2011 (accessed on 12 February 2018).
11. Gornitz, V.; White, T.W.; Cushman, R.M. Vulnerability of the East coast, U.S.A. to future sea level rise. *J. Coast. Res.* **1990**, *9*, 201–237.
12. Gornitz, V. Global coastal hazards from future sea level rise. *Glob. Planet. Chang.* **1991**, *89*, 379–398.
13. Thieler, E.R.; Hammar-Klose, E. *National Assessment of Coastal Vulnerability to Sea-Level Rise: Preliminary Results for U.S. Atlantic Coast*; Open-File Report 99-593; U.S. Geological Survey: Reston, VA, USA, 1999; 1 Sheet. Available online: http://pubs.usgs.gov/of/1999/of99-593/ (accessed on 12 February 2018).
14. Thieler, E.R.; Hammer-Klose, E.S. *National Assessment of Coastal Vulnerability to Sea-Level Rise: Preliminary Results for the US Pacific Coast*; Open File Report 00-178; United States Geological Survey (USGS): Woods Hole, MA, USA, 2000; 1p. Available online: https://www2.usgs.gov/science/cite-view.php?cite=1442 (accessed on 12 February 2018).
15. Thieler, E.R.; Williams, S.J.; Beavers, R. *Vulnerability of U.S. National Parks to Sea-Level Rise and Coastal Change*; U.S. Geological Survey Fact Sheet FS 095-02; U.S. Geological Survey: Reston, VA, USA, 2002; 2p. Available online: http://pubs.usgs.gov/fs/fs095-02/ (accessed on 12 February 2018).
16. Pendleton, E.A.; Thieler, E.R.; Williams, S.J. Coastal Vulnerability Assessment of War in the Pacific National Historical Park to Sea-Level Rise. U.S. Geological Survey Open-File Report 2005a-1056. 2005. Available online: https://pubs.usgs.gov/of/2005/1056/ (accessed on 12 February 2018).
17. Pendleton, E.A.; Thieler, E.R.; Williams, S.J. Coastal Vulnerability Assessment of National Park of American Samoa to Sea-Level Rise. U.S. Geological Survey Open-File Report 2005b-1055. 2005. Available online: https://pubs.usgs.gov/of/2005/1055/ (accessed on 12 February 2018).
18. Pendleton, E.A.; Thieler, E.R.; Williams, S.J. Coastal Vulnerability Assessment of Channel Islands National Park (CHIS) to Sea-Level Rise. U.S. Geological Survey Open-File Report 2005-1057. 2005. Available online: https://pubs.usgs.gov/of/2005/1057/ (accessed on 12 February 2018).
19. Abuodha, P.A.; Woodroffe, C.D. *Assessing Vulnerability of Coasts to Climate Change: A Review of Approaches and Their Application to the Australian Coast*; Australian National Centre for Ocean Resources and Security University of Wollongong: Wollongong, Australia, 2007; 458p, Available online: http://citeseerx.ist.psu.edu/viewdoc/download?doi=10.1.1.461.2048&rep=rep1&type=pdf (accessed on 12 February 2018).
20. Srinivasa Kumar, T.; Mahendra, R.S.; Nayak, S.; Radhakrishnan, K.; Sahu, K.C. Coastal Vulnerability Assessment for Orissa State, East Coast of India. *J. Coast. Res.* **2010**, *26*, 523–534. [CrossRef]
21. Mohd Fauzi, M.; Lee Hin, L.; Mohd Kamarul Huda, S. Coastal Vulnerability Assessment towards Sustainable Management of Peninsular Malaysia Coastline. *Int. J. Environ. Sci. Dev.* **2014**, *5*, 533–538.
22. Loinenak, F.; Hartoko, A.; Muskananfola, M.R. Mapping of coastal vulnerability using the coastal vulnerability index and geographic information system. *Int. J. Technol.* **2015**, *5*, 819–827. [CrossRef]

23. Ozyurt, G. Vulnerability of Coastal Areas to Sea Level Rise: A Case of Study on Göksu Delta. Master's Thesis, Graduate School of Natural and Applied Sciences of Middle-East Technical University, Ankara, Turkey, 2007. Available online: http://etd.lib.metu.edu.tr/upload/12608146/index.pdf (accessed on 12 February 2018).
24. Ozyurt, G.; Ergin, A.; Esen, M. Indicator Based Coastal Vulnerability Assessment Model to Sea Level Rise. In Proceedings of the Seventh International Conference on Coastal and Port Engineering in Developing Countries COPEDEC VII "Best Practices in the Coastal Environment", Dubai, UAE, 24–28 February 2008; Available online: https://icce-ojs-tamu.tdl.org/icce/index.php/icce/article/view/1905/pdf_144 (accessed on 12 February 2018).
25. Palmer, B.J.; Van der Elst, R.; Mackay, F.; Mather, A.A.; Smith, A.M.; Bundy, S.C.; Thackeray, Z.; Leuci, R.; Parak, O. Preliminary coastal vulnerability assessment for KwaZulu-Natal, South Africa. *J. Coast. Res.* **2011**, *64*, 1390–1395.
26. Kantamaneni, K.; Du, X.; Aher, S.; Singh, R.M. Building Blocks: A Quantitative Approach for Evaluating Coastal Vulnerability. *Water* **2017**, *9*, 905. [CrossRef]
27. Giorgi, F.; Lionello, P. Climate change projections for the Mediterranean region. *Glob. Planet Chang.* **2008**, *63*, 90–104. [CrossRef]
28. Marcos, M.; Jorda, G.; Gomis, D.; Perez, B. Changes in storm surges in southern Europe from a regional model under climate change scenarios. *Glob. Planet Chang.* **2011**, *77*, 116–128. [CrossRef]
29. Krestenitis, Y.N.; Androulidakis, Y.S.; Kontos, Y.N.; Georgakopoulos, G. Coastal inundation in the north-eastern Mediterranean coastal zone due to storm surge events. *J. Coast. Conserv.* **2011**, *15*, 353–368. [CrossRef]
30. Jordà, G.; Gomis, D.; Álvarez-Fanjul, E.; Somot, S. Atmospheric contribution to Mediterranean and nearby Atlantic sea level variability under different climate change scenarios. *Glob. Planet Chang.* **2012**, *80–81*, 198–214. [CrossRef]
31. Conte, D.; Lionello, P. Characteristics of large positive and negative surges in the Mediterranean Sea and their attenuation in future climate scenarios. *Glob. Planet Chang.* **2013**, *111*, 159–173. [CrossRef]
32. Vousdoukas, M.I.; Voukouvalas, E.; Mentaschi, L.; Dottori, F.; Giardino, A.; Bouziotas, D.; Bianchi, A.; Salamon, P.; Feyen, L. Developments in large-scale coastal flood hazard mapping. *Nat. Hazards Earth Syst. Sci.* **2016**, *16*, 1841–1853. [CrossRef]
33. Vousdoukas, M.I.; Voukouvalas, E.; Annunziato, A.; Giardino, A.; Feyen, L. Projections of extreme storm surge levels along Europe. *Clim. Dyn.* **2016**, *47*, 3171–3190. [CrossRef]
34. European Environmental Agency (EEA). Mediterranean Sea Region Briefing—The European Environment—State and Outlook 2015. Available online: https://www.eea.europa.eu/soer-2015/countries/mediterranean (accessed on 12 February 2018).
35. Doukakis, E. Coastal vulnerability and risk parameters. *Eur. Water* **2005**, *11/12*, 3–7.
36. Gaki-Papanastassiou, K.; Karymbalis, E.; Serafim, P.; Zouva, C. Coastal vulnerability assessment to sea-level rise based on geomorphological and oceanographical parameters: The case of Argolikos Gulf, Peloponnese, Greece. *Hell. J. Geosci.* **2010**, *45*, 109–122.
37. Karymbalis, E.; Chalkias, C.; Chalkias, G.; Grigoropoulou, E.; Manthos, G.; Ferentinou, M. Assessment of the Sensitivity of the Southern Coast of the Gulf of Corinth (Peloponnese, Greece) to Sea-level Rise. *Cent. Eur. J. Geosci.* **2012**, *4*, 561–577. [CrossRef]
38. Ojeda Zújar, J.; Álvarez Francoso, J.I.; Martín Cajaraville, D.; Fraile Jurado, P. El uso de las Tecnologías de la información geográfica para el cálculo del índice de vulnerabilidad costera (CVI) ante una potencial subida del nivel del mar en la costa Andaluza (España). *GeoFocus* **2009**, *9*, 83–100.
39. Hereher, M.E. Coastal Vulnerability Assessment for Egypt's Mediterranean Coast. *Geomat. Nat. Hazards Risk* **2013**, *6*, 342–355. [CrossRef]
40. UNEP/MAP. *State of the Mediterranean Marine and Coastal Environment*; UNEP/MAP–Barcelona Convention: Athens, Greece, 2012; Available online: http://www.undp.org/content/dam/turkey/docs/Publications/EnvSust/akdenizrapor.pdf (accessed on 12 February 2018).
41. Sánchez-Arcilla, A.; Mosso, C.; Sierra, J.P.; Mestres, M.; Harzallah, A.; Senouci, M.; El Raey, M. Climatic drivers of potential hazards in Mediterranean coasts. *Reg. Environ. Chang.* **2011**, *11*, 617–636. [CrossRef]
42. Tomasicchio, G.R.; Sanchez Arcilla, A.; D'Alessandro, F.; Ilic, S.; James, M.; Fortes, C.J.E.M.; Sancho, F.; Schüttrumpf, H. Large-scale flume experiments on dune erosion processes. *J. Hydraul. Res.* **2011**, *49*, 20–30. [CrossRef]

43. Tomasicchio, G.R.; D'Alessandro, F.; Barbaro, G. Composite modelling for large-scale experiments on wave-dune interactions. *J. Hydraul. Res.* **2011**, *49*, 15–19. [CrossRef]
44. Sancho, F.; Abreu, T.; D'Alessandro, F.; Tomasicchio, G.R.; Silva, P.A. Surf hydrodynamics under collapsing coastal dunes. *J. Coast. Res.* **2011**, *64*, 144–148.
45. D'Alessandro, F.; Tomasicchio, G.R.; Musci, F.; Ricca, A. Dune Erosion Physical, Analytical and Numerical Modelling. In Proceedings of the 33rd International Conference on Coastal Engineering, Santander, Spain, 1–6 July 2012; Printed Online, Paper n. Sediment.32, 2012. Available online: https://icce-ojs-tamu.tdl.org/icce/index.php/icce/article/view/6900/pdf (accessed on 12 February 2018).
46. D'Alessandro, F.; Tomasicchio, G.R. Wave-dune interaction and beach resilience in large-scale physical model tests. *Coast. Eng.* **2016**, *116*, 15–25. [CrossRef]
47. Pergent, G.; Pergent-Martini, C.; Boudouresque, C.F. Utilisation de l'herbier à Posidonia oceanica comme indicateur biologique de la qualité du milieu littoral en Méditerranée: État des connaissances. *Mesogée* **1995**, *54*, 3–27.
48. Boudouresque, C.F.; Jeudy de Grissac, A. L'herbier à Posidonia oceanica en Méditerranée: Les interactions entre le plant et le sédiment. *J. de Recherche Océanographique* **1983**, *8*, 99–122.
49. Jeudy de Grissac, A.; Boudouresque, C.F. Rôles des herbiers de phanérogames marines dans les mouvements des sédiments côtiers: Les herbiers à Posidonia oceanica. In Proceedings of the Colloque franco-japonais Oceanographie, Marseille, France, 16–21 September 1985; Volume 1, pp. 143–151.
50. Blanc, J.J.; Jeudy de Grissac, A. Reflexion sur la regression des herbiers a Posidonies (Departements du var e des bouches di Rhone). In *International Workshop on Posidonia Oceanica Meadows*; GIS Posidonie Publisher: Carry-le-Rouet, France, 1989; Volume 2, pp. 273–285.
51. Basterretxea, G.; Orfila, A.; Jordi, A.; Casas, B.; Lynett, P.; Liu, P.L.F.; Duarte, C.M.; Tintor, J. Seasonal Dynamics of a Microtidal Pocket Beach with Posidonia oceanica Seabeds (Mallorca, Spain). *J. Coast. Res.* **2004**, *20*, 1155–1164. [CrossRef]
52. Cavazza, W.; Immordino, F.; Moretti, L.; Peirano, A.; Pironi, A.; Ruggiero, F. Sedimentological parameters and seagrass distributions as indicators of anthropogenic coastal degradation at Monterosso Bay (Ligurian Sea, NW Italy). *J. Coast. Res.* **2000**, *16*, 295–305.
53. De Falco, G.; Molinaroli, E.; Baroli, M.; Bellacicco, S. Grain size and compositional trends of sediments from Posidonia oceanica meadows to beach shore, Sardinia, Western Mediterranean. *Estuar. Coast. Shelf Sci.* **2003**, *58*, 299–309. [CrossRef]
54. Di Carlo, G.; Badalamenti, F.; Jensen, A.C.; Kock, E.W.; Riggio, S. Colonisation process of vegetative fragments of Posidonia oceanica (L.) Delile on rubble mounds. *Mar. Biol.* **2005**, *147*, 1261–1270. [CrossRef]
55. Pasqualini, V.; Pergent-Martini, C.; Clabaut, P.; Pergent, G. Mapping of Posidonia oceanica using aerial photographs and side scan sonar: Application off the island of Corsica. *Estuar. Coast. Shelf Sci.* **1998**, *47*, 359–368. [CrossRef]
56. De Falco, G.; Baroli, M.; Murru, E.; Piergallini, G.; Cancemi, G. Sediment analysis evidences two different depositional phenomena influencing seagrass distribution in the Gulf of Oristano (Sardinia—Western Mediterranean). *J. Coast. Res.* **2006**, *22*, 1043–1050. [CrossRef]
57. De Falco, G.; Simeone, S.; Baroli, M. Management of Beach-Cast Posidonia oceanica Seagrass on the Island of Sardinia (Italy, Western Mediterranean). *J. Coast. Res.* **2008**, *24*, 69–75. [CrossRef]
58. Boudouresque, C.F.; Bernard, G.; Bonhomme, P.; Charbonnel, E.; Diviacco, G.; Meinesz, A.; Pergent, G.; Pergent-Martini, C.; Ruitton, S.; Tunesi, L. *Protection and Conservation of Posidonia Oceanica Meadows*; RaMoGe and RAC/SPA: Tunis, Tunisia, 2012.
59. Vacchi, M.; De Falco, G.; Simeone, S.; Montefalcone, M.; Morri, C.; Ferrari, M.; Bianchi, C.N. Biogeomorphology of the Mediterranean Posidonia oceanica seagrass meadows. *Earth Surf. Process. Landf.* **2017**, *42*, 42–54. [CrossRef]
60. Sit-Apulia. Available online: http://www.sit.puglia.it/ (accessed on 22 January 2018).
61. Pendleton, E.A.; Thieler, E.R.; Williams, S.J. Coastal Vulnerability Assessment of Golden Gate National Recreation Area to Sea-Level Rise. U.S. Geological Survey Open-File Report 2005-1058. Available online: https://pubs.usgs.gov/of/2005/1058/ (accessed on 12 February 2018).
62. NOAA. Available online: https://www.star.nesdis.noaa.gov/sod/lsa/SeaLevelRise (accessed on 26 February 2018).
63. EEA Database. Available online: https://www.eea.europa.eu (accessed on 26 February 2018).

64. Report Regione Apulia. Inventario e Cartografia delle Praterie di Posidonia nei Compartimenti Marittima di Manfredonia, Molfetta, Bari, Brindisi, Gallipoli e Taranto. 2005. Available online: http://www.pugliamare.it/studi/Inventario%20e%20cartografia%20delle%20praterie%20di%20posidonia/RELAZIONE%20ANALITICA.pdf (accessed on 12 February 2018).
65. Hammar-Klose, E.S.; Pendleton, E.A.; Thieler, E.R.; Williams, S.J. Coastal Vulnerability Assessment of Cape Cod National Seashore to Sea-Level Rise. USGS Open-File Report 02-233. Available online: https://pubs.usgs.gov/of/2002/of02-233/caco.htm (accessed on 12 February 2018).
66. Eurosion Project. Available online: http://www.eurosion.org/ (accessed on 27 February 2018).
67. Tragaki, A.; Gallousi, C.; Karymbalis, E. Coastal hazard vulnerability assessment based on geomorphic, oceanographic and demographic parameters: The case of the Peloponnese (Southern Greece). *Land* **2018**, *7*, 56. [CrossRef]
68. Tomasicchio, G.R.; Francone, A.; D'Alessandro, F.; Medellin, G.; Torres-Freyermuth, A.; Barbaro, G. Morphodynamic model to simulate shoreline evolution at any coastal mound. In Proceedings of the 36th International Conference on Coastal Engineering, Baltimore, MD, USA, 30 July–3 August 2018. in press.
69. Luijendijk, A.; Hagenaars, G.; Ranasinghe, R.; Baart, F.; Donchyts, G.; Aarninkhof, F. The State of the World's Beaches. *Sci. Rep.* **2018**, 1–11. [CrossRef] [PubMed]
70. Liu, H.; Jezek, K.C. Automated extraction of coastline from satellite imagery by integrating Canny edge detection and locally adaptive thresholding methods. *Int. J. Remote Sens.* **2004**, *25*, 937–958. [CrossRef]
71. García-Rubio, G.; Huntley, D.; Kingston, K.; Esteves, L. Shoreline identfication using satellite images. In *Proceedings of Coastal Dynamics 2009*; World Scientific: Singapore, 2009; pp. 1–10. [CrossRef]
72. Goncalves, G.; Duro, N.; Sousa, E.; Figueiredo, I. Automatic extraction of tide-coordinated shoreline using open source software and landsat imagery. In Proceedings of the ISPRS—The International Archives of the Photogrammetry, Remote Sensing and Spatial Information Sciences, Berlin, Germany, 11–15 May 2015; Volume XL-7/W3, pp. 953–957.
73. Pardo Pascual, J.E.; Almonacid Caballer, J.; Ruiz Fernández, L.Á.; Palomar-Vázquez, J. Automatic extraction of shorelines from Landsat TM and ETM+ multi-temporal images with subpixel precision. *Remote Sens. Environ.* **2012**, *123*, 1–11. [CrossRef]
74. Giardino, A.; Nederhoff, K.; Vousdoukas, M. Coastal hazard risk assessment for small islands: Assessing the impact of climate change and disaster reduction measures on Ebeye (Marshall Islands). *Reg. Environ. Chang.* **2018**, 1–12. [CrossRef]
75. Roelvink, J.A.; Reniers, A.; Van Dongeren, A.; Van Thiel de Vries, J.; McCall, R.; Lescinksi, J. Modelling storm impacts on beaches, dunes and barrier islands. *J. Coast. Eng.* **2009**, *56*, 1133–1152. [CrossRef]

water

MDPI

Article

A Systematic Review of Coastal Vulnerability Assessment Studies along Andhra Pradesh, India: A Critical Evaluation of Data Gathering, Risk Levels and Mitigation Strategies

Komali Kantamaneni [1,*], N.N.V. Sudha Rani [2], Louis Rice [3], Koyel Sur [4,5], Menaha Thayaparan [6], Udayangani Kulatunga [6], Rajshree Rege [7], Komali Yenneti [8] and Luiza C. Campos [9]

1 Warsash School of Maritime Science and Engineering, Solent University, East Park Terrace, SO14 OYN Southampton, UK
2 The Interdisciplinary Programme in Climate Studies Centre, Indian Institute of Technology, Mumbai 400076, India; sudha.nnvs@gmail.com
3 Centre for Architecture and Built Environment Research, University of the West of England, BS16 1QY Bristol, UK; Louis.Rice@uwe.ac.uk
4 Space Applications Centre, Indian Space Research Organisation (ISRO), Gujarat 380015, India; koyelsur3@aau.in
5 Anand Agriculture University, Anand 388110, India
6 Department of Building Economics, Faculty of Architecture, University of Moratuwa, Katubedda 10400, Sri Lanka; mthayaparan@uom.lk (M.T.); ukulatunga@uom.lk (U.K.)
7 Physics, maths and Science-Physics and Astronomy group, Tarrant County College, NW campus, Fort worth, TX 76179, USA; rv_rajshree@yahoo.co.in
8 Faculty of Built Environment, University of New South Wales, Sydney 2052, Australia; k.yenneti@unsw.edu.au
9 Department of Civil, Environment & Geomatic Engineering, University College London (UCL), Gower Street, WC1E 6BT London, UK; l.campos@ucl.ac.uk
* Correspondence: Komali.kantamaneni@solent.ac.uk

Received: 3 October 2018; Accepted: 6 February 2019; Published: 23 February 2019

Abstract: The establishment and alteration of any coastal feature is largely dependent upon complex hydrological and geomorphologic processes. Therefore, understanding hazard factors and threat risk level is crucial for mitigating risk in coastal zones. This study examines coastal vulnerability factors and their influence along the Coastal Andhra Pradesh (CAP) region in India. CAP has been exposed to frequent hydrological and meteorological hazards due to variations in the geographical, geological, and bathymetric characteristics. Despite substantial vulnerabilities, the risk to the coastline of Andhra Pradesh has not been rigorously evaluated. The current research systematically reviews the drivers and effects of hazards and vulnerabilities in CAP. Findings indicate that urban cities have a considerably higher risk of cyclones and floods due to their locations on the Bay of Bengal tectonic plate, the topology of this coastal region, and higher population density. The study revealed that the Coastal Vulnerability Index (CVI) data along CAP is mostly gathered using low-resolution satellite data and/or field observation surveys. The study further revealed that there are very few existing mitigation strategies developed or discussed within the obtained results. However, more accurate data gathering techniques for coastal vulnerability factors are available such as Unmanned Aerial Vehicles (UAVs): Air-borne and LiDAR sensors, which provide very high resolution data and low-cost accessibility to physically inaccessible places, making them suitable for vulnerability data collection in coastal locations. These findings are useful for stakeholders seeking to reduce or ameliorate the impact of coastal disasters and their impacts on the CAP economy, environment, and population. The study further helps to reduce the existing shortcomings in the assessment techniques used previously.

Keywords: coastal vulnerability; natural hazards; disasters; coastal Andhra Pradesh; Unmanned Aerial Vehicles

1. Introduction

Coastal environments are particularly vulnerable to the effects of disasters; partly due to denser urbanization and human populations and related economic activities such as agriculture, aquaculture, tourism, industries, trades and transportation in these locations [1–5]. Coastal areas tend to be relatively highly urbanized with higher concentrations of human populations [6,7]. Approximately 40% of the global human population lives within 60 km of the coast [8] and more than 25% of the human population in India lies within 50 km of the coast [9]. Natural processes, particularly those affected by climate change, combined with human activities, are becoming significant factors for coastal vulnerabilities. Risks at coastal regions in India are rapidly increasing; major threats for vulnerability are mean sea level rise, extreme events like flash floods, and cyclone induced storm surges [10–13]. Increased green-house gas emissions, predominantly from human activities, are predicted to expedite climate change and sea level rises that are already making coastal zones vulnerable in the 21st century [14–18]. Warming oceans lead to sea level rises, which in turn provide a source of thermal expansion of sea water, contributing further to intensified and more frequent cyclone-induced storm surges [19–22] addressed the potential climate change impacts owing to drought, floods and cyclone events with more intensity and asserted that India has very high risk levels (class 9 among risk class 0–10). As a result of the changes related to climate change, there are significant challenges for assessing coastal vulnerability and subsequent adaptation and mitigation strategies [23,24]. Systematic coastal vulnerability assessments [25–28] are essential for managing coastal threats. According to Indian Coastal Zone (CRZ) regulations, the buffer zones, also known as active zones, usually cover a region of 500 m from the shoreline; these zones are more usually focused on coastal vulnerability studies. However, coastal areas up to 100 km perpendicular from the shoreline should be considered as vulnerable areas for assessment studies. Coastal vulnerability resistance developments and strategies done to date are limited; they are only applicable to the coastal buffer zone (yet the effects of disasters impact on areas beyond this zone) and mostly concern coastal geomorphology issues but do not pertain to other extreme events such as cyclones and inundation [29–31].

Comprised of different geological, ecological, biological, urban and socio-economic features, the Indian coast is subject to varying degrees of exposure to multiple hazards. In the current scenario, the environmental stresses on coastal zones are increasing significantly in terms of social, physical and economic variables. Social parameters such as population density, drinking water demands owing to over-pumping of fresh-water, communication networks, roads with transportation, drainage, infrastructure, agriculture, aquaculture and industrialization are the prime factors that are providing harmful feedback loops to existing coastal vulnerability. Physical parameters such as sea level rise, slope, cyclones and storm surges are the natural contributing factors (or hazards) for increasing vulnerabilities along the coast. Sand mining is another crucial parameter in terms of coastal vulnerability along this coast. The increasing levels of sand mining along CAP will further result in ecosystem damage in terms of habitat, erosion, riverine system changes, and increased destruction due to extreme events. Therefore, coastal vulnerability assessment techniques require a multi-hazard assessment methodology dependent upon the location at a regional level integrated with global level parameters. Several studies of coastal vulnerability have been conducted at various geographical locations along the Indian coast using a variety of approaches to mapping and data collection methods. Kumar et al. (2006) [32] used mapping techniques to evaluate the potential vulnerability implications, due to sea level variations, along the Cochin coast. Kanakara et al. (2007) [33] used an integrated MIKE-21 model for the estimation of oil spill vulnerability for different locations along the Gulf of Katchchh by using land-use, land cover changes, shoreline changes, rates of erosion and accretion, and sediment transport parameters during pre-monsoon, post-monsoon seasons. Chandrasekar et al. (2011) [34] studied coastal geomorphological vulnerability along the Tamil Nadu coast and revealed that intensive sand mining and coral mining in barrier coral islands are adversely changing beach morphology. Mani Murali et al. (2013) [35] considered a Physical Vulnerability Index (PVI), Social Vulnerability Index (SVI) and calibrated Coastal Vulnerability Index (CVI) for the Puducherry coast. Saxena et al. (2013) [36] developed a composite vulnerability index for coastal hazards along the Cuddlore coast, Tamil Nadu. The parameters considered for this study were erosion mapping, flood hazard mapping, and sea level rise combined together and then integrated with socio-economic influences. Appelquist et al. (2015) [37] proposed a new methodology (adapted from a method initially proposed by Appelquist (2013) [38] to assess and manage the Karnataka coast with a new multi-hazard technique known as the Coastal Hazard Wheel (CHW). The prime focus of this methodology is for screening the coastal area at local, regional and national levels of hazard, mainly in regions where available data is inadequate. These vulnerability assessment studies were mostly conducted using low resolution spatial data; and there were very few mitigation strategies evident. Therefore, it is necessary to adopt a more holistic and integrated approach to overcome these limitations. The concepts' vulnerability, hazard and risks are interrelated, but address distinct issues. According to Gilard (2016) [39], hazards are mostly natural phenomena, whereas vulnerability is associated with socioeconomic parameters affected by hazards and natural processes. For instance, when a vulnerable region is impacted by a hazard, there exists a potential risk that needs to be addressed. The risks are given a rating to estimate the vulnerability level (for example: high, medium, low). Therefore, when a particular coastal vulnerability is estimated, an integrated approach is required that can distinguish between these three concepts.

The main aim of this work is to review the existing coastal vulnerability assessment studies along CAP to identify the vulnerable locations with its main focus on CVI methodology approaches. A systematic literature review of CVI was carried out along the coastal Andhra Pradesh. The literature review for this region was organized according to each of the CVI parameters; including the data-collection technique, the magnitude of the threat, drivers for vulnerability, and the prevalence of mitigation strategies. Furthermore, the current study also identified gaps in knowledge and some recommendations for future research.

2. Snapshot of Coastal Vulnerability Methodologies

The primary technique used to monitor coasts for vulnerability assessments is spatial data with Geographic Information System (GIS). The method for gathering spatial GIS data is satellite imagery; this is due to its relative accessibility, economic affordability, and regular repetitive coverage. GIS-based spatial data is used in a number of different analytical methodologies, which include 'Decision Support Systems' (DSS) and index based techniques. DSS approaches include 'Community Vulnerability Assessment Tool' (CVAT), DINAS-coast, Dynamic Interactive Vulnerability Assessment (DIVA), Digital Shoreline Assessment System (DSAS) and vulnerability assessment [4]. Index-based techniques such as the Coastal Vulnerability Index (CVI) are also used widely across the world [28,40–50] 'Integrated Valuation of Ecosystem Services and Tradeoffs' (InVEST) is an open source software model that has a wide range of models to analyze a range of coastal vulnerabilities, including social, geographical, biological and economic factors [27]. Indian coasts have been studied mostly through the use of CVI methods for physical assessments [51–54] most of these studies required a certain amount of field data for the evaluation, due to limitations in the available spatial and temporal satellite data resolution [55] Remote-sensing approaches evident in the existing literature range from air-born to space-borne data gathering techniques; however, coastal vulnerability studies along the Indian coast were mostly restricted to the use of earth observing space-borne sensors [56–62]. Table 1 illustrates the range of parameters used across India to evaluate coastal vulnerability in diverse circumstances.

Table 1. Vulnerability assessment and parameters considered for different Indian geographic locations (modified from Rani et al. 2015). [3]

Vulnerability Method	Geographical Location	Parameters Considered	Remarks	Reference
CVI due to erosion	Coast between Kanyakumari and Tuticorin, Tamil Nadu	Geomorphology, shoreline change rate, coastal slope, sea level change, mean wave height, mean tidal range	Physical and human intervention processes are major causes of erosion	[63]
CVI due to erosion	Coast between Rasulpur (Midnapur) and Subarnarekha (Balasore), Orissa	Shore line change rate, land use and human activities, population density	Assessment of CVI using socio-economic parameters	[59]
Multi hazard vulnerability	Cuddalore, Pondicherry and Villupuram districts, Tamil Nadu	Probability of maximum storm surge height during the return period, future sea level rise, coastal erosion and high resolution coastal topography	Used multi hazard vulnerability technique	[64]
CVI	Orissa State, East Coast of India	Shoreline change rate, significant wave height, sea-level change rate, tidal range, coastal regional elevation, coastal slope, tsunami run-up and coastal geomorphology	-	[53]
CVI	From Talapady to Surathkal along Mangalore Coast	Geomorphology, regional coastal slope, shoreline change rates, population	Assessment of CVI using socio-economic parameters	[51]
CVI	Udupi coast in Karnataka	Geomorphology, shoreline erosion/accretion, coastal slope, mean tide range, mean significant wave height, mean sea level rise	Considered the CVI due to future SLR	[52]
CRI	Coastal stretch from Kattivakkam to Kovalam, Tamil Nadu	Environmental Vulnerability Index (EVI), Social Vulnerability Index (SVI), Hazard Potential Index (HPI), Mitigation Capacity Index (MCI)	Computed integrated coastal risk index. It is a multi-scale approach. Considered cyclones as indicator coastal risk	[65]
Potential Vulnerability Implications	Cochin	Sea-level variations	Used sea-level-rise scenario as an indicator for vulnerability of Cochin coast	[32]
Gulf of Kachchh, India	Gulf of Kachchh, India	Oil pollution, social and cultural values, scientific values, environmental, and economic values	Used an integrated numerical simulation modelling integrated with GIS	[33]
Coastal geomorphological vulnerability	Coastline between Kallar and Vembar lies in the Gulf of Mannar, Tamil Nadu	Land use/land cover changes, shoreline changes over the years, rate of erosion and accretion, sediment transport during pre-monsoon, monsoon, post-monsoon seasons	Beach morphological changes are influenced by intensive sand mining along the coast and coral mining in the barrier coral islands	[34]

3. Study Area

Coastal Andhra Pradesh (CAP) is located along the Eastern Indian coastline towards the Bay of Bengal at latitude 13°24′ N to 19°54′ N and longitude 80°02′ E to 86°46′ E (Figure 1) and is approximately 966 km in length. CAP has a hot and humid climate and consists of nine coastal districts with an area of 95,442 km^2, which is 60% of the total state area [66]. The Andhra Pradesh (Figure 1) coastline is the second largest shoreline after Gujarat in India. The population is 34 million, which is 70% of the total state population; the major cities are Vijayawada, Visakhapatnam, Guntur, Rajahmundry, Kakinada, Eluru, Nellore, and Ongole [66]. CAP has many hectares of fertile agricultural land due to the Krishna and Godavari deltas. Agriculture and fishing, which are the main fiscal activities in this region, are often disrupted by monsoonal variations, changes in climate, rapid changes in coastal geomorphology, local and sub-local elevations, and the rate of shoreline changes [67].

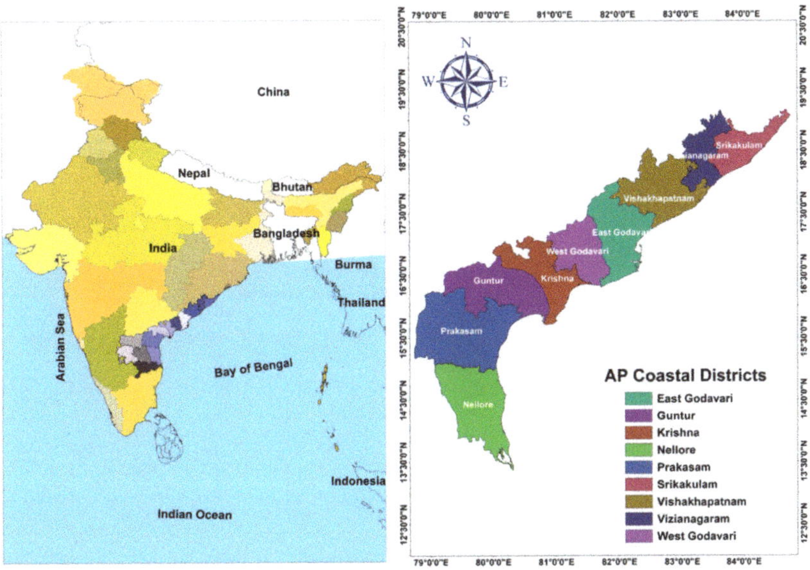

Figure 1. A spatial map of India with CAP indicated.

The rainfall is influenced by both South-West and North-East monsoons and the average annual rainfall along CAP is 1078.0 mm [68]. Andhra Pradesh, particularly coastal Andhra, is prone to cyclones, storm surges and floods. Moderate to high intensity cyclones occur every two to three years. Figure 2 shows the detailed spatial map for Andhra Pradesh coastal districts with the total number of cyclones occurring during the time period of 1951 to 2010. According to UNEP [6], more than 40% of the state is vulnerable to tropical storms, hurricane winds, severe floods and other associated natural hazards. The Godavari and Krishna River floods often contribute to serious disasters in the East and West Godavari and Krishna districts. The death toll and infrastructure damage due to these disasters is higher than other Indian coastal areas. According to the National Disaster Management Authority (NDMA) report (2016) [69], Andhra Pradesh coast is the second-most cyclone affected Indian region after Odisha and second-largest vulnerable region to floods after Kerala. The Ministry of Environment & Forests, Govt. of India (2012) [70] reported more frequent and intensified cyclonic activity and associated storm surges in CAP as a result of the impacts of climate change on air and sea-surface temperatures.

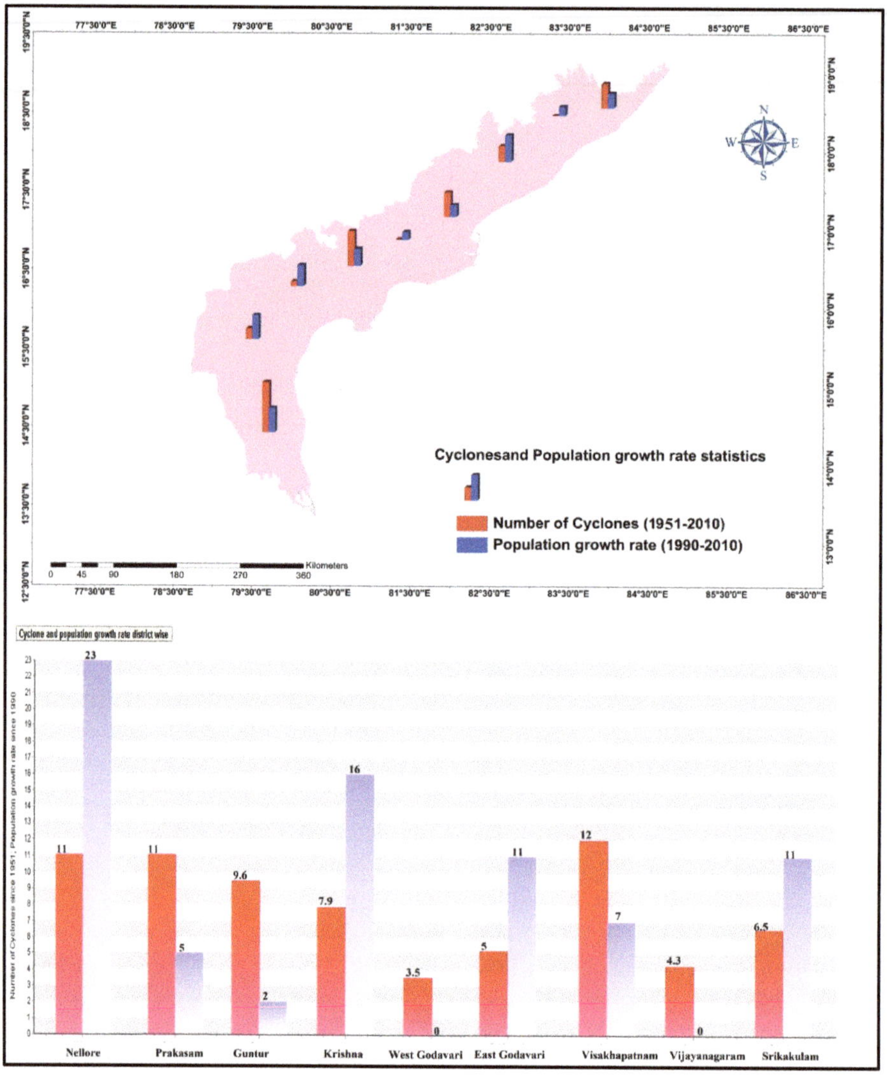

Figure 2. Total number of cyclones and population growth rate. Developed based upon the statistics of Revenue Disaster Management Andhra Pradesh (2017) [71].

4. Methodology

The current study is a systematic review of existing literature on coastal vulnerability assessments of the Indian coastline with emphasis on coastal Andhra Pradesh. The review systematically searched the following databases: ScienceDirect, Google Scholar, Web of Science (WoS), and Scopus to recognise the relevant literature using the keywords *coastal vulnerability, India, coastal Andhra Pradesh, erosion, GIS, spatial analysis*. Before finalising the existing studies to carry out the analysis, the current study performed a scoping exercise to recognise main themes, gaps and trends in the literature on coastal vulnerability studies. The review process comprised three main steps: searching (1), screening (2) finalization and analysis (3) (Figure 3). Initially, the first the review identified 125 articles and documents that met the search criteria. During the screening procedure, the current study applied exclusion and inclusion criteria to the search results to categorize the selected studies. Accordingly, the current study appraised the article titles and abstracts to evaluate their relevance to the study and, consequently excluded 75 articles deemed to be irrelevant. At this stage, 50 articles were selected for the full-text assessment. However, after full text analysis, a further 25 more articles were excluded due to irrelevance. Accordingly, only 25 articles from 125 articles have been finalized for the analysis. The omitted papers discussed coastal vulnerability in general, but their emphasis was not on the Coastal Andhra Pradesh region. Furthermore, other excluded papers examined CVI procedures at some point, but from integrated models' viewpoints, which are not relevant to the current study. In addition, the study also explored various government websites such as AP State of Portal, Government of India National Disaster Management Authority (NDMA), Indian Metrological Department (IMD), Regional Metrological Centre (RMC) and the Mumbai and National Cyclone Risk Mitigation Project (NCRMP) for grey literature. This grey literature was searched and evaluated according to the same criteria used for the selection of journal articles and other associated documents. Accordingly, 32 of these reports have been included for analysis. In total, 57 documents regarding coastal vulnerability studies have been analyzed.

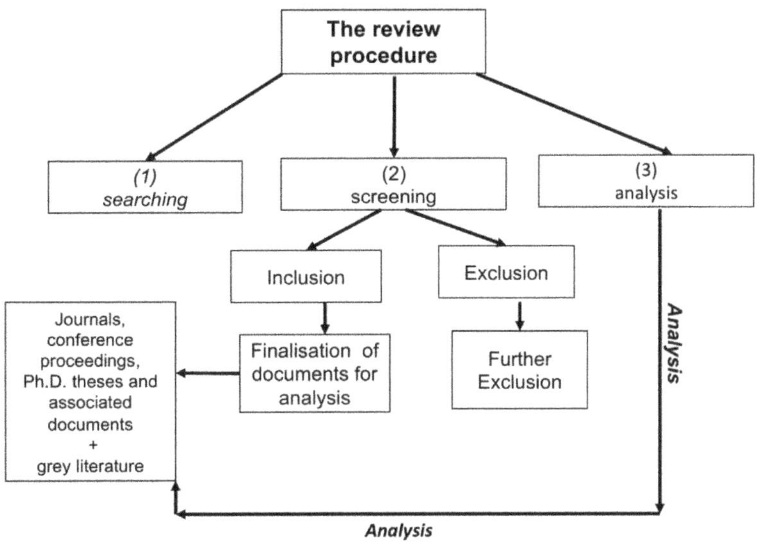

Figure 3. Methodological procedure.

Coastal Vulnerability Index Formulations and Parameters

Several CVI approaches, formulations and parameters are used to evaluate coastal vulnerability. Listed below are the main methodological frameworks and parameters. The first application of CVI (Equation (1)) was developed by Gornitz in 1990 with a focus on sea level rise, which considered the following parameters:

(a) Geomorphology
(b) Shoreline change-rate
(c) Coastal-slope
(d) Relative sea-level rise
(e) Mean significant wave-height
(f) Mean tidal-range

The principal formula for calculating CVI is given as follows:

$$CVI = \frac{\sqrt{a \times b \times c \times d \times e \times d \times e \times f \times g}}{7} \quad (1)$$

The above equation is the standard way to assess coastal vulnerability; however, it is not the only possible way to estimate the CVI quantitatively. Gornitz (1990) [72] proposed various ways to calculate the CVI, though the above equation is most widely used, as the geometric average is the most sensitive to individual parameters. Most of the studies along CAP used the above equation for vulnerability assessment. However, Rao et al. 2010 [73] used a different approach by using the summing of variables with prescribed weightage given to each parameter. For this assessment, five parameters were considered and the equation is as follows (Equation (2)):

$$CVI = 4a + 4c + 2b + f + e \quad (2)$$

where the variables a, c, b, f and e have their usual meanings defined as above.

The majority of studies [32,50–53] focused on basic factors as major drivers for coastal vulnerability along the Indian coast. Krishnan et al. (2018) [74] proposed a cumulative vulnerability index (CuVI) framework to map the coastal vulnerability along the Maharashtra coast. The CuVI is a function of exposure, sensitivity and adaptive capacity and uses the exposure-index (EI), sensitivity-index (SI) and adaptive-capacity-index (ACI). It covers all physical, social, and economic variables cumulatively for the preparation of a spatial decision-making map. Table 2 sets out the data-gathering approaches (in terms of conventional and spatial data) used for various CVI assessments as well as others that can be used. Here, conventional data is defined as point level observations or in-situ data that is collected throughout the year, whereas spatial data is satellite/aircraft based. There are various social parameters such as population, population growth rate, transportation with road networks, mobile-television facilities for each home for hazard warning awareness, age, gender, etc. CVI analyses require the data to be classified into different risk ranges; these are illustrated in Table 3.

Table 2. Different types of data that can be used for CVI studies in India.

Parameter	Data
	Spatial
Shoreline change	Satellite or Aircraft or UAVs
Coastal slope	DEM generated from SRTM or any SAR data
Bathymetry	GEBCO
Sea level change	Satellite altimeter data from TOPEX/Poseidon
Land use land cover change	Satellite data (like LANDSAT 8)
	Conventional
Significant wave height	JASON-1 data or Wintidex software generated data
Cyclones and storm surge	IMD cyclone data
Coastal slope	Elevation measured from point observations
Bathymetry	National hydrographic charts
Sea level change	Tide gauge data from Permanent Service for Mean Sea Level (PSMSL)
Historical floods, water quality	Disaster management reports
Social parameters	Census reports

Table 3. Coastal Vulnerability Index ranges with their risk values considered for CAP. Source: [9,58].

CVI (Coastal Vulnerability Index) Ranking Criteria					
Parameters	**Very Low (1)**	**Low (2)**	**Medium (3)**	**High (4)**	**Very High (5)**
Geomorphology	Rocky coasts	Indented coasts	Beach ridge, high dunes and vegetation	Low fore dunes (<3 m), estuaries and lagoons	Mudflats, mangroves, beaches and barriers/spits
Slope (%)	>1.00	0.50–1.00	0.10–0.50	0.05–0.10	<0.05
Mean sea level change rate (deg)	<1.80	1.8–2.5	2.50–3.00	3.00–3.40	>3.40
Significant mean wave height (m)	<0.55	0.55–0.55	0.85–1.05	1.05–1.25	<1.25
Mean tidal range (m)	<1.00	1.0–2.0	2.00–4.00	4.00–6.00	>6.00

5. Results

Based on the literature reviewed, several of the existing studies evaluated Indian coastal vulnerability. However, very few studies were performed along CAP and most of these studies were based on the conventional CVI technique developed by Gornitz et al. (1990) [72]; however, these studies differ from each other in the selection of parameters.

5.1. Coastal Geomorphology

The coastline of Andhra Pradesh is segregated into three distinct types: a rocky coast starting from North of Godavari delta, a vegetation coast in the Krishna-Godavari deltaic region, and a sandy coast from south of Krishna delta to Pulikat lake (SAC, ISRO report, 2011); a more detailed breakdown of land-use cover is shown in Figure 3, which revels that the entire CAP is predominantly agricultural. The geomorphology of the CAP is very diverse and economically important due to several features such as the Krishna-Godavari river delta, which plays a huge role in agriculture activities; developed/urbanized coastal areas along Visakhapatnam and Machilipatnam port; and ecologically sensitive places like mangroves and mudflats. Pramanik et al. (2016) [75] described the region as sensitive to muddy and soft sand and categorised the entire coast into two risk rates (higher and lower). Rao et al. (2010) [73] used IRS P6 AWiFS (The Advanced Wide Field Sensor) data to map geomorphology and its risk rankings, showing that the Krishna-Godavari deltaic regions are highly vulnerable due to the presence of mud flats, mangroves and the beach ridge complex. Overall, the mangroves and mudflat dominated areas of Kakinada, Mummidivaram, Avanigadda and Repalle were categorised as very high risk.

5.2. Shoreline Change-Rate

Shoreline change rate can be defined as the rate at which shore gets eroded or accreted due to wave-action, sea level-rise or other hazards and processes that affect the land. Although much of this is a natural process, added urbanization, infrastructure development and increased population can exacerbate the process. The CAP region is particularly prone to erosion due to its location and spatial distribution along the ocean. Factors such as sand mining, dredging and hard engineered protection for CAP are making it more susceptible to erosion. Almost all of the studies for the CAP to assess the shoreline erosion used Landsat data. For example, Basheer Ahammed et al. (2016) [76] used Landsat satellite data for the time-period of 1972–2015, and showed that most erosion was observed along the Krishna-Godavari delta and highest accretion was confined to the estuary outlet of the Krishna-Godavari deltaic region. The reason for this accretion could be mostly due to sediment transport from the estuary rather than that of coastal accretion, whereas sand mining could be one of the parameters for extensive erosion patterns. Vivek et al. (2016) [9] used the Landsat data for the period of 1973–2015 and found that some regions, such as Bheemunipatnam, central Chebrolu,

Veerappa Konda and south of Tuni, experienced a (maximum) exposure rate of almost 10 meters per year. Rani et al. (2018) [67] used 20 years (1997–2017) of spatial data (Landsat 4 to Landsat OLI) as well as conventional data (such as PSMSL, Survey of India toposheets) and showed that an approximately 2.5 km shoreline shift, in terms of erosion and 1.82 km shift in terms of accretion, was observed along the Vizianagaram–Srikakulam coast. These studies are useful to measure the shoreline erosion statistically; however, there is a need to improve the spatial, as well as temporal resolutions, for more accurate assessment.

5.3. Coastal Slope

Coastal elevation is the degree of steepness of the region with respect to the ocean; its measurement helps to identify the vulnerability of the coast. Rao et al. (2010) [73] used SRTM DEM data to evaluate the coastal slope range and its risk rate. Basheer Ahammed et al. (2016) [76] extracted coastal slope data from GEBCO (General Bathymetric Chart of Oceans) to identify that the lower southern part of CAP has steeper slopes, which leads to higher vulnerability as well as other risks. The available literature revealed different forms of data for coastal slope and its vulnerability mapping: Rani et al. (2018) [67] used ASTER DEM (Advanced Spaceborne Thermal Emission and Reflection Radiometer Digital Elevation Model) data, while SRTM DEM was used by Pramanik et al. (2016) [75] (both SRTM and ASTER DEM have a resolution of 30m for the Indian region). These studies further noted that CAP steep slopes are at high risk due to shoreline erosion and inundation due to storms. One more parameter that can classify the risk along the coastal zone is bathymetry, particularly in relation to wave height. The higher the bathymetry, the higher will be the risk to the coast owing to sea level rise and surge height. Rani et al. (2018) [64] used SRTM DEM with Directorate General of Hydrocarbons (DGH). While Pramanik et al. (2016) [75] used DGH data and revealed that Mummidivaram and Avanigadda zones are under very high risk followed by Pithapuram, Kakinada, Amalapuram and Machilipatnam, lower risk rates were identified for Tuni, Razole and Narsapur.

5.4. Relative Sea Level Rise

According to Unnikrishnan et al. (2007) [77], sea level changes can be due to two phenomena—one is global (mean sea level change) while the other is regional (extreme sea level changes). Therefore, sea level rise is a major phenomenon that combines global and regional processes that affects climate change induced coastal vulnerability along any coast. There are several ways to retrieve data for sea level change, for example from satellites and observations. Vivek et al. (2016) [9] brought together data from GLOSS (Global Sea Level Observing System) and found that the highest sea level rise was 5 cm/year along the Nellore coast and the lowest sea level rise of 0.8 cm/year was recorded in Pulicat coast. Pramanik et al. (2016) [75] used the IPCC global average sea level rise data and found 40% of CAP's population vulnerable to potential flooding by mid-century.

5.5. Significant Wave Height

Significant wave height, when associated with a tropical cyclone-induced storm surge, creates disaster at higher levels and CAP is highly prone to tropical cyclones. CAP wave height records from wave rider buoys were taken from the National Institute of Oceanography (NIO) to calculate the significant wave height [75]. They considered the significant wave height values of the pre- and post-monsoon seasons for 2013, which were found in the range of 0.5 m to 2.5 m. Rao et al. (2010) [73] used the MIKE-21 (Spectral Wave) model to measure the wave height from the European Center for Medium Range Weather Forecast (ECMWF) wind velocity component data. There are other more accurate ways to measure significant wave height, such as Wavewatch III, but this has not been utilized for the CAP region so far.

5.6. Mean Tidal Range, Tropical Cyclone and Induced Storm Surge

The tidal range is the vertical difference between high and low tide. For the CAP region, Rao et al. (2010) [73] used hydrographic chart data. According to the study, the tidal range for CAP varies between 0.7 to 1.4 m, which would be classified as a low vulnerability region. The influence of cyclones and the related induced storm surges are more prevalent along the East coast of India, particularly along Andhra Pradesh and Odisha [78]. Severe cyclones since 1977 are listed in Table 4, while Figure 4 shows land use land cover distribution along CAP. As shown in Figure 4, CAP predominantly suffered from agricultural and the cyclone-induced storm surge and its inundation, which resulted in socio-economic losses. There has been a reduction of risk and loss of life in CAP since 1977, due to improved early warning protection systems that give accurate predictions for cyclonic landfall; however, economic loss has been increasing (Table 4). Surge inundation damages the soil system through penetration of saltwater into the coastal upper aquifer and damages agricultural and biophysical ecosystems. Rao et al. (2010) [73] created a vulnerability map for CAP in the context of storm surges and cyclone winds and determined that Nellore district is at highest risk followed by Srikakulam and Visakhapatnam districts; the lowest vulnerable risk locations are located at Krishna and Guntur districts. Using Expert Decision Support System (EDSS), Matta et al. (2015) [79] concluded that existing CAP cyclone shelters can only accommodate 8% of the rural population, meaning that 92% of the rural populace is still at risk from cyclones. None of the CVI-based systems (that included other parameters) used tropical cyclones as a vulnerability parameter.

Tropical cyclone induced storm surge is a significant threat to CAP, but it has not been rigorously accounted for in CVI assessments previously. One exception is the research by Vivek et al. (2016) [9] who used tsunami run up as one of the parameters for CVI along CAP. Land subsidence, depletion of groundwater, and saltwater intrusion into groundwater are additional parameters that can be incorporated into coastal vulnerability assessments for better results. The resultant CVI categorized Nellore and West Godavari as highly vulnerable. In contrast, low vulnerability ranked areas were observed along the Visakhapatnam region [9,76]. The studies conducted by Vivek et al. (2016) [9] differed because of the inclusion of the parameter 'tsunami' into the CVI studies. The slight differences in the resultant maps from available literature can be accounted for by variations in the parameters and the parameter observation time. Rao et al. (2010) [73] produced a detailed range of risks along CAP, and the analysis shows that over 70% of CAP is vulnerable. Diverse studies evaluated CAP vulnerability by implementing various parameters and techniques under CVI formation; this is shown in Table 5.

Table 4. Severe cyclonic storm and their impact since 1977 along CAP.

Number	Severe Cyclonic Storm/Factor Effected	Maximum Wind (km/h)	Lowest Pressure (mbh)	Fatalities	Economic Loss (million-USD)
1	1977	165	919	14,204	499
2	1990	230	920	967	600
3	1996	145	988	1077	602
4	2014	185	215	124	3400

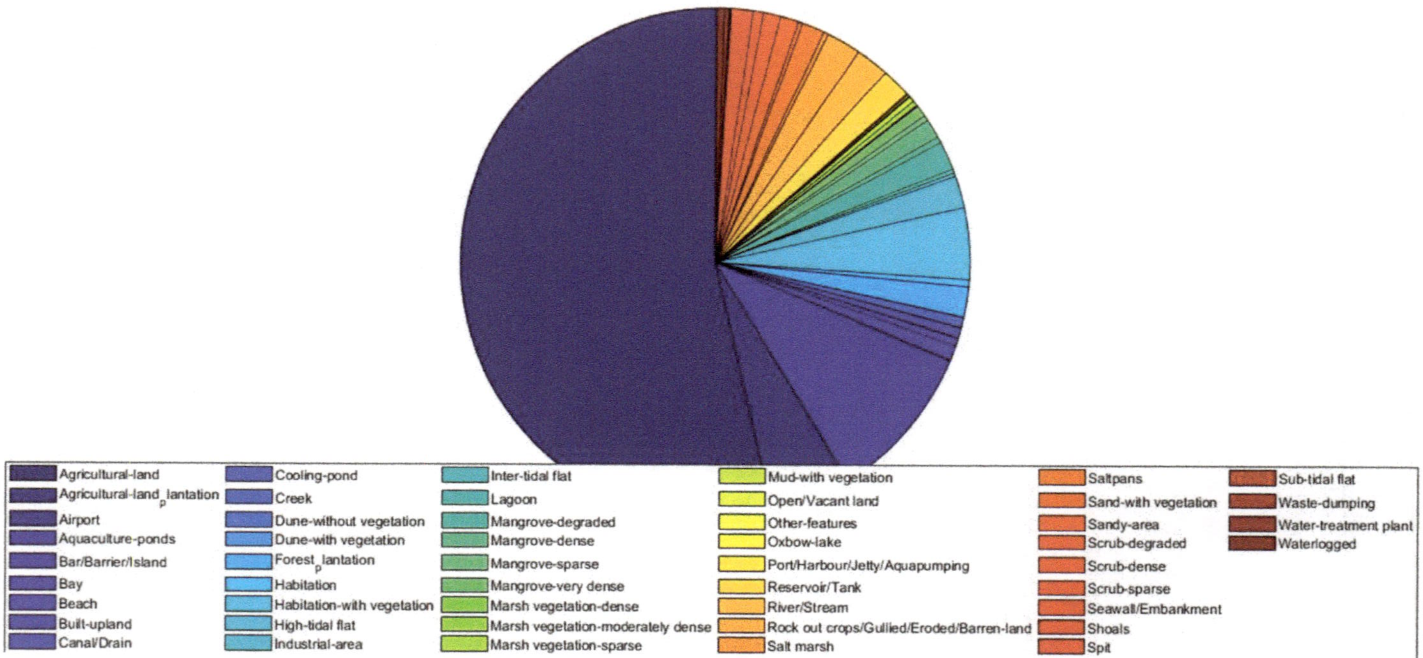

Figure 4. Land use and land cover distribution along the coastal Andhra Pradesh (redeveloped from NRSC, ISRO report 2012–2013 [80].

Table 5. Vulnerability assessment and parameters considered for different geographic locations of Andhra Pradesh.

Vulnerability Hazard	Location	Vulnerability Parameters	Technique	Remarks	Reference
CVI–Square Root of Product (SRoP)	Vijayanagaram-Srikakulam coast, AP	Shoreline change rate, coastal elevation, bathymetry, sea level rise, significant wave height	Remote sensing (space-borne) and GIS	-	[67]
CVI–SRoP	North East-coast of AP	Shoreline change rate, coastal slope, regional elevation, sea level change rate, significant wave height, tidal range, geomorphology, tsunami run-up	Remote sensing and GIS	Parts of Bheemavaram and Visakhapatnam are highly vulnerabble; Nakkapalli and Kancherpulem are at medium risk; Thallapalem is at low risk	[9]
CVI—SRoP	Entire AP coast	Shoreline change, coastal slope, coastal regional elevation, mean sea level, mean tide range, and geomorphology	Remote sensing and GIS	Small stretches in Nellore have very high vulnerability; Nellore, Kothapatnam and Kakinada Southern part are highly vulnerable; Most parts of Sompeta, Koduru and Kothapatnam are moderately vulnerable; Vishakhapatnam and Vizianagaram coasts are very low vulnerability	[76]
CVI (summing of variables with weightage)	AP	Geomorphology, coastal slope, shoreline change, mean spring tide, significant wave height	Remote sensing and GIS	Considered sea-level rise as an indicator for coastal vulnerability. The Krishna–Godavari delta coast is notably very highly vulnerable due to its low flat terrain and mud flats	[73]
CVI–SRoP	Krishna–Godavari delta region, AP	Shoreline change rate, coastal elevation, slope, geomorphology, sea level rise, significant wave height, mean tidal range	Remote sensing and GIS	Tuni, Kundukur, Narsapur, Razole at low risk; Pithapuram, Amalapuram, Chirala at High risk; Kakinada, Machilipatnam, Bapatla, Ongole at medium risk; Mummidivaram, Avanigadda & Repalle at very high risk	[75]

Table 5. *Cont.*

Vulnerability Hazard	Location	Vulnerability Parameters	Technique	Remarks	Reference
Vulnerability from storm surges and cyclone wind fields on the coast	AP	Vulnerability measured for the situations—frequent (return period of 10 years), 50-year return period, global warming-likely scenario, global warming-extreme case	-	-	[77]
Cyclone Vulnerability and Risk Analysis	AP	Vulnerable locations, casualties' prediction and shelters for cyclone induced vulnerability mitigation	GIS mapping	Prime focus is on socio-economical assessment for mitigation purposes.	[80]
Vulnerability of Indian Coastal Districts to Sea-Level Rise and Climate Extremes	Entire India	Demographic: Population density, annual growth rate of population, population at risk due to sea-level rise. Physical: Coast length, insularity, frequency of cyclones based on historic data, probable maximum surge height, area at risk of inundation due to SLR, number of vulnerable houses Economic: Agricultural dependency, income and/or infrastructure index. Social: Literacy, spread of institutional set-up Life-Loss: Storms and human casualties	Available historic statistical data	The study finely described the adaptation and mitigation strategies along the Indian coast with policy issues	[78]

6. Discussion

Coastal management along CAP is very important due to its agricultural productivity and socioeconomic significance. However, the importance of coastal zone management and work towards adaptation and mitigation strategies along CAP is still in its infancy unlike other southern states of India. Objective coastal vulnerability assessments are necessary [73] due to the influence of a range of parameters along the coast.

6.1. CVI Analysis

The current study revealed the importance of coastal zone risks, management, and the primary challenges associated with it. Coastal Andhra Pradesh is mostly vulnerable to cyclones with their induced storm surges, floods and coastal erosion [81,82]. The importance is evidenced by historical hazards like the devastating 1997 Andhra Pradesh cyclone, 1999 Orissa cyclone, 2004 tsunami or most recent 2017 Chennai and 2018 Kerala floods. However, cyclone hazard assessments were not included in CVI assessments for CAP. For the CAP region, CVI criteria included basic parameters like shoreline-change assessment, sea-level-rise, coastal elevation, coastal geomorphology, bathymetry, significant wave height. Most of them were missing the following factors: tropical cyclone, tsunami and storm-surge run-up. However, even if these additional parameters are included in the evaluation, the studies are further limited as a result of the exclusion of social and economic factors. Disasters are often associated with human activities and the resultant stresses. Saxena et al. (2013) [36] described that the magnitude and risk of coastal disasters due to hazards should be addressed with the sensitivity and resilience of the exposed population, rather than the intensity of natural hazard. The protection of an area is deemed vulnerable if the area is sufficiently important to economic, cultural and environmental (ecological and biological) aspects [3,59]. Therefore, the CAP region needs better understanding of the human and economic damage/loss associated with it and studies should focus on a more integrated approach that combines natural, anthropogenic and climate change-induced vulnerabilities.

Rao et al. 2010 [73] showed that the risk levels were very high at Penna and Krishna-Godavari delta regions. These are very low-lying and almost flat areas with mudflats, mangrove swamps, and lagoons/backwaters. Whereas, a study by Basheer Ahammed et al. 2016 [76] showed that the Nellore region and Narsapur area are highly vulnerable. The reason for this difference in the risk rate evaluation can be the difference in the years of data as well as the approach. The CVI assessment by Rao et al. 2010 [69] was performed using a summation of parameters with weightage given to each factor; further SLR was considered as an indicator to give weightages to the parameters; whereas, for Ahammad et al. 2016, CVI analysis was based on the basic Gornitz 1998 formula. Rao et al. 2008, 2010 [58,73] provided the vulnerability maps for the storm surges and cyclone wind effect along CAP. Surge inundation is a very serious issue, as CAP has a history of inundation that lead to huge loss of life and property during the 1977 cyclone. His study showed that the East Godavari to Guntur stretch and Southern Nellore coast are under the widest surge inundation, while wind speeds are highest in Visakhapatnam, though the entire CAP stretch is not far behind. Cyclone vulnerability and risk analysis performed by Matta et al. 2015 [80] studied the population and land region exposed to cyclone and wind speed during the cyclones. These Mandals level maps revealed that out of 430 mandals, 190 are highly exposed to cyclone winds with speeds of 235 km/h. Such studies can be very useful for policy makers to mitigate future damage at a social level. Given the importance of cyclones and their influences on the CAP region, there exists a large gap in CVI estimations with cyclones as there were very few studies reported in this context.

Figure 5 provides some of the district-wise social parameters along CAP. From the figure, it is notable that almost all the districts are equally exposed to socioeconomic parameters. Nevertheless, households without proper drainage (West Godavari district) are at a higher risk of cyclone-induced storm surges and their inundation. Drinking water facility is not fully covered in any of the districts as most of the coastal villages depend upon wells. Additionally, over-pumping of well water along the coastal regions might further lead to salt water intrusion into ground water and land subsidence.

Population density accelerates the vulnerability risk rate further. In terms of population density per square km, the highest risk was observed for Krishna district, followed by East Godavari, Srikakulam and Guntur districts.

6.2. Mitigation Strategies and Approaches

Andhra Pradesh coastal management has developed some adaptive strategies for stresses such as provision of cyclone shelters, disaster response force for responsive activities, cyclone and tsunami warning centers [83]; likewise, protection works for rivers and seacoasts were implemented to improve resilience to flooding (National Disaster Management Authority, 2016 [69]. Mitigation of coastal land loss due to erosion is mostly through hard engineering interventions such as the construction of groynes, dykes and seacoasts (Bhattacharyya et al. 2016 [84]. However, the literature revealed no such mitigation strategies employed over the CAP region, instead most focused on the rate of vulnerability with respect to the parameters considered. Nevertheless, these engineering structures can help to reduce the erosion in the short-term due to longshore drift; however, in the longer-term, these engineering interventions may increase erosion rates further down the coast. Furthermore, maintenance of these artificial engineering structures is relatively expensive [85]. Therefore, a more effective and more affordable alternative might be required, such as the use of ecosystem-based green-infrastructure solutions, to improve resiliency and to reduce the impact of some coastal activities such as sand mining.

6.3. Data Gathering Techniques

The most common weakness of the many studies of hazards on the Indian coast is the relatively poor resolution of data. The major focus of existing research has mostly used data gathered from field observations or satellites. CAP, in particular has less data gathered compared to states like West Bengal, Odisha, Tamil Nadu, Kerala and Karnataka. The studies focusing on mitigation and adaptation strategies to CAP hazards are very few to none. Furthermore, temporal variations of coastal processes vary from a daily time period to many years at a climatic level; however, the current data gathering techniques do not accurately capture this temporality sufficiently. At a global level, there have been significant improvements in approaches to data gathering since the majority of these studies were undertaken, particularly using new digital technologies. There is a need to study CAP with better data and constant observation; emerging technologies such as Unmanned Aerial Vehicles (UAVs) with airborne and LiDAR sensors can help to achieve a better understanding of CAP. Such digital technologies can provide spatial data at a much higher level of resolution than existing approaches, allowing data specificity at a much more local level. Moreover, the cost of these technologies has reduced recently and is now a cost-effective means for gathering data.

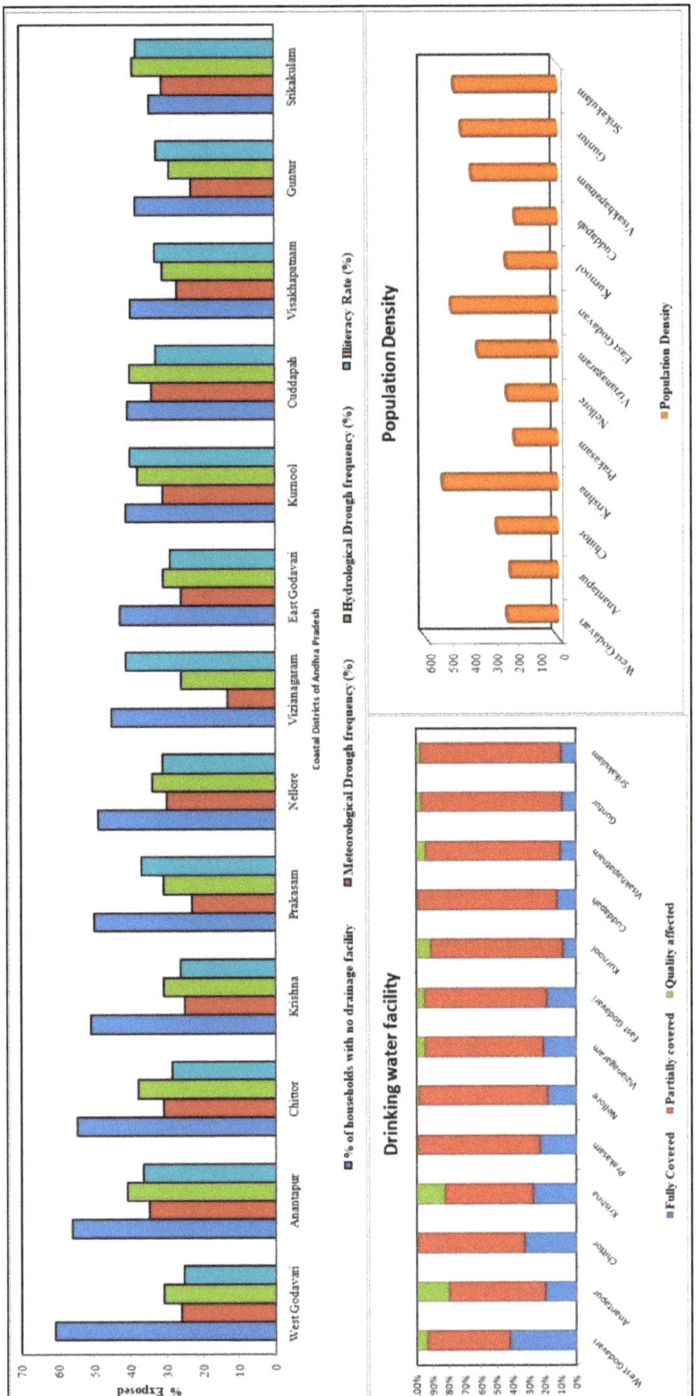

Figure 5. Social components exposed due to coastal vulnerability.

7. Conclusions

The current study research aims to generate up-to-date knowledge for use in the development and implementation of local, regional and national policies, and disaster and coastal risk management procedures. The current study systematically reviews existing literature to examine the occurrence, magnitude and impacts of the natural hazards affecting the Indian coast with special focus on Coastal Andhra Pradesh. Data collection methods and data sources are critically appraised as well as existing mitigation strategies proposed as a result of the analyses. The most recurrent natural hazards within the coastal Andhra Pradesh region are tropical cyclones, coastal inundation, monsoon-floods along the rivers and torrential rains. The present study highlights that, due to its geographical location on the Bay of Bengal Tectonic Plate, Coastal Andhra Pradesh is highly prone to natural coastal disasters. This study also distinguishes the probable long-term influence of sea level rise due to human-induced global warming and rapid urbanization on Coastal Andhra Pradesh in specific areas. The predicted sea level rise and temperatures are likely to result in the loss of residential and agricultural land, an upsurge in salinity and the decline of water quality, especially in agricultural fields, with predominantly negative impacts on the economy and population of coastal Andhra Pradesh.

The review reveals that the majority of existing data is collected from either satellite data or field observations that are very low in resolution, and are not continuous. This research can be used as a primary assessment for the design of a resiliency assessment framework for natural disaster management for the Andhra Pradesh region, and beyond. This resiliency assessment framework should play a vital role in decision-making within the reduction of coastal area vulnerability to natural hazards such as cyclones and floods, and also might improve its resilience capacity. Its application is likely to enable positive results regarding effective disaster management and control, by reducing harm to economies and safeguarding the security of local and regional communities. The current review of the coastal vulnerability of coastal Andhra Pradesh will be of interest to researchers, coastal engineers, environmentalists and government officials, who want to evaluate the prospective consequences of multiple disasters for coastal geographical areas across the world. The study shows that very little has been done to study the overall coastal vulnerability along CAP with only a few of the CVI parameters being accounted for, and based on relatively low resolution data. More research is required to observe, measure, and assess adaptation and mitigation measures in this area to improve resiliency.

Author Contributions: K.K., N.N.V.S.R. and L.R. designed the study. K.K. designed the methodology. K.K., N.N.V.S.R. and L.C.C. wrote the paper; L.R., K.S., M.T., U.K., K.Y and R.R. modified the manuscript and advised on physical and socio-economic coastal vulnerability studies.

Funding: This research was funded by the GCRF (Global Challenges Research Fund), United Kingdom.

Acknowledgments: Particular acknowledgement is made to Muhammad Kamran Naeem (Solent University) for his assistance in generating the pie chart using MetLab.

Conflicts of Interest: The authors declare no conflict of interest.

References

1. Sekovski, I.; Newton, A.; Dennison, W.C. Megacities in the coastal zone: Using a driver-pressure-state-impact-response framework to address complex environmental problems. *Estuar. Coast. Shelf Sci.* **2012**, *96*, 48–59. [CrossRef]
2. Gari, S.R.; Newton, A.; Icely, J.D. A review of the application and evolution of the DPSIR framework with an emphasis on coastal social-ecological systems. *Ocean Coast. Manag.* **2015**, *103*, 63–77. [CrossRef]
3. Rani, N.S.; Satyanarayana, A.; Bhaskaran, P.K. Coastal vulnerability assessment studies over India: A review. *Nat. Hazards* **2015**, *77*, 405–428. [CrossRef]
4. Paula, J. Overall assessment of the state of the coast in the Western Indian. *Ocean Reg. State Coast Rep.* **2016**, 500–527. [CrossRef]
5. Kantamaneni, K.; Gallagher, A.; Du, X. Assessing and mapping regional coastal vulnerability for port environments and coastal cities. *J. Coast. Conserv.* **2019**, *23*, 59–70. [CrossRef]

6. Brown, S.; Nicholls, R.J.; Lázár, A.N.; Hornby, D.D.; Hill, C.; Hazra, S.; Addo, K.A.; Haque, A.; Caesar, J.; Tompkins, E.L. What are the implications of sea-level rise for a 1.5, 2 and 3 °C rise in global mean temperatures in the Ganges-Brahmaputra-Meghna and other vulnerable deltas? *Reg. Environ. Chang.* **2018**, *18*, 1829–1842. [CrossRef]
7. Sahoo, B.; Bhaskaran, P.K. Multi-hazard risk assessment of coastal vulnerability from tropical cyclones—A GIS based approach for the Odisha coast. *J. Environ. Manag.* **2018**, *206*, 1166–1178. [CrossRef] [PubMed]
8. UNEP. Global Environment Outlook 3: Past, Present and Future Perspectives. UNEP/Earthscan: Nairobi, Kenya, 2012. Available online: https://wedocs.unep.org/bitstream/handle/20.500.11822/8609/GEO-3%20REPORT_English.pdf?sequence=7&isAllowed=y (accessed on 12 September 2014).
9. Vivek, G.; Mahendra, R.S.; Mohanty, P.C.; Srinivasa Kumar, T. Sachikanta Nanda Coastal Vulnerability Assessment for North East Coast of Andhra Pradesh, India. *Int. J. Remote Sens. Geosci. (IJRSG)* **2016**, *5*, 1–7.
10. Patwardhan, A.; Semenov, S.; Schnieder, S.; Burton, I.; Magadza, C.; Oppenheimer, M.; Pittock, B.; Rahman, A.; Smith, J.; Suarez, A. Assessing key vulnerabilities and the risk from climate change. *Clim. Chang.* **2007**, 779–810.
11. Few, R. Flood hazards, vulnerability and risk reduction. In *Flood Hazards and Health*; Routledge: London, UK, 2013; pp. 20–39. [CrossRef]
12. Elliott, M.; Cutts, N.D.; Trono, A. A typology of marine and estuarine hazards and risks as vectors of change: A review for vulnerable coasts and their management. *Ocean Coast. Manag.* **2014**, *93*, 88–99. [CrossRef]
13. Ramesh, R.; Chen, Z.; Cummins, V.; Day, J.; D'Elia, J.; Dennison, B.; Forbes, D.L.; Glaeser, B.; Glaser, M.; Glavovic, B.; et al. Land–ocean interactions in the coastal zone: Past, present & future. *Anthropocene* **2015**, *12*, 85–98.
14. Nicholls, R.J.; Wong, P.P.; Burkett, V.R.; Codignotto, J.O.; Hay, J.E.; McLean, R.F.; Ragoonaden, S.; Woodroffe, C.D. Coastal systems and low-lying areas. In *Climate Change 2007: Impacts, Adaptation and Vulnerability: Contribution of Working Group II to the Fourth Assessment Report of the Intergovernmental Panel on Climate Change*; Parry, M.L., Canziani, O.F., Palutikof, J.P., van der Linden, P.J., Hanson, C.E., Eds.; Cambridge University Press: Cambridge, UK, 2017; pp. 315–356.
15. Schuur, E.A.G.; McGuire, A.D.; Schädel, C.; Grosse, G.; Harden, J.W.; Hayes, D.J.; Hugelius, G.; Koven, C.D.; Kuhry, P.; Lawrence, D.M.; et al. Climate change and the permafrost carbon feedback. *Nature* **2015**, *520*, 171–179. [CrossRef] [PubMed]
16. Dow, K.; Downing, T.E. *The Atlas of Climate Change: Mapping the World's Greatest Challenge*; University of California Press: Brighton, UK, 2016.
17. Lal, M. Global climate change: India's monsoon and its variability. *J. Environ. Stud. Policy* **2003**, *6*, 1–34.
18. Lioubimtseva, E.; Henebry, G.M. Climate and environmental change in arid Central Asia: Impacts, vulnerability, and adaptations. *J. Arid Environ.* **2009**, *73*, 963–977. [CrossRef]
19. Karim, M.F.; Mimura, N. Impacts of climate change and sea-level rise on cyclonic storm surge floods in Bangladesh. *Glob. Environ. Chang.* **2008**, *18*, 490–500. [CrossRef]
20. Knutson, T.R.; McBride, J.L.; Chan, J.; Emanuel, K.; Holland, G.; Landsea, C.; Held, I.; Kossin, J.P.; Srivastava, A.K.; Sugi, M. Tropical cyclones and climate change. *Nat. Geosci.* **2010**, *3*, 157–163. [CrossRef]
21. Vousdoukas, M.I.; Mentaschi, L.; Voukouvalas, E.; Verlaan, M.; Jevrejeva, S.; Jackson, L.P.; Feyen, L. Global probabilistic projections of extreme sea levels show intensification of coastal flood hazard. *Nat. Commun.* **2018**, *9*, 2360. [CrossRef] [PubMed]
22. The South Asian Disaster Report (SADR), Changing Limits, Impeding Risks, Emerging Perspectives. 2010. Available online: http://www.ndma.gov.pk/Publications/South%20Asia%20Disaster%20Report%202010%20unisdr.pdf (accessed on 12 March 2016).
23. McCarthy, J.J.; Canziani, O.F.; Leary, N.A.; Dokken, D.J.; White, K.S. *Climate Change 2001: Impacts, Adaptation, and Vulnerability: Contribution of Working Group II to the Third Assessment Report of the Intergovernmental Panel on Climate Change*; Cambridge University Press: Cambridge, UK, 2001; Volume 2.
24. Moss, R.H.; Edmonds, J.A.; Hibbard, K.A.; Manning, M.R.; Rose, S.K.; van Vuuren, D.P.; Carter, T.R.; Emori, S.; Kainuma, M.; Kram, T.; et al. The next generation of scenarios for climate change research and assessment. *Nature* **2010**, *463*, 747–756. [CrossRef] [PubMed]
25. Neumann, B.; Vafeidis, A.T.; Zimmermann, J.; Nicholls, R.J. Future coastal population growth and exposure to sea-level rise and coastal flooding-a global assessment. *PLoS ONE* **2015**, *10*, e0118571. [CrossRef] [PubMed]

26. Chen, W.-Y.; Suzuki, T.; Lackner, M. *Sea-Level Rise and Hazardous Storms: Impact Assessment on Coasts and Estuaries*; Springer: New York, NY, USA, 2016.
27. Arkema, K.K.; Griffin, R.; Maldonado, S.; Silver, J.; Suckale, J.; Guerry, A.D. Linking social, ecological, and physical science to advance natural and nature-based protection for coastal communities. *Ann. N. Y. Acad. Sci.* **2017**, *1399*, 5–26. [CrossRef] [PubMed]
28. Kantamaneni, K.; Phillips, M.; Thomas, T.; Jenkins, R. Assessing coastal vulnerability: Development of a combined physical and economic index. *Ocean Coast. Manag.* **2018**, *158*, 164–175. [CrossRef]
29. Blaikie, P.; Cannon, T.; Davis, I.; Wisner, B. *At Risk: Natural Hazards, People's Vulnerability and Disasters*, 1st ed.; Routledge: London, UK, 1994. [CrossRef]
30. Parry, M.; Parry, M.L.; Canziani, O.; Palutikof, J.; Van der Linden, P.; Hanson, C. *Climate Change 2007-Impacts, Adaptation and Vulnerability: Working Group II Contribution to the Fourth Assessment Report of the IPCC*; Cambridge University Press: Cambridge, UK, 2007; Volume 4.
31. Nguyen, T.T.; Bonetti, J.; Rogers, K.; Woodroffe, C.D. Indicator-based assessment of climate-change impacts on coasts: A review of concepts, methodological approaches and vulnerability indices. *Ocean Coast. Manag.* **2016**, *123*, 18–43. [CrossRef]
32. Kumar, P.D. Potential vulnerability implications of sea level rise for the coastal zones of Cochin, southwest coast of India. *Environ. Monit. Assess.* **2006**, *123*, 333. [CrossRef] [PubMed]
33. Kanakara, R.; Subramanian, B. Oil spill sensitivity analysis and risk assessment for Gulf of Kachchh, India, using integrated modeling. *J. Coast. Res.* **2007**, *14*, 1251–1258. [CrossRef]
34. Chandrasekar, N.; Joevivek, V. *John Prince Soundaranayagam DC Geospatial Analysis of Coastal Geomorphological Vulnerability along Southern Tamil Nadu Coast*; Geospatial World Forum: Hyderabad, India, 2011.
35. ManiMurali, R.; Ankita, M.; Amrita, S.; Vethamony, P. Coastal Vulnerability Assessment of Puducherry Coast, India, using the Analytical Hierarchical Process. *Nat. Hazards Earth Syst. Sci.* **2013**, *13*, 3291–3311. [CrossRef]
36. Saxena, S.; Purvaja, R.; Suganya, G.M.D.; Ramesh, R. Coastal hazard mapping in the Cuddalore region, South India. *Nat. Hazards* **2013**, *66*, 1519–1536. [CrossRef]
37. Appelquist, L.R.; Balstrøm, T. Application of a new methodology for coastal multi-hazard-assessment & management on the state of Karnataka, India. *J. Environ. Manag.* **2015**, *152*, 1–10.
38. Appelquist, L.R. Generic framework for meso-scale assessment of climate change hazards in coastal environments. *J. Coast. Conserv.* **2013**, *17*, 59–74. [CrossRef]
39. Gilard, O. Hazards, Vulnerability and Risk. In *Climate Change and Agriculture Worldwide*; Springer: New York, NY, USA, 2016; pp. 19–29.
40. Thieler, E.R.; Hammar-Klose, E.S. *National Assessment of Coastal Vulnerability to Sea-level Rise: Preliminary Results for the US Gulf of Mexico Coast*; Geological Survey Open-File Report; U.S. Geological Survey: Woods Hole, MA, USA, 2000.
41. Boruff, B.J.; Emrich, C.; Cutter, S.L. Erosion hazard vulnerability of US coastal counties. *J. Coast. Res.* **2005**, *21*, 932–942. [CrossRef]
42. Szlafsztein, C.; Sterr, H. A GIS-based vulnerability assessment of coastal natural hazards, state of Pará, Brazil. *J. Coast. Conserv.* **2007**, *11*, 53–66. [CrossRef]
43. Abuodha, P.A.; Woodroffe, C.D. Assessing vulnerability to sea-level rise using a coastal sensitivity index: A case study from southeast Australia. *J. Coast. Conserv.* **2010**, *14*, 189–205. [CrossRef]
44. Torresan, S.; Critto, A.; Rizzi, J.; Marcomini, A. Assessment of coastal vulnerability to climate change hazards at the regional scale: The case study of the North Adriatic Sea. *Nat. Hazards Earth Syst. Sci.* **2012**, *12*, 2347–2368. [CrossRef]
45. Ahsan, M.N.; Warner, J. The socioeconomic vulnerability index: A pragmatic approach for assessing climate change led risks—A case study in the south-western coastal Bangladesh. *Int. J. Disaster Risk Reduct.* **2014**, *8*, 32–49. [CrossRef]
46. Cui, L.; Ge, Z.; Yuan, L.; Zhang, L. Vulnerability assessment of the coastal wetlands in the Yangtze Estuary, China to sea-level rise. *Estuar. Coast. Shelf Sci.* **2015**, *156*, 42–51. [CrossRef]
47. Satta, A.; Snoussi, M.; Puddu, M.; Flayou, L.; Hout, R. An index-based method to assess risks of climate-related hazards in coastal zones: The case of Tetouan. *Estuar. Coast. Shelf Sci.* **2016**, *175*, 93–105. [CrossRef]

48. Coletti, A.; De Nicola, A.; Villani, M.L. Building climate change into risk assessments. *Nat. Hazards* **2016**, *84*, 1307–1325. [CrossRef]
49. Ciccarelli, D.; Pinna, M.S.; Alquini, F.; Cogoni, D.; Ruocco, M.; Bacchetta, G.; Sartic, G.; Fenu, G. Development of a coastal dune vulnerability index for Mediterranean ecosystems: A useful tool for coastal managers? *Estuar. Coast. Shelf Sci.* **2017**, *187*, 84–95. [CrossRef]
50. Ghoussein, Y.; Mhawej, M.; Jaffal, A.; Fadel, A.; El Hourany, R.; Faour, G. Vulnerability assessment of the South-Lebanese coast: A GIS-based approach. *Ocean Coast. Manag.* **2018**, *158*, 56–63. [CrossRef]
51. Vittal Hegde, A.; Radhakrishnan Reju, V. Development of coastal vulnerability index for Mangalore coast, India. *J. Coast. Res.* **2007**, *23*, 1106–1111. [CrossRef]
52. Dwarakish, G.S.; Vinay, S.A.; Natesan, U.; Asano, T.; Kakinuma, T.; Venkataramana, K.; Pai, B.J.; Babita, M.K. Coastal vulnerability assessment of the future sea level rise in Udupi coastal zone of Karnataka state, west coast of India. *Ocean Coast. Manag.* **2009**, *52*, 467–478. [CrossRef]
53. Kumar, T.S.; Mahendra, R.; Nayak, S.; Radhakrishnan, K.; Sahu, K. Coastal vulnerability assessment for Orissa State, east coast of India. *J. Coast. Res.* **2010**, *26*, 523–534. [CrossRef]
54. Kumar, A.A.; Kunte, P.D. Coastal vulnerability assessment for Chennai, east coast of India using geospatial techniques. *Nat. Hazards* **2012**, *64*, 853–872. [CrossRef]
55. Klemas, V.V. Coastal and environmental remote sensing from unmanned aerial vehicles: An overview. *J. Coast. Res.* **2015**, *31*, 1260–1267. [CrossRef]
56. Prabhakara Rao, P.; Nair, M.; Raju, D. Assessment of the role of remote sensing techniques in monitoring shoreline changes: A case study of the Kerala coast. *Int. J. Remote Sens.* **1985**, *6*, 549–558. [CrossRef]
57. Pavri, F. Urban expansion and sea-Level rise related flood vulnerability for Mumbai (Bombay), India using remotely sensed data. In *Geospatial Techniques in Urban Hazard and Disaster Analysis*; Springer: Dordrecht, The Netherlands, 2009; pp. 31–49.
58. Rao, K.N.; Subraelu, P.; Rao, T.V.; Malini, B.H.; Ratheesh, R.; Bhattacharya, S.; Rajawat, A. Sea-level rise and coastal vulnerability: An assessment of Andhra Pradesh coast, India through remote sensing and GIS. *J. Coast. Conserv.* **2008**, *12*, 195–207.
59. Jana, A.; Bhattacharya, A.K. Assessment of coastal erosion vulnerability around Midnapur-Balasore Coast, Eastern India using integrated remote sensing and GIS techniques. *J. Indian Soc. Remote Sens.* **2013**, *41*, 675–686. [CrossRef]
60. Kaliraj, S.; Chandrasekar, N.; Ramachandran, K.; Srinivas, Y.; Saravanan, S. Coastal landuse and land cover change and transformations of Kanyakumari coast, India using remote sensing and GIS. Egypt. *J. Remote Sens. Space Sci.* **2017**, *20*, 169–185.
61. Thakur, S.; Dey, D.; Das, P.; Ghosh, P.; De, T. Shoreline change detection using Remote Sensing in the Bakkhali Coastal Region, West Bengal, India. *Indian J. Geosci.* **2017**, *71*, 611–626.
62. Venkataraman, C.; Mishra, T.; Ghosh, S.; Karmakar, S. *Climate Change Signals and Response: A Strategic Knowledge Compendium for India*; Springer: Singapore, 2018.
63. Mujabar, P.S.; Chandrasekar, N. Coastal erosion hazard and vulnerability assessment for southern coastal Tamil Nadu of India by using remote sensing and GIS. *Nat. Hazards* **2013**, *69*, 1295–1314. [CrossRef]
64. Mahendra, R.; Mohanty, P.; Bisoyi, H.; Kumar, T.S.; Nayak, S. Assessment and management of coastal multi-hazard vulnerability along the Cuddalore–Villupuram, east coast of India using geospatial techniques. *Ocean Coast. Manag.* **2011**, *54*, 302–311. [CrossRef]
65. Poompavai, V.; Ramalingam, M. Geospatial analysis for coastal risk assessment to cyclones. *J. Indian Soc. Remote Sens.* **2013**, *41*, 157–176. [CrossRef]
66. AP (Andhra Pradesh). State Portal. AP at a Glance. 2017. Available online: https://www.ap.gov.in/?page_id=30 (accessed on 25 March 2018).
67. Rani, M.; Rehman, S.; Sajjad, H.; Chaudhary, B.; Sharma, J.; Bhardwaj, S.; Kumar, P. Assessing coastal landscape vulnerability using geospatial techniques along Vizianagaram–Srikakulam coast of Andhra Pradesh, India. *Nat. Hazards* **2018**, *94*, 711–725. [CrossRef]
68. NRSC (National Remote sensing Centre). AP at a Glance. 2014. Available online: https://nrsc.gov.in/uim_2014_proceedings/papers_ppts/UIM2014_US5_EVinodKumar.pdf (accessed on 24 June 2016).
69. National Disaster Management Authority (NDMA). Annual Report for 2016–2017-Government of India. 2016. Available online: https://ndma.gov.in/images/annreport/ENG-2016-17-AR.pdf (accessed on 25 March 2018).

70. The Ministry of Environment & Forests, Govt. of India. Coastal Zones of India. 2012. Available online: http://www.moef.nic.in/sites/default/files/Coastal_Zones_of_India.pdf (accessed on 26 June 2017).
71. AP State Disaster Management Authority. Revenue Department. 2017. Available online: http://www.apsdma.ap.gov.in (accessed on 22 March 2018).
72. Gornitz, V. Vulnerability of the East Coast, USA to future sea level rise. *J. Coast. Res.* **1990**, *9*, 201–237.
73. Rao, A.; Babu, S.; Prasad, K.; Murty, T.R.; Sadhuram, Y.; Mahapatra, D. Investigation of the generation and propagation of low frequency internal waves: A case study for the east coast of India. *Estuar. Coast. Shelf Sci.* **2010**, *88*, 143–152. [CrossRef]
74. Krishnan, P.; Ananthan, P.S.; Purvaja, R.; Joyson Joe Jeevamani, J.; Amali Infantina, J.; Srinivasa Rao, C.; Anand, A.; Mahendra, R.S.; Sekar, I.; Kareemulla, K.; et al. Framework for mapping the drivers of coastal vulnerability and spatial decision making for climate-change adaptation: A case study from Maharashtra, India. *Ambio* **2018**, *48*, 192–212. [CrossRef] [PubMed]
75. Pramanik, M.K.; Biswas, S.S.; Mondal, B.; Pal, R. Coastal vulnerability assessment of the predicted sea level rise in the coastal zone of Krishna–Godavari delta region, Andhra Pradesh, east coast of India. *Environ. Dev. Sustain.* **2016**, *18*, 1635–1655. [CrossRef]
76. Basheer Ahammed, K.; Mahendra, R.; Pandey, A. Coastal Vulnerability Assessment for Eastern Coast of India, Andhra Pradesh by Using Geo-Spatial Technique. *Geoinform. Geostat. Overv.* **2016**, *4*. [CrossRef]
77. Unnikrishnan, A.; Shankar, D. Are sea-level-rise trends along the coasts of the north Indian Ocean consistent with global estimates? *Glob. Planet. Chang.* **2007**, *57*, 301–307. [CrossRef]
78. Kumar, K.K.; Tholkappian, S. Relative vulnerability of Indian coastal districts to sea-level rise and climate extremes. *Int. Rev. Environ. Strat.* **2006**, *6*, 3–22.
79. Matta, S.; Tejaswi Baddam, S. Cyclone Vulnerability and Risk Analysis for Coastal Districts of Andhra Pradesh. 2015. Available online: http://imetsociety.org/wp-content/pdf/vayumandal/2015/2015_6.pdf (accessed on 26 March 2017).
80. NRSC-ISRO. NATURAL RESOURCES CENSUS Ninth Cycle (2012-13) NATIONAL LAND USE AND LAND COVER MAPPING USING MULTI-TEMPORAL AWiFS DATA (LULC-AWiFS) Remote Sensing Applications Area & Regional Remote Sensing Centres NATIONAL REMOTE SENSING CENTER & North Eastern Space Applications Centre (NESAC) Indian Space Research Organisation. 2013. Available online: https://bhuvan-noeda.nrsc.gov.in/gis/thematic/tools/document/LULC250/1213.pdf (accessed on 23 February 2019). [CrossRef]
81. Kavya, A.; Bhavani, A.G.; Rao, P.J.; Chandrudu, V.B. Impact of Land Use/Land Cover and Mangrove Degradation on Coastal Erosion in Godavari Delta Region, Andhra Pradesh—A Geospatial Approach. In *Proceedings of International Conference on Remote Sensing for Disaster Management*; Springer Series in Geomechanics and Geoengineering; Springer: Cham, Switzerland, 2019; pp. 413–421.
82. Kumari, K.P.; Srinivas, K. Applications of RS and GIS Techniques for Disaster Studies in East Godavari District, Andhra Pradesh, India. In *Proceedings of International Conference on Remote Sensing for Disaster Management*; Springe Series in Geomechanics and Geoengineering; Springer: Cham, Switzerland, 2019; pp. 199–214.
83. Topno, P.N. Multi-stakeholder Support in Disaster Risk Governance in India. In *Disaster Risk Governance in India and Cross Cutting Issues*; Springer: Singapore, 2018; pp. 127–142.
84. Bhattacharyya, R.; Ghosh, B.N.; Dogra, P.; Mishra, P.K.; Santra, P.; Kumar, S.; Fullen, M.A.; Mandal, U.K.; Anil, K.S.; Lalitha, M.; et al. Soil conservation issues in India. *Sustainability* **2016**, *8*, 565. [CrossRef]
85. FAOUN (The Food and Agriculture Organization of the United Nations). Chapter-4: Protection from Coastal Erosion. 2016. Available online: http://www.fao.org/docrep/010/ag127e/AG127E09.htm (accessed on 10 August 2017).

water

MDPI

Article

Long-Term GIS Analysis of Seaside Impacts Associated to Infrastructures and Urbanization and Spatial Correlation with Coastal Vulnerability in a Mediterranean Area

Salvador Garcia-Ayllon

Department of Civil Engineering, Technical University of Cartagena, Cartagena 30203, Spain; salvador.ayllon@upct.es; Tel.: +34-968-325-768

Received: 10 October 2018; Accepted: 9 November 2018; Published: 13 November 2018

Abstract: The relationship between the impacts of coastal perimeter transformations derived from human activity and coastal vulnerability is not easy to assess. The impacts associated with coastal dynamics are phenomena that usually develop very extensively over a considerable time. These are transformations that cause significant environmental damage in vulnerable coastal areas, but whose results are very often not really visible until 10, 20, or even 40 years have elapsed. In addition, the analysis and quantification of the current context in complex territories is particularly difficult, since the spatial feedback of various issues and its consequences can generate an uncertain scenario with many interrelated variables. In this field, the use of GIS tools can be of great help to objectively analyze the relationship between coastal anthropization and its impact on its vulnerability in order to correct wrong inertias in vulnerable coastal areas. To this end, a long-term GIS analysis has been carried out of the impacts from urbanization and seaside infrastructures suffered by a complex Mediterranean coastal area in Spain. This territory, with singular elements such as dune ridges, beaches located in protected areas, and a coastal lagoon, will be evaluated using GIS spatio-temporal indicators over the last 90 years and geostatistical correlation methods. This approach will allow us to better understand the relationship between territorial transformations on the coast and the current coastal vulnerability of this area.

Keywords: seaside impacts; coastal vulnerability; Mar Menor; long-term GIS analysis; marine infrastructures impact; coastal urbanization impact

1. Introduction

1.1. The Concept of Coastal Vulnerability: Causes and Assessment

The concept of coastal vulnerability presents a very broad field of study that has been analyzed from different points of view for some time [1]. There are several approaches that focus on, for example, their analysis of the environmental [2], economic [3], physical [4], or social impact [5], and some that even consider the combination of several of them [6,7]. This problem can affect various environments such as large cities [8], natural areas [9], tourist beaches [10], developing countries [11], or small islands [12], among others, and is often associated with phenomena such as urbanization and the construction of infrastructures [13,14], hazards [15,16], and in recent years, the global climate change of the planet [8,17]. However, there are many other less common reasons that may bring about this problem, such as land subsidence [4], hydrology [18], or soil erosion [19] for example, with a complex set of these causes ultimately being the final origin of the vulnerability of a coastal territory in many cases.

Monitoring this phenomenon for the performance of strategies for its mitigation is not straight forward [20,21]. The scientific bibliography provides several analysis tools such as the classical study of sedimentary transport [22], the development of risk management matrixes [18] or the analysis of the rise in sea level [23] of current scenarios. In this field, it is evident that the physical transformations of the territory derived from human activity inevitably modify the coastal vulnerability of an area. That is why it is very important to understand the negative repercussions of human activities in order to avoid undesirable tendencies that accentuate its coastal vulnerability. Nevertheless, how can we determine the extent to which the constructions or land transformations increase the coastal vulnerability of a large-scale territory? The effects appear slowly and are very often difficult to measure [1]. Furthermore, there is usually no clear interconnection between causes and effects in this matter. The question becomes even more complex when several variables are interwoven in the analysis.

A habitually complex context in this matter is the determination of coastal vulnerability in urbanized areas. This phenomenon is particularly challenging if we seek to determine the origin of the current effects in areas that have been subjected to fast urbanization processes, because of the difficulty of evaluating multiple long-term impacts in short periods of time [24]. In this case, integrated analysis such as the DPSIR (Drivers forces, Pressures, State, Impact & Response) method [25] or multi-level nested frameworks for socioecological systems [26,27], for example, are also recommended. Due to that, despite being a more recent technique in this field, the implementation of multivariable GIS analysis is today becoming more and more widespread, thanks to the progressive improvement in the method's accuracy [28]. This tool allows the spatial variable in the analysis to be introduced and is transversal to the diverse origins of the issues.

A problem usually faced by this methodology is the difficulty in modeling the different physical phenomena that contribute to impacts in the coastal area [20]. It should also be noted that current issues associated with coastal vulnerability require a precise analysis of subjects whose time frame in reality extends over a period of years or even decades [1,29]. The effects of buildings and infrastructures, hazards or climate change in coastal dynamics, the transformation of the territory, or the alteration of the soil usually require a precise evaluation over 10, 30, or even 50 years to be reliably appreciated.

In this context, GIS analysis can prove very interesting if we have the necessary geo-referenced information [30–32]. In this field, several interesting studies are available that parameterize the coastal vulnerability of a territory through some spatial indicator [8,33,34]. Nevertheless, the retro-historic GIS type of analysis in this matter from an integrated perspective is a very scarcely studied field of research [35]. The implementation of GIS indicators through historical spatial information allows modeling the phenomenon from a numerical approach that can additionally be analyzed from a statistical perspective [36]. This approach becomes even more infrequent in scientific studies if we are faced with territorially complex areas in which there are several interrelated causes whose effects feed on each other. In this context, the use of geo-statistical tools can be very useful and innovative since it enables spatially cause-effect relationships to be correlated in these complex environments.

1.2. The Mar Menor Case Study: A Coastal Complex and Antropized Territory

In this study, the relationship between coastal transformations and the current vulnerability for a complex territory such as the Mar Menor, a coastal lagoon located in the Southeast of Spain, will be evaluated. This is a coastal territory of high environmental value [37–39] and with a 20 km long dune cord that separates two seas with very different characteristics (Figure 1). The area has been subjected to the impact of an intense human activity linked to tourism [40,41] in recent decades, which has resulted in massive urbanization, the construction of marine infrastructures, and the land transformation of its coastal perimeter [42].

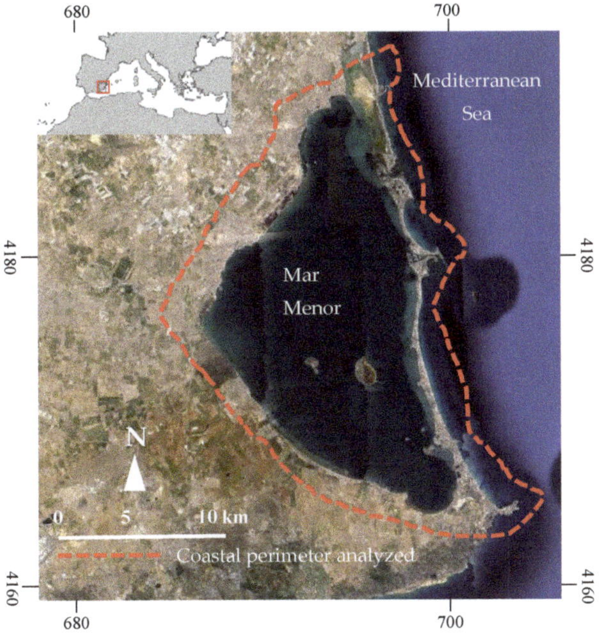

Figure 1. Scope of the study (source: SITmurcia [43]).

Several studies have examined the impact of these and other activities in the area at a biological and ecological level [44–46]. However, the assessment of the physical vulnerability of the coastal territory as a result of human activity remains scarcely investigated since it is a very long-term impact that is not easy to visualize at a glance, requiring large-scale and long-term analysis. In this sense, the extent to which the different constructions and transformations of the territory have affected its current coastal vulnerability is not known, and has been the subject of social controversy on different occasions. Consequently, through the development of a retrospective GIS analysis, the physical impact of different phenomena in the coastal perimeter will be evaluated from the 1970's (moment when mass tourism began in the area) to the present day. This impact will be correlated at a spatial level with the different current coastal risks by using geo-statistical tools in order to determine the extent to which human actions have contributed positively or negatively to the coastal vulnerability of the area during the last decades (analyzed area included in a KML file as supplementary material).

2. Materials and Methods

The selected territory is a very interesting case in which to propose this new methodology of long-term GIS retro-historic geo-statistical analysis of coastal vulnerability for several reasons. In the first place, it is a complex case of evaluation in which the different parameters of analysis may have certain interrelationship of feedback, generating a framework of difficult identification of the cause-effect relationships. Secondly, it deals with a series of impacts and physical transformations in the coastal perimeter whose overall effects are not easy to evaluate, since they are phenomena whose incidence emerges over decades. Finally, it is a territory in which geo-referenced spatial information is available for almost 90 years (the first aerial photographs were taken in 1929), thus enabling a wide retro-historic GIS analysis with numerous data.

The analysis of its coastal vulnerability will be carried out from an evaluation perspective of its physical support against the different impacts of direct and indirect human activity. This evaluation will focus on how the human transformations of the territory during the last decades have increased, or not, its current coastal vulnerability. It should be noted that this territory has been subjected to human action since Roman times; since then some small fishing population settlements have existed for example [47]. Nevertheless, it was with the arrival of tourism from the 1960's [40] that the current main coastal impacts started (urbanization, construction of ports, and marine infrastructures, artificial widening of the communication channels between the Mar Menor and the Mediterranean, land fillings of beaches, etc.).

Therefore, based on all the above, a diagnosis of the long-term impacts of physical transformations on this coastal area will be performed using GIS retro-historic indicators; the following three global phenomena are used as evaluation variables: The urbanization of the coastal perimeter, the incidence of ports and marine infrastructures, and the alteration of coastal edges uses. Later, the geo-statistical correlation between these GIS indicators and current coastal vulnerabilities will be spatially assessed. Below, the modeling criteria of the retro-historic GIS indicators, vulnerability evaluation, and geo-statistical methodology are presented.

2.1. Retrohistoric GIS Indicators of Impact

2.1.1. Analysis of the Impact of Urbanization on the Coastal Perimeter: UTD Index

The phenomena of fast urbanization are one of the elements most commonly associated with coastal vulnerability. On the one hand, this fast urbanization usually introduces an important process of physical transformation of the coast that generates important environmental impacts on the original status of the territory. These impacts may cause alterations in coastal dynamics whose effects may not become apparent for years or even decades. The situation became especially interesting in this complex territorial context where we have a coastal perimeter that is highly urbanized as a consequence of tourism in the last decades and two seas separated by an ancient dune cord now strongly urbanized since the middle of the 1960's (Figure 2).

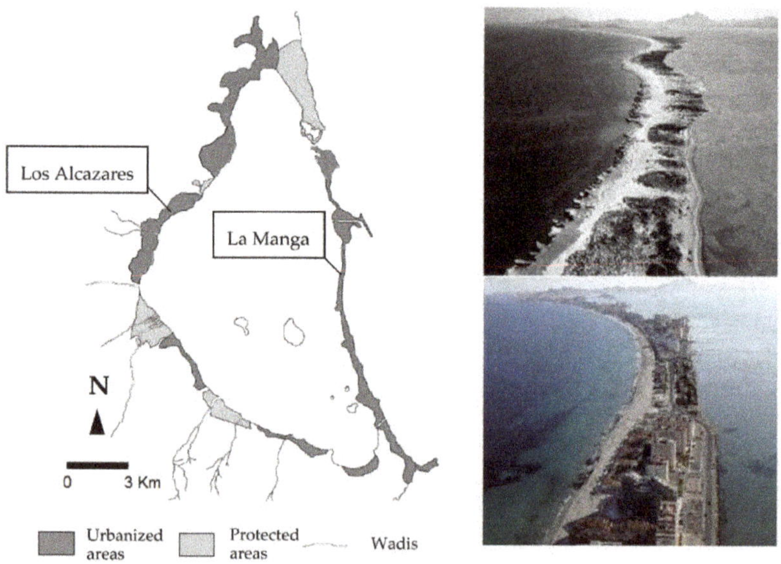

(a) **(b)**

Figure 2. (**a**) Current situation of the Mar Menor: Urbanized areas, protected salt marshes and wadis. (**b**) La Manga ancient dune cord in 1950 (**up**) and now (**down**, source: Paisajes Españoles S.A.).

On the other hand, fast and uncontrolled urbanization processes tend to notably increase the vulnerability risks associated with natural hazards such as floods. In this case, the coastal perimeter is strongly threatened as it is the mouth of numerous wadis in its Eastern and Southern perimeters. This context, coupled with the accelerated urban growth on the coast and the Mediterranean weather in which the rains are scarce but torrential, form a dangerous cocktail. In this sense, we can find several recent episodes in which these runoff waterways (through which water usually does not circulate), have become the natural floods channels, devastating important coastal settlements such as Los Alcazares (Figure 3).

(a) (b)

Figure 3. Floods occurred in the coastal town of Los Alcazares in (**a**) 1953 and (**b**) 2016 (source: Town Council of Los Alcazares).

To evaluate the impact of the urbanization processes in the coastal perimeter in the last decades, a long-term analysis will be carried out using a GIS indicator of urban transformation density UTD (1). This indicator will make a comparison between the state of a territory during a period $t_1 - t_2$ and the intensity of urbanization detected during a similar previous period of time $t_0 - t_1$. In this way, the indicator will take into account both the processes of transformation of natural areas into urbanized ones, as well as the transformation of areas already urbanized by new buildings. This will give us at the surface level not only an idea of the intensity of transformation of the coastal territory, but also a map of its building density. The indicator is formulated as follows:

$$UTD_{t_1-t_0}^{t_2-t_1} = \frac{\iiint F(x,y,z)_{t_1}^{t_2}}{\iiint H(x,y,z)_{t_0}^{t_1}} \quad (1)$$

with $F(x,y,z)_{t_1}^{t_2}$ being the distribution of building density increase between t_1 and t_2 and $H(x,y,z)_{t_0}^{t_1}$ the building density increase between t_0 and t_1 for a coastal buffer of 2 km.

2.1.2. Analysis of the Impact of the Construction of Marine Infrastructures (CIR)

The Mar Menor and the Mediterranean Sea have in these coastal perimeters a varied catalog of marinas (11) and seaside infrastructures such as dikes and breakwaters (>20, Figure 4). All these elements have been built mainly during the 1970's, 1980's, and 1990's with the arrival of tourism.

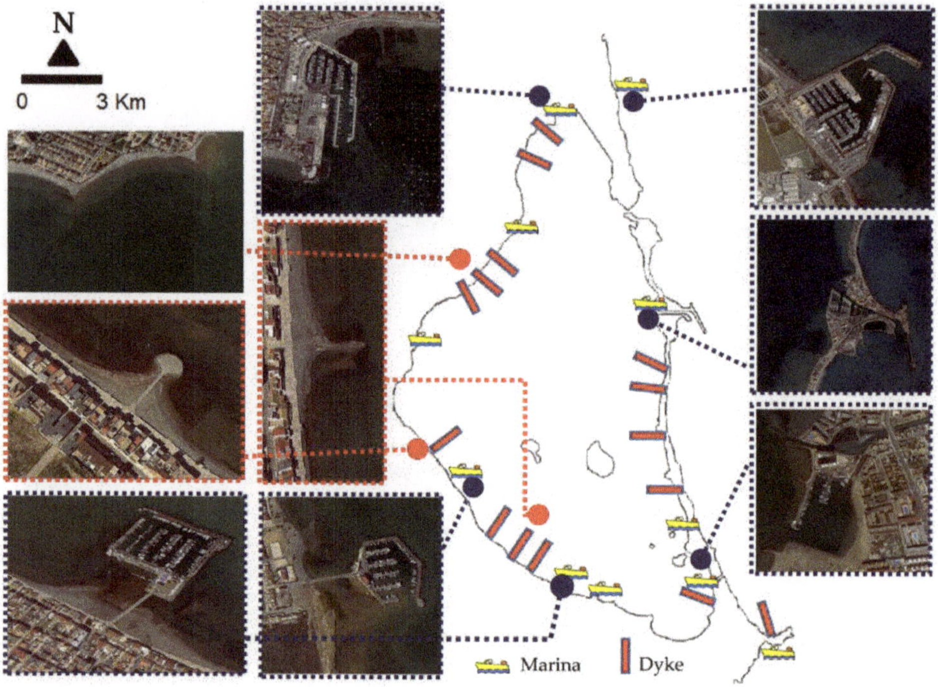

Figure 4. Marinas and coastal infrastructures built in the Mar Menor and the Mediterranean Sea since the 1950's (data source: SITmurcia [43]).

To evaluate the impact of these infrastructures, the evolution of the changes in the coastal perimeter over time associated with their construction will be analyzed spatially. The analysis will be carried out through the use of a GIS indicator. This indicator will measure the surface area of the seaside perimeter altered derived from the construction of each infrastructure during a period of time, in relation to the alteration observed during an analogous period on the same stretch of coastline before the construction of the infrastructure. As the area of influence of each of the infrastructures, a length of 200 m buffer will be taken upstream and downstream on the coastline perimeter from the location of the infrastructure. This length of analysis will be reduced if unchanged behavior is detected during 50 consecutive meters, since the coastal area of influence of the infrastructure is understood as minor. The GIS indicator will be called the Index of Coastal Infrastructure Repercussion (CIR) and it is obtained, as shown in Reference (2):

$$CIR_{t_1-t_0}^{t_2-t_1} = \frac{\iint_{t_1}^{t_2} F(x,y)}{\iint_{t_0}^{t_1} H(x,y)} \tag{2}$$

with $F(x,y)$ being the function that spatially quantifies the neighboring coastal surface area modified from the construction of an infrastructure from the time of its construction (t_1) to the present (t_2) and $H(x,y)$ the function that spatially quantifies the evolution of this coastal surface into a period

$t_0 - t_1$ analogous in time and prior to the construction of the infrastructure (see null-hypothesis in Section 2.1.4).

2.1.3. Analysis of Direct Land Transformations (DLT) in the Coastal Edge

There are other factors with major impacts that may generate an artificial alteration of the coastal configuration by direct anthropization not linked to building constructions or marine infrastructures. This is the case, for example, of channel dredging, agricultural land use modifications, coastal landfills, or simply alterations of the coastal space that cannot be attributed to a single cause or whose origin is unknown. In the case analyzed, we have a varied catalog of anthropic activities in this field (Figure 5), to which the territorial complexity of the existence of the ancient dune cord that separates the Mar Menor from the Mediterranean Sea must also be added. These two water masses are connected through 5 channels called "golas", whose configuration is quite heterogeneous. On the one hand we find three such channels in the Northern area, which, being located in an environmentally protected area, remain theoretically unaltered. On the other hand, the other two channels are further South and there is also an interior dredged area in the shape of a clover in the North, whose natural configurations were altered, with its bottom having been widened and dredged to facilitate maritime navigation of large boats. This last phenomenon has given rise to controversies about a possible "mediterraneanisation" process of the Mar Menor, which will be discussed later.

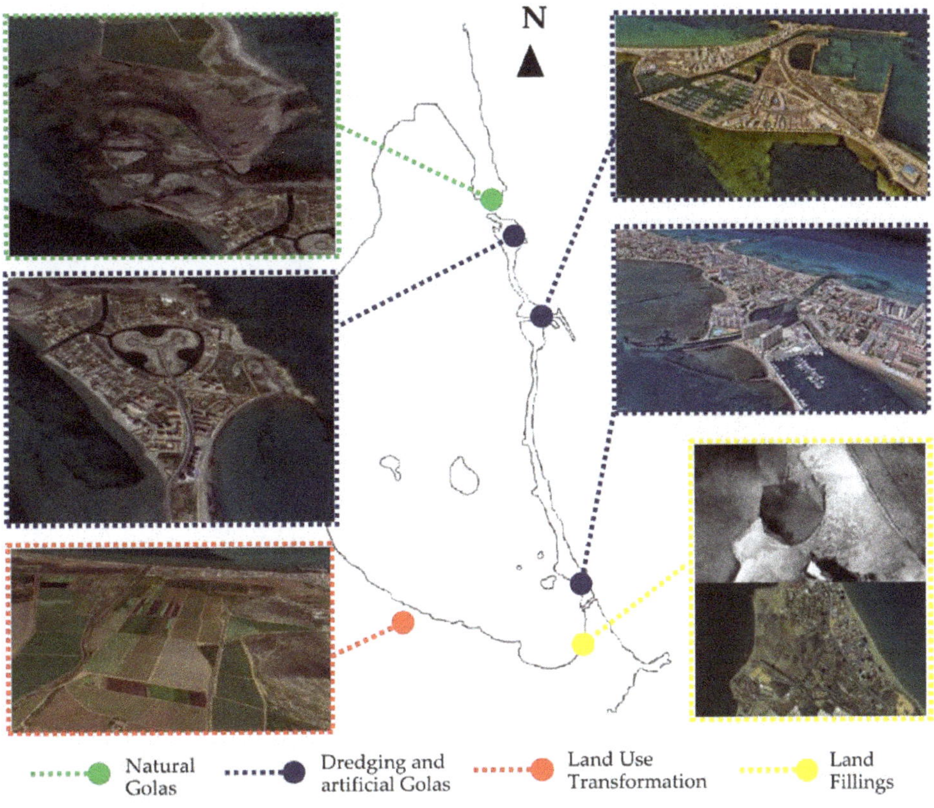

Figure 5. Marinas and coastal infrastructures of the Mar Menor and the Mediterranean Sea (data source: SITmurcia [43]).

Another important issue in this section is the alteration of the coastal territory through land fillings or land use change. In the first case, we find important alterations of the original profile of the beaches with sand landfills or land gained from the sea, especially in La Manga area. In the second case, the transformation of land for agricultural activities stands out especially in the innermost perimeter of the Mar Menor. This transformation affects the urbanized and non-urbanized areas of the coast by modifying the natural relief of the soil, influencing the arrival of water and sludge as a result of flooding.

All these concepts represent different forms of anthropization of the coastal perimeter through different processes of direct transformation of land that will be grouped in the index of direct land transformation (DLT):

$$DLT_{t_2-t_0}^{t_2-t_1} = \iiint \frac{G_1(x,y,z)_{t_1}^{t_2}}{J_1(x,y,z)_{t_3}^{t_0}} + \iiint \frac{G_2(x,y,z)_{t_1}^{t_2}}{J_2(x,y,z)_{t_3}^{t_0}} + \iiint \frac{G_3(x,y,z)_{t_1}^{t_2}}{J_3(x,y,z)_{t_3}^{t_0}} \quad (3)$$

with G_i being the function that quantifies direct land transformations by the dredging actions (1), coastal landfills (2), and non-urban land use changes (3) for a coastal strip 2 km wide between t_1 and t_2, and J_i the function that spatially quantifies the same coastal area transformed because of these three phenomena during a period $t_0 - t_1$ analogous in time and prior to $t_1 - t_2$ (see null-hypothesis in Section 2.1.4).

2.1.4. The Null Hypothesis: Zero-Effect Reference Sample

One of the basic aspects of this GIS retro-historical evaluation methodology of human impacts in the territory is inter-temporal comparative analysis [35]. In this context, an analogous concept to a control sample of a laboratory test or the null hypothesis of a statistical analysis is necessary. In this sense, the case analyzed is very interesting because it has spatially geo-referenced information of all its territory with aerial photography in the years 1929, 1932, 1945, and 1956. This information is of great interest since this period between 1929 and the mid-1960s predates the arrival of tourism and the urbanization process of the coast, allowing a geo-referenced spatial comparative analysis.

We should also take into account that the level of accuracy in the geo-referenced information of both periods (1929–1956 and 1956–2017) is not the same. However, as can be seen in Table 1, the information available is sufficiently accurate to enable a temporary comparative analysis to be performed using GIS indicators of the impact of the urbanization process, the construction of ports and coastal infrastructures, and the development of large transformation of lands in the coastal perimeter.

Table 1. Technical characteristics of geo-referenced data used.

Mapping Data	Pixel Size Projected on the GSD Ground (cm)		Planimetric Accuracy (X,Y) Mean Squared Error (m)	Altimetric Accuracy (Z) Mean Squared Error (m)	Mesh Step
	Flight	Orthophoto			
1929–1956	45	50	<1.00	<2.00	5 × 5
1957–2017	22	25	<0.50	<1.00	5 × 5

2.2. *Spatial Coastal Vulnerability Assessment: Methodological Considerations to the Model*

The assessment of coastal vulnerability must be carried out through contrasted and homogeneous scientific methodologies at a spatial level. An approximate estimation of a coastal global vulnerability Π_z of the different areas of a territory can be obtained by means of an indicator that groups together the most common existing risks. This approximation Π_z can be estimated as the homogenous sum of the different existing vulnerabilities as a result of natural hazards modeled through GIS indices δ_i and weighted by corrective coefficients λ_i to statistically assess their probability. The format of global index (4), partial indexes (5), and weighting coefficients (6) will be as follows:

$$\Pi_z = \sum_n \lambda_i \delta_i \quad \text{with } \Pi \in [0,1] \tag{4}$$

$$\delta = \{\delta_1; \delta_2; \delta_i; \delta_n\} \quad \text{with } \delta \in [0,1] \tag{5}$$

$$\lambda = \{\lambda_1; \lambda_2; \lambda_i; \lambda_n\} \quad \text{with } \lambda \in [0,1] \tag{6}$$

where partial indexes of vulnerability risks δ are detected in a territory and the weighting coefficients λ are obtained as follows:

$$\delta_i = \Phi\left[\frac{T_j}{U_M}\right] \quad \text{with } \delta \in [0,1] \tag{7}$$

$$\lambda_i = \Psi\left(\frac{\lambda_i}{\lambda_m}\right) \quad \text{with} \sum_i \lambda_i = 1 \tag{8}$$

The function Φ will model the risk maps for each hazard variable δ_i in a dimensionless way. This function is made up of a density map of average values T for each one of the j areas of the territory divided by the maximum values U of the territory. The function Ψ will establish the values of the weighting coefficients λ_i as a function of a probability ratio of each of the variables in relation to an average value λ_m. This probabilistic assessment should be carried out based on a statistical evaluation criterion that can be modeled in a common way for all the risk variables δ_i (in this case, the measurement will be based on the different return periods for flooding for example). It should be noted that the different GIS indicators may be modeling phenomena of a very different nature (floods of different origin, earthquakes, hurricanes, impacts of climate change, fire risk, etc.). Nevertheless, it is important to remember that the effect derived from these elements must be implemented in a homogeneous manner by generating as output units that can be added in a dimensionless way to perform the global coastal vulnerability index.

In this study, the flooding of land and marine origin have been selected as major risks to analyze the global assessment of the coastal vulnerability index of this territory. Other natural hazards, such as those derived from the risk of fire, seismic movements, or hurricanes, have not been taken into account since they are not significant cases with a negligible historical occurrence rate. The methodology for the evaluation of both flood risks has been developed as follows.

2.2.1. Flood Risk of Marine Origin

To face the complexity of the different aspects that make up the calculation of the flood throughout this coastal territory, a three-phase methodology based on the criteria of the Methodological Guide of the Spanish National Flood Mapping System [48] has been followed.

In the first phase, the entire coastline is flooded only by the dynamics of the sea level (derived from the effects of astronomical and meteorological tides) without surf. With this approach, there are valid results in the areas where the waves have no relevance (inside estuaries or sheltered from external infrastructures). To do this, the extreme regime of flood elevation from the series of sea level data (available for more than 60 years of data) is adjusted for each position along the coast, calculating the level of flood associated with the return period T = 100 years with the Peaks Over Threshold (POT) technique [49] and adjusting the distribution function by means of the Generalized Extreme Value (GEV) statistical distribution. As a result, the sea level is obtained for each position, from which the corresponding Digital Terrain Model (hereinafter, DTM) benchmark will have to be subtracted from the coast to obtain the openwork on the ground.

In the second phase, the coastal areas where the waves hit directly are corrected, which is the coast line which is not protected from the waves. In this second approach, terrain profiles are drawn to resolve the flood in two dimensions, profile by profile, incorporating the combined effect of waves and sea level. The effect of the swell is evaluated by means of the two-dimensional numerical model IH-2VOF [50], which solves the Navier-Stokes equations, by using the Downscaled Ocean Waves (DOW) wave database [51], obtained from the data series of the C3E project [52]. This allows us to correctly characterize the wave propagated to the coast with a spatial resolution of at least 200 m.

Finally, in the third phase, the flood envelope is obtained by the sum of the flood zone by level and the flood zone by swell.

2.2.2. Flood Risk of Land Origin

The delimitation of the flood zones is carried out by defining the so-called Significant Potential Flood Hazard Areas (SPFHAs). These areas are obtained from the Preliminary Flood Risk Assessment (hereinafter, PFRA) in accordance with Directive 2007/60 of the European Commission [53] in several ways:

- From a hydrological study in which the flow rates are determined associated to the corresponding Return Period considered in the PFRA, in this case 100 years. Once the flows have been defined, a hydraulic study determines the levels reached by the sheet of water and with them the extension of the flooded area associated with that frequency.
- From geomorphological-historical studies that allow the delimitation of areas with a low probability of flooding, based on historical evidence, and identified geo-morphologically.
- Based on a mixed methodology, which includes the two previous methods, allowing more reliable results.

In the case of considering structures of rolling or derivation of flows in the hydrological calculation it is considered that the flows are in an altered regime; otherwise they are deemed to be in a natural regime. For the analysis of the flood risks of land origin, the protocols established in the Methodological Guide for the development of the Spanish National Flood Mapping System [48] to model cartographic, hydrologic, geomorphologic and hydraulic boundary conditions have been used. In this case, since the most affected area has a more complex orography (see the results section below), Laser Imaging Detection and Ranging (LIDAR) tools have been used to generate the DTM. This higher level of precision is justified by the need for modeling the surfaces of the watersheds in a reliable way with the actual drainage directions. The cells used are at least 25×25 in agricultural or natural land and 5×5 in urban areas.

For this method, in relation to the concept of return period, it is important to make certain clarifications. In numerical terms, it is equivalent to the probability of having an equal or higher avenue flow in a given year, that is, the probability of exceeding the flow in a year. For example, for a return period of 100 years, that probability $F(x) = 1/T = 1/100 = 0.01 = 1\%$. Thus, there is a 1% probability that one year this flow value will be exceeded and a 99% probability that it will not be exceeded. However, this does not imply that two or more avenues of such or higher intensity cannot occur within the same year, since the return period is a statistical concept and depends on the duration of the interval considered. Should we wish to calculate the probability of equaling or exceeding this value during a period of N years (statistical concept of Risk) for a return period T, it would be calculated by means of the following expression (9):

$$1 - [1 - (1/T)]^{N} \tag{9}$$

Thus, according to Table 2, an area affected by flooding in a period of 100 years return zone has a probability of 22.2% of being flooded in a period of 25 consecutive years and 39.5% to be flooded in 50 consecutive years.

Table 2. Probability of occurrence for T = 100 years.

Return Period T = 100	Consecutive Years					
	1	2	5	25	50	100
Probability of occurrence (%)	1	2	4.9	22.2	39.5	63.4

On the other hand, it should also be noted that the calculated flood areas have important limitations regarding the flood that would occur in a specific event. Current techniques, although they are very precise, have important restrictions that make the actual flood of an event vary significantly from what was calculated. In this sense, we must highlight two important limitations:

- The first is related to the effects of erosion, landslides, sedimentations, etc. Herein they are only taken into account by means of geomorphological criteria, and in a specific situation, especially in high slopes, the geomorphological changes can be very important and the results differ significantly.
- The second is that the existing mathematical models cannot take into account the effects that certain elements can produce by obstructions and derivations of the flow. For example, fallen trees, vehicles, etc., can obstruct bridges, streets, etc., thus deriving the flow in any direction, which is unpredictable. In this same context, we have the cases of large flood areas in large (concentrated or dispersed) urban settlements. The difficulties to adequately represent all the artificial elements that influence the characteristics of the flow in the hydraulic model, and the computational limitations derived from a large amount of processed data, may also cause the flood results to differ from the model forecasts.

2.3. Geostatistical Correlation between Human Actions and Coastal Vulnerability

Once the distributions at the spatial level of the impact and coastal vulnerability indexes have been obtained, we can evaluate the possible spatial correlation between them by using geo-statistical methods. This analysis will allow us to assess to what extent the transformations made by human activity in the coastal perimeter of a territory have influenced the current coastal vulnerability existing in it. The spatial relationships will be parameterized and assessed through the use of Global Moran's I [54] and Anselin Local Moran's I [55] bivariate statistics, both are geo-processing tools from ArcGIS Pro 10.5.0 (ESRI Corporation, Redlands, CA, USA).

Bivariate global spatial autocorrelation will allow us to assess the statistical correlation of a set of geo-located data obtained spatially and the sign of this autocorrelation (positive or negative). Bivariate Global Moran's I statistic formula is given as I (9):

$$I = \frac{n}{S_0} \frac{\sum_{i=1}^{n} \sum_{j=1}^{n} w_{i,j} z_i z_j}{\sum_{i=1}^{n} z_i^2} \tag{10}$$

where z_i is the deviation of an attribute for feature i from its mean $\left(x_i - \overline{X}\right)$, $w_{i,j}$ is the spatial weight between feature i and j, n is equal to the total number of features, and S_0 is the aggregate of all the spatial weights of (11):

$$S_0 = \sum_{i=1}^{n} \sum_{j=1}^{n} w_{i,j} \tag{11}$$

The z_I-score for the statistic is computed, as in Reference (12):

$$z_I = \frac{I - E[I]}{\sqrt{V[I]}} \tag{12}$$

where $E[I]$ and $V[I]$ can be calculated as follows:

$$E[I] = -1/(n-1) \tag{13}$$

$$V[I] = EI^2 - E[I]^2 \quad (14)$$

Global spatial GIS autocorrelation will return three values: the Moran's I Index, z-score, and p-value. Given a series of spatial features and an associated attribute, bivariate Global Moran's I statistic indicates whether the pattern expressed is clustered, dispersed, or random and its degree of statistical correlation. When the z-score or p-value indicates statistical significance, a positive Moran's I index value indicates a tendency toward clustering, while a negative Moran's I index value indicates tendency toward dispersion. The z-score and p-value are measures of statistical significance which inform us whether or not to reject the null hypothesis. For this analysis, the null hypothesis states that the values associated with features do not have any statistical correlation.

From this information, we will be able to implement, in a geo-located way, the so-called hot and cold points in the mapping through the Local Indicators of Spatial Association (LISA) from Anselin [55]. Each Anselin Local Moran's I statistic of spatial association I is given as:

$$I_i = \frac{x_i - \overline{X}}{S_i^2} \sum_{j=1, j=i}^{n} w_{i,j}\left(x_j - \overline{X}\right) \quad (15)$$

where x_i is an attribute for feature i, $\overline{X}$ is the mean of the corresponding attribute, $w_{i,j}$ is the spatial weight between feature i and j, and:

$$S_i^2 = \frac{\sum_{j=1, j=i}^{n} \left(x_j - \overline{X}\right)^2}{n-1} \quad (16)$$

with n equating to the total number of features. The z_I-score for the statistic is computed as:

$$z_I = \frac{I - E[I]}{\sqrt{V[I_i]}} \quad (17)$$

where $E[I]$ and $V[I]$ can be calculated as follows:

$$E[I] = -\frac{\sum_{j=1, j=i}^{n} w_{i,j}}{n-1} \quad (18)$$

$$V[I] = EI^2 - E[I_i]^2 \quad (19)$$

For this analysis, the null hypothesis states that the values correlation of two elements are randomly distributed. Thus, the higher (or lower) the z-score, the stronger the intensity of the clustering of these values. A z-score near zero indicates no apparent clustering within the study area. A positive z-score indicates clustering of high values. A negative z-score indicates clustering of low values. This numerical evaluation will be implemented through GIS mapping to distinguish configuration patterns of High-High clusters (high levels of impact associated with high levels of vulnerability), Low-Low clusters (low levels of impact associated with low levels of vulnerability), and spatial outliers, either High-Low (high levels of impact associated with low levels of vulnerability) or Low-High (low levels of impact associated with high levels of vulnerability).

Therefore, the bivariate statistical correlation analysis between the distributions of different GIS indicators will help us to understand, spatially, the extent to which the impacts produced by human action affect coastal vulnerability.

3. Results

The exposed methodology has been applied to the surface area detailed above with the following results. In the first place, the retro-historic GIS analysis of the impacts on the coastal edge has been carried out. Secondly, the coastal vulnerability of each area has been spatially evaluated. Finally, the spatial correlation of both phenomena has been evaluated by geo-statistical methods.

3.1. GIS Retrohistoric Analysis of the Anthropization Impacts in the Coastal Perimeter

A long-time spatial analysis of the seaside impacts in the area has been carried out through GIS retro-historic methods from the 1960's to the present. The results obtained in the analysis have been differentiated according to the structure detailed in the methodology section.

3.1.1. Urbanization Impacts

The transformation density of the urbanization processes can be observed in a summarized way in Figure 6. To simplify the spatial representation of the GIS indicators for the transformation of the coastal perimeter as a result of urbanization, outputs have been tessellated using the ArcGIS Pro 10.5.0 program (ESRI Corporation, Redlands, CA, USA). The tessellated polygons have a size of 25 × 25 m to allow an understandable visualization of the results at a large scale (in case a tile has a surface area in two or more categories, it is assigned to the category with the most surface area present). At the intensity level, it is observed how the largest and fastest-growing population densities have been generated in the ancient dune cord called La Manga. At a quantitative level of surface transformation, it can be observed that the greatest results in absolute values are found in the urban sprawl from the settlements of the inner perimeter (San Javier and Los Alcazares coastal towns).

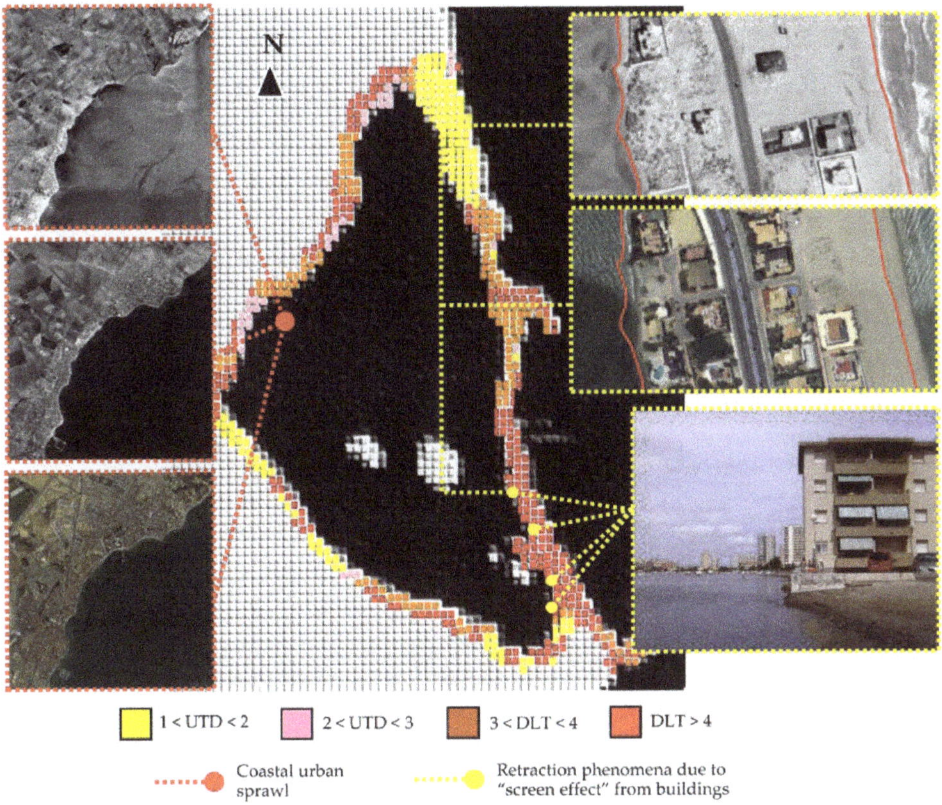

Figure 6. Mapping of $UTD_{1929-1973}^{1973-2017}$ index for seaside impacts associated to urbanization. The accelerated evolution of urban sprawl in the population of Los Alcazares in 1956–1981–2017 is detailed in red on the left. The most significant cases of dune shrinkage phenomena on the beaches of the Mar Menor in La Manga are indicated in yellow on the right (below) and the detailed evolution of the dune profile of one of them from 1981 (marked with a red line) until 2017 (above).

It is interesting to observe in the area of La Manga how sometimes there is no correspondence between the initial natural surface and the actual urbanized surface at the two-dimensional level. The incidents detected correspond mainly to retraction phenomena on the beaches of the ancient dune cord of the Mar Menor. These beaches were formerly fed by the sand from the beaches of the Mediterranean thanks to the prevailing winds from the East. The current "screen effect" generated by the massive construction of buildings in the old dune belt has caused retraction effects in a generalized manner in the dune profile of the Mar Menor beaches, with alarming cases being observed in which the disappearance of the beaches reaches the very foundations of some houses.

It is interesting to observe how the phenomenon of dune shrinkage of the beaches occurs mainly in the lower half of the old dune cord. This question is explained by La Manga being chronologically urbanized progressively from the South to the North, from the mid-1960's to the late 1990's. As a consequence of this, the phenomena of dune shrinkage (which are deferred in time and often require an average of 10–15 years to significantly emerge) appear mainly in the buildings of the Southern half that were built in the decades of the 1970's and 1980's. In addition, this duality is accentuated by the fact that the so-called "screen effect" is more intense in the Southern half, given that the urbanization in that area is denser. However, this does not mean that the same situation will not occur in the northern half in the future, when the construction in that area (which is not yet fully built) has been consolidated and the dune shrinkage phenomena has taken enough time to bring out its first consequences.

3.1.2. Ports and Coastal Infrastructure Impacts

The detected impacts derived from marinas and coastal infrastructures are widely distributed spatially (Figure 7). The distribution of the impacts has also been tessellated in meshes of 25 × 25 m to make the characterization of the phenomenon on a large scale more understandable.

Its impact on the coastal perimeter extends over the last decades and it is observed that it varies depending on two main parameters. On the one hand, it is necessary to take the geographical situation of each element into account. Variables such as the proximity of golas, tall buildings, or the mouths of the wadis can influence the global impact. Obviously, the behavior of a coastal infrastructure in the Mar Menor will be very different from that of one in the Mediterranean Sea. However, even the situation of the infrastructure within the Mar Menor shows influence in its effects on the coast.

On the other hand, the type of infrastructure built exerts the main influences. In the case of breakwater dikes, there is no great variability, with only their length being the differentiating variable. Nevertheless, in the case of ports, regardless of their size (those of greater size may have a greater impact), we find three distinct impact typologies. First, we have the traditional ports whose dock is on land reclaimed from the sea through sheltered dikes. Secondly, we find ports built with interior marinas inland, or naturally sheltered in bays. These ports are usually located in the urban plot of the coastal towns. As the third and last case, we have so-called "island ports". This mixed case includes the ports that are separated from the coastline and are linked to it through an element that allows the passage of water and sedimentary dynamics.

In the case of the dikes or breakwaters we find elements that do have an impact on the coastal dynamics, as well as elements that do not, the length of each element being a determining factor of the intensity from the impact. Among the elements that have an impact, it should be noted that there are positive, negative, and mixed consequences. Positive impacts are understood as those that contribute to stabilize the dune profile for the maintenance of the beach surface or those that correct the negative impacts of an area derived from the action of ports or dikes from neighboring areas. Negative impacts are understood to be those that negatively affect the dune stability of the area in which they are located, causing generalized phenomena of dune retraction. Finally, mixed elements are those that simultaneously generate results that can be cataloged as negative and as positive, such as increasing the beach surface on one side at the cost of reducing it on another, or maintaining the dune stability of one area, generating instability in another nearby area.

In the case of marinas, their size does seem to be directly correlated with their coastal impact. For example, the Tomas Maestre marina (numbered as 7 in Figure 4), the largest marina in Spain (and one of the largest in Europe with 1700 moorings) located in the North of La Manga and crossed by one of the golas, does not offer a remarkable behavior at the level of alteration of the dune profile. Nevertheless, we find common differentiated behaviors based mainly on the location and typology.

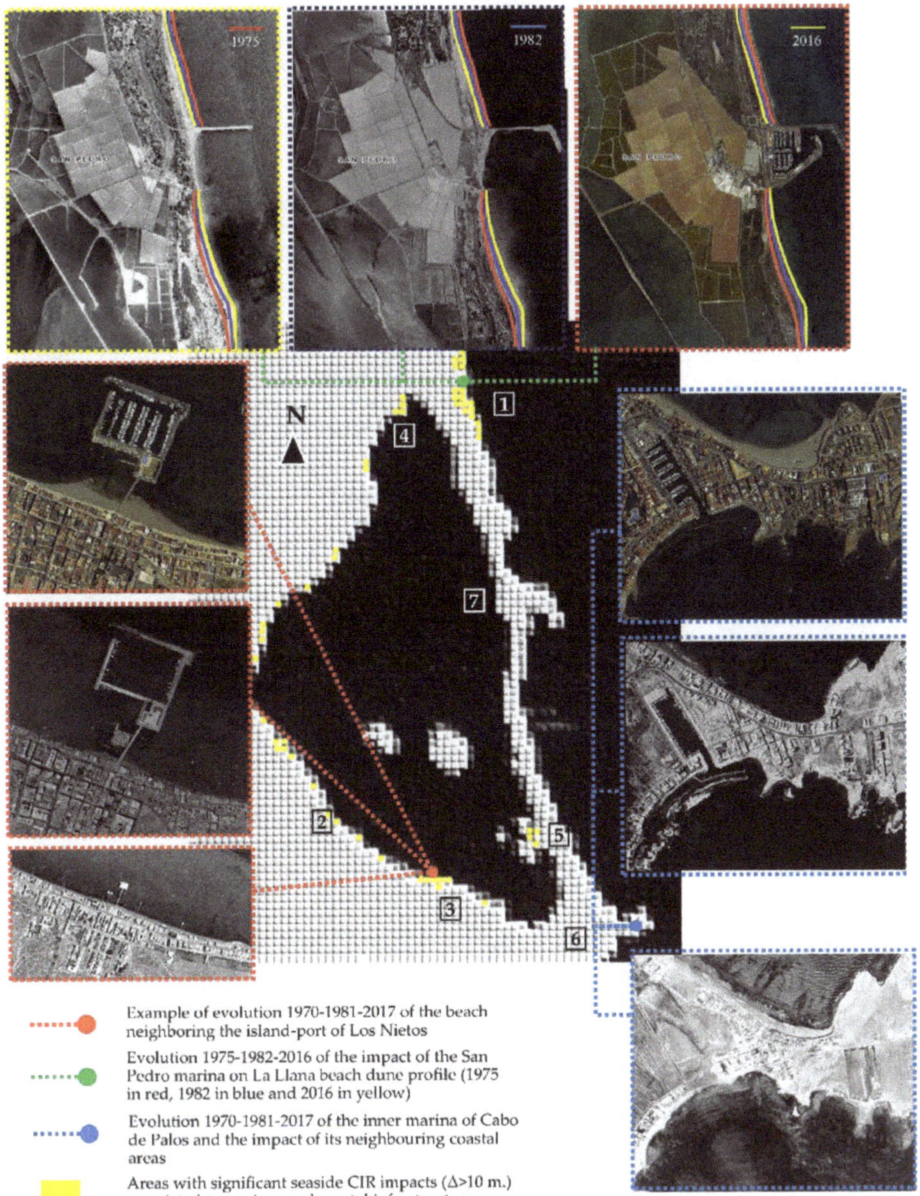

Figure 7. Tessellation of significant structural CIR impacts in the seaside edge associated to coastal alterations caused by marinas and coastal infrastructures between 1970 and 2017.

In the case of ports located on land reclaimed from the sea and harboring the port dock through breakwaters, we can find usual coastal behavior in these infrastructures in the Mediterranean. The beaches located to the North of the port increase their surface area while those located to the South experience important retraction phenomena over the years. In this section, the case of the San Pedro del Pinatar marina (numbered as 1 in Figure 7), located on the Mediterranean side, is especially interesting, and it has made almost 80% of the well-known La Llana Beach disappear within a period of 20 years (see the evolution in the upper part of Figure 7).

In the case of exempt ports called "island ports", the detected behavior is different. The area in front of the port experiences a growth of the dune surface according to the port-coast union axis. This growth has constituted, in the ports of this typology, what we would call a half-tombolo after 20–30 years of impact and it is foreseeable that it will end up eventually forming a complete tombolo (a bar of sand or shingle joining the port to the mainland). This dune growth also tends to be controversial, since when it is generated in the bottom of the sea currents, it tends to accumulate sludge on the beach that turns out to be very annoying for bathers. In addition, this growth takes place at the expense of the neighboring beaches North and South of the port, which tend to shrink and even in some cases disappear.

Finally, in the case of the marinas located in the interior of the land zone, the casuistic is more varied and depends more on the geographical position and the physical conditions that surround the port. In the case of the port of Cape of Palos (numbered as 6 in Figure 7) in the Southern Mediterranean end of La Manga, a pure inner land port, it can be seen in Figure 7 that there is no effect on the neighboring beaches (when, on the contrary, there are alterations on the beaches on the other side of the cape as a result of a dike construction, as can be seen in the Northern part of the photos). Nevertheless, we also find other cases such as the port of Lo Pagan (numbered as 4) and the Dos Mares yacht club (numbered as 5), rather half-inner land infrastructures, in which there are hard accumulations around the port. These growths are related to the boundary conditions of each port infrastructure and also lead to the accretion of sludge that damages neighboring beaches. The most significant set of impacts are detailed in Table 3.

Table 3. CIR and main impacts detected of the marinas on their neighboring beaches.

Beach Area	Location (Figure 7)	Infrastructure Associated	Average Wide Variations (Δ)	Period Evaluated	*CIR*
La Llana	Mediterranean (1)	Marina of San Pedro del Pinatar	+51.67 m. (north) −84.42 m. (south)	1978–2017	7.12
Los Urrutias	Mar Menor (2)	Los Urrutias island-type port	+22.56 (max)/0 (min) −18.12 (average)	1975–2017	2.54
Los Nietos	Mar Menor (3)	Los Nietos island-type port	+59.17 (max)/0 (min) −10.12 (average)	1970–2017	2.78
Villananitos	Mar Menor (4)	Marina Lo Pagan half-innerland type	+14.27 m. (north) +26.97m. (south)	1966–2016	1.95
El Ciervo	Mar Menor (5)	Marina Dos Mares half-innerland type	+17.17 m. (max) −7.64 m. (min)	1966–1999	1.89

3.1.3. Coastal Land Alterations Impacts

The analysis of the impacts derived from the direct alterations of the coastal space denotes a heterogeneous distribution associated to specific actions with diverse results. The area most subjected to this type of impact is mainly the old dune cord currently urbanized called La Manga (Figure 8). This area has various actions such as the direct alteration of its surface geometry, the artificial modification of two of the so-called "golas" (natural communication channels between the Mar Menor and the Mediterranean Sea, numbered as 1), the internal dredging of some areas such as the artificial clover called Veneziola (numbered as 3), or the landfilling of large marine areas such as the one called El Vivero (numbered as 2), or the road to connect the inland area with the Ciervo Island (numbered as 5).

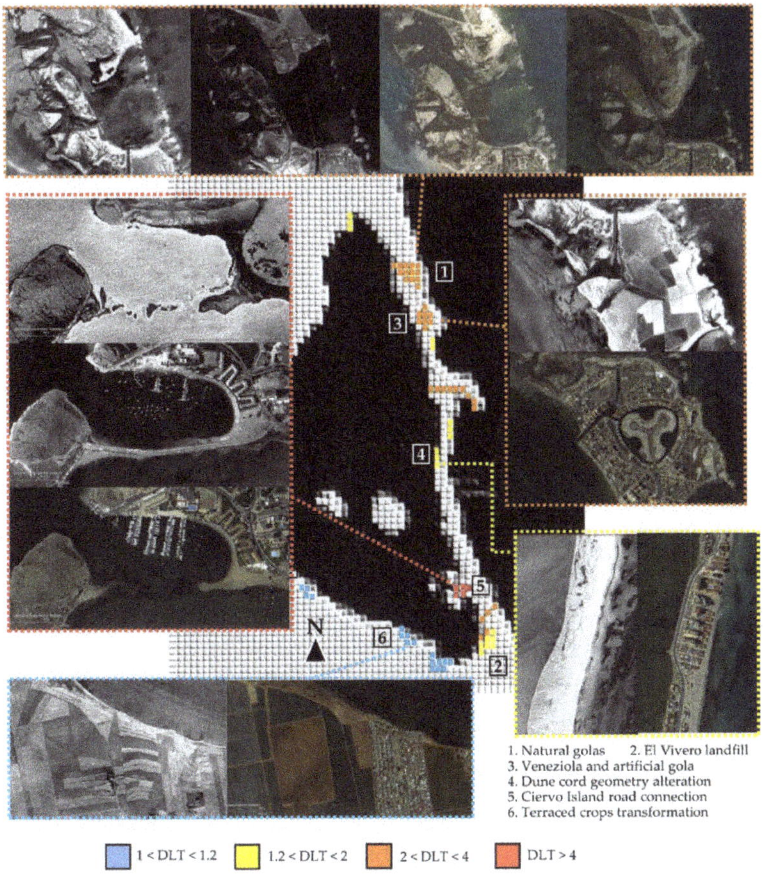

Figure 8. Tessellation of the $DLT_{1974-1929}^{2017-1974}$ mapping for significant impacts in the seaside edge associated to land alteration transformations between 1974 and 2017.

The behavior of the GIS evolution for all these elements is not homogeneous either. On the one hand, we find those whose new physical configuration has consolidated over time reaching a static equilibrium situation. This is the case of alterations in the layout of the dune cord consolidated by road infrastructures and buildings or the landfilling of El Vivero. On the other hand, we find cases in which there is a certain dynamic equilibrium (in which a static equilibrium position is not reached) or in which nature always directly opposes with forcefulness in the same sense as human action in a reiterative way. The first case can be found in the two so-called artificial "golas" or the dredging of Veneziola. In these cases, the artificial dredging made to reach a larger draft thus enabling large vessels to navigate it is not maintained at stable depths due to coastal dynamics, forcing periodical dredging. The other three natural golas also warrant special mention, whose surface and depth have moved in different directions over the last decades. However, in this case, a clear and direct cause cannot be established, since there are several nearby elements with a certain impact capacity such as sports ports, dredging, or alterations to the coastal surface. Finally, we found situations whose completely unbalanced configuration generated important alterations and whose impacts were not acceptable either environmentally or socially by the population, forcing the restoration of them to their original situation. One such case is of the road that connected the Ciervo island, whose obstruction of the sea currents generated such a "dam effect" to the North of the same that forced its dismantling.

For the mapping of the distribution of the DLT transformation indicator during the period 1967-2107, four different levels have been established to categorize the four most representative casuistic of the phenomenon. In the first place we have those cases in which the transformations are limited to the change of land use, with which the functions G_i and J_i barely have differences of less than 20%. This category fundamentally includes transformations from natural to agricultural land or significant changes in the type of agricultural use (urbanization transformations are assessed in the 3.1.1 indicator). Within this first case, the substitution of the old terraced crops by horticultural plantations of intensive agriculture stands out in the area. In the second category, we find those cases in which there has been an alteration of the land-sea configuration, but which has been consolidated in a static equilibrium over time without new alterations. This is the case of the mentioned coastal landfills and geometry changes consolidated by the road infrastructures in the old dune cord. In the third level we find those transformations of the land-sea geometry that have not reached a static equilibrium and are currently in what we have called a dynamic equilibrium. This is the case of artificially-dredged golas whose draft tends to be reduced by the effect of coastal dynamics, or of the golas that still remain natural, whose surface and draft has been changing in the last decades in different directions as a side effect of several nearby impacts. Finally, in the fourth category we find land-sea configurations that have undergone important transformations in different ways as a result of the generation of a dynamic equilibrium that is "not stable". This is the case of the execution of coastal infrastructures such as the road that connected the old dune cord with Ciervo Island, whose serious alteration of coastal dynamics forced the infrastructure to be dismantled.

In Figure 8, the distribution of the impact has been tessellated in meshes of 25 × 25 m to make the characterization of the phenomenon on a large scale more understandable. For this analysis, we must highlight that in the case of the golas, the z parameter is not obtained by geo-referencing. It has been obtained based on the abundant bathymetric documentation of the nautical charts of the area that exist for many dates. It should be noted that the level of detail of these charts for navigation is lower than in the case of the rest of the information; with the information basically being lines of homogeneous depth. However, this can be understood as an acceptable simplification of the model, since in artificial golas the depth is the result of dredging to facilitate navigation, so the bottom tends to be of a homogeneous depth. Likewise, in the so-called natural golas, since these are fairly shallow waters located between the two seas, the slope of the terrain is usually flat.

3.2. GIS Analysis of Coastal Vulnerability

A spatial analysis of the coastal vulnerability for land and marine flooding has been carried out based on the geo-referenced data available in the Ministry of Environment, Agriculture, and Water of Spain [56]. A return period of T = 100 years has been chosen for the forecast simulation, since it provides spatially significant results, whilst at the same time it results in an order of magnitude close to the total period evaluated, making the results more interesting in order to establish conclusions later.

3.2.1. Coastal Vulnerability Associated to Risk of Marine Flooding

A forecast simulation of coastal vulnerability in the territory analyzed was carried out for a return period of T = 100 years according to the criteria set out in the methodology section. The envelope of worst case scenarios for maximum events indicates an impact mainly concentrated in the Mediterranean side of the old dune cord of La Manga (Figure 9). The inner coastal perimeter of the Mar Menor scarcely suffers from the flooding of the beaches, apart from occasionally some single-family homes located on the front line of the coast. In the case of La Manga, the most affected part surpassed by a possible flood is the area corresponding to the natural golas, located in the far North (which is logical). Regarding the urbanized areas, it is interesting to observe how the effects in the Northern area are much greater than in the Southern zone. The urban configuration of the Southern area is more resistant to the phenomena of flooding of marine origin due to its greater density and more compact structure.

The most vulnerable urban areas are found in the Northern section (especially in the points with the smallest width of the dune cord and with low density urban configurations or with partial building) and in areas with artificial "golas", which are completely flooded by water. Given that a homogeneous distribution of coastal vulnerability is not found, this will be discussed in the discussion section, based on the results of the geo-statistical analysis as to whether the action of man has contributed to worsen or improve the existing situation in this section.

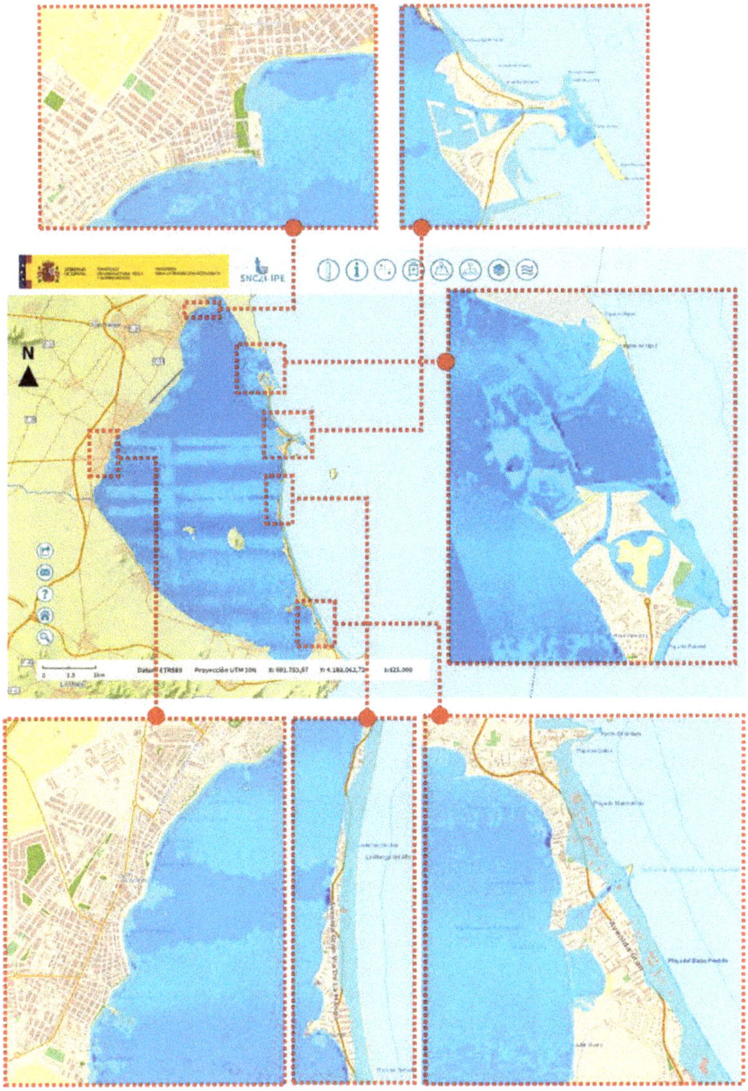

Figure 9. Forecast modeling of the danger by marine flooding on a simplified MDT of the coastal strip for a return period T = 100 years. Several detailed examples of the impact in different areas are attached (data source for GIS analysis: see Reference [56]).

3.2.2. Coastal Vulnerability Associated to Risk of Land Flooding

The analysis of coastal vulnerability associated with flood risk of river origin is even more interesting. In this case, the reverse phenomenon occurs, with the interior coastal perimeter of the Mar Menor being the area where the impact is concentrated (Figure 10). This is a consequence of the existing orographic distribution with numerous wadis that flow into that stretch of coast. In this case, the natural areas are not actually the most affected (despite the existence of crypto-wetlands theoretically within these protected areas), but the areas of greatest concentration are urban and agricultural ones. Due to this, a detailed DTM has been incorporated into these urban areas using LIDAR technology to obtain more precise results.

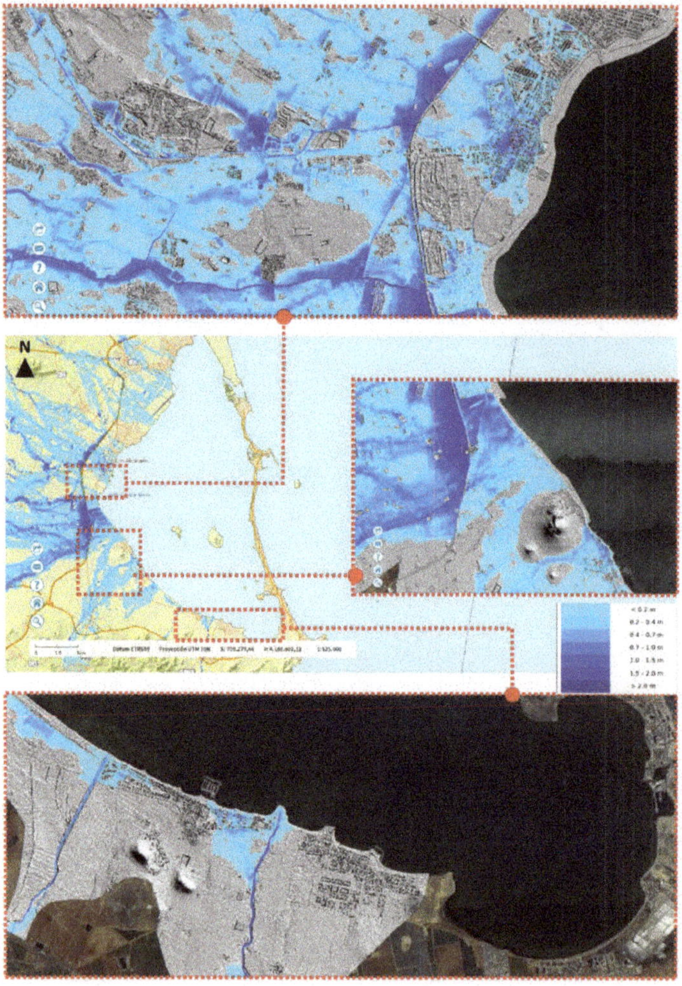

Figure 10. Modeling of the danger by fluvial flooding on a LIDAR DTM of the coastal strip for a return period T = 100 years. Three detailed examples of the impact are attached: on the urban plot in the western urban settlements (**top**), in the protected natural areas (**middle**) and in the urbanized area of the southern perimeter (**below**), (data source for GIS analysis: see Reference [56]).

In the West side we find large urban areas subject to an intense phenomenon of flooding throughout its urban surface (particularly close to the coastal town of Los Alcazares). This phenomenon is mainly a consequence of the urban configuration and the growth patterns of the town during recent decades, which have not taken the risks associated with the orography of the land into account. This growth pattern and urban structure of medium-low density have only aggravated existing problems. However, it should be noted that the highway (built in the 1990's) that surrounds the town exerts a "dam effect". This new local configuration greatly protects the town, being possibly related to the fact that the recent floods in the town have been less catastrophic than the last ones of 60 years ago. Even so, this configuration is a double-edged sword, since the possibility of occurrence of a maximal event of greater dimensions (for example, with a return period T = 500 years) could suppose the blockage of the drainage elements of the highway, thus making this dam overflow directly into the population, which would significantly aggravate the consequences for the town.

In the Southern area, we find a very different problem. In this case, the agricultural areas are the most affected. However, the current orographic configuration of the agricultural lands allows easy access of the waters to several small coastal towns, flooding them. As we have already seen, this orographic configuration has not always been thus, but has been subjected to various land processes of transformation of use and structure over the last decades. We find transformations throughout these last decades that may be susceptible to worsening as well as improving the status quo of the territory analyzed. Therefore, in the last section the extent to which human action has contributed to worsen or improve the existing situation in this field will also be discussed, based on the results of the geo-statistical analysis.

3.3. Geostatistical Bivariate Analysis of GIS Indicators

In the first place, we have made the two GIS indicators of the distribution of flood risks of marine and land origin dimensionless, as explained in the methodology section. Then, we have integrated them into a single statistically homogenous GIS indicator, called index of global coastal vulnerability I_{GCV}. This indicator has been spatially correlated with the different UTD, CIR, and DLT impact indicators through the Global Moran's I statistic. This bivariate analysis assesses the relationship two to two between the indicators at the level of two-dimensional autocorrelation in the spatial plane. The aggregate result of this first global correlation can be summarized for the scope of study in Table 4.

Table 4. Bivariate Global Moran's I statistics for spatial autocorrelation between I_{GCV} global coastal vulnerability index and the three GIS impact indices UTD, CIR, and DLT.

Bivariate Global Moran's I	$UTD–I_{GCV}$	$CIR–I_{GCV}$	$DLT–I_{GCV}$
Global Moran's Index	0.57	0.41	0.60
z-score	52.8	37.4	55.8
p-value	>0.01	>0.01	>0.01
F-Statistic	75.2	70.4	78.5
R-Squared	0.18	0.20	0.19
Adjusted R-Squared	0.19	0.22	0.20
Number of observations	1063	839	1291

The results show positive global statistical autocorrelation between indices, but higher levels of correlation for the $DLT\text{-}I_{GCV}$ and $UTD\text{-}I_{GCV}$ couples than the $CIR\text{-}I_{GCV}$ one. This difference may be due to the more direct incidence at a spatial level of the processes of urbanization and direct transformation of the coastal edge, rather than the execution of marine and port infrastructures. The incidence of the latter can be derived indirectly at a spatial level far from the focus of the impact through the coastal dynamics, so a more local geo-statistical analysis is necessary in order to analyze this relationship.

The level of local correlation between the different pairs of indices was evaluated by means of the Anselin Local Moran's I (LISA) statistic. This statistic allows us to transpose at a local spatial level the so-called hot spots (high impact-high vulnerability), cold spots (low impact–low vulnerability) and significant outliers (Low-High and High-Low cross-links) to avoid the limitations of global geo-statistical analysis. A varied catalog of different situations can be observed; the most significant results of this statistic are summarized in the tessellated mesh distribution of Figure 11.

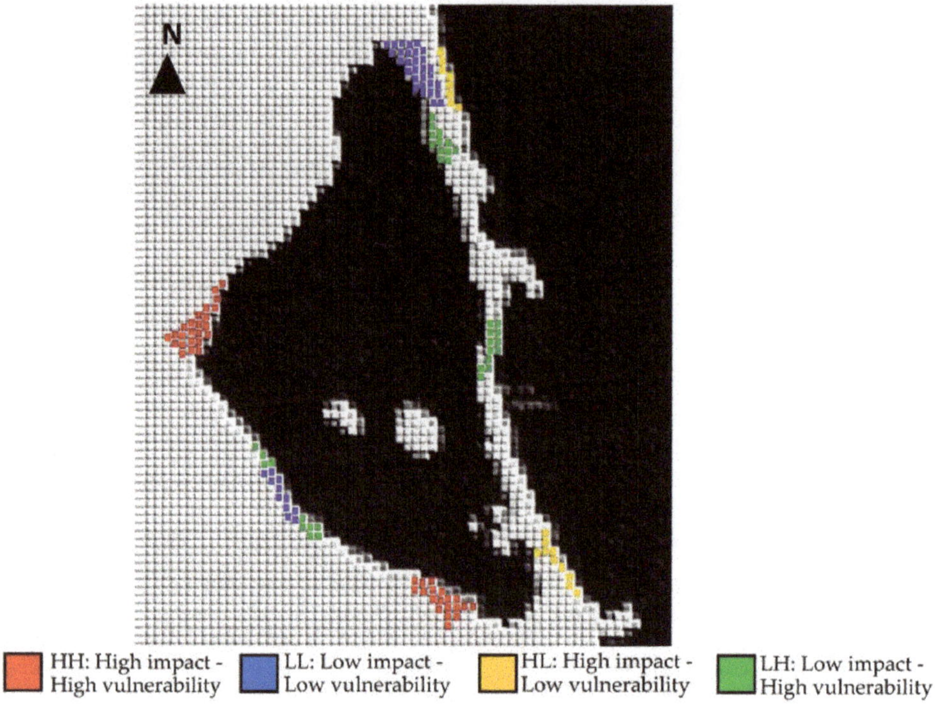

Figure 11. Tessellated mesh distribution of Local Anselin Moran's I statistic for aggregated impacts and vulnerability correlation analysis with hot (HH) and cold (LL) spots and outliers (LH and HL).

We can here observe how the processes of urbanization in the inner perimeter of the Mar Menor have notably increased the coastal vulnerability of land origin in the territory, despite the theoretical beneficial impact of some infrastructures such as the highway. Similarly, the transformation of agricultural orography in the Southern perimeter has favored this type of vulnerability, increasing the risk of flooding of land origin. Nevertheless, it is paradoxical in this case to observe how the impact of the so-called island-ports in the Mar Menor has increased the protection against the marine flooding of the Mar Menor in some villages such as Los Nietos.

On the other hand, the progressive dislocation of beaches in the Northern area of La Manga as a result of artificial dredging and the development of port infrastructures has increased the coastal vulnerability of the Northern half of La Manga due to flooding of marine origin coming from the Mediterranean. A different situation is found in the Southern half. There, the same dense and compact urbanization process that mitigates the impact of the coastal vulnerability of marine origin from the Mediterranean on the urban plot is responsible for the indirect vulnerability that occurs on the side of the Mar Menor by the disappearance of several beaches due to the effect on coastal dynamics.

4. Discussion

The Mar Menor, La Manga and its neighboring coastal territory have long been a common object of social and scientific controversy. As mentioned before, there are numerous studies in the scientific bibliography related to the Mar Menor [57–64] and the problems derived from its unique configuration and high environmental value. Most of them converge in associating the environmental impacts of the territory individually, mainly with the intensive urbanization of the coastal perimeter, the construction of marine infrastructures and the anthropization of the environment with actions such as intensive agriculture or the artificial widening of the "golas". Nevertheless, practically none of them address the problems of this territory in a global way from the perspective of the physical impact on the territory. They focus instead on analyzing the existing segmented consequences from the point of view of water quality [65,66], marine fauna [37,61], seabed vegetation cover [38,60], birds [39,67], landscape [47], etc.

In addition, the main biological, chemical, or ecological approach of most of the existing studies in the area has traditionally focused on variables with little connection with the coastal vulnerability of the territory as impact factors (nitrates from agriculture, heavy metals from old mining areas, etc.). Therefore, except for a number of specific and segmented studies in the field of coastal hydrodynamics [68,69], sea level [70], or territorial anthropization [34,47] there was no analysis from the physical perspective of the coastal territory that allows a diagnosis of the global vulnerability in this area.

This study presents a different approach to all the existing analyses to date. The different factors related to the current coastal vulnerability of the territory at a global level have been analyzed spatially. Additionally, through an innovative methodology based on the retrohistoric GIS analysis of anthropization processes derived from human activity, it has been possible to geo-statistically correlate the link between the impacts of phenomena such as the construction of ports or the change of land use, with the current coastal vulnerability. This type of diagnosis may prove to be of great interest in order to develop strategies for mitigating coastal vulnerability in a complex territory, since it not only analyzes existing risks through hazard maps, but takes into account how human activity contributes to current problems.

In this sense, to correctly implement these strategies to mitigate the existing vulnerability, it is very important to know which elements have a negative impact and which have a positive or negative impact, both in the current scenario and in their trend trajectory. In addition, it must be taken into account that complex environments such as the one analyzed can have cross-linked impacts, generating a negative effect in one place, while generating a positive effect in another. This mapping context with different scenarios of the couple human impact-vulnerability (HH, LL, HL, and LH) would allow us to implement a segmented mitigation strategy with different sub-strategies depending on how different impacts of human activity affect the current coastal vulnerability at a spatial level.

For example, in the case study analyzed, in the inland Western area of the Mar Menor, one should preferably act on the orography of the wadis, making it compatible with the current urban layout and preventing future urban developments from being carried out in the areas of greatest risk. In the Southern perimeter, attention must also be paid to the flood risk of land origin. However, where it is really necessary to act is in the configuration of the agricultural areas, since the current urban settlements have a fairly stabilized growth and it is the transformation of the agricultural orography that fosters the greatest risk of flooding at present. On the other hand, in the La Manga area, the need to implement at least two different sub-strategies is observed. In the Northern zone, it would be necessary to act in the urban plot configuration (since it is observed that it is continuing to be urbanized at present) against the risk of flooding of marine origin from the Mediterranean (in this case aggravated by the impact of the sedimentary dynamics of the area linked to the port infrastructures located further North). A very different strategy would be needed in the Southern area, where the configuration of the urban plot mitigates the risk of flooding of marine origin from the Mediterranean, while indirectly promoting negative effects on coastal vulnerability on the Mar Menor side. In this case it would be necessary to implement a strategy to rebalance the current situation between the two sides of the ancient dune cord.

This approach is especially interesting for a coastal vulnerability study given that in territories such as this one, issues like the impact of rising sea levels as a result of climate change have been socially very controversial. In fact, approaches in this territory have traditionally existed that lack a solid scientific basis; these have generated some social alarm about the impact of the risk of a sea level rise due to climate change (Figure 12). In complex environments such as this one, approaches with great scientific rigor, such as can be seen in Reference [70], may even be insufficient to assess coastal vulnerability at a comprehensive level, since it is also necessary to take into account the impacts and cross-links of other variables related to the anthropization processes of human activity and to develop detailed digital models of the terrain and urban plot able of simulating the physical reality at the local level.

Figure 12. Simulation carried out in 2007 of the estimated sea level rise in different sites of La Manga for the year 2050 as a result of climate change. Source: Greenpeace.

In this sense, possible lines of future research in this area for this territory to formulate more precise and segmented mitigation strategies could be carried out to deepen the study at the local level of certain elements whose performance was positive or negative depending on the context of analysis. For example, retrospective GIS analysis and geo-statistical correlation could be used to determine in greater detail which seaside dikes are detrimental or beneficial to the stability of sedimentary dynamics, or to assess the long-term effect of dredging on the golas at an environmental level. Both issues have traditionally generated important social debates in the area: In the first case in relation to the need to eliminate or not the marine dikes to improve the quality of beaches, and in the second, on the possibility of further expanding the golas with the aim of improving the quality of the waters in the Mar Menor, despite the risk of deepening the process of "mediterraneanisation" of the lagoon and increasing the risk of marine flooding.

Another important issue concerns the limitations of the methodology used. In the analyzed case, a very accurate geo-referenced cartography was available for different dates, as well as a DTM model that enables the orographic natural and urban reality to be represented in a reliable way. Additionally, the existence of an orographically simple terrain and an urban configuration without great constructive singularities have both considerably facilitated the work. It was only necessary to calibrate the mathematical models to represent the "dam effect" of the AP-7 highway, and the marine flooding in the golas as a result of the dredging presents some minor uncertainties. However, the model at the global level can be considered very robust, with the data used as input to the geo-statistical analysis being considered very reliable. Even so, this statement clearly cannot be extended to every study, and it is thus necessary to maximize the precision in the DTM model and take caution in mathematical models when we are faced with complex natural orography or urban configurations with several constructive singularities.

Consequently, it is clear that this diagnostic methodology can be very beneficial to improve current coastal vulnerability assessment systems and implement more accurate mitigation strategies. It is true that it is a methodology that requires an important geo-referenced database of historical character and with a great deal of precision. In this context, it cannot be said that this new technique can currently be applied in a generalized manner for analysis of coastal vulnerability in any territory. Nevertheless, we must bear in mind that national GIS databases and international Spatial Data Infrastructures (SDI) systems are becoming generalized throughout the world and perfecting their level of accuracy. Therefore, it may be an approach with a promising future that in the coming years will allow for the opening of increasingly sophisticated new lines of research in the field of coastal vulnerability.

5. Conclusions

This study has presented an innovative methodology for analyzing the coastal vulnerability of a territory based on the GIS evaluation of the spatial statistical correlation of long-term anthropic impacts and the distribution of current risks. The geo-statistical analysis carried out for the case of the Mar Menor Mediterranean lagoon reveals that the urbanization processes being developed in the last decades have generated imbalances. On the one hand, they provoke the retraction of the coastline in the old dune cord called La Manga, increasing coastal vulnerability due to flooding of marine origin. On the other hand, the inadequate urban sprawl has notably increased the vulnerability in the interior coastal perimeter due to flooding of fluvial origin in some coastal towns.

In the case of infrastructures, we find a catalog of very heterogeneous situations. The construction of some ports (although not all) have severely affected the balance of beaches, making them expand or almost disappear (and therefore increasing the risk of marine flooding). The role of several motorways, whose barrier effect theoretically mitigates the risk of terrestrial flooding in some coastal towns, should also be highlighted. This situation is actually a double-edged sword because, in the event of a flood that overcomes that dam effect, the consequences could be catastrophic. Finally, land use transformations or dredging and earthmoving in the coastal strip have also caused different impacts on the vulnerability of the territory. Changes in agricultural use in the Southern fringe have increased the risk of land-based flooding, while dredging in the La Manga golas increases that of marine origin.

Supplementary Materials: The KML file including the area of analysis is available online at http://www.mdpi.com/2073-4441/10/11/1642/s1.

Funding: This research received no external funding.

Conflicts of Interest: The author declares no conflict of interest.

References

1. Bevacqua, A.; Yu, D.; Zhang, Y. Coastal vulnerability: Evolving concepts in understanding vulnerable people and places. *Environ. Sci. Policy* **2018**, *82*, 19–29. [CrossRef]
2. Denner, K.; Phillips, M.R.; Jenkins, R.E.; Thomas, T. A coastal vulnerability and environmental risk assessment of Loughor Estuary, South Wales. *Ocean Coast. Manag.* **2015**, *116*, 478–490. [CrossRef]
3. Kantamaneni, K. Counting the cost of coastal vulnerability. *Ocean Coast. Manag.* **2016**, *132*, 155–169. [CrossRef]
4. Rimba, A.B.; Osawa, T.; Parwata, I.N.S.; As-syakur, A.R.; Kasim, F.; Astarini, I.A. Physical assessment of coastal vulnerability under enhanced land subsidence in Semarang, Indonesia, using multi-sensor satellite data. *Adv. Space Res.* **2018**, *61*, 2159–2179. [CrossRef]
5. Su, S.; Pi, J.; Wan, C.; Li, H.; Xiao, R.; Li, B. Categorizing social vulnerability patterns in Chinese coastal cities. *Ocean Coast. Manag.* **2015**, *116*, 1–8. [CrossRef]
6. Kantamaneni, K.; Phillips, M.; Thomas, T.; Jenkins, R. Assessing coastal vulnerability: Development of a combined physical and economic index. *Ocean Coast. Manag.* **2018**, *158*, 164–175. [CrossRef]
7. Tragaki, A.; Gallousi, C.; Karymbalis, E. Coastal Hazard Vulnerability Assessment Based on Geomorphic, Oceanographic and Demographic Parameters: The Case of the Peloponnese (Southern Greece). *Land* **2018**, *7*, 56. [CrossRef]
8. Mavromatidi, A.; Briche, E.; Claeys, C. Mapping and analyzing socio-environmental vulnerability to coastal hazards induced by climate change: An application to coastal Mediterranean cities in France. *Cities* **2018**, *72*, 189–200. [CrossRef]
9. Lee, Y.-J. Relationships among Environmental Attitudes, Risk Perceptions, and Coping Behavior: A Case Study of Four Environmentally Sensitive Townships in Yunlin County, Taiwan. *Sustainability* **2018**, *10*, 2663. [CrossRef]
10. Toubes, D.R.; Gössling, S.; Hall, C.M.; Scott, D. Vulnerability of Coastal Beach Tourism to Flooding: A Case Study of Galicia, Spain. *Environments* **2017**, *4*, 83. [CrossRef]
11. Ahmed, B.; Kelman, I.; Fehr, H.K.; Saha, M. Community Resilience to Cyclone Disasters in Coastal Bangladesh. *Sustainability* **2016**, *8*, 805. [CrossRef]
12. Becker, P. Dark Side of Development: Modernity, Disaster Risk and Sustainable Livelihoods in Two Coastal Communities in Fiji. *Sustainability* **2017**, *9*, 2315. [CrossRef]
13. Brown, J.M.; Morrissey, K.; Knight, P.; Prime, T.D.; Almeida, L.P.; Masselink, G.; Bird, C.O.; Dodds, D.; Plater, A.J. A coastal vulnerability assessment for planning climate resilient infrastructure. *Ocean Coast. Manag.* **2018**, *163*, 101–112. [CrossRef]
14. Kantamaneni, K.; Du, X.; Aher, S.; Singh, R.M. Building Blocks: A Quantitative Approach for Evaluating Coastal Vulnerability. *Water* **2017**, *9*, 905. [CrossRef]
15. Sahoo, B.; Bhaskaran, P.K. Multi-hazard risk assessment of coastal vulnerability from tropical cyclones—A GIS based approach for the Odisha coast. *J. Environ. Manag.* **2018**, *206*, 1166–1178. [CrossRef] [PubMed]
16. Pérez-Maqueo, O.; Martínez, M.L.; Sánchez-Barradas, F.C.; Kolb, M. Assessing Nature-Based Coastal Protection against Disasters Derived from Extreme Hydrometeorological Events in Mexico. *Sustainability* **2018**, *10*, 1317. [CrossRef]
17. Santos-Lacueva, R.; Clavé, S.A.; Saladié, Ò. The Vulnerability of Coastal Tourism Destinations to Climate Change: The Usefulness of Policy Analysis. *Sustainability* **2017**, *9*, 2062. [CrossRef]
18. Amuzu, J.; Jallow, B.P.; Kabo-Bah, A.T.; Yaffa, S. The Climate Change Vulnerability and Risk Management Matrix for the Coastal Zone of The Gambia. *Hydrology* **2018**, *5*, 14. [CrossRef]
19. Lins-de-Barros, F.M. Integrated coastal vulnerability assessment: A methodology for coastal cities management integrating socioeconomic, physical and environmental dimensions–Case study of Região dos Lagos, Rio de Janeiro, Brazil. *Ocean Coast. Manag.* **2017**, *149*, 1–11. [CrossRef]
20. Di Luccio, D.; Benassai, G.; Di Paola, G.; Rosskopf, C.M.; Mucerino, L.; Montella, R.; Contestabile, P. Monitoring and Modelling Coastal Vulnerability and Mitigation Proposal for an Archaeological Site (Kaulonia, Southern Italy). *Sustainability* **2018**, *10*, 2017. [CrossRef]
21. Ruol, P.; Martinelli, L.; Favaretto, C. Vulnerability Analysis of the Venetian Littoral and Adopted Mitigation Strategy. *Water* **2018**, *10*, 984. [CrossRef]

22. Ouillon, S. Why and How Do We Study Sediment Transport? Focus on Coastal Zones and Ongoing Methods. *Water* **2018**, *10*, 390. [CrossRef]
23. Panpeng, J.; Ahmad, M.M. Vulnerability of Fishing Communities from Sea-Level Change: A Study of Laemsing District in Chanthaburi Province, Thailand. *Sustainability* **2017**, *9*, 1388. [CrossRef]
24. Li, Y.; Zhang, X.; Zhao, X.; Ma, S.; Cao, H.; Cao, J. Assessing spatial vulnerability from rapid urbanization to inform coastal urban regional planning. *Ocean Coast. Manag.* **2016**, *123*, 53–65. [CrossRef]
25. Newton, A.; Weichselgartner, J. Hotspots of coastal vulnerability: A DPSIR analysis to find societal pathways and responses. *Estuar. Coast. Shelf Sci.* **2014**, *140*, 123–133. [CrossRef]
26. Ostrom, E. A General Framework for Analyzing Sustainability of Social-Ecological Systems. *Science* **2009**, *325*, 419–422. [CrossRef] [PubMed]
27. Garcia-Ayllon, S. The Integrated Territorial Investment (ITI) of the Mar Menor as a model for the future in the comprehensive management of enclosed coastal seas. *Ocean Coast. Manag.* **2018**, *166*, 82–97. [CrossRef]
28. Lathrop, R.; Auermuller, L.; Trimble, J.; Bognar, J. The Application of WebGIS Tools for Visualizing Coastal Flooding Vulnerability and Planning for Resiliency: The New Jersey Experience. *ISPRS Int. J. Geo-Inf.* **2014**, *3*, 408–429. [CrossRef]
29. Williams, S.J.; Ismail, N. Climate Change, Coastal Vulnerability and the Need for Adaptation Alternatives: Planning and Design Examples from Egypt and the USA. *J. Mar. Sci. Eng.* **2015**, *3*, 591–606. [CrossRef]
30. García-Ayllón, S. Integrated management in coastal lagoons of highly complexity environments: Resilience comparative analysis for three case-studies. *Ocean Coast. Manag.* **2017**, *143*. [CrossRef]
31. Hou, D.; O'Connor, D.; Nathanail, P.; Tian, L.; Ma, Y. Integrated GIS and multivariate statistical analysis for regional scale assessment of heavy metal soil contamination: A critical review. *Environ. Pollut.* **2017**, *231*, 1188–1200. [CrossRef] [PubMed]
32. Rahayuningsih, T.; Muntasib, E.K.S.H. Nature Based Tourism Resources Assessment Using Geographic Information System (GIS): Case Study in Bogor. *Procedia Environ. Sci.* **2016**, *33*, 365–375. [CrossRef]
33. Zanetti, V.B.; de Sousa Junior, W.C.; De Freitas, D.M. A Climate Change Vulnerability Index and Case Study in a Brazilian Coastal City. *Sustainability* **2016**, *8*, 811. [CrossRef]
34. García-Ayllón, S. GIS Assessment of Mass Tourism Anthropization in Sensitive Coastal Environments: Application to a Case Study in the Mar Menor Area. *Sustainability* **2018**, *10*, 1344. [CrossRef]
35. García-Ayllón, S. Retro-diagnosis methodology for land consumption analysis towards sustainable future scenarios: Application to a mediterranean coastal area. *J. Clean. Prod.* **2018**, *195*, 1408–1421. [CrossRef]
36. García-Ayllón, S. Diagnosis of complex coastal ecological systems: Environmental GIS analysis of a highly stressed Mediterranean lagoon through spatiotemporal indicators. *Ecol. Indic.* **2017**, *83*. [CrossRef]
37. García-Sánchez, M.; Pérez-Ruzafa, I.M.; Marcos, C.; Pérez-Ruzafa, A. Suitability of benthic macrophyte indices (EEI, E-MaQI and BENTHOS) for detecting anthropogenic pressures in a Mediterranean coastal lagoon (Mar Menor, Spain). *Ecol. Indic.* **2012**, *19*, 48–60. [CrossRef]
38. Pérez-Ruzafa, A.; Marcos, C.; Bernal, C.M.; García-Sánchez, M.; Pérez-Ruzafa, I.M. Cymodocea nodosa vs. Caulerpa prolifera: Causes and consequences of a long term history of interaction in macrophyte meadows in the Mar Menor coastal lagoon (Spain, southwestern Mediterranean). *Estuar. Coast. Shelf Sci.* **2012**, *110*, 101–115. [CrossRef]
39. Robledano, F.; Esteve, M.A.; Martínez-Fernández, J.; Farinós, P. Determinants of wintering waterbird changes in a Mediterranean coastal lagoon affected by eutrophication. *Ecol. Indic.* **2011**, *11*, 395–406. [CrossRef]
40. Garcia-Ayllon, S. Geographic Information System (GIS) Analysis of Impacts in the Tourism Area Life Cycle (TALC) of a Mediterranean Resort. *Int. J. Tour. Res.* **2016**, *18*, 186–196. [CrossRef]
41. Perez-Ruzafa, A.; Marcos-Diego, C.; Ros, J.D. Environmental and biological changes related to recent human activities in the Mar Menor (SE of Spain). *Mar. Pollut. Bull.* **1991**, *23*, 747–751. [CrossRef]
42. García-Ayllón, S. La Manga case study: Consequences from short-term urban planning in a tourism mass destiny of the Spanish Mediterranean coast. *Cities* **2015**, *43*, 141–151. [CrossRef]
43. Region de Murcia SIT Murcia: Visor del Sistema de Informacion Territorial. Available online: http://sitmurcia.es/visor/? (accessed on 14 August 2018).
44. De Pascalis, F.; Pérez-Ruzafa, A.; Gilabert, J.; Marcos, C.; Umgiesser, G. Climate change response of the Mar Menor coastal lagoon (Spain) using a hydrodynamic finite element model. *Estuar. Coast. Shelf Sci.* **2012**, *114*, 118–129. [CrossRef]

45. Velasco, A.M.; Pérez-Ruzafa, A.; Martínez-Paz, J.M.; Marcos, C. Ecosystem services and main environmental risks in a coastal lagoon (Mar Menor, Murcia, SE Spain): The public perception. *J. Nat. Conserv.* **2018**, *43*, 180–189. [CrossRef]
46. Robledano, F.; Esteve, M.A.; Calvo, J.F.; Martínez-Paz, J.M.; Farinós, P.; Carreño, M.F.; Soto, I.; Avilés, M.; Ballesteros, G.A.; Martínez-Baños, P.; et al. Multi-criteria assessment of a proposed ecotourism, environmental education and research infrastructure in a unique lagoon ecosystem: The Encañizadas del Mar Menor (Murcia, SE Spain). *J. Nat. Conserv.* **2018**, *43*, 201–210. [CrossRef]
47. García-Ayllón, S.; Miralles, J.L. The environmental impacts of land transformation in the coastal perimeter of the mar menor lagoon (SPAIN). *Int. J. Des. Nat. Ecodyn.* **2014**, *9*, 109–128. [CrossRef]
48. MARM. *Guía Metodológica para el Desarrollo del Sistema Nacional de Cartografía de Zonas Inundables*; MARM: Madrid, Spain, 2011; ISBN 978-84-491-1136-5.
49. Laface, V.; Malara, G.; Romolo, A.; Arena, F. Peak over threshold vis-à-vis equivalent triangular storm: Return value sensitivity to storm threshold. *Coast. Eng.* **2016**, *116*, 220–235. [CrossRef]
50. Iturrioz, A.; Guanche, R.; Armesto, J.A.; Alves, M.A.; Vidal, C.; Losada, I.J. Time-domain modeling of a fixed detached oscillating water column towards a floating multi-chamber device. *Ocean Eng.* **2014**, *76*, 65–74. [CrossRef]
51. Camus, P.; Mendez, F.J.; Medina, R.; Tomas, A.; Izaguirre, C. High resolution downscaled ocean waves (DOW) reanalysis in coastal areas. *Coast. Eng.* **2013**, *72*, 56–68. [CrossRef]
52. Universidad de Cantabria Proyecto C3E. Available online: http://www.c3e.ihcantabria.com/ (accessed on 18 August 2018).
53. European Union. *Directive 2007/60/EC of the European Parliament and of the Council of 23 October 2007 on the Assessment and Management of Flood Risks*; European Union: Brussels, Belgium, 2007.
54. Wu, C. Handbook of Applied Spatial Analysis: Software Tools, Methods and Applications edited by Manfred, M. Fischer and Arthur Getis. *J. Reg. Sci.* **2012**, *52*, 386–388. [CrossRef]
55. Anselin, L. Local Indicators of Spatial Association—LISA. *Geogr. Anal.* **1995**, *27*, 93–115. [CrossRef]
56. Ministry of Environment Agriculture and Water of Spain National System of Cartography of Floodplains. Available online: https://sig.mapama.gob.es/snczi/visor.html?herramienta=DPHZI (accessed on 14 August 2018).
57. Stefanova, A.; Hesse, C.; Krysanova, V. Combined Impacts of Medium Term Socio-Economic Changes and Climate Change on Water Resources in a Managed Mediterranean Catchment. *Water* **2015**, *7*, 1538–1567. [CrossRef]
58. Hesse, C.; Stefanova, A.; Krysanova, V. Comparison of Water Flows in Four European Lagoon Catchments under a Set of Future Climate Scenarios. *Water* **2015**, *7*, 716–746. [CrossRef]
59. García-Ayllón, S.; Miralles, J.L. The feedback between a tourism model and urban planning in La Manga, a mass tourist destination on the Spanish coast. *WIT Trans. Ecol. Environ.* **2014**, *187*, 339–349. [CrossRef]
60. García-Sánchez, M.; Korbee, N.; Pérez-Ruzafa, I.M.; Marcos, C.; Domínguez, B.; Figueroa, F.L.; Pérez-Ruzafa, Á. Physiological response and photoacclimation capacity of Caulerpa prolifera (Forsskål) J.V. Lamouroux and Cymodocea nodosa (Ucria) Ascherson meadows in the Mar Menor lagoon (SE Spain). *Mar. Environ. Res.* **2012**, *79*, 37–47. [CrossRef] [PubMed]
61. Pérez-Ruzafa, A.; García-Charton, J.A.; Barcala, E.; Marcos, C. Changes in benthic fish assemblages as a consequence of coastal works in a coastal lagoon: The Mar Menor (Spain, Western Mediterranean). *Mar. Pollut. Bull.* **2006**, *53*, 107–120. [CrossRef] [PubMed]
62. González-Wangüemert, M.; Giménez-Casalduero, F.; Pérez-Ruzafa, Á. Genetic differentiation of Elysia timida (Risso, 1818) populations in the Southwest Mediterranean and Mar Menor coastal lagoon. *Biochem. Syst. Ecol.* **2006**, *34*, 514–527. [CrossRef]
63. García-Oliva, M.; Pérez-Ruzafa, Á.; Umgiesser, G.; McKiver, W.; Ghezzo, M.; De Pascalis, F.; Marcos, C. Assessing the Hydrodynamic Response of the Mar Menor Lagoon to Dredging Inlets Interventions through Numerical Modelling. *Water* **2018**, *10*, 959. [CrossRef]
64. Pérez-Ruzafa, A.; De Pascalis, F.; Ghezzo, M.; Quispe-Becerra, J.I.; Hernández-García, R.; Muñoz, I.; Vergara, C.; Pérez-Ruzafa, I.M.; Umgiesser, G.; Marcos, C. Connectivity between coastal lagoons and sea: Asymmetrical effects on assemblages' and populations' structure. *Estuar. Coast. Shelf Sci.* **2018**. [CrossRef]

65. García-Pintado, J.; Martínez-Mena, M.; Barberá, G.G.; Albaladejo, J.; Castillo, V.M. Anthropogenic nutrient sources and loads from a Mediterranean catchment into a coastal lagoon: Mar Menor, Spain. *Sci. Total Environ.* **2007**, *373*, 220–239. [CrossRef] [PubMed]
66. Moreno-González, R.; Campillo, J.A.; León, V.M. Influence of an intensive agricultural drainage basin on the seasonal distribution of organic pollutants in seawater from a Mediterranean coastal lagoon (Mar Menor, SE Spain). *Mar. Pollut. Bull.* **2013**, *77*, 400–411. [CrossRef] [PubMed]
67. Campillo, J.A.; Sevilla, A.; Albentosa, M.; Bernal, C.; Lozano, A.B.; Cánovas, M.; León, V.M. Metabolomic responses in caged clams, Ruditapes decussatus, exposed to agricultural and urban inputs in a Mediterranean coastal lagoon (Mar Menor, SE Spain). *Sci. Total Environ.* **2015**, *524–525*, 136–147. [CrossRef] [PubMed]
68. Baudron, P.; Cockenpot, S.; Lopez-Castejon, F.; Radakovitch, O.; Gilabert, J.; Mayer, A.; Garcia-Arostegui, J.L.; Martinez-Vicente, D.; Leduc, C.; Claude, C. Combining radon, short-lived radium isotopes and hydrodynamic modeling to assess submarine groundwater discharge from an anthropized semiarid watershed to a Mediterranean lagoon (Mar Menor, SE Spain). *J. Hydrol.* **2015**, *525*, 55–71. [CrossRef]
69. García, G.; Muñoz-Vera, A. Characterization and evolution of the sediments of a Mediterranean coastal lagoon located next to a former mining area. *Mar. Pollut. Bull.* **2015**, *100*, 249–263. [CrossRef] [PubMed]
70. Martínez-Graña, A.; Gómez, D.; Santos-Francés, F.; Bardají, T.; Goy, J.L.; Zazo, C. Analysis of Flood Risk Due to Sea Level Rise in the Menor Sea (Murcia, Spain). *Sustainability* **2018**, *10*, 780. [CrossRef]

water

MDPI

Article

Coastal Flooding Hazard Due to Overflow Using a Level II Method: Application to the Venetian Littoral

Chiara Favaretto *, Luca Martinelli and Piero Ruol

ICEA Department, Padua University, v. Ognissanti 39, 35129 Padova, Italy; luca.martinelli@unipd.it (L.M.); piero.ruol@unipd.it (P.R.)

* Correspondence: chiara.favaretto@dicea.unipd.it; Tel.: +39-049-827-7995

Received: 19 November 2018; Accepted: 21 December 2018; Published: 13 January 2019

Abstract: In recent years, marine flooding and its impacts have become a question of growing interest, since coastal areas are the most heavily populated and developed land zones in the world. This paper presents a rapid tool for mapping at regional scale the hazard associated with coastal flooding due to overflow. The tool merges a recently developed numerical model that solves a simplified form of the Shallow-Water Equations and is suited for Graphic Processing Unit (GPU) acceleration, with a Level II reliability method that allows producing hazard maps of inland flooding propagation. The procedure was applied to two stretches of the Venetian littoral, i.e., Valle Vecchia and Caorle, located in the northern Adriatic Sea. The application includes the site descriptions and the resulting hazard maps that show the probability of failure in each point of the coast for a given inland inundation level.

Keywords: marine inundation; FORM; reliability analysis; Veneto coast

1. Introduction

The EU water flood directive (2007/60/EC) points out the importance of evaluating coastal flooding hazard maps. Under climate change, some authors predicted that sea-levels and subsidence will rise [1] and storm surge intensity will increase [2,3]. Therefore, it is possible that, in the near future, many human infrastructures will be affected by marine flooding and, assuming the coastal population will grow [4], tools for risk mitigations are deemed necessary.

Coastal flooding may be triggered by many causes, possibly in combination: (i) high sea level (and wave run-up) overtopping the artificial dikes/barriers or natural dunes; (ii) breaching of cliffs, dunes or more generally erosion of the coastal defence; (iii) river overflow; and (iv) rain excess or insufficient drainage system. However, it is very complex to account for all these mechanisms. In particular, breaching is very difficult to predict, since it reasonably depends on the (spatially distributed) geotechnical characteristics of the coastal dunes and, for instance, on the presence of vegetation or type of revetment.

In the absence of a complete framework for the assessment of coastal flooding hazard, the maps are frequently based on simple indexes that typically combine topographic information (usually available to the coastal managers within GIS database) to sea level and wave run-up. Many specific open-source tools and approaches have been developed to provide an assessment of the potential flood risk for coastal zone and to support decision-making processes [5,6]. For instance, the National Oceanic and Atmospheric Administration [7] developed a GIS awareness tool to display the resulting inundation from storm surges along the US coastal states vulnerable to hurricanes. Wadey et al. [8] presented a methodology for integrating existing models for the rapid simulation of coastal flood events across a large and varied case study area on the UK south coast. Spaulding et al. [9] developed flood inundation maps for Charlestown (RI) using state of the art, fully coupled high-resolution surge and wave models. In Italy, Aucelli et al. [10] recently studied the inundation hazard and risk in Volturno coastal plain in

Campania (Italy). Di Luccio et al. [11] focused on the effect of different run-up lumped formulas on the index for coastal vulnerability assessment of a microtidal beach located along the southern Ionian coast of Calabria (Italy). Di Risio et al. [12] assessed coastal hazard to wave-induced flooding at the national scale. In other cases, the mapping of areas potentially exposed to inundation is carried out by a chain of models (e.g., [13,14]), solving the wave transformation problem in the sea area and the flow propagation in the inland area.

The issue is particularly relevant for the Venetian littoral, where local managers require (possibly GIS-integrated) rapid tools to simulate the coastal flood by means of wave overtopping in an urban area at regional scale. The northern Adriatic coast is subject to rapidly evolving pressures from a range of drivers, including natural and anthropogenic ones: for example, rapid morphological evolution of Po River Delta [15], human-induced subsidence caused by fluid withdrawal [16], and changing wave climate [2]. The study here presented originates from the practical need to fill a gap in the coastal flooding assessment in the Veneto region, until now studied with simplified approaches. The objective of this study was to develop a flood propagation model and to establish a methodology that has the ambition to help the coastal managers and stakeholders in producing flooding hazard maps required under the EU water floods directive.

For the evaluation of the coastal hazard, Lerma et al. [14] compared a traditional approach, based on the simulation of events with given return periods, to a novel empirical method that tries to better define, by analyzing the uncertainties of the hydraulic drivers, the real return period of the flooding event. In the work of Martinelli et al. [17], the failure probability is found following the Level II methodology. They defined in each location the combination of loads that induces a flooding depth larger than a given threshold. The final result is the actual probability that a specific failure mechanism (e.g., overflow) occurs, for an in-erodible topography (e.g., deterministic resistance). The results produced by such models are certainly more rigorous than the traditional approach for the selected failure mode, since they are statistically based. However, they cannot include the expert judgement of the coastal managers. It should be stressed that the maps proposed by this work require an integration before they may be used for coastal management purposes, for instance other failure mechanism and information on the exposed value. Mitigation measures against marine inundation include alert services [18], evacuation procedures [19,20], flood insurance programs [21], structural defences (e.g., dikes and breakwaters [22]), planning options [23], community information and participation [21,24,25], and protection of natural buffer zones (e.g., dunes [26,27]).

This paper includes two main sections and a concluding paragraph. First, the dynamic reduced-complexity model of coastal flooding is briefly described together with the procedure for mapping coastal flooding hazard. The method's application to two stretches of coast belonging to the Venetian littoral is then presented (Valle Vecchia and Caorle). Lastly, conclusions are drawn.

2. Methodology

2.1. GPU-Based Swallow Water Equation Model

The propagation model is a raster-based inundation model, presented by Favaretto et al. [28,29], that solves for each cell of the domain a simplified form of the Shallow-Water Equations (Equations (1) and (2)) applied in the x and y directions to simulate two-dimensional flow over a raster grid.

$$\frac{\partial w}{\partial t} + \frac{\partial q_x}{\partial x} + \frac{\partial q_y}{\partial y} = 0 \tag{1}$$

$$\frac{\partial q_\xi}{\partial t} + \frac{\partial (q_\xi^2/w)}{\partial \xi} = -gh(\frac{\partial h}{\partial \xi} + j) \tag{2}$$

where $\xi = x$ or y. In Equations (1) and (2), $h = w + z$ (w = water depth, z = bed elevation) is the water surface elevation and q (m^2/s) is the discharge per unit width assuming a rectangular channel.

In the momentum equation (Equation (2)), to simplify the equation and to speed up the model, only the more relevant terms are taken into account. The inertial terms are included to avoid overestimation of the fluxes between adjacent cells and to account for the flooding duration. The friction term j is important for the correct simulation of the flow propagation; however, it is linearized, with a coefficient K_H that accounts for the linearized turbulent friction flow.

$$K_H = \frac{8g}{3\pi} \frac{Q_{MAX}}{K_\xi^2 w_F^{1/3}} \tag{3}$$

where w_F is the maximum available depth through which water can flow between two adjacent cells and it is defined, following [30], as the difference between the highest water free surface in the two cells and the highest bed elevation. Q_{MAX} is the maximum discharge during the sinusoidal oscillation. In practice, the value of Q_{MAX} is assessed as the maximum discharge flowing through the cells.

Forces induced by the advection terms would dominate in presence of small scale features (which generate significant velocity spatial derivatives), but, in absence, bed friction dominates over the advection terms that may therefore be neglected [31].

The final formulation of the momentum equation, only valid where the flow advection is relatively unimportant, becomes:

$$\frac{\partial q_\xi}{\partial t} + g w_F \frac{\partial h}{\partial \xi} + K_H \frac{q_\xi}{w_F^2} = 0 \tag{4}$$

The discretization method implemented is first order in space and time (Euler scheme minimizes memory requirements), and a semi-implicit treatment is used for the friction term to improve stability. The domain coincides with the DTM (Digital Terrain Model), usually subdivided into square cells (area equal to $\Delta x \Delta y$) forming a regular grid. The numerical model is subject to the Courant–Freidrichs–Levy (CFL) condition: $CFL = V\Delta t/\Delta x < 1$. To avoid non-physical discontinuities in the flow and to reduce the dispersion error in the numerical scheme, a relaxation technique was used. The discharge q at time t is evaluated in two steps: first, q is obtained on the basis of the balance equation and then q is updated as a weighted average that accounts for the discharge of the neighbouring points. To preserve positivity, the maximum discharge that flows between adjacent cells is limited by a two-step approach. The positive preserving property is crucial when in part of the domain no water is present, or when the water depth is very small, and little oscillations may lead to negative depths, eventually resulting in the simulation to fail [32]. Specifically, if the depth variation Δh evaluated in a cell is negative (namely, the cell is emptying too much due to a large time step), its absolute value must be less or equal to the current available depth. If it is not, the "negative" volume is subtracted from the spatially connected cells with positive balance to ensure continuity.

Computational efficiency is one of the key issues in the application of raster-based models (or more refined hydraulic models) with Δx of order 1–5 m. Computational speed depends mainly on grid size and, in fact, dynamic models are often applied to coarse grid only ($\Delta x > 100$ m), whereas, in this study, the grid cell dimension is $\Delta x = 1$ m. Furthermore, the inundated area evolves throughout the simulation in a manner that is rarely known a priori, meaning that the model domain must be very large to incorporate the flood uncertain extent.

To reduce the model runtime (as suggested in [33,34]), it was decided to write a code suitable for a GPU card to reduce the model run time through parallelization. The code was written in MATLAB and, using the Parallel Computing Toolbox with minimal code changes, the simulation can be run in parallel.

The GPU used to execute the simulation was a Nvidia Tesla K80 (4992 core, 12 GByte memory). GPUs achieve high performance by calculating many results in parallel since each computation is processed by a different core. For very large domains (10^8 cells, typical of a regional map), it was found that the GPU time is 2–3% of the CPU (Central Processing Unit) time [28,29].

The overall model consistency was successfully validated through a number of analytical and experimental benchmarks [28,29]. The behavior of the model for very large domains was checked against the analytical solution of the Shallow-Water Equations regarding an oscillatory flow in a parabolic bowl with friction. The numerical ability to simulate wet/dry transitions and the importance of the convective terms were evaluated through a set of comparisons with the experimental investigation of Synolakis [35] for a solitary wave run-up on a simple beach. The flow propagation in a three-dimensional domain was analyzed through a comparison with the experimental investigation of Briggs et al. [36] for solitary wave interaction around a circular island. To prove the model ability to simulate a real case of coastal flooding, an event that occurred at Caorle (VE) in December 2008, documented by videos and reports, was also analysed.

2.2. Procedure for the Assessment of Coastal Flooding Hazard Maps

In the following, a single mechanism is analyzed as responsible for coastal flooding: the overflow over an in-erodible bathymetry. Clearly, since the topography is in-erodible, the breaching mechanism is not possible and the methods is best suited to cases where coastal flood defence is artificial (e.g., road, dike). The approach of this work is based on the Level II methodology that is described in [37,38]. In short, the procedure aims at finding the p_f of a system, i.e., the probability that it fails under certain conditions. In the coastal inundation analysis, it is the probability that a portion of inland is flooded under certain extreme conditions.

Mathematically, p_f is the probability that the random variables $\mathbf{X} = (X_1, X_2, ..., X_N)$ are in the unsafe region, defined by $g(\mathbf{X}) < 0$ [39] ($g(\mathbf{X})$ is the performance function and $g(\mathbf{X}) = 0$ is the limit state, see also Figure 1a). If the joint probability density function (joint pdf) of $\mathbf{X}$ is $f_x(\mathbf{X})$, the probability of failure is evaluated as:

$$p_f = P\{g(\mathbf{X}) < 0\} = \int_{g(\mathbf{X})<0} f_x(\mathbf{X}) dX \quad (5)$$

One of the most commonly used analysis methods is the First Order Reliability Method (FORM [40]). The basic idea is to ease the computational difficulties through simplifying the integrand of $f_x(\mathbf{X})$, achieved through transforming the random variables $\mathbf{X}$ to an equivalent independent standard normal random variable space $\mathbf{U} = (U_1, U_2, ..., U_N)$, and approximating the performance function $g(\mathbf{X})$ with a linear approximation (Figure 1b). The point that has the highest probability density on the performance $g(\mathbf{U})$, called "design value", is the one with the shortest distance from the limit state to the origin in the standard space. The minimum distance β is called reliability index and it is shown in Figure 1b. In conclusion, the probability of failure is given by $p_f = 1 - \Phi(\beta)$.

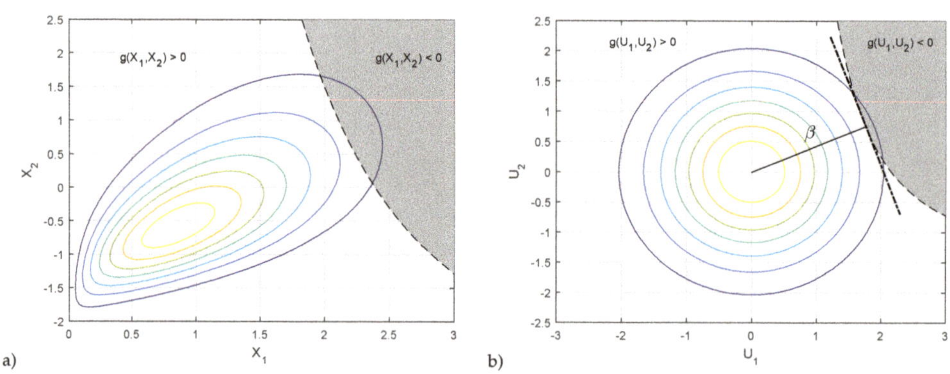

Figure 1. (**a**) Joint probability density function in X-plane and performance function $g(X_1, X_2) = 0$; and (**b**) joint pdf in U-plane and performance function $g(U_1, U_2) = 0$.

The application of this method was carried out in an unconventional way. Only the uncertainties of two variables were considered, namely sea level and wave height, whereas the other variables were deterministically selected equal to the expected value, given that sea level and wave height are at "design value". Some variables, such as wave period and direction, have a low sensitivity, i.e., their uncertainties do not significantly affect the final result. Other variables, such as the storm duration, were not deeply analyzed, since their joint statistics is uncertain.

In the following, the entire procedure for the assessment of coastal flooding hazard maps is described in four steps. First, the offshore marine loads **X** that contribute to the rise of the water level at the domain are identified. They need to be transformed into the standard space **U**. The second step shows how to select and transfer a set of offshore wave conditions (X_i) to the boundary of the inundation domain. The third step shows how the coastal flood model, described in Section 2.1, is applied to find the value of the flood level, i.e., the performance function $g(X_i)$ in each point of the map. Finally, a method is described to define the limit state in each point of the domain.

(a) Definition of the offshore marine loads. The coastal flooding may be forced by a combination of several variables that contribute to the rise of the water level at the domain boundary. The rise of water level is due to: (1) the sea level ζ formed by the astronomical tide (ζ_A) and the meteorological contribution (storm surge, ζ_S) caused by wind and pressure effects; and (2) the wave contribution, i.e., breaking waves contribute to the water level rise through wave run up and wave set-up Z, which mainly depend on wave height, period, and direction.

Extreme sea conditions **X** can be studied in terms of significant wave height (Hs), peak period (Tp), mean wave direction, sea level ζ (tide + surge) and storm duration. This information can be derived from direct measures (e.g., from buoys) or computed dataset (from numerical model such as WAM [41]). From these data, a homogeneous and independent sample can be derived by identification of sea storms.

The definition of "sea storm" is not unique and standardized since many methods can be applied to identify an extreme event from a wave dataset. The differences in each method are relative to the variables used for the analysis, the threshold fixed and the geographical configurations of the basin [42]. The common definition defines a storm event as a sequence of sea states during which the significant wave height is above a given threshold H_{crit} and does not fall below it for a predefined time interval d_{crit} [43,44]. Usually, the threshold H_{crit} is related to the average significant wave height Hs_{Mean} calculated from its time series in the considered zone, so that it depends on the characteristics of the recorded sea states (e.g., $\sim 1.5\ Hs_{Mean}$ [43,45]). Moreover, to guarantee the independence of the selected storms, a minimum time interval between two storms has to be set (e.g., 72 h proposed in [44]), hence two storms with time interval smaller than the minimum are considered as one storm event [46]. After the storm identification, all variables **X** can be analyzed to find their joint statistics. Note that the variations of some input variables, such as the wave period (or, better, the wave steepness) and the storm duration, induce low variability on the model output and the associated sensitivity is very low. Their mean value can be considered as model input. Therefore, the joint statistics is limited to wave height Hs and sea level ζ. The final step is to transform these random variables $\mathbf{X} = [Hs, \zeta]$ to equivalent standard normal random variables, finding the transformation function $\mathbf{U} = F_T(\mathbf{X})$.

(b) Definition of the boundary conditions at the shoreline. A wave transformation model from offshore to onshore needs to be applied to a set of offshore conditions to estimate the value of the set-up Z and the residual wave height at the shoreline refereed to the Mean Sea Level (MSL). The wave transformation model used is the "Dally, Dean and Dalrymple model" [47] for breaker decay that is capable of describing wave transformation across beaches of irregular profile shape. It considers that the wave breaking starts when $H > 0.78d$ (d is the water depth) and continues until some stable wave height is attained (usually $H > 0.4d$). An example of its application is shown in Figure 2.

The final boundary condition is therefore composed by a constant value $\zeta_{TOT} = \zeta + Z$ and an irregular impulsive signal with height equal to H_{RES} and period equal to the offshore period Tp (example in Figure 3).

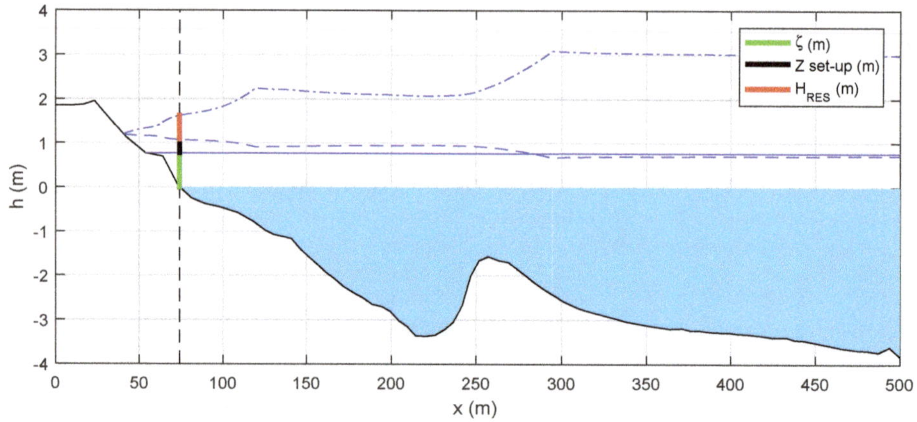

Figure 2. Example of the wave transformation model results and definition of the boundary conditions at the isobath 0 m.

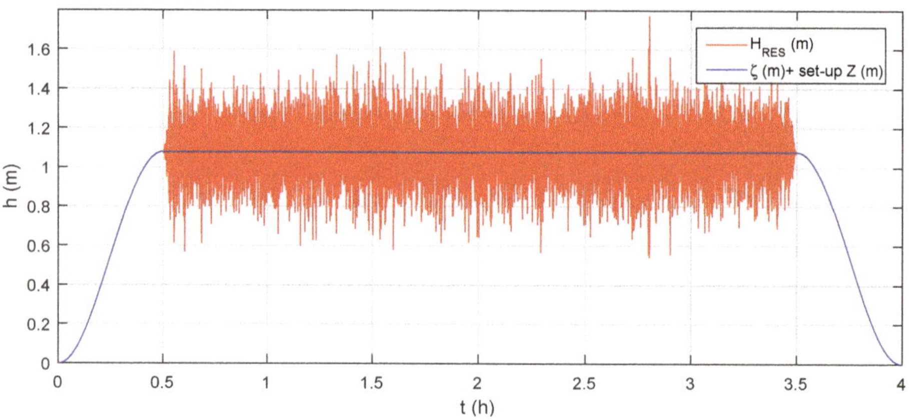

Figure 3. Example of final boundary condition at the shoreline refereed to the Mean Sea Level (MSL).

(c) Coastal inundation modeling. The model for coastal flooding propagation, presented in Section 2.1, can be applied introducing the following essential data: the position of the shoreline refereed to the MSL (or the shoreline position in MSL condition); the inland topography; and the roughness K_s. For each of the boundary conditions selected, the propagation model is run. The maximum water depth reached in each grid cell of the domain is saved and the obtained maps are flooding maps relative to each couple of Hs and ζ.

(d) Reliability analysis. The probability of failure is defined as the probability that a portion of inland is flooded under certain values of wave height and sea level. Therefore, the limit state $g(\mathbf{X}) = 0$ is evaluated through an interpolation with the obtained results. In each cell of the domain, the maximum water level reached during the simulation for a fixed value of ζ is plotted against different values of Hs (Figure 4a). Then, two couples that correspond to a water level in the cell equal to h_F (in Figure 4a h_F = 0.5 m) are extrapolated. Finally, the transformation of the two couples from the physical space to the standard space is applied and the minimum distance β (and consequently the p_f) from the limit state to the origin of the standard space is computed (Figure 4b).

The last step is the fulfilment of the hazard maps in terms of return period $T_R = 1/(N_y \times p_f)$, where N_y are the number of extreme events recorded in each year of observation (based on the data available). The results are the exceedance probability values of a given inundation level for each pixel of the domain (i.e. DTM).

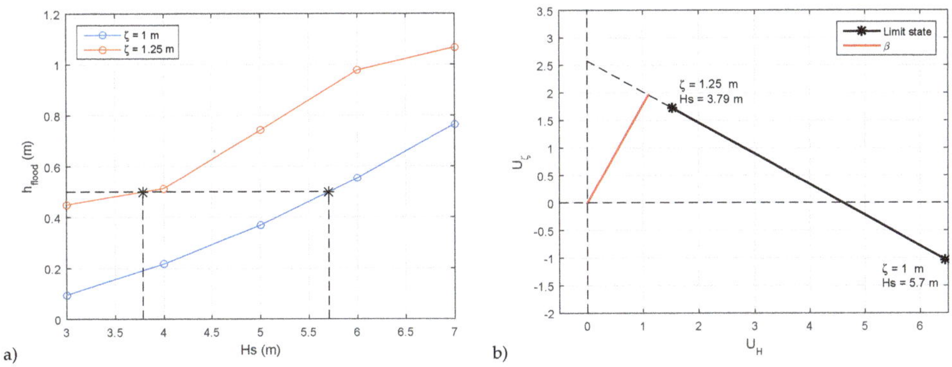

Figure 4. (**a**) Example of evaluation of the limit state in the physical space; and (**b**) limit state in standard space g(**U**) and evaluation of the distance β.

3. An Application to Two Stretches of the Venetian Littoral

The aim of this section is to present local inundation maps produced for two stretches of the Venetian littoral (Figure 5), which show the annual probability of exceedance for a given flood level. A detailed description of this littoral is given in [48] where, for a better understanding of sediment transport patterns, the coast of the Veneto region is subdivided into 20 homogeneous littoral cells separated by lagoon inlets or river mouths.

The Adriatic coast is plagued by a combination of high waves and storm surges, which are responsible for the flooding of coastal areas, in particular, Venice and its lagoon. The north Adriatic Sea is characterized by two main wind (and correspondingly wave) regimes, which are primarily influenced by local orography. The prevailing winds along the Venetian coastline are the Bora and the Scirocco, which blow from the northeast and southeast, respectively. In [49], information and data for the Venetian littoral are collected, harmonized and stored in a single geographical information system (called Coastal GIS).

This information comprises topographic and bathymetric surveys over a range of time (bathymetry: 2005, 2007/2008, 2010, and 2012/2014; DTM and DEM: 2008and 2012/2013) that are essential for both the wave transformation model and for the flooding model. The latest available information was used as input in the models. More in detail, DTM representing the inland topography have a grid size equal to 1 m. A common frame for vertical reference data was considered.

A wave dataset measured (1987–2017) at the oceanographic tower "Acqua Alta" situated on 16 m of water depth. (MLLW) in the Gulf of Venice (Lat 45°18′51.27″ N, Lon: 12°30′29.93″ E) allowed statistically analyzing the marine climate along this coast. Different wave gauges have been used since the start of the measurements at the Acqua Alta research tower and the instrument system has been progressively upgraded and repositioned during maintenance operations. The registered data include: significant wave height Hs (m), maximum wave height H_{MAX}, mean period Tm and peak period Ts (s), mean wave direction (°N), and sea level ζ (m ZMPS, where ZMPS is a reference level for Venice named Zero Mareografico Punta della Salute). Figure 6 shows the time series of the recorded significant wave height (black line).

The data sample, selected from the measurements following the aforementioned procedure and selecting $H_{crit} \sim 1.5Hs_{Mean}$ = 1 m and d_{crit} = 3 h, includes 974 wave storms recorded from 1987 to 2017 and it was also used to find some correlation among variables. The dots in Figure 6, together with the wave series, represent the maximum significant wave heights of the 974 storms; green dots are the peaks over threshold 1 m, while red dots are the peaks over threshold 2.5 m.

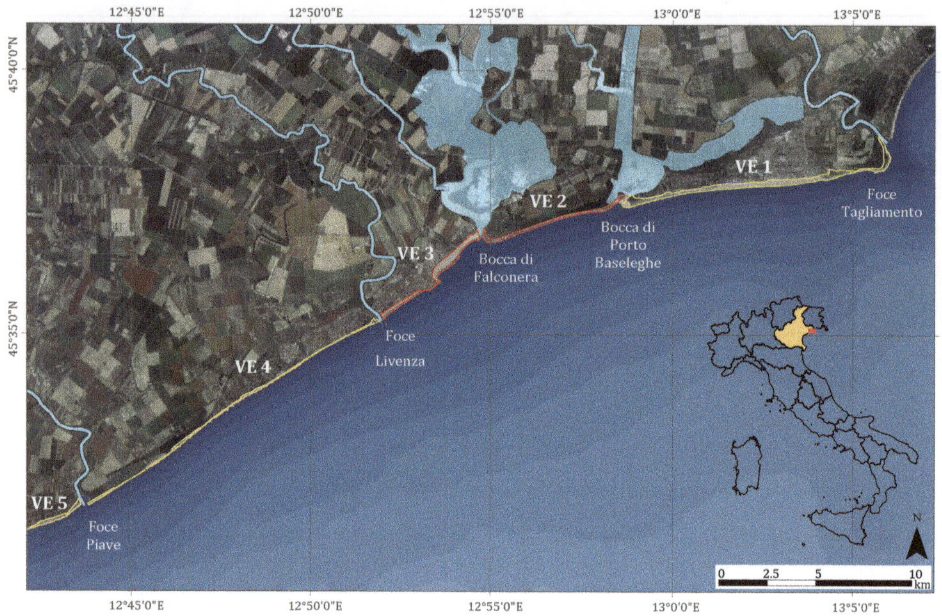

Figure 5. The northern part of the Venetian littoral and its subdivision into coastal cells: VE1, Bibione; VE2, Valle Vecchia; VE3, Caorle; VE4, Porto Santa Margherita–Duna Verde–Eraclea; VE5, Jesolo.

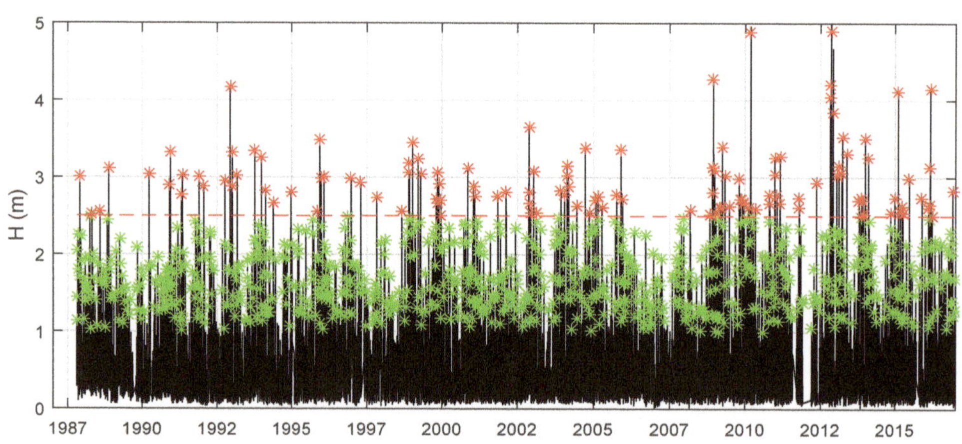

Figure 6. Time series of significant wave height Hs (black line), where dots represent the 974 storms: green dots are peak over threshold 1 m, while red dots are peak over threshold 2.5 m.

In Figure 7a, the wave heights Hs of each storm are plotted against their corresponding wave periods and classified on the basis of their wave steepness (Hs/L_0), allowing to evaluate the dependency law between this two variable. Figure 7b shows the duration of the storm d, defined as the time interval in which the significant wave height exceeds the 50% of Hs_{MAX} and includes the maximum measured wave height Hs_{MAX} of the storm. Moreover, the transformation from the space of physical variables to the standardized variable space was applied to the data sample. The transformation applied (in the following $F_T(Hs, \zeta)$) was based on the Nataf transformation [50], which describes the joint probability density function of random variables based on their individual marginal distributions and the coefficients of correlation using a Gaussian copula.

Finally, twenty couples of wave heights H_S and sea levels ζ were chosen as offshore input, as reported in Figure 8. The twenty resulting maps were considered sufficient to define the failure domain in detail. The wave steepness was chosen equal to 0.04, therefore the wave period Tp was estimated as $Tp = 3.75Hs^{0.554}$ and the typical total storm duration was assumed equal to the typical value of 4 h, with 30 min of ramping up and 30 min of ramping down.

Since maps of the land uses are not available, for all simulations, a uniform friction coefficient was applied K_s = 33 $m^{1/3}s^{-1}$.

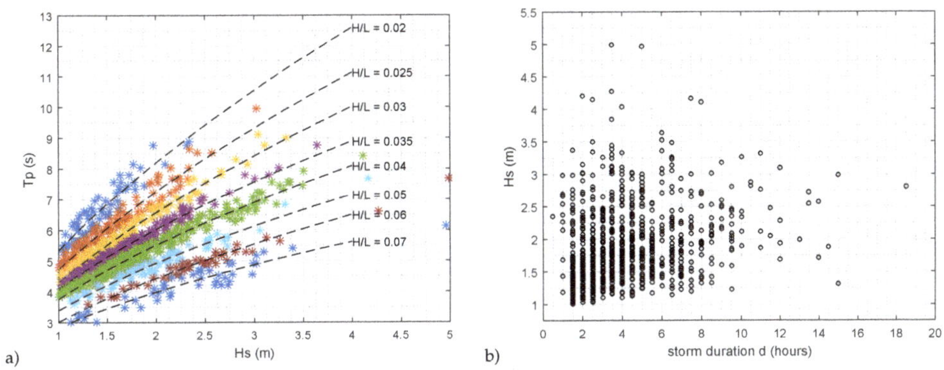

Figure 7. (**a**) Hs, Tp and steepness Hs/L correlation; and (**b**) storm duration d vs. Hs.

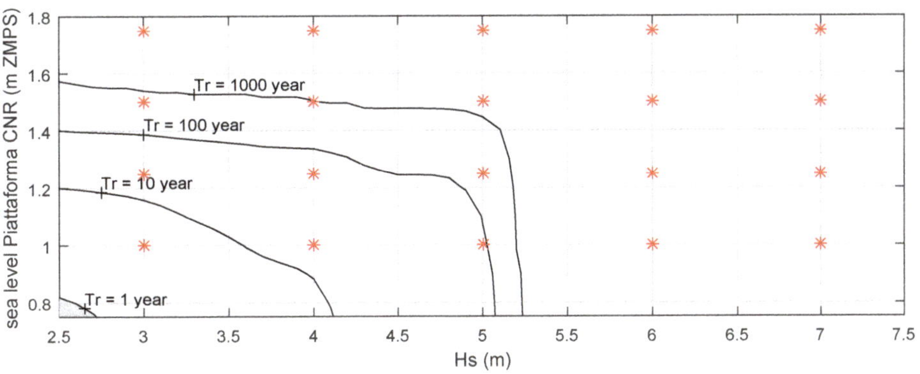

Figure 8. Red dots are the couple Hs–ζ chosen as input for the models (ZMPS is a reference level for Venice).

The following subsections present the results obtained by applying the proposed methodology in two different littoral cells (visible in Figure 5): the Valle Vecchia coastline (littoral cell No. VE2) and the Caorle coastline (littoral cell No. VE3), characterized by very different land uses.

3.1. Valle Vecchia Littoral Cell

The Valle Vecchia coastline (VE2 cell) is 5.5 km long and confined by lagoon's inlets of Baseleghe at the northeast and Falconera at the southwest. The entire cell is a major environmental area, protected and designated as Natura 2000 sites (SCI IT3250033 and SPA IT3250041), free of urban and tourist settlements and without any coastal defence structure. In the back-shore, a system of dunes is present. Behind the dunes, there are valleys and crop fields. The long-shore sediment transport, coming from the northern cell (VE1) is equal to ~50,000 m^3/year. Analyzing the recent evolution of the shoreline, it is possible to verify that the cell is substantially in accretion, even if dunes are subject to local erosion. In fact, the risk of coastal flooding is high and sometimes inundation occurred in the back-shore valley. The water has always entered through some gaps (mainly for pedestrian or vehicle paths) on the dunes. The topographic data highlight different dunes, with crest height in the range 2–5 m.

The aforementioned steps for the coastal flooding risk assessment were applied to the Valle Vecchia coastline. Figure 9 shows an example of flooding maps for Hs = 6 m and ζ = 1.75 m ZMPS: some areas are inundated, mainly close to the lagoon's inlet of Falconera.

From the 20 maps created, it is possible to extract, for every pixel, which couples of Hs and ζ caused a level of inundation equal to 50 cm (limit state). The transformation $F_T(Hs, \zeta)$, previously defined, was applied to every couple and, following the FORM methods, the probability of failure p_f was evaluated in the whole littoral cell. Figure 10 shows this result, i.e., the present hazard map arranged with a chosen limit state equal to 0.5 m. The area characterized by a return period lower than 10 years is ~0.1 km^2.

Figure 11a shows a zoom of the hazard map and highlights that a portion of the valleys and crop fields located behind the dunes have a T_R < 10 years. Therefore, the probability that these parts are inundated is very high.

3.2. Caorle Littoral Cell

The Caorle coastline (VE3 cell in Figure 5) is 5 km long, its borders are the mouth of the lagoon's channel Falconera to the north and the mouth of the River Livenza to the south, both armored with jetties. The economy is mainly based on tourism (~4,500,000 visitors in 2017) and fishing.

The cell can be subdivided into three main parts: (i) "Spiaggia di Levante" at northeast; (ii) "Murazzi" in the central part; and (iii) "Spiaggia di Ponente" at southwest. The first stretch of this coast at northeast (named "Spiaggia di Levante") has a normal shoreline direction equal to 140° N, very different from the adjacent ones. This coast is characterized by very fine sediments, with silty fraction that causes drainage problems and occasionally, during the most intense precipitation, the formation of puddles. The emerged beach is very wide and the submerged beach is characterized by gentle slopes. The long-shore sediment transport is ~20,000 m^3/year, directed from northeast to southwest.

The historic centre of Caorle is located in the central part. This stretch of coast (800 m long) is bordered at north by a cusp, where a church named "Chiesa della Madonna dell'Angelo" is located. To mitigate the risk to human health, economic activities and cultural heritage, the shoreline position in the central area is stabilized by a sea wall (named "Murazzi").

In the southern part, the long-shore sediment transport remains approximately equal to 20,000 m^3/year, again directed from northeast to the southwest. This sand partly nourishes the beach named "Spiaggia di Ponente", partly deposits in the area next to the jetty of the mouth of the River Livenza and only a few thousand cubic meters go to the southern cell (VE4). Some portions of the town have ground elevations lower than 0.5 m and no system of dunes is present.

The aforementioned steps for the coastal flooding hazard assessment were applied to the Caorle coastline. Figure 12 shows the flooding maps for Hs = 6 m, ζ = 1.5 m ZMPS: the historic town is partly flooded and the overtopping occurs at the southern bound of the seawall.

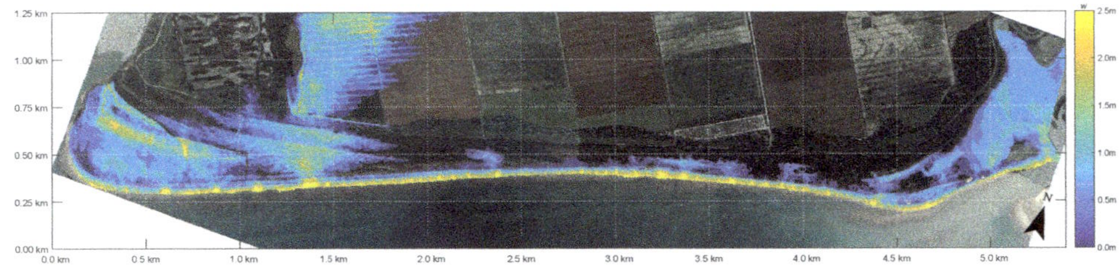

Figure 9. Flooding map (cell VE2): $Hs = 6$ m, $\zeta = 1.75$ m ZMPS.

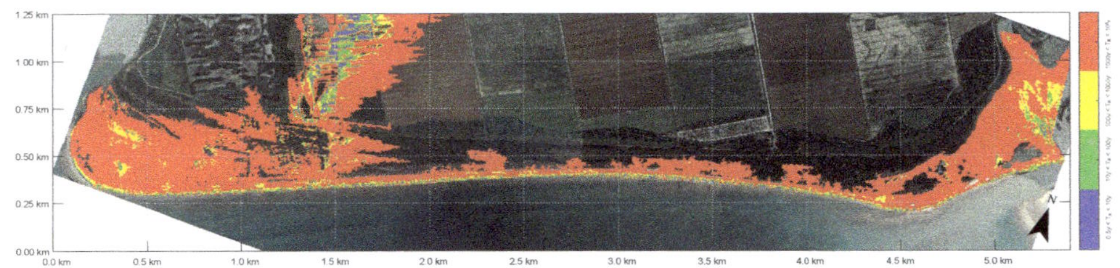

Figure 10. Hazard map (cell VE2): Probability that flooding level 0.5 m is exceeded, expressed as return period T_R.

As for the previous stretch of coast, following the FORM methods, present hazard map was arranged with a chosen limit state equal to 0.5 m (Figure 13). The town centre, as expected, has some portions with a high probability of failure (T_R < 10 years). Moreover, the zones next to the two mouths (Falconera and Livenza) are partially exposed to coastal flooding hazard. More in general, the area characterized by a return period lower than 10 years is only ~70,000 m^2.

Figure 11b shows a zoom of the hazard map and highlights that a portion of the historic city have a T_R < 10 years.

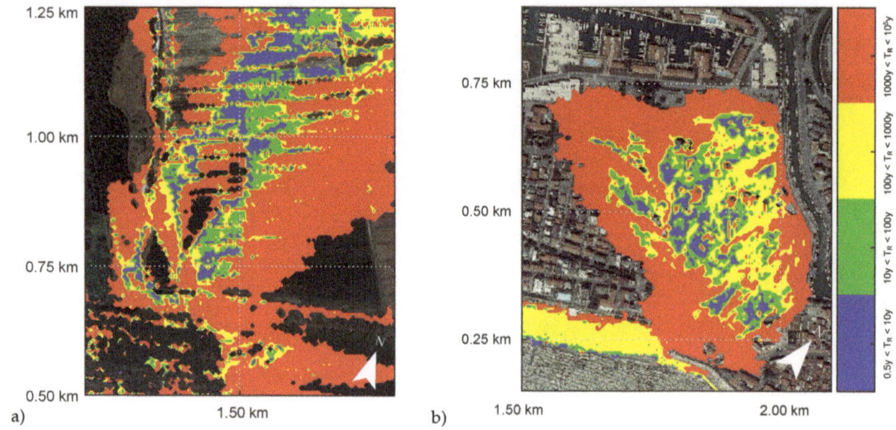

Figure 11. (**a**) Zoom of the Valle Vecchia hazard map shown in Figure 10; and (**b**) zoom of the Caorle hazard map shown in Figure 13.

4. Discussion

The obtained maps suffer from a number of limitations and simplifications of the hydrodynamic model used. However, the proposed probabilistic methodology has a wider perspective, since it allows drawing maps of coastal flooding vulnerability on the basis of a general flood inundation model.

The methodology can be easily coupled with other models to include other key aspects of the coastal flooding assessment. For example, the boundary conditions can be linked with meteorological models to include the effects of hurricanes, typhoons and cyclones and/or with riverine models. In fact, the combinations of multiple phenomena, not only related to sea conditions, frequently exacerbate the effect of marine inundation. For instance, the simultaneous occurrence of intense precipitations or river overflows (in areas such as deltas and estuaries) with extreme waves and sea levels could contribute to coastal flooding. Similarly, the presence of a drainage system could be less efficient than expected during sea storms, generating critical scenarios.

A single failure mechanism is taken into account in this study, i.e., the overtopping. This mechanism is suited to represent coastal flooding in areas with "rigid boundaries", for example an area with concrete sea walls or with paved coastal roads that typically are not subject to local damage or collapse. In other cases, however, other features, can influence failure in the long term, for instance, local erosion and, more generally, morphological changes in the topography (for example, the breaching of a dune system or of a fragile coastal defence structure). The inclusion of this mechanism is not straightforward since it involves a deep knowledge of sea–dike/dune resistance and, in the long term, its maintenance strategy (e.g., nourishment and dunes reinforcement).

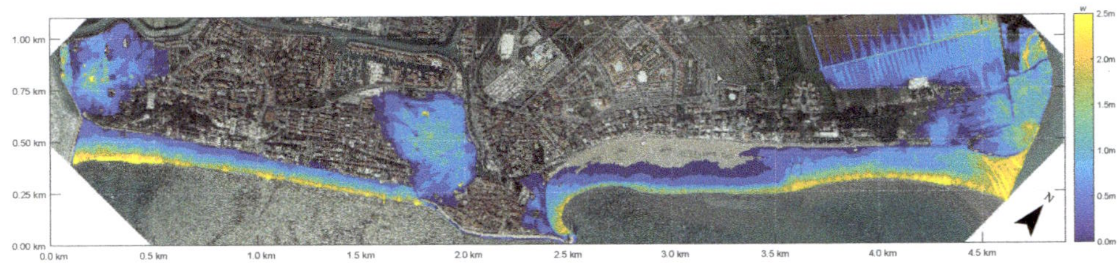

Figure 12. Flooding map (cell VE3): $Hs = 6$ m, $\zeta = 1.5$ m ZMPS.

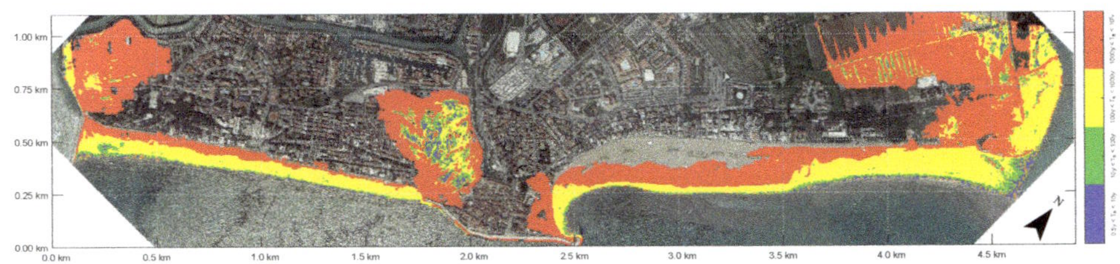

Figure 13. Hazard map (cell VE3): Probability that flooding level 0.5 m is exceeded, expressed as return period T_R.

5. Conclusions

This paper presents a numerical model for the inland flood propagation and an approach for the assessment of coastal flooding vulnerability. The proposal is a novel 2D model that solves the shallow-water equations using a linearized friction term and assumes negligible advective accelerations. The formulation takes advantage of an appropriate vectorization method and a positivity preserving scheme. The key feature of this model is the fast computational speed obtained by means of a formulation suited to GPU cards, thus making it ideal for handling high-resolution maps on a regional scale.

An application to the Venetian coast was carried out, thanks to a wide geomorphological and hydraulic knowledge of this regional area. The statistical analysis of the main marine drivers acting on this coast was carried out to force the wave transformation model and to obtain the boundary condition to be imposed to the shoreline. Finally, a Level II reliability analysis was applied to the results gained through the flooding propagation model, establishing the desired hazard maps. Two hazard maps were produced, showing the probability that any point/pixel is flooded by at least 50 cm.

The methodology seems promising and meets the requirements of local stakeholders for a flexible tool that predicts coastal flooding and highlights the more exposed and vulnerable areas, as the hardware required (GPU) to run the model is certainly well within the financial and technical means of local administrations, unlike the large cluster of CPUs needed to run the existing operational dynamic models.

Finally, the produced hazard maps could be easily integrated with information on the exposed values in the coastal area (e.g., indicative number of inhabitants potentially affected, type of economic activity of the area potentially affected, and installation which might cause accidental pollution in case of flooding [51]) to prepare flood risk maps that show the potential adverse consequences for human health, the environment, cultural heritage and economic activity associated with coastal flooding.

Author Contributions: Conceptualization and Methodology, C.F., L.M. and P.R.; Investigation, C.F., L.M. and P.R.; Writing—Review and Editing, C.F., L.M. and P.R.; Supervision, C.F., L.M. and P.R.

Funding: The Agreement between Regione Veneto and University of Padova entitled "Gestione Integrata della Zona Costiera. Progetto per lo studio ed il monitoraggio della linea di costa per la definizione degli interventi di difesa dei litorali dall'erosione nella Regione Veneto" is gratefully acknowledged.

Conflicts of Interest: The authors declare no conflict of interest.

References

1. Cazenave, A.; Cozannet, G.L. Sea level rise and its coastal impacts. *Earth's Future* **2014**, *2*, 15–34. [CrossRef]
2. Pomaro, A.; Cavaleri, L.; Lionello, P. Climatology and trends of the Adriatic Sea wind waves: Analysis of a 37-year long instrumental data set. *Int. J. Climatol.* **2017**, *37*, 4237–4250. [CrossRef]
3. Rahmstorf, S. Rising hazard of storm-surge flooding. *Proc. Natl. Acad. Sci. USA* **2017**, *114*, 11806–11808. [CrossRef] [PubMed]
4. Nicholls, R.J.; Cazenave, A. Sea-level rise and its impact on coastal zones. *Science* **2010**, *328*, 1517–1520. [CrossRef] [PubMed]
5. Vafeidis, A.T.; Nicholls, R.J.; McFadden, L.; Tol, R.S.; Hinkel, J.; Spencer, T.; Grashoff, P.S.; Boot, G.; Klein, R.J. A new global coastal database for impact and vulnerability analysis to sea-level rise. *J. Coast. Res.* **2008**, *24*, 917–924. [CrossRef]
6. Zanuttigh, B.; Simcic, D.; Bagli, S.; Bozzeda, F.; Pietrantoni, L.; Zagonari, F.; Hoggart, S.; Nicholls, R.J. THESEUS decision support system for coastal risk management. *Coast. Eng.* **2014**, *87*, 218–239. [CrossRef]
7. Zachry, B.C.; Booth, W.J.; Rhome, J.R.; Sharon, T.M. A national view of storm surge risk and inundation. *Weather Clim. Soc.* **2015**, *7*, 109–117. [CrossRef]
8. Wadey, M.P.; Nicholls, R.J.; Hutton, C. Coastal flooding in the Solent: An integrated analysis of defences and inundation. *Water* **2012**, *4*, 430–459. [CrossRef]
9. Spaulding, M.L.; Grilli, A.; Damon, C.; Fugate, G.; Oakley, B.A.; Isaji, T.; Schambach, L. Application of state of art modeling techniques to predict flooding and waves for an exposed coastal area. *J. Mar. Sci. Eng.* **2017**, *5*, 10. [CrossRef]

10. Aucelli, P.P.C.; Di Paola, G.; Incontri, P.; Rizzo, A.; Vilardo, G.; Benassai, G.; Buonocore, B.; Pappone, G. Coastal inundation risk assessment due to subsidence and sea level rise in a Mediterranean alluvial plain (Volturno coastal plain–southern Italy). *Estuarine Coast. Shelf Sci.* **2017**, *198*, 597–609. [CrossRef]
11. Di Luccio, D.; Benassai, G.; Di Paola, G.; Rosskopf, C.; Mucerino, L.; Montella, R.; Contestabile, P. Monitoring and modelling coastal vulnerability and mitigation proposal for an archaeological site (Kaulonia, Southern Italy). *Sustainability (Switzerland)* **2018**, *10*, 2017. [CrossRef]
12. Di Risio, M.; Bruschi, A.; Lisi, I.; Pesarino, V.; Pasquali, D. Comparative Analysis of Coastal Flooding Vulnerability and Hazard Assessment at National Scale. *J. Mar. Sci. Eng.* **2017**, *5*, 51. [CrossRef]
13. Gallien, T. Validated coastal flood modeling at Imperial Beach, California: Comparing total water level, empirical and numerical overtopping methodologies. *Coast. Eng.* **2016**, *111*, 95–104. [CrossRef]
14. Lerma, A.; Bulteau, T.; Elineau, S.; Paris, F.; Durand, P.; Anselme, B.; Pedreros, R. High-resolution marine flood modelling coupling overflow and overtopping processes: Framing the hazard based on historical and statistical approaches. *Nat. Hazards Earth Syst. Sci.* **2018**, *18*, 207–229. [CrossRef]
15. Simeoni, U.; Corbau, C. A review of the Delta Po evolution (Italy) related to climatic changes and human impacts. *Geomorphology* **2009**, *107*, 64–71. [CrossRef]
16. Carbognin, L.; Teatini, P.; Tosi, L. Eustacy and land subsidence in the Venice Lagoon at the beginning of the new millennium. *J. Mar. Syst.* **2004**, *51*, 345–353. [CrossRef]
17. Martinelli, L.; Zanuttigh, B.; Corbau, C. Assessment of coastal flooding hazard along the Emilia Romagna littoral, IT. *Coast. Eng.* **2010**, *57*, 1042–1058. [CrossRef]
18. Pescaroli, G.; Magni, M. Flood warnings in coastal areas: How do experience and information influence responses to alert services? *Nat. Hazards Earth Syst. Sci.* **2015**, *15*, 703–714. [CrossRef]
19. Wolshon, B.; Urbina, E.; Wilmot, C.; Levitan, M. Review of policies and practices for hurricane evacuation. I: Transportation planning, preparedness, and response. *Nat. Hazards Rev.* **2005**, *6*, 129–142. [CrossRef]
20. Jia, X.; Morel, G.; Martell-Flore, H.; Hissel, F.; Batoz, J.L. Fuzzy logic based decision support for mass evacuations of cities prone to coastal or river floods. *Environ. Model. Softw.* **2016**, *85*, 1–10. [CrossRef]
21. Kunreuther, H.; Pauly, M. Rules rather than discretion: Lessons from Hurricane Katrina. *J. Risk Uncertain.* **2006**, *33*, 101–116. [CrossRef]
22. Burcharth, H.F.; Zanuttigh, B.; Andersen, T.L.; Lara, J.L.; Steendam, G.J.; Ruol, P.; Sergent, P.; Ostrowski, R.; Silva, R.; Martinelli, L.; et al. Chapter 3—Innovative Engineering Solutions and Best Practices to Mitigate Coastal Risk. In *Coastal Risk Management in a Changing Climate*; Butterworth-Heinemann: Boston, MA, USA, 2015; pp. 55–170.
23. Vanderlinden, J.P.; Baztan, J.; Coates, T.; Dávila, O.G.; Hissel, F.; Kane, I.O.; Koundouri, P.; McFadden, L.; Parker, D.; Penning-Rowsell, E.; et al. Chapter 5—Nonstructural Approaches to Coastal Risk Mitigations. In *Coastal Risk Management in a Changing Climate*; Butterworth-Heinemann: Boston, MA, USA, 2015; pp. 237–274.
24. Kusumo, A.; Reckien, D.; Verplanke, J. Utilising volunteered geographic information to assess resident's flood evacuation shelters. case study: Jakarta. *Appl. Geogr.* **2017**, *88*, 174–185. [CrossRef]
25. Coastal Flooding Hazard—Community information and participation. Available online: https://coastalfloodinghazard.wordpress.com/community-information (accessed on 7 January 2019).
26. Hoggart, S.; Hawkins, S.J.; Bohn, K.; Airoldi, L.; van Belzen, J.; Bichot, A.; Bilton, D.T.; Bouma, T.J.; Colangelo, M.A.; Davies, A.J.; et al. Chapter 4–Ecological Approaches to Coastal Risk Mitigation. In *Coastal Risk Management in a Changing Climate*; Butterworth-Heinemann: Boston, MA, USA, 2015; pp. 171–236.
27. Gómez-Pina, G.; Muñoz-Pérez, J.J.; Ramírez, J.L.; Ley, C. Sand dune management problems and techniques, Spain. *J. Coast. Res.* **2002**, *36*, 325–332. [CrossRef]
28. Favaretto, C.; Martinelli, L.; Ruol, P. A Model of Coastal Flooding Using Linearized Bottom Friction and its Application to a Case Study in Caorle, Venice Italy. *Int. J. Off. Polar Eng.* **2019**, *29*, 1–9.
29. Favaretto, C.; Martinelli, L.; Ruol, P. Raster Based Model of Inland Coastal Flooding Propagation Using Linearized Bottom Friction and Application to a Real Case Study in Caorle, Venice (IT). In Proceedings of the 28th International Ocean and Polar Engineering Conference, Sapporo, Japan, 10–15 June 2018.
30. Bates, P.D.; Horritt, M.S.; Fewtrell, T.J. A simple inertial formulation of the shallow water equations for efficient two-dimensional flood inundation modelling. *J. Hydrol.* **2010**, *387*, 33–45. [CrossRef]
31. Hunter, N.M.; Bates, P.D.; Horritt, M.S.; Wilson, M.D. Simple spatially-distributed models for predicting flood inundation: A review. *Geomorphology* **2007**, *90*, 208–225. [CrossRef]

32. Khorshid, S.; Mohammadian, A.; Nistor, I. Extension of a well-balanced central upwind scheme for variable density shallow water flow equations on triangular grids. *Comput. Fluids* **2017**, *156*, 441–448. [CrossRef]
33. Brodtkorb, A.R.; Sætra, M.L.; Altinakar, M. Efficient shallow water simulations on GPUs: Implementation, visualization, verification, and validation. *Comput. Fluids* **2012**, *55*, 1–12. [CrossRef]
34. Liang, W.Y.; Hsieh, T.J.; Satria, M.T.; Chang, Y.L.; Fang, J.P.; Chen, C.C.; Han, C.C. A GPU-based simulation of tsunami propagation and inundation. In *Proceedings of the International Conference on Algorithms and Architectures for Parallel Processing*; Springer: Berlin/Heidelberg, Germany, 2009; pp. 593–603.
35. Synolakis, C. The runup of solitary waves. *J. Fluid Mech.* **1987**, *185*, 523–545. [CrossRef]
36. Briggs, M.J.; Synolakis, C.E.; Harkins, G.S.; Green, D.R. Laboratory experiments of tsunami runup on a circular island. *Pure Appl. Geophys.* **1995**, *144*, 569–593. [CrossRef]
37. Kottegoda, N.; Rosso, R. *Probability, Statistics, and Reliability for Civil and Environmental Engineers*; McGraw-Hill: New York, NY, USA, 1997.
38. Madsen, H.O.; Krenk, S.; Lind, N.C. *Methods of Structural Safety*; Courier Corporation: North Chelmsford, MA, USA, 2006.
39. Du, X. First order and second reliability methods. In *Probabilistic Engineering Design*; John Wiley & Sons Inc.: Hoboken, NJ, USA, 2005; pp. 1–33.
40. Hasofer, A.M.; Lind, N.C. Exact and invariant second-moment code format. *J. Eng. Mech. Div.* **1974**, *100*, 111–121.
41. Group, T.W. The WAM model—A third generation ocean wave prediction model. *J. Phys. Oceanogr.* **1988**, *18*, 1775–1810. [CrossRef]
42. Armaroli, C.; Ciavola, P.; Perini, L.; Calabrese, L.; Lorito, S.; Valentini, A.; Masina, M. Critical storm thresholds for significant morphological changes and damage along the Emilia-Romagna coastline, Italy. *Geomorphology* **2012**, *143*, 34–51. [CrossRef]
43. Boccotti, P. On coastal and offshore structure risk analysis. *Excerpta Ital. Contrib. Field Hydraul. Eng.* **1986**, *1*, 19–36.
44. Mendoza, E.; Jimenez, J.; Mateo, J.; Salat, J. A coastal storms intensity scale for the Catalan sea (NW Mediterranean). *Nat. Hazards Earth Syst. Sci.* **2011**, *11*, 2453–2462. [CrossRef]
45. Archetti, R.; Paci, A.; Carniel, S.; Bonaldo, D. Optimal index related to the shoreline dynamics during a storm: The case of Jesolo beach. *Nat. Hazards Earth Syst. Sci.* **2016**, *16*, 1107–1122. [CrossRef]
46. Lin-Ye, J.; García-León, M.; Gràcia, V.; Ortego, M.I.; Stanica, A.; Sánchez-Arcilla, A. Multivariate hybrid modelling of future wave-storms at the northwestern Black Sea. *Water* **2018**, *10*, 221. [CrossRef]
47. Dally, W.R.; Dean, R.G.; Dalrymple, R.A. Wave height variation across beaches of arbitrary profile. *J. Geophys. Res. Oceans* **1985**, *90*, 11917–11927. [CrossRef]
48. Ruol, P.; Martinelli, L.; Favaretto, C. Vulnerability Analysis of the Venetian Littoral and Adopted Mitigation Strategy. *Water* **2018**, *10*, 984. [CrossRef]
49. Ruol, P.; Martinelli, L.; Favaretto, C. *Gestione Integrata Della Zona Costiera. Studio e Monitoraggio per la Definizione Degli Interventi di Difesa dei Litorali Dall'Erosione Nella Regione Veneto—Linee Guida*; Edizioni Libreria Progetto; University of Padua: Padova, Italy, 2016.
50. Nataf, A. Determination des Distribution don't les marges sont Donnees. *C. R. Acad. Sci.* **1962**, *225*, 42–43.
51. Prinos, P.; Kortenhaus, A.; Swerpel, B.; Jiménez, J.A. *Review of Flood Hazard Mapping*; Floodsite Report, No. T03-07-01,54; University of Athens: Athens, Greece, 2008.

water

MDPI

Article

Coupled Wave-2D Hydrodynamics Modeling at the Reno River Mouth (Italy) under Climate Change Scenarios

Maria Gabriella Gaeta [1,*], Davide Bonaldo [2], Achilleas G. Samaras [3], Sandro Carniel [2] and Renata Archetti [1,2]

1 Department of Civil, Chemical, Environmental, and Materials Engineering (DICAM), University of Bologna, Bologna 40136, Italy; renata.archetti@unibo.it
2 Institute of Marine Sciences, National Research Council (CNR-ISMAR), Venice 30122, Italy; davide.bonaldo@ve.ismar.cnr.it (D.B.); sandro.carniel@ismar.cnr.it (S.C.)
3 Department of Civil Engineering, Aristotle University of Thessaloniki, Thessaloniki 54124, Greece; asamaras@civil.auth.gr
* Correspondence: g.gaeta@unibo.it; Tel.: +39-051-2090508

Received: 4 July 2018; Accepted: 28 September 2018; Published: 2 October 2018

Abstract: This work presents the results of the numerical study implemented for the natural area of Lido di Spina, a touristic site along the Italian coast of the North Adriatic Sea, close to the mouth of River Reno. High-resolution simulations of nearshore dynamics are carried out under climate change conditions estimated for the site. The adopted modeling chain is based on the implementation of multiple-nested, open-source numerical models. More specifically, the coupled wave-2D hydrodynamics runs, using the open-source TELEMAC suite, are forced at the offshore boundary by waves resulting from the wave model (SWAN) simulations for the Adriatic Sea, and sea levels computed following a joint probability analysis approach. The system simulates present-day scenarios, as well as conditions reflecting the high IPCC greenhouse concentration trajectory named RCP8.5 under predicted climate changes. Selection of sea storms directed from SE (Sirocco events) and E–NE (Bora events) is performed together with Gumbel analysis, in order to define ordinary and extreme sea conditions. The numerical results are here presented in terms of local parameters such as wave breaking position, alongshore currents intensity and direction and flooded area, aiming to provide insights on how climate changes may impact hydrodynamics at a site scale. Although the wave energy intensity predicted for Sirocco events is expected to increase only slightly, modifications of the wave dynamics, current patterns, and inland flooding induced by climate changes are expected to be significant for extreme conditions, especially during Sirocco winds, with an increase in the maximum alongshore currents and in the inundated area compared to past conditions.

Keywords: climate changes; sea-level rise; TELEMAC; natural beach; flooded area

---

1. Introduction

Strong anthropic pressures, together with the effects of a changing climate, are contributing to the recent increase in the vulnerability of coastal areas. This is particularly significant in the case of extreme events, which, even in the case of a possible decrease of the storminess, are expected to be superimposed to an increased sea level, with an overall intensification of the flooding hazard in coastal regions (for the Mediterranean Sea see for instance [1,2]).

The development of efficient tools to accurately represent nearshore wave and current-induced dynamics is essential for today's operational and forecasting applications in coastal zones [3,4]. High-resolution wave and hydrodynamics modeling offer an extensive range of capabilities to support coastal planning and decision assessment, accounting for typical features at a coastal engineering scale, such as nonlinear processes of wave propagation and interactions between offshore and coastal structures and the inclusion of inshore boundary conditions, such as river run-off.

In recent years, research efforts have focused on the development of methodological frameworks based on advanced numerical modeling [5,6], which can be used to study the effects of future climate change scenarios affecting both the intensity and frequency of storm-surge events, wave climate, currents, sea-level rise, and riverine sediment discharge. Since the above-mentioned phenomena may increase the flood risk for coastal areas, the understanding of their dynamics at coastal scale becomes essential for the design of climate-change resilience protection and, in general, spatial planning activities.

Thus, the development of multipurpose measures mitigating erosion and inundation and increasing coastal defense efficiency requires a challenging prediction of sea forcings variation induced by the estimated effects of climate change.

Recently, regional future scenarios accounting for the Intergovernmental Panel on Climate Change (IPCC) sea-level projections at 2100 characterized by Representative Concentration Pathways equal to +8.5 W/m^2 (hereafter RCP8.5) were applied along the Italian peninsula at DTM (Digital Terrain model) scale [7,8], in order to define the most vulnerable coastal areas in the country and to assess coastal mitigation and adaptation strategies in response to climate change [9]. This static GIS approach can be applied to indicate flood-prone areas at regional [10] or national [11] scales, which can potentially become inundated under future sea-level conditions [12]. However, the implementation of coupled wave-hydrodynamics models, in addition to this static approach, may be advisable on a local scale to obtain more accurate results in terms of nearshore dynamics connected to vulnerability levels and to account for processes of wave propagation, littoral drift, and coastal flooding.

The objective of the present paper is the description of an operational strategy for the development of a multiple-nesting system based on the implementation of open-source numerical models at a high resolution. The natural beach of Lido di Spina, in the vicinity of River Reno mouth, along the eastern coast of Northern Italy (Section 2), is selected to simulate nearshore dynamics induced by the projected effects of climate-change scenarios.

In the specific area of study presented here, we recall the efforts carried out in a 3D context [13–15] and in integrated 2D as well [16], the latter representing a good trade-off between computational costs and results. The proposed study aims to investigate on a site scale how climate changes may impact local parameters such as wave breaking position, alongshore currents intensity and direction and flooded area.

Section 3 describes materials and methods adopted in the present study: In Section 3.1, the framework of the multi-model approach is introduced by specifying the modeling chain of the implemented open-source models, namely, SWAN and TELEMAC. The results of two thirty-year numerical sea state simulations of past (1971–2000) and future (2071–2100) scenarios forced by the climate model COSMO-CLM, in a severe emission future scenario (RCP8.5 in [17]), are presented in Section 3.2, under ordinary and extreme sea events. The implementation of the high-resolution coastal model at the natural beach of Lido di Spina is described in Section 3.3. The model results for both simulated scenarios are discussed in terms of wave nearshore dynamics, current patterns, and flooded areas in Section 4. Conclusions derived from this study are presented in Section 5.

2. Description of the Study Area

Figure 1 shows the area selected as the representative study case: The natural beach of Lido di Spina and adjacent low-lying zones, located along the coast of Emilia Romagna Region, in the Northern Adriatic Sea. The zone was declared a National (Italian) Nature reserve in 1981, as part of the Po Delta Regional Park, and the Sites of European Community Importance under Directive 92/43/EEC. The site was an extremely dynamic area, under erosion in the last decade and exposed to several climate change related effects, e.g., higher occurrence of flooding and beach retreat [18,19].

The modeled site includes, from North to South, the coastal area between the south-side of Porto Garibaldi and the Reno River mouth (Figure 1), and comprises the mouth of Logonovo channel (zone *a*) at Lido degli Estensi, where sediments transported northwards from the Reno outflow are intercepted by the jetty of Porto Garibaldi, the touristic shoreline of Lido di Spina (zone *b*), and the old outlet plain of Gobbino channel (zone *c*) with low-lying plains.

As is common in all the Northern Adriatic Sea, the wave climate in the study site is characterized by storms mainly generated by northeasterly and southeasterly winds, named Bora and Sirocco, respectively [20,21], the latter inducing the highest surge levels [9].

The investigated area is subject both to a high rate of coastal erosion, with a beach retreat of about 3 km over the last 70 years and a total of about 75 ha of land lost, and to high, human-induced land subsidence due to fluid extraction and off-shore gas platforms activity [22,23]. Moreover, the coastal area presents highly urbanized touristic resorts, known for the wide and sandy open beaches and the natural areas of the Po Delta Park.

The recent local management strategies to protect the area from flooding and erosion have been based on frequent beach nourishments, using depositional materials from the north of the study area, and on the construction of defense structures, such as stone revetments (especially in the area northern the mouth of the River Reno, see [18]). Sea banks along the coastal area strongly preserved it from inundation, although these defense structures may turn out to be rather ineffective in case of poor maintenance and frequently occurring storm events have rendered these defense structures ineffective [24]. In addition, important morphological variations at the mouth of the River Reno, as those related to the deconstruction of the final part of its right riverside, partly compromised the inland safety.

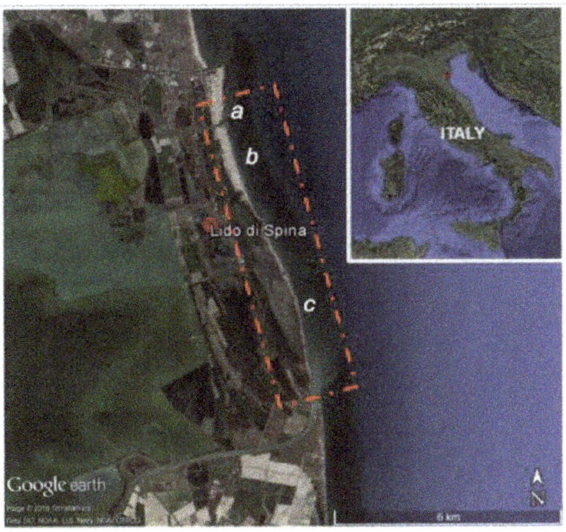

Figure 1. Wider view of the study area of Lido di Spina (images from Google Earth [25]; privately processed). The plot also shows the location of areas of interest: Logonovo channel at Lido degli Estensi (zone *a*), Lido di Spina (zone *b*), and Gobbino channel (zone *c*).

3. Materials and Methods

3.1. Adopted Methodology and Numerical Modelling Chain

The proposed multiple-nesting methodology is based on a progressive down-scaling, from global climate to coastal models, as described in the following.

a. Wind fields in the Adriatic Sea are obtained every 6 h at 8 km horizontal resolution from the Regional Climate Model (RCM) COSMO-CLM [26], a climate version of the operational weather forecast model COSMO-LM [27]. The simulation generating this dataset [28] encompasses the Italian peninsula and its marine regions, describing the climate evolution under the IPCC RCP8.5 scenario [29,30] in the period 1971–2100. The model is initialized by ERA-Interim Reanalysis [31], and the same model provided boundary conditions for the period 1971–2005, and by the coupled atmosphere-ocean general circulation model CMCC-CM [32] in the period 2006–2100. Compared to previous implementations [33], the validation of the COSMO-CLM wind fields based on in-situ observations along the Italian coast [34] exhibits a particular skill in capturing the wind directional distribution. In turn, results on climatological projections suggest a tendency towards an overall wind energy decrease (in the range of 0–10% across the whole Adriatic Sea), with a stronger decrease along the Bora jets patterns.

b. The spectral properties of the sea states in the Adriatic basin are described by means of the phase-averaged wave model SWAN [35] forced by the climatological wind fields from RCM COSMO-CLM [34]. Wave model simulations are carried out under the severe climate change conditions envisaged by the IPCC RCP8.5 scenario in the period 2071–2100 (henceforth referred to as SCE—SCEnarios—dataset) and, for an overall assessment of the model performances, under control conditions in the period 1971–2000 (henceforth referred to as CTR—ConTRol—dataset). The geographical SWAN domain, encompassing the whole Adriatic Sea, is discretized into a curvilinear orthogonal grid with a resolution ranging approximately from nearly 10 km in the southeastern regions to 2 km along the northern Italian coasts. The spectral domain is divided into 25 logarithmically-distributed values between 0.05 and 0.5 Hz and 36 directional sectors. Since the focus of the model is on the northern basin and in the absence of regional wave dataset consistent with the wind forcing, following the same approach described by previous studies [36,37], no waves are prescribed entering the southern boundary, whereas internally-generated energy is allowed to radiate outward.

c. The coupled wave-2D hydrodynamics model of the natural beach of Lido di Spina (Figure 1) is implemented by means of the TELEMAC suite [38]. The above-described SWAN results of wave field are extrapolated at the offshore boundary nodes of the detailed mesh performed for TELEMAC runs. The spectral module TOMAWAC (henceforth denoted as TOM [38]) is forced under control and climatological scenarios, and wave-induced currents are taken into account coupling the 2D hydrodynamics module TELEMAC2D (henceforth denoted as TEL2D [39]).

Table 1 shows the numerical characteristics of each of the different modeling chain level of the proposed nesting approach, i.e., the model extension, the mesh-size range, the simulated period, the type of run, and the forcings inducing atmospheric and sea dynamics.

Table 1. Numerical characteristics of the proposed nesting approach for each of the different chain levels, i.e., the model extension, the mesh size range, the simulated period, the type of run, and the forcings inducing atmospheric and sea dynamics.

Chain Level	Model	Extension	Mesh Size Range	Period	Run Type	Forcings
a	RCM COSMO-CLM	Italian marine regions [28]	0.0715° (~8 km)	1971–2100	Atmospheric circulation	SST, atmospheric conditions from [31,32]
b	SWAN	Adriatic Sea	2–10 km	1971–2000	Non-stationary wave fields	From (a), wind at 10-m height
				2071–2100	Non-stationary wave fields	
c	TELEMAC	Lido di Spina (4 km long and 2 km wide)	25–50 m	1971–2000	Stationary wave fields	From (b), after statistics: wave spectra, sea-level rise
				2071–2100	Stationary wave fields	

The level (a) of the modeling chain, although described in the table, is not discussed in the present study, since its implementation and results are largely described and examined in [28]. Its numerical outputs, such as wind fields under different scenarios, are used to force the wave model SWAN, developed by [34] for the Adriatic Sea, namely, the chain level (b).

The numerical results of SWAN offshore the study site are (1) statistically elaborated and compared to the transposed observations and under CTR and SCE scenarios, (2) analyzed in order to identify extreme events and estimate Gumbel distributions for each simulated scenario, and (3) used to force the TELEMAC model, implemented at high resolution for the beach of Lido di Spina. Details on the statistics applied on the SWAN outputs and on the sea state conditions simulated by TELEMAC are given in the following sections.

3.2. Past and Projected Future Wave Climate Scenarios

3.2.1. Statistics on Wave Forcings

The 10-year-long data acquired by the regional NAUSICAA buoy, offshore the near coastline of Cesenatico, at a water depth of 10 m (data available at http://www.smr.arpa.emr.it/dext3r), are analytically transposed offshore the study site. Since the buoy data are not measured at waters deep enough to neglect the refraction effect on the wave directions [40]—that are supposed to be actually altered by refraction—the recorded waves at NAUSICAA are already slightly rotated towards the perpendicular to isobaths lines, ranging approximately between 55 and 60° N.

The wave rose showing the distribution of the significant wave height and the mean wave directions at the study area and as a result of the transposed observations is presented in Figure 2, showing that:

- the most energetic waves, up to 3.7 m of height, propagate from sectors 50–60° N;
- the most frequent conditions, with wave height up to 1 m, come from 100° N;
- calm sea conditions comprise 20.45% of annual frequency.

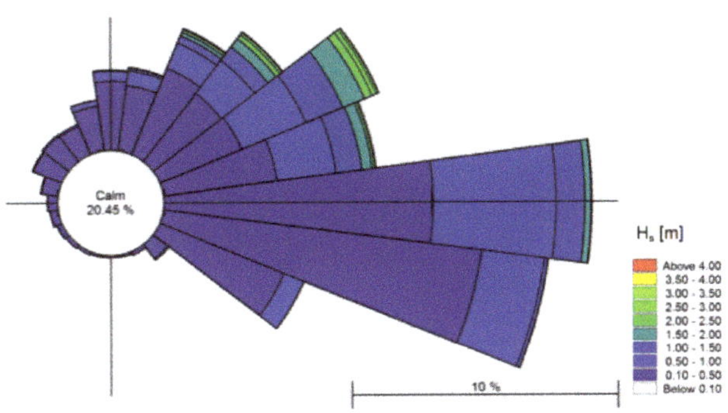

Figure 2. Wave rose at the study area of Lido di Spina from the transposed observations.

For the simulations carried out in this work, SWAN outputs are retrieved offshore the beach of Lido di Spina, at the nearest grid nodes to the offshore boundary of the implemented TELEMAC.

The seasonal variability of the monthly average of the significant wave height for Sirocco and Bora events is reported in Figure 3, both for the 30-year-long numerical simulations by the SWAN model results under CTR and SCE scenarios and for the transposed measurements from the 10-year-long NAUSICAA buoy records.

A slightly local increase in wave energy intensity is forecasted for events from Sirocco and especially during winter, while Bora energy tends to decrease on average. By comparing with CTR runs, the available transposed observations show good agreement in case of Sirocco events, while, during Bora events, the numerical results underestimate the monthly average of the significant wave height in winter and fall seasons, despite a slight overestimation during summer events.

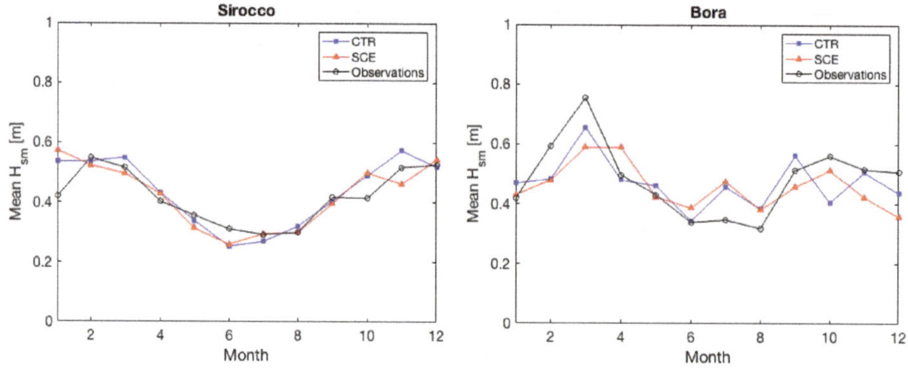

Figure 3. Monthly averages of the significant wave height $H_{s,m}$ offshore of the study area of Lido di Spina that resulted from CTR (1971–2000) and SCE (2071–2100) scenarios performed by SWAN implementation [34] and from NAUSICAA—transposed observations (2007–2017) for Sirocco (**left panel**) and Bora (**right panel**) events.

3.2.2. Extreme Event Analysis: Joint Probability of Significant Wave Height, Period, and Sea-Level Rise

Storms are identified following the procedure proposed by [41] and classified based on their generating wind direction (Sirocco or Bora). For each of the two dominant directions, extreme events are analyzed considering the return period (henceforth referred to as T_R) associated with their maximum significant wave height and fitting the modeled storms with a return period greater than 1 year by means of Gumbel distributions.

In comparison with the control scenario, the analysis in Figure 4 reveals a significant increase in the extreme events coming from Sirocco for the future projection, which is also expected to be important for return periods greater than 50 years; on the other hand, extreme Bora events are expected to reduce their occurrence at the study site.

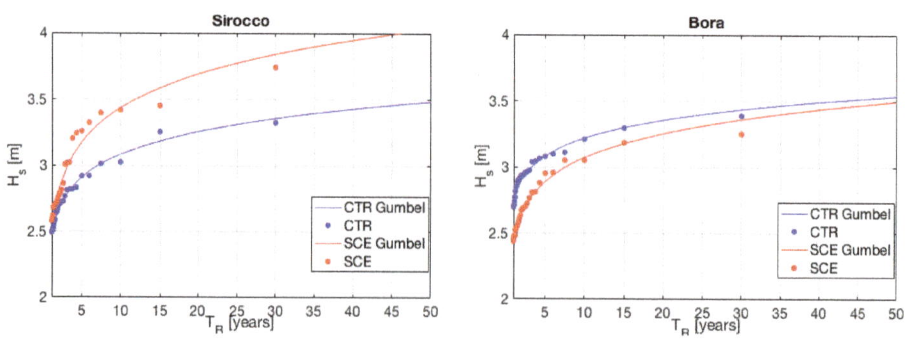

Figure 4. Statistics of extreme events for CTR (blue) and SCE (red) runs for Sirocco (**left panel**) and Bora (**right panel**) events. Filled dots show extreme values from simulations; tendency curves following Gumbel distribution are plotted in solid lines.

The joint distributions of the significant wave height H_s and the peak wave period T_p along the offshore boundary of TELEMAC model are obtained from the transposed observations and plotted in Figure 5 for Sirocco (left panel) and Bora (right panel) events.

The two variables can be directly correlated for extreme events (i.e., $H_s > 2$ m) with an exponential law (black lines in Figure 5), giving a correlation coefficient of 0.92 and 0.97 for Sirocco and Bora, respectively.

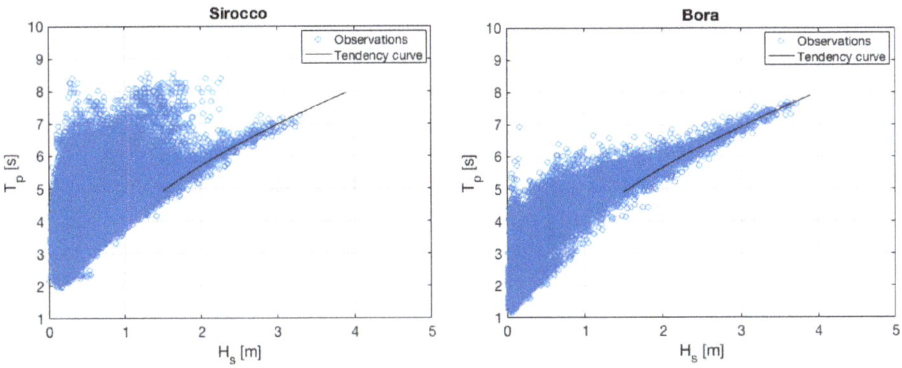

Figure 5. Relation between H_s and T_p for Sirocco (**left panel**) and Bora (**right panel**) events for the NAUSICAA-transposed observations at the study site. Tendency curves for the extreme events are also plotted with solid black lines.

The effects of the peak water level, including high tidal level and storm surge, are included in the numerical simulations of the study site, following the accurate analysis performed by [40]. They applied a joint probability analysis by means of a copula-based approach [42] to the registered significant wave heights at NAUSICAA buoy and to the contemporary peak water levels measured at Porto Corsini, located 20 km south of Lido di Spina.

Thanks to the proposed cumulated density functions for the two contemporary variables, peak water levels with return periods T_R equal to 1 year and 25 years are extracted for the present study, under the assumption that extreme sea storms are generated and forced by same meteorological events at the two locations, i.e., Cesenatico and Lido di Spina.

The authors of [1,43] suggested the existence of strong correlation in the Northern Adriatic Sea between winds and maximum sea levels induced by meteorological events, and this aspect is particularly evident during Sirocco events. Under future scenario conditions, their validated simulations showed how inter-annual evolution of maximum sea levels is expected to be essentially stable from 2051 to 2100, as well as how unchanged extreme statistics for events characterized by return period of less than 100 years [44]. Following these future projections, no modifications in the extreme statistics connecting peak sea-water levels and significant wave heights are reasonably assumed; therefore, also for SCE runs, the peak water levels contemporary to extreme significant wave heights are also extrapolated by the authors of [40].

Finally, sea-level rise in the SCE runs is taken into account by uniformly increasing the bottom depth by +0.70 m: This is a basin-scale bulk estimate based on the recent projections provided by [8] for the Northern Adriatic Sea, based on the authors of [17,45] adopting RCP8.5 scenarios of climate change and adjusting them to the projected rates of vertical land movements (isostasy and tectonics).

Table 2 summarizes all the conditions used in this work for the TELEMAC simulations, in terms of sets of significant wave height H_s, peak period T_p, mean wave direction Dir, peak water level PWL (comprising higher tidal level and storm surge), and the expected sea-level rise SLR values.

Table 2. Characteristics of the simulated scenarios, with the significant wave height H_s, the peak period T_p, the mean wave direction Dir, the peak water level PWL, and the sea-level rise SLR.

Scenario	Period	Storm Direction	Statistics	T_R (year)	Run Name	H_s (m)	T_p [(1)] (s)	Dir (° N)	PWL [(2)] (m)	SLR (m)
CTR	1971–2000	Sirocco	Ordinary	1	CTR-S1	2.50	6.56	90	0.79	0.00
		Sirocco	Extreme	25	CTR-S25	3.31	7.55	90	1.05	0.00
		Bora	Ordinary	1	CTR-B1	2.76	6.64	45	0.83	0.00
		Bora	Extreme	25	CTR-B25	3.4	7.38	45	1.10	0.00
SCE	2071–2100	Sirocco	Ordinary	1	SCE-S1	2.57	6.65	90	0.81	+0.70
		Sirocco	Extreme	25	SCE-S25	3.77	8.06	90	1.18	+0.70
		Bora	Ordinary	1	SCE-B1	2.46	6.27	45	0.79	+0.70
		Bora	Extreme	25	SCE-B25	3.31	7.28	45	1.05	+0.70

Note: (1) The values of T_p were estimated according to the exponential curves in Figure 5. (2) The joint probability analysis presented in [40] was adopted to estimate the peak water levels, as described in the text.

3.3. Set-up of the Coupled Wave-2D Hydrodynamics Numerical Model

Approaching the coastal region, waves generated offshore are influenced by shoaling, refraction, and loss of energy either due to bottom friction or wave breaking [46]. To simulate all these physical processes, including wave-induced currents, the present study is carried out using the wave and hydrodynamics models of the TELEMAC-MASCARET suite that is distributed under a General Public License and is available at TELEMAC [47].

The suite comprises finite-element-based solvers to simulate shallow water hydrodynamics and wave propagation, and is able to model inshore water levels and wave spectra under different forcings. The different included modules can simulate wind wave propagation, ground water flows, tracer transport, sediment transport, and morphodynamics.

In the proposed approach, the wave and 2D hydrodynamics modules of TELEMAC are implemented in order to propagate offshore waves and currents, and reproduce nearshore dynamics and flood processes.

TOM module is a third-generation spectral wave model and solves a simplified equation for the spectral-angular density of wave action by means of a finite-element type method, in order to describe wave propagation and dynamics in coastal areas [38]. The model takes into account bathymetric wave breaking, bottom friction, non-linear wave-wave interactions, wind wave generation, and white-capping. The authors of [48] showed that it is also able to represent the spread wave fields induced by the presence of sandbanks in the nearshore region.

TEL2D module solves the 2D shallow water equations (also referred to as Saint-Venant equations [39]), derived by integrating the Reynolds-averaged Navier-Stokes equations over the flow depth. Several options for the horizontal diffusion terms (depth-averaged k-ε, or constant eddy viscosity models) and source terms (atmospheric pressure gradients, Coriolis force, etc.) could be chosen in the model setup, while the numerical discretization adopts classical methods for the advection terms, such as characteristics, and distributive schemes. Recently, the implementation of implicit schemes enabling relaxation of the CFL limitation on time steps made TEL2D applicable to the treatment of tidal flats, ensuring positive water depth and mass conservation without extra limitation of the time step [49,50].

Since using the same horizontal discretization with a series of Delaunay triangular unstructured elements, the two described modules can be directly coupled (two-way coupling) to account for the effects of waves on the mean coastal circulation reproducing wave-induced currents leading to littoral drift. The gradients of the radiation stress induced by waves are computed using the theory of [51] as part of the hydrodynamics equations in TEL2D.

The implementation of the numerical models in TELEMAC suite, after a series of preliminary tests on the optimal edge dimension for the mesh by using the freely-available pre-processing tool Blue Kenue™ [52], results in a variable density unstructured mesh, consisting of two density areas, the characteristic triangular element size being set to 25 m close to the shore and on the inland areas, in order to better capture the presence of sandbanks and nearshore seabed profiles, and to 50 m for the rest of the computational domain; the total number of nodes is equal to 40,350.

LIDAR topo-bathymetrical data collected by Emilia Romagna Region in 2012 with a spatial resolution up to 1 m are interpolated on the implemented mesh as presented in Figure 6, allowing a detailed analysis of wave-induced hydrodynamics and floods under different sea-level rise scenarios.

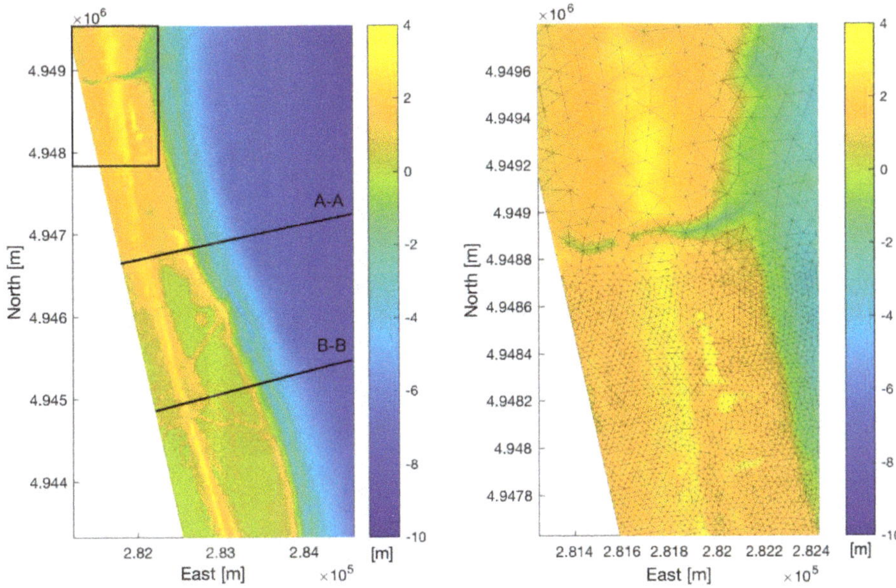

Figure 6. Interpolated bathymetry map in UTM-WGS84 coordinates used for the TELEMAC simulations (**left panel**), as well as the position of the two selected transects A-A and B-B (solid lines). Zoom on the implemented variable density unstructured mesh for the study area ((**right panel**) corresponding to the black box in (**left panel**)).

TOM and TEL2D are properly set up for the studied area based on previous experience on coupled wave-2D hydrodynamics runs for the representation of nearshore processes, as presented in [53,54]. The processes included in the wave model simulations are (a) energy dissipation due to wave breaking according to [55], (b) energy dissipation due to bottom friction according to [56], and (c) nonlinear transfer of energy due to triad (three-wave) interactions according to [57]. No movable seabed, no defense breaching, and no past subsidence-induced movements are assumed in the study.

The simulated conditions as described in Table 2 are performed for the duration necessary to reach the equilibrium states for all the spectral and hydrodynamics variables, i.e., 3 h, and the presented results are extracted at the final time step of the computations.

4. Results and Discussion

Numerical outputs from TELEMAC runs under CTR and SCE scenarios and for Sirocco and Bora events are presented in the following sections, in terms of wave nearshore dynamics, current patterns, and flooded areas.

4.1. Wave and Hydrodynamics

Model results for SCE runs described in Table 2 are presented in the following figures. For Sirocco and Bora events, the significant wave height distribution and mean wave direction vectors that resulted from TOM are shown in Figures 7 and 8, respectively. The increment of water levels in SCE runs, including sea-level rise induced by climate changes, leads to wave propagation over the initially dry zone of the beach, especially in case of T_R = 25 years, revealing an expected increase of sea flooding events in the future.

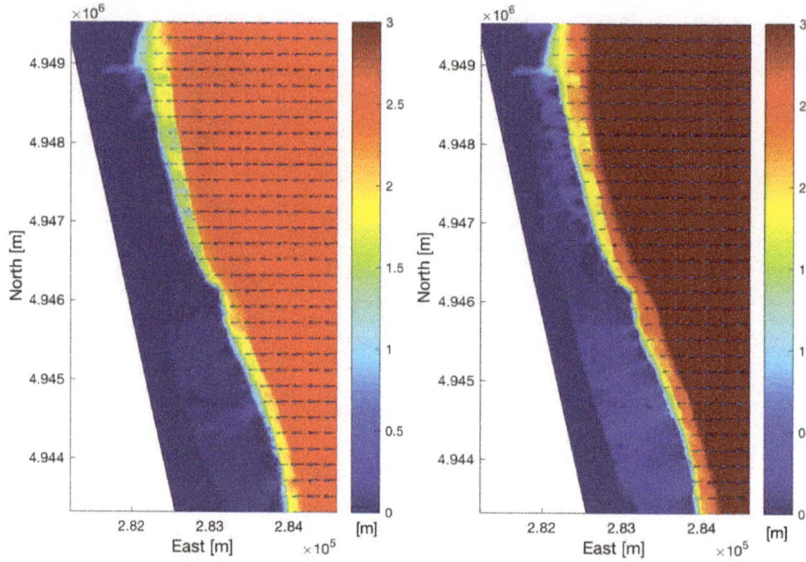

Figure 7. Sirocco events. Distribution of significant wave height and mean wave direction vectors for SCE—S1 (**left panel**) and SCE—S25 (**right panel**) runs, respectively.

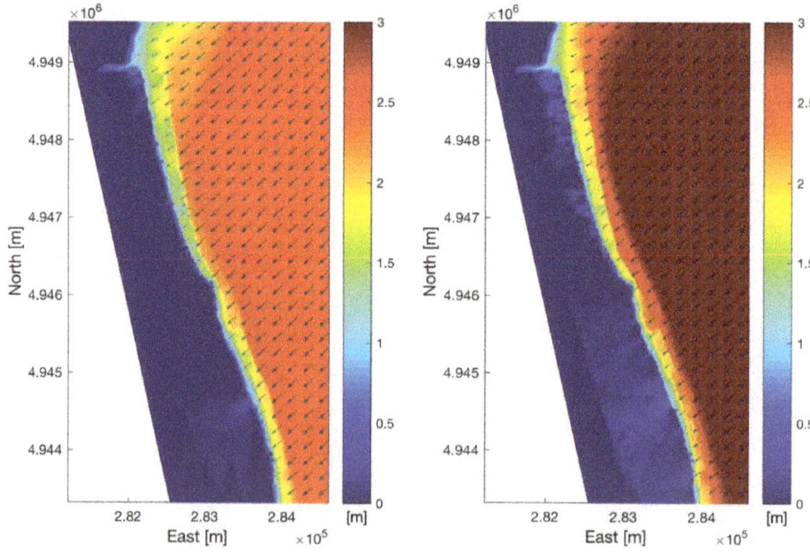

Figure 8. Bora events. Distribution of significant wave height and mean wave direction vectors for SCE—B1 (**left panel**) and SCE—B25 (**right panel**) runs, respectively.

Since differences between runs may appear less significant than in reality on a scale of representation of the global model domain; a comparison of the results is performed along linear transects from the offshore computational boundary to the shoreline (as represented in the left panel of

Figure 6), where the distance is taken from the original shoreline position at 0.0 m of mean sea level (hereafter MSL).

The first transect A-A is located in the zone *a* (in Figure 1) of the beach of Lido di Spina and is characterized by a natural submerged bar at a 160 m offshore distance from the shoreline, at a water depth of around 1.8 m, and by an emerged beach at +1.4 m level with a 2% slope. The transect B-B is instead located in the low-lying area of the old outlet of Bellocchio channel (zone *c* in Figure 1) and presents a submerged natural sand bar at a distance of 220 m from the shoreline, at a water depth of 3 m. The spatial evolution of the significant wave height along these two transects is reported in Figures 9 and 10 for Sirocco and Bora events, respectively.

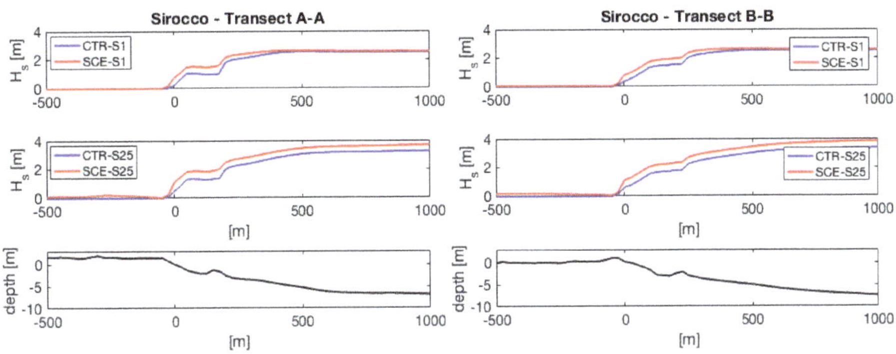

Figure 9. Sirocco events. Cross-shore distribution of the significant wave height H_s as calculated by TOM along the transect A-A (**left panels**) and transect B-B (**right panels**).

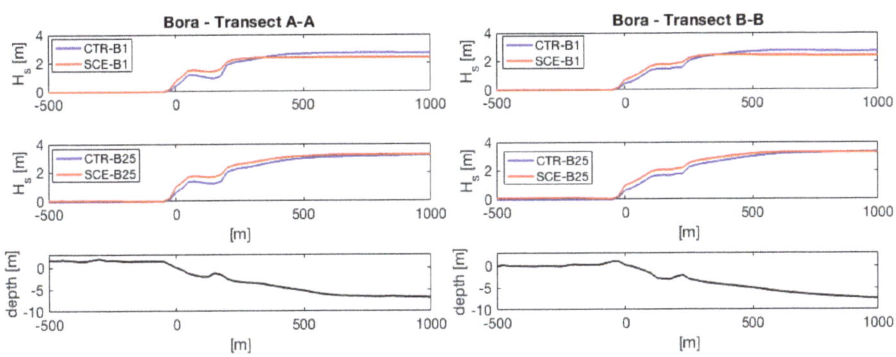

Figure 10. Bora events. Cross-shore distribution of the significant wave height H_s as calculated by TOM along the transect A-A (**left panels**) and transect B-B (**right panels**).

Wave dissipation occurs as a result of both breaking and bottom friction. TOM model simulates the evolution of the wave height, frequency, and direction as the waves propagate towards the beach. In the presence of sand bars, the cross-shore distribution of significant wave height usually has a typical evolution, influenced by the composite bathymetry. Indeed, wave steadily propagates towards the breaking point, beyond which it starts to decrease in height up to submerged bar crest. Then, it remains constant until water depth reaches again the bar crest depth; then, it slowly decreases up to reach the shoreline.

Simulations of CTR and SCE scenarios produce evident variation in wave breaking position and height, denoting a shift to higher wave heights closer to the shoreline in case of SCE runs (especially in case of Sirocco events) and a significant shift of the breaking line position towards the shoreline.

The wave radiation stresses give rise to the alongshore velocity profiles, which are responsible for inter-annual sediment transport, as reported in Figures 11 and 12, showing Sirocco and Bora (with opposite ordinate axis) events, respectively, where positive values denote currents directed from South to North. The maximum value of alongshore velocity occurs within the breaking zone of each simulation.

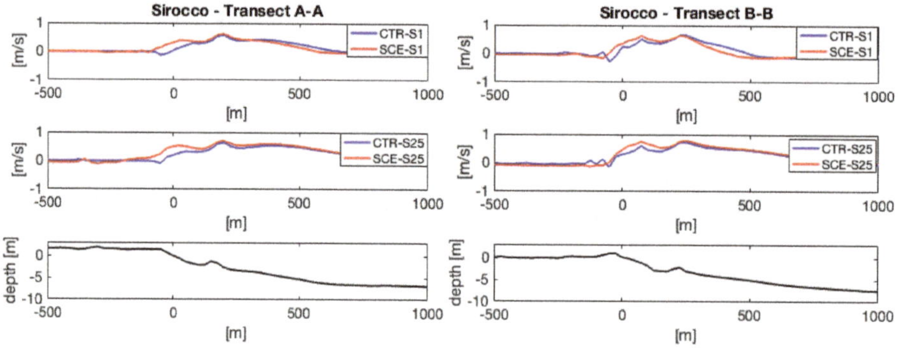

Figure 11. Sirocco events. Cross-shore distribution of the alongshore velocity as calculated by TEL2D for ordinary S1 and extreme S25 events and of the water depth along transect A-A (**left panels**) and transect B-B (**right panels**).

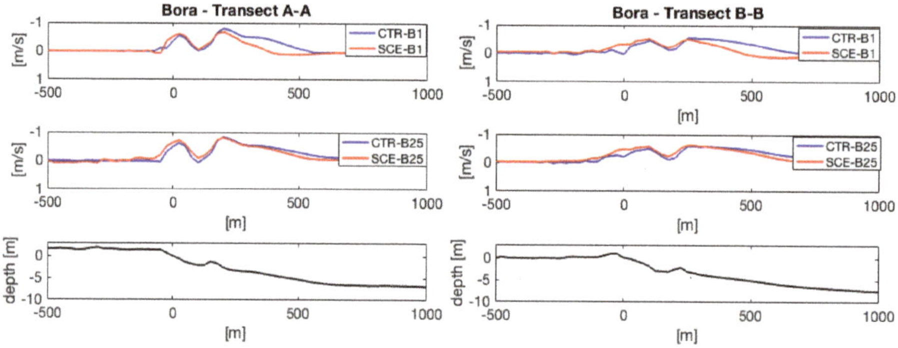

Figure 12. Bora events. Cross-shore distribution of the alongshore velocity as calculated by TEL2D for ordinary B1 and extreme B25 events, and of the water depth along transect A-A (**left panels**) and transect B-B (**right panels**).

In presence of submerged bars, as in the case of the adopted bathymetry, the distribution of alongshore current usually has two peaks, each corresponding to the peaks of wave breaking-points [58,59], with the maximum on the bar crest and the minimum much closer to the shoreline. In the critical part of the surf zone beneath the alongshore-current jet, the intense wave-current stresses occur, while prior to breaking and in the inner surf zone, the current strength is smaller.

The combined effects of the expected sea-level rise and the intensity increase of Sirocco events bring a modest shift in nearshore wave-induced currents, resulting from the respective change in wave dynamics. This is evident for both the analyzed transects in case of Sirocco events but only for the low-lying area of Bellocchio channel in case of Bora events.

4.2. Inundated Areas

Figures 13 and 14 show the comparative effects of changing climate scenarios on the inundated areas that result from Sirocco and Bora runs, respectively: Black lines indicate the initial shoreline at a water depth equal to ±0 m MSL, while blue and red lines represent the upper limits of the inundated area at the end of CTR and SCE simulations.

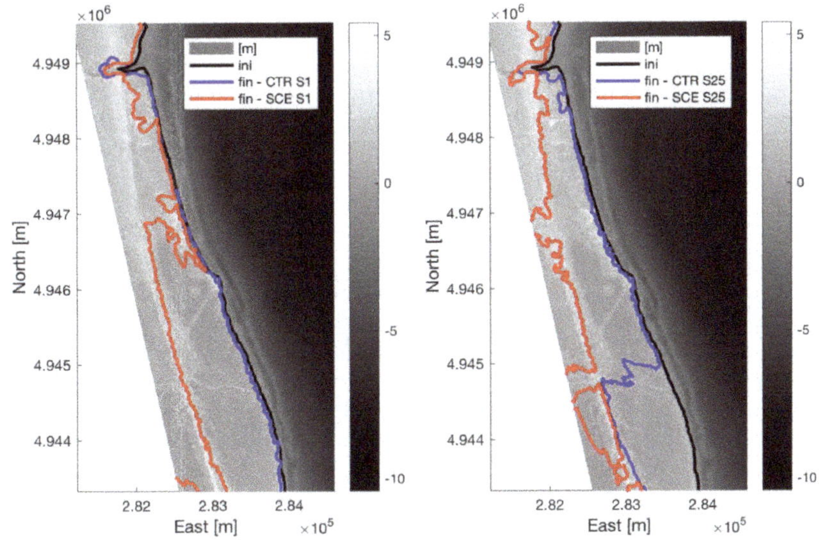

Figure 13. Sirocco events. Inundated areas for ordinary (**left panel**) and extreme (**right panel**) events and for CTR (blue lines) and SCE (red lines) scenarios. Black lines indicate the initial shoreline.

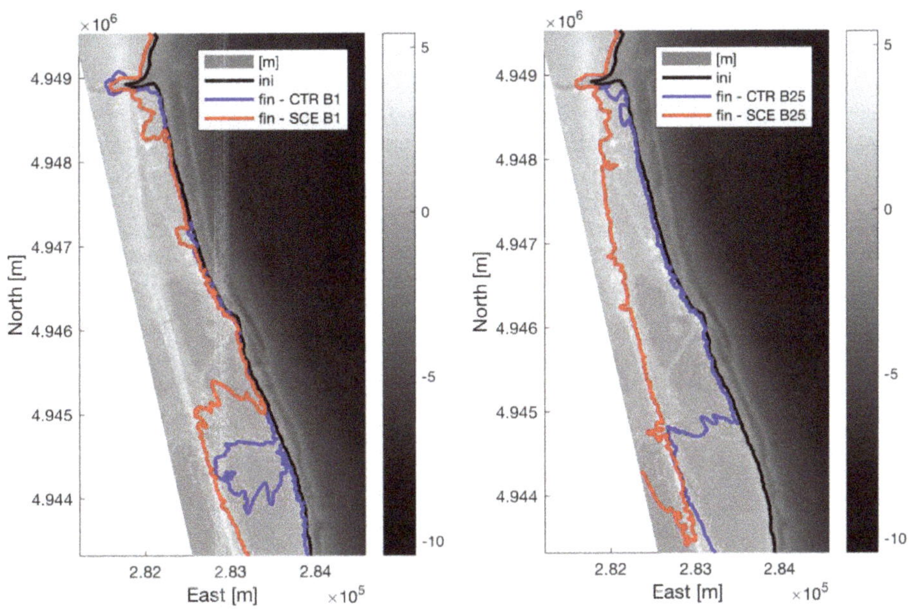

Figure 14. Bora events. Inundated areas for ordinary (**left panel**) and extreme (**right panel**) events and for CTR (blue lines) and SCE (red lines) scenarios. Black lines indicate the initial shoreline.

Differences between CTR and SCE scenarios are the most noticeable ones among the presented results on the scale of the representation; they highlight the utility and effectiveness of the implemented numerical approach in the simulation of the respective phenomena, and they are used to make a comparative analysis.

The areas mainly suffering flooding in the scenarios of climate changes are the northern mouth of Logonovo channel (zone *a* in Figure 1) and the southern low-lying plain of the older outlet of Bellocchio channel (zone *c* in Figure 1), which are expected to be the most vulnerable and exposed to flooding hazards induced by sea-level rise, and during ordinary events.

Indeed, looking at the numerical results for the three different zones in which the studied site is divided in Figure 1:

- In the northern area, the channel outlet is subject to a general increment of flooded area extension in all the simulated scenarios and despite the absence of river discharges, revealing the high vulnerability of these flood prone areas to climate changes.
- The natural beach of Lido di Spina seems not to be significantly inundated in case of ordinary events both from Sirocco and Bora, as well as in future scenarios as resulted after SCE runs; the area, instead, is much more strongly threatened by the predicted future sea-level rise in case of extreme events (Figures 13 and 14, right panels).
- The older outlet of Bellocchio channel seems not to be particularly affected in terms of inundated area by climate change scenarios under extreme conditions (T_R = 25 years), both for Sirocco and Bora events (Figures 13 and 14, right panels). While in case of ordinary event generated by Bora wind (Figure 14, left panel), and already for conditions typical of the CTR scenario, the area experiences a sea inundation that significantly increases in SCE scenario; sea-level rise is also responsible of a significant increase in the flooded area extension also in case of ordinary event generated by Sirocco event (Figure 13, left panel).

4.3. Discussion of the Effects of Expected Climate Changes

Changes in deep-water wave climate drive littoral hydrodynamics and morphological variation, such as the patterns of nearshore wave propagation and alongshore currents over local shelf bathymetry.

A quantitative comparison between the different scenarios and events is proposed in Table 3, in which differences between the simulated CTR and SCE scenarios are reported in terms of characteristics of wave-current-induced dynamics with the aim of estimating the effects of expected climate changes at the study site.

Table 3. Computed variation induced by climate changes between SCE and CTR runs of wave breaking line position Δx_B, maximum alongshore velocity $\Delta U_{ls, MAX}$, and percentage of inundated area $\Delta\Omega$.

Event	Δx_B (m)	$\Delta U_{ls, MAX}$ (%)	$\Delta\Omega$ (-)
Sirocco T_R = 1 year	−20	+2	9
Sirocco T_R = 25 years	±0.0	+8	4
Bora T_R = 1 year	−200	−8	3.5
Bora T_R = 25 years	−70	±0	3

In particular, the following features are calculated from the numerical results:

- Variation of the wave breaking line position Δx_B (m) calculated as

$$\Delta x_B = x_{B, SCE} - x_{B, CTR} \tag{1}$$

in which $x_{B,SCE}$ and $x_{B,CTR}$ are the breaking line positions for SCE and CTR runs, respectively. This data would allow estimating wave energy in the surf zone and impacting the shoreline.

- Changes in the maximum alongshore currents calculated $\Delta U_{ls,\,MAX}$ (%) as

$$\Delta U_{ls,\,MAX} = (U_{ls,\,MAX\,SCE} - U_{ls,\,MAX\,CTR})/U_{ls,\,MAX\,CTR} \times 100 \quad (2)$$

in which $U_{ls,\,MAX\,SCE}$ and $U_{ls,\,MAX\,CTR}$ are the maximum alongshore velocity computed in SCE and CTR runs, respectively. This value could give evidence of the estimated variation induced by climate changes in morphological patterns driving the inter-annual area evolution.

- The computed percentage of inundated area variation $\Delta\Omega$ (-) is calculated as

$$\Delta\Omega = (\Omega_{SCE} - \Omega_{CTR})/\Omega_{CTR} \quad (3)$$

in which Ω_{SCE} and Ω_{CTR} are the estimated inundated areas for SCE and CTR runs, respectively; this information could help in evaluating the flooding hazard increment connected to climate changes.

According to computed results reported in Table 3, modifications of the wave dynamics, currents patterns, and inland water propagation induced by climate changes are expected to be mainly significant for extreme conditions, and especially for Sirocco events, with increases in the maximum alongshore currents and in the flooded area dimensions. Decreases or negligible variations are instead estimated for Bora events in terms of alongshore currents, again revealing how long-term effects of a changing climate along the North Adriatic shoreline may be stronger related to Sirocco events.

Expected effects of increments in wave height and in sea level in case of Sirocco events during SCE runs (as reported in Table 2) may induce no evident variations in the breaking line position, while wave energy impacting the shoreline and inland area may be strongly affected by wave energy flux, which is related to wave height and storm duration. An onshore shift of the breaking line in case of Bora events is instead connected to a balance between wave height decrement and sea-level rise during SCE runs (as reported in Table 2).

An overall increase of the flooded area extension ranging from 3 to 9 times the CTR values is observed in the results of all the simulated scenarios, showing that the greatest threat from future climate changes, especially during ordinary sea storms generated by Sirocco winds, would be the relative rise of sea level, while slight modifications of the littoral currents would drive morphological dynamics in the shoreline stability and evolution.

5. Conclusions

Wave-current dynamics represent a feature of paramount importance in the northern Adriatic basin, both for triggering local circulation [13,60] influencing connections in terms of amount of dense water produced [59] and driving sediment re-suspension and mobilization [14,61,62]. Modifications induced by climate changes in wave-current dynamics need to be carefully evaluated, since they can be regarded as among the main threats to coastal protection and stability, with important implications for the assessment of littoral management plans and for mitigation efforts [4,63].

In the present study, coupled wave-2D hydrodynamics simulations carried out by means of the open-source TELEMAC suite were implemented in the coastal area of the River Reno mouth, along the eastern coast of Northern Italy. A multiple-nesting approach was developed by adopting COSMO-CML and SWAN results for the Adriatic Sea in order to run past (1971–2000), and future climate change (2071–2100) scenarios were developed at high-resolution coupled wave-2D-hydrodynamics TELEMAC model.

Presented results showed modifications in the area dynamics as a response to expected climate change scenarios and previously integrated carried out studies, limited to sea level dynamics. A slight local growth in the wave energy intensity for Sirocco events, however, inducing an increase in alongshore currents and inundated areas, is observed. On the other hand, during Bora events, negligible changes in wave-induced coastal dynamics are forecasted, while flooded area dimensions again increased under the simulated sea-level rise effects.

The presented numerical results showed that flooding hazards and changes in littoral hydrodynamics at the selected site are nowadays already significant, especially during Sirocco extreme events [24], and are expected to further increase in the future. It is also found that, according to the basin sea level projections elaborated by IPCC [8,17], the highest contribution to the coastal vulnerability of the studied beach is due to the relative rise of sea level, especially when this is combined with extreme sea storms.

Overall, the described methodology is deemed to be of general interest for ocean and coastal modelers involved in the development of procedures for offline multiple-nested high-resolution simulations from global circulation models, the adaptation of coastal models to platforms of operational oceanography, and the implementation of the above in coastal planning and design actions at the light of climate change scenarios.

Future studies of the selected site will focus on the implementation of a validated morphological model to perform erosional risk analysis and the running of long-term scenarios in order to estimate the shoreline evolution and seasonal variability of coastal dynamics in the area.

Author Contributions: For the presented research, D.B. performed the climatologic simulations with SWAN, M.G.G. and D.B. analyzed the wave climate and performed statistics, M.G.G. and A.G.S. performed the numerical simulations set-up and analyzed the results from TELEMAC runs, and R.A. and S.C. conceived the research aim and contributed to the discussion on the study outcomes.

Funding: This work was supported by the Flagship Project RITMARE—The Italian Research for the Sea, coordinated by the Italian National Research Council and funded by the Italian Ministry of Education, University, and Research within the National Research Program 2011–2013.

Acknowledgments: Geological Unit from Emilia Romagna Region is acknowledged to provide DTM data of the study area. D.B. and S.C. acknowledge the support from the UE H2020 program under grant agreement n. 730030 ("CEASELESS" Project) and the INTERREG MED "CO-EVOLVE" Project. The authors thankfully acknowledge Edoardo Bucchignani (CMCC and CIRA, Capua, Italy) for providing the COSMO-CLM wind fields used in the wave model simulations.

Conflicts of Interest: The authors declare no conflict of interest.

References

1. Androulidakis, Y.S.; Kombiadou, K.D.; Makris, C.V.; Baltikas, V.N.; Krestenitis, Y.N. Storm surges in the Mediterranean Sea: Variability and trends under future climatic condition. *Dyn. Atmos. Ocean.* **2015**, *71*, 56–82. [CrossRef]
2. Lionello, P.; Conte, D.; Marzo, L.; Scarascia, L. The contrasting effect of increasing mean sea level and decreasing storminess on the maximum water level during storms along the coast of the Mediterranean Sea in the mid 21st century. *Glob. Planet. Chang.* **2017**, *151*, 80–91. [CrossRef]
3. Carisi, F.; Domeneghetti, A.; Gaeta, M.G.; Castellarin, A. Is anthropogenic land-subsidence a possible driver of riverine flood-hazard dynamics? A case study in Ravenna, Italy. *Hydrol. Sci. J.* **2017**, *62*, 1–16. [CrossRef]
4. Bonaldo, D.; Antonioli, F.; Archetti, R.; Bezzi, A.; Correggiari, A.; Davolio, S.; de Falco, G.; Fantini, M.; Fontolan, G.; Furlani, S.; et al. Integrating multidisciplinary instruments for assessing coastal vulnerability to erosion and sea level rise: Lessons and challenges from the Adriatic Sea, Italy. *J. Coast. Conserv.* **2018**, *22*, 1–19. [CrossRef]
5. Samaras, A.G.; Koutitas, C.G. Modeling the impact of climate change on sediment transport and morphology in coupled watershed-coast systems: A case study using an integrated approach. *Int. J. Sediment Res.* **2014**, *29*, 304–315. [CrossRef]
6. Tsoukala, V.K.; Chondros, T.M.; Kapelonis, Z.G.; Martzikos, N.; Lykou, A.; Belibassakis, K.; Makropoulos, C. An integrated wave modelling framework for extreme and rare events for climate change in coastal areas—The case of Rethymno, Crete. *Oceanologia* **2016**, *58*, 71–89. [CrossRef]

7. Lambeck, K.; Antonioli, F.; Anzidei, M.; Ferranti, L.; Leoni, G.; Scicchitano, G.; Silenzi, S. Sea level change along the Italian coast during the Holocene and projections for the future. *Quat. Int.* **2011**, *232*, 250–257. [CrossRef]
8. Antonioli, F.; Anzidei, M.; Amorosi, A.; Presti, V.L.; Mastronuzzi, G.; Deiana, G.; De Falco, G.; Fontana, A.; Fontolan, G.; Lisco, S.; et al. Sea-level rise and potential drowning of the Italian coastal plains: Flooding risk scenarios for 2100. *Quat. Sci. Rev.* **2017**, *158*, 29–43. [CrossRef]
9. Perini, L.; Calabrese, L.; Luciani, P.; Olivieri, M.; Galassi, G.; Spada, G. Sea-level rise along the Emilia-Romagna coast (Northern Italy) in 2100: Scenarios and impacts. *Nat. Hazards Earth Syst. Sci.* **2017**, *17*, 2271–2287. [CrossRef]
10. Perini, L.; Calabrese, L.; Salerno, G.; Ciavola, P.; Armaroli, C. Evaluation of coastal vulnerability to flooding: Comparison of two different methodologies adopted by the Emilia-Romagna region (Italy). *Nat. Hazards Earth Syst. Sci.* **2016**, *16*, 181–194. [CrossRef]
11. Di Risio, M.; Bruschi, A.; Lisi, I.; Pesarino, V.; Pasquali, D. Comparative analysis of coastal flooding vulnerability and hazard assessment at national scale. *J. Mar. Sci. Eng.* **2017**, *5*, 51. [CrossRef]
12. Seenath, A.; Wilson, M.; Miller, K. Hydrodynamic versus GIS modelling for coastal flood vulnerability assessment: Which is better for guiding coastal management? *Ocean Coast. Manag.* **2016**, *120*, 99–109. [CrossRef]
13. Carniel, S.; Warner, J.C.; Chiggiato, J.; Sclavo, M. Investigating the impact of surface wave breaking on modelling the trajectories of drifters in the Northern Adriatic Sea during a wind-storm event. *Ocean Model.* **2009**, *30*, 225–239. [CrossRef]
14. Boldrin, A.; Carniel, S.; Giani, M.; Marini, M.; Bernardi Aubry, F.; Campanelli, A.; Grilli, F.; Russo, A. The effect of Bora wind on physical and bio-chemical properties of stratified waters in the Northern Adriatic. *J. Geophys. Res.* **2009**, *114*, C08S92. [CrossRef]
15. Carniel, S.; Sclavo, M.; Archetti, R. Towards validating a last generation, integrated wave-current-sediment numerical model in coastal regions using video measurements. *Ocean. Hydrobiol. Stud.* **2011**, *40*, 11–20. [CrossRef]
16. Archetti, R.; Zanuttigh, B. Integrated monitoring of the hydro-morphodynamics of a beach protected by low crested detached breakwaters. *Coast. Eng.* **2010**, *57*, 879–891. [CrossRef]
17. Stocker, T.F.; Qin, D.; Plattner, G.-K.; Tignor, M.; Allen, S.K.; Boschung, J.; Nauels, A.; Xia, Y.; Bex, V.; Midgley, P.M. Summary for policy makers. In *Climate Change 2013: The Physical Science Basis*; Cambridge University Press: Cambridge, UK, 2013; pp. 3–29.
18. Villatoro, M.; Silva, R.; Méndez, F.J.; Zanuttigh, B.; Pan, S.; Trifonova, E.; Losada, I.J.; Izaguirre, C.; Simmonds, D.; Reeve, D.E.; et al. An approach to assess flooding and erosion risk for open beaches in a changing climate. *Coast. Eng.* **2014**, *87*, 50–76. [CrossRef]
19. Sanuy, M.; Duo, E.; Jäger, W.S.; Ciavola, P.; Jiménez, J.A. Linking source with consequences of coastal storm impacts for climate change and risk reduction scenarios for Mediterranean sandy beaches. *Nat. Hazards Earth Syst. Sci.* **2018**, *18*, 1825–1847. [CrossRef]
20. Artegiani, A.; Barale, V.; Cushman-Roisin, B.; Gačić, M.; Kourafalou, V.H.; Lascaratos, A.; Malačić, V.; Manca, B.B.; Mantziafou, A.; Orlić, M.; et al. Northern Adriatic Sea. In *Physical Oceanography of the Adriatic Sea Past, Present and Future*; Cushman-Roisin, B., Gačić, M., Poulain, P.M., Artegiani, A., Eds.; Kluwer Academic Publishers: Dordrecht, The Netherlands, 2001; pp. 143–165.
21. Lionello, P.; Cavaleri, L.; Nissen, K.M.; Pino, C.; Raicich, F.; Ulbrich, U. Severe marine storms in the Northern Adriatic: Characteristics and trends. *Phys. Chem. Earth Parts A/B/C* **2012**, *40–41*, 93–105. [CrossRef]
22. Aguzzi, M.; Bonsignore, F.; De Nigris, N.; Morelli, M.; Paccagnella, T.; Romagnoli, C.; Unguendoli, S. *Stato Del Litorale Emiliano—Romagnolo Al 2012. Erosione e Interventi di Difesa, Quaderni di ARPAE—Regione Emilia Romagna*; Arpae Emilia-Romagna: Bologna, Italy, 2012; ISBN 978-88-87854-41-1. (In Italian)
23. Martinelli, L.; Zanuttigh, B.; Corbau, C. Assessment of coastal flooding hazard along the Emilia Romagna Littoral, IT. *Coast. Eng.* **2010**, *57*, 1042–1158. [CrossRef]
24. Archetti, R.; Gaeta, M.G. Design of multipurpose coastal protection measures at the Reno river mouth (Italy). In Proceedings of the 28th International Ocean and Polar Engineering Conference, Sapporo, Japan, 10–15 June 2018; pp. 1343–1348.
25. Google Earth (Image ©2017 TerraMetrics, Data SIO, NOAA, U.S. Navy, NGA, GEBCO, 2017). Available online: https://www.google.com/earth/ (accessed on 5 April 2018).

26. Rockel, B.; Will, A.; Hense, A. The regional Climate Model COSMO-CLM (CCLM). *Meteorol. Z.* **2008**, *17*, 347–348. [CrossRef]
27. Steppeler, J.; Doms, G.; Schattler, U.; Bitzer, H.W.; Gassmann, A.; Damrath, U.; Gregoric, G. Meso-gamma scale forecasts using the nonhydrostatic model LM. *Meteorol. Atmos. Phys.* **2003**, *82*, 75–96. [CrossRef]
28. Bucchignani, E.; Montesarchio, M.; Zollo, A.L.; Mercogliano, P. High-resolution climate simulations with COSMO-CLM over Italy: Performance evaluation and climate projections for the 21st century. *Int. J. Climatol.* **2016**, *36*, 735–756. [CrossRef]
29. Moss, R.; Edmons, J.; Hibbard, K.; Manning, M.; Rose, S.; van Vuuren, D.P.; Carter, T.; Emori, S.; Kainuma, M.; Kram, T.; et al. The next generation of scenarios for climate change research and assessment. *Nature* **2010**, *463*, 747–756. [CrossRef] [PubMed]
30. Church, J.A.; Clark, P.U.; Cazenave, A.; Gregory, J.M.; Jevrejeva, S.; Levermann, A.; Merrifield, M.A.; Milne, G.A.; Nerem, R.S.; Nunn, P.D.; et al. Sea level change. In *Climate Change 2013: The Physical Science Basis. Contribution of Working Group I to the Fifth Assessment Report of the Intergovernmental Panel on Climate Change;* Cambridge University Press: Cambridge, UK; New York, NY, USA, 2013.
31. Dee, D.; Uppala, S.; Simmons, A.; Berrisford, P.; Poli, P.; Kobayashi, S.; Andrae, U.; Balmaseda, M.; Balsamo, G.; Bauer, P.; et al. The era-interim reanalysis: Configuration and performance of the data assimilation system. *Q. J. R. Meteorol. Soc.* **2011**, *137*, 553–597. [CrossRef]
32. Scoccimarro, E.; Gualdi, S.; Bellucci, A.; Sanna, A.; Fogli, P.; Manzini, E.; Vichi, M.; Oddo, P.; Navarra, A. Effects of tropical cyclones on ocean heat transport in a high resolution coupled general circulation model. *J. Clim.* **2011**, *24*, 4368–4384. [CrossRef]
33. Bellafiore, D.; Bucchignani, E.; Gualdi, S.; Carniel, S.; Djurdjevic, V.; Umgiesser, G. Assessment of meteorological climate models as inputs for coastal studies. *Ocean Dyn.* **2012**, *62*, 555–568. [CrossRef]
34. Bonaldo, D.; Bucchignani, E.M.; Ricchi, A.; Carniel, S. Wind storminess in the Adriatic Sea in a climate change scenario. *Acta Adriat.* **2017**, *58*, 195–208.
35. Booij, N.; Ris, R.C.; Holthuijsen, L.H. A third-generation wave model for coastal regions: 1. Model description and validation. *J. Geophys. Res.* **1999**, *104*, 7649–7666. [CrossRef]
36. Benetazzo, A.; Fedele, F.; Carniel, S.; Ricchi, A.; Bucchignani, E.; Sclavo, M. Wave climate of the Adriatic Sea: A future scenario simulation. *Nat. Hazards Earth Syst. Sci.* **2012**, *12*, 2065–2076. [CrossRef]
37. Benetazzo, A.; Bergamasco, A.; Bonaldo, D.; Falcieri, F.M.; Sclavo, M.; Langone, L.; Carniel, S. Response of the Adriatic Sea to an intense cold air outbreak: Dense water dynamics and wave-induced transport. *Prog. Oceanogr.* **2014**, *128*, 115–138. [CrossRef]
38. Benoit, M.; Marcos, F.; Becq, F. Development of a third generation shallow-water wave model with unstructured spatial meshing. In Proceedings of the 25th International Conference on Coastal Engineering, Orlando, FL, USA, 2–6 September 1996.
39. Hervouet, J.M. *Hydrodynamics of Free Surface Flows: Modelling with the Finite Element Method;* John Wiley & Sons, Ltd.: London, UK, 2007; p. 360.
40. Masina, M.; Lamberti, A.; Archetti, R. Coastal flooding: A copula based approach for estimating the joint probability of water levels and waves. *Coast. Eng.* **2015**, *97*, 37–52. [CrossRef]
41. Boccotti, P. *Wave Mechanics for Ocean Engineering;* Elsevier: New York, NY, USA, 2000.
42. Husler, J.; Reiss, R.-D. Maxima of normal random vectors: Between independence and complete dependence. *Stat. Probab. Lett.* **1989**, *7*, 283–286. [CrossRef]
43. Conte, D.; Lionello, P. Characteristics of large positive and negative surges in the Mediterranean Sea and their attenuation in future climate scenarios. *Glob. Planet. Chang.* **2013**, *111*, 159–173. [CrossRef]
44. Vousdoukas, M.I.; Voukouvalas, E.; Annunziato, A.; Giardino, A.; Feyen, L. Projections of extreme storm surge levels along Europe. *Clim. Dyn.* **2016**, *47*, 3171–3190. [CrossRef]
45. Rahmstorf, S. A semi-empirical approach to projecting future sea-level rise. *Science* **2007**, *315*, 368–370. [CrossRef] [PubMed]
46. Cavaleri, L.; Abdalla, S.; Benetazzo, A.; Bertotti, L.; Bidlot, J.-R.; Breivik, Ø.; Carniel, S.; Jensen, R.E.; Portilla-Yandun, J.; Rogers, W.E.; et al. Wave modelling in coastal and inner seas. *Prog. Oceanogr.* **2018**. [CrossRef]
47. TELEMAC. The Mathematically Superior Suite of Solvers 2017. Available online: http://www.opentelemac.com (accessed on 10 April 2018).

48. Kuang, C.P.; Stansby, P. Modelling directional random wave propagation inshore. *Proc. Inst. Civ. Eng. Marit. Eng.* **2004**, *157*, 123–131. [CrossRef]
49. Hervouet, J.-M.; Razafindrakoto, E.; Villaret, C. Dealing with dry zones in free surface flows: A new class of advection schemes. In Proceedings of the 34th IAHR World Congress, Brisbane, Australia, 26 June–1 July 2011.
50. Villaret, C.; Hervouet, J.-M.; Kopmann, R.; Merkel, U.; Davies, A. Morphodynamic modeling using the Telemac finite-element system. *Comput. Geosci.* **2013**, *53*, 105–113. [CrossRef]
51. Longuet-Higgins, M.S.; Stewart, R.W. Radiation stresses in water waves; a physical discussion, with applications. *Deep Sea Res. Oceanogr. Abstr.* **1964**, *11*, 529–562. [CrossRef]
52. CHC—Canadian Hydraulics Centre, National Research Council. *Blue Kenue. Reference Manual*; Canadian Hydraulics Centre, National Research Council: Ottawa, ON, Canada, 2010.
53. Samaras, A.G.; Gaeta, M.G.; Miquel, A.M.; Archetti, R. High-resolution wave and hydrodynamics modelling in coastal areas: Operational applications for coastal planning, decision support and assessment. *Nat. Hazards Earth Syst. Sci.* **2016**, *16*, 1499–1518. [CrossRef]
54. Gaeta, M.G.; Samaras, A.G.; Federico, I.; Archetti, R.; Maicu, F.; Lorenzetti, G. A coupled wave-3D hydro-dynamics model of the Taranto Sea (Italy): A multiple-nesting approach. *Nat. Hazards Earth Syst. Sci.* **2016**, *16*, 2071–2083. [CrossRef]
55. Battjes, J.A.; Janssen, J.P.F.M. Energy loss and set-up due to breaking of random waves. In Proceedings of the 16th International Conference on Coastal Engineering, Hamburg, Germany, 27 August–3 September 1978; pp. 569–587.
56. Hasselmann, K.; Barnett, T.P.; Bouws, E.; Carlson, H.; Cartwright, D.E.; Enke, K.; Ewing, J.A.; Gienapp, H.; Hasselmann, D.E.; Kruseman, P.; et al. Measurements of wind-wave growth and swell decay during the Joint North Sea Wave Project (JONSWAP). *Ergänz. Dtsch. Hydrogr. Z. Reihe* **1973**, *A 8*, 95.
57. Eldeberky, Y.; Battjes, J.A. Parameterisation of triads interactions in wave energy models. In Proceedings of the Coastal Dynamics Conference '95, Gdansk, Poland, 4–8 September 1995; pp. 140–148.
58. Ruessink, B.; Miles, J.; Feddersen, F.; Guza, R.; Elgar, S. Modeling the alongshore current on barred beaches. *J. Geophys. Res.* **2001**, *106*, 22451–22463. [CrossRef]
59. Feddersen, F.; Guza, R. Observation of nearshore circulation: Alongshore uniformity. *J. Geophys. Res.* **2003**, *108*, 6-1–6-10. [CrossRef]
60. Sclavo, M.; Benetazzo, A.; Carniel, S.; Bergamasco, A.; Falcieri, F.M.; Bonaldo, D. Wave-current interaction effect on sediment dispersal in a shallow semi-enclosed basin. *J. Coast. Res.* **2013**, *65*, 1587–1592. [CrossRef]
61. Carniel, S.; Bonaldo, D.; Benetazzo, A.; Bergamasco, A.; Boldrin, A.; Falcieri, F.M.; Sclavo, M.; Trincardi, F.; Langone, L. Off-shelf fluxes across the southern Adriatic margin: Factors controlling dense-water-driven transport phenomena. *Mar. Geol.* **2016**, *375*, 44–63. [CrossRef]
62. Archetti, R. Quantifying the evolution of a beach protected by low crested structures using video monitoring. *J. Coast. Res.* **2009**, *25*, 884–899. [CrossRef]
63. Bonaldo, D.; Benetazzo, A.; Sclavo, M.; Carniel, S. Modelling wave-driven sediment transport in a changing climate: A case study for northern Adriatic Sea (Italy). *Reg. Environ. Chang.* **2015**, *15*, 45–55. [CrossRef]

water

MDPI

Article

Vulnerability Analysis of the Venetian Littoral and Adopted Mitigation Strategy

Piero Ruol *, Luca Martinelli and Chiara Favaretto

ICEA Department, Padua University, v. Ognissanti 39, 35129 Padova, Italy; luca.martinelli@unipd.it (L.M.); chiara.favaretto@dicea.unipd.it (C.F.)

* Correspondence: piero.ruol@unipd.it; Tel.: +39-049-827-7905

Received: 30 May 2018; Accepted: 24 July 2018; Published: 26 July 2018

Abstract: This paper discusses the key aspects of the recent Coastal Plan of the Veneto Region (IT). Its aim is to propose a single mitigation strategy for coastal erosion that is valid for the whole Veneto Region, and possibly elsewhere, as well as a method to assign a priority level to any action. The suggested mitigation action against erosion depends on urbanization level, beach width, as well as cross-shore and long-shore sediment transport. The criterion used to give a priority level to mitigation actions is based on a vulnerability index that takes into account erosive tendency, existing coastal flooding hazards, coast value, environmental relevance, tourist pressure, urbanization level, the presence of production activities, and cultural heritage. A sample case featuring the littoral of Rosolina is also provided and includes a site description, the sediment budget, critical issues and possible mitigation measures.

Keywords: coastal plan; erosion, coastal flooding; sediment budget; mitigation strategies guidelines; littoral cell

1. Introduction

In the recent past, in Italy, when coastal managers at the Regional Authorities were faced with coastal erosion problems, they were frequently guided by an empirical approach based on simple transport equations, or even solely by their intuition. More recently, however, the perspective has changed somewhat and it has become evident that the complexity of coastal systems needs to be studied through a homogenous and multidisciplinary approach with reference to large spatial and temporal scales that take into account a wide range of phenomena and topics. Coastal zones must be analysed from several points of view (e.g., geophysical, biological, socioeconomic, political, cultural, historical) and with different approaches (e.g., research, planning, operational purposes). Successful management requires a thorough understanding of the physical processes impacting the coast to create a strategic vision of the future, establishing a framework to guide future actions [1–3]. In light of this, local authorities are expected to produce a medium-term coastal plan that can effectively reduce and manage the risks that natural processes (e.g., storms, floods and erosion) pose to human health, the environment, cultural heritage, and businesses. They are also required to produce a flooding vulnerability map that includes coastal inundation in compliance with EU Directive (2007/60/EC). As a result, many of the available studies (e.g., Hinkel et al. [4], Weisse et al. [5], Toimil et al. [6]) that include the most recent IPCC Assessment [7] show that European coasts are exposed to erosion, rising sea-levels and climate change and discuss how to manage this threat at large scale.

Guidance for shoreline management has been provided by several recent research projects. The following studies comprise a demonstration site on the North-Eastern coast of Italy, and were carried out in collaboration with local stakeholders: RITMARE Flagship Project (Antonioli et al. [8], Bezzi et al. [9]), CAMP Italy Project, COASTANCE (Montanari and Marasmi [10]), DELOS, (Zanuttigh et al., [11]) COASTGAP-MED, MEDSANDCOAST-ENPI, COASTAL Mapping-DG

MARE, EUROSION [12], THESEUS (Zanuttigh [13]), MICORE (Ciavola et al. [14]) or RISC-KIT (Armaroli et al. [15]). Specific open-source tools and approaches have been developed to support decision-making processes (Zanuttigh et al. [16], Torresan et al. [17]; Vafeidis et al. [18], van Dongeren et al. [19]; Stelljes at al. [20]).

In order to implement the monitoring and coastal plans, that can be adopted by the local authorities, and/or in order to draw the vulnerability maps required by the Flood Directive, the valuable results obtained by tools such as the THESEUS DSS require an engineering synthesis, e.g., studies like those described in Preti et al. [21], Petrillo et al. [22], Ruol et al. [23]. Such studies provide simple guidelines for the design of erosion and coastal flooding mitigation measures that follow coherent and rational criteria.

This paper hence aims to propose erosion mitigation criteria that are valid for the whole Veneto Region, and possibly for adjoining coasts. It also aims to provide practical guidelines on how to interpret coastal monitoring analysis, select when, where and what mitigation measures should be adopted, and suggest a methodology for assigning a priority level to any action.

The analysis here presented originates from the practical need to investigate the Venetian littoral and to establish a coastal plan based on a balanced combination of scientific rigorous approach and expert, discussion-based, assumptions. The Northern Adriatic coast is subject to rapidly evolving pressures from a range of drivers, including natural and anthropogenic ones (e.g., rapid morphological evolution of Po River Delta [24], human-induced subsidence caused by fluid withdrawal [25], changing wave climate [26]), all of which require an integrated approach. The study also outlines a detailed monitoring plan and shows that the mitigation measures adopted depend on the monitoring results.

In addition to this introduction, this paper includes two main sections and a concluding paragraph. First, the analysis methods are described, which include a set of available mitigation options (whose effectiveness is based on the behaviour of previous works on the Venetian littoral) and the rational criteria for their selection. The method is applied to the whole Venetian littoral, together with an example of the mitigation measures adopted on a short stretch of coast. Lastly, conclusions are drawn.

2. Methodology

The geography of the Venetian littoral is presented in Section 3.1. The Coastal Plan of the Veneto Region (IT) is divided into a descriptive phase and a design phase.

The descriptive phase involves collecting all of the available coastal data, possibly integrated or obtained with numerical tools. These data may then be used to extend knowledge and reduce uncertainties. The coastal plan includes a sediment budget assessment that provides essential information for the subsequent steps.

The design phase carries out a critical interpretation of the sediment budget results, selects the most appropriate mitigation option based on a univocal criterion and prioritizes mitigation actions. All major issues along the littoral zone were discussed with a wide stakeholder group.

The core of the method (common to both the descriptive and design phase) involves subdividing the regional coast into littoral cells and organising all of the information and results into a single geographical information system (Coastal GIS). The concept of "sediment cells" [12,27] allows for a better understanding of sediment transport patterns. These cells are stretches of coast with similar characteristics bordered by morphological features, such as river mouths, inlets and port dams, meaning that, in the absence of major obstacles to long-shore currents, sediment is relatively free to move inside the morphological feature.

In addition, each cell is divided in half to form two semi-cells (S-C). Dividing these cells into two half (not necessarily equal in length) has practical advantages for the sediment budget balance, as the boundary conditions between them are morphologically continuous. Therefore, the long-shore sediment transport is continuous here and easy-to-compute by using wave climate and local beach orientation, and a new set of independent equations is added, allowing a more reliable assessment of the entire budget.

The Coastal GIS is a large database that is easy to access and update, and could be shared with all the stakeholders and local managers. The use of GIS for coastal management has expanded rapidly during the past decade (Bartlett and Smith [28], Wright and Bartlett [29]), and is suited to the Williams [30] approach, based on "getting", "reordering" and "refining" the information. The collected information is suitable to be integrated within a CoastalME type framework [31].

Figure 1 shows the general structure of the proposed methodology, whereby the information gathered from both the descriptive and design phases is stored in the Coastal GIS.

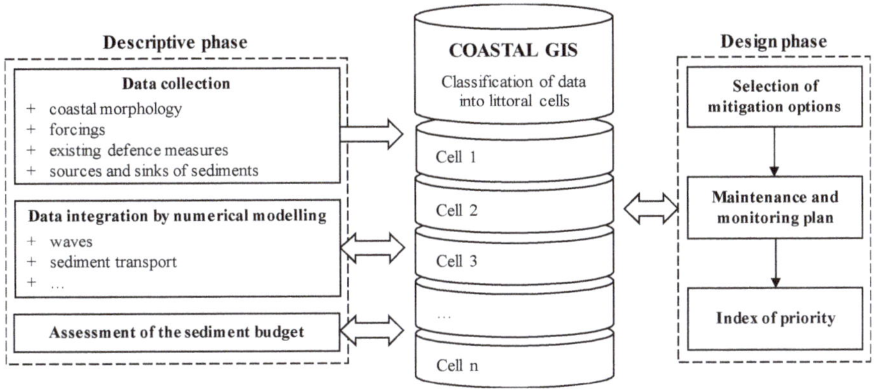

Figure 1. Breakdown of coastal plan activities. The Coastal GIS includes both data (measured or computed) gathered in the descriptive phase and the results following the design phase.

2.1. Descriptive Phase

2.1.1. Data Collection and Harmonisation

Data and available measurements are collected, harmonised and stored in the Coastal GIS. Data for sediment budget analysis consist of all the variables that characterise local coastal morphology (i.e., shorelines, bathymetries, DTMs, grain size distribution, dune characterisation, subsidence, etc.), the forcing (i.e., wave, wind, tide, surge, currents, etc.) together with existing defence measures, and the sources and sinks of sediments (i.e., fluvial sediment transport, littoral sediment transport, a detailed history of past nourishments and dredgings).

Other information is required to define the constraints (e.g., areas with special legal protection or regulation, urban planning), and the value of the area in general (i.e., environmental relevance, urban and tourist pressure, local economy, cultural heritage, etc.). Additional data may be relevant to conducting flood-risk and vulnerability assessment (e.g., inland mapping, land use, emergency plans).

2.1.2. Data Integration by Numerical Modelling

Numerical models are frequently used in coastal planning, since they are useful for integrating any available information, especially on forcing, coastal flooding and sediment transport, e.g., [13,21–23,27,31].

Models such as ECMWF, AdCirc, Wavewatch III help provide a detailed description of the wind, currents and wave climate [32], both in terms of extremes and average values. In some cases, it is useful to statistically analyse the forcing provided by extensive databases (e.g., NOAA's Historical Hurricane Tracks [33]).

Evaluation of the coastal flooding risk is a specific task that may be integrated using raster images included in the Coastal GIS as described in [34]. The flooding probability derived by the model is stored in form of vulnerability maps [35].

After the data collection phase, other numerical models (e.g., Mike21, X-Beach, LitPack, Gencade) are used to achieve a complete and homogeneous description of the sediment transport discharge for each of the sediment cells identified, taking into account the grain size distribution. The simulations evaluate the river sediment supply and long-shore and cross-shore sediment transport at regional level during the predefined time interval selected for the sediment balance. These data are essential for the assessment of the sediment budget and hence for the design phase.

2.1.3. Assessment of Sediment Budget

The sediment budget is essentially a mass balance equation applied to a specified time interval. It is convenient to subdivide the coastline into a number of small stretches and apply the balance to each stretch *i*:

$$\varepsilon_i = \partial V_i - Q_i \Delta t = 0, \tag{1}$$

where ∂V_i is the volume of accretion or erosion estimated by a comparison of the bathymetries surveyed at the beginning and at the end of the time range Δt, and Qi is the net sediment discharge shown in Figure 2 and given by:

$$Q_i = -Q_{LS(North)i} + Q_{LS(South)i} + Q_{Fi} - Q_{CRi} + Q_{Ni} - Q_{Di}, \tag{2}$$

where Q_{LS} is the long-shore sediment transport (*North* identifies the discharge at the northern boundary and *South* at the southern boundary in an ideal beach aligned in the North-South direction); Q_F is the additional river sediment supply; Q_{CR} is the cross-shore sediment transport; and Q_N and Q_D are the volumes added or subtracted due to nourishment or dredging respectively. Subsidence and sea level rise do not affect the sediment balance directly but they have the same effect as generalized erosion as they alter the marine accommodation space. In order to preserve the equilibrium profile, the accommodation space must be balanced by beach nourishment.

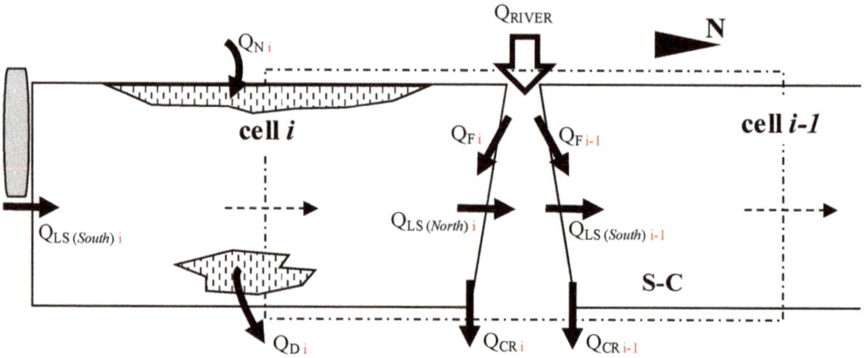

Figure 2. Sediment balance diagram. Littoral cells are limited by morphological features (continuous line). The inclusion of river sediment transport in the balance equations would be straightforward if the cell control volume extended to the dash-dot box.

The balance in Equation (1) forms a system of $2n - 1$ equations, where each stretch *i* corresponds to a semi-cell (S-C). As anticipated at the beginning of Section 2, boundary conditions between the two halves of the same littoral cell are morphologically continuous, whereas the littoral cell boundaries are placed in correspondence with morphological elements, e.g., river mouths, that can complicate the assessment of the long-shore sediment transport.

River mouths need to be considered with special attention. The inclusion of river sediment transport in the balance equations would be straightforward if the control volume extended to the two S-Cs adjacent to the mouth (i.e., the dash-dot box in Figure 2) since river supply is equivalent to nourishment. However, Equation (1) is applied to each S-C. In the adopted scheme, river sediment transport is divided and allocated to both the adjacent S-Cs, and the fine fraction losses are treated as additional off-shore sediment transport. The approach is similar to the one described by Samaras and Koutitas [36,37]. To provide a single procedure for all of the cell boundaries, the long-shore transport Q_{LS} is calculated even where the boundary is a groin, a port, a river mouth, etc. However, in presence of the latter, further calculations are required to establish Q_F. Q_F is given by the total river supply (Q_{RIVER} in Figure 2) minus the fine fraction Q_{CR} that will be transferred seaward, minus the previously calculated value of Q_{LS}. Obviously, in the trivial case of a stable river mouth, the river supply is equal to the long-shore sediment transport, and the resulting value Q_F is zero. It is also immediately clear that an insufficient river supply results in a negative value of Q_F.

The system in Equation (1) is coupled since long-shore sediment transport represents a mixed term, e.g., the $Q_{LS(South)}$ of the Northern S-C is equal to the $Q_{LS(North)}$ of the Southern one (dotted arrow in the middle of the cell *i* showed in Figure 2). It is solved using a compensation of error technique based on a matrix of uncertainties given a priori. Error compensation by least squares adjustment is obtained by solving an overdetermined system of equations based on the principle of least squares of observation residuals. It is used extensively in the disciplines of surveying, geodesy, and photogrammetry [38]. Guidelines on the evaluation of the a priori uncertainties of each term can be found, for instance, in [39]. Support of the stakeholders is essential for this phase.

2.2. Design Phase

2.2.1. Mitigation Options and Selection Criteria

Initially, based on the descriptive phase, the specific causes inducing erosion should be found, e.g., reduction of river sediment transport, increased subsidence rate, etc. Appropriate action should then be geared towards reducing these causes, e.g., providing sediment bypass in the presence of river dams and limiting extractions from the soil, etc. Similarly, reintegration of damaged structures/environmental areas shall be considered, with dune restoration being a typical measure that provides a reserve of sand in the event of storms as well as a safety against coastal flooding.

Engineering solutions for mitigation of flood and erosion risks described in [39] include both active methods, based on the reduction of the incident wave energy, such as the use of wave energy converters, floating breakwaters and artificial reefs, and passive methods, consisting of increase in overtopping resistance of dikes, improvement of resilience of breakwaters against failures, and the use of beach nourishment (possibly with innovative layout [40]) as well as tailored dredging operations. Suggestions on design optimization, optimal placement, and efficiency from an ecological perspective are outlined in [41].

Non-structural mitigation options are discussed in [42] where it is pointed out that they should be considered as part of a potential portfolio in which their combination transcends their sum, rather than standalone measures. Also, mobilization of stakeholders in the implementation process is an important issue to achieve effective results.

This sub-section puts forward the more common options and discuss their suitability for the coast of the Veneto region, which is mainly low and sandy.

Low sandy beaches can be subdivided into two categories: (1) "urbanized coasts", which are intensely developed and have a high economic value, for which the main goal is to defend urban and tourist activities, possibly preserving a large beach; (2) "non-urbanized coasts", natural littoral, for which the aim is to preserve the environmental value.

The proposed mitigation options follow this classification.

1. Urbanized low sandy coasts: The criterion is based on two main physical characteristics: the net long-shore sediment transport (*net LST*) and the gross long-shore sediment transport (*gross LST*). The rate of cross-shore sediment transport (*CST*) is also taken into account when *LST* values are low. Figure 3 shows a scheme of the guidelines for this type of coast. In all the cases, nourishment is the main strategy adopted (e.g., [43,44]), possibly together with structures (e.g., submerged barrier, low-crested structure, groins) or/and other measures (e.g., down-drift maintenance, up-drift bypass).

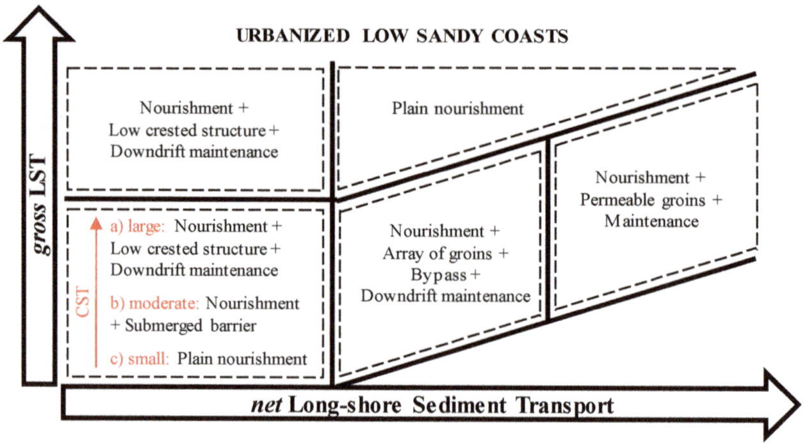

Figure 3. Guidelines for erosion mitigation strategies (*div(LST)* = divergence of long-shore sediment transport, *CST* = cross-shore sediment transport) for urbanized low sandy coasts.

In particular, for very low *LST* (i.e., both *net* and *gross*), obstacles to *LST* are considered fairly ineffective at intercepting sediments. When the *CST* is also negligible (case "c" in Figure 3), causes of erosion are independent of coastal dynamics, and plain nourishment is the most suitable action [45]. However, when *CST* is appreciable (case "b" in Figure 3), the building of a submerged barrier is suggested. This structure is placed at the breaking line, deeply submerged. The expected piling up [46] and wave transmission [47] are small. It is considered effective in stabilising the cross-section profile, since breaking always occurs at the same point, and therefore the bar is stabilised, minimising the offshore *CST* associated with offshore bar migration [48]. For very large *CST* (case "a" in Figure 3), a more effective solution is required. The candidate is the low crested structure (*LCS*) scheme with small gaps [11], which essentially confines the sediment in the protected area, i.e., the area bordered by the shoreline, the lateral low crested groins and the offshore barrier. Note that a large piling up is expected to occur in this type of protected area, forming an obstacle to coastal dynamics and possibly resulting in some down-drift maintenance being needed [49]. The same solution is appropriate even when the *gross LST* is large but the *net LST* is small.

For large values of *net* and *gross LST*, the best solution is probably a battery of groins, as it tends to stabilise the shoreline in a saw-tooth shape and reduces *LST*, with benefits for the protected area [50]. However, the downdrift area (at the end of the battery of groins) must be properly maintained, and a bypass system may help reduce the sediment resource needed. In the event of extreme *net LST* (which only occurs with extreme *gross LST*), permeable groins may be considered. By reducing, but not blocking, the littoral current, the velocity differential between the velocity seaward and in the pile-groin fields is smaller than with impervious groins [51]. The primary objective of the design is to reduce the littoral current velocity to an extent that rip currents and large-scale circulations in the groin field are minimised.

When *net LST* is appreciable but *gross LST* extreme, groins are evidently not very effective [52] and do not compensate the minor *CST* caused by the increased wave reflection as it interacts with the structure. Plain nourishment is, therefore, the solution suggested.

In areas where the shoreline is expected to move significantly, such as in proximity of a river mouth, for which exceptional riverine sediment transport may occur, fixed structures should not be built (they may become unsuited to the modified shoreline).

2. Non-urbanized low sandy coasts: The criterion for natural beaches (Figure 4) is based on two physical characteristics, i.e., *net LST* and beach width. According to Van der Nat et al. [53], who classify the degree to which coastal designs are nature-based using criteria for ecosystem-based management, the "building with nature" approach is particularly suited to this category. Three soft stabilization measures are taken into account: two measures envisage supplying sand to the beach in the form of distributed or concentrated nourishment, and one contemplates increasing sand-dune volume.

When the beach is wide and *net LST* large, i.e., sediment transport direction is well-defined, concentrated nourishment updrift is the least impacting solution for reducing erosion in this location, as it limits the impact of the works on a single spot and relies on littoral dynamics to redistribute sand downdrift (an extreme example is the "Sand Engine" project in the Netherlands [54]). For small *net LST*, the ability to redistribute nourishment is limited, so nourishment should cover the whole of the eroding coastline. Wherever the beach is narrow (i.e., may be completely eroded within a short time period), and a drastic modification of the natural environment is not viable, sand may be accumulated creating new dune systems, thus reinforcing the existing beach [55,56].

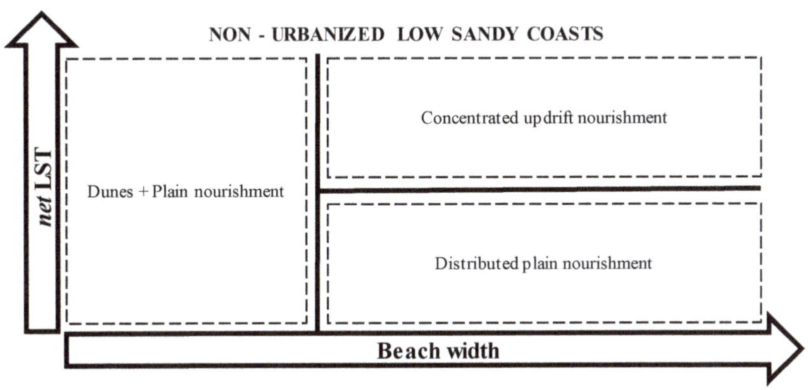

Figure 4. Guidelines for erosion mitigation strategies (*div (LST)* = divergence of long-shore sediment transport) for non-urbanized low sandy coasts.

All of the aforementioned mitigation options require some degree of nourishment, making resource availability a critical issue. It is, therefore, necessary that the plan includes the location of any sediment resource along the coast, offshore and/or inland. In most cases, the sediment balance is very helpful for finding the position and size of any coastal stocks [44]. Clearly, this sediment stock makes for excellent nourishment since it belongs to the same environment and is the most economical to derive. Some care must be taken when planning the annual volumes to be dredged from these areas, and a permanent monitoring plan is necessary.

The aforementioned measures mitigate the risk of erosion. In the event of flood risk, however, strategies must be integrated with other measures [57]. Dune restoration may be considered the first candidate in order to defend the inland from marine inundation [16]. Seawalls, that induce a local erosion due to reflection, are cost-effective solutions to limit the overtopping, especially when the parapet is appropriately shaped [58,59]. *LCS* and other detached parallel structure, by reducing the wave energy incident directly to the coast, may have a significant impact in the reduction of run-up.

Similarly, large nourishments have a very beneficial effect, both favouring energy dissipation, due to a reduced water depth, and reducing run-up, due to milder foreshore slope.

In addition to these measures, other anti-erosion and anti-flooding actions are available for both urbanized and natural coasts: managed realignment; urban planning; stabilization of river mouth (embankments to confine river discharge, plus bypass or sediment stock for updrift nourishment); lagoon dredging for environmental purposes; revetments (only where the unprotected area has no environmental and tourist interests); beach dewatering systems [60,61]; artificial reefs for habitat restoration; wave energy converter farms acting as coastal defence [62,63], seasonal interventions (e.g., sand accumulation, sand bags). It is also recommended that innovative solutions be lab-tested to ensure that they are effective in cost/benefit terms before they can be safely used on-site [16].

2.2.2. Maintenance and Monitoring Plan

The Coastal GIS should be constantly updated and all new events recorded in the database. Updates must, therefore, include results of both periodic and occasional monitoring, all the activities authorized by Regional Authorities, and an ex-post description of any coastal works carried out. It is especially important that the GIS is updated with:

1. Co-ordinates and details of the methodology used to integrate information on waves, fluvial sediment transport, bathymetric profiles, etc. The plan must have a regional scale and be planned over a broad period, possibly using a common approach; it must also focus closely on the expected results.
2. Results of monitoring. Each mitigation option is typically associated with a monitoring plan designed to check maintenance effectiveness and verify whether the option selected has achieved the objectives. It is highly advisable to plan an annual topo-bathymetric survey for the nourishment and additional (less frequent but specific) surveys for dunes, barriers and groins etc. These surveys should then be analysed and compared with maintenance plan predictions so that remedial action can be taken.
3. Activities authorized by Coastal Authorities, such as dredging and nourishment, are to be recorded so that it is clear on which cell they were carried out, i.e., by subdividing the total dredged and nourished volumes according to the semi-cells shown in the plan. This is of great help for overall coastal management since it reduces uncertainty about total volumes.
4. Any non-coastal-defence-related activities that may still contribute to the general knowledge framework.

2.2.3. Index of Priority

In addition to establishing anti-erosion and coastal flooding measures, work must be prioritized on the basis of resources and needs, and a chronological list of operations must be drawn up. The index of priority for each cell is defined as:

$$IP = \sum_i V_{M_i} \times \sum_i V_{SE_i}, \tag{3}$$

where the morphological vulnerability index (V_M) is the sum of the erosive tendency index and the risk of coastal flooding index; the socioeconomic vulnerability index (V_{SE}) is the sum of six indexes specifying the coast value (Table 1). The scale used for every index goes from 1 to 4 (min IP = 12, max IP = 192). The minimum and maximum value of each index are based on the lowest and highest qualitative functional response among all the analysed cells with respect to the specific aspect.

Table 1. Description of terms V_M and V_{SE} in Equation (3).

Morphological Vulnerability (V_M)	Socioeconomic Vulnerability (V_{SE})
Erosive Trend Coastal Flooding Risk	Value of coastal defence action (frequency and financial investments) Environmental relevance (presence of natural areas, Natura2000 sites, etc.) Tourist pressure (number of tourist/year, bathing facilities, tourist facilities, etc.) Urbanization (presence of cities, type of inland, etc.) Production activities (fishing, mussel/clam farming, etc.) Cultural heritage (monuments, archaeological sites, etc.)

Note that the index of priority is equivalent to the "risk" defined by Benassai [64] as the product between morphological vulnerability and socioeconomic vulnerability.

3. The Venetian Littoral

3.1. Description of the Area and Assessment of Sediment Budget

The Coastal Plan of the Veneto Region (IT) [23] (carried out according to the methods described in the previous Section) investigates the Venetian coastline (Figure 5), which is 160 km long and faces the Northern Adriatic Sea. The coastline's borders are the mouth of the River Tagliamento to the North and the mouth of the River Po di Goro to the South. It is subdivided into two provinces and ten coastal municipalities.

The Adriatic Sea is rectangular-shaped, is about 750 km long and 200 km wide, and is connected to the Mediterranean Sea at its Southern end by the Strait of Otranto, which is about 80 km wide. Its depth is rather limited in the Northern part, where the bottom descends south-eastwards with a 1 in 1000 slope. The Adriatic coast is plagued by a combination of high waves and storm surges, which are responsible for the flooding of coastal areas, in particular, Venice and its lagoon. The highest surge was on 4 November 1966 when the sea level rose approximately 180 cm above the mean sea level (MSL) and persisted above the 100 cm mark for more than 15 h (Canestrelli [65]).

The North Adriatic Sea is characterized by two main wind (and correspondingly wave) regimes, which are primarily influenced by local orography. The prevailing winds along the Venetian coastline are the Bora and the Scirocco, which blow from the North-East and South-East respectively.

The Venetian coastline is characterized by low beaches, lagoons (i.e., Caorle, Venice and Po River Delta) and the mouths of seven rivers: the Tagliamento, Livenza, Piave, Sile, Brenta, Adige, and Po.

Along the 100 km stretch of coast from the mouth of the River Tagliamento to the Porto Caleri inlet [23] lie a vast number of areas with a high tourist value (e.g., Bibione, Caorle, Jesolo, Lido di Venezia, Sottomarina). Many of them are protected by coastal structures (e.g., groins, seawalls, breakwaters), and few are free of urban settlements (e.g., Valle Vecchia). Only a few, mainly discontinuous, dune systems can be found along the coast because they were destroyed at various times in the past.

The remaining Venetian littoral comprises the Po Delta, which covers 610 km^2 and has 60 km of coast stretching from the Porto Caleri inlet to the mouth of the River Po di Goro. The active river branches of the River Po are (from North to South) Po di Maistra, Po di Pila, Po di Tolle, Po di Gnocca and Po di Goro. The coastal fringe is characterized by a sequence of low sandy and vulnerable barrier islands, beaches and spits that separate lagoons, fishing valleys, bays, tidal flats and marshes from the sea. Inland, ground elevation is almost completely below sea level (locally $-2.5/-3.0$ m.s.l.), and consequently the risk of coastal flooding is very high. The morphological characteristics of the Po Delta make it Italy's largest wetland, as well as particularly unstable and very fragile when subjected to human pressure.

The Venetian coast is subdivided into 20 homogeneous littoral cells separated by inlets or the mouths of rivers (from North to South) Tagliamento, Bocca di Porto Baseleghe, Bocca di Falconera, Livenza, Piave, Sile, Bocca di Lido, Bocca di Malamocco, Bocca di Chioggia, Brenta, Adige, Bocca di Caleri, Bocca di

Porto Levante, Po di Maistra, Busa Tramontana, Busa Dritta, Busa di Scirocco, Busa Storiona, Po di Tolle, Po di Gnocca, and Po di Goro. These 22 limits are shown in Figure 5 and listed in Table 2.

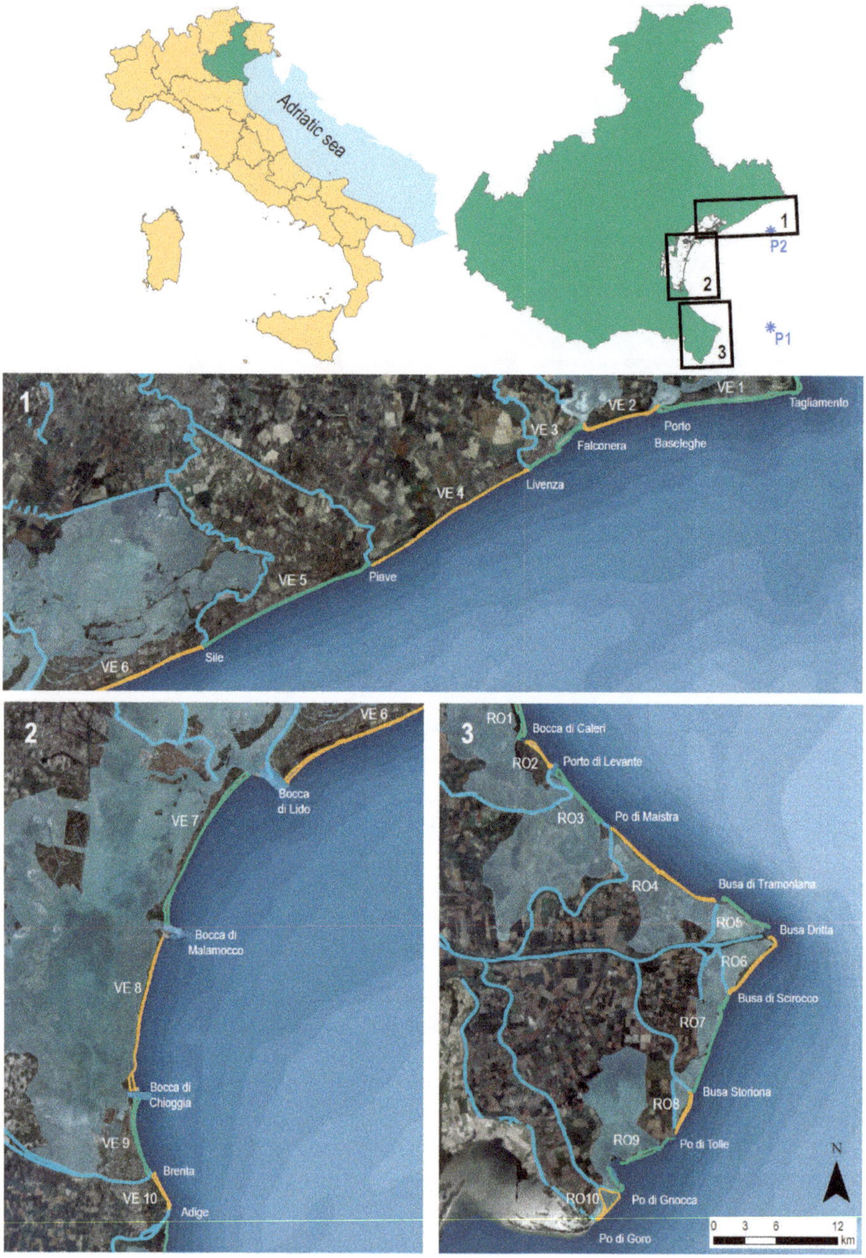

Figure 5. Venetian littoral and its subdivision into coastal cells.

Table 2. Littoral cell boundary, sediment diameter (d_{50}, mm) sampled at different depths and evaluated with the Dean rule (d_{EQ}, mm).

Cell	North Bound	South Bound	d_{50} 0 m/−2 m	d_{50} −2 m/−4 m	d_{50} −4 m/−6 m	d_{50} <−6 m	d_{EQ}
VE1	Tagliamento	Porto Baseleghe	0.323	-	-	-	0.14
VE2	Porto Baseleghe	Falconera	0.173	-	-	-	0.13
VE3	Falconera	Livenza	0.175	0.235	0.118	-	0.16
VE4	Livenza	Piave	0.205	0.173	0.143	-	0.16
VE5	Piave	Sile	0.153	0.268	0.118	-	0.15
VE6	Sile	Bocca di Lido	-	0.235	0.213	-	0.14
VE7	Bocca di Lido	Bocca di Malamocco	0.181	-	0.186	-	0.12
VE8	Bocca di Malamocco	Bocca di Chioggia	-	-	-	-	0.20
VE9	Bocca di Chioggia	Foce Brenta	-	0.144	0.168	-	0.15
VE10	Brenta	Adige	-	-	-	-	0.24
RO1	Adige	Bocca di Caleri	-	0.190	0.191	0.129	0.18
RO2	Bocca di Caleri	Porto di Levante	-	0.139	0.126	0.110	0.11
RO3	Porto di Levante	Po di Maistra	-	0.143	0.185	0.142	0.11
RO4	Po di Maistra	Busa di Tramontana	-	0.173	0.208	0.133	0.13
RO5	Busa di Tramontana	Busa Dritta	-	0.234	0.230	0.101	0.17
RO6	Busa Dritta	Busa di Scirocco	-	0.273	0.258	0.207	0.17
RO7	Busa di Scirocco	Busa Storiona	-	0.167	0.148	0.132	0.17
RO8	Busa Storiona	Po di Tolle	-	0.068	0.197	0.100	0.15
RO9	Po di Tolle	Po di Gnocca	-	0.144	0.146	-	0.10
RO10	Po di Gnocca	Po di Goro	-	-	-	-	0.13

The plan [23] is based on the information and data available for the Venetian littoral; they comprise offshore wave characteristics, sediment grain size, topographic and bathymetric surveys over a range of time (bathymetry: 2005, 2007/2008, 2010, 2012/2014, DTM: 2008, 2012/2013), subsidence rate (1992–2000 and 2002–2010), shoreline position (1983, 2000, 2003, 2012), flooding risk maps (from 2007/60/EC directive) and a catalogue of existing shore protection structures and nourishment/dredging carried out.

Nearshore wave conditions were evaluated using the SWAN model (Simulating WAve Nearshore, [66]), developed by Delft University of Technology (NL) and based on offshore wave data. Unfortunately, a spatially refined evaluation of the offshore wave statistics obtained by oceanographic models was not available, as it is still an ongoing project. Wave information was therefore obtained by existing WAM simulations forced by data from the European Centre for Medium-Range Weather Forecasts (ECMWF) between June 1992 and December 2008, and were restricted to two points in the Northern Adriatic sea (P1: Longitude 13°00′ Latitude 45°00′, P2: Longitude 13°00′ Latitude 45°30′, wave roses in Figure 6).

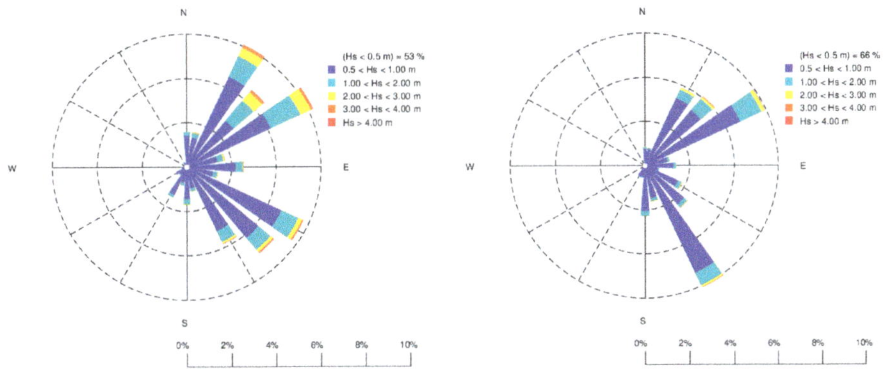

Figure 6. Offshore wave climate in the Northern Adriatic Sea: point P1 (left), point P2 (right).

The nearshore wave climate (−10 m depth) was obtained with the SWAN model. The SWAN transforms the directional wave spectrum, which cannot be fully described by a single small plot inside a regional map. For this reason, the energy has been integrated in frequency and its directional distribution is given in the figure. For instance, Figure 7 presents the energy distribution for the point P1 and P2, whose wave climate is fully characterised by the wind rose in Figure 6. Figure 8 presents the energy distribution obtained through the SWAN model, propagating the waves from P1 and P2. The Northern part of the Venetian littoral is mainly subject to waves from the South-East, with the Scirocco blowing along the main axis of the basin and acting on a much longer fetch in this zone. Due to shoreline orientation, the Northern part of the Po Delta is exposed to the Bora, which causes high waves, although fetch is limited, with wave periods ranging between 5 and 7 s. In the Southern part of the Po Delta, sea conditions are governed by both Bora and Scirocco wind and waves. Figure 8 clearly shows that the Po Delta is characterized by larger wave energy.

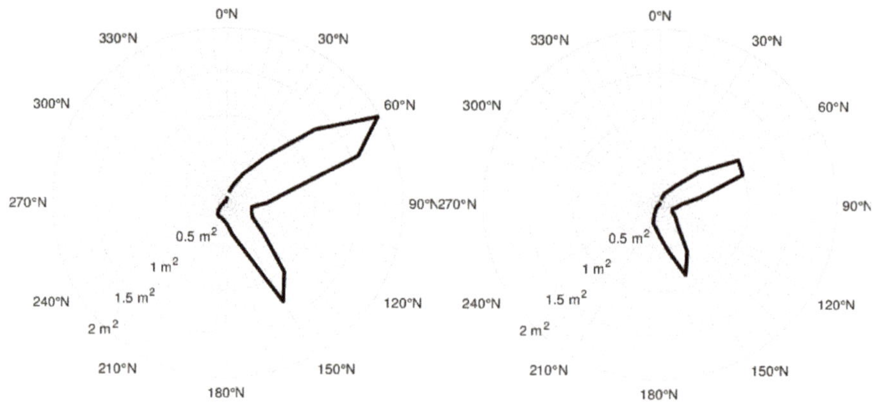

Figure 7. Offshore Energy polar plot relative to point P1 (left), point P2 (right) summarising the wave rose in Figure 6.

Sediment surveys for the Northern Adriatic Sea have been re-organized and stored in a Coastal GIS. Table 2 shows the average sediment size for each littoral cell at different depths. In general, sediment on the Veneto coast is fine sand, with grain diameter ranging between 0.12 and 0.25 mm. As expected, grain size is coarser near the shoreline and decreases seawards. Deviations occur is some places, e.g., RO8, where rivers transport fine sediment that may deposit in the nearshore zone.

Dean [67] proposed an equilibrium profile, $y = Ax^{2/3}$, giving a relationship between water depth (y) and the distance from the shoreline (x) via parameter A that, according to Hanson and Kraus [68] ($A = 0.41 d_{50}{}^{0.94}$ for d_{50} <0.4 mm), is a function of the median diameter d_{50} (mm). The formulation can be inverted to start from the bathymetric profile, so that parameter A can be adapted, and the corresponding "equilibrium diameter" d_{EQ} can be found. The result for each littoral cell is shown in the last column of Table 2 and can be compared with actual grain size at different depths. The value of d_{EQ} is very useful since it provides an average bed-profile shape immediately.

Assessment of the river sediment transport is complicated by the almost complete absence of systematic hydrographic surveys. Therefore, numerical models based on what little information was available were used instead. In order to evaluate river sediment discharge at the mouth, Lanzoni [69] proposed a one-dimensional numerical model using topographic surveys, the annual hydrological regime, and a medium grain size. Considering steady forcing conditions, the model estimates a "formative discharge" that produces the river topography observed and the corresponding sediment transport capacity. This approach was applied to the main rivers in the Veneto Region (Tagliamento, Piave, Brenta, Adige, and Po).

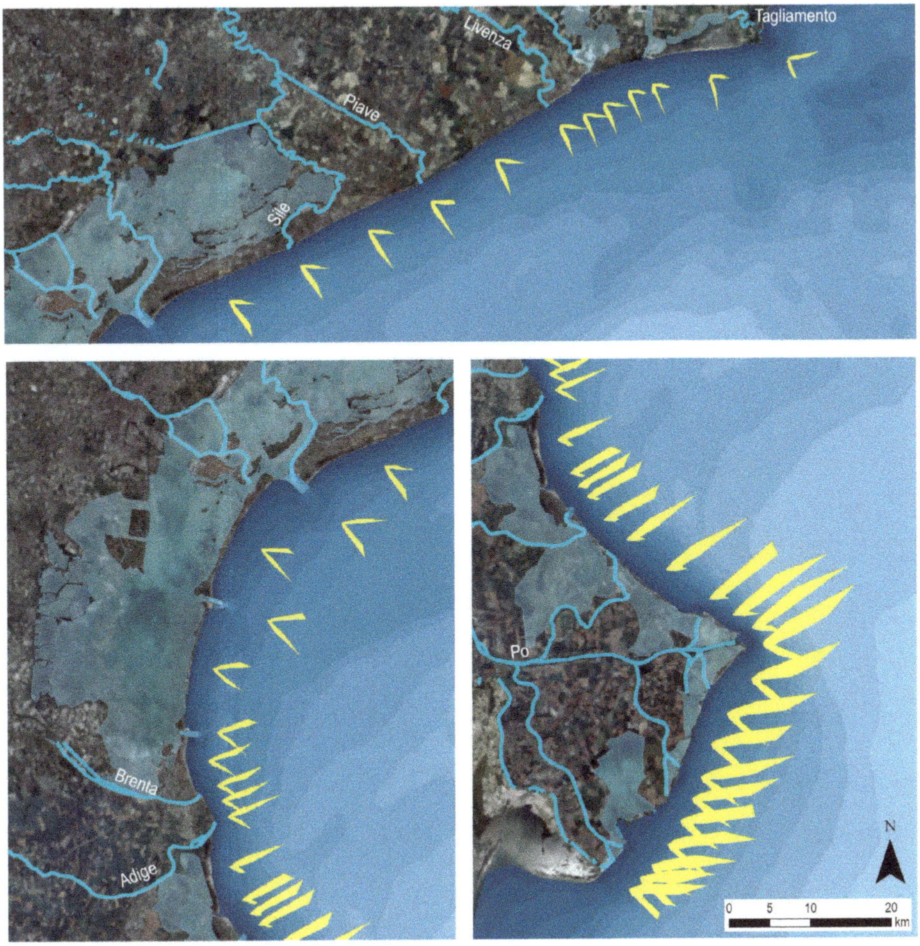

Figure 8. Nearshore wave climate (energy polar plot) for the Venetian littoral. The scale of the energy plot is common to the three images, allowing a qualitative comparison.

The rate of long-shore sediment transport was based on the local computation of the wave and currents for each wave state with the formula proposed by Bijker [70], following the procedure pointed out in [71], integrating across the profile and averaging. Results are shown in Figure 9 and the method is described in [23]. The spatial pattern of the simulated net transport contains divergence and convergence areas that separate areas with oppositely directed net sediment fluxes. Divergence points are located in front of the mouths of the two main rivers (Adige and Po). Convergence points are at Bocca di Lido and Bocca di Caleri. The latter, placed between cell RO1 and RO2, is a highly persistent point of convergence for net transport and thus, as observed, a deposition area (volume ~100,000 m^3/year). The most dynamic zone is the Po Delta area, which has a symmetric morphology with a divergence net sediment transport of ~200,000 m^3/year.

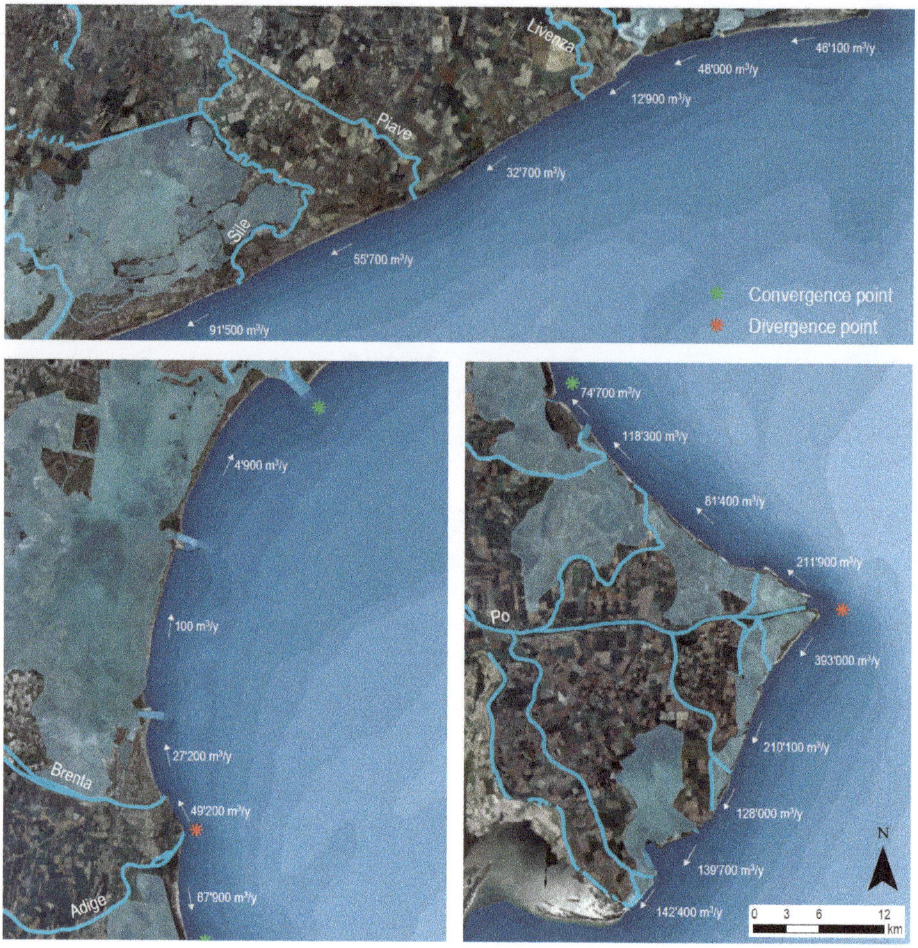

Figure 9. Long-shore sediment transport for the Venetian littoral.

It was evaluated, by comparing few methods [72], that the cross-shore sediment transport Q_{CR} was heading offshore and approximately equal to 1 m^3/km/year, except where the cell boundary element is a river mouth, where an additional contribution is assessed to simulate the sediment plume losses.

Subsidence along the coast is associated with natural causes related to the area's geological history (e.g., sediment consolidation) and with anthropogenic activities, mainly fluid withdrawal. An innovative technique called Advanced Differential Interferometric SAR (A-DInSAR [73,74]) was applied in order to measure the deformation of the Earth's surface. The subsidence in the Northern part is equal to 1–2 mm/year and is mainly related to natural causes. The subsidence in the Po Delta is much larger and ranges from 3–5 mm/year, with it being linked to both natural and anthropogenic causes.

Accumulation and erosion were measured up to the depth of closure to compare successive bathymetric profiles and calculate the volume of accretion and erosion. Details and methodology are given in [75].

After the evaluation of every source term, the sediment balance for each littoral cell was calculated with Equation (1). Each homogenous littoral cell was divided into two parts in order to better appreciate the coastal processes involved. Note that the long-shore sediment transport in the middle of the cell must be the same for each semi-cell. The long-shore sediment transport at the boundary between adjacent cells may differ on account of potential depositional or erosive areas surrounding inlets or river mouths.

The balance is solved by using a compensation of error technique based on a matrix of a priori uncertainties. The accuracy of each variable is weighted with a specific coefficient and the mass unbalance for each cell is subdivided among the estimated variables forming the budget based on their weight. The final sediment budget is summarized in Table 3: 19 semi-cells (total length ~58 km) have a depositional behaviour with a volume >10,000 m^3/year; 15 semi-cells (total length ~62 km) have an erosive behaviour with a volume ≤10,000 m^3/year; and 6 semi-cells (total length ~19 km) are almost stable, with a volume in the range of ±10,000 m^3/year.

Table 3. Sediment balance (thousand m^3/year).

Cell	Part	Q_{LS} (1) Long-shore	Q_{LS} (2) Long-shore	Q_{CR} Cross-shore	Q_F Fluvial	Q_N Nourished	Q_D Dredged	∂V
VE1	N	−30.9	−46.1	22.7	11.0	51.0	49.0	−24.9
VE1	S	−46.1	−48.9	7.1	65.9	-	-	55.9
VE2	N	−67.9	−48.0	4.4	-	-	-	15.6
VE2	S	−48.0	−36.4	3.0	-	22.8	17.2	14.1
VE3	N	−20.3	−12.9	4.6	-	24.0	0.0	26.9
VE3	S	−12.9	−10.0	3.8	-	22.4	5.1	16.3
VE4	N	−10.0	−32.7	8.9	-	64.4	-	32.8
VE4	S	−32.7	−57.3	79.8	13.0	29.9	-	−61.5
VE5	N	−62.6	−55.7	183.7	117.3	40.0	40.0	−59.6
VE5	S	−55.7	−91.2	19.7	-	35.6	-	−19.6
VE6	N	−75.6	−91.5	22.1	-	20.1	-	−17.9
VE6	S	−91.5	−24.3	17.2	-	-	20.2	29.8
VE7	N	−0.7	4.9	9.3	-	10.0	-	6.4
VE7	S	4.9	2.1	9.9	-	10.0	-	−2.7
VE8	N	2.0	0.1	18.8	-	-	-	−20.7
VE8	S	0.1	2.3	18.0	-	-	-	−15.8
VE9	N	−0.6	27.2	4.6	-	-	-	23.2
VE9	S	27.2	16.2	48.0	19.8	39.9	20.0	−19.3
VE10	N	48.2	49.2	43.6	8.3	34.9	20.0	−19.3
VE10	S	49.2	58.0	93.2	62.5	35.0	-	13.1
RO1	N	−64.9	−87.9	100.1	80.4	30.0	-	−12.6
RO1	S	−87.9	−65.0	6.0	-	-	-	16.8
RO2	N	79.2	74.7	2.8	-	-	-	−7.2
RO2	S	74.7	62.9	3.6	-	-	-	−15.4
RO3	N	75.4	118.3	7.9	-	-	-	35.0
RO3	S	118.3	86.3	31.6	60.1	-	-	−3.5
RO4	N	86.3	81.4	39.1	71.4	-	-	27.4
RO4	S	81.4	207.6	20.8	-	-	59.3	46.1
RO5	N	195.6	211.9	6.7	-	-	-	9.6
RO5	S	211.9	359.8	368.8	240.4	-	-	19.4
RO6	N	−409.6	−393.0	370.5	534.0	-	-	180.2
RO6	S	−393.0	−305.3	8.7	-	-	-	78.9
RO7	N	−234.7	−210.1	9.4	-	-	-	15.2
RO7	S	−210.1	−252.5	9.0	-	-	-	−51.3
RO8	N	−230.5	−128.0	6.2	-	-	-	96.3
RO8	S	−128.0	−103.5	96.9	99.1	-	-	26.8
RO9	N	−172.9	−139.7	97.3	64.4	-	-	0.3
RO9	S	−139.7	−190.5	78.6	106.2	-	-	−23.2
RO10	N	−95.8	−142.4	5.0	-	-	-	−51.6
RO10	S	−142.4	−155.5	119.5	120.8	-	-	−11.8

3.2. Design Phase

Based on the sediment balance and on the criteria used to select the mitigation options (see Section 2), a coastal management proposal for every stretch of coast was carried out. For the sake of brevity, only a summary of the mitigation measures is presented here, and a detailed example of the RO1 cell is described in the next paragraph.

In the Northern zone, the goal is to "hold" the shoreline position and to protect shore-based activities using protection measures with minimum impact (e.g., avoiding seawalls, detached barriers, etc.). Management in this zone includes building a series of groins in cells VE4, VE5 and VE10 and adding large nourishments to cells VE4, VE7, VE8 and VE10 (volume = 3,650,000 m^3). The global volume of sand needed for maintenance is approximately 385,000 m^3/year.

A "Building with Nature" methodology is applied to the Po Delta area. Localized sand nourishment and dune reinforcement are nature-based defences that provide several ecosystem services, including flood/erosion risk mitigation and environmental conservation. The volume of sand needed for maintenance (nourishments and dunes) is approximately 145,000 m^3/year.

A scheduled monitoring program was established across the entire Regional littoral in order to collect the data and information necessary.

The proposed index of prioritization was also applied to each cell to assess a chronological list of the operations based on the available resources. Figure 10 (top) shows the morphological and socioeconomic vulnerabilities along the littoral. The indexes reflect the urban essence of the area between the mouths of the rivers Tagliamento and Adige (where the main tourist activities—5,000,000 visitor/month in summer—are concentrated); the main cultural heritage sites (Venice, Caorle); and the natural essence of the Po Delta (an area with one of Italy's highest environmental values).

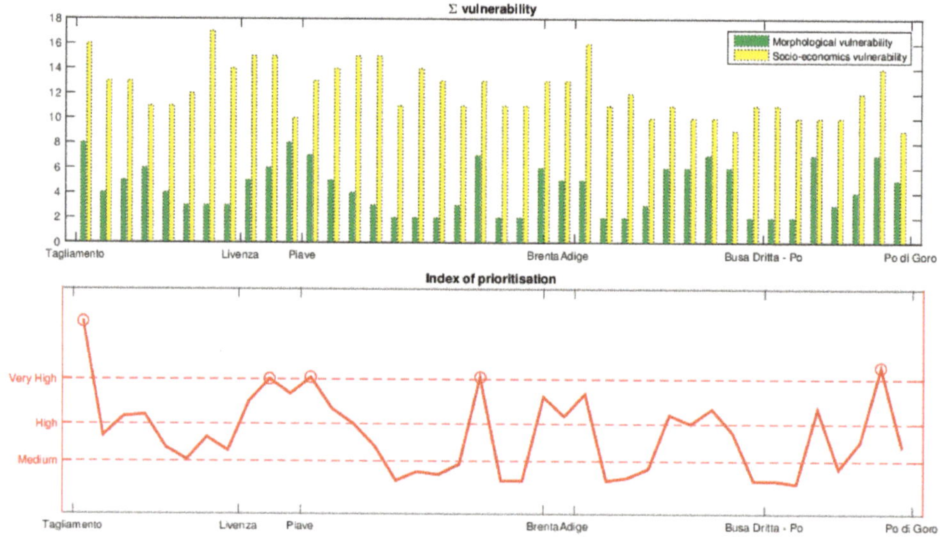

Figure 10. Vulnerability and index of priority.

Figure 10 (bottom) shows the index of priority, with the circles highlighting the 5 stretches of coast with the highest index of priority. For each stretch, the main issues are presented below.

1. Coast right of the mouth of the River Tagliamento (VE1): intensive erosive processes due to reduced river supply and a system of attached breakwaters that trap the long-shore sediment transport.
2. VE4 cell (Porto Santa Margherita, Duna Verde and Eraclea-Venice): erosive processes due to a decreased river sediment transport supply and the presence of reflective structures along the coast.
3. Coast right of the mouth of the River Piave (VE5): intensive erosive processes due to decreased river sediment supply.
4. VE 8 cell (Pellestrina-Venice): huge nourishment was conducted in the late 1990s. The effectiveness of this measure is now reduced, and intensive erosive processes are affecting this littoral. This cell is a thin barrier/island separating the Venice lagoon from the sea.
5. Scardovari spit (VE9), which confined the Scardovari lagoon (in the South of the Po Delta): this vulnerable sand formation is subject to erosion processes and to episodic overwashing, which endangers the lagoon environment.

3.3. *Example for RO1 Littoral Cell*

The Rosolina coastline (RO1 cell) is 8 km long and the normal shoreline direction is 80° N. The economy in the northern and central part is based mainly on tourism (1,100,000 visitors in 2016) and fishing in the backshore Caleri lagoon. The lagoon and the Southern part of the cell are major environmental areas, protected and designated as Natura 2000 sites (SCI IT3270017 and SPA IT3270023).

The cell is delimited to the North by the mouth of the River Adige which, in the final stretch, flows parallel to the beach and is confined by a weak dike that closes an old branch of the river mouth. The sediment budget analysis (Table 3) put the total fluvial sediment transport at ~265,000 m^3/year, of which ~145,000 m^3/year was directed toward the cell being studied (the remaining volume is directed northward toward the adjacent cell VE10). The fine sediment is lost and only ~45,000 m^3/year contributes to cell advancement. Until 2007, there was a narrow beach on the sea side of the weak dike, but it has now completely disappeared, as a result of the imbalance between potential long-shore transport (~65,000 m^3/year) and the actual river contribution. The crest height of this dike is very low (+1.5 m above sea level) and some waves overtop it, creating a depositional area inside the river mouth and obstructing river outflow.

Further South along the right side of the mouth of the River Adige, the beach is protected by a system of 5 groins and a detached submerged breakwater. Erosion (~13,000 m^3/year) also predominates in this area due to the limited river sediment supply mentioned above, therefore an insufficient volume of 30,000 m^3 is nourished every year in a bid to balance the long-shore transport directed to the Southern semi-cell (~88,000 m^3/year). The interaction between nearshore hydrodynamics and the submerged barrier caused the formation of a deep channel (~3 m) in the breakwater's seaward zone. The channel obstructs natural circulation and sand deposition from the River Adige towards the beach.

No structures were built in the southern semi-cell, as it is characterized by deposition phenomena (shoreline accretion equal to 4 m/year, volume of accretion equal to ~15,000 m^3/year) since long-shore sediment transport in the Southern boundary of RO1 is reduced to ~65,000 m^3/year.

The seabed appears steeper in the northern and central part than in the southern part (Caleri inlet). The −5 m isobath is 400 m from the shoreline in the northern zone, while it is 1000 m from the shoreline in the southern zone (Figure 11, top).

The Caleri inlet is a convergence point between two adjacent cells since the sand comes from the River Adige to the North and from cell RO2 to the South, making the inlet a potential dredging area (available volume ~140,000 m^3/year).

The mitigation planned (Figure 11, bottom) follows on from the criteria in Figure 3. Given the large LST and the significant divergence of the LST along the northern semi-cell, the existing groins are considered appropriate, with them being reinforced and their number slightly increased. The annual nourishment volume is obtained from the sediment balance results.

Sediment resources can be derived from dredging the Caleri inlet, and it is sufficient to provide sediment elsewhere, too. A cautious dredging of only half of the forecasted annual increase in stock volume is addressed in the initial plan, and the monitoring programme will check the actual potential.

In practice, mitigation measures involve:

1. Constructing a series of groins (min. 2) in the central zone in order to reduce erosion on this stretch of coast.
2. Maintenance dredging the Caleri inlet (~70,000 m^3/year) in order to nourish the northern and central areas and to ensure that boats can continue to navigate the lagoon inlet.
3. A monitoring program.

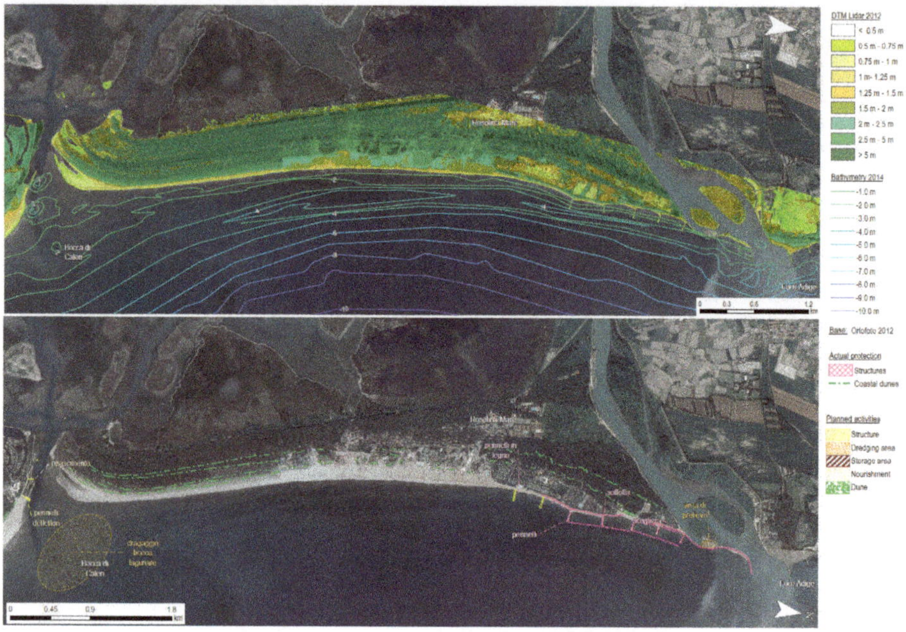

Figure 11. Topo-bathymetry (top) and adopted mitigation for the RO1 cell.

4. Conclusions

The paper presents the method used to set up the Coastal Plan for the Venetian littoral. This method was based on a homogeneous approach at regional level consisting of a descriptive and a design phase and focused on an intensive use of a Coastal GIS database that stored both information and results, which were then organised into littoral cells and semi-cells.

The GIS also contains the sediment balance assessment obtained by analysing collected and modelled data, the accuracy of which is based on a compensation criterion. The ensuing erosion and coastal-flooding hazard, integrated with information on socioeconomic vulnerability, resulted in a priority index that highlighted the most critical areas. The information on relative sea level rise due to climate change and geodynamics are stored as they may have a relevant role for a correct design of the mitigation measures characterized by a long lifetime.

A noteworthy contribution of this paper is given in Section 2.2.1, which provides an example of a single mitigation strategy against erosion and coastal flooding for the region's valuable urbanized and non-urbanized low sandy coasts. The consistent application of this method to the Veneto Region produced measures that were globally accepted by stakeholders, and the plan was adopted by the Regional Authorities with regional decree DGR no. 898-14/06/2016. It is conjectured that the same approach may be adopted, with due modifications, in other low sandy beaches, as well.

After a general description of the main results for the Venetian coastal plan [23], the Rosolina (RO) coastline is described in detail and provided as an example case.

Author Contributions: All the authors contributed equally in conceiving, developing and writing the article.

Acknowledgments: The Agreement between Regione Veneto and University of Padova entitled "Gestione Integrata della Zona Costiera. Progetto per lo studio ed il monitoraggio della linea di costa per la definizione degli interventi di difesa dei litorali dall'erosione nella Regione Veneto" is gratefully acknowledged.

Conflicts of Interest: The authors declare no conflict of interest.

References

1. French, J.; Payo, A.; Murray, B.; Orford, J.; Eliot, M.; Cowell, P. Appropriate complexity for the prediction of coastal and estuarine geomorphic behaviour at decadal to centennial scales. *Geomorphology* **2016**, *256*, 3–16. [CrossRef]
2. Haasnoot, M.; Kwakkel, J.H.; Walker, W.E.; ter Maat, J. Dynamic adaptive policy pathways: A method for crafting robust decisions for a deeply uncertain world. *Glob. Environ. Chang.* **2013**, *23*, 485–498. [CrossRef]
3. Van Maanen, B.; Nicholls, R.J.; French, J.R.; Barkwith, A.; Bonaldo, D.; Burningham, H.; Brad Murray, A.; Payo, A.; Sutherland, J.; Thornhill, G.; et al. Simulating mesoscale coastal evolution for decadal coastal management: A new framework integrating multiple, complementary modelling approaches. *Geomorphology* **2016**, *256*, 68–80. [CrossRef]
4. Hinkel, J.; Lincke, D.; Vafeidis, A.T.; Perrette, M.; Nicholls, R.J.; Tol, R.S.J.; Marzeion, B.; Fettweis, X.; Ionescu, C.; Levermann, A. Coastal flood damage and adaptation costs under 21st century sea-level rise. *Proc. Natl. Acad. Sci. USA* **2014**, *111*, 3292–3297. [CrossRef] [PubMed]
5. Weisse, R.; Bellafiore, D.; Menéndez, M.; Méndez, F.; Nicholls, R.J.; Umgiesser, G.; Willems, P. Changing extreme sea levels along european coasts. *Coast. Eng.* **2014**, *87*, 4–14. [CrossRef]
6. Toimil, A.; Losada, I.J.; Camus, P.; Díaz-Simal, P. Managing coastal erosion under climate change at the regional scale. *Coast. Eng.* **2017**, *128*, 106–122. [CrossRef]
7. Stocker, T.F.; Qin, D.; Plattner, G.-K.; Tignor, M.; Allen, S.K.; Boschung, J.; Nauels, A.; Xia, Y.; Bex, V.; Midgley, P.M. IPCC Climate Change 2013: The Physical Science Basis, WG1. In *The Fifth Assessment Report of the Intergovernmental Panel on Climate Change*; Cambridge University Press: Cambridge, UK, 2013.
8. Antonioli, F.; Anzidei, M.; Amorosi, A.; Presti, V.L.; Mastronuzzi, G.; Deiana, G.; De Falco, G.; Fontana, A.; Fontolan, G.; Lisco, S.; et al. Sea-level rise and potential drowning of the Italian coastal plains: Flooding risk scenarios for 2100. *Quat. Sci. Rev.* **2017**, *158*, 29–43. [CrossRef]
9. Bezzi, A.; Pillon, S.; Martinucci, D.; Fontolan, G. Inventory and conservation assessment for the management of coastal dunes, Veneto coasts, Italy. *J. Coast. Conserv.* **2018**, *22*, 503–518. [CrossRef]
10. Montanari, R.; Marasmi, C. *New Tools for Coastal Management in Emilia-Romagna*; Technical Report: Bologna, Italy, 2012; Available online: http://www.coastance.eu (accessed on 4 April 2018).
11. Zanuttigh, B.; Martinelli, L.; Lamberti, A.; Moschella, P.; Hawkins, S.; Marzetti, S.; Ceccherelli, V.U. Environmental design of coastal defence in lido di dante, italy. *Coast. Eng.* **2005**, *52*, 1089–1125. [CrossRef]
12. Doddy, P.; Ferreria, M.; Lombardo, S.; Luicus, I.; Misdorp, R.; Niesing, H.; Smallegange, M. *Living with Coastal Erosion in Europe–Sediment and Space for Sustainability*; Results from the Eurossion Study; European Commission Publication: Luxembourg, Luxembourg, 2004.
13. Zanuttigh, B. Coastal flood protection: What perspective in a changing climate? The theseus approach. *Environ. Sci. Policy* **2011**, *14*, 845–863. [CrossRef]
14. Ciavola, P.; Ferreira, O.; Haerens, P.; Van Koningsveld, M.; Armaroli, C. Storm impacts along european coastlines. Part 2: Lessons learned from the MICORE project. *Environ. Sci. Policy* **2011**, *14*, 924–933. [CrossRef]
15. Armaroli, C.; Duo, E. Validation of the coastal storm risk assessment framework along the Emilia-Romagna coast. *Coast. Eng.* **2018**, *134*, 159–167. [CrossRef]

16. Zanuttigh, B.; Simcic, D.; Bagli, S.; Bozzeda, F.; Pietrantoni, L.; Zagonari, F.; Hoggart, S.; Nicholls, R.J. Theseus decision support system for coastal risk management. *Coast. Eng.* **2014**, *87*, 218–239. [CrossRef]
17. Torresan, S.; Critto, A.; Rizzi, J.; Zabeo, A.; Furlan, E.; Marcomini, A. Desyco: A decision support system for the regional risk assessment of climate change impacts in coastal zones. *Ocean Coast. Manag.* **2016**, *120*, 49–63. [CrossRef]
18. Vafeidis, A.T.; Nicholls, R.J.; McFadden, L.; Tol, R.S.J.; Hinkel, J.; Spencer, T.; Grashoff, P.S.; Boot, G.; Klein, R.J.T. A new global coastal database for impact and vulnerability analysis to sea-level rise. *J. Coast. Res.* **2008**, *244*, 917–924. [CrossRef]
19. Van Dongeren, A.; Ciavola, P.; Viavattene, C.; de Kleermaeker, S.; Martinez, G.; Ferreira, O.; Costa, C.; McCall, R. Risc-kit: Resilience-increasing strategies for coasts-toolkit. *J. Coast. Res.* **2014**, *70*, 366–371. [CrossRef]
20. Stelljes, N.; Martinez, G.; McGlade, K. Introduction to the RISC-KIT web based management guide for DRR in European coastal zones. *Coast. Eng.* **2018**, *134*, 73–80. [CrossRef]
21. Preti, M. *Stato del litorale Emiliano-Romagnolo All'anno 2007 e Piano Decennale di Gestione*; Technical Report; I quaderni di ARPA Emilia Romagna: Bologna, Italy, 2008.
22. Petrillo, A.F.; Bruno, M.F.; Francioso, R.; Nobile, B.; Tomasicchio, R.; D'alessandro, F.; Di Pace, P.; Gencarelli, R. Studi Propedeutici per la Predisposizione del Piano Stralcio della Dinamica delle Coste. AdB Puglia, Politecnico di Bari, Università del Salento, Allegato 3.1 & 3.2, 2010. Available online: old.regione.puglia.it/web/files/demaniomarittimo/06__All.3.2__Strutture_per_la_difesa_delle_coste.pdf;old.regione.puglia.it/web/files/demaniomarittimo/05__All.3.1__Strutture_di_mitigazione_del_rischio.pdf (accessed on 4 April 2018).
23. Ruol, P.; Martinelli, L.; Favaretto, C. *Gestione Integrata della zona Costiera. Studio e Monitoraggio per la Definizione degli Interventi di Difesa dei Litorali Dall'erosione nella Regione Veneto—Linee Guida*; Edizioni Libreria Progetto: Padova, Italy, 2016.
24. Simeoni, U.; Corbau, C. A review of the delta Po evolution (Italy) related to climatic changes and human impacts. *Geomorphology* **2009**, *107*, 64–71. [CrossRef]
25. Carbognin, L.; Teatini, P.; Tosi, L. Eustacy and land subsidence in the Venice lagoon at the beginning of the new millennium. *J. Mar. Syst.* **2004**, *51*, 345–353. [CrossRef]
26. Pomaro, A.; Cavaleri, L.; Lionello, P. Climatology and trends of the Adriatic Sea wind waves: Analysis of a 37-year long instrumental data set. *Int. J. Climatol.* **2017**, *37*, 4237–4250. [CrossRef]
27. California Coastal Sediment Management Workgroup. *California Coastal Sediment Master Plan Status Report; Draft for Public Review and Comment*; California Geological Survey: Santa Rosa, CA, USA, 2006.
28. Bartlett, D.; Smith, J.L. *GIS for Coastal Zone Management*; CRC Press: Boca Raton, FL, USA, 2004.
29. Wright, D.; Bartlett, D. *Marine and Coastal Geographical Information Systems*; Taylor and Francis: London, UK, 2000.
30. Williams, A.; Rangel-Buitrago, N.G.; Pranzini, E.; Anfuso, G. The management of coastal erosion. *Ocean Coast. Manag.* **2017**. [CrossRef]
31. Payo, A.; Favis-Mortlock, D.; Dickson, M.; Hall, J.W.; Hurst, M.D.; Walkden, M.J.A.; Townend, I.; Ives, M.C.; Nicholls, R.J.; Ellis, M.A. Coastal Modelling Environment version 1.0: A framework for integrating landform-specific component models in order to simulate decadal to centennial morphological changes on complex coasts. *Geosci. Model Dev.* **2017**, *10*, 2715–2740. [CrossRef]
32. Cavaleri, L.; Abdalla, S.; Benetazzo, L.; Bertotti, J.-R.; Bidlot, Ø.; Breivik, S.; Carniel, R.E.; Jensen, J.; Portilla-Yandun, W.E.; Rogers, A.; et al. Wave modelling in coastal and inner seas. *Prog. Oceanogr.* **2018**. [CrossRef]
33. Roy, C.; Kovordányi, R. Tropical cyclone track forecasting techniques—A review. *Atmos. Res.* **2012**, *104*, 40–69. [CrossRef]
34. Favaretto, C.; Martinelli, L.; Ruol, P. Raster Based Model of Inland Coastal Flooding Propagation Using Linearized Bottom Friction and Application to a Real Case Study in Caorle, Venice (IT). In Proceedings of the Twenty-Eight International Ocean and Polar Engineering Conference (ISOPE2018), Sapporo, Japan, 10–15 June 2018.
35. Martinelli, L.; Zanuttigh, B.; Corbau, C. Assessment of coastal flooding hazard along the Emilia Romagna littoral, IT. *Coast. Eng.* **2010**, *57*, 1042–1058. [CrossRef]
36. Samaras, A.G.; Koutitas, C.G. The impact of catchment management on coastal morphology. The case of Fourka in Greece. *J. Coast. Res.* **2009**, 1686–1690.

37. Samaras, A.G.; Koutitas, C.G. Comparison of three longshore sediment transport rate formulae in shoreline evolution modeling near stream mouths. *Ocean Eng.* **2014**, *92*, 255–266. [CrossRef]
38. Mikhail, E.D.; Gracie, G. *Analysis & Adjustment of Survey Measurements*; Van Nostrand Reinhold: New York, NY, USA, 1981.
39. Burcharth, H.F.; Zanuttigh, B.; Andersen, T.L.; Lara, J.L.; Steendam, G.J.; Ruol, P.; Sergent, P.; Ostrowski, R.; Silva, R.; Martinelli, L.; et al. Innovative Engineering Solutions and Best Practices to Mitigate Coastal Risk. *Coast. Risk Manag. Chang. Clim.* **2014**, 55–170. [CrossRef]
40. Ruol, P.; Martinelli, L.; Favaretto, C.; Scroccaro, D. Innovative Sand Groin Beach Nourishment with Environmental, Defense and Recreational Purposes. In Proceedings of the Twenty-Eight International ocean and polar engineering conference (ISOPE2018), Sapporo, Japan, 10–15 June 2018.
41. Hoggart, S.; Hawkins, S.J.; Bohn, K.; Airoldi, L.; van Belzen, J.; Bichot, A.; Bilton, D.T.; Bouma, T.J.; Colangelo, M.A.; Davies, A.J.; et al. Ecological approaches to coastal risk mitigation. *Coast. Risk Manag. Chang. Clim.* **2015**, 171–236. [CrossRef]
42. Vanderlinden, J.P.; Baztan, J.; Coates, T.; Dávila, O.G.; Hissel, F.; Kane, I.O.; Koundouri, P.; McFadden, L.; Parker, D.; Penning-Rowsell, E.; et al. Nonstructural approaches to coastal risk mitigations. *Coast. Risk Manag. Chang. Clim.* **2015**, 237–274. [CrossRef]
43. Bergillos, R.J.; López-Ruiz, A.; Principal-Gómez, D.; Ortega-Sánchez, M. An integrated methodology to forecast the efficiency of nourishment strategies in eroding deltas. *Sci. Total Environ.* **2018**, *613*, 1175–1184. [CrossRef] [PubMed]
44. Stronkhorst, J.; Huisman, B.; Giardino, A.; Santinelli, G.; Santos, F.D. Sand nourishment strategies to mitigate coastal erosion and sea level rise at the coasts of Holland (The Netherlands) and Aveiro (Portugal) in the 21st century. *Ocean Coast. Manag.* **2018**, *156*, 266–276. [CrossRef]
45. Dean, R.G. *Beach Nourishment: Theory and Practice*; World Scientific Publishing Company: Toh Tuck Link, Singapore, 2003.
46. Zanuttigh, B.; Martinelli, L.; Lamberti, A. Wave overtopping and piling-up at permeable low crested structures. *Coast. Eng.* **2008**, *55*, 484–498. [CrossRef]
47. Buccino, M.; Calabrese, M. Conceptual approach for prediction of wave transmission at low-crested breakwaters. *J. Waterw. Port Coast. Ocean Eng.* **2007**, *133*, 213–224. [CrossRef]
48. Martinelli, L.; Zanuttigh, B.; De Nigris, N.; Preti, M. Sand bag barriers for coastal protection along the emilia romagna littoral, northern adriatic sea, Italy. *Geotext. Geomembr.* **2011**, *29*, 370–380. [CrossRef]
49. Hawkins, S.J.; Burcharth, H.F.; Zanuttigh, B.; Lamberti, A. *Environmental Design Guidelines for Low Crested Coastal Structures*; Elsevier: Oxford, UK, 2010.
50. Basco, D.R.; Pope, J. *Groin Functional Design Guidance from the Coastal Engineering Manual*; Coastal Education & Research Foundation, Inc.: Collier County, FL, USA, 2004; pp. 121–130.
51. Raudkivi, A.J. Permeable pile groins. *J. Waterw. Port Coast Ocean Eng.* **1996**, *122*, 267–272. [CrossRef]
52. Kristensen, S.E.; Drønen, N.; Deigaard, R.; Fredsoe, J. Impact of groyne fields on the littoral drift: A hybrid morphological modelling study. *Coast. Eng.* **2016**, *111*, 13–22. [CrossRef]
53. Van der Nat, A.; Vellinga, P.; Leemans, R.; Van Slobbe, E. Ranking coastal flood protection designs from engineered to nature-based. *Ecol. Eng.* **2016**, *87*, 80–90. [CrossRef]
54. Stive, M.J.F.; de Schipper, M.A.; Luijendijk, A.P.; Aarninkhof, S.G.J.; van Gelder-Maas, C.; van Thiel de Vries, J.S.M.; de Vries, S.; Henriquez, M.; Marx, S.; Ranasinghe, R. A new alternative to saving our beaches from sea-level rise: The sand engine. *J. Coast. Res.* **2013**, *290*, 1001–1008. [CrossRef]
55. Hanley, M.E.; Hoggart, S.P.G.; Simmonds, D.J.; Bichot, A.; Colangelo, M.A.; Bozzeda, F.; Heurtefeux, H.; Ondiviela, B.; Ostrowski, R.; Recio, M.; et al. Shifting sands? coastal protection by sand banks, beaches and dunes. *Coast. Eng.* **2014**, *87*, 136–146. [CrossRef]
56. Gracia, C.A.; Rangel-Buitrago, N.; Oakley, J.A.; Williams, A. Use of ecosystems in coastal erosion management. *Ocean Coast. Manag.* **2018**, *156*, 277–289. [CrossRef]
57. Villatoro, M.; Silva, R.; Méndez, F.J.; Zanuttigh, B.; Pan, S.; Trifonova, E.; Losada, I.J.; Izaguirre, C.; Simmonds, D.; Reeve, D.E.; et al. An approach to assess flooding and erosion risk for open beaches in a changing climate. *Coast. Eng.* **2014**, *87*, 50–76. [CrossRef]
58. McCabe, M.V.; Stansby, P.K.; Apsley, D.D. Random wave runup and overtopping a steep sea wall: Shallow-water and boussinesq modelling with generalised breaking and wall impact algorithms validated against laboratory and field measurements. *Coast. Eng.* **2013**, *74*, 33–49. [CrossRef]

59. Castellino, M.; Sammarco, P.; Romano, A.; Martinelli, L.; Ruol, P.; Franco, L.; De Girolamo, P. Large impulsive forces on recurved parapets under non-breaking waves. A numerical study. *Coast. Eng.* **2018**, *136*, 1–15. [CrossRef]
60. Damiani, L.; Aristodemo, F.; Saponieri, A.; Verbeni, B.; Veltri, P.; Vicinanza, D. Full-scale experiments on a beach drainage system: Hydrodynamic effects inside beach. *J. Hydraul. Res.* **2011**, *49*, 44–54. [CrossRef]
61. Damiani, L.; Vicinanza, D.; Aristodemo, F.; Saponieri, A.; Corvaro, S. Experimental investigation on wave set-up and nearshore velocity field in presence of a BDS. *J. Coast. Res.* **2011**, *SI 64*, 55–59.
62. Ruol, P.; Zanuttigh, B.; Martinelli, L.; Kofoed, P.; Frigaard, P. Near-shore floating wave energy converters: Applications for coastal protection. *Coast. Eng. Proc.* **2011**, *1*. [CrossRef]
63. Mendoza, E.; Silva, R.; Zanuttigh, B.; Angelelli, E.; Lykke Andersen, T.; Martinelli, L.; Nørgaard, J.Q.H.; Ruol, P. Beach response to wave energy converter farms acting as coastal defence. *Coast. Eng.* **2014**, *87*, 97–111. [CrossRef]
64. Benassai, G.; Chirico, F.; Corsini, S. Una metodologia sperimentale per la definizione del rischio da inondazione costiera. *Stud. Costieri* **2009**, *16*, 51–72.
65. Canestrelli, P.; Mandich, M.; Pirazzoli, P.A.; Tomasin, A. Wind, depression and seiches: Tidal perturbations in Venice (1951–2000). Centro Previsioni e Segnalazioni Maree. *Comune Venezia* **2001**, 1–104.
66. Booij, N.; Ris, R.C.; Holthuijsen, L.H. A third-generation wave model for coastal regions: 1. Model description and validation. *J. Geophys. Res. Ocean.* **1999**, *104*, 7649–7666. [CrossRef]
67. Dean, R.G. *Equilibrium Beach Profiles: US Atlantic and Gulf Coasts*; Department of Civil Engineering and College of Marine Studies; Technical Report 12; University of Delaware: Newark, DE, USA, 1977.
68. Hanson, H.; Kraus, N.C. *GENESIS: Generalized Model for Simulating Shoreline Change*; Report 1. Technical Reference; Technical Report CERC-89-19; Coastal Engineering Research Center: Vicksburg, MS, USA, 1989.
69. Lanzoni, S.; Luchi, R.; Bolla Pittaluga, M. Modeling the morphodynamic equilibrium of an intermediate reach of the po river (Italy). *Adv. Water Resour.* **2015**, *81*, 95–102. [CrossRef]
70. Bijker, E.W. Some Considerations about Scales for Coastal Models with Movable Bed. Ph.D. Thesis, Delft Hydraulics Laboratory Publication, Delft, The Netherlands, 1967.
71. Bayram, A.; Larson, M.; Miller, H.C.; Kraus, N.C. Cross-shore distribution of longshore sediment transport: Comparison between predictive formulas and field measurements. *Coast. Eng.* **2001**, *44*, 79–99. [CrossRef]
72. Schoonees, J.S.; Theron, A.K. Evaluation of 10 cross-shore sediment transport/morphological models. *Oceanogr. Lit. Rev.* **1996**, *2*, 136. [CrossRef]
73. Ferretti, A.; Prati, C.; Rocca, F. Permanent scatterers in sar interferometry. *IEEE Trans. Geosci. Remote. Sens.* **2001**, *39*, 8–20. [CrossRef]
74. Lanari, R.; Mora, O.; Manunta, M.; Mallorqui, J.J.; Berardino, P.; Sansosti, E. A small-baseline approach for investigating deformations on full-resolution differential sar interferograms. *IEEE Trans. Geosci. Remote Sens.* **2004**, *42*, 1377–1386. [CrossRef]
75. Fontolan, G.; Bezzi, A.; Martinucci, D.; Pillon, S.; Popesso, C.; Rizzetto, F. Sediment budget and management of the Veneto beaches, Italy: An application of the modified Littoral Cells Management System (SICELL). Proceedings of Coastal and Maritime Mediterranean Conference, Ferrare, Italy, 25–27 November 2015.

water

MDPI

Article

The Study of the Coastal Management Criteria Based on Risk Assessmeant: A Case Study on Yunlin Coast, Taiwan

Wei-Po Huang [1,2,*], Jui-Chan Hsu [1], Chun-Shen Chen [3] and Chun-Jhen Ye [1]

[1] Department of Harbor and River Engineering, National Taiwan Ocean University, Keelung 20224, Taiwan; 20652005@mail.ntou.edu.com.tw (J.-C.H.); j925372guy@gmail.com (C.-J.Y.)
[2] Center of Excellence for Ocean Engineering, National Taiwan Ocean University, Keelung 20224, Taiwan
[3] Water Resources Planning Institute, Water Resources Agency, Ministry of Economic Affairs, Taichung 41350, Taiwan; csc@wrap.gov.tw
* Correspondence: a0301@mail.ntou.edu.com.tw; Tel.: +886-2-2462-2192 (ext. 6154)

Received: 18 June 2018; Accepted: 25 July 2018; Published: 26 July 2018

Abstract: In this study, we used the natural and anthropogenic characteristics of a coastal region to generate risk maps showing vulnerability and potential hazards, and proposed design criteria for coastal defense and land use for the various kinds of risks faced. The Yunlin coast, a first-level protection area in mid-west Taiwan, was then used as an example to illustrate the proposed design criteria. The safety of the present coastal defenses and land use of the Yunlin coastal area was assessed, and coastal protection measures for hazard prevention were proposed based on the generated risk map. The results can be informative for future coastal management and the promotion of sustainable development of coastal zones.

Keywords: coastal defense; risk maps; non-engineering measure; coastal vulnerability

1. Introduction

Like most developing countries, Taiwan's coast has been alternatively used for settlement, agriculture, trade, industry, and recreation without careful and thorough planning in the development stage since 70s. The continuous expansion and diversity of urbanization together with the accumulation of deleterious effects on the coastal zone has intensified natural disasters in certain areas and their consequences for coastal residence. The conflicts between coastal exploitation and restriction continued because the Coastal Zone Management Act had not yet become a statutory law. This made it difficult to draw up strategies for Integrated Coastal Zone Management (ICZM), as there was no legal ground for the planning of land use in coastal areas. The Act came into effect in February 2015. It regulates the classification of coastal areas as first- or second-level coastal protection areas; and specifies which authorities are responsible for drawing up the respective coastal protection plans. Huang et al. (2016) (see also Chien et al., 2016) showed the zoning principles of the coastal protection areas which were based on the severity level of coastal hazards, however, neglecting the vulnerability [1,2]. The possible risks that a coastal area faces are not clearly identified. Thus, the most important step in facilitating coastal management is to generate risk maps in order to develop robust adaptation strategies and measures for different levels of protection areas respectively.

The concept of hazard risk analysis proposed by the United Nations Disaster Relief Organization (UNDRO, 1980) [3] involves a comprehensive examination of the relationship between hazard and vulnerability (risk = hazard × vulnerability). The hazard potential factor refers to the variability of hazards. In general, an increase in hazard intensity and frequency causes more serious damage and loss. There are five kinds of hazards for coastal areas in Taiwan, namely, storm surge, coastal erosion, flooding, ground subsidence, and tsunami (Chien et al., 2012) [4]. While the first four hazards are most frequent, the last one is relatively rare. The Coastal Zone Management Act therefore listed the first four hazards as the 'primary concerns' of coastal risks.

Coastal vulnerability can be defined as a measure of the degree to which natural hazards can affect coastal residents (McCarthy et al., 2001; van der Veen and Logtmeije, 2005; Parkinson and McCue, 2011) [5–7]. Possible losses increase when vulnerability increases (Cutter, 1996) [8]. Different variables have been used to evaluate coastal vulnerability according to the research orientation and perspective, for example the effect of sea-level rise (Khouakhi et al., 2013; Özyurt and Ergin, 2010), coastal erosion (Fitton et al., 2016; Merlotto, et al., 2016; Tarragoni et al., 2014), and sustainable development (Schernewski et al., 2014) [9–14]. Generally speaking, the variables can be categorized into geophysical, social, and socio-environmental contexts (Zanetti et al., 2016) [15]. The geophysical vulnerability focused on the relationships between physical features and coastal hazards. Hammar-Klose and Thieler (2001) used the six physical variables proposed by Gornitz et al. (1994) and Shaw et al. (1998) to assess the vulnerability of coasts in the United States [16–18]. Social vulnerability identifies the characteristics of coastal communities that enable them to respond to and recover from hazards (Cutter et al., 2003) [19]. In socio-environmental vulnerability the combined effects of both social and environmental vulnerabilities were taken into consideration (Wang et al., 2014) [20]. The United Nations Office for Disaster Risk Reduction (UNISDR, 2004) assessed vulnerability by grading physical, environmental, social and economic variables [21]. As social and economic conditions were taken into consideration, the potential threats of coastal hazards to natural surroundings and coastal residents can be clearly evaluated.

The combinations of the potential hazards and coastal vulnerability can be used to generate risk maps. Chien et al. (2013) used the risk maps to assess existing hazard prevention and coastal management measures in Taiwan [22]. Wang et al. (2014) conducted a comprehensive risk assessment strategy based on the risk matrix approach (RMA), which consisted of a probability phase and a severity phase [20]. Note that a relationship between the hazard map and the coastal protection criteria has not been clearly identified in any of these studies. Carrasco et al. (2012) and Ward et al., (2014) pointed out that flood hazard maps based on return periods could be useful in evaluating the physical damage to infrastructure, economy, and ecological resources of a region [23,24]. The risk maps with classified grades can be helpful in understanding the possible risks that a coastal area faces and determining the criteria for coastal protection.

Traditionally, coastal hazard management has concentrated on providing protection against floods through coastal defenses. However, severer marine conditions including sea level rise and the increasing intensity of storms due to climate change appear to be unavoidable and will inevitably threaten the residents of coastal areas (Van Vuren et al., 2004) [25]. To strengthen coastal defenses unlimitedly against climate change effects may lead to significant ecological losses and high cost (Penning-Rowsell et al., 2014) [26]. Nowadays, the coastal hazard management has moved away from engineering-dominated approaches to formulate adaptation strategies (Dinh et al., 2012; ESCAP/UNISDR, 2012; Luo et al., 2015; Salik et al., 2015) [27–30]. Non-engineering measures can be incorporated, for those hazards beyond the design criteria of coastal defenses. The non-engineering measures should be encouraged since, now the "Coast Zone Management Act" has come into effect, the government will have a legal basis to take action. Regulations to limit or even ban further exploitation of hazard-prone coastal areas can thus now be enacted and climate change adaptation strategies enforced.

Coastal risk maps are important for both decision-makers and engineers. They contain essential information for the former to make policies, for the latter to assess the design criteria of the coastal defenses, and are also required to implement the Coastal Zone Management Act. In this study, the coastal risk map was drawn to assess the current coastal areas whether coastal areas should be managed by mitigation and/or adaptation. We believe that this could be helpful in achieving integrated coastal management in Taiwan.

2. Background

2.1. Coastal Protection Areas in Taiwan

To promote the sustainable development of coastal areas, coastal zone planning under the Act consists of designating conservation and protection areas with the aim of establishing a spatial development strategy. Conservation areas afford legal protection to natural resources, whereas protection areas are aimed at preventing coastal hazards and protecting the lives and assets of residents. Coastal areas, consisting of shore land areas and offshore areas, were demarcated and declared by the Construction and Planning Agency of the Ministry of the Interior (CPAMI). Shore areas were defined as extending from the mean high tide line to the nearest provincial highway, major coastal road, or ridgeline. Relevant assessments were restricted to the areas defined in this manner, that is, conservation or protection areas cannot be demarcated beyond these coastal areas. The two areas were demarcated and announced in the "Integrated Coastal Zone Management Plan" (CPAMI, 2017) [31]. This study focuses on the management of coastal protection areas.

Coastal protection areas were designated as first- and second-grade according to the severities of aforementioned four hazards. Table 1 lists the severity of the hazards and zoning principles. Two types of inundations, storm surge and flood, are considered to be hazardous to coastal areas. The surge hazard is defined as the coastal area inundated by sea water. It should be noted that most of the sea walls in Taiwan are higher than the 100-year surge water levels. Most of the flooding by seawaters is due to run-ups and overtopping of waves, where the duration is very short and flooded areas are relatively small. The "Integrated Coastal Zone Management Plan" thus treats the potential of surge hazard as the elevation difference between the heights of shore land and storm surge, ignoring the function of seawalls. Flooding is due to overland flows. Coastal erosion takes erosion rates and potential effects into consideration. Ground subsidence has an added effect to the other three hazards. However, it is considered as to pose no immediate threats to coastal residences alone. Thus, areas having ground subsidence were demarcated into protection areas when overlaid with areas having one or more of the other three hazards.

Table 1. Grading of coastal protection areas.

Coastal Hazard	High Potential	Medium Potential	Zoning Principle	
			First-Grade	Second-Grade
Storm surge	Shore land areas with an elevation lower than the level of a 50-year storm surge height, and a flooding depth of 1 m or more	Shore land areas with an elevation lower than the level of a 50-year storm surge height, and a flooding depth of less than 1 m	(1) Coastal sectors categorized as high potential coastal hazard (2) Areas with severe ground subsidence and comprising coastal sectors with single medium potential coastal hazard	Coastal sectors with one or more medium potential compound hazards.
Coastal erosion	Shore land areas identified as susceptible to coastal erosion and its possible effects within 10 years	Shore land areas identified as susceptible to coastal erosion and its possible effects within 10–30 years		
Flood	Shore land areas within the range of a 50-year flood, and with a flooding depth of 1 m or more	Shore land areas within the range of a 50-year flood, and with a flooding depth between 0.5–1 m		
Ground subsidence	Areas that encounter severe ground subsidence, as determined by the Water Resources Agency	-		

After determination of the potential severity of coastal hazards, the "Coastal Protection Sectors" can be delimited following the zoning principles. First-grade coastal protection sectors are those with high potential coastal hazard. Sectors with severe ground subsidence and having areas with one other potential coastal hazard are also categorized as the first grade coastal protection sectors. The second-grade coastal protection sectors are those with one or more medium potential compound hazards. Furthermore, coastal sectors with similar natural hazards and protection requirements were incorporated in the same grade and their extents were zoned according to the appropriate administrative boundaries or landmarks.

Figure 1 presents the distributions of four kinds of hazard prone areas. The total coastline of the first- and second-level protection areas was 478.3 km and 181 km, respectively. Most of the first-level protection areas were distributed in the southwestern region of Taiwan, in the counties of Changhua, Yunlin, Kaohsiung, and Pingtung. All these areas suffer from severe ground subsidence induced by the excessive pumping of groundwater for aquaculture. We selected Yunlin, one of these counties, for further discussion.

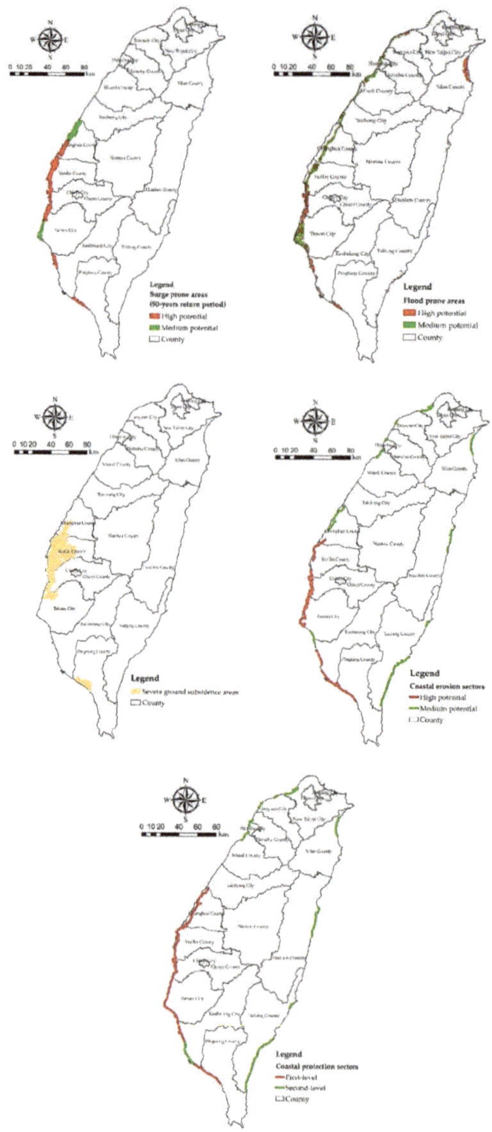

Figure 1. Distribution of hazard prone areas, and first- and second-level coastal protection areas.

2.2. Coastal Protection Strategies

In the past, coastal protection measures were formulated based on the Seawall Management Regulations, which stipulate that the measures can only be implemented within the extent of seawall areas. The zoning of seawall areas was greatly restricted because of inflexible coastal protection measures due to peripheral social and economic developments. The need to protect coastal areas from the impact of tides and waves when other auxiliary protection measures were lacking led to the construction of hard engineering structures (e.g., seawalls) with relatively strict design criteria and resulting in the construction of rather massive structures. However, extreme climate events have become more frequent and severe (Webster and Holland, 2005; Landsea et al., 2006; Elsner et al., 2008) [32–34]. For example, in recent decades, the annual minimum typhoon pressure of typhoons that have made landfall in Taiwan has gradually decreased, while the maximum wind speed has increased (Lan et al., 2013) [35]. The conventional mode of using a single protection strategy for coastal areas has thus become outdated. To depend on conventional protection modes, current protection structures must be reinforced to respond to the unpredictable trends of environmental changes. Nevertheless, the use of a single protection measure is limited with regard to economics, environmental impact, and protection effectiveness. Global practice for coastal protection remedies has gradually demarcated setback lines for coastal areas with high hazard risks. In other words, there is no longer a complete reliance on protection defense when facing unpredictable natural hazards. Instead, the conventional conception of zero disaster is discarded, hazards are allowed to occur to an acceptable extent, and attempts are made to reduce hazard-induced damage through risk management. It is suggested that both engineering and non-engineering measures should be conducted to facilitate the goals of ICZM.

3. Methodology

Coastal areas have different characteristics and degrees of exploitation; therefore, a single set of protection design criteria cannot satisfy the aim of sustainable coastal development. Environmental characteristics should be taken into consideration when formulating design criteria for the coastal defense of different regions. The process should be based on hazard risk classes to provide references for developing corresponding design criteria and formulating hazard protection and management measures.

As stated earlier, coastal protection areas in Taiwan were designated according to the criteria set out in the Coastal Zone Management Act, which did not include tsunami hazards. Wave gauging stations around Taiwan also indicate that there has been no tsunami that has caused any casualties in the last century (Central Weather Bureau, CWB; Chen and Chen, 2011; Kontar et al., 2014) [36–38]. Nevertheless, Taiwan is located in the Circum-Pacific seismic zone, and the threat of tsunamis is not negligible. Particularly, the potential threat of tsunamis originating from the Manila Trench, the East Luzon trench, and the Ryukyu trench is of concern (Lin et al., 2015; Wu and Huang, 2009; Wu et al., 2015) [39–41]. Therefore, the tsunami hazards are included in consideration.

The present study aims to establish methods for assessing the design criteria of coastal defense and land use management in various coastal areas. These methods were mainly based on a set of systematic assessment principles, from which relevant indicators were selected for further management.

Based on the risk management policy proposed by the Executive Yuan, Taiwan, the hazard risk can be defined in terms of the product of potential hazards and vulnerability. Coastal hazards were classified into the five coastal hazard types discussed above. While the hazard index can be quantified based on hazard severity, the coastal vulnerability index (CVI) must be developed relative to specific combinations of different objectives, processes, and spatial and temporal scales (Özyurt and Ergin, 2010) [10]. Furthermore, both natural and anthropogenic factors should be considered. The objective of a protection area is to minimize the impacts of coastal hazards on residents, which means that the focus is on socio-environmental concerns; in contrast, natural factors are incorporated into the hazard indices. Chien et al. (2012) suggested that vulnerability in protection areas should refer to the possibility of life-threatening events or property loss induced by potential hazard factors in a given hazard-prone area [42]. We therefore chose socio-environmental indicators for the assessment of CVI.

First, we decided on the spatial units used to estimate vulnerability in our analysis. Although adopting large-scale units may allow easy and rapid operations and high data accessibility, the resulting failure to reflect local or regional characteristics may lead to their underrepresentation during analysis of the results. To ascertain coastal characteristics accurately, we adopted townships/villages as the analytical and statistical spatial unit, and used currently accessible data in this study. According to the spatial overlay, coastal areas in Taiwan were comprised of 110 townships, which were further subdivided into 898 villages.

Second, we selected indicators for grading before conducting risk analyses and assessments. From a statistical perspective, adopting more indicators generates results that are more representative of the characteristics of analyzed targets. However, in practice, the information required for indicators frequently fails to satisfy analytical requirements for spatial units and accuracy, and relevant survey data may even be completely lacking. This study proposed the following principles for selecting indicators:

- Adopt indicators that can be obtained from accessible databases, or use simple statistical analyses that satisfy the spatial unit and accuracy requirements.
- To ensure data impartiality, data or research projects announced or published by public institutions or government authorities are to be prioritized.

Accordingly, the indicators chosen to assess vulnerability in this study were as follows: population density, annual comprehensive income, and land use.

Regarding indicator weights, expert consensus (e.g., the Analytic Hierarchy Process or the Delphic Hierarchy Process) has generally been relied upon in previous studies, albeit still modified by the personal approaches of the experts involved and the number of survey samples (Ward, 2014) [24]. For this reason, this study still focused on establishing a methodology and assessing its feasibility. Hazard and vulnerability factors were given equal weights in the calculation.

Table 2 graded the scores of each hazard factors and CVI for further risk analysis. Each factor is independent. Hazard potential was defined as the ratio of the hazard-prone area to the shore land area; the greater of these two indicators defined the score of the hazard factor.

Table 2. Classification of hazard factors and CVI.

Hazard Factor Grading			Cvi Grading		
Score	Hazard Type	Hazard Potential [1]	Population Density	Comprehensive Income	Land Use
5	5 types of hazards	66–100%	80–100%	80–100%	Residential, commercial, educational, and medical areas
4	4 types of hazards	-	60–80%	60–80%	Industry, port activity, and public infrastructure areas
3	3 types of hazards	33–66%	40–60%	40–60%	Productive area (agriculture, aquaculture, and livestock breeding)
2	2 types of hazards	-	20–40%	20–40%	Non-productive areas (mining, salt, sandstone, funerary, artificial lakes and channels)
1	Single hazard	0–33%	0–20%	0–20%	Natural areas
0	No hazard	-	-	-	-

[1] Hazard potential is defined as ratio of the hazard-prone area to shore land area.

Hazard-prone areas were demarcated following the criteria for "High potential" (Table 1). Areas prone to flooding due to storm surges and floods were estimated through numerical simulations. Coastal erosion and ground subsidence areas were demarcated based on survey data sourced from the Water Resource Agency of Taiwan. Furthermore, potential tsunami threats for coasts around Taiwan were taken from the results of the National Science and Technology Center for Disaster Reduction (NCDR, 2015) [43]. The detailed procedure can be found in the following case study.

Vulnerability was scored on a scale of 1 to 5, with 1 indicating the least vulnerability, and 5 indicating greatest vulnerability. The population density and comprehensive income of the 898 villages within coastal areas in Taiwan were divided into five classes, by ranking them in 20% increments in ascending order (Figure 2). Five classes were also used to score land use, with vulnerability referring to the impact on human life and property (Table 2). The level of vulnerability estimated in the risk matrix is the average score of the three indicators.

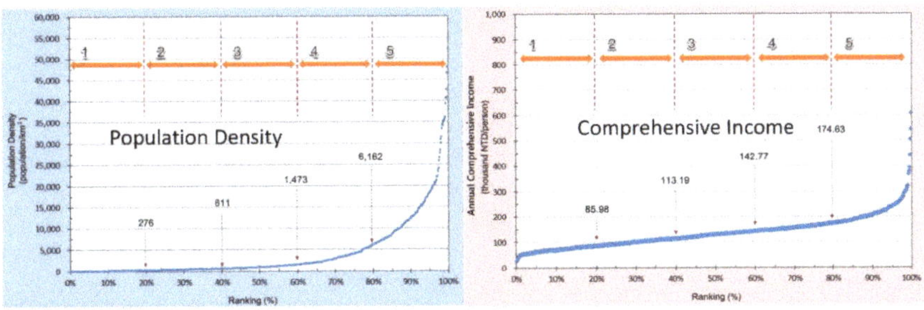

Figure 2. Population density and comprehensive income within coastal areas.

The hazard and vulnerability factors were multiplied in a 6 × 5 risk matrix, generating six risk classes ranging from A to F that denoted high, high-intermediate, moderate, low-intermediate, low, and minimal protection levels, respectively. These risk classes were subsequently used to determine the appropriate design criteria. The assessment procedure and framework for this method are presented in Figure 3.

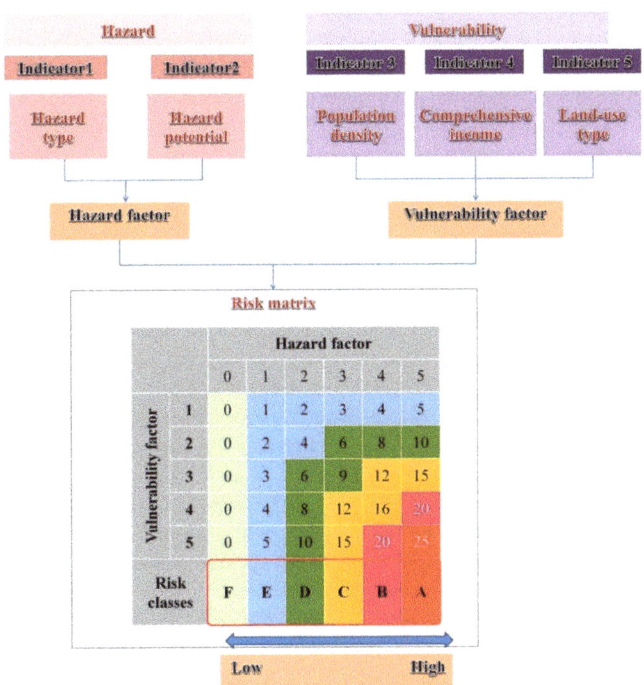

Figure 3. Procedure and framework for coastal risk assessment.

4. Results

4.1. Formulation of Design Criteria

It should be noted that most of the coastal protective facilities in Taiwan were completed within the last three decades. The design criteria were based on the marine climate, including the impact of waves and surges, for a return period of 50–100 years. However, land use in the protection area was not taken into consideration. It is not economically viable to use the same design criteria to protect shore land areas that have different land uses; for example, residential use should assume a larger protective return period than agricultural use. For sustainable use in a coastal zone, the design criteria should be formulated based on the requirements of coastal protection and hazard management. In this study, shore land areas with different natural and anthropogenic environmental characteristics were objectively investigated and provided with distinct protection strategies and suitable design criteria, upon which subsequent comprehensive protection strategies were planned and designed.

It is noted that ordinary coastal defenses are not designed to withstand tsunami impacts, but normal wave climates and the effects of typhoons. Given the high return period of major tsunami events in the area, it is arguably better to mitigate their potential effects using non-engineering measures.

The design criteria were divided into two categories. One category focused on regulating coastal defenses, and a set of coastal protection structural design criteria were formulated based on the marine climate. These design criteria specified that the protection capabilities of coastal defenses must fulfill the safety standards formulated based on the wave and water level conditions of a certain return period. The other category of design criteria highlighted the need to ensure the safety of coastal social and economic environments, and a set of design criteria for protection was formulated for hazard management.

4.2. Design Criteria for Coastal Defense and Land Use

Coastal erosion and ground subsidence are long-term hazards which pose no immediate threat to human lives. In contrast, waves, storm surge, and tsunamis are major destructive events that can constitute life-threatening hazards. Consequently, the relevant coastal defenses must be capable of controlling the tides and waves caused by extreme conditions, preventing them from severely affecting protected coastal areas, while reducing coastal hazards. This principle was a crucial factor in the safety validation conducted in the design of coastal defenses. The design criteria proposed here is mainly concerned with the safety of coastal defenses, while against waves and storm surges. However, tsunami forces were not considered when assessing the engineering measures, since these typically have a high design criterion that is probably able to withstand low-level tsunami events. To be on the safe side, possible inundation through tsunami was set as 1 m. Higher than 1 m, it is proposed that tsunami threats be treated with non-engineering measures. Regions with potential inundation through tsunami of less than 1 m are suggested to follow the measures against surges.

The design criteria for different types of land use were selected according to risk levels. It is suggested that building foundations should be elevated above the proposed design criteria to prevent inundation caused by floods or surges. Agriculture and aquafarming areas situated at elevations lower than specified in the design criteria can experience constant inundation, and should be reported. In such cases, we suggest modification of the land use of these areas.

Integrated coastal protection should be realized through a combination of engineering and non-engineering measures. The design criteria for these two categories were flexibly formulated based on actual combinations of protection measures. The proposed principles for formulating coastal protection structural design criteria and land use refer to various risk levels are presented in Table 3 and as follows:

- High risk level (level A): marine conditions (including waves and surges) in a 100-year return period were adopted as the design criteria.

- High-intermediate risk level (level B): a 50–100 year return period was adopted as the design criterion. However, to avoid any negative environmental impact caused by design criteria upgrades, the original design criterion is still considered applicable for coastal defenses meeting the criteria of a 50-year return period, if modifying the coastal defenses (structural measures) or extending the buffer zone (nonstructural measures) enables the coastal defense to reduce external impact sufficiently for the original design criterion to withstand it.
- Moderate risk level (level C): a 50-year return period was adopted as the design criterion.
- Low-intermediate risk level (level D): a 25–50 year return period was adopted as the design criterion. As above, to prevent any negative environmental impact caused by upgrading design criteria, the original design criterion was still considered applicable if other supportive measures can reduce external impact sufficiently for the original design criterion to withstand it.
- Low risk level (level E): a 25-year return period was adopted as the design criterion.
- Minimal risk level (level F): no protection facility is required.

Both coastal defenses and planning of land use can be reevaluated at a later point.

Table 3. Grading of risk levels and coastal protection design criteria.

Risk Levels		Coastal Protection Structural Design Criteria	Protection Design Criteria for Land Use Management				
		Coastal Structural Protection Facilities	Residential Areas or Crucial Social and Economic Areas	Industrial Land	Public Evacuation Facilities	Agriculture, Fishery, and Animal Husbandry	Nonproductive Land
High	Level A	100-year return period				≤25	-
High intermediate	Level B	50–100 year return period.					
Moderate	Level C	50-year return period					
Low intermediate	Level D	25–50 year return period.					
Low	Level E	25-year return period					
Minimal	Level F	-					
Considering factor		Wave and storm surge level	Storm surge level and regional flood potential				

5. Discussion

This study used Yunlin County as case study for further discussion to verify the proposed assessment principles on design criteria and risk analyses. It is hoped that the results can be used for future coastal management. The data used in this study were obtained from the databases listed below:

- Storm surges: The 50-year return period of storm surge height along the Yunlin coast is +2.64 m, based on "The Assessment on Coastal Protection of Sea Dikes" plan (Water Resources Planning Institute (WRPI), 2014) [44]. The design codes proposed in this plan were applied by the River Management Offices to assess the safety and capability of currently existing sea walls. Areas with potential inundation depths greater than 1 m were estimated based on storm surge water levels and land elevation. The criterion of having an inundation depth of 1 m or more is given in Table 1, which defines the severity of coastal hazards where inundation may cause danger to life. As the height of the sea wall throughout the areas studied is greater than the potential storm surge heights, the extent of inundation-prone areas was estimated with the assumption that there were no sea walls in the area. The Taixi and Kouhu Townships were estimated to be the areas that most prone to surge hazards (Figure 4a).
- Floods: A GIS layer of flood-prone areas with an inundated depth of over 1 m for a 50-year return period was acquired from the Water Hazard Mitigation Center (WHMC, 2014). (WHMC, 2014) [45]. Flood hazard-prone areas are shown in Figure 4b.

- Coastal erosion: The erosion coastlines of Yunlin County were estimated using data from the historical bathymetry survey. The coastline of Kouhu Township was found to be subject to coastal erosion (Figure 4c).
- Ground subsidence: The ground subsidence area was acquired from the Water Resource Agency (2014) [46]. The entire area of the Yunlin coast was found to be affected by ground subsidence (Figure 4d).
- Tsunami: Areas under potential tsunami threats were studied by the NCDR (2015) [43]. The results were derived from the numerical simulations of 600 scenarios. Eighteen possible epicenters of earthquakes distributed in the oceans around Taiwan were considered. These include the Manila trench, one of the most hazardous tsunami source regions. The maximum possible earthquake scenarios within these trenches and troughs were considered. The areas that can be potentially affected by tsunamis in Yunlin County have a similar distribution to storm surge inundation areas. Essentially, Taixi and Kouhou Townships were estimated to be most prone to tsunami hazards (Figure 4e).

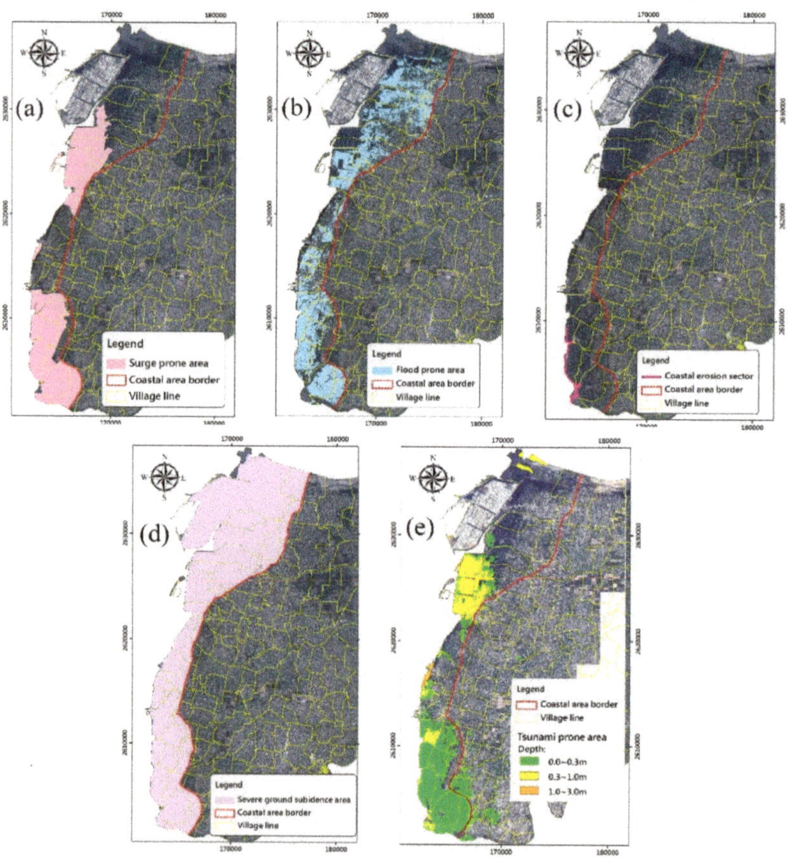

Figure 4. GIS layers of the coastal hazard-prone areas in Yunlin County, consisting of (**a**) surge, (**b**) flood, (**c**) erosion, (**d**) ground subsidence, and (**e**) tsunami prone area layers.

The data for the five hazards were all acquired from official sources, in accordance with the stated requirement of data impartiality in this study. After overlaying the resulting five coastal hazard-prone area layers, we found that Kouhu Township was subject to a combination of all five hazard types.

This township made up over 66% of the total hazard-prone area, resulting in a hazard factor score of 5. The rest of the villages had scores of 3 or above. The hazard factor score of each village is presented in Figure 5.

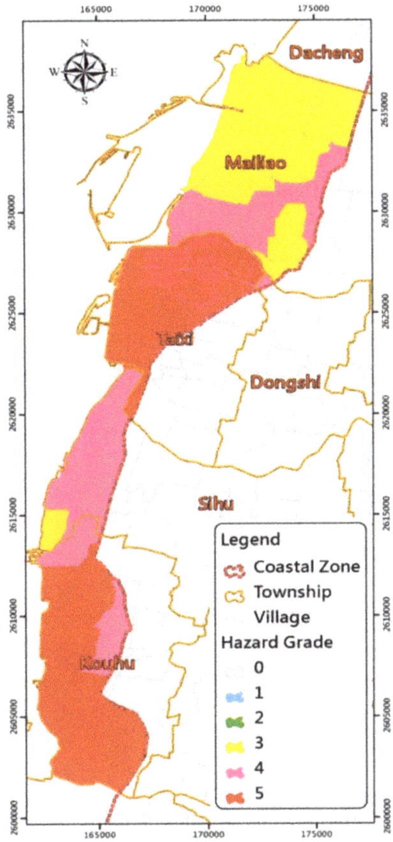

Figure 5. Hazard grading of the coastal villages in Yunlin County.

Population density, comprehensive income, and land use data were acquired from county government statistics, and were used to estimate vulnerability. Population densities and comprehensive incomes of the villages within the coastal area of Yunlin range from 1 to 4, and 1 to 3, respectively. Land use scores are 2 or 3, with most of the villages constituting productive areas, although a few non-productive areas also exist. The vulnerability grading was derived from the average score of the three indicators for each village (Figure 6).

Risk maps for each village (Figure 7) were constructed by multiplying their respective hazard (Figure 5) and vulnerability (Figure 6) scores. The results show that most villages were classified as Level D (low-intermediate risk). Level C (moderate risk) was the highest risk class reached, and among these villages, only Yongfeng (in the Taixi Township) was adjacent to the shore.

Coastal areas that are adjacent to others with similar natural and anthropogenic environmental characteristics can be incorporated into a single protection area unit, and assigned an appropriate protection level. Table 4 presents the assessment of the coastal defense of Yunlin's coastal areas. The results show an overall coastal protection level of D (low-intermediate risk). The run-up heights of waves with a 25- and 50-year return period were adopted as the coastal protection structural design criterion.

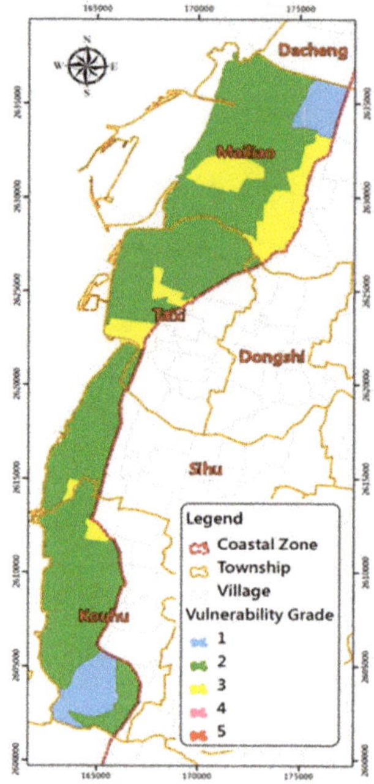

Figure 6. Vulnerability grading of the coastal villages in Yunlin County.

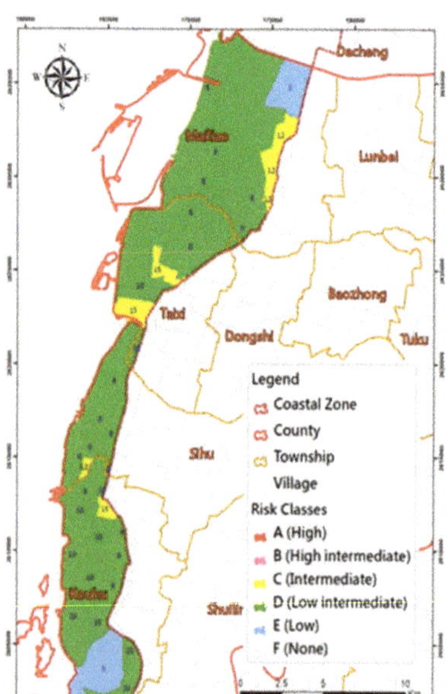

Figure 7. Risk classes of the coastal areas in Yunlin County.

Table 4. Risk level assessment and suggested design criteria in Yunlin County.

Coastal Administrative Division	Risk Assessment	Suggested Design Criteria (Return Period)	Protection Facility
Mailiao Township: Houan Village Haifeng Village	C, D	25–50	Hoanliao seawall
Taixi Township: Wengang Village Hefeng Village Wugang Village Haikou Village Xiding Village	C, D	25–50	Xinxing Tidal Land seawall Taisi Tidal Land seawall Haikou seawall Haixin seawall
Sihu Township: Lincuo Village Lunbei Village Guanggou Village Bozi Village	C, D	25–50	Lincuoliao seawall Santiaolun seawall Boziliao seawall
Kouhu Township: Xialun Village Qinghan Village Gangxi Village Taizi Village	C, D, E	25–50	Xialun seawall Yanchoucuo seawall Qinghan seawall Xingang seawall Taizi Village seawall

The designed and surveyed height of the coastal defenses is shown in Table 5. The height is referred to mean sea water level. Waves reaching a run-up height equal to the expected 25- and 50-year return period events are also listed in the two columns to the far right for comparison. These data were estimated using DHI MIKE 21 numerical models, including the effects of waves and tides. It is apparent that all the existing coastal defenses in Yunlin met the safety standards.

Table 5. Assessment of coastal defenses in Yunlin County.

Protection Facility	Constructed Year	Design Height (m)	Surveyed Height (m) (2012)	Subsidence Rate (m/year) [3]	Run-Up Height (m)	
					25-Year Return Period	50-Year Return Period
Hoanliao seawall	2000	6.50	6.16	0.03	[1]	
Taixi tide barrier	2004	5.00	4.68	0.05	2.98	3.59
Haikou seawall	1994	6.00	4.42	0.10	2.55	2.64 [2]
Xinxing tide barrier	1996	6.00	4.39	0.11	2.55	2.64 [2]
Lincuoliao sea-wall	1997	6.50	5.85	0.41	2.55	2.64 [2]
Santiaolun seawall	2010	5.50	5.41	0.04	3.25	3.70
Boziliao seawall	1998	6.50	5.89	0.05	3.51	4.00
Xialun seawall	2000	6.50	5.78	0.07	3.28	3.75
Yanchoucuo seawall	1996	6.50	5.58	0.07	2.55	2.64 [2]
Qinghan seawall	1993	5.00	4.06	0.05	2.55	2.64 [2]
Xingang seawall	1998	6.50	6.00	0.04	2.55	2.64 [2]
Taizi Village seawall	1999	6.50	5.99	0.05	3.58	4.00

[1] Due to land reclamation, this seawall was located inland and was not subjected to any marine force. [2] The seawall did not face the sea directly; the effect of waves can be neglected and assessed by the design surge height. [3] It was estimated from the elevation change of the control points that set on the crest of sea wall. The subsidence rates are the average values from 2007 to 2012 except Santiaolun seawall from 2010 to 2012 [44].

However, it should be noted that ground subsidence is still active in the coastal areas of Yunlin, with subsidence rates around coastal defenses in the order of 0.03–0.11 m/year. In addition, global sea level rise rates were estimated at 0.0028–0.0036 m/year between 1993 and 2010 (Church et al., 2013) [47]. Lan et al. (2013) simulated sea-level-rise scenarios in the range of 0.10 to 0.15 m during the time period ranging from 2020 to 2039 for the case of the southwest coast of Taiwan [35]. Thus, the level of protection offered by the breakwater/revetment is likely to be downgraded in the near future, given the influence that sea level rise and ground subsidence have on the long-term stability of structure. The data also shows that the coastal facilities located in Taixi Township are experiencing a higher subsidence rate than the others, and are thus categorized as having the highest risk level. We suggest that periodical monitoring projects be conducted for further assessment. According to our risk assessment, once the protection level of the existing breakwater/revetment goes below the standard 25-year return period, modification of the defense measures should be conducted immediately.

5.1. Non-Engineering Measures

Considering Yunlin's relatively high hazard score, both engineering and non-engineering measures should be conducted simultaneously, as part of ICZM. We suggest the following non-engineering measures.

- Delimiting the buffer zone

The buffer zone in Yunlin took into consideration the dominant coastal hazards of storm surges, ground subsidence and tsunamis. Areas with intermittent or even frequent flooding were given special attention. The townships of Taixi and Kouhu have the highest hazard scores and delimiting the buffer zones is suggested as their first priority. However, regulations in the buffer zone may hinder coastal development; two levels of buffer zones were therefore suggested. The first one was delimited based on the more frequent hazards, such as storm surges and ground subsidence, with strict limitations. The second one considers the very rare hazard, such as tsunami, and the regulations there are more flexible. The first level is marked 50 m further landward of the 25-year return period storm surge water level. No further exploitation within this buffer zone should be permitted. The latter one is delimited based on the tsunami potential risk area which was shown in Figure 4e. A restrictions and permissions system should be implemented where the tsunami potential inundation depths are more than 1 m. Any activity that may lead to explosion or toxic pollution should be forbidden. The development of residential communities in highly populated areas is to be regulated. Low density developments are to be encouraged.

- Construction of the hazard maps

A hazard database should be established and maintained, and a risk management and economic analysis should be carried out to develop a coastal protection policy and regime.

- Land-use modification

The overuse of groundwater is the main factor causing ground subsidence in this region. A large amount of groundwater is being extracted and used for aquaculture; therefore, the use of land for new aquaculture applications should be restricted. Local government agencies should be given a portion of grant funds intended to reduce the development of fish farms. For maintaining groundwater balance, river water and rainwater can be used, and wastewater from industrial plants may be treated and then discharged into the rivers or artificial lake to increase the infiltration.

It is suggested that land-use changes into detention basins or is used for building solar power systems. These regions have suffered constant inundation and consequently yielded below-average production benefits. The proposed change of land use is a win-win solution to reduce the impact of inundation and provide new resources.

- Building renovations

Due to the threat of flooding, buildings should be modified to prepare for flood hazards. Some suggested facilities or modifications include waterproof gates, foundations on stilts, and temporary polder dykes. The entire coastal area in Yunlin is affected by ground subsidence, which resulted in low-lying lands. Buildings with low foundations experienced frequent inundation on the lower floors, including the basement and first floor. Hence, limiting the use of low floors to decrease the loss of assets is another option. Following such limitations, reductions in housing and land taxes can be used to encourage residency. The Taixi and Kouhu townships are especially encouraged to deploy these measures. The foundation elevation of any new buildings should be higher than the 25-year design criterion. It can also reduce potential losses induced by tsunamis.

Any changes in land use patterns that fall under scores 4 and 5 in Table 2 should be made tsunami resistant. This would typically require buildings to be made using concrete, and preferably allow vertical evacuation under the worst case scenario high-return tsunami events. Evacuation shelters should be designated in the regions where the tsunami potential inundation depths exceed more than 1 m. The public should be made aware of emergency evacuation routes through the use of mass media and public address systems.

- Improvement of forecasting and warning

The main purposes of the defense system are the improvement of forecasting and warning for storm surge, flood, and tsunami. Disaster prevention training and education to make people aware of coastal hazards and what to do in case of hazards should be conducted on community basis. These exercises and education on precautions and preparedness against hazards will reduce potential losses.

5.2. Public Participation

Responsible authorities are now preparing drafts of coastal protection plans with respect to different regions in Taiwan, and should be announced in 2020 following the Act. During this period, public participations in establishing coastal defense and management policies are to be encouraged. Three-stage of public participations, including opinion polls, public hearings, and consensus conferences, should be carried out before the draft of coastal protection area planning are to be completed. Risk maps generated by this study can be a useful tool to show the public that the threats of flooding are both realistic and imminent, demonstrating possible coastal risks and their effects. At the same time, non-engineering measures proposed in our study can be used to offer alternative measures for mitigation, improving acceptance and understanding by the public. Furthermore, when necessary, factors and weights proposed in this study can be easily adjusted in future studies in accordance with public interests.

6. Conclusions

Decision of adaptation or mitigation strategies for coastal defenses is made on a national scale in Taiwan. For this, policy makers should be provided with as much information as possible. On the other hand, too much information at one time often causes confusion, leaving it hard to make correct decisions. Furthermore, the dynamic coastal processes are complicated, often with severe socio-economic consequences, making the correct decision is therefore of vital importance. As the authorities are formulating costal protection plans in Taiwan at the moment, coastal risks each region are facing should be clearly identified in the first place. Coastal protection sectors have been announced by CPAMI based on the severity level of coastal hazards, but the risks to coastal residents have yet to be assessed [31]. In this paper, we propose a scheme for a quick but informative risk assessment. It is based on the method of the United Nations Disaster Relief Organization [3], which uses both hazard and vulnerability indicators. It can be readily to be applied since the graded coastal hazard severities are used. In the process of formulating a coastal protection plan, if necessary, the vulnerability factors or weights can be altered according to the public interests. In this way, coastal management criteria will be formulated according to coastal risks. Both residents and authorities can be made aware of the risks of the area, and the officials can draw a prioritized list based on the criteria and make defense plans accordingly. Rational use of resources for adaptation or mitigation measures can be also be achieved.

Yunlin County, with its existing coastal defenses and the present status of land use, was used for the case study to test the proposed assessment principles of coastal management criteria. It was found that the risk class of the coastal areas in Yunlin County can be categorized as low-intermediate. The coastal hazards are comprised mainly of storm surge and ground subsidence. We then proposed implementation of both engineering and non-engineering measures to promote sustainable management of coastal zones, and to reduce the impact of coastal hazards.

Author Contributions: W.-P.H. designed the study and developed the methodology. J.-C.H. and C.-S.C. collected the data and analyzed the results. C.-J.Y. primarily made figures and tables of the manuscript.

Acknowledgments: This paper is partly aided by a project of the Ministry of Science and Technology, Taiwan, Project No. MOST 106-2621-M-019-003-MY2. Their support is deeply appreciated.

Conflicts of Interest: The authors declare no conflict of interest.

References

1. Huang, W.P.; Chien, L.K.; Lee, L.H. Risk Maps and Coastal Defense Criteria in Taiwan. *J. Mar. Sci. Technol.* **2016**, *24*, 1105–1114.
2. Chien, L.K.; Hsu, C.H.; Huang, W.P.; Wang, I.C. The Study of Designation of Coastal Protection Zone and Management in Taiwan. In Proceedings of the Twenty-Sixth International Ocean and Polar Engineering Conference, Rhodes, Greece, 26 June–2 July 2016; pp. 1287–1293.
3. United Nations Disaster Relief Organization (UNDRO). *Natural Disasters and Vulnerability Analysis*; Report of Experts Group Meeting; UNDRO: Geneva, Switzerland, 1980.
4. Chien, L.K.; Chiu, S.Y.; Tseng, W.C.; Hsu, C.H. The Study of Coastal Risk Assessment of Planning and Management for Coastal Area in Taiwan. In Proceedings of the Twenty-Second International Offshore and Polar Engineering Conference, Rhodes, Greece, 17–22 June 2012; pp. 17–22.
5. McCarthy, M.A.; Possingham, H.P.; Day, J.R.; Tyre, A.J. Testing the accuracy of population viability analysis. *Conserv. Biol.* **2001**, *15*, 1030–1038. [CrossRef]
6. Van der Veen, A.; Logtmeijer, C. Economic hotspots: Visualizing vulnerability to flooding. *Natl. Hazards* **2005**, *36*, 65–80. [CrossRef]
7. Parkinson, R.W.; McCue, T. Assessing municipal vulnerability to predicted sea level rise: City of Satellite Beach, Florida. *Clim. Chang.* **2011**, *107*, 203–223. [CrossRef]
8. Cutter, S.L. Vulnerability to environmental hazards, Progress in Human. *Geography* **1996**, *20*, 529–539.
9. Khouakhi, A.; Snoussi, M.; Niazi, S.; Raji, O. Vulnerability assessment of Al Hoceima bay (Moroccan Mediterranean coast): A coastal management tool to reduce potential impacts of sea-level rise and storm surges. *J. Coast. Res.* **2013**, *65*, 968–973. [CrossRef]
10. Özyurt, G.; Ergin, A. Improving coastal vulnerability assessments to sea-level rise: A new indicator based methodology for decision makers. *J. Coast. Res.* **2010**, *26*, 265–273. [CrossRef]
11. Fitton, J.M.; Hansom, J.D.; Rennie, A.M. A national coastal erosion susceptibility model for Scotland. *Ocean Coast. Manag.* **2016**, *132*, 80–89. [CrossRef]
12. Merlotto, A.; Bértola, G.R.; Poccolo, M.C. Hazard, vulnerability and coastal erosion risk assessment in Necochea Municipality, Buenos Aires Province, Argentina. *J. Coast. Conserv.* **2016**, *20*, 351–362. [CrossRef]
13. Tarragoni, C.; Bellotti, P.; Davoli, L.; Raffi, R.; Palmieri, E.L. Assessment of coastal vulnerability to erosion: The case of Tiber River Delta (Tyrrhenian Sea, Central Italy), Italian. *J. Eng. Geol. Environ.* **2014**, *14*, 5–16.
14. Schernewski, G.; Schönwald, S.; Kataržytė, M. Application and evaluation of an indicator set to measure and promote sustainable development in coastal areas. *Ocean Coast. Manag.* **2014**, *101*, 2–13. [CrossRef]
15. Zanetti, V.B.; De Sousa Junior, W.C.; De Freitas, D.M. A Climate Change Vulnerability Index and Case Study in a Brazilian Coastal City. *Sustainability* **2016**, *8*, 811. [CrossRef]
16. Hammar-Klose, E.S.; Thieler, E.R. *Coastal Vulnerability to Sea Level Rise: A Preliminary Database for the U.S. Atlantic, Pacific, and Gulfof Mexico Coasts*; U.S. Geological Survey Digital Data Series DDS-68 1CD-ROM; U.S. Geological Survey: Reston, VA, USA, 2001.
17. Gornitz, V.M.; Daniels, R.C.; White, T.W.; Birdwell, K.R. The development of a coastal risk assessment database: Vulnerability to sea-level rise in the U.S. southeast. *J. Coast. Res.* **1994**, 327–338.
18. Shaw, J.; Taylor, R.B.; Forbes, D.L.; Ruz, M.H.; Solomon, S. *Sensitivity of the Canadian Coast to Sea-Level Rise*; Geological Survey of Canada Bulletin; Geological Survey of Canada Offices: Ottawa, ON, Canada, 1998.
19. Cutter, S.L.; Boruff, B.J.; Shirley, W.L. Social Vulnerability to Environmental Hazards. *Soc. Sci.* **2003**, *84*, 242–261. [CrossRef]
20. Wang, G.G.; Liu, Y.J.; Wang, H.B.; Wang, X.Y. A comprehensive risk analysis of coastal zones in China Estuarine. *Coast. Shelf Sci.* **2014**, *140*, 22–31. [CrossRef]
21. United Nations Office for Disaster Risk Reduction (UNISDR). *Living with Risk: A Global Review of Disaster Reduction Initiatives*; United Nations Publication: Geneva, Switzerland, 2004.

22. Chien, L.K.; Lei, R.J.; Chiu, S.Y.; Hung, L.J.; Lee, B.C. Coastal Vulnerability and Risk Assessment in Complex Disasters and Adaptation Strategies Exploration. In Proceedings of the Twenty-Third International Offshore and Polar Engineering Conference, Anchorage, AK, USA, 30 June–5 July 2013; pp. 617–624.
23. Carrasco, A.R.; Ferreira, Ó.; Matias, A.; Freire, P. Flood hazard assessment and management of fetch-limited coastal environments. *Ocean Coast. Manag.* **2012**, *65*, 15–25. [CrossRef]
24. Ward, T.J. The condition of Australia's marine environment is good but in declined—An integrated evidence based national assessment by expert elicitation. *Ocean Coast. Manag.* **2014**, *100*, 86–100. [CrossRef]
25. Van Vuren, S.; Kok, M.; Jorissen, R.E. Coastal Defence and Societal Activities in the Coastal Zone: Compatible or Conflicting Interests? *J. Coast. Res.* **2004**, *20*, 550–561. [CrossRef]
26. Penning-Rowsell, E.C.; De Vries, W.S.; Parker, D.J.; Zanuttigh, B.; Simmonds, D.; Trifonova, E.; Hissel, F.; Monbaliu, J.; Lendzion, J.; Ohle, N.; et al. Innovation in coastal risk management: An exploratory analysis of risk governance issues at eight THESEUS study sites. *Coast. Eng.* **2014**, *87*, 210–217. [CrossRef]
27. Dinh, Q.; Balica, S.; Popescu, I.; Jonoski, A. Climate change impact on flood hazard, vulnerability and risk of the Long Xuyen Quadrangle in the Mekong Delta. *Int. J. River Basin Manag.* **2012**, *10*, 103–120. [CrossRef]
28. ESCAP/UNISDR. Reducing Vulnerability and Exposure to Disasters. The Asia-Pacific Disaster Report. 2012. Available online: http://www.unisdr.org/ (accessed on 1 June 2018).
29. Luo, S.; Feng, C.; Liu, H.; Lei, G.; Qi, H.; Su, X. Adaptive measures adopted for risk reduction of coastal erosion in the People's Republic of China. *Ocean Coast. Manag.* **2015**, *103*, 134–145. [CrossRef]
30. Salik, K.M.; Jahangir, S.; Wu, Z.; Su, H. Climate change vulnerability and adaptation options for the coastal communities of Pakistan. *Ocean Coast. Manag.* **2015**, *112*, 61–73. [CrossRef]
31. Construction and Planning Agency of the Ministry of the Interior (CPAMI). Integrated Coastal Zone Management Plan. 2017. Available online: https://www.cpami.gov.tw/%E6%9C%80%E6%96%B0%E6%B6%88%E6%81%AF/%E6%A5%AD%E5%8B%99%E6%96%B0%E8%A8%8A/18529-%E6%95%B4%E9%AB%94%E6%B5%B7%E5%B2%B8%E7%AE%A1%E7%90%86%E8%A8%88%E7%95%AB.html (accessed on 1 June 2018).
32. Webster, P.J.; Holland, G.J.; Curry, J.A.; Chang, H.R. Changes in tropical cyclone number, duration, and Intensity in a warming environment. *Science* **2005**, *309*, 1844–1846. [CrossRef] [PubMed]
33. Landsea, C.W.; Harper, B.A.; Hoarau, K.; Knaff, J.A. Can We Detect Trends in Extreme Tropical Cyclones? *Science* **2006**, *313*, 452–454. [CrossRef] [PubMed]
34. Elsner, J.B.; Kossin, J.P.; Jagger, T.H. The increasing intensity of the strongest tropical cyclones. *Nature* **2008**, *455*, 92–95. [CrossRef] [PubMed]
35. Lan, Y.J.; Hsu, T.W.; Lin, Y.C.; Huang, C.J. An Adaptation Due to Climate Change in Southwest Coast of Taiwan. *Coast. Manag.* **2013**, *41*, 172–189. [CrossRef]
36. Central Weather Bureau (CWB). Available online: http://scweb.cwb.gov.tw/Twenty.aspx?ItemId=4&loc=tw (accessed on 1 June 2018).
37. Chen, G.Y.; Chen, Y.Y. *Tsunami Affected Area and Flooding Possibility on the Coast of Taiwan*; Institute of Transportation Ministry of Transportation and Communications: Taichung, Taiwan, 2011.
38. Kontar, Y.A.; Korotenko, K.A.; Santiago-Fandiño, V. Chapter 2 in Tsunami events and lessons learned: Environmental and societal significance. In *Assessment and Modeling of Dispersal Contamination Incoming with Submarine Groundwater Discharge (SGD) in Tsunami Affected Coastal Areas*; Springer: Dordrecht, The Netherlands, 2014; pp. 55–88.
39. Lin, S.C.; Wu, T.R.; Yen, E.; Chen, H.Y.; Hsu, J.; Tsai, Y.L.; Lee, C.J.; Liu, P.L.F. Development of a tsunami early warning system for the South China Sea. *Ocean Eng.* **2015**, *100*, 1–18. [CrossRef]
40. Wu, T.R.; Huang, H.C. Modeling Tsunami Hazards from Manila Trench to Taiwan. *J. Asia Earth Sci.* **2009**, *36*, 21–28. [CrossRef]
41. Wu, B.R.; Ke, S.S.; Wu, T.R.; Liu, K.S. Tsunami Inundation Map and its Application on Evacuation Planning in Taiwan. *J. Earthq. Tsunami* **2015**, *9*, 1540004-1–1540004-8. [CrossRef]
42. Chien, L.K.; Tseng, W.C.; Chiu, S.Y.; Hsu, C.H.; Huang, W.P. Coastal vulnerability applied on prevention and management under climate change impact. *Adv. Mater. Res.* **2012**, *356–360*, 791–800. [CrossRef]
43. National Science and Technology Center for Disaster Reduction (NCDR). 2015. Available online: http://satis.ncdr.nat.gov.tw/Dmap/102Catalog-Map.aspx (accessed on 1 June 2018).
44. Water Resources Planning Institute. *The Assessment on Coastal Protection of Sea Dikes*; Water Resources Agency, Ministry of Economic Affairs: Taichung, Taiwan, 2014. (In Chinese)

45. Water Hazard Mitigation Center. Flood Disaster Prevention Plan. 2014. Available online: http://www.dprc.ncku.edu.tw/download/index2.html (accessed on 1 June 2018).
46. Water Resource Agency. Severe Land Subsidence Area in Taiwan. 2014. Available online: http://gic.wra.gov.tw/gic/HomePage/Index.aspx (accessed on 1 June 2018).
47. Church, J.A.; Clark, P.U.; Cazenave, A.; Gregory, J.M.; Jevrejeva, S.; Levermann, A.; Merrifield, M.A.; Milne, G.A.; Nerem, R.S.; Nunn, P.D.; et al. Sea Level Change. In *Climate Change 2013: The Physical Science Basis. Contribution of Working Group I to the Fifth Assessment Report of the Intergovernmental Panel on Climate Change*; Cambridge University Press: Cambridge, UK; New York, NY, USA, 2013.

water

MDPI

Article

A Regional Application of Bayesian Modeling for Coastal Erosion and Sand Nourishment Management

Alessio Giardino [1,*], Eleni Diamantidou [2], Stuart Pearson [1,3], Giorgio Santinelli [1] and Kees den Heijer [1,3]

1. Deltares, Unit Marine and Coastal Systems, Boussinesweg 1, 2629 HV Delft, The Netherlands; S.G.Pearson@tudelft.nl (S.P.); Giorgio.Santinelli@deltares.nl (G.S.); c.denheijer@tudelft.nl (K.d.H.)
2. Van Oort, Schaardijk 211, 3063 NH Rotterdam, The Netherlands; diamantidou_el@hotmail.com
3. Civil Engineering and Geoscience, Delft University of Technology, 2628 CN Delft, The Netherlands

* Correspondence: Alessio.giardino@deltares.nl; Tel.: +31-(0)-6-1582-1874

Received: 26 September 2018; Accepted: 22 December 2018; Published: 1 January 2019

Abstract: This paper presents an application of the Bayesian belief network for coastal erosion management at the regional scale. A "Bayesian ERosion Management Network" (BERM-N) is developed and trained based on yearly cross-shore profile data available along the Holland coast. Profiles collected for over 50 years and at 604 locations were combined with information on different sand nourishment types (i.e., beach, dune, and shoreface) and volumes implemented during the analyzed time period. The network was used to assess the effectiveness of nourishments in mitigating coastal erosion. The effectiveness of nourishments was verified using two coastal state indicators, namely the momentary coastline position and the dune foot position. The network shows how the current nourishment policy is effective in mitigating the past erosive trends. While the effect of beach nourishment was immediately visible after implementation, the effect of shoreface nourishment reached its maximum only 5–10 years after implementation of the nourishments. The network can also be used as a predictive tool to estimate the required nourishment volume in order to achieve a predefined coastal erosion management objective. The network is interactive and flexible and can be trained with any data type derived from measurements as well as numerical models.

Keywords: BERM-N; coastal erosion; sea level rise; sand nourishments; Bayesian belief network; JarKus data; coastal state indicators; dune foot; momentary coastline; Holland coast

---

1. Introduction

Coastline retreat is a worldwide phenomenon caused by an imbalance between sediment supply and demand. The authors of [1] showed that 24% of the world's sandy beaches are eroding, with erosion rates exceeding 0.5 m/year, while 28% are accreting and 48% are stable. Causes of erosion may be either anthropogenic (e.g., construction of river dams, ports or coastal protection works, sediment mining) or natural (e.g., natural gradients in alongshore sediment transport, storms, presence of submarine canyons near the coastline) [2]. Additionally, coastal retreat can be exacerbated by the effects of sea level rise and soil subsidence. The Holland coast is a typical example of an erosive coastline, along which coastal erosion is managed by applying sand nourishments.

The management of coastal erosion relies on the continuous development of new and flexible solutions and tools, which can account for changes in natural hazard conditions and anthropogenic interventions and which are based on long-term observations [3,4]. Coastal erosion is, however, a process characterized by high temporal and spatial variability. Several examples are available of small-scale data or numerical modeling studies with long-term datasets (e.g., [5–7]). However, the use of data analysis or numerical modeling at the local scale often provides information which is site-dependent and only reliable for a short time period for which they are validated. This may not be sufficient for decision makers and coastal managers to assess the effectiveness of coastal erosion strategies at larger spatial and temporal scales.

The use of a Bayesian belief network (BBN) can provide a very powerful tool to bridge the existing gap between the needs of coastal managers and the currently available data and numerical models [8]. BBN is a method of reasoning using probabilistic relationships between causes and effects. Bayesian statistics have been applied to very diverse fields. In the field of coastal engineering, they have been used for predicting coastal dune and beach erosion and overwash [9–13], assessing wave height evolution in the surf zone [14], and estimating offshore wave heights and depth given limited offshore information [15]. Recent studied have used BBN to evaluate the coastal vulnerability due to sea level rise [16], the relationships between observed damages and multiple hazard indicators due to hurricanes [17], and the estimation of flooding hazards at coral reef islands [18]. They have also been used to assess the applicability of parametric morphodynamic classification of beach states [19]. BBN can also be an effective tool for data mining where large datasets are available [20,21].

The studies in which BBN was applied to evaluate the effectiveness of disaster risk reduction (DRR) measures are still rather limited in number and are restricted to local study cases. The authors of [22] applied BBN to assess the effectiveness of coastal protection measures (i.e., a seawall) and vulnerability reduction strategies at a small town in North Norfolk (UK). The effectiveness of exposure reduction strategies (i.e., partial house removal) and beach replenishment, at one coastal sector in South Portugal was also assessed using BBN [23]. The authors of [24] applied BBN to assess the effectiveness of measures against sea-level rise (i.e., salt marsh restoration and beach nourishments) in a lagoon in northeast Italy.

This paper describes the implementation and application of a "Bayesian ERosion Management Network" (BERM-N) to be used for coastal erosion management at a larger (regional) scale. The network is based on data collected over more than 50 years along the entire Holland coast, one of the most data-rich environments in the world. The large data availability allows to capture the spatial and temporal variability related to coastal erosion processes and human interventions. BERM-N is used in this manuscript to assess the effectiveness of different adaptation solutions in the form of beach and shoreface nourishments, and also as a predictive tool to achieve a predefined coastal erosion management objective.

2. Bayesian Modeling for Coastal Erosion Management

A Bayesian network is a method of reasoning using probabilities, where the nodes of the network represent variables and the arrows indicate the cause–effect relationships between the nodes. The advantage of using this approach is that by combining the information from multiple variables, it makes it possible to make robust data-driven or model-driven forecasts.

At the heart of Bayesian networks lies the Bayes rule, generally expressed as:

$$p(F_i|O_j) = p(O_j|F_i)p(F_i)/p(O_j), \quad (1)$$

where the left-hand term is the updated conditional probability (or 'posterior probability') of a forecast F_i, given a particular set of observations, O_j (Pearl, 1988). The first term on the right-hand side is the likelihood of observations O_j given that the forecast F_i is true. The second term on the right is the prior probability distribution of F_i (e.g., the probability of a given forecast based on the entire training dataset, in the absence of any additional observations). The denominator on the right side is the prior probability distribution of O_j.

In this specific case, the posterior probability is described by the distribution of a chosen coastal state indicator (e.g., changes in shoreline or dune foot position), in response to a nourishment strategy (e.g., nourishment type or nourishment volume), representing the observations. Constructing a network first involves defining the variables which best describe the process to be studied. In this study, the main variables are selected through the identification of measurable indicators describing the coastal erosion process and the relation between adaptation measures and coastal response. The selected variables are represented by nodes in the BBN. The states of these nodes must be mutually exclusive and collectively exhaustive. A BBN is a directed acyclic graph, where the direct cause–effect relationship between two variables in a BBN is illustrated by an arrow. The feeding of information into the Bayesian network in order to construct the conditional probability tables is called "training".

An advantage of BBN is that it is interactive and fast to operate. Once the network is trained, it can be used to carry out quick simulations based on the data included in the network. Moreover, being a probabilistic method, it can account for uncertainties by specifying probability distributions for input variables and generating such distributions as output.

3. Study Area

The Holland coast is a sandy, microtidal, wave-dominated coast located in the central part of the Netherlands (Figure 1). The region is divided in three subregions: Noord-Holland, Rijnland, and Delfland. The coastline is bounded in the north by a tidal inlet named the Marsdiep, connecting the Wadden Sea to the North Sea, and in the south by the long jetty near Hook of Holland, which allows ships to access the Port of Rotterdam. The coastal defense mainly consists of a sandy dune system 117 km long.

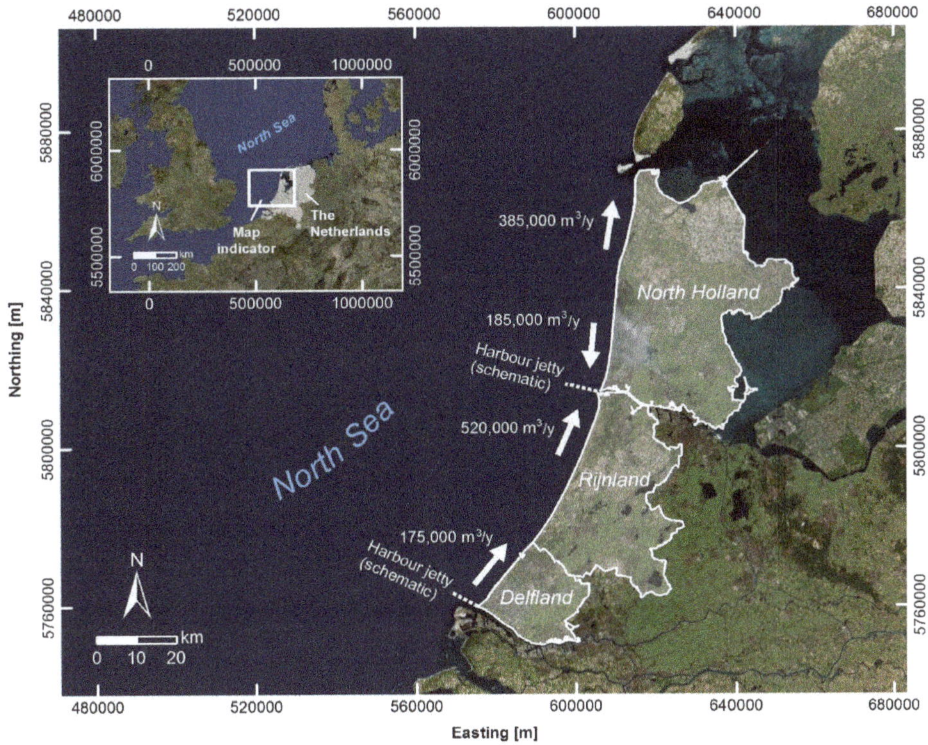

Figure 1. Map of Holland, the Netherlands, including the three subregions considered in this study: Delfland, Rijnland, and North Holland. Net yearly alongshore sediment transport rates are also shown [25].

A detailed sediment budget analysis for the entire country has been presented by several authors (see for example: [25,26]). Net alongshore sediment transport is mostly northward directed, with yearly rates ranging approximately between 200,000 and 500,000 m^3/year (Figure 1). In general, the natural supply of sediment to the coast is very limited and, consequently, the coastline is retreating [4,27].

Protection against flooding is traditionally the primary objective of coastal policy in the Netherlands. However, since 1990, additional objectives have been included: the sustainable preservation of safety against flooding and of values and functions in the dune area [28]. To fulfill these objectives, the yearly volume of sand for nourishments along the entire Dutch coast was first increased from about 3 million m^3 to 6 million m^3 of sand in 1990, and then to 12 million m^3 in 2001 [29]. Even higher volumes might be necessary in the future to cope with the more severe predicted sea level rise scenarios. The total nourishment volumes implemented along the Holland Coast between 1965–1990, 1991–2000, and 2001–2016, and divided for different nourishment types (beach nourishments, shoreface nourishments, dune nourishments, and others), are shown in Figure 2. Beach and dune nourishments are generally implemented directly on the beach or dunes. Shoreface nourishments are implemented in proximity of the breaker bars (≈3 to 5 m water depth). As shown in Figure 2, the total volume of shoreface nourishments has been increasing since 1990, due to their lower cost (relative to beach nourishments) and lower interference with the coastal environment.

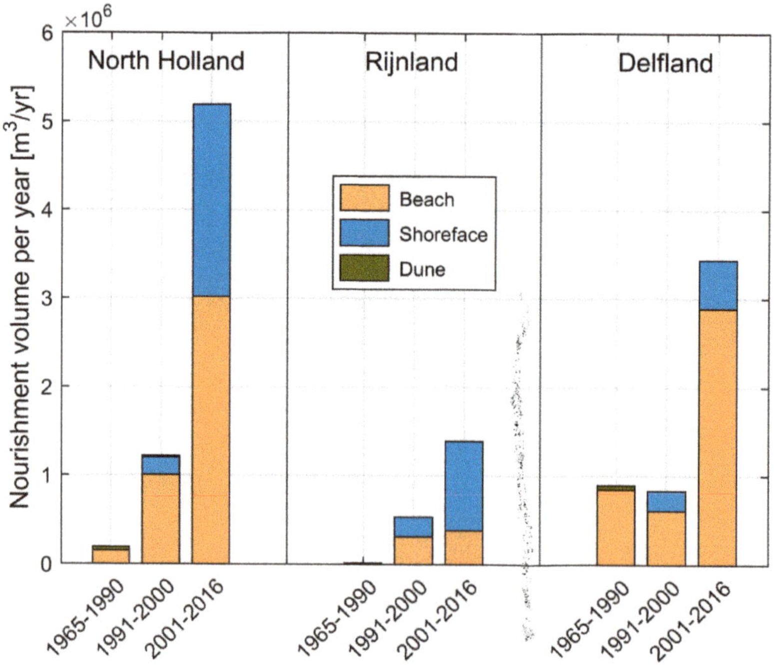

Figure 2. Nourishment volumes (millions m^3/year) at the Holland coast for the three periods: 1965–1990, 1991–2000, and 2001–2016. Note that the North Holland coast is ≈55 km long, Rijnland ≈41 km, and Delfland ≈21 km.

As a result of these large nourishment works, the generally erosive trends which characterized the shoreline before 1990 have changed into generally accretive trends. This was confirmed for example by analyzing volume changes within different water depths (−8 ÷ −3 m NAP and −3 ÷ +3 m NAP) for different time periods [30].

4. Material and Methods

4.1. Data Availability

Two main data types were used to build BERM-N: (a) Information on nourishment types and volumes and (b) changes in coastal state indicators to quantify the coastal erosion processes and the effects of different nourishment schemes.

The nourishment database supplied by Rijkswaterstaat (Ministry of Transport, Public Works, and Water Management) was used, which includes information on nourishment type, year and month when construction started/ended, location, and volume. For simplicity, the nourishment volumes were converted to m^3 of sand nourishment per linear meter of coastal length where the nourishment was built.

The coastal state indicators were derived based on the JarKus profiles. The JarKus ("Jaarlijkse Kustmeting", Annual Coastal Measurement) program was started in 1963 by Rijkswaterstaat. Coastal elevation profiles are measured yearly, during spring, with a longshore spacing of 250 m. Measurements above land were carried out until the 1990s using a differential GPS (dGPS), while measurements are currently carried out using LIDAR. Measurements below water are carried out using multibeam surveying techniques.

Thus, in total, coastal indicators were derived from 31,408 cross-shore profile measurements (604 transects $\times$ 52 years). In the cross-shore direction, measurements begin at approximately the first dune row and continue until the -8 m $\approx$ -12 m contour with respect to MSL (mean sea level), depending on the transect and the year. The cross-shore resolution of the measurements increases from 20 m offshore up to 5 m at the coast. An example of a series of JarKus profile measurements at a single location is given in Figure 3.

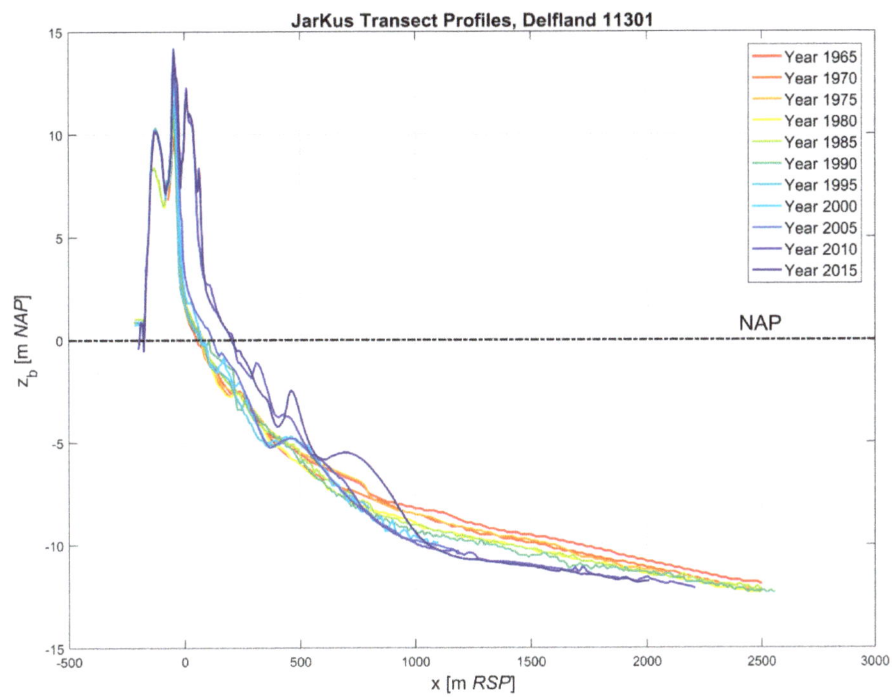

Figure 3. Example of the morphological development of a single cross-shore JarKus profile (transect 11,301) located at Delfland, between 1965 and 2015.

The authors of [31] defined suitable coastal indicators to assess the morphological development of the Holland coastline due to natural and anthropogenic pressure factors. In particular, the following indicators were used for setting up BERM-N:

- Changes in MCL (momentary coastline) position, defining the position of the coastline as a function of the volumes of sand in the near shore zone, approximately between the dune foot (+3 m NAP, where NAP $\approx$ mean sea level) and -5 m NAP (Figure 4) [32]. Positions are given with respect to predefined reference points at each transect (i.e., the RSP points "RijkStrandPalen" = "Beach Poles").
- Changes in DF (dune foot) position, defining the position of the dune foot, and estimated as the most seaward intersection of the +3 m NAP line and the cross-shore profile.

4.2. Bayesian Network

To construct the Bayesian belief network, the Netica software package [33] was chosen. Netica uses the probabilistic inference algorithms of Reference [34], and it is widely used for coastal applications (e.g., [12,14,18,20]). The standard network construction and updating procedures outlined in [33] were followed to create BERM-N.

BERM-N, after construction and training in Netica, is shown in Figure 5. In the network, nodes have been grouped in three categories, namely:

- Time interval and spatial characterization of the study area (in yellow);
- nourishment type and volume (in purple);
- effects on the morphological indicators (in green).

These are discussed in greater detail below.

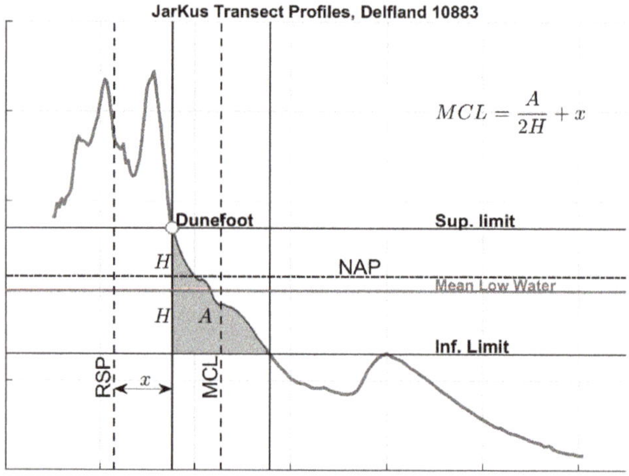

Figure 4. Computation of the momentary coastline (MCL) volume for a given JarKus transect. *A* is the area used to compute the Momentary Coastline position. *A* is delimited by an upper boundary, corresponding to the dune foot position, and a lower boundary, at a distance equal to 2 x *H* from the dune foot position. *H* is defined as the distance between the dune foot position and the mean low water line. RSP is the reference point from which distances are computed ("rijksstrandpalen"). Therefore, the MCL position can be estimated as MCL = (A/2H) + x, with x being the distance between the RSP line and the dune foot position.

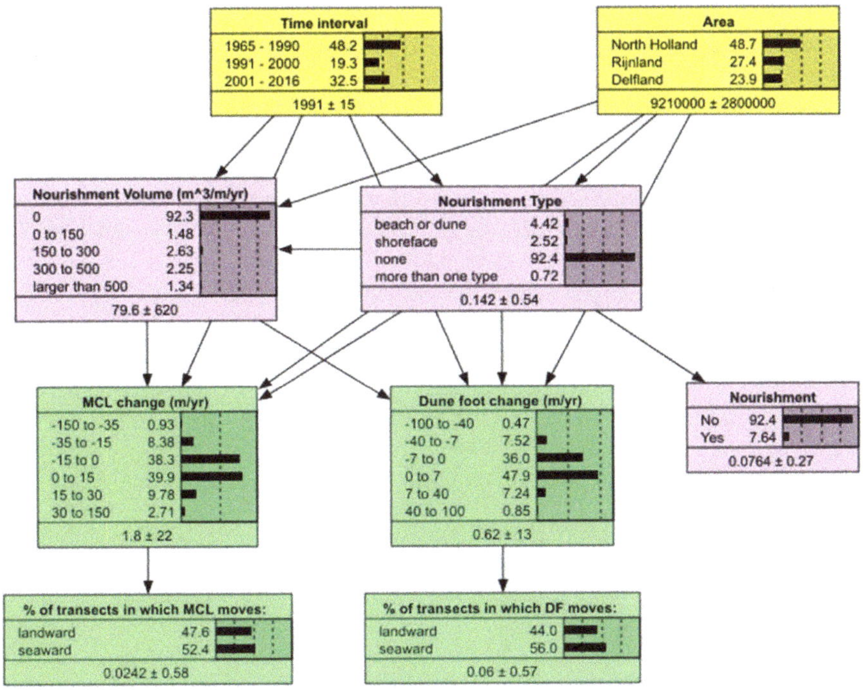

Figure 5. Visualization of BERM-N. Nodes have been grouped in three categories according to the color. Yellow is used for the nodes describing the spatial characterization of the study area and the time interval; purple for the nodes describing the nourishment types and volumes; and green for the nodes describing the effects on the morphological indicators. At each node, the first column indicates the chosen discretization intervals, while the second column (adjacent to the histogram) is the percentage of the prior distribution in each bin. The last line indicates the mean of the prior distribution ± one standard deviation.

4.2.1. Time Interval and Spatial Characterization of the Study Area

It includes the following nodes:

- **Time interval**: 1965–1990; 1991–2000; 2001–2016. Time intervals have been chosen in order to discriminate different periods in which the nourishment policy has been adapted (i.e., in 1990 and 2000).
- **Area**: North Holland, Rijland, and Delfland. These are the three coastal sections in which the Holland coast is divided.

4.2.2. Nourishment Type and Volume

It includes the following nodes:

- **Nourishment volume** (in $m^3/m/year$): Yearly nourishment volume divided by the length of the nourishment.
- **Nourishment type**: Beach or dune nourishment; shoreface nourishment; no nourishment; more than one type of nourishment at the same transect. It describes the nourishment type.
- **Nourishment**: Yes; no. To discriminate transects which have been nourished at least once during the entire period (1965–2016) from the ones which have never been nourished.

4.2.3. Effects on the Morphological Indicators

- **MCL change** (m/year): To quantify changes in coastline (MCL) position.
- **Dune foot (DF) change** (m/year): To quantify changes in dune foot position.
- **Percentage of transects in which the momentary coastline (MCL) moves**: Landward; seaward. To quantify the percentage of transects in which MCL has a positive (seaward), negative (landward) shift as a result of the effects of natural morphological changes and nourishments.
- **Percentage of transects in which the dune foot (DF) moves**: Landward; seaward. To quantify the percentage of transects in which the dune foot has a positive (seaward), negative (landward) shift as a result of the effects of natural morphological changes and nourishments.

5. Results

In this section, a number of possible applications of BERM-N and results are described. The predictive skill of the network was evaluated by means of confusion matrixes (similarly to References [12,18]). This resulted in predictive skills of 77% and 87%, respectively, for MCL and DF changes.

5.1. Prior Probability Distributions

The prior probability distribution describes the baseline situation, in which no node is constrained, and the network is trained based on all data from all the available 604 transects at the 52 years in which measurements were available (Figure 5). Some relevant information can be depicted from this baseline situation.

First of all, the node "Area" shows that most of the observations are available for the North Holland coastal section, in view of its wider extension, considering that JarKus transects are almost equally spaced.

Of all data, 92.4% fall in the category no nourishment, meaning that when considering all the transects for all the years, 92.4% of them do not contain a nourishment at a given year (see node "Nourishment Type"). Of all data, 4.42% include a beach or a dune nourishment, 2.52% a shoreface nourishment and only 0.72% more than one nourishment type per year. As a consequence, most of the data indicate a nourishment volume equal to 0 (see node "Nourishment Volume").

The nodes "MCL change" and "dune foot change" show that the indicators MCL and dune foot, based on all data, tend to move, in average, seaward 1.8 m/year and 0.6 m/year, respectively. This also suggests that beaches are becoming wider since the MCL position is moving seaward at a faster rate than the dune foot position, as a result of the volume of sediments in the nearshore zone.

The fact that most of the data show a seaward migration is confirmed by the nodes "percentage of transects in which MCL moves" and "percentage of transects in which DF moves". In particular, 52.4% of data show a seaward migration of MCL and 56.0% of the data a seaward migration of DF. Nevertheless, there is also a consistent percentage of transects which are characterized by a landward trend (i.e., erosion).

5.2. Assessment of the Effectiveness of Sand Nourishments Against Erosion

This section illustrates the effectiveness of past nourishments in addressing the coastal erosion along the Holland coast. In order to do so, the node "Nourishment" was constrained to a "No" then to a "Yes" value, to consider, respectively, transects which have not been nourished and transects which have been nourished (see Figure 6). The nodes: "Percentage of transects in which MCL moves" and "percentage of transects in which DF moves" show, respectively, the percentages of transects in which an erosive and accretive trend has to be expected. In particular, in case of no nourishment, about half of the cases are characterized by an erosive trend (i.e., 49.3% if considering MCL and 45.2% if considering the dune foot position as coastal indicator). The effect of nourishment is to decrease the number of erosive transects to 26.6%, when considering MCL as indicator, and 29.5% when considering dune foot as indicator.

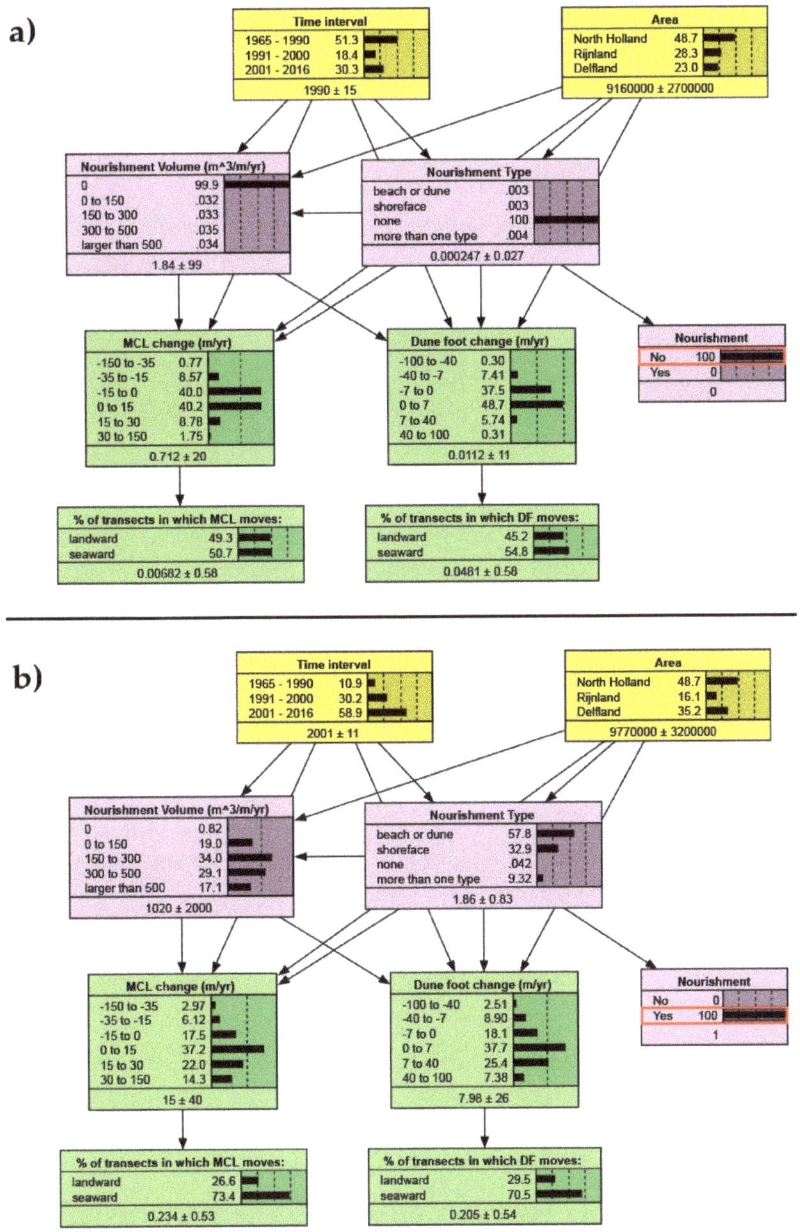

Figure 6. BERM-N application to assess the effectiveness of sand nourishments. Panel above (**a**): BERM-N constrained in order to consider only transects which have not been nourished (see red box). Panel below (**b**): BERM-N constrained in order to consider only transects which have been nourished (see red box).

5.3. Assessment of the Effectiveness of Different Nourishment Designs

In this section, the effectiveness of the two most common nourishment designs in the Netherlands (i.e., beach and shoreface nourishments) has been assessed using BERM-N. In order to do so, the node "nourishment type" was constrained first to "beach or dune" and then to "shoreface". To facilitate the visualization, the resulting changes in MCL and DF position (i.e., landward or seaward) are shown using a pie chart in Figure 7.

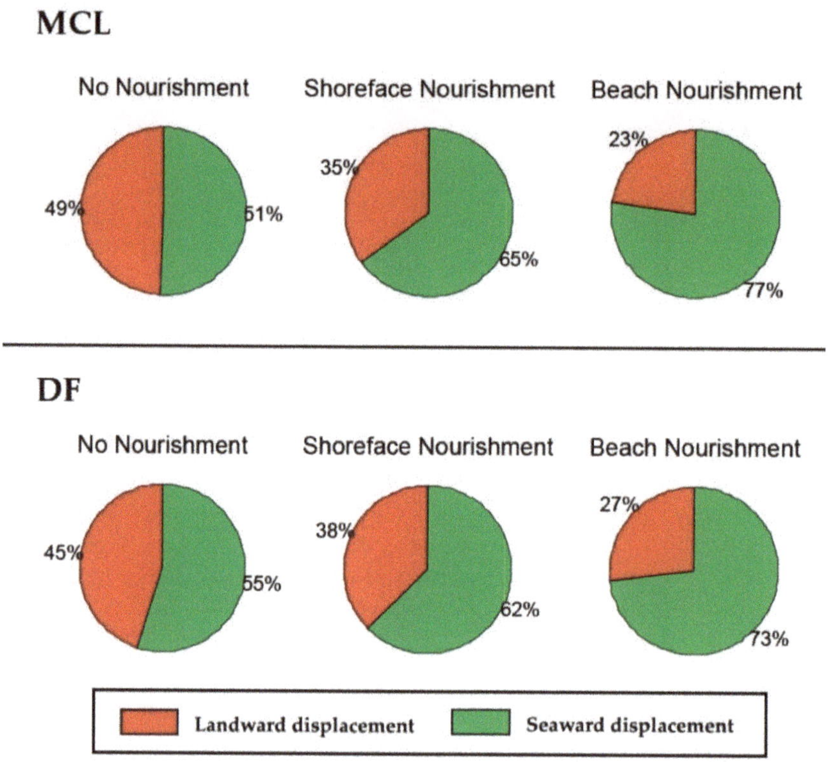

Figure 7. Effects of no nourishments, shoreface, and beach nourishments on MCL (**upper panel**) and dune foot (DF) (**lower panel**) indicators. The red color represents the probability of a landward displacement of the indicator, whereas a green color represents the probability of a seaward displacement. Values indicate the mean values of the distributions.

The figure confirms the positive effect of nourishments on counteracting coastal erosion on both indicators of MCL and DF position, as indicated by the green color (i.e., seaward displacement). Among the two types of nourishments, the effect of beach nourishments appears more pronounced than the effect of shoreface nourishments.

To shed light on the effect of different nourishment type in time, after implementation of a nourishment, the same network was trained with three different datasets. These three datasets were derived considering the effects on the indicators, respectively, one year, five years, and ten years after implementation of a nourishment. In case of multiple nourishments taking place within the considered time window, only the first nourishment was taken into account.

The results are shown in Figure 8, respectively, for the MCL indicator (upper panel) and DF indicator (lower panel). A first, important observation which can be derived from the figure is that even in the case of no nourishment, the percentage of transects characterized by indicators experiencing a landward trend (erosion) decreases when moving from a one-year time window to a five- or ten-year time window. This is the result of the large nourishment volumes implemented along the entire Holland coast. As nourished sand tends to move in an alongshore direction (Figure 1), even transects which have not been nourished tend to benefit from the sand nourished at neighboring transects.

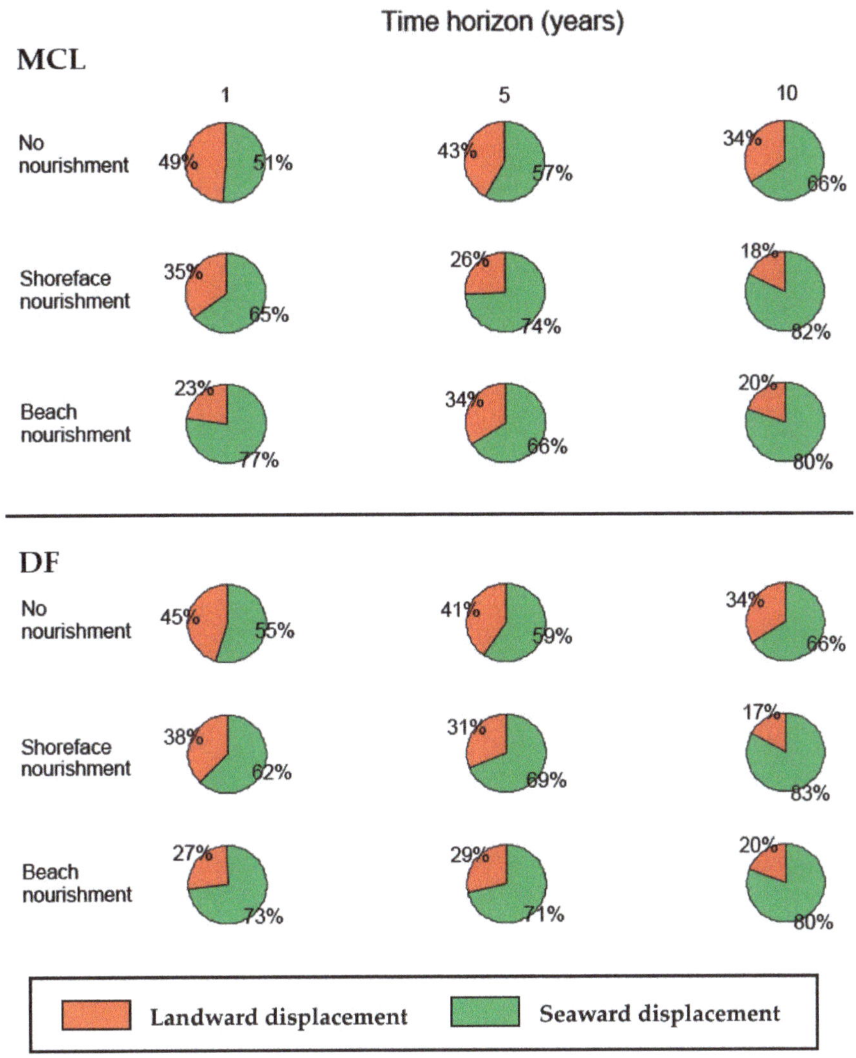

Figure 8. Effects of no nourishments, shoreface, and beach nourishments on MCL (**upper panel**) and DF (**lower panel**) indicators. The effects are shown in three different columns, to indicate the effects one year, five years and ten years after implementation of a nourishment.

The effect of nourishments is to further decrease the number of transects experiencing erosion with a similar qualitative effect on both indicators. When comparing the effect of shoreface and beach nourishments on the two indicators, one can see that the effect of shoreface nourishments is less evident than that of beach nourishments one year after construction. In the case of shoreface nourishments, 35% of the transects at which MCL was computed still experience a landward trend one year after implementation of the nourishment, compared with 23% in the case of beach nourishments. Similarly, for the DF position, 38% of the cases show a landward trend one year after implementation of a shoreface nourishment, against just 27% in case of a beach nourishment. However, the positive effect of shoreface nourishments on the indicators continues in time and becomes more noticeable 5 years after implementation of the nourishment, with a similar (i.e., for DF indicator) or even higher (i.e., for MCL indicator) impact than for beach nourishments.

The comparison of the effect of beach and shoreface nourishment after 10 years requires careful interpretation. In particular, the figure seems to suggest that the effectiveness of beach nourishments starts increasing again 10 years after implementation of the nourishments, after a drop at 5 years. In particular, the number of cases receiving a positive effect from beach nourishments after 10 years is 80% for both MCL and DF indicators, and only 66% for MCL and 71% for DF after 5 years. This is, however, rather due to the effect of sand coming from neighboring transects than the result of sand from the nourished transects. Beach nourishments have lifetimes in the order of 1–5 years, depending on the size of the nourishment and the local wave climate [30]. The effect of shoreface nourishments 10 years after implementation is consistent with the observation at 5 years, with a larger effect of shoreface nourishments on the indicators compared to that of beach nourishments.

5.4. *Application of BERM-N as Predictive Tool to Achieve a Predefined Coastal Management Objective*

In this last application, BERM-N was used as predictive tool to quantify the sand volumes required in order to reach a specific objective. The objective which is tested in this example is a coastline at which any erosion trend (i.e., described by the MCL and DF indicators) is stopped and replaced by accretive trends over the entire region. In order to do so, the two nodes ("% of transects in which MCL/DF moves") are constrained to a value equal to "seaward".

Figure 9 shows that in order to achieve this objective, an average nourishment volume of 161 $m^3/m/year$ would be required, distributed over 13.7% of the transects. Considering a total length of the coastline equal to 117 km, this would correspond to a yearly volume of about 2.5 million m^3 of sand/year (i.e., 161 $m^3/m/year \times 117{,}000\ m \times 0.137$). This is less than the current nourishment volume applied along the Holland coast, which currently results in an average accreting coastline as a result of the large nourishment volumes applied yearly. These nourishments are applied not just to preserve the current position of the coastline, but to maintain the entire coastal foundation in future sea level rise scenarios, as well as to provide the boundary conditions for the developments of additional functions, such as wide beach and dunes for nature and recreation (Section 3).

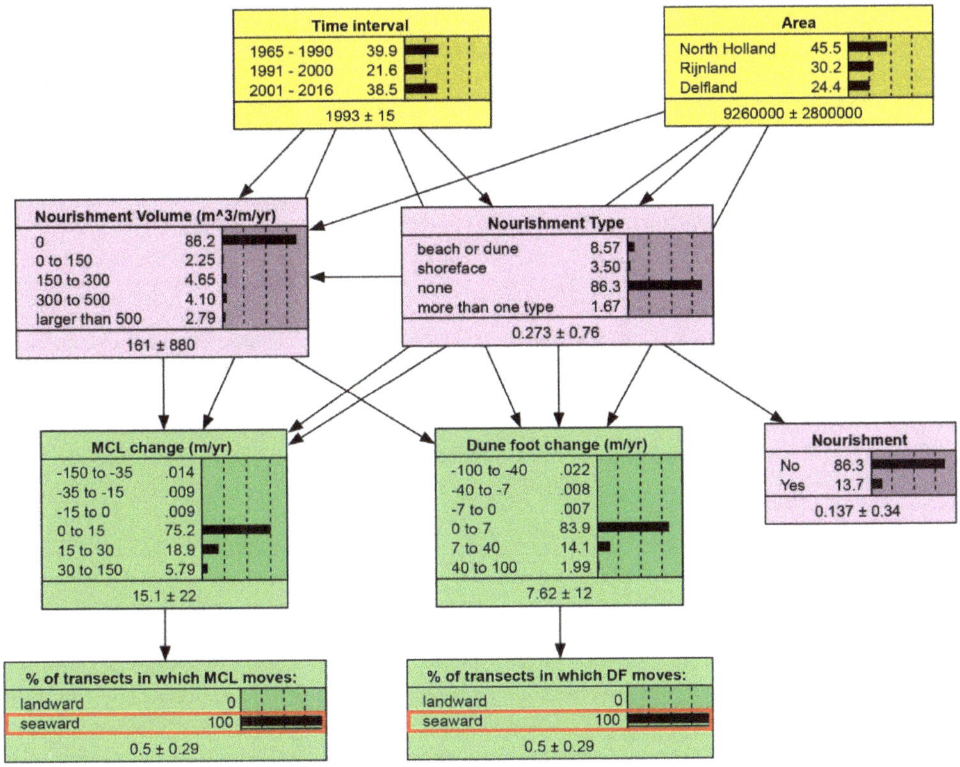

Figure 9. BERM-N application as predictive tool to assess the required nourishment volume in order to reach a predefined coastal erosion management objective: Reduce to zero the erosion over the entire coastline, only allowing for accretion. The red boxes show the two nodes constrained to a condition of "seaward" movement at the two coastline indicators.

6. Discussions

For clarity, discussion points have been grouped under three main topics.

6.1. General Assumptions Related to the Construction of the BERM-N

In order to develop BERM-N, a number of general assumptions have been adopted. First of all, the indicators chosen to describe the morphological development of the region (i.e., MCL and DF positions) are arbitrary. In principle, other indicators could also be selected, such as beach width (i.e., useful for recreational purposes) or the probability of dune breaching (i.e., important for safety). However, we have given preference to these indicators, as they are widely used for similar studies in the region [31,32] as well as by the local government.

The choice of processes to be included in the network is also arbitrary. In this paper, we have chosen to focus on the effects of anthropogenic forcing (i.e., nourishments) as the major driver for coastal development along the Holland coast. On the other hand, natural forcing has been omitted from the network. This choice is due to the fact that the large-scale and multi-year development of this coastline in the recent years are mainly the result of the large nourishment volumes which have been applied to manage coastal erosion. However, this is not the case at most coastlines around the world, where the effect of natural forcing is dominant. Another assumption relates to the discretization (i.e., number of bins and size of each bin) [35] of the continuous variables such as MCL change. We followed the guidance of the authors of [15], who suggest that each bin should be as wide as possible to limit the computational expense and capture multiple data points, but narrow enough to give meaningful forecasts and resolve uncertainty. Hence, node discretization presents a tradeoff between the precision and accuracy of predictions. In this study, the bin width was chosen in such a way that each bin would contain a sufficient number of data points, so we have avoided using bins with a very small size. Similarly, there should be a sufficient number of bins in order to properly describe the statistical distribution of the variable under consideration. BBN cannot extrapolate, only interpolate within the bounds of the data with which they have been trained, so the extremes of the training dataset determine the upper and lower limits of each node.

A final assumption relates to how to account for the effects of multiple nourishments in subsequent years, when considering time windows of several years (i.e., 5 or 10), and specifically at the transects where multiple nourishments occurred. This effect is currently included in BERM-N by only looking at the first nourishment within the considered time window. Although other options may be possible and were explored while setting up the network (i.e., adding the total volumes or correcting volumes with functions which would account for the exact timing when a nourishment occurred), it should be recognized that each of these approaches contains a bias. It was finally decided to adopt the approach used in the paper, because of its simplicity and because it provided the highest predictive skill.

6.2. *Alongshore Effects of Sand Nourishments*

The analysis as presented in this study fully relies on data measured at cross-shore transects along the entire region (i.e., nourishment volumes, types and spatial displacement of the indicators resulting from the nourishments). The basic assumption is that the effect of nourishments can be described by cross-shore processes and that a nourishment will only affect the coastal indicators at the transect in which it has been applied. In reality, sand nourishments, and in particular shoreface nourishments, also have an alongshore effect, as discussed in Section 5.3. This is a simplification, as it is not straightforward to link the displacement of a coastal indicator resulting from a sand nourishment applied at a different location.

This alongshore effect is also visible as a background signal in the displacements of the coastal indicators. For example, Figures 7 and 8 show that, even without nourishments, the coastal indicators are characterized by accretive trends, which is most likely related to the positive effect of sand nourishments applied at neighboring transects.

6.3. *BERM-N Tool for Coastal Erosion Management of Past and Future Conditions*

BERM-N, as presented in this study, is fully trained based on data from past observations, collected over more than 50 years. This means that the response and outcome of the network is derived based on these past observations. This further implies that in case of future changes to the physical system (e.g., accelerated sea level rise) or development of new anthropogenic interventions which may interfere with the coastal systems, the model may have insufficient data on which to base its predictions. Bayesian networks can interpolate but not extrapolate beyond the data with which they have been trained. A possibility to overcome this is by complementing and retraining the BERM-N with data derived from numerical modeling simulations (e.g., [36]), which may also include information on possible future scenarios.

Similarly, the network has been trained with data from all past nourishment projects carried out in the region, i.e., one of the most nourished region in the world, per kilometer length of coastline (see, e.g., [37]). This is a very extensive database, including nourishments with different sizes and volumes. Nevertheless, this also means that the network can only be used as a predictive tool to simulate cases which are included within the statistical distribution of these past nourishment projects. In case, for example, the effects of a very different type of nourishment should be simulated (e.g., much larger in size than the ones included in the database), it may be required to extend the database with data derived from numerical modeling simulations, which include these new nourishments as synthetic cases.

7. Conclusions

In this study, an application of Bayesian statistics for coastal erosion management at the regional scale has been presented. A tool based on the use of a Bayesian network (BERM-N) was constructed using measured data collected along the entire Holland coast for a period of over 50 years at 604 cross-shore transects. The effects of different sand nourishment designs on two pre-identified coastal indicators (i.e., dune foot and momentary coastline position) were assessed at these transects.

Different applications of BERM-N have been described. A first application was used to quantify the issue of coastal erosion along the Holland coast region. Additionally, the tool has shown how sand nourishments have been effective in counteracting this issue. In a second application, the tool was used to assess the effectiveness of different nourishment designs (i.e., shoreface vs. beach nourishments) in counteracting coastal erosion. The results have shown how beach nourishments have an immediate effect on the chosen indicators. On the other hand, shoreface nourishments become more effective with time, reaching a similar effect to the one of beach nourishments after about five years. For coastal managers, this is important, as beach nourishments may be more suitable for achieving short-term safety objectives (≈months/years), whereas shoreface nourishment may be more appropriate to counteract erosion at medium- and long-term (≈decades) temporal scale. In a last application, the tool was used as a predictive tool to assess the nourishment volumes required in order to achieve a specific coastal management objective (e.g., extending the coastline in seaward direction of a given value).

The advantage of the tool is that it is flexible, fast, and interactive and can accommodate the use of measured data (i.e., as shown in this study) as well as synthetic data derived from numerical modeling simulations. This makes it suitable for discussions with policy makers and other stakeholders, following a collaborative modeling approach [38].

Author Contributions: A.G. and E.D. conceptualized the research; E.D. implemented the methodology; E.D. and G.S. contributed to data analyses and figures; A.G. prepared the original draft of the manuscript and contributed to further revisions; S.P. and K.d.H. supervised the research; S.P. reviewed and edited the original manuscript and further revisions.

Funding: This research received no external funding.

Acknowledgments: We would like to thank Dirk Knipping, who has helped developing the first setup of the Bayesian network. We are also thankful to Matthijs Kok, Saskia van Vuren, and Sierd de Vries for their suggestions in developing this research. Finally, we would like to thank Björn Röbke for his help in preparing Figure 1.

Conflicts of Interest: The authors declare no conflict of interest.

References

1. Luijendijk, A.; Hagenaars, G.; Ranasinghe, R.; Baart, F.; Donchyts, G.; Aarninkhof, S. The State of the World's Beaches. *Sci. Rep.* **2018**, *8*, 6641. [CrossRef] [PubMed]
2. Giardino, A.; Schrijvershof, R.; Nederhoff, C.M.; de Vroeg, H.; Brière, C.; Tonnon, P.K.; Caires, S.; Walstra, D.J.; Sosa, J.; van Verseveld, W.; et al. A quantitative assessment of human interventions and climate change on the West African sediment budget. *Ocean Coast. Manag.* **2018**, *156*, 249–265. [CrossRef]
3. Giardino, A.; De Boer, W.; Den Heijer, K.; Huisman, B.; Mulder, J.; Walstra, D.-J. Innovative approaches and tools for erosion control and coastline management. In Proceedings of the 10th Global Congress on ICM: Lessons Learned to Address New Challenges, EMECS 2013—MEDCOAST 2013 Joint Conference, Marmaris, Turkey, 30 October–3 November 2013; Volume 2.
4. Stronkhorst, J.; Huisman, B.; Giardino, A.; Santinelli, G.; Santos, F.D. Sand nourishment strategies to mitigate coastal erosion and sea level rise at the coasts of Holland (The Netherlands) and Aveiro (Portugal) in the 21st century. *Ocean Coast. Manag.* **2018**, *156*, 266–276. [CrossRef]
5. Robinet, A.; Idier, D.; Castelle, B.; Marieu, V. Environmental Modelling & Software A reduced-complexity shoreline change model combining longshore and cross-shore processes: The LX-Shore model. *Environ. Model. Softw.* **2018**, *109*, 1–16. [CrossRef]
6. Ranasinghe, R.; Callaghan, D.; Stive, M.J.F. Estimating coastal recession due to sea level rise: Beyond the Bruun rule. *Clim. Chang.* **2012**, *110*, 561–574. [CrossRef]

7. Callaghan, D.P.; Nielsen, P.; Short, A.; Ranasinghe, R. Statistical simulation of wave climate and extreme beach erosion. *Coast. Eng.* **2008**, *55*, 375–390. [CrossRef]
8. Callaghan, D.P.; Baldock, T.E.; Shabani, B.; Mumby, P.J. Communicating physics-based wave model predictions of coral reefs using Bayesian belief networks. *Environ. Model. Softw.* **2018**, *108*, 123–132. [CrossRef]
9. Hapke, C.; Plant, N. Predicting coastal cliff erosion using a Bayesian probabilistic model. *Mar. Geol.* **2010**, *278*, 140–149. [CrossRef]
10. Den Heijer, C.K.; Baart, F.; van Koningsveld, M. Assessment of dune failure along the Dutch coast using a fully probabilistic approach. *Geomorphology* **2012**, *143–144*, 95–103. [CrossRef]
11. Wilson, K.E.; Adams, P.N.; Hapke, C.J.; Lentz, E.E.; Brenner, O. Application of Bayesian Networks to hindcast barrier island morphodynamics. *Coast. Eng.* **2015**, *102*, 30–43. [CrossRef]
12. Poelhekke, L.; Jäger, W.S.; van Dongeren, A.; Plomaritis, T.A.; McCall, R.; Ferreira, Ó. Predicting coastal hazards for sandy coasts with a Bayesian Network. *Coast. Eng.* **2016**, *118*, 21–34. [CrossRef]
13. Plant, N.G.; Stockdon, H.F. Probabilistic prediction of barrier-island response to hurricanes. *J. Geophys. Res.* **2012**, *117*. [CrossRef]
14. Plant, N.G.; Holland, K.T. Prediction and assimilation of surf-zone processes using a Bayesian network. Part I: Forward models. *Coast. Eng.* **2011**, *58*, 119–130. [CrossRef]
15. Plant, N.G.; Holland, K.T. Prediction and assimilation of surf-zone processes using a Bayesian network. Part II: Inverse models. *Coast. Eng.* **2011**, *58*, 256–266. [CrossRef]
16. Gutierrez, B.T.; Plant, N.G.; Thieler, E.R. A Bayesian network to predict coastal vulnerability to sea level rise. *J. Geophys. Res. Earth Surf.* **2011**, *116*. [CrossRef]
17. van Verseveld, H.C.W.; van Dongeren, A.R.; Plant, N.G.; Jäger, W.S.; den Heijer, C. Modelling multi-hazard hurricane damages on an urbanized coast with a Bayesian Network approach. *Coast. Eng.* **2015**, *103*, 1–14. [CrossRef]
18. Pearson, S.G.; Storlazzi, C.D.; van Dongeren, A.R.; Tissier, M.F.S.; Reniers, A.J.H.M. A Bayesian-Based System to Assess Wave-Driven Flooding Hazards on Coral Reef-Lined Coasts. *J. Geophys. Res. Oceans* **2017**, *122*, 10099–10117. [CrossRef]
19. Loureiro, C.; Ferreira, Ó.; Cooper, J.A.G. Applicability of parametric beach morphodynamic state classification on embayed beaches. *J. Mar. Geol.* **2013**, *346*, 153–164. [CrossRef]
20. Yates, M.L.; Le Cozannet, G. Brief communication "evaluating European Coastal Evolution using Bayesian Networks". *Nat. Hazards Earth Syst. Sci.* **2012**, *12*, 1173–1177. [CrossRef]
21. Bulteau, T.; Baills, A.; Petitjean, L.; Garcin, M.; Palanisamy, H.; Le Cozannet, G. Gaining insight into regional coastal changes on La Réunion island through a Bayesian data mining approach. *Geomorphology* **2015**, *228*, 134–146. [CrossRef]
22. Jäger, W.S.; Christie, E.K.; Hanea, A.M.; den Heijer, C.; Spencer, T. A Bayesian network approach for coastal risk analysis and decision making. *Coast. Eng.* **2018**, *134*, 48–61. [CrossRef]
23. Plomaritis, T.A.; Costas, S.; Ferreira, Ó. Use of a Bayesian Network for coastal hazards, impact and disaster risk reduction assessment at a coastal barrier (Ria Formosa, Portugal). *Coast. Eng.* **2018**, *134*, 134–147. [CrossRef]
24. Catenacci, M.; Giupponi, C. Integrated assessment of sea-level rise adaptation strategies using a Bayesian decision network approach. *Environ. Model. Softw.* **2013**, *44*, 87–100. [CrossRef]
25. Van Rijn, L.C. Sediment transport and budget of the central coastal zone of Holland. *Coast. Eng.* **1997**, *32*, 61–90. [CrossRef]
26. Van Der Spek, A.J.F.; Lodder, Q. A new sediment budget for The Netherlands; the effects of 15 years of nourishing (1991–2005). In *The Proceedings of the Coastal Sediments 2015*; Word Scientific Publishing: Singapore, 2015.
27. Beets, D.J.; Van Der Spek, A.J.F. The Holocene evolution of the barrier and the back-barrier basins of Belgium and the Netherlands as a function of late Weichselian morphology, relative sea-level rise and sediment supply. *Neth. J. Geosci.* **2000**, *79*, 3–16. [CrossRef]
28. Mulder, J.P.M.; Hommes, S.; Horstman, E.M. Implementation of coastal erosion management in the Netherlands. *Ocean Coast. Manag.* **2011**, *54*, 888–897. [CrossRef]
29. Giardino, A.; Mulder, J.; De Ronde, J.; Stronkhorst, J.; Giardino, A. Sustainable Development of the Dutch Coast: Present and Future. *J. Coast. Res.* **2011**, *61*, 166–172. [CrossRef]

30. Van Rijn, L.C. Coastal erosion and control. *Ocean Coast. Manag.* **2011**, *54*, 867–887. [CrossRef]
31. Giardino, A.; Santinelli, G.; Vuik, V. Coastal state indicators to assess the morphological development of the Holland coast due to natural and anthropogenic pressure factors. *Ocean Coast. Manag.* **2014**, *87*, 93–101. [CrossRef]
32. van Koningsveld, M.; Mulder, J.P.M. Sustainable Coastal Policy Developments in The Netherlands. A Systematic Approach Revealed. *J. Coast. Res.* **2004**, *202*, 375–385. [CrossRef]
33. Norsys Software Corp. *Netica TM, Application for Belief Networks and Influence Diagrams: User's Guide*; Norsys Software Corp: Vancouver, BC, Canada, 2003; pp. 1–91.
34. Spiegelhalter, D.J.; Dawid, A.P.; Lauritzen, S.L.; Robert, G.; Spiegelhalter, D.J.; Dawid, A.P.; Lauritzen, S.L.; Cowell, R.G. Bayesian Analysis in Expert Systems Linked references are available on JSTOR for this article. *Bayesian Anal. Expert Syst.* **1993**, *8*, 219–247.
35. Beuzen, T.; Splinter, K.D.; Marshall, L.A.; Turner, I.L.; Harley, M.D.; Palmsten, M.L. Bayesian Networks in coastal engineering: Distinguishing descriptive and predictive applications. *Coast. Eng.* **2018**, *135*, 16–30. [CrossRef]
36. Giardino, A.; den Heijer, K.; Santinelli, G.; van der Werf, J. *Tools for Medium- and Long-Term Prediction of Nourishments*; Deltares: Delft, The Netherlands, 2013.
37. Hamm, L.; Capobianco, M.; Dette, H.H.; Lechuga, A.; Spanhoff, R.; Stive, M.J. A summary of European experience with shore nourishment. *J. Coast. Eng.* **2002**, *47*, 237–264. [CrossRef]
38. Basco-Carrera, L.; Warren, A.; van Beek, E.; Jonoski, A.; Giardino, A. Collaborative modelling or participatory modelling? A framework for water resources management. *Environ. Model. Softw.* **2017**, *91*, 95–110. [CrossRef]

water

MDPI

Article

Source Apportionment Assessment of Marine Sediment Contamination in a Post-Industrial Area (Bagnoli, Naples)

Sara Giglioli [1], Loris Colombo [1], Pasquale Contestabile [2,3], Luigi Musco [4], Giovanna Armiento [5], Renato Somma [6,7], Diego Vicinanza [2,4,*] and Arianna Azzellino [1,3]

1 Department of Civil and Environmental Engineering, Politecnico di Milano, 20133 Milano, Italy; sara.giglioli@polimi.it (S.G.); loris.colombo@polimi.it (L.C.); arianna.azzellino@polimi.it (A.A.)
2 Department of Engineering, University of Campania, 81031 Caserta, Italy; pasquale.contestabile@unicampania.it
3 CoNISMa Inter-University National Consortium for Marine Sciences, 00196 Roma, Italy
4 Stazione Zoologica Anton Dohrn, 80121 Napoli, Italy; luigi.musco@szn.it
5 Agenzia Nazionale per le Nuove Tecnologie, L'energia e lo Sviluppo Economico Sostenibile, 00196 Roma, Italy; giovanna.armiento@enea.it
6 Osservatorio Vesuviano, Istituto Nazionale di Geofisica e Vulcanologia, 80124 Naples, Italy; renato.somma@ingv.it
7 CNR IRISS-Istituto di Ricerca su Innovazione e Servizi per lo Sviluppo, 80124 Napoli, Italy
* Correspondence: diego.vicinanza@unicampania.it

Received: 18 June 2020; Accepted: 31 July 2020; Published: 3 August 2020

Abstract: The area of Bagnoli (Gulf of Naples, central Tyrrhenian Sea) has been heavily exposed to pollution for over a century due to the presence of industrial sites along its coastline. The aim of this study is to analyze contaminant concentrations (i.e., heavy metals and hydrocarbons) in seabed sediments through a statistical multivariate approach. Multivariate methods permit us to describe the pollution dynamics affecting the area and distinguish between anthropogenic and natural pollution sources. Additionally, the association between contamination patterns and the wave climate characteristics of the gulf (i.e., wave period, direction, height, power, and energy) is investigated. The study confirms that the main contamination source in the Bagnoli bay is anthropogenic activities (i.e., former steel plant and sewage discharges) for the majority of investigated pollutants. It also provides evidence, however, for the potential co-existence of multiple anthropogenic and geogenic sources of arsenic and other metals that may be originating also from the water-rock interaction and submarine volcanic emissions in the Phlegraean area.

Keywords: arsenic; hydrocarbons; heavy metals contamination; marine pollution; multivariate analysis; Bagnoli; Naples

1. Introduction

The contamination of marine environments represents an increasing global concern because of the potential risks to both human health and along the coast heavily affecting the marine ecosystems. The Mediterranean Sea, due to reduced circulation and the presence of multiple industrial inputs along the coastline, is particularly vulnerable to environmental impacts and risks. Moreover, there is evidence that contamination may persist long after the end of industrial activities [1]. The historical industrial district and metallurgical production at the Bagnoli steel factory (ILVA), active for roughly a century, has exposed the marine sediments of the Gulf of Pozzuoli (GoP) to pollution by heavy metals and polycyclic aromatic hydrocarbons (PAHs). This area became a key site for twentieth-century Italian economic growth through industrial plants that produced steel and cement using iron ore and coal transported from other locations and processed on site. These activities, now recognized as detrimental to the environment and economically impractical, prompted the industrial area to be dismantled starting in the mid-1990's [2]. The impact of heavy industries was never completely remediated, however, and this negligence results in high concentrations of PAHs and trace metals in marine sediments [3]. Sediment contamination is still evident in the vicinity of the industrial sites but also widespread in neighboring areas due to re-suspension of sediments and to water currents [4–7]. Furthermore, sewage loss from wastewater treatment plants or accommodation facilities along the western Bagnoli coastline increased the magnitude of the marine environment contamination [8,9]. Bagnoli has served as the subject of numerous scientific studies aimed at verifying the current condition of the water and marine sediments present in the GoP. Recent work has focused on bioaccumulation and biomarkers investigations to better understand the toxic effects and mechanism of action of contaminants [10–12]. Recently, an integrated assessment called sediment quality triad (SQT) was used to consider chemical analyses and biological effects as different lines of evidence (LOEs) to describe environmental quality of marine sediments [2,13]. A multidisciplinary approach (the weight of evidence approach, WoE) permits researchers to interpret various environmental conditions compared to a univariate analysis, such as a chemical approach, and circumvent overestimated and costly management decisions [11]. The WoE approach, integrating five lines of evidence (LOEs: sediment chemistry, bioavailability of chemicals, subcellular effects, toxicity at organism level and at the community level) was successfully applied to the Bagnoli case study and revealed a clearly polluted area but less critical levels of pollution when compared to the results of sediment chemistry alone [2].

This study, developed in the framework of the ABBaCo project which started in 2017, is designed to (a) update and improve the characterization of the environmental quality of the Bagnoli industrial site, (b) identify contamination sources, and (c) propose suitable remediation strategies. In the Bagnoli area, the determination of polluting sources is particularly challenging due to the cohabitation between natural processes and anthropic activities, which assesses sources apportionment more difficult. Moreover, some contaminations are nowadays almost ubiquitous (e.g., PAHs contamination is present in urban areas [14], rural areas [15], and coastal areas [5]).

The GoP is an area characterized by intense volcanic activity due to the presence of a large caldera—the Phlegraean fields—which is one of the most densely populated active volcanoes on Earth. It is characterized by a strong record of historical unrest and eruption events that date back to 2.2 ka BP. Since the 1950's, the Phlegraean fields' area has undergone four episodes of caldera-wide uplift and seismicity, which have raised the coastal town of Pozzuoli, near the center of unrest by 4.5 m and triggered the repeated evacuation of some 40,000 people. During the last 20 years of subsidence, following the uplift peak reached in 1984, the caldera started a new, low-rate uplift episode accompanied by low-magnitude seismicity and marked geochemical changes in fumaroles [16]. For this reason, some elements, such as arsenic (As), represent a natural enrichment. In some Italian areas, such as the Po plain, it is recognized that arsenic originates from the reductive dissolution of Fe oxides [17]. Therefore, in the present study, the presence of arsenic is assessable as a characteristic natural background level of the study area [6]. This natural presence of heavy metals in the sea sediments of the gulf is attributed to an active system of submarine thermal springs near the Bagnoli coastline that constantly release volcanic gasses [18,19]. PAHs also originate, however, from percolation through the soils or landfills contaminated by industrial activities [7].

For the assessment of the source apportionment in a highly polluted area, the use of statistical tools is well recognized. Some studies report the advantages of using multivariate statistical analysis [20–27] to interpret the contaminant distributions and the pollutants patterns. For the marine environment, some studies use principal component analysis (PCA) [7,15,28–30] and bivariate correlation analysis (Pearson coefficient) [31,32], nonparametric multivariate multiple regression analyses [33,34], canonical analysis of principal coordinates [33], multivariate linkage tree analysis [34] and randomized analysis of variance (PERMANOVA) [8,33,34] to investigate the co-occurrence of a suite of pollutants in sediments and to assess the related response on the biological assemblages that inhabit the seabed. Some correlation analyses were carried out between pollutant concentrations and sediment granulometry [31]. Other studies examined the distribution patterns of the meiofauna and the diversity and abundance of microorganisms inhabiting the sediments of GoP in relation to environmental variation and chemical pollution [8,33,34]. Most PAH inputs in the environment are linked to anthropogenic activity (e.g., wastes from industrialized and urbanized areas, off-shore petroleum hydrocarbons production or petroleum transportation) [35]. One of the main issues is the connection between pollutant concentrations and their possible source (natural or anthropic) combined with the influence of wave climate on their concentration patterns. Within this project, the present research aims to apply a robust statistical approach (PCA/FA) to demonstrate its practical application for assessing the main contamination sources, distinguishing among natural or anthropic/industrial contributions, and finding correlation between the contaminant concentrations and wave hydrodynamics in the area. The workflow framework of the study is shown in Figure 1.

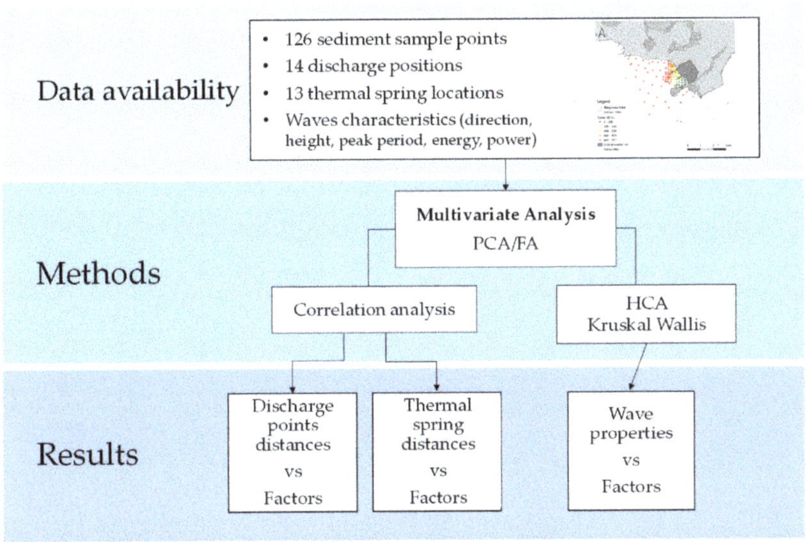

Figure 1. Workflow scheme of the study. This work basically generates three main results boxes: the first and the second box include bivariate correlations analysis, respectively between PC/Fs and distances from sewage discharge and PC/Fs and distances from thermal spring, while the third box includes the hierarchical cluster analysis (HCA) and Kruskal Wallis test.

2. Materials and Methods

2.1. Study Area and Geological Settings

The study area, Bagnoli, is located within the Bay of Pozzuoli, in the western sector of the urban territory of Naples. Close to the Bagnoli site, the already mentioned Phlegraean fields is a large "super volcano" situated to the west of Naples, Italy. Declared a regional park in 2003, the area of the caldera consists of 24 craters and volcanic edifices, most of them submerged under the bay with very high hydrothermal activity and effusive gaseous manifestations [19] (Figure 2). From a geological point of view, this large caldera is in a state of quiescence. The Phlegraean fields experienced extreme volcanic activity in the last 39,000 years. The two main events occurred 35,000 years ago (the 'Campanian ignimbrite eruption'), followed by caldera collapse, and 15,000 years ago (the 'Neapolitan Yellow Tuff', NYT eruption). The NYT eruption formed the central and eastern part of the Bagnoli–Fuorigrotta plain, which is constituted by a sequence of pyroclastic and volcanic material that thickens along the Agnano crater. Agnano volcanic products dominates the western part of the plain. The last eruptive event occurred in A.D. 1538 and gave rise to the Mt. Nuovo cone. From then on, only bradyseismic and hydrothermal activity (in the Solfatara and Pisciarelli area) are present in the area. The ground deformation in this area was recognized as caused mainly by the actions of long periods of uplift, eruptive activity, and possible deep magma movements [36–38]. According to some studies [39], the Phlegraean fields activity has been predominantly explosive and characterized by an interaction between magma and water. There is evidence of both interaction of the magma with sea water as well as intra-caldera lake water and deep-seated aquifers in some Plinian volcanic events. High concentrations of metals and metalloids, such as As, Mn and Fe are found in groundwater and in the sea sediments near the main tectonic and hydrothermal activity [3,18,31,40,41].

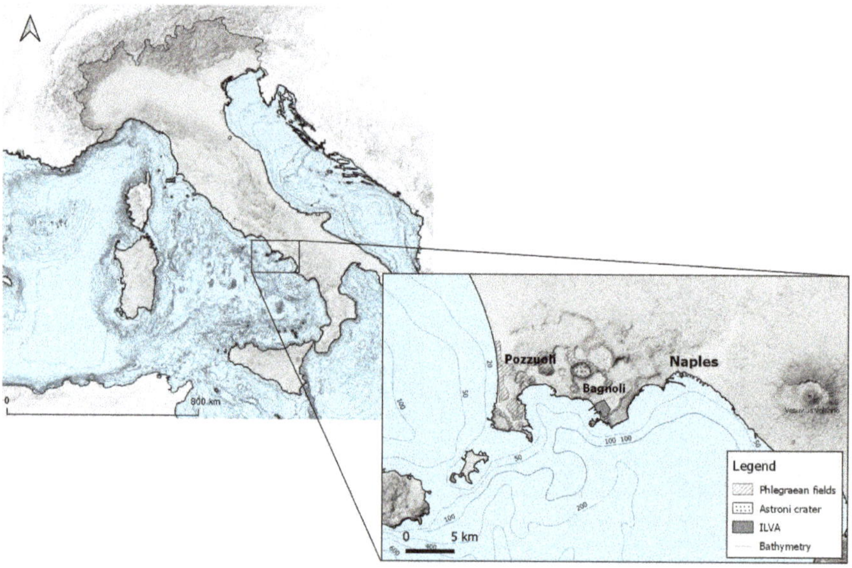

Figure 2. Location of the study area, Bagnoli bay, Naples, Italy.

Since the early 1900's Bagnoli has been affected by the presence of various industries. The most important industrial activity was through ILVA steelworks, formerly Italsider. The ILVA steel plant was located along the Bagnoli coastline and was characterized by two long piers (still present now) which served as a berth for large ships that carried raw materials such as coal, iron ore and limestone to the steelworks. In the early 1960's, in order to enlarge the plant, the area between the two piers was filled with industrial waste coming from the steelworks. A new coastline was designed, giving more space for the industrial activities, but causing consistent damage to the marine ecosystems. Furthermore, the presence of nine discharge points along the coastline, combining both industrial and civil sewage, may have contributed to the deterioration of the quality of the environment. Some of these collectors are no longer in use since the steelworks stopped its production in 1990.

The Bagnoli Bay is also characterized by the presence of some submarine thermal springs, mainly located close to the former industrial area. Some offshore thermal springs are also present in the GoP.

2.2. Data Availability: Sampling, Sewage Discharge, and Wave Information

In this study, a total of 126 sampling points were considered as shown in Figure 3 (black dot points). These samples were part of an explorative campaign carried out in 2017 in the GoP by the Stazione Zoologica Anton Dohrn [42,43]. On the sea bottom, 94 sediment samples were collected using drilling (0-50 cm) and 32 sediment samples were collected by bucket (surface layer). In each point, the concentrations of heavy metals and PAHs were measured. A total of 11 heavy metals (i.e., Al, As, Cd, Cr, Cu, Fe, Hg, Ni, Pb, V, and Zn) and 18 PAHs (i.e., naphthalene, anthracene, phenanthrene, acenaphthylene, acenaphthene, fluorene, fluoranthene, pyrene, benzo(a)anthracene, chrysene, benzo(b)fluoranthene, benzo(a)pyrene, benzo(k)fluoranthene, indeno(1,2,3-cd)pyrene, benzo(ghi)perylene, dibenz(ah)anthracene, benz(j)fluoranthene, benz(e)pyrene) indicated by Environmental Protection Agency (EPA) as important toxicological contaminants were considered. All the samples and the analytical determination derive from the series of studies conducted under the framework of the ABBaCo project [4,33]. Contaminants general statistics are reported in Table 1.

Table 1. General statistics of the sampled contaminants in Bagnoli sea sediments. Heavy metals (in mg/kg) and polycyclic aromatic hydrocarbons (PAHs, in µg/kg) are divided by bold line.

Contaminant	*N*	Minimum	Maximum	Mean	Std. Deviation
Aluminum (Al)	126	27827.37	92240.68	65268.54	16565.95
Arsenic (As)	126	18.66	136.66	64.32	24.59
Cadmium (Cd)	126	0.26	14.65	2.03	2.96
Chromium (Cr)	126	11.16	1022.02	51.62	108.50
Copper (Cu)	126	6.64	209.58	48.38	42.17
Iron (Fe)	126	21578.56	209372.34	76045.46	45464.22
Mercury (Hg)	126	0.01	7.51	0.73	1.02
Nickel (Ni)	126	4.15	94.14	14.20	9.50
Lead (Pb)	126	25.29	1425.65	306.64	334.94
Vanadium (V)	126	42.03	360.13	107.69	33.89
Zinc (Zn)	126	93.55	3132.46	706.18	741.38
Naphthalene	126	0.50	308169.70	7055.02	33307.90
Anthracene	126	4.69	147085.93	9281.80	21470.44
Phenanthrene	126	9.80	427669.32	19897.58	55329.63
Acenaphthylene	125	1.93	97037.60	3570.29	12733.90
Acenaphthene	126	0.50	261079.10	6093.03	28521.31
Fluorene	125	1.72	243499.32	6815.46	29311.06
Fluoranthene	126	22.34	384779.23	38409.73	65196.81
Pyrene	126	20.04	314505.67	33190.47	55808.17
Benzo(a)anthracene	126	11.05	143895.53	14480.05	23670.28
Chrysene	126	10.39	123533.76	13286.81	21443.57
Benzo(b)fluoranthene	126	8.62	133453.01	16081.11	25336.33
Benz(a)pyrene	126	10.21	160764.77	20325.76	32175.75
Benz(k)fluoranthene	126	9.76	77963.48	8639.16	14020.82
Indeno(1,2,3,c,d)pyrene	126	17.13	91995.93	11367.30	18115.95
Benz(g,h,i)perylene	126	22.03	109180.22	13201.81	21859.16
Dibenz(a,h)anthracene	126	4.03	33074.18	3255.55	5528.15
Benz(j)fluoranthene	126	7.32	77596.95	8227.36	13993.61
Benz(e)pyrene	126	6.86	125705.41	15362.81	24770.26

A total of 14 sewage discharge points were mapped and selected in the study area (from 1 to 9C, in Figure 3). Some of these discharge points were supposed to be inactive (e.g., Points 4, 7 and 8). However, it is well recognized that due to poor maintenance and some illegal discharges in sewage disposal these points, especially when heavy rains occur, release water carrying waste and sewage to the coastline. For this reason, all the discharge points shown in the figure have been considered as active. It should also be noted that the discharge points labeled as "C" in Figure 3 are control points chosen as the nearest to discharge point (i.e., 9C is the control point of discharge point 9).

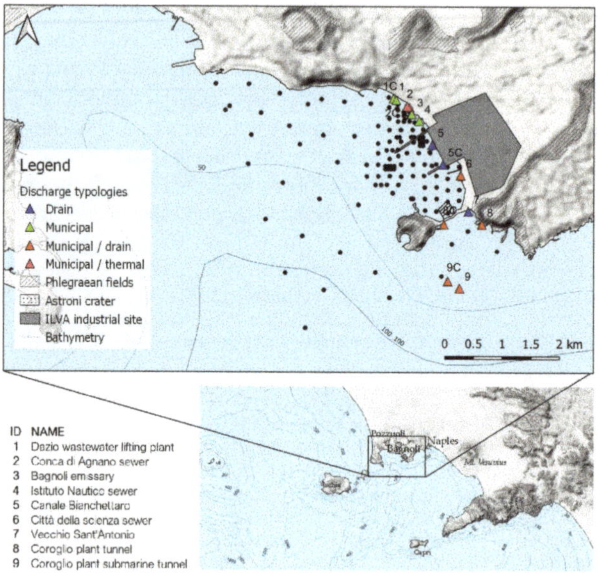

Figure 3. Location and typology of sewage discharges located along the Bagnoli coastline. The capital letter "C" indicates the discharges which control the nearest discharge point (i.e., 9C is the control point of Discharge Point 9).

Representing natural contaminant sources spread along the Bagnoli bay seabed, 13 thermal springs were identified [18] (Figure 4).

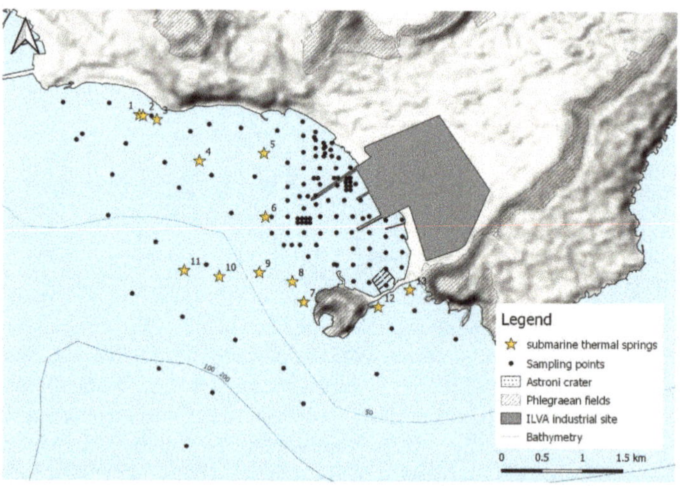

Figure 4. Location of natural sources inside the Bagnoli bay [12,18].

In addition to the presence of sewage discharges and thermal springs, concentration patterns can be influenced by marine dynamics that remobilize sediments and accumulate them in places where there are steady-state conditions. In order to include the analysis, the main effects of the wave-induced hydrodynamics inside the GoP area in the analysis, we incorporated wave properties generated by a numerical model of the gulf.

2.3. Numerical Wave Modelling

The wave climate analysis was carried out using the hindcast data from the European Centre for Medium-Range Forecasts (ECMWF). Significant wave height (H_s), mean period (T_m), and mean direction (Θ_m) for the time period ranging from January 1979 to December 2018 were extracted from the wave model (WAM) of the ERA-interim archive (ERA: ECMWF Atmospheric Reanalysis), available for download online [44]. Both the offshore energetic patterns and the nearshore water conditions have been studied by means of the MIKE 21 SW coastal propagation model [45]. The model has been previously calibrated and validated applying a multi-collocation-based estimation approach as described in [46]. Such studies perform an optimization procedure for:

1. The WAM offshore hindcast data, consisting in an amplification of each value of WAM dataset by means of an "enhancement factor". The enhancement factor has been obtained by comparison of WAM hindcast dataset at a point located offshore the study area and the time series obtained by transposition of the available data from an offshore wave buoy record. Then, the fictitious WAM time series has been used as input for the numerical model.
2. The nearshores wave propagation model output, applying a two-step calibration strategy by comparison with measurements from acoustic Doppler current profilers and a set of innovative low-cost drifter-derived GPS-based wave buoys [47] located both inside and outside the GoP.

It is worth noting that spreading and dispersion studies and studies of water quality or ecological systems in marine areas are generally carried out by means of coastal hydrodynamic models coupled with sediment transport and particle tracking models. Generally, such modelling is computationally demanding. Therefore, in the perspective to undertake multi-year wave hindcast studies more quickly, a spectral modeling of wave propagation has been used. In order to assess the hypothesis that the bay's hydrodynamics is a factor of influence on pollution patterns, a single wave scenario able to represent the yearly average condition in the bay in terms of wave energy flux, had to be selected. Such a sea state, here defined as "energy equivalent", was obtained considering the wave power content of each wave from the whole dataset of 40-year wave records. Operatively, from the 6-h data of H_s, T_m, θ_m provided by the fictious WAM model, a 6-h wave power and a wave energy dataset were obtained. For a specific sea state, the average wave energy per unit area is proportional to the square of the significant wave height, H_s, according to the known relationship [48]:

$$E_{wave} = \frac{1}{16}\rho g H_s^2 \left[\frac{J}{s}\right] \quad (1)$$

where ρ is the sea water density, assumed equal to 1025 kg/m^3, g is the gravity acceleration (equal to 9.81 m/s^2). For a given spectrum, the significant wave height computation is based on zero-order moment of the spectral function and readily estimated as follows:

$$H_s = H_{m0} = 4\sqrt{m_0} \quad (2)$$

and the wave characteristic/statistic periods can be defined as:

$$T_e = 2\pi \frac{m_{-1}}{m_0} \quad (3)$$

$$T_m = T_{01} = 2\pi \frac{m_0}{m_1} \quad (4)$$

where T_e, T_{01} are the energy period and the spectral mean period, respectively, with the spectral moment being defined as:

$$m_n = \int_0^\infty S(f) f^n df \quad (5)$$

where $n = -2, -1, 0, 1, 2, \ldots$

The wave power in irregular waves can be computed as:

$$P = \frac{\rho \times g^2 \times H_s^2 \times T_e}{64 \times \pi} \tag{6}$$

Following a conservative approach, according to [49,50], the energy period in the present study has been assumed as 1.14 T_m. Equation (6) can be written in the approximate deep water expression:

$$P = 0.459 \times H_s^2 \times T_e \tag{7}$$

For each j-th year, an average wave power ($\overline{P_J}$) has been computed. Considering that the energy period can be computed directly using the approximate formula:

$$T_e = 4.5 \times \sqrt{H_s} \tag{8}$$

then, an "energetic" yearly significant wave height, H_e, can be calculated as:

$$H_{e,j} = \sqrt[2.5]{\frac{P_j}{0.459 \times 4.5}} \tag{9}$$

for $j = 1, \ldots, 40$, and a correspondent energy period can be estimated as follows:

$$T_{ee,j} = 4.5 \times \sqrt{H_{e,j}} \tag{10}$$

The 40-year average for these parameters gives $H_e = 0.93$ m and $T_{ee} = 4.3$ s. Regarding the representative direction, D_e, a vectoral analysis about the energy flows (carried out comparing records of ADCP installed in the bay and correspondent offshore waves) indicate as the offshore wave direction can be considered 217° N. Finally, the "energy equivalent scenario" with H_e, T_{ee} and D_e has been propagated applying the MIKE 21 SW model. The seabed was performed by interpolating at the grid nodes the information provided by the General Bathymetric Chart of the Oceans (GEBCO) database [51]. A gross verification of the 10-m isobath, as representative of the nearshore region in a wave propagation model, has been applied by comparison with a recent bathymetric campaign [19].

2.4. Statistical Analysis: Multivariate Analysis and Bivariate Correlation

All the statistical computations were implemented using the statistical software IBM SPSS Statistics® (version 26.0), while spatial queries, distance calculation and data mapping were managed through the open source environmental software QGIS (version 3.8.3, https://qgis.org/en/site/).

PCA [52] is a powerful technique to investigate patterns of correlations among a set of variables. PCA in fact allows to extract a set of new variables, uncorrelated, called principal components (herein after PCs), each aggregating the variables more strongly correlated with the PC and among themselves. PCA seeks in fact a linear combination of input variables, which extracts the maximum variance of the bivariate correlation matrix. Following this, with a second linear combination, orthogonal from the first one, PCA extracts the maximum variance of the remaining variance, and so on. PCs represent all the linear combinations of the original variables weighted by their contribution in explaining the variance, in a specific orthogonal dimension. PCA was turned into a factor analysis (FA) to reduce the contribution of the less significant parameters within each component, by extracting a new set of vari-factors through rotating the axes defined by the PCA extraction. The Varimax rotation criterion was used to rotate the PCA axes allowing us to maintain the axes, orthogonality. The number of factors to be retained was chosen based on the "eigenvalue higher than 1" criterion (i.e., all the factors that explained less than the variance of one of the original variables were discarded).

To assess the variability within the whole set of contaminants, PCA/FA was performed considering all 11 metals, and all 18 PAHs (Table 1). For each of the N sampling points, PCA enabled to calculate principal component scores aggregating the information of the different parameters' concentration values. Moreover, in order to assess correlations between the sampled contaminants concentrations and the position of sewage discharges and thermal springs, a distance matrix was created through the QGIS vector toolbox, containing the pairwise distance between each sampling point and each discharge point. The measurements-discharges distance matrix was then associated to the component loadings and a bivariate Pearson's linear correlation analysis was performed. Finally, in order to assess the possible effects of the waves' dynamics on the concentration patterns found with PCA, a Kruskal Wallis test was performed to test the significance of hydrodynamics' effect on the contaminant concentrations, likewise similar studies [53–56]. Five wave characteristics were considered, such as the direction, the height, the mean peak period, the energy and the resulting power of the wave motion inside the bay. These characteristics were subdivided into classes, identified based on a hierarchical cluster analysis (HCA) [52] and tested as factors on the vari-factors assumed as dependent variable in the Kruskal Wallis test. Therefore, the KW null hypothesis was the equality of the vari-factor medians across wave characteristics classes. The level of significance assumed was P-value lower than 0.05.

3. Results

3.1. PCA and PCs Extraction

PCA/FA allowed us to extract four rotated PCs accounting for about 86% of cumulative variance from the original 29 variables. Table 2 shows the matrix of the Factor loadings, which represent the correlation of each variable with each rotated PCs.

Table 2. Factor loadings matrix: the factor loading values higher than |0.5| are highlighted in bold and indicated as the significant loadings for the corresponding factor. Total variance explained by the rotated components (Fs) are shown.

Rotated Component matrix				
Contaminant	Factor			
	1	2	3	4
Al	**−0.660**			
As				**0.881**
Cd	**0.746**	0.440		
Cr			**0.789**	
Cu	**0.646**		0.475	
Fe	**0.568**			0.446
Hg	**0.683**			
Ni			**0.551**	
Pb	**0.820**			
V			**0.710**	
Zn	**0.809**			
Naphthalene		**0.915**		
Anthracene	0.516	**0.845**		
Phenanthrene		**0.903**		
Acenaphthylene		**0.928**		
Acenaphthene		**0.92**		
Fluorene		**0.942**		
Fluoranthene	0.670	**0.722**		
Pyrene	**0.705**	0.688		
Benz(a)anthracene	**0.708**	0.688		
Chrysene	**0.723**	0.668		
Benz(b)fluoranthene	**0.788**	0.597		
Benz(a)pyrene	**0.793**	0.584		

Table 2. *Cont.*

Rotated Component matrix				
Contaminant	**Factor**			
	1	**2**	**3**	**4**
Benz(k)fluoranthene	**0.768**	0.619		
Indeno(1,2,3,c,d)pyrene	**0.786**	0.595		
Benzo(g,h,i)perylene	**0.790**	0.584		
Dibenz(a,h)anthracene	**0.743**	0.636		
Benz(j)fluoranthene	**0.763**	0.63		
Benz(e)pyrene	**0.788**	0.59		
% of Variance	38.314	35.447	6.872	5.565
Cumulative %	38.314	73.760	80.543	86.108

Based on factor loadings, the Factors can be interpreted as follows:

1. F1 accounts for 38.3% of the variance and it is loaded mostly by PAHs and by some heavy metals (i.e., Cd, Cu, Fe).
2. F2 accounts for 35.4% of the variance and it is loaded by the remaining PAHs with a lower molecular weight and higher vapor pressure.
3. F3 accounts for 6.8% of the variance and it is loaded by Cr, Ni and V.
4. F4 accounts for 5.5% of the variance and it is loaded by As.

PCA/FA highlights that PAHs concentrations are split into two different pollution components and this is consistent with their mobility characteristics. The molecular weight is in fact considered as an index of the dispersion of a pollutant: the heavier a substance is, in terms of molecular weight, the lower their environmental mobility is (Supplementary Materials: Table S1 shows the PAH molecular weights).

3.2. Factors vs. Sewage Discharges

After the factors identification, a bivariate correlation analysis with the distances from the existent discharges was performed. In Table 3, the Pearson's linear correlation coefficients are shown. It can be observed that the significant correlations are all negative, outlining that higher is the distance of the sampling point from the discharge point, the lower is the pollutant concentration. It is also worthwhile to observe that the points labeled as C, control points of the specific discharge points, show the same pattern of correlation of actual discharge points.

Table 3. Correlation coefficients between the vari-factors obtained by principal component analysis (PCA)/factor analysis (FA) and their distances from the sewage discharges. The correlations highlighted in bold and with a superscript are significant. The number of data are 125.

	F1		F2		F3		F4	
Discharge Points	**Pearson's Correlation**	**Sign. (Two-Tailed)**	**Pearson's Correlation**	**Sign. (Two-Tailed)**	**Pearson's Correlation**	**Sign. (Two-Tailed)**	**Pearson's Correlation**	**Sign. (Two-Tailed)**
1	−0.081	0.368	−0.070	0.438	0.175	0.050	**−0.536 ****	0.000
1C	−0.083	0.357	−0.068	0.450	0.169	0.060	**−0.546 ****	0.000
2	−0.100	0.268	−0.058	0.524	0.135	0.133	**−0.588 ****	0.000
2C	−0.096	0.289	−0.060	0.508	0.142	0.114	**−0.581 ****	0.000
3	−0.118	0.191	−0.055	0.540	0.120	0.180	**−0.595 ****	0.000
4	−0.144	0.108	−0.047	0.602	0.093	0.300	**−0.606 ****	0.000
5	**−0.203 ***	0.024	−0.030	0.746	−0.006	0.945	**−0.590 ****	0.000
5C	**−0.195 ***	0.030	−0.017	0.854	−0.086	0.339	**−0.517 ****	0.000
6	−0.149	0.098	−0.007	0.938	−0.105	0.242	**−0.458 ****	0.000
7	−0.132	0.142	−0.001	0.990	−0.160	0.075	**−0.294 ****	0.001
8	−0.118	0.190	0.003	0.974	−0.162	0.072	**−0.256 ****	0.004
8C	−0.174	0.052	−0.005	0.952	**−0.218 ***	0.015	**−0.210 ***	0.018
9	−0.136	0.130	0.011	0.905	**−0.242 ****	0.006	−0.039	0.667
9C	−0.147	0.101	0.009	0.924	**−0.252 ****	0.005	−0.030	0.738

* Correlation is significant at level 0.05 (two-tailed). ** Correlation is significant at level 0.01 (two-tailed).

The results show that the pollutants included in F1 are inversely correlated with the distance from the discharges 5 and 5C (i.e., the discharge and the control point of the former industrial area). No significant correlations were instead found between the pollutants loaded on F2 and the distances from the discharges, while F3, including Cr, Ni and V, correlates inversely with 8C, 9, 9C discharge points. These discharges are located on the southern side of the Nisida isthmus and are the discharges of the Coroglio Plant 8 and 9 (Figure 3). Finally, F4 showed significant correlations with distances from all discharges except 9 (and 9C). Being F4 loaded only by As, it may be speculated that As concentrations did not depend on specific discharge points, but it may rather derive from the whole coastal area and it is distributed along the whole gulf. The geology of the land surrounding the gulf (i.e., pyroclastic rocks) had in fact characteristics that were consistent with a geogenic origin of the As. Moreover, this analysis suggested that the presence of As in sediments is not necessarily due to the industrial spillage, but it may have a possible natural origin related to the geology of the area.

Based on these results, F2 was the only pollution component, which was not found correlated with any discharge point. This can be a consequence of the higher mobility of these compounds and their tendency to disperse.

3.3. Factors vs. Thermal Springs

As mentioned before, F4 (representative of As contamination pattern) resulted to be inversely correlated to the majority of the discharge points located along the Bagnoli coastline, suggesting that the As contamination in the gulf might not be due to a single source, but it may be caused by a presence of multiple sources (i.e., discharged into the gulf through groundwater). In addition, As might be released through submarine thermal springs that were clustered on the Pozzuoli Bay seabed. In order to investigate this possibility, another correlation analysis was performed. Being As loaded only on F4, the other three components were excluded from this analysis (Table 4).

Table 4. Correlations of As with respect to the position of natural thermal springs present on Bagnoli bay seabed. Significant correlations are highlighted in bold. The number of data are 125.

Discharge Points	F4 (Arsenic)	
	Pearson Correlation	Sig. (Two-Tailed)
T1	0.088	0.329
T2	0.081	0.370
T3	0.051	0.572
T4	−0.036	0.687
T5	**−0.382** **	0.000
T6	**−0.254** **	0.004
T7	−0.067	0.457
T8	−0.110	0.221
T9	−0.001	0.995
T10	**0.233** **	0.009
T11	**0.345** **	0.000
T12	**−0.190** *	0.034
T13	**−0.268** **	0.002

* Correlation is significant at the 0.05 level (2-tailed). ** Correlation is significant at the 0.01 level (2-tailed).

Some positive correlations between As and the distance from Thermal Springs 10 and 11 were found, whereas inverse correlations were found with T5, T6, T12 and T13. Such an ambiguous pattern might be attributed to the fact that only some of those hydrothermal springs were fully active as was shown by recent studies [16,18,19]. As some of those springs are located in the open sea, the positive correlations might be the effect of the bathymetry of the bay [19], where the seabed sinks rapidly from the coastal area, determining the coastal accumulation of the sediments enriched with As.

It is also worth noting that these springs were in the proximity of discharge points (e.g., Thermal Springs 5 and 6 which are close to the "Conca d'Agnano", Discharge 2, or Thermal springs 12 and 13, which are close to Discharges 8 and 9, shown in Figure 3). All these drains are likely to be affected by groundwater and thermal water infiltrations which may possibly contain As [57].

3.4. Pollution Patterns and Wave Hydrodynamics

In order to assess whether the wave hydrodynamics have an influence on pollutant concentrations in sediments, the wave characteristics estimated at each sampling point were interpolated and pollutant vari-factors were tested against five classes of wave characteristic (i.e., wave height, peak period and mean direction) through the Kruskal Wallis test. Figure 5 shows the bathymetry implemented in the wave numerical model, the significant wave height and the energy period as results of the wave propagation of the 40-year averaged energy equivalent scenario.

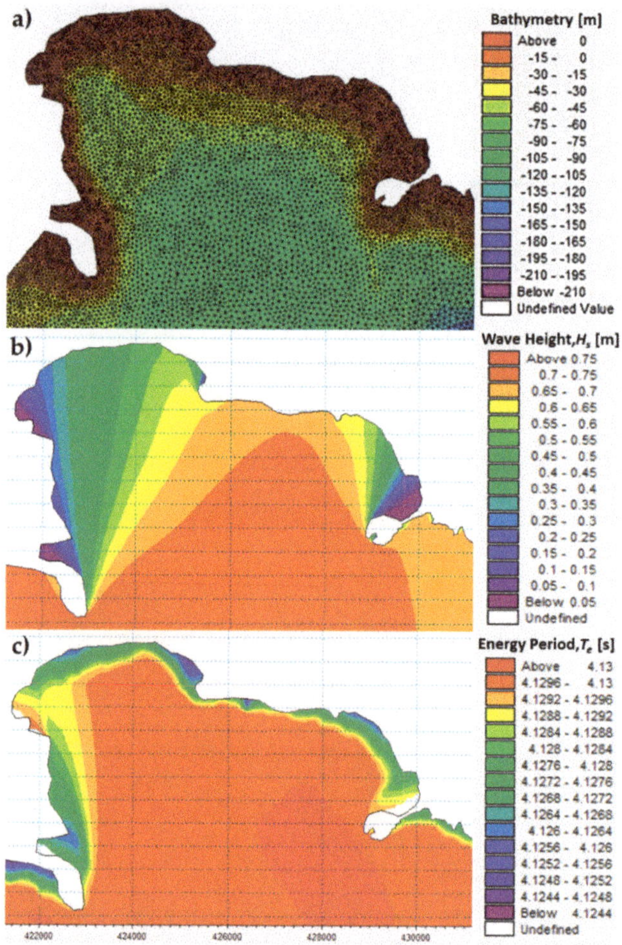

Figure 5. (**a**) Bathymetry implemented in the wave propagation model; (**b**) resulting significant wave height in the bay after propagation of the 40-year averaged energy equivalent sea state; (**c**) the same of (**b**) but in terms of energy period.

Besides wave height, peak period and mean direction, the energy content and the energy flux were also calculated. Figure 6 showed the distribution of the wave energy and wave power at each sampling point.

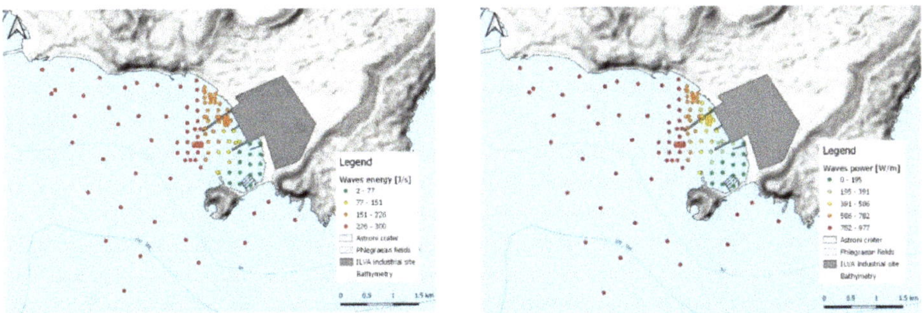

(a) (b)

Figure 6. Distribution of the waves' energy (**a**) and the waves' power (**b**) at each sampling point.

HCA results suggested to subdivide the wave hydrodynamics profiles of the Gulf into four classes of increasing hydrodynamics/energy (see Table 5). These classes have been tested as fixed factors in the Kruskal Wallis test design.

Table 5. Characteristics of the four wave hydrodynamics classes.

Wave Hydrodynamics Classes		Direction	Wave Height	Period	Energy	*P* Wave
1	Mean	224.00	0.14	4.13	14.98	48.34
	Std. Deviation	0.00	0.06	0.00	8.77	28.30
	Median	224.00	0.18	4.13	20.42	65.92
	Minimum	224.00	0.06	4.13	2.27	7.32
	Maximum	224.00	0.18	4.13	20.42	65.93
	N	10	10	10	10	10
2	Mean	207.45	0.54	4.13	190.24	614.10
	Std. Deviation	7.37	0.08	0.00	50.93	164.44
	Median	204.00	0.57	4.13	204.79	660.86
	Minimum	192.00	0.33	4.13	68.64	221.49
	Maximum	224.00	0.63	4.13	250.17	807.78
	N	73	73	73	73	73
3	Mean	193.95	0.67	4.13	283.91	916.66
	Std. Deviation	2.03	0.03	0.00	23.69	76.52
	Median	192.00	0.69	4.13	300.10	968.97
	Minimum	192.00	0.63	4.13	250.17	807.39
	Maximum	196.00	0.69	4.13	300.10	969.02
	N	37	37	37	37	37
4	Mean	196.00	0.69	4.16	300.10	977.18
	Std. Deviation	0.00	0.00	0.00	0.00	0.00
	Median	196.00	0.69	4.16	300.10	977.18
	Minimum	196.00	0.69	4.16	300.10	977.18
	Maximum	196.00	0.69	4.16	300.10	977.18
	N	3	3	3	3	3

The different pollution components (i.e., F1, F2, F3 and F4) showed different results:

1. F1, loaded by the heavier PAH compounds and by some heavy metals and F4 loaded by arsenic, were both found significantly influenced by wave hydrodynamics (test results were respectively H: 12.9; df: 3, $P < 0.01$; and H: 51; df: 3, $P < 0.01$).

2. F2, loaded by the lighter PAHs, and F3, loaded by chromium, nickel and vanadium were not found influenced by the wave hydrodynamics ($P > 0.05$).

Moreover, pairwise comparisons where significance values have been adjusted by the Bonferroni correction for multiple tests allowed to assess that the F1 pollution component had the highest concentration values associated with classes of intermediate hydrodynamics ($P < 0.05$, see Figure 7a.) whereas F4 (and therefore arsenic) was found mostly associated with a low hydrodynamics ($P < 0.05$, see Figure 7b.).

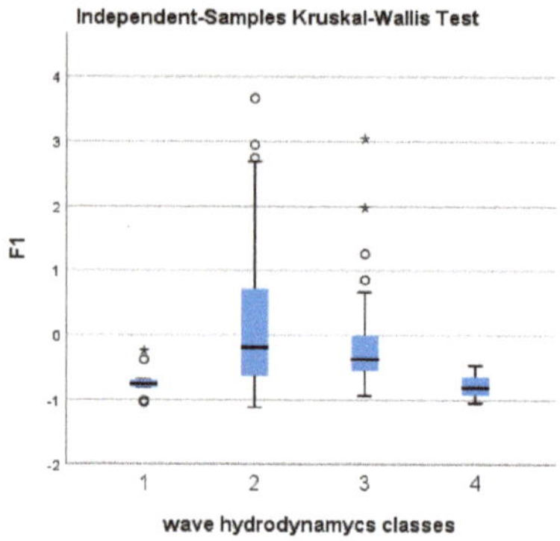

(a)

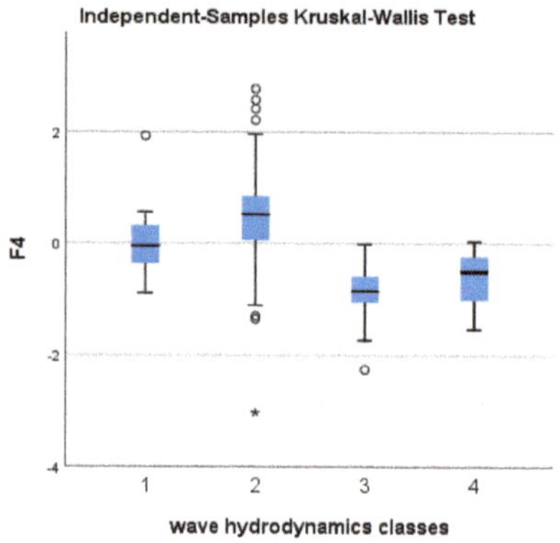

(b)

Figure 7. Box plots showing (**a**) F1 concentration values (heavy PAHs) vs. wave hydrodynamics classes; (**b**) F4 concentration values (As) vs. wave hydrodynamics classes.

Therefore, the PAHs' F1 component appears to be affected by the wave-induced currents, being the concentration pattern of these pollutants in sediments dependent also on the shallow water hydrodynamic parameters in the bay. On the other hand, the wave hydrodynamics was not found to influence the PAHs' F2 component whose higher mobility had probably contributed to homogenize their contamination level in the bay. Likewise, F3 pollutants (i.e., chromium, nickel and vanadium) do not any the wave-induced pattern in the bay. Arsenic (F4) instead was found strongly associated to the areas closest to the coast and characterized by a low wave hydrodynamics.

4. Discussion

This study confirms that a multivariate statistical analysis (PCA/FA) approach can be extremely effective in assessing the apportionment of contaminant sources, even in a site with a complex geological characteristic and a long historical industrial development. With respect to the previous studies (e.g., [3,4,6,7,18]) which used similar methods to investigate the sediments or biota ([31,33,34]), this study considers a larger portion of the Gulf of Bagnoli, and it investigates the relationship of metals/PAHs with both the hydrodynamics of the Gulf and potential discharges/sources. Additionally, while studies in literature considered mostly the sum of total PAHs, in this study 18 PAH compounds are considered individually as their contamination pattern in the gulf. This aspect of the analysis allows us to determine that PAHs of higher molecular weight follow the pattern of metals such as Cd, Cu, Hg, Pb, Zn and Fe and are mainly located near the former discharges of the ILVA steelwork plant. PAHs of lower molecular weight are more dispersed in the gulf. Moreover, while heavier compounds seemed to be influenced by the wave climate of the bay, lighter compounds were apparently much less influenced by it.

Heavy metals such as chromium, nickel and vanadium were found more concentrated near the Nisida northwestern coast and in offshore waters (Figure 8) and did not appear to be strongly influenced by the wave hydrodynamics of the outer gulf. These findings were coherent with the findings of other studies ([4,6,58,59]) even though the measurements in this study derived mostly from drilling and did not include sediment cores.

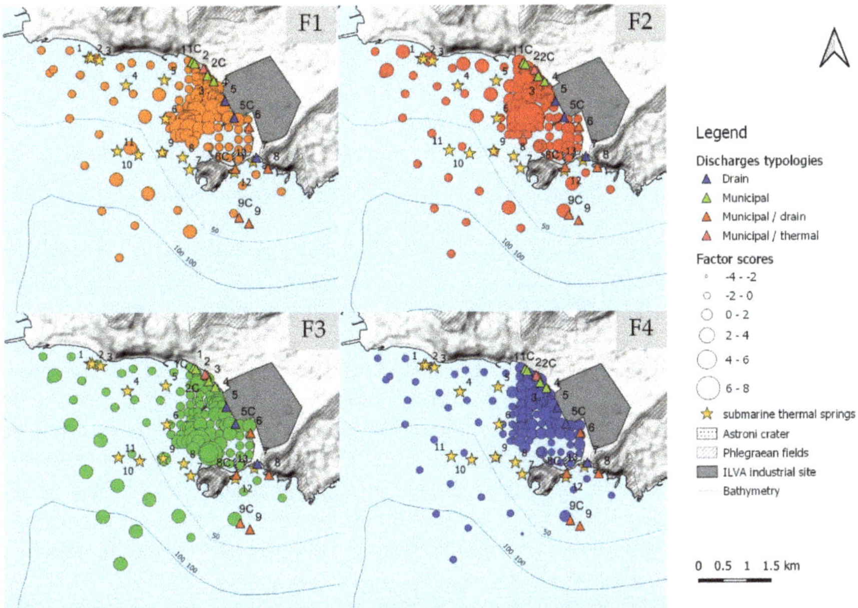

Figure 8. Representation of factor scores inside the bay. The interpolation of factor scores was made with the IDW (inverse distance weighting) method.

Armiento et al. [4] within the same ABBaCo framework had analyzed both drilling (i.e., top layer sediments or TL) and sediment cores and found that that TL samples were characterized by a lower contamination than sediment cores. It is important to note that the correlation patterns found in TL samples reflected what was found in studies that analyzed historical contamination patterns through sediment cores.

The association of Cd, Cu, Hg, Pb, Zn and PAHs was described by Romano et al. through the analysis of sediment cores in a study [6] that determined the historical contamination pattern from industrial activity. The same study found that As, Cr, Ni and V were associated with deeper levels (209–299 cm) of the cores having with the highest percentages of clay fraction, suggesting a prevalent natural contribution for these elements. The only contaminant that is almost completely uncorrelated with the other pollutants is arsenic. This metalloid is significantly present in the bay showing a gradient from offshore to the inshore zones. Arsenic variability was found significantly correlated with both the distance from discharge sites located along the coastline and with some thermal springs present on the seabed. This pattern suggests that it is impossible to attribute the origin to a single source, even when the anthropogenic or the geogenic origin is slightly dominant.

Finally, our study confirms that arsenic is not correlated with other elements and highlights that the spatial pattern of its contamination might be dependent on both a land-driven origin (e.g., As-enriched groundwater, see [58]) and the presence of subaerial/submarine geothermal springs [59]).

The correlation of arsenic with the sewage discharge sites suggests a potential effect due to multiple anthropic activities: the former ILVA steelworks used arsenic in their production cycle but glass factories, located along the coastline in the past, may have contributed to the contamination due to their use of arsenic in glass production. Furthermore, the correlation with thermal springs might account for a "natural" arsenic source and conditions are not infrequent where discharges are characterized by a mixture of anthropogenic and geogenic sources. One example of this is the Conca d'Agnano discharge point which releases a mix of water from Agnano lake (i.e., thermal) and municipal wastewaters. Sewage discharge might also contain As-enriched groundwaters due to gas-water-rock interactions inside the aquifer [57,60]. The diffusion of arsenic in the bay was found to be influenced by the wave hydrodynamics, suggesting that arsenic dispersion in sediments might be attributed to various sources (anthropic, natural, or "mixed") and further diffused due to the particular marine dynamics in the Bagnoli Bay.

5. Conclusions

The results of this study contribute to the reconstruction of the contamination history of the Bagnoli coastal zone, determined by the presence and past activity of industrial sites. Due to the high urbanization of the Neapolitan coastal zone it is impossible to find an unpolluted area-characterized by the same natural environmental framework of the impacted industrial site—to use as a control area. In the present study, a PCA/FA assessment was carried out in the Bagnoli Bay in order to demonstrate the utility of the robust statistical tool in a complex environmental/industrial scenario to investigate the concentrations' variability and their relationship with the locations of sewers, industrial sites, thermal springs, and the wave action inside the GoP. This study confirms the importance of performing statistically robust multidimensional analysis to support the source apportionment assessment of marine sediment contamination. PCA/FA, considering several parameters, allowed better discrimination among the many contamination components affecting the Bagnoli Gulf area and proved to be a very powerful tool in a complex environment with a mixture of effects due to anthropogenic and natural sources. The results of the analyses confirm that the main contamination source is anthropogenic activities (i.e., former steel plant and sewage discharges) but it also suggests the existence of multiple anthropogenic and geogenic sources of arsenic and other metals that might be originating from the volcanic rocks present in the Phlegraean area.

These findings suggest the need to define a "natural background level" (NBL) for the area for arsenic and other heavy metals to distinguish the natural from the anthropogenic component of the contamination. The source apportionment assessment presented here may help define such NBLs, and also facilitate decisions on the contamination control and the remediation management of the area, permitting public authorities to apply knowledge-based management actions.

Supplementary Materials: The following are available online at http://www.mdpi.com/2073-4441/12/8/2181/s1, Table S1: The molecular weights of PAHs, split into to their principal components.

Author Contributions: Conceptualization and methodology: all authors contributed equally. Data interpretation, A.A., R.S., S.G.; resources, D.V., L.M., P.C., G.A.; writing—original draft preparation, S.G. and L.C., A.A.; writing—review and editing, all authors contributed equally. All authors have read and agreed to the published version of the manuscript.

Funding: This research was funded by the Italian Ministry for Education, University and Research (MIUR) through the ABBaCo project, grant number C62F16000170001.

Acknowledgments: We thank "Anton Dohrn" Zoological Station, ISPRA (Institute for the Environmental Protection and Research), CNR (National Research Center), INGV (National Institute of Geophysics and Volcanology) and INVITALIA for providing the information about Bagnoli Bay and about available datasets PAHs and heavy metals concentrations. We thank also Università degli Studi Luigi Vanvitelli for providing the characterization of the Bagnoli Bay wave climate. Further, we gratefully acknowledge Karen Holmberg for having read the paper, giving useful suggestions and correcting the English version. We also thank two anonymous reviewers for useful suggestions, which helped to clarify and improve the paper.

Conflicts of Interest: All authors declare no conflicts of interest.

References

1. Richir, J.; Salivas-Decaux, M.; Lafabrie, C.; Lopez y Royo, C.; Gobert, S.; Pergent, G.; Pergent-Martini, C. Bioassessment of trace element contamination of Mediterranean coastal waters using the seagrass Posidonia oceanica. *J. Environ. Manag.* **2015**, *151*, 486–499. [CrossRef] [PubMed]
2. Morroni, L.; d'Errico, G.; Sacchi, M.; Molisso, F.; Armiento, G.; Chiavarini, S.; Rimauro, J.; Guida, M.; Siciliano, A.; Ceparano, M.; et al. Integrated characterization and risk management of marine sediments: The case study of the industrialized Bagnoli area (Naples, Italy). *Mar. Environ. Res.* **2020**, *160*, 104984. [CrossRef]
3. Romano, E.; Ausili, A.; Zharova, N.; Celia Magno, M.; Pavoni, B.; Gabellini, M. Marine sediment contamination of an industrial site at Port of Bagnoli, Gulf of Naples, Southern Italy. *Mar. Pollut. Bull.* **2004**, *49*, 487–495. [CrossRef] [PubMed]
4. Armiento, G.; Caprioli, R.; Cerbone, A.; Chiavarini, S.; Crovato, C.; De Cassan, M.; De Rosa, L.; Montereali, M.R.; Nardi, E.; Nardi, L.; et al. Current status of coastal sediments contamination in the former industrial area of Bagnoli-Coroglio (Naples, Italy). *Chem. Ecol.* **2020**, *36*(6), 579–597. [CrossRef]
5. Arienzo, M.; Donadio, C.; Mangoni, O.; Bolinesi, F.; Stanislao, C.; Trifuoggi, M.; Toscanesi, M.; Di Natale, G.; Ferrara, L. Characterization and source apportionment of polycyclic aromatic hydrocarbons (pahs) in the sediments of gulf of Pozzuoli (Campania, Italy). *Mar. Pollut. Bull.* **2017**, *124*, 480–487. [CrossRef]
6. Romano, E.; Bergamin, L.; Celia Magno, M.; Pierfranceschi, G.; Ausili, A. Temporal changes of metal and trace element contamination in marine sediments due to a steel plant: The case study of Bagnoli (Naples, Italy). *Appl. Geochem.* **2018**, *88*, 85–94. [CrossRef]
7. Trifuoggi, M.; Donadio, C.; Mangoni, O.; Ferrara, L.; Bolinesi, F.; Nastro, R.A.; Stanislao, C.; Toscanesi, M.; Di Natale, G.; Arienzo, M. Distribution and enrichment of trace metals in surface marine sediments in the Gulf of Pozzuoli and off the coast of the brownfield metallurgical site of Ilva of Bagnoli (Campania, Italy). *Mar. Pollut. Bull.* **2017**, *124*, 502–511. [CrossRef]
8. Bertocci, I.; Dell'Anno, A.; Musco, L.; Gambi, C.; Saggiomo, V.; Cannavacciuolo, M.; Lo Martire, M.; Passarelli, A.; Zazo, G.; Danovaro, R. Multiple human pressures in coastal habitats: Variation of meiofaunal assemblages associated with sewage discharge in a post-industrial area. *Sci. Total Environ.* **2019**, *655*, 1218–1231. [CrossRef]
9. Jenkins, S.H. *Mediterranean Coastal Pollution: Proceedings of a Conference Held in Palma*; Pergamon Press: New York, NY, USA, 1980.
10. Regoli, F.; Giuliani, M.E. Oxidative pathways of chemical toxicity and oxidative stress biomarkers in marine organisms. *Mar. Environ. Res.* **2014**, *93*, 106–117. [CrossRef]
11. Regoli, F.; d'Errico, G.; Nardi, A.; Mezzelani, M.; Fattorini, D.; Benedetti, M.; Di Carlo, M.; Pellegrini, D.; Gorbi, S. Application of a weight of evidence approach for monitoring complex environmental scenarios: The case-study of off-shore platforms. *Front. Mar. Sci.* **2019**, *6*, 377. [CrossRef]
12. Menghan, W.; Stefano, A.; Annamaria, L.; Claudia, C.; Antonio, C.; Wanjun, L.; Marco, S.; Angela, D.; Benedetto, D.V. Compositional analysis and pollution impact assessment: A case study in the Gulfs of Naples and Salerno. *Estuar. Coast. Shelf Sci.* **2015**, *160*, 22–32. [CrossRef]
13. Chapman, P.M. Determining when contamination is pollution—Weight of evidence determinations for sediments and effluents. *Environ. Int.* **2007**, *33*, 492–501. [CrossRef] [PubMed]
14. Colombo, L.; Alberti, L.; Mazzon, P.; Formentin, G. Transient Flow and Transport Modelling of an Historical CHC Source in North-West Milano. *Water* **2019**, *11*, 1745. [CrossRef]
15. Stevenazzi, S.; Masetti, M.; Beretta, G. Pietro Groundwater vulnerability assessment: From overlay methods to statistical methods in the Lombardy Plain area. *Acque Sotter.-Ital. J. Groundw.* **2017**, *6*, 1–11.
16. Troise, C.; De Natale, G.; Schiavone, R.; Somma, R.; Moretti, R. The Campi Flegrei caldera unrest: Discriminating magma intrusions from hydrothermal effects and implications for possible evolution. *Earth-Sci. Rev.* **2019**, *188*, 108–122. [CrossRef]
17. Rotiroti, M.; Sacchi, E.; Fumagalli, L.; Bonomi, T. Origin of Arsenic in Groundwater from the Multilayer Aquifer in Cremona (Northern Italy). *Environ. Sci. Technol.* **2014**, *48*, 5395–5403. [CrossRef]
18. Albanese, S.; De Vivo, B.; Lima, A.; Cicchella, D.; Civitillo, D.; Cosenza, A. Geochemical baselines and risk assessment of the Bagnoli brownfield site coastal sea sediments (Naples, Italy). *J. Geochem. Explor.* **2010**, *105*, 19–33. [CrossRef]

19. Somma, R.; Iuliano, S.; Matano, F.; Molisso, F.; Passaro, S.; Sacchi, M.; Troise, C.; De Natale, G. High-resolution morpho-bathymetry of Pozzuoli Bay, southern Italy. *J. Maps* **2016**, *12*, 222–230. [CrossRef]
20. Idowu, O.; Carbery, M.; O'Connor, W.; Thavamani, P. Speciation and source apportionment of polycyclic aromatic compounds (PACs) in sediments of the largest salt water lake of Australia. *Chemosphere* **2020**, *246*, 125779. [CrossRef]
21. Yuanan, H.; He, K.; Sun, Z.; Chen, G.; Cheng, H. Quantitative source apportionment of heavy metal(loid)s in the agricultural soils of an industrializing region and associated model uncertainty. *J. Hazard. Mater.* **2020**, *391*, 122244. [CrossRef]
22. Henriksson, S.; Hagberg, J.; Bäckström, M.; Persson, I.; Lindström, G. Assessment of PCDD/Fs levels in soil at a contaminated sawmill site in Sweden—A GIS and PCA approach to interpret the contamination pattern and distribution. *Environ. Pollut.* **2013**, *180*, 19–26. [CrossRef] [PubMed]
23. Meng, L.; Zuo, R.; Wang, J.-S.; Yang, J.; Teng, Y.-G.; Shi, R.-T.; Zhai, Y.-Z. Apportionment and evolution of pollution sources in a typical riverside groundwater resource area using PCA-APCS-MLR model. *J. Contam. Hydrol.* **2018**, *218*, 70–83. [CrossRef] [PubMed]
24. Rahman, M.S.; Hossain, M.B.; Babu, S.M.O.F.; Rahman, M.; Ahmed, A.S.S.; Jolly, Y.N.; Choudhury, T.R.; Begum, B.A.; Kabir, J.; Akter, S. Source of metal contamination in sediment, their ecological risk, and phytoremediation ability of the studied mangrove plants in ship breaking area, Bangladesh. *Mar. Pollut. Bull.* **2019**, *141*, 137–146. [CrossRef] [PubMed]
25. Azzellino, A.; Colombo, L.; Lombi, S.; Marchesi, V.; Piana, A.; Merri, A.; Alberti, L. Groundwater diffuse pollution in functional urban areas: The need to define anthropogenic diffuse pollution background levels. *Sci. Total Environ.* **2019**, *656*, 1207–1222. [CrossRef]
26. Alamdar, R.; Kumar, V.; Moghtaderi, T.; Naghibi, S.J. Groundwater quality evaluation of Shiraz City, Iran using multivariate and geostatistical techniques. *SN Appl. Sci.* **2019**, *1*, 1367. [CrossRef]
27. Sheikhy Narany, T.; Ramli, M.F.; Aris, A.Z.; Sulaiman, W.N.A.; Fakharian, K. Spatiotemporal variation of groundwater quality using integrated multivariate statistical and geostatistical approaches in Amol-Babol Plain, Iran. *Environ. Monit. Assess.* **2014**, *186*, 5797–5815. [CrossRef]
28. Lattuada, M.; Albrecht, C.; Wilke, T. Differential impact of anthropogenic pressures on Caspian Sea ecoregions. *Mar. Pollut. Bull.* **2019**, *142*, 274–281. [CrossRef]
29. Bajt, O.; Ramšak, A.; Milun, V.; Andral, B.; Romanelli, G.; Scarpato, A.; Mitrić, M.; Kupusović, T.; Kljajić, Z.; Angelidis, M.; et al. Assessing chemical contamination in the coastal waters of the Adriatic Sea using active mussel biomonitoring with Mytilus galloprovincialis. *Mar. Pollut. Bull.* **2019**, *141*, 283–298. [CrossRef]
30. Stefania, G.A.; Rotiroti, M.; Buerge, I.J.; Zanotti, C.; Nava, V.; Leoni, B.; Fumagalli, L.; Bonomi, T. Identification of groundwater pollution sources in a landfill site using artificial sweeteners, multivariate analysis and transport modeling. *Waste Manag.* **2019**, *95*, 116–128. [CrossRef]
31. Romano, E.; Bergamin, L.; Ausili, A.; Pierfranceschi, G.; Maggi, C.; Sesta, G.; Gabellini, M. The impact of the Bagnoli industrial site (Naples, Italy) on sea-bottom environment. Chemical and textural features of sediments and the related response of benthic foraminifera. *Mar. Pollut. Bull.* **2009**, *59*, 245–256. [CrossRef]
32. Sinaei, M.; Zare, R. Polycyclic aromatic hydrocarbons (PAHs) and some biomarkers in the green sea turtles (Chelonia mydas). *Mar. Pollut. Bull.* **2019**, *146*, 336–342. [CrossRef] [PubMed]
33. Gambi, C.; Dell'Anno, A.; Corinaldesi, C.; Lo Martire, M.; Musco, L.; Da Ros, Z.; Armiento, G.; Danovaro, R. Impact of historical contamination on meiofaunal assemblages: The case study of the Bagnoli-Coroglio Bay (southern Tyrrhenian Sea). *Mar. Environ. Res.* **2020**, *156*, 104907. [CrossRef] [PubMed]
34. Tangherlini, A.; Corinaldesi, M.; Rastelli, C.; Musco, E.; Armiento, L.; Danovaro, G.; Dell'Anno, R. Chemical contamination can promote turnover diversity of benthic prokaryotic assemblages: The case study of the Bagnoli-Coroglio bay (Southern Tyrrhenian Sea). *Mar. Environ. Res.* **2020**, in press. [CrossRef]
35. Soclo, H.H.; Garrigues, P.; Ewald, M. Origin of polycyclic aromatic hydrocarbons (PAHs) in coastal marine sediments: Case studies in Cotonou (Benin) and Aquitaine (France) Areas. *Mar. Pollut. Bull.* **2000**, *40*, 387–396. [CrossRef]
36. Del Gaudio, C.; Aquino, I.; Ricciardi, G.P.; Ricco, C.; Scandone, R. Unrest episodes at Campi Flegrei: A reconstruction of vertical ground movements during 1905–2009. *J. Volcanol. Geotherm. Res.* **2010**, *195*, 48–56. [CrossRef]

37. Lima, A.; De Vivo, B.; Spera, F.J.; Bodnar, R.J.; Milia, A.; Nunziata, C.; Belkin, H.E.; Cannatelli, C. Thermodynamic model for uplift and deflation episodes (bradyseism) associated with magmatic-hydrothermal activity at the Campi Flegrei (Italy). *Earth-Sci. Rev.* **2009**, *97*, 44–58. [CrossRef]
38. Isaia, R.; Vitale, S.; Marturano, A.; Aiello, G.; Barra, D.; Ciarcia, S.; Iannuzzi, E.; Tramparulo, F.D.A. High-resolution geological investigations to reconstruct the long-term ground movements in the last 15 kyr at Campi Flegrei caldera (southern Italy). *J. Volcanol. Geotherm. Res.* **2019**, *385*, 143–158. [CrossRef]
39. Rosi, M.; Sbrana, A.; Principe, C. The phlegraean fields: Structural evolution, volcanic history and eruptive mechanisms. *J. Volcanol. Geotherm. Res.* **1983**, *17*, 273–288. [CrossRef]
40. Sellerino, M.; Forte, G.; Ducci, D. Identification of the natural background levels in the Phlaegrean fields groundwater body (Southern Italy). *J. Geochemical Explor.* **2019**, *200*, 181–192. [CrossRef]
41. De Vivo, B.; Lima, A. *Characterization and Remediation of a Brownfield Site. The Bagnoli Case in Italy*; Elsevier B.V.: Amsterdam, The Netherlands, 2008.
42. ABBaCo Project, 2018. Sperimentazioni Pilota Finalizzate al Restauro Ambientale e Balneabilità del SIN Bagnoli-Coroglio. 2018. Available online: http://www.szn.it/index.php/en/research/integrative-marine-ecology/research-projects-emi/abbaco (accessed on 8 July 2020).
43. Musco, L.; Bertocci, I.; Buia, M.C.; Cannavacciuolo, M.; Conversano, F.; Gallo, A.; Gambi, M.C.; Ianora, A.; Iudicone, D.; Margiotta, F.; et al. Restauro ambientale e balneabilità a del SIN di Bagnoli Coroglio-Progetto ABBaCo Workshop SiCon2017. Siti contaminati. In Proceedings of the Esperienze Negli Interventi di Risanamento. Roma, Facoltà Ingegneria Civile ed Industriale. La Sapienza, Roma, Italy, 8–10 February 2017.
44. European Centre for Medium-Range Weather Forecasts. Available online: http://www.ecmwf.int/ (accessed on 8 July 2020).
45. Dhi Water and Environment. Available online: https://www.dhigroup.com/ (accessed on 8 July 2020).
46. Contestabile, P.; Conversano, F.; Centurioni, L.; Golia, U.M.; Musco, L.; Danovaro, R.; Vicinanza, D. Multi-collocation-based estimation of wave climate in a non-tidal bay: The case study of Bagnoli-Coroglio bay (Tyrrhenian Sea). *Water* **2020**, *12*, 1936. [CrossRef]
47. Centurioni, L.; Braasch, L.; Di Lauro, E.; Contestabile, P.; De Leo, F.; Casotti, R.; Franco, L.; Vicinanza, D. A New strategic wave measurement station off Naples Port main breakwater. *Coast. Eng. Proc.* **2017**, *1*, 36. [CrossRef]
48. Holthuijsen, L.H. Waves in oceanic and coastal waters. *Waves Ocean. Coast. Waters* **2007**, *9780521860*, 1–387.
49. Contestabile, P.; Vicinanza, D. Coastal defence integratingwave-energy-based desalination: A case study in Madagascar. *J. Mar. Sci. Eng.* **2018**, *6*, 64. [CrossRef]
50. Contestabile, P.; Di Lauro, E.; Galli, P.; Corselli, C.; Vicinanza, D. Offshore Wind and Wave Energy Assessment around Malè and Magoodhoo Island (Maldives). *Sustainability* **2017**, *9*, 613. [CrossRef]
51. General Bathymetric Chart of the Oceans. Available online: http://www.gebco.net/ (accessed on 8 July 2020).
52. Afifi, A.; May, S.; Clark, V.A. Cluster analysis. In *Computer-Aided Multivariate Analysis*; Chapman & Hall-CRC Press: London, UK, 2003.
53. Baltas, H.; Sirin, M.; Dalgic, G.; Bayrak, E.Y.; Akdeniz, A. Assessment of metal concentrations (Cu, Zn, and Pb) in seawater, sediment and biota samples in the coastal area of Eastern Black Sea, Turkey. *Mar. Pollut. Bull.* **2017**, *122*, 475–482. [CrossRef]
54. Safakhah, N.; Ghanemi, K.; Nikpour, Y.; Batvandi, Z. Occurrence, distribution, and risk assessment of bisphenol A in the surface sediments of Musa estuary and its tributaries in the northern end of the Persian Gulf, Iran. *Mar. Pollut. Bull.* **2020**, *156*, 111241. [CrossRef]
55. Kim, S.; Lee, Y.S.; Moon, H.B. Occurrence, distribution, and sources of phthalates and non-phthalate plasticizers in sediment from semi-enclosed bays of Korea. *Mar. Pollut. Bull.* **2020**, *151*, 110824. [CrossRef]
56. Renzi, M.; Romeo, T.; Guerranti, C.; Perra, G.; Italiano, F.; Focardi, S.E.; Esposito, V.; Andaloro, F. Temporal trends and matrix-dependent behaviors of trace elements closed to a geothermal hot-spot source (Aeolian Archipelago, Italy). *Procedia Earth Planet. Sci.* **2011**, *4*, 10–28. [CrossRef]
57. Valentino, G.M.; Stanzione, D. Source processes of the thermal waters from the Phlegraean Fields (Naples, Italy) by means of the study of selected minor and trace elements distribution. *Chem. Geol.* **2003**, *194*, 245–274. [CrossRef]

58. Aiuppa, A.; Avino, R.; Brusca, L.; Caliro, S.; Chiodini, G.; D'Alessandro, W.; Favara, R.; Federico, C.; Ginevra, W.; Inguaggiato, S.; et al. Mineral control of arsenic content in thermal waters from volcano-hosted hydrothermal systems: Insights from island of Ischia and Phlegrean Fields (Campanian Volcanic Province, Italy). *Chem. Geol.* **2006**, *229*, 313–330. [CrossRef]
59. Celico, P.; Dall'Aglio, M.; Ghiara, M.R.; Stanzione, D.; Brondi, M.; Prosperi, M. Geochemical monitoring of the thermal fluids in the Phlegeran Fields from 1970 to 1990. *Boll. Soc. Geol. Ital.* **1992**, *111*, 409–442.
60. De Vivo, B.; Lima, A. *The Bagnoli-Napoli Brownfield Site in Italy: Before and After the Remediation*, 2nd ed.; Elsevier B.V.: Amsterdam, The Netherlands, 2018.

water

MDPI

Review

Coastal Erosion Management at Callao (Peru) in the 17th and 18th Centuries: The First Groin Field in South America?

Carlos Mestanza [1,2], Marco Piccardi [3] and Enzo Pranzini [3,*]

1 Facultad de Ciencias de la Vida, Universidad Estatal Amazonica, Sucumbos EC210150, Ecuador; cmestanza@ug.uchile.cl
2 Research Center, Green Amazon, Nueva Loja EC110150, Ecuador
3 Dipartimento di Scienze della Terra, Università di Firenze, 50121 Firenze, Italy; m.piccardi@tin.it
* Correspondence: epranzini@unifi.it; Tel.: +39-055-275-6547

Received: 6 June 2018; Accepted: 27 June 2018; Published: 4 July 2018

Abstract: Analysis of written documents, projects and historical cartography of the 17th and 18th centuries allows reconstruction of the shore protection works performed at Callao (Peru) to defend the settlements and its boundary wall from storm waves and tsunamis. Groins appeared for the first time in early 18th century maps, together with gently sloping revetments in an unrealized project of the same period. Comparisons between Callao projects and those performed in Europe in the same centuries show a uniformity in design and construction materials that overpasses the environmental differences of the sites. Successes and failures followed each other and, although an understanding of coastal dynamics and the positive and negative effects of the various works were known, it was not possible to guarantee the stability of the walls and consequently the safety of the city from sea attack. A strategic retreat was even considered.

Keywords: historical cartography; shore protection structures history; groins; gently sloping revetment; earthquake; tsunami; strategic retreat

---

1. Introduction

Coastal erosion, because of sea level rise, subsidence, reduced river sediment input and, more recently, anthropogenic interventions on the watershed, is a natural process that humans have tended to oppose ever since they settled near the shoreline [1]. At sites, the construction of ports, docks, piers, breakwaters along the coast itself induced or increased the process [2]. In some areas, hurricanes e.g., [3] and tsunamis e.g., [4] transformed this slowly-evolving process into a catastrophic one. Nevertheless, coastal settlements increased and expanded during the centuries, thanks to their advantageous position for climate, resources access, movements and—recently—recreation [5].

Where it was not possible to retreat, shore protection structures had to be built to hold the current position, either for the need to stay close to water (commercial, military, industrial purpose) or because safer land was unavailable; frequently, it was investment on reclamations done in that specific area that discouraged retreat.

Different materials were used in the past (rock, wood, fagots) in a variety of configurations, such as revetments, rip-raps, seawalls, detached breakwaters, and groins. Most ancient structures were subjected to continuous and expensive maintenance, and now are lost, either because land was abandoned to the sea (generally rural areas) or because they have been continuously replaced with stronger ones.

Elsewhere, military defences (e.g., boundary walls) also served as shore protection structures, such as seen in Roman and pre-Roman maritime cities. To reconstruct the evolution of these defence structures, written documents, drawings, and old maps are the main sources of information. It is known that earth and stones were used for seawall construction in 713 AD at Quintang Estuary (China), further replaced with bamboo baskets filled with stones in 910 AD [6], and rock seawalls were built in China in the 14th century [7]. Due to the widespread coastal accretion that characterized Europe until the 18th century (generally as a consequence of deforestation), protection works of the coast were carried out in limited areas, mostly along the North Sea coast. Nieuwhof [8] referring to the excavations carried out by Bazelmans [9], and by Bazelmans et al. [10] describes the northwest Friesland dikes (no more than 70 cm high, composed of neatly-stacked peat sods against a core of loose bulk material) of Peins-Oost (1st century B.C.) and Dongjum-Heringa (2nd century A.D.).

Dykes have been present since the medieval period: in Germany since the 11th century to oppose storm surges [11], the 12th century in the Netherlands—frequently associated with land reclamation [12], the 14th century in Belgium [13], and the 16th century in Denmark [14]. These were frequently made by materials found in situ: sand plus clay and peat reinforced with wood or fagots, and these materials were used, in different combinations, when rocks became unavailable. Importing rocks and stones from Scandinavia was expensive and its use in coastal protection was parsimonious. Building with fagots (*fascinage*) is a long-lasting tradition in protection of river banks and coasts. As *fascinage* works are nowadays used in several undeveloped countries, one may infer that many centuries ago they were present all over the world's coasts, where erosion was threatening settlements or cultivated areas. More sophisticated, but still primitive structures include gabions made of twigs and branches, in the manner of basket-work filled with pebbles and laid one on top of the other when they had to reach a greater thickness (Figure 1).

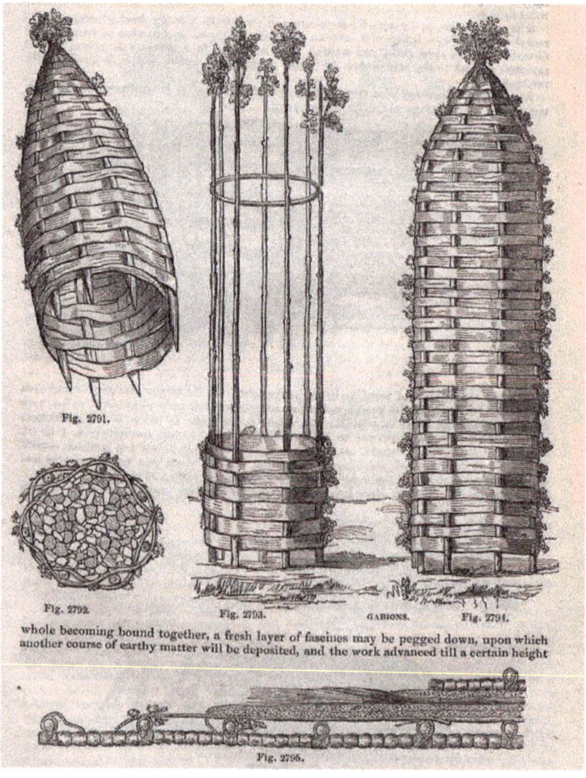

Figure 1. Gabions construction scheme by Cresy [15].

Groins built with these "technologies" were located on the sea side of fortified cities in the 16th–18th centuries, as shown in several ancient maps, e.g., Flissingen (Figure 2) and Ostende (Figure 3). Their functioning was to prevent scouring (and therefore collapsing), affecting structures directly exposed to wave actions due to longshore transport, backwash and, to a lesser extent, direct cross-shore wave action [16]. On account of this, for any tower, fortress and walled town located on the beach, it was of outmost importance to prevent such a process, and this was frequently achieved by constructing groins, which intercept longshore transport. On the other hand, vertical breakwaters were built by the Romans just to prevent harbor entrance siltation; but in this case a stout rock foundation was laid at the base [17].

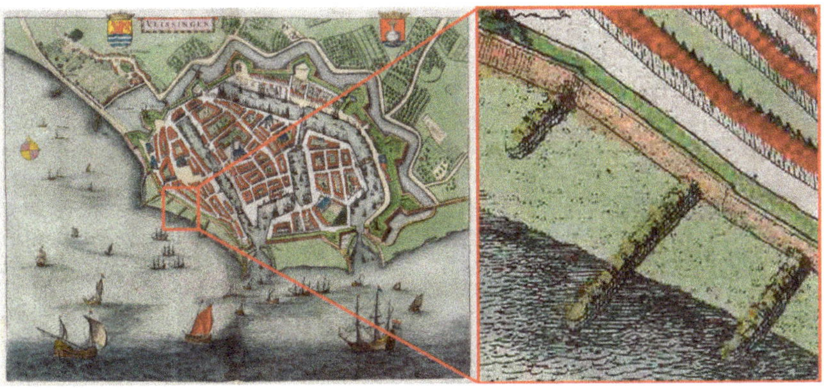

Figure 2. Groins at Flissingen (The Netherlands) in a 1649 map (ed. 1652) by Joan Blaeu [18]. Courtesy Universiteitsbibliotheek Utrecht.

Maps do not always allow identification of the design of individual groins and the material used, but in some cases, pictures can be extremely detailed and annotations give further information. For example, Figure 2 shows several shore protection structures made up of wood piles boxes, but it is impossible to see what lies inside. More explicative is the 1734 map of Ostende (Figure 3), where groins are better drawn and writing in Latin *Cistae ligneae lapidibus repletae* and in Frisian *Holtzerne Kuften mit steunen ange fillet* explains that they are formed by wood baskets filled with stones [19]. Siltation, in this case and in that of Flissingen, was not a problem for navigation and beaching, since access to the town was guaranteed through the river mouth or channels entrances protected by jetties.

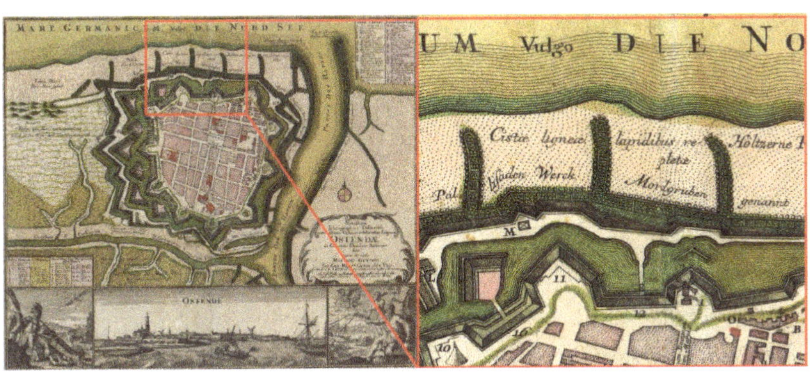

Figure 3. Ostende in a 1734 map by Seutter [20] (detail on the right).

These vernacular structures, known from antiquity, are still present in many countries where local coastal populations struggle against the sea. Being recently adopted by "green engineering", the knowledge of their design and efficiency could help future projects.

2. Materials and Methods

For the present study, approximately 50 maps, plans and views drawn or printed from the 17th to 18th centuries, representing the port of Callao and its surroundings, were analyzed. Further, several texts were analyzed, both accompanying the previous documents and independently. They were the result of European expeditions to the South Seas, organized by different navies (Spanish, Dutch, French, English, etc.) and their content cannot be ignored for a full comprehension of the historical iconography. In addition, there are drawings apparently isolated from any context, whose collocation is more difficult.

On this matter, a comparative, synchronic and diachronic analysis was performed to identify original documents and locate them in true time, discarding late copies or variants, which frequently constitute a trap for hasty scholars.

In this analysis, one must be aware of the fact that maps and drawings had a strategic importance, and copyists in the service of the various countries were reproducing documents, frequently stolen, to enemies of competing fleets. On the other hand, publishers and booksellers were deleting original dates and adding a more recent one to make the book more inviting and to sell out remnants [21]; similarly, maps were "updated" only changing their age [22].

In this case, dating is even harder. Evidences of this state of things comes from the scene at Callao published in 1729 within a collection of views by van der Aa [23] (Figure 4), being nothing else, but a reproduction of a 1671 Dutch view [24]. These older documents actually were probably inspired by the view of the naval blockade of Callao in May 1624 by the 11 ships of the "Nassau Fleet" under Jacques l'Hermite [25], and Figure 5. In this case, the state of the wall and of the beach, which is our interest, is very likely to be that of 1624 and not of 1729.

Figure 4. The first wall of Callao in a detail of a view published in a collection dated 1729 by van der Aa [23], which is actually a reproduction of a 1671 Dutch view. Courtesy: Bibliothèque nationale de France.

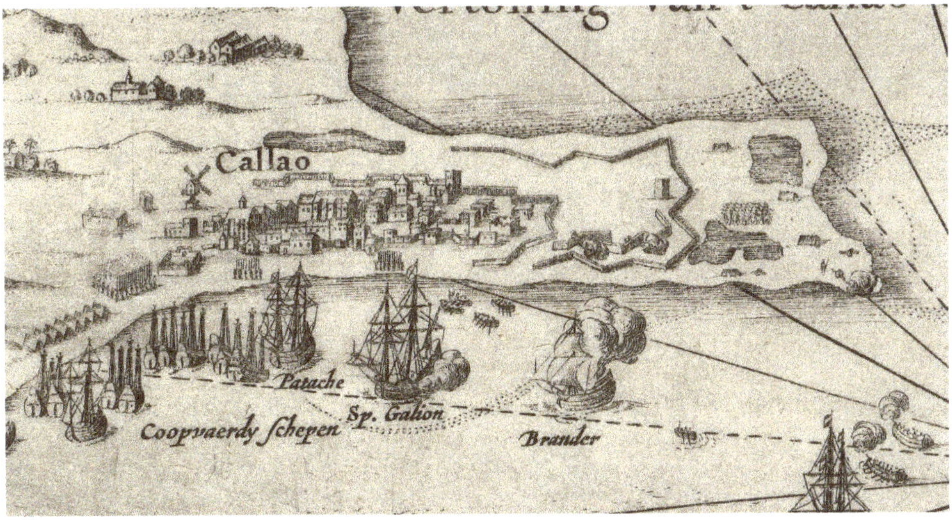

Figure 5. Callao in a 1624. Detail of view by an unknown author published in ca. 1663 [26]. Courtesy: Rijksmuseum, Amsterdam.

As far as accuracy is concerned, Callao maps and views used in this study, although not suitable to quantify beach evolution, can give accurate information on the presence/absence of coastal structures and on their effect on coastal morphology (i.e., silting side), especially when flanked by written descriptions.

Shore Protection at Callao

The Peruvian coast stretches along the north-west South America continental margin, a very tectonically active area [4] exposed to earthquakes, landslides, floods, sea storms and tsunamis. Callao, the most important harbour in Peru (Figure 6), had to face these catastrophic events since its foundation in 1537. Its first boundary wall, built at the beginning of the 17th century, was replaced by new walls between 1640 and 1647 [27]. These primarily had a military function, but on its sea side they also had to protect the settlement from wave attack, which proved to be stronger than that of any enemy's fleets, as written in chronicles and shown in Figure 7.

The history of the town and of its walls has been recounted by Melo in 1899–1900 [29], Arrũs in 1905 [27] and later by Lohmann Villena in 1963 [30]. They, and mostly the latter, gave a detailed account of the evolution of the settlement and its defensive structures from an historical perspective, but also provided sound technical information on the shore protection structures built to defend the town from the sea; these stimulate a comparison with present day knowledge of coastal engineering.

The area where Callao was built in 1537 has been inhabited since the pre-ceramic period (Cotton pre-ceramic, 2500–1750 B.C. [31], but no buildings are drawn on a 1624 view, except those forming a small colonial settlement (Figure 4), which quickly became the most important harbour on the South America Pacific coast.

A detailed analysis of 17th century events is provided by Lohmann Villena [30]. According to his reconstruction, the fence present on the 1624 map was not completed, and between 1640 and 1647 new walls were built, but their seaward side segment soon had structural problems, mostly due to overtopping by storm wave attacks in 1647 and 1651.

Anyhow, the coup de grace to the city and walls occurred with the 1655 earthquake (estimated magnitude 7.4 with epicentre in front of Isla de San Lorenzo [32]) and the following tsunami, which destroyed part of the walls, the Jesuit Church and the Viceroy residence. A further storm in 1656 created additional damages in the unprotected town. According to Lohman-Villena [30], to replace the collapsed wall segment, a breakwater (18.50 m long, 3.36 m wide and 4.20 high) was built. However, breaches created by further earthquakes are visible in the map copied by Lemoine [33] and attributable to the years 1724–1727 (Figure 7).

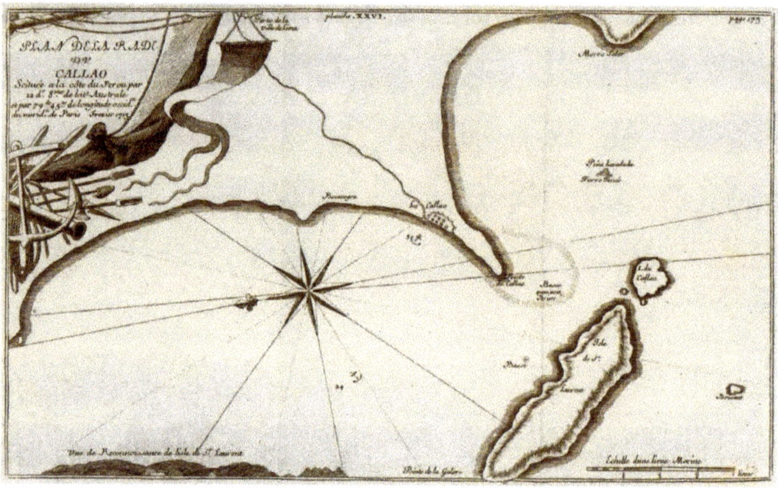

Figure 6. The bay of Callao and Isla San Lorenzo in a map surveyed by Frézier [28] in 1713 and published in 1716. Courtesy: Zentralbibliothek, Zürich.

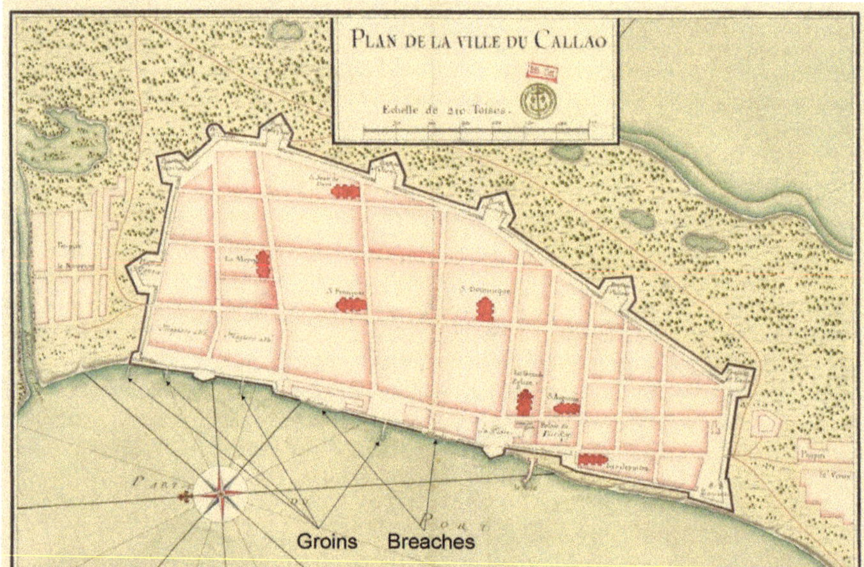

Figure 7. Groins in Callao (1724–1727) in a city plan copied by Lemoine [33]. Courtesy: Bibliothèque nationale de France.

Earthquakes continued to hit Callao even in the following decades, and the continuous interventions did not stop wall deterioration. After the 1687 earthquake, repairs were done on the seaside wall placing masonry and planting a crescent-shaped fence with three rows of stakes filled with stones [30] (p. 132). In 1694, a pier was built, which extended it some 60 m further (Figures 7–9) with stones brought from the neighboring San Lorenzo island (Figure 6). This pier (stonekey) worsened the wall stability condition, as realized by Frézier during his visit in 1713: *This Port was in a bad Condition in the Year 1713; there were five Breaches in it, and the Sea daily ruins the Wall, fince there has been a Stone Key built, the Situation whereof stops the S. W. Surf, and occasions a Return of the Water, which saps the Walls of the Town* [28] English edition [34] (p. 196).

Between the end of the 17th and the beginning of the 18th century, several gabions were built using piles from straight trees coming from Guayaquil (Ecuador), each 7 to 9 m long and 0.5 m in diameter, locked with ribbons and nailed. Using a pile driver, they were hammered to a depth of 2 to 2.5 m, their interstices being filled with large stones, as reported by Lohmann Villena [30] (p. 134).

In the 18th century, the rise of the Bourbon dynasty in Spain and the *pacto de familia* with France increased the presence of French military engineers in Spanish colonies. From 1707 to 1711, Jean Baptiste de Rosmain was in charge as Inspector of Callao fortifications. To prevent enemies from climbing over the wall, he eliminated sand that had accumulated against it [30] (p. 137).

Maybe protection of the wall from sea attack was deputed to a *glacis qui règne le long de la cité nouvelle muraille bien pavè et entrelassè de gros madriers jusques au bord de la mer* (gently sloping revetment running along the wall and well paved with sheet interlaced with large planks up to the shore line; as in Figure 9 map cartouche).

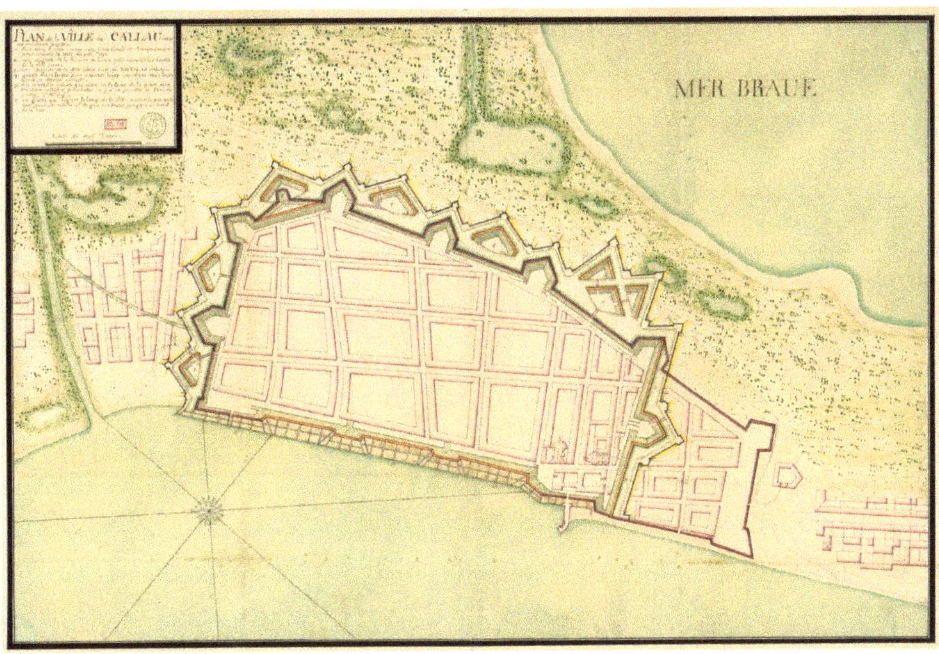

Figure 8. Callao: project of the walls and of their protection on the sea side formed by wooden board boxes filled with well-arranged stones [35]. Courtesy: Bibliothèque nationale de France.

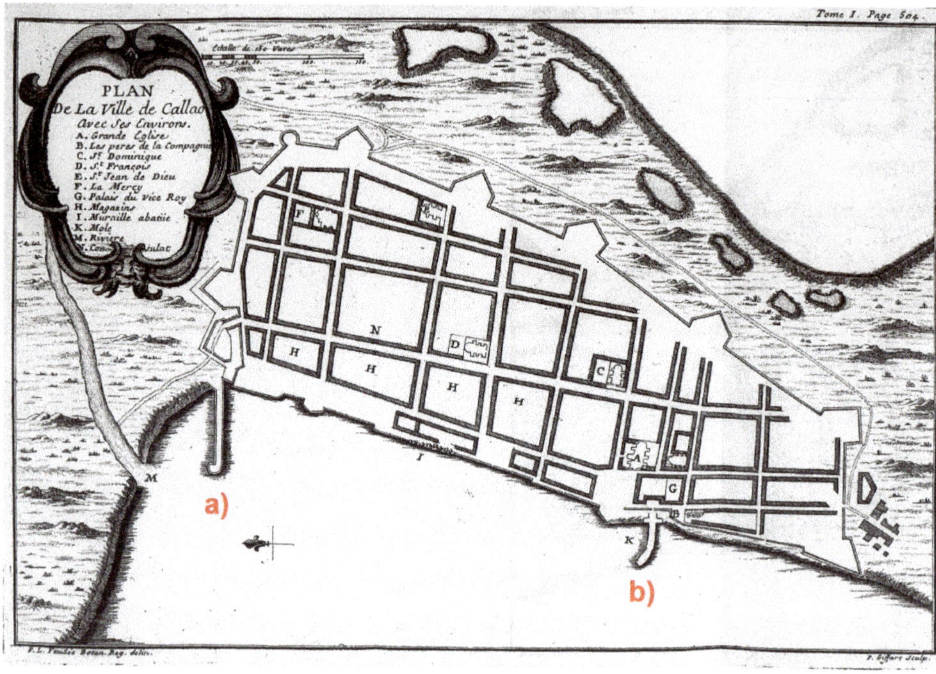

Figure 9. Callao in a city plan drawn by Feuillée [36], in 1710 (1714 ed). On the eastern side of the town, the jetty designed by de Rosmain and never realized (**a**); on the western side the 1694 jetty (**b**). Courtesy: ETH-Bibliothek Zürich.

This unrealized project is known thanks to a drawing by Lemoine [35] (Figure 8)—probably the cartographer François-Pierre Le Moyne, 1713–1795—employed at the *Dépot des cartes et plans* of the French navy, where he worked also as a map copyist at reproducing, inter alia, many maps of South America. De Rosmain also designed a jetty to be built on the eastern side of the town to host water supply boats, which is visible on a map by Feuillée dated 1710 [36] (Figure 9), but never realized. In any case, no intervention had the desired effect, so that in the 1720s the hypothesis to rebuild Callao further from the sea was advanced—a forerunner project of strategic retreat.

After de Rosmain's death, the responsibility for Callao's defensive structures was inherited by Pedro de Peralta Barnuevo, mathematician and astronomer, later assisted by engineer Nicols Rodriguez. The latter, as previously done by Frézier, paid attention to wall scouring induced by the 1694 jetty. To prevent this process, they proposed to build groins to intercept longshore transport in order to create a beach in front of the walls. They were placed 'at proportional distances so that the effect of each reached to the next' [30] (p.143). Construction started in 1724 and Figure 7 is very likely related to the first construction step, with four out of the eight designed groins, which appear in Figure 10.

Here, the groin length/spacing ratio is approximately 1/2–1/4, similar to values found in most groin fields today [38]. Four shore parallel gabions lines were built between the groins with the purpose of retaining sediments dragged by waves. No information was given on the construction material, but from the drawing they seem very similar to the *cistae ligneae* present at Ostenda (Figure 3).

The analysis of hundreds of maps representing 17th–18th century coastal settlements in South America shows that no other groin-based shore protection project was performed before 1724. We wait till 1779 to see the construction of a groin field at Cartagena de Indias (Colombia), but inside a detached breakwater [39].

In Figure 10, groins are silted on the west side, demonstrating an eastward longshore transport, confirmed by the fact that Rio Rimac, the main feeder of this coastal segment [40] empting to the east of the town, does not create a beach in adherence to the breakwater of the present city port. In addition, the delta formed during the March 2017 catastrophic flood (El Niño event, 10 times the average rainfall of that month) could not reach the breakwater. Wave reflection today on that oblique structure can be thought of as an additional process to limiting river sediment dispersion to the west.

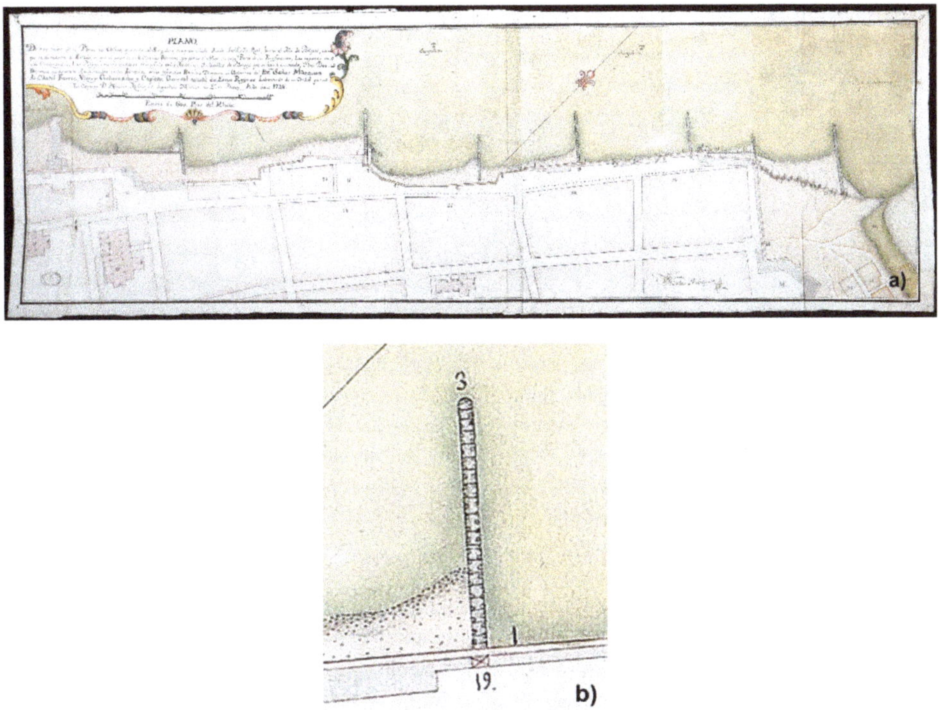

Figure 10. (**a**) Groins at Callao in 1728; (**b**) Detail on a groin [37]. In the cartouche: Map of a part of Callao (. . .) in which the state of the breaches done by the sea in this part of the wall is shown, together with the repairs done, and the beach created, thanks to the positive effects of the constructed groins (. . .). Courtesy: Archivo general de Indias, Sevilla.

On 28 October 1746, another catastrophic earthquake (estimated magnitude 8.4 with epicentre on the northern coast of the Lima department [32]) followed by a tsunami, completely razed the town; a belt of 5 km of land was flooded and 4800 out of the 5000 inhabitants of the town, died [29]. A surely minor loss was the destruction of any shore protection structure that might have been built on that coast, probably the oldest groin field in South America.

For years, the various viceroys, well aware of the area's hazards, gave up reconstructing the town, again, indirectly interpreting the idea of relocation proposed a few decades previously. Defence Military protection of the landing place was commissioned to Real Felipe fort (Figure 11), whose construction started immediately after the earthquake. Nevertheless, an aggregation of spontaneous buildings near the fort recreated a town, although without a master plan.

During the 18th century, only a short extension in L shape of the 1694 jetty was performed to give shelter to the boats, but in an 1865 map, an oblique breakwater protects both a mooring area and a landfill where the railway arrives. Callao was becoming an industrial settlement with dockyards, carpentry, ice and sugar factories. Today, this segment of coast is almost fully armoured, with harbours, marinas, docks and revetments, whereas the areas behind hosts industries and storage yards, with residential settlements on the southern stretch only, where a narrow beach still exists. Approximately 800,000 inhabitants live in this suburb of Lima.

Figure 11. A 1768 copy by Ordoñez [41] of a map of Lima and surroundings (1747–1768), with Real Felipe fort (see red arrow) built on the site where Callao was present (courtesy: Biblioteca Nacional de Espana).

3. Conclusions

The history of shore protection structures is a pillar within coastal engineering science, not only to reconstruct the evolution of concepts still in progress, but also to discover forgotten experiences that may help in soft shore protection strategies. What happened at Callao is still mostly hidden, but what appears from ancient maps and written documents traces a straight line with what was happening in Europe in those centuries [42]. This is not surprising, the designers being Spanish or French. However, noteworthy is the fact that structures used in Europe were replicated in a very different environment. It is evident that Callao's groins were designed within a colonialist mind set, therefore following European technology, but the presence of similar structures (for shape and construction materials) along the coasts worldwide, including rural beaches of developing countries, demonstrates the plurality of their birth site.

However, building and maintaining a groin field needed the backing of wealthy investors, an unusual occurrence in pre-17th century South America. This supports the hypothesis that the groin field at Callao, the first gold and silver exporting harbour in South America, is the oldest in that continent. In addition, a current study based on approximately 400 17th–18th maps of coastal settlement in that continent did not find similar defences older than those shown at Callao.

Reading Callao's shore protection structures' history put in evidence that most of what we know today regarding coastal dynamics and shore protection was just known in the 17th to 19th centuries. The main difference is that new materials to build structures and physical and numerical models to design them are now available.

However, within the forecasted sea level rise and increased storminess [43], strategic retreat is the most sustainable solution, at least where it is possible [44]. The same solution was proposed for Callao, just at the beginning of the 18th century.

Author Contributions: Historical and environmental analysis, C.M.; historical cartography, M.P.; coastal engineering, E.P.

Funding: This research received no external funding.

Acknowledgments: We are grateful to Allan T. Williams for advice relevant to improve this paper.

Conflicts of Interest: The authors declare no conflict of interest.

References

1. Pranzini, E. Coastal erosion and shore protection. In *Encyclopedia of Vernacular Architecture*; Vellinga, M., Ed.; Bloomsbury Publishing: London, UK, 2018; in press.
2. Jayakumar, K.; Malarvannan, S. Assessment of shoreline changes over the Northern Tamil Nadu Coast, South India using Web GIS techniques. *J. Coast. Conserv.* **2016**, *20*, 477–487. [CrossRef]
3. Morton, R.A.; Barras, J.A. Hurricane Impacts on Coastal Wetlands: A Half-Century Record of Storm-Generated Features from Southern Louisiana. *J. Coast. Res.* **2011**, *27*, 27–43. [CrossRef]
4. Young, K.R.; León, B. Natural Hazards in Peru: Causation and Vulnerability. *Dev. Earth Surf. Process.* **2009**, *13*, 165–180.
5. Nicholls, R.J. Rising sea levels: Potential impacts and responses. In *Global Environmental Changes*; Hester, R.E., Harrison, R.M., Eds.; Royal Society of Chemistry: Cambridge, UK, 2002; Volume 17, pp. 83–107.
6. Jiang, W.; Tao, C. The seawall in Quintang estuary. In *Engineered Coasts*; Chen, J., Eisma, D., Hotta, K., Walker, H.J., Eds.; Kluwer Academic Publishers: Dordrecht, The Netherlands, 2002; pp. 139–150.
7. Wang, L.; Xie, Y.; Wu, Y.; Guo, Z.; Cai, Y.; Xu, Y.; Zhu, X. Failure mechanism and conservation of the ancient seawall structure along Hangzhou Bay, China. *J. Coast. Res.* **2012**, *28*, 1393–1403. [CrossRef]
8. Nieuwhof, A. Living in a dynamic landscape: Prehistoric and proto-historic occupation of the northern-Netherlands coastal area. In *Science for Nature Conservation and Management: The Wadden Sea Ecosystem and EU Directives, Proceedings of the 12th International Scientific Wadden Sea Symposium, Wilhelmshaven, Germany, 30 March–3 April 2009*; Marencic, H., Eskildsen, K., Farke, H., Hedtkamp, S., Eds.; Common Wadden Sea Secretariat: Wilhelmshaven, Germany, 2010.

9. Bazelmans, J.; Gerrets, D.A.; de Koning, J.; Vos, P.C. Zoden aan de dijk: Kleinschalige dijkbouw in de late prehistorie en protohistorie van noordelijk Westergo. *De Vrije Fries* **1999**, *79*, 7–73.
10. Bazelmans, J. Die Wurten von Dongjum-Heringa, Peins-Oost und Wijnaldum-Tjitsma: Kleinmaßstäblicher Deichbau in ur- und frügeschichtlicher Zeit des nördlichen Westergo. In *Kulturlandschaft Marsch. Natur-Geschichte-Gegenwart*; Landesmuseum für Natur und Mensch: Oldenburg, Germany, 2005; pp. 68–84.
11. Niemeyer, H.D.; Eiben, H.; Rohde, H. History and heritage of German coastal engineering. In *History and Heritage of Coastal Engineering*; Krauss, N.C., Ed.; ASCE: New York, NY, USA, 1996; pp. 169–213.
12. Van der Meulen, F.; van der Valk, B.; Arens, B. The Netherlands. In *Coastal Erosion and Protection in Europe*; Pranzini, E., Williams, A.T., Eds.; Routledge: London, UK, 2013; pp. 136–157.
13. Charlier, R.H. Belgian coastal erosion. *Prof. Geogr.* **1955**, *7*, 10–12. [CrossRef]
14. Kraus, N.C. *History and Heritage of Coastal Engineering*; American Society of Civil Engineers: New York, NY, USA, 1996; p. 45.
15. Cresy, E. *An Encyclopedia of Civil Engineering*; Longman: Harlow, UK, 1847.
16. Kraus, C.N.; McDougal, W.G. The Effects of Seawalls on the Beach: Part I, An Updated literature Review. *J. Coast. Res.* **1996**, *12*, 691–701.
17. Franco, L. *History of Coastal Engineering in Italy*; American Society of Civil Engineers: New York, NY, USA, 1996; pp. 275–335.
18. Blaeu, J. Vlissingen in Toonneel der Steden van de Vereenighde Nederlanden, Met Hare Beschrijvingen, 1652. Available online: http://objects.library.uu.nl/reader/index.php?obj=1874-237713&lan=en#page//52/21/31/52213138426633615601451169782851892476.jpg/mode/1up (accessed on 29 June 2018).
19. Pranzini, E. Coastal erosion and shore protection: A brief historical analysis. *J. Coast. Conserv.* **2017**, 1–4. [CrossRef]
20. Seutter, M. Novissima Ichnographica Delineatio Munitissimae Urbis et Celeberrimi Emporii Ostendae. In *Comitatu Flandriae Austriacae Sitae, cur et Coela Matthaei Seutteri*; Private Collection E.P.; Seutter, Mathaeus: Augsburg, Germany, 1734.
21. Baldacci, O. Le carte nautiche e il portolano di Bartolomeo Crescenzio. In *Rendiconti della Classe di Scienze Morali, Storiche e Filologiche*; Accademia Nazionale dei Lincei: Roma, Italy, 1949; Volume 4, pp. 601–635.
22. Lepore, F.; Piccardi, M.; Pranzini, E. Latitudine senza latitudine. Determinazioni astronomiche e carte per navigare dell'Europa tardo medievale: Le innovazioni di Francesco Beccari. *Studi Costieri* **2017**, *27*, 33–108.
23. Van der Aa, B. Le Port de Callao de Lima dans l'Amerique in La Galerie Agréable du Monde, où L'on Voit en un Grand Nombre de Cartes Très Exactes et de Belles Tailles Douces Les Principaux Empires, Roïaumes, Républiques, Provinces, Villes, Bourgs et Forteresses, Tome Second d'Amerique 1729. Available online: http://gallica.bnf.fr/ark:/12148/btv1b10509759w.r=btv1b10509759w?rk=21459;2 (accessed on 29 June 2018).
24. Montanus, A.; van Meurs, J. De Nieuwe en Onbekende Weereld of Beschryving van America en 't Zuid-Land, Vervaetende D'oorsprong der Americaenen en Zuid-Landers, Gedenkwaerdige Togten Derwaerds, Gelegendheid der Vaste Kusten, Eilanden, Steden, Sterkten, Dorpen, Tempels, Bergen, Fonteinen, Stroomen, Huisen, De Natuur Van Beesten, Boomen, Planten en Vreemde Gewasschen, Gods-Dienst en Zeden, Wonderlijke Voorvallen, Vereeuwde en Nieuwe Oorloogen: Verciert Met af-Beeldsels na 't Leven in America Gemaekt Amsterdam, 1671. Available online: https://archive.org/stream/denieuweenonbeke00mont#page/n5/mode/2up (accessed on 29 June 2018).
25. Saeghman, G.J. *Journael van de Nassausche Vloot, Zijnde een Beschryvingh van de Voyagie om den Gantschen Aerdt-Kloot, Gedaen Met Elf Schepen, Onder 't Beleydt Vanden Admirael Jaques l'Heremite, Ende Vice-Admirael Geen Huygen Schapenham, Inde Jaren 1623, 1624, 1625, en 1626*; Gillis Joosten Saeghman: Amsterdam, The Netherlands, 1663.
26. Unknown, Vertoning van't Callao de Lima (De Vloot Voor Anker bij Callao, 1624), [ed.] 1663 ca. Available online: https://www.rijksmuseum.nl/nl/collectie/RP-P-OB-80.316 (accessed on 29 June 2018).
27. Arrús, D. El Callao en La Época del Coloniaje Los Antes y Despues de la Catástrofe de 1746 Posicion Real y Efectiva de la Ciudad Real, en Relación Con el Antiguo Presidio. Fundación de Bellavista. Terremotos Piratas Y, Datos Histórico-Cronológicos de los Virreyes y Gobernadores del Perú y de los Sucesos Más Reservas Notables Desde 1530 á 1829, Imprenta de El Callao, Callao 1904. Available online: https://archive.org/details/elcallaoenlapoc00arrgoog (accessed on 29 June 2018).

28. Frézier, A.F. Plan de la Rade du Callao in Relation Du Voyage De La Mer Du Sud Aux Côtes Du Chily Et DuPerou, A Paris, 1716. Available online: https://www.e-rara.ch/zuz/doi/10.3931/e-rara-54648 (accessed on 29 June 2018).
29. Melo, R. El Callao: Geographic Historical Monograph, Librería e Imprenta Gil ed.; 1899–1900. Available online: http://bdh-rd.bne.es/viewer.vm?id=0000042274&page=1 (accessed on 29 June 2018).
30. Lohmann Villena, G. *Las Defensas Militares de Lima y Callao Hasta 1746*; C.S.I.C: Sevilla, Spain, 1964.
31. Engels, F. Early Sites on the Peruvian Coast. *Southwest. J. Anthropol.* **1957**, *13*, 54–68. [CrossRef]
32. Silgado, F.E. Historia de los Sismos Mas Notables Ocurridos en el Peru: 1513–1974. Primera Edicion Digital, Víctor López Guzmán, Lima. 2012. Available online: http://www.guzlop-editoras.com/web_des/ing01/ingsismica/pld0579.pdf (accessed on 29 June 2018).
33. Lemoine Fecit, Plan de la Ville du Callao, (1724–1727). Available online: http://gallica.bnf.fr/ark:/12148/btv1b53122271w.r=btv1b53122271w?rk=21459;2 (accessed on 29 June 2018).
34. Frézier, A.F. *Relation Du Voyage De La Mer Du Sud Aux Côtes Du Chily Et Du Perou*; Printed for Jonah Bowyer: London, UK, 1717; Available online: https://archive.org/details/voyagetosouthse00frzi (accessed on 29 June 2018).
35. Lemoine (Attr.) Plan de la Ville de Callao Avec ses Nouveaux Projets, (1724–1727). Available online: http://gallica.bnf.fr/ark:/12148/btv1b53123330h.r=btv1b53123330h?rk=21459;2 (accessed on 29 June 2018).
36. Feuillée, L. Plan de la Ville de Callao Avec Ses Environs in Journal des Observations Physiques, Mathématiques et Botaniques, Faites par L'ordre du Roy sur les Côtes Orientales de l'Amérique Méridionale, et Dans les Indes Occidentales, Depuis L'année 1707, Jusques en 1712 et Dans un Autre Voiage Fait par le Même Ordre à la Nouvelle Espagne, & Aux Isles de l'Amérique, 1725, Vol I. Available online: http://www.e-rara.ch/zut/doi/10.3931/e-rara-1565 (accessed on 29 June 2018).
37. Rodríguez, N. Plano de Una Parte de la Plaze de Callao, Que Mira al Surgidereo, Comprebendida Desde el Muelle Real; Hasta el Rio de Pitipiti, en que se Demuestra el Estado en que al Presente se Hallan las Brechas que Avriò el Mar en Esta Parte de su Fortifacion, los Reparos en ella Costruydos, Las Playas Criadas Mediante el Beneficio de los Redientes o Muelles di Pioltage que se Han Exexutado [...]1728, Sevilla, Archivo General de Indias. Available online: http://pares.mcu.es/ParesBusquedas/servlets/Control_servlet?accion=3&txt_id_desc_ud=22499&fromagenda=N (accessed on 29 June 2018).
38. Silvester, R.; Hsu, J.R.C. *Coastal Stabilization. Innovative Concepts*; PRT Prentice Hall: Englewood Cliffs, NJ, USA, 1993; p. 539.
39. Segovia, S.R.; Mogollón, H.; Mayer, B. *The Fortifications of Cartagena de Indias: Strategy and History*; El Ancora Editores: Santafé, Bogotá, 1998.
40. Taype Ramos, V.; Bird, E.C.F. Peru. In *Encyclopedia of the World's Coastal Landforms*; Bird, E.C.F., Ed.; Springer: Dordrecht, The Netherlands, 2010; pp. 207–211.
41. Ordoñez, A. Descrypcyon de la Cyudad de Lyma y Puerto del Callao, 1768 (Copy). Available online: http://bdh.bne.es/bnesearch/detalle/bdh0000032258 (accessed on 29 June 2018).
42. Pranzini, E.; Williams, A.T. (Eds.) *Coastal Erosion and Protection in Europe*, 1st ed.; Routledge: London, UK, 2013; p. 457.
43. Douglas, B.C.; Kearney, M.S.; Leatherman, S.P. *Sea Level Rise: History and Consequences*; Academic Press: San Diego, CA, USA, 2001; p. 232.
44. Williams, A.T.; Rangel-Buitrago, N.; Pranzini, E.; Anfuso, G. The management of coastal erosion. *Ocean Coast. Manag.* **2018**, *156*, 4–20. [CrossRef]

water

MDPI

Article

Shore Protection Structures Increase and Evolution on the Northern Tuscany Coast (Italy): Influence of Tourism Industry

Enzo Pranzini [1], Giorgio Anfuso [2,*], Irene Cinelli [1], Marco Piccardi [1] and Giovanni Vitale [3]

1 Dipartimento di Scienze della Terra, Università di Firenze, 50125 Firenze, Italy; enzo.pranzini@unifi.it (E.P.); irene.cinelli@hotmail.it (I.C.); m.picche@gmail.com (M.P.)
2 Departamento de Ciencias de la Terra, Universidad de Cádiz, 11510 Puerto Real, Spain
3 CNR Ibimet—Consorzio LaMMA, 50125 Firenze, Italy; giovanni.vitale79@gmail.com
* Correspondence: giorgio.anfuso@uca.es

Received: 4 October 2018; Accepted: 7 November 2018; Published: 13 November 2018

Abstract: Analysis of a large shoreline database (from 1878 to 2017) and recompilation of information on type/age of shore protection structures along the Northern Tuscany, allowed a deep insight of the progressive armouring of this coastal sector. The area experienced beach erosion since the end of the 19th century due to reduced sediment inputs from rivers and harbour constructions. Shore protection structures started to develop at the beginning of the 20th century, first to protect settlements and coastal roads, later to maintain a beach for tourist activity. The changing of the goal and the increasing awareness of the negative impact of some structures resulted in an evolution of coastal defence projects: initially, seawalls and revetments, later detached breakwaters and, more recently, groins. Today, a reduction in hard structures is perceived by removing or lowering detached breakwaters and groins below mean sea level. The forcing function of the growing tourism industry is producing a demand for projects and their design is detailed in this paper: results will be of use in the correct design of a long-term, general, erosion management plan to restore the natural sediment circulation patterns.

Keywords: coastal erosion; shore protection; coastal armouring; 3S tourism; beach economy

1. Introduction

Beaches are of outmost importance for coastal tourism economy in several countries [1], for example, USA coastal states receive *circa* 85% of tourist related revenues [2]. In 2011 international and national visitors spent US $5.7 billion at Miami [3], where the beach is fully maintained via nourishment works according to a strategy that is extremely cost effective [4]. On the Pacific coast, Venice Beach (California) has 16 million annually tourist visits, more than the combined visits (*circa* 12 million) to Yellowstone, Yosemite and Grand Canyon National parks. Tourism revenues are even more important in small countries and islands, for example, in Hawaii, beaches supply one third of all jobs in the State [5] and, in the Mediterranean, the small island (8336 km^2) of Crete (Greece), is visited by 2.8 million tourists annually [6]; Balearic Islands (4992 km^2), Spain, recorded almost 20 million visitors in 2017 with an associated amount of revenues of 16 M€—an increase of 12% respect to 2016 [7]. Sea side tourism is a relevant part of the Italian economy too, with 118.6 million visits in 2016, 29.5% of the whole national tourism [8]. Bathing establishments produce an Economic Value Added (EVA) of 800 M€ per year, more than 3% of the Italian EVA and give work to 300,000 persons [9]. Coastal tourism is strongly affected by beach erosion [4,10,11]. For example, at the Hoi, an World Heritage site in Vietnam, the total annual revenue losses for beach erosion in 2020 are estimated in 29.6 M US dollars [12].

A The Bahamas, the economic losses in properties and accommodation offer for the next 20 years could be up to US $44 billion [13]. In Jamaica coastal erosion affects beaches, properties and coral reef: present annual economic losses (valued at 19 M US $) could record an additional US $1.3 million per year increase if the reef continues to degrade—because of the loss of tourism related activities [14].

Erosion trend acquires a great relevance in Italy too, where 41.9% of the beaches are eroding, that is, 1657 km out of 3951 km [8]. This process started in Northern Italy and progressively expanded to the south, following the economic development of the country [15] and was immediately opposed with shore protection structures. Only recently important beach nourishment projects have been carried out but hard shore protection is still the most frequent option, thanks to the familiarity that administrators, technicians and stakeholders have with them, notwithstanding the negative impact they can have on the coastal system.

Shore protection structures are known to have been built at least since the 1st century B.C. in Germany [16,17] and The Netherlands [18]; the 6th century A.D. in Venice [19], the 8th century in China [20] and the beginning of the 16th century in Belgium [21]. Most of these ancient structures aimed at protecting short coastal segments, frequently where villages or military structures were located [22] but the need to prevent storm surge flooding in subsiding coastal lowlands along the North Sea brought construction of long dykes, for example, in Germany as early as the 11th century [23].

Nineteenth century coastal development (human settlements, industrial areas, communications) and the emergence of widespread coastal erosion (due to sediment input reduction, coastal structure construction and sea level rise) caused an increase of shore protection structures. These extended further where coastal tourism started to be a social fashion as well as an economic business. Buildings, generally second houses and hotels, stretched along the coast, at places on the foredunes or even on the beaches themselves, reducing enormously coastal resilience [24].

However, in Spain [25] and in many Mediterranean countries, coastal protection structures really mushroomed only post WWII due to the low level of human occupation, the moderate tourist demand for beaches and the low intensity of erosion problems [26]. In Italy and France, the coast was mostly depopulated [27,28] and only a few tourist destinations existed in well-established localities. Protection of isolated buildings, coastal roads and railways and promenades was mostly performed with revetments but this system was gradually abandoned after the 1940s.

Examples of this trend have been reported for many countries, from Italy [25] to Spain [27], which is now the European premier beach holiday tourist destination [29,30]. In the Tropics, Mexico, with international tourism [31,32] and Colombia, with a national one [33], are following the Italian bad example. Similar evolution has been experienced by countries not in any seaside destination list, such as, Québec, Canada [34] and Washington, USA [35].

Most of the first Italian projects utilized revetments but beaches disappeared in sedimentary deficient areas. The introduction of detached breakwaters produced limited results and the structures proved to be unsuitable for bathing activities; groin fields followed and were expanded with the backing of beach concessionaires.

However, coastal tourism has come into conflict with traditional coastal defence works, inducing their transformation into submerged structures, like happened in the study area. Similar projects have been carried out in other places in Italy, for example, at Follonica [36] and Igea Marina [37], where detached breakwaters have been widened and lowered below sea level. Examples can also be found in other countries: in Spain, at Marbella and Benalmádena, groins were greatly reduced in numbers [38] and, at Estepona (Spain), they were removed and submerged shore-parallel structures substituted; a similar project was carried out in Greece, at Katerini beach [39]. Demolition, or abandonment of 19 shore-parallel seawalls present at Sandy Hook Spit, New Jersey (USA), was proposed along 10 km of coastline including both the ocean and bay shores to allow natural shoreline processes to prevail as part of a strategy of adaptation to sea level rise associated with climate change [40]. Beach nourishment often accompanied these works.

This evolution is understandable only if the economic value of these beaches is considered, not only for beach activities but also the spin-off on several components of the tertiary economy. Research carried out on the northern coastal segment on beach value perception by the managers of different economic activities (bathhouse, hotels, restaurants, fashion shops, news sellers, food stores, etc.) has proved this statement [41]. Most interviewees answered that the loss of beach could induce closure of their activity. Interviews were done within 2 km from the shoreline and the perception of beach impact on economic activities slowly reduced inland but remained relevant.

In many cases it is not possible to trace coastal erosion development together with shore protection proliferation, due to the lack of both shoreline data and structures construction and modification through time. For the present case study, both shoreline position and structure construction time are available from 1878, when the beach was stable or accreting; in addition, along the investigated coast, urban settlements, tourist places, commercial/recreational harbours and natural parks are present, allowing identification as to how shore protection structures are designed to counteract beach erosion, which recently has been coming under increasing pressure from the tourism industry.

2. Study Area

A continuous 63.5 km-long sand barrier connects Monte Marcello promontory to the Livorno rocky salient (Figure 1), where the largest harbour of Tuscany was constructed in the 16th century. Three main rivers empty on this coast, R. Magra, close to Monte Marcello, R. Serchio, 39 km to the south and R. Arno, 11 km from the southern limit of the physiographic unit. The latter is the main feeder river of this coast, with approximately 1,524,000 t/year of total load [42], R. Magra contributes with 632,000 t/year to the coastal sediment budget [43], whereas R. Serchio, whose watershed is in large parts on limestone gives a limited input (approx. 23,000 t/year, [43]). River Magra, coming down from high mountains close to the coast, brings mixed sand and gravel sediments, whereas R. Arno, crossing the Florence and Pisa plains, deposits all its coarse sediments before reaching the sea. Present input is far lower than that of the previous centuries: for example, R. Arno sediment yield is only 37% of that estimated for the 1500–1800 AD period [44].

Other small water courses (e.g., Cinquale, Versilia and Calambrone), although their bedload has never been assessed, are considered by the authors to be insignificant to the sediment input.

Wave climate is characterized by a limited wave angular dispersion, with storms coming exclusively from the SW (Figure 1). Longshore transport diverges at the River Arno delta apex and sediments from this river feed the coast as far as Livorno and up to Marina di Pietrasanta, where a convergence with sediments coming from the R. Magra was identified via morphological, sedimentological, petrographic and numerical models [44,45]. The littoral unit is divided into four main littoral cells [46] (Figure 1), from north to south these are:

(i) Monte Marcello—Marina di Pietrasanta;
(ii) Marina di Pietrasanta—R. Arno mouth (these two are limited by a convergence zone);
(iii) R. Arno mouth—R. Calambrone mouth (limited by the previous by a divergence zone);
(iv) the small cell in the R. Calambrone—Livorno area, limited by a convergence zone and produced by wave reflection on the oblique breakwaters of the harbour [47,48].

Two harbours intercept longshore transport: the first is Marina di Carrara, whose breakwaters have been extended to a 10 m water depth allowing only finer sediments to overpass the port structure; the second is Viareggio harbour, whose entrance originally was at −5.0 m and now—after the updrift beach reached the breakwater tip—allows natural sediments bypass.

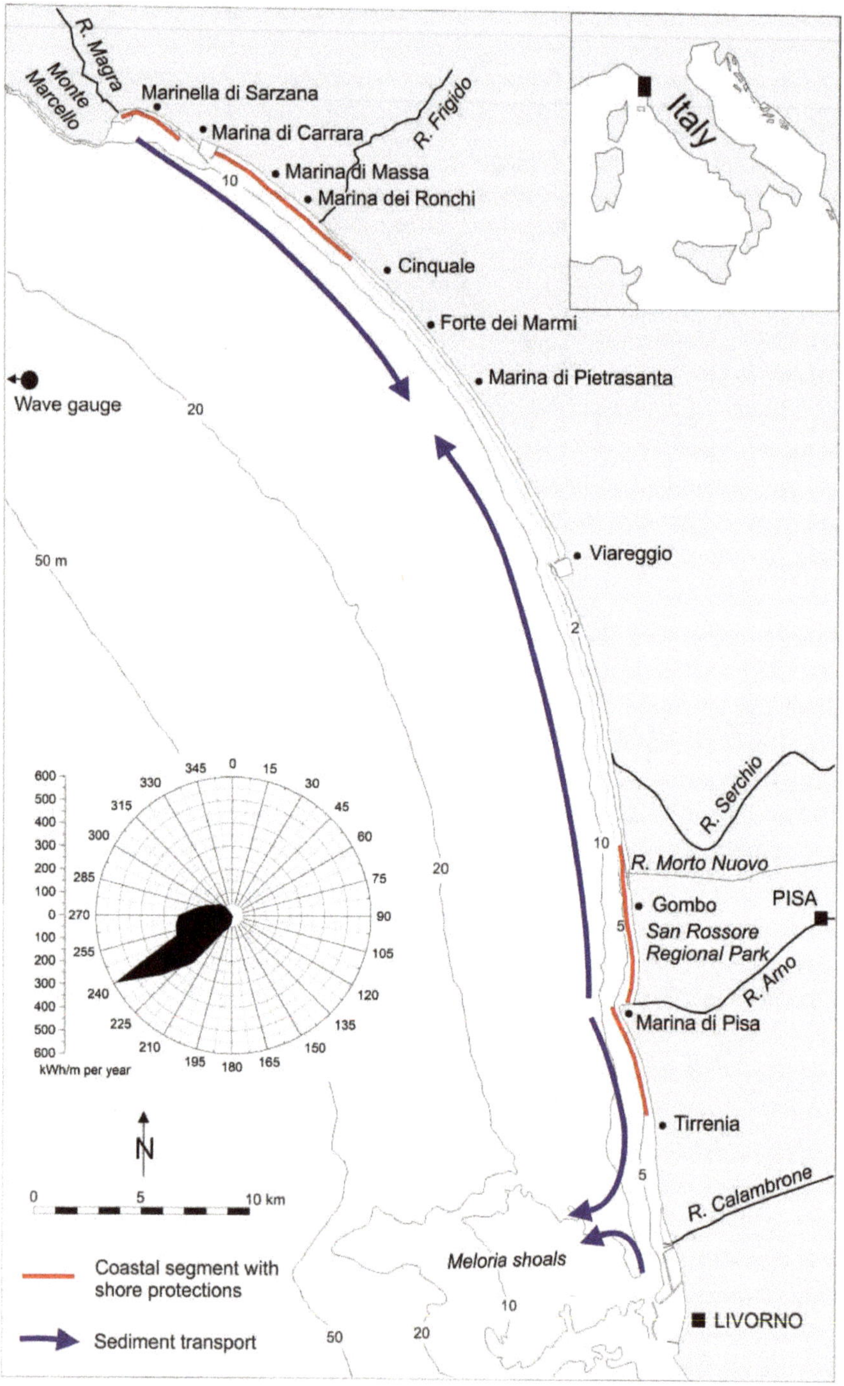

Figure 1. Location map of the study area with longshore transport directions [46]; wave data from "Rete Ondametrica Nazionale"—La Spezia gauge December 2009–December 2014.

Coastal tourism started very early on this coast with the first bathing facilities built at Viareggio in 1827, where Liberty bathhouses, kiosks and hotels still increase the tourist attractiveness of the site. Bathers from Pisa were crowding at Gombo, where a bathhouse has been active since 1830s and ended when it was moved in 1869 to the southern side of the R. Arno delta because the land owner, King Vittorio Emanuele II, did not like so many visitors. Three years later (1872), on the same side of R. Arno, Marina di Pisa was founded after a masterplan drawn by the Pisa Municipality to create a tourist "new town": a careless decision, since the delta of the River Arno, where the settlement was located, was just showing the first signs of an erosion trend that, on the unprotected northern delta lobe, resulted in 1.3 km of coastal retreat in the following 120 years. The area became Presidential property in 1956 but was then given to the Tuscany Region in 1999 and now is a regional park.

In the northern sector, industrial activity—at first connected with Carrara marble quarries—was gradually flanked and later overpassed by tourism activity; marble loading piers, replaced by the Marina di Carrara harbour, now are tourist attractions. Viareggio harbour, originally a small dock along a drainage channel, which grew with several expansions protruding into the sea to reach the present configuration, now hosts few fishing boats but several recreational boats and is bordered by dockyards building mega yachts sailing on the world seas.

It is difficult to evaluate beach tourism economy since available data on tourist revenues for the investigated zone are aggregated within administrative areas, that is, provinces. Hence, they include revenues linked to both tourism in towns (in this case Pisa, Lucca and Livorno) and beach tourism related activities. However, a proof of importance of the latter is provided by the number of bathing establishments located along this coast: excluding the Regional Park, where this activity is not allowed, 645 concessions are given (15 for each km of coast); here concessionaires not only rent cabanas and beach umbrellas but also manage bars and restaurant. The average cost for a cabana plus one beach umbrella and two sun chairs is about 40 €/day in this area and umbrella density is approximately 1/10 m^2 of beach.

3. Materials and Methods

Shoreline evolution was studied starting from a dataset developed at the Earth Science Department of the University of Florence and based on the ortho-rectification of 1938, 1954, 1967, 1978 and 1985 aerial photographs, integrated in 2005 with direct DGPS surveys. The late 1800s shoreline position was taken from the first edition of the 1:25,000 topographic maps of the Istituto Geografico Militare (IGM) [49] produced between 1878 and 1881, from now on referred as the 1878 shoreline. The 2017 shoreline was extracted from high resolution satellite images (Pleiades) by the "Consorzio LaMMA" within a commitment conferred by the Tuscany Region government. Shoreline position accuracy was estimated to be 30 m for the 1878 shoreline, 10 m for the 1938 to 1985 shorelines and 1 m for the 2005 and 2017 ones [50,51]. Beach erosion and accretion values are considered significant if they exceed the accuracy of the figures given. For the present study, the 63.5 km long coast was divided into 246 sectors; each one being approximately 250 m long and representative of a uniform (natural or artificial) coastal sector.

For each sector and time interval, beach surface variation (m^2) was measured [52] via Geographic Information System (QGIS) and mean shoreline displacement value (m) was computed. Coastal structures type, position and length were acquired from the University of Florence dataset and from geo-referenced aerial photographs; in addition, several later IGM map editions were used. To each structure (or modification) the construction time was considered as the year of its first appearance in the maps/air photos, unless more specific information were available from written reports; this gives a "stepped shape" to the lines tracing the growth of the structure length. Old photographs and postcards were helpful in this research.

Coastal armouring is the result of the construction of different structures, some addressed to oppose beach erosion (seawalls, revetments, groins, detached breakwaters, etc.), others to stabilize the entrance of rivers and channels or as part of harbours. The latter structures often cause downdrift beach erosion that requires emplacement of protection structures. Although this paper deals with the temporal evolution of coastal structures (i.e., time is the first variable), their description will follow a geographic sequence (from north to south). However, for reasons of clarity, the analysis will start from three erosion hotspots: the two main river mouths and the harbour of Marina di Carrara.

4. Coastal Erosion

Since sea level stabilization (approx. 6000 year ago) this coast was accreting due to the huge amount of sediments delivered by the two main rivers (Arno and Magra). Only two significant erosion phases have been recorded on the River Arno delta, the first at the Fall of the Roman Empire, the second consequent to the 14th cent. Black Death; in both the cases a tremendous demographic crisis occurred accompanied by the abandonment of many cultivated areas where forest grew back [48].

The present coastal erosion phase, just like the previous, started at river mouths and gradually extended to lateral beaches [53]. The main causes were river sediment input reduction due to reforestation, dam construction and river bed quarrying. Further, harbour construction and the first shore protection structures increased the intensity of erosion processes. Following the longshore transport direction, erosion expanded north and south of the R. Arno mouth and south of the R. Magra one (the coast to the north is rocky).

On the southern lobe of the R. Arno, the "new town" of Marina di Pisa was under construction when beach erosion started. When approximately 150 m of beach was lost, shore protection projects were initiated; on the delta uninhabited northern side nature was left alone and 1300 m of land disappeared (Figure 2). Only a small coastal sector located 4 km to the north, where the Presidential Villa is located, was in the early 1960s protected with 5 detached breakwaters that increased downdrift erosion reflected by an evident log-spiral plan form [54].

Approximately 1.5 km to the north occurs the Fiume Morto Nuovo (New Dead River) outlet, an artificial mouth of the Fiume Morto draining the northern Pisa plain, which was diverted to the south in 1926–1930 to shorten its course and favour water discharge. In 1933, jetties were built, which were further extended with several projects. This had a strong feeding effect and now the downdrift beach is some 260 m narrower than the updrift one (Figure 2).

Proceeding to the north, the continuous seaward expansion of Viareggio harbour, carried out to guaranty deep water at its entrance, induced downcoast erosion until a by-pass system was constructed in 1954 [55]. This proved to be expensive and became useless when sediments started to naturally bypass the structure. However, the bar overpassing the harbour entrance is frequently dredged and sediments deposited to the north.

On the northern side of the coast (Figure 3), the R. Magra never built a real delta but the Istituto Idrografico della Marina Italiana (IMI) 1881 nautical chart shows a salient in correspondence with the 5 and 10 m isobaths [56]. In 1938 approximately 3 km of coast to the south side of the river mouth were eroded, with a recession of more than 500 m; on the IIMI 1954 nautical chart the salient is no longer present [56].

When erosion nearly reached Marina di Carrara (Figure 3), harbour construction began. The project was redacted in 1920 but work started in 1922 [57] and the beach of this locality—otherwise destined to be eroded—started to expand so much that today it is approximately 300 m wider than in 1920s (Figures 2 and 3). This favoured and accelerated erosion processes on southern coastal sectors (i.e., downdrift of the harbour, Figure 2), where now 10 km of intensively developed coast is severely eroding and a variety of shore protection structures have been built.

Over the whole coast (Monte Marcello—Livorno), available data shows that the beach surface grew until 1954 with a mean shoreline displacement of approximately 80 m (Figure 4). The 1967 survey was performed to evaluate damages done by intense storms, which concurrently occurred with the November 1966 flood in Florence and a mean shoreline retreat of 8.3 m was registered. However, in the following years, part of this loss was recuperated and the total beach surface in 1985 was approximately the same as in 1954. Post this, a gradual surface reduction took place and, in 2017, the total beach surface was 9.6 m narrower than in 1954 but remained 69.8 m larger than in 1881 and 6.1 m larger than in 1938. Things are not so beneficial as these figures would let it be imagined: this positive budget is the result of very wide beach progradation and severe diffuse erosion. Whereas beach expansion gives limited gains, especially where sand surface exceeds the needs of tourists [58], beach erosion prevents, or strongly compromises this economy [59].

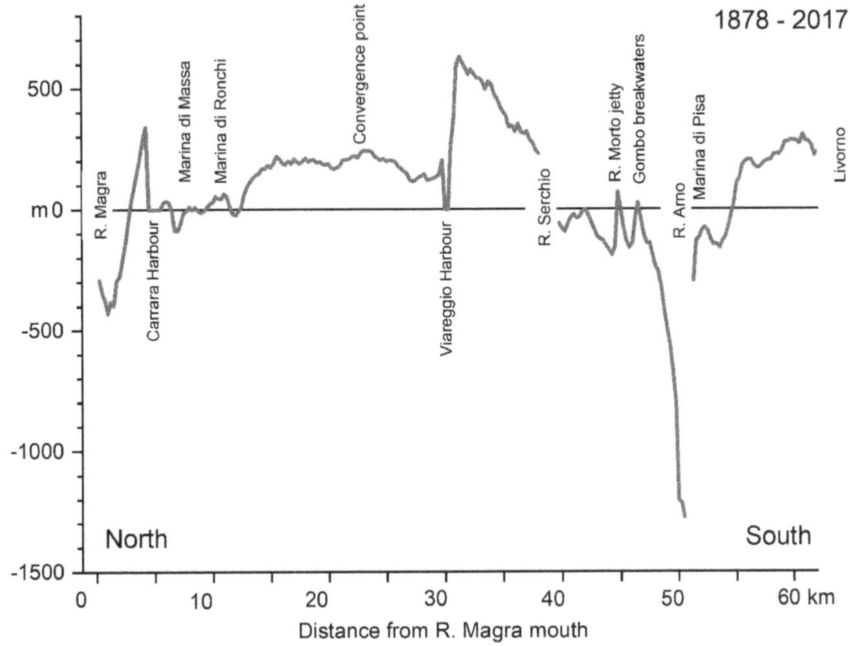

Figure 2. Shoreline displacement along the study area from 1878 to 2017 (from south of Marina di Pisa to Livorno the period is 1881–2017).

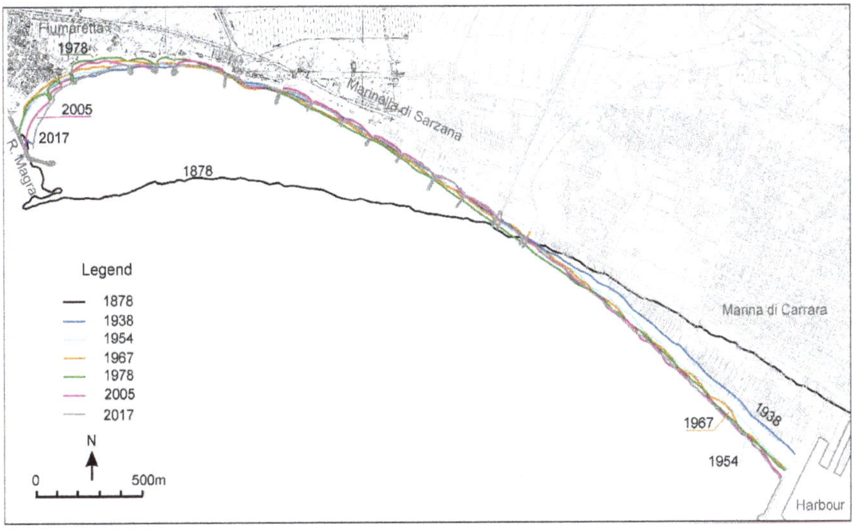

Figure 3. Shoreline evolution from 1878 to 2017 between the R. Magra mouth and Marina di Carrara harbour.

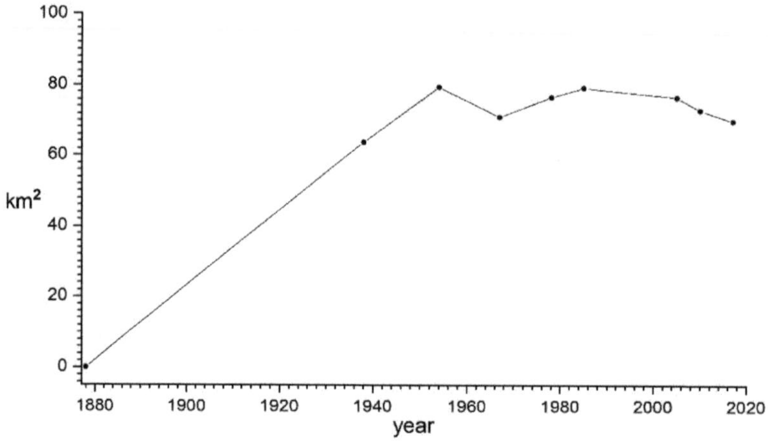

Figure 4. Mean shoreline trend of the study area compared to 1878.

In the last years (2005–2017; Figure 5) some projects gave positive results: south of the Magra River sediment bypassing from the river terminal course to the beach allowed beach expansion (Figure 3). This sediment movement was possible because, after several decades of river dredging prohibition, the natural sedimentary dynamics was restored and sediments were able to arrive to the mouth.

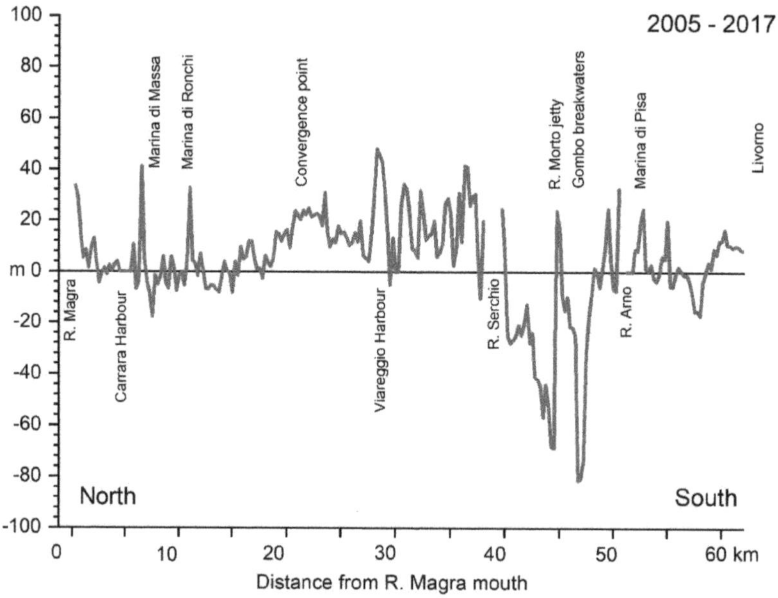

Figure 5. Shoreline displacement along the study area from 2005 to 2017.

At Marina di Ronchi emerged and submerged groins were built and a few nourishment works carried out: this allowed to revert a decennial negative trend. Similar structures stopped coastal erosion north of R. Arno mouth where erosion recorded values of 10 m/yr during the 1997–2000 period. Gravel nourishment at Marina di Pisa returned the beach to this town after one century. The opposite is still the situation updrift of the R. Morto Nuovo jetty and between Marina di Pisa and Tirrenia but the largest erosion is observed at Gombo, where a project based on an artificial tombolo built with a geotextiles failed and up to 80 m of shoreline retreat was registered during the 2005–2017 period (Figure 5).

4.1. Coastal Structures Number, Type and Length

In addition to the two harbours whose breakwaters extend for about 3420 m and occupy a coastal length of 1800 m, 160 individual structures (155 still existing) were mapped for a total length of 29,995 m (Table 1; Figure 6). There are 16 jetties at river/channel mouths, which extend for 2534 m; some of which were originally groins but the jetties later built on the other side of the channel flanked the river mouth (e.g., Brugiano R.); others act as real groins with a relevant updrift and downdrift effect (e.g., R. Morto Nuovo).

Table 1. Structures length along the study area in 2017.

Structure Type	Length (m)
Harbours	3420
Seawalls & Revetments	5405
Groins ("I" and "T" shaped, emerged)	3652
Groins (with submerged extension)	3559
Groins (submerged in rocks or concrete elements)	105
Groins (submerged in geotextiles)	757
Detached breakwaters (emerged)	5415
Detached breakwaters (submerged)	5098
Artificial islands	50
Jetties	2534
Total	29,995

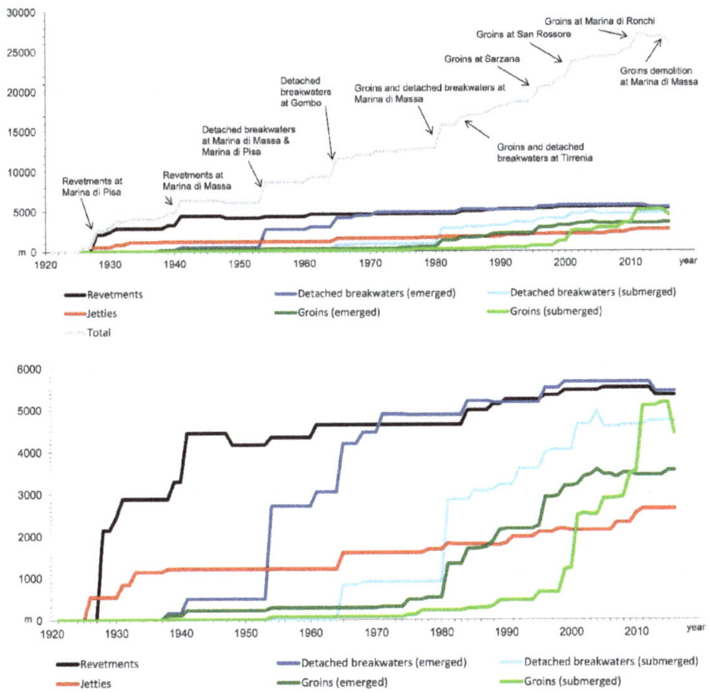

Figure 6. Cumulative length of the various shore protection structures from 1920 to 2016. Main interventions are indicated. **Top**: with total length; **Bottom**: without it to better show each structure types length.

Seawalls and revetments extend for 5405 m, at some places faced by emerged detached breakwaters, for example, at Marina di Pisa. The latter are 34 in number for 5415 m, whereas 15 are submerged and extend for 5098 m.

Traditional groins are 47 in number for a total length of 3652 m and 31 in number (4421 m) are completely or partially submerged, in rocks, geotextiles or precast concrete. There are also three rocky artificial islands; but this figure does not consider other similar structures that after construction have been connected to the land and now are considered as groins (e.g., south of R. Magra mouth). To these structures, four fishing piers for a total length of 1180 m are to be added: at Marina di Massa (215 m), Forte dei Marmi (300 m), Marina di Pietrasanta (380 m) and Lido di Camaiore (285 m). The first two were built in the last quarter of the 19th century for the shipping of marble, the latter two were constructed between 2005 and 2008 for tourist use only. Since they do not interact with coastal dynamics, they will not be considered further.

4.2. Coastal Armouring Development

On the 1878 IGM map (1:25,000 scale) no coastal structures are present along the investigated area, except for two 300 m long jetties at the Canale di Burlamacca, the entrance to the Viareggio dock. They were built in the 16th century to allow safe boat entrance to the channel and continuously extended to cope with the accreting beach (since they were later incorporated into Viareggio harbour, they are not considered in Figure 6). Beach asymmetry is well evident, with the southern side (updrift) *circa* 125 m wider than the northern one (downdrift).

On the 1938 IGM map a detached breakwater is present on the northernmost sector, where the road connecting Sarzana with Marina di Carrara turns to follow the coast (Figure 7); this structure was built after 1928 and, although deteriorated, still exists. On the same map a *circa* 300-m-long jetty is present on the R. Magra outfall's southern side; it was built in 1931 to prevent river mouth siltation both to favour river discharge during floods and to allow boat access to shipyards located along the river's terminal stretch.

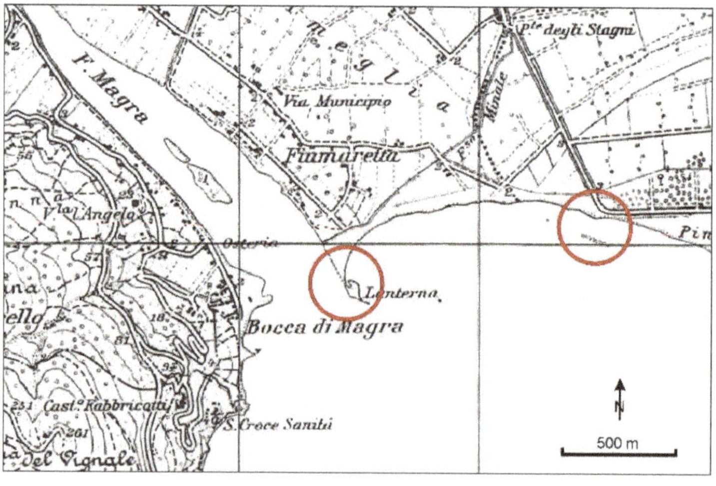

Figure 7. The first two structures (red circles) built on the northern coast and present on I.G.M. 1:25,000 map surveyed in 1938: post-1928 detached breakwater and 1931 jetty.

Reduction in river sediment input and offshore dispersion induced by this jetty, trigged downdrift erosion but as this sector was almost undeveloped, shore protection was limited to the above-mentioned detached breakwater. It was only in the 1970s that the beach at Fiumaretta (at the river mouth) was defended by five rocky round islands and some of these were further joined by submerged breakwaters or connected to the shore with short groins (Figures 3 and 8). Further changes to the structures were carried out in the 1980s, adding and reshaping elements and in 2005 an artificial island was extended seaward with a submerged groin. In 1982 a gap was opened in the jetty at the river mouth to allow sediments to feed the beach but due to induced sea water pollution and stakeholders' complaints, it was closed in 1990.

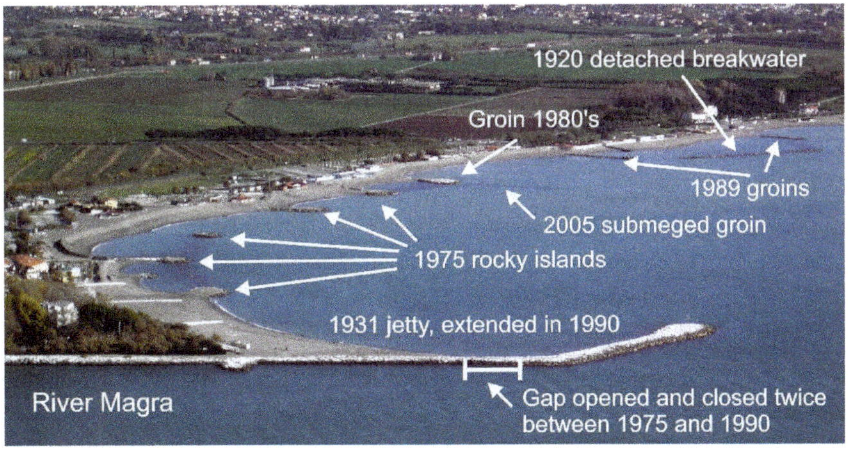

Figure 8. River Magra jetty and Fiumaretta beach (November 2005).

To the south, at Marinella di Sarzana (Figure 3), eight groins were built between 1989 and the mid-1990s and three of them were extended with submerged segments in 2005, within a major defence project which was never finished. It is in these years that orthogonal structures exceed revetments in length (Figure 6).

Even more complex is the evolution of the structures at Marina di Massa, where shore protection projects registered several failures leading to demolition or modification of previous works, whose history is almost impossible to be fully reconstructed. The most detailed account for this area was given by [59], which analysed all available projects at the local office of Genio Civile per le Opere Maritime (Coastal Engineering Department of the Public Work Ministry).

As previously stated, coastal erosion was approaching Marina di Carrara beach when harbour construction started in 1922 (Figure 3); the updrift coast inverted its trend but the downdrift sector was severely eroded. Here, 300 m of coast was protected with a revetment and detached breakwater in 1930, both being progressively further expanded to the south to reach respectively 500 and 800 m in 1939 (Figure 6).

At the beginning of the following decade shore protection expanded more to the south for approximately 2 km, with revetments and detached breakwaters and at the end of the 1950s the coast south of the harbour was defended for more than 3 km (Figure 9). Not only revetments and detached breakwaters were extended but they also needed several maintenance works and strengthening.

Further south, where the beach was accreting until 1938, in 1954 it was approximately 100 m narrower. Here coastal road and houses were far from the shoreline but the back beach was occupied by bathhouses fronting a beach surface on which to locate sun umbrellas. This surface gradually reduced and structures, at that time built with timber, were continuously set back, until the road prevented any further retreat. Permeable groins with timber poles—but also with iron bars—were built but no information remains except from some old photographs (Figure 10).

Contrary to what was carried out in the sector immediately south of the harbour where the road was protected with revetments and the beach was lost, here the stakeholders request was to prevent further retreat and maintain the beach. In the late 1970s–early 1980s—an attempt was done by emplacement of four (out of the six designed) artificial round islands [59] but their ineffectiveness soon suggested to build a landward connection with groins and a submerged breakwater connecting the islands.

Figure 9. Marina di Carrara: area downdrift of the harbour with different defence structures (Photo Provincia di Livorno).

Figure 10. Marina di Massa: permeable groins in a picture taken in 1968 (the pier in the background).

Coastal armouring proceeded to the south with nine very closely spaced groins (average Length/Distance = 1) unable to intercept longshore transport (see [60] for a functional design of groin systems) and the beach continued to be eroded even if a submerged breakwater was constructed to connect their tips. Only rising it to the mean sea level favoured effectiveness of the defences and the beach expanded for approximately 16 m on average [61]: very fine sand enters in suspension into the protected area during storms producing a not very attractive tourist beach but one which is very appreciated by concessionaires that are able to rent more sun umbrellas (Figure 11).

Downdrift erosion was soon evident at Marina di Ronchi [62], where a similar project was required by the concessionaires to stop erosion and, possibly, to enlarge the beach surface. Awareness that a further erosion expansion could occur, lead authorities to test different solutions and submerged groins constructed with experimental sand bags were emplaced [63]. After positive results, a more complete project started in 2009–2010 with seven submerged groins formed by geo-containers extending to a 4 m water depth. In 2012–2015 the inshore part was replaced by rock groins with a submerged extension, which partly covered the geotextile elements. A jetty with an "L" shaped submerged extension was built on the southern side of the Magliano R. mouth, which limits Ronchi beach to the north. It worthwhile to report that artificial beach nourishments accompanied most of these works, especially the most recent ones.

Figure 11. Groins with tips connected by submerged breakwaters. On the left the Brugiano R. with jetties; on the centre the Marina di Massa pier (Photo Provincia di Livorno). Summer shot captures sun umbrella density.

But it is the coast at Marina di Pisa, which gave origin to the first structure proliferation in this littoral unit. Maps dated 1878, 1907 and 1920 (Figure 12) showed no structures in front of the town but analysing old photos and postcards, it is possible to observe perishable structures (permeable wooden fences and groins) on the beach fronting the recently built settlement, which can be dated to the very early 20th century (Figure 13).

The 1928 IGM map shows several short groins on the first kilometre of coast south of the R. Arno. Ten years later a detached breakwater connected to an older groin appeared immediately south of the river mouth and in 1954 another detached breakwater was evident. At this time, the first kilometre of coast had a continuous revetment (Figure 6).

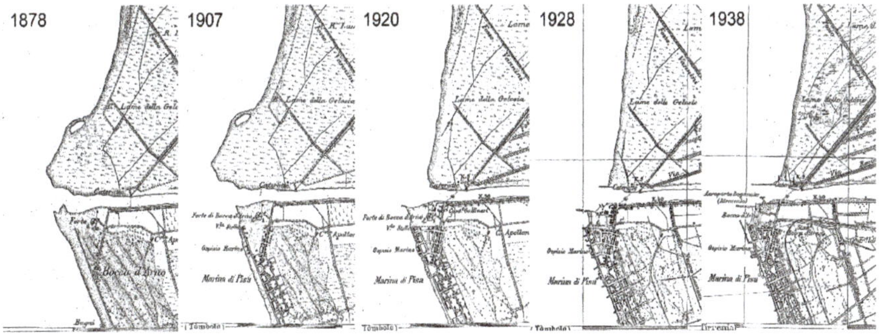

Figure 12. Coastal occupation, armouring and progressive beach disappearance at Marina di Pisa on IGM maps between 1878 and 1938 (in 1907 an update was performed on the southern coast only).

These defences gradually evolved into a more structured project performed in the 1960s to arrive at ten detached breakwaters, some of which were connected to the coast with groins: each kilometre of coast was protected by 2.3 km of rock. It is in this period that detached breakwater length overpassed that of revetments in the study area (Figure 6).

The shoreline position was fixed but erosion continued on the submerged profile: at the external foot of the detached breakwater a depth of 7 m was reached, both for the general sedimentary deficit and for basal scouring, as in Reference [64]. These breakwaters become inadequate to protect the settlement and their elevation from 2.0 m to 3.5 m was postulated, when the need for softer shore protection structures was growing in Tuscany under the motto "Back to the beach" [65]. A new project started with lowering of the old breakwaters to 0.50 m below mean sea level and widening their crest;

a 30 m wide gravel beach was created to absorb the overpassing wave energy. Currently four of the ten breakwaters have been converted by transforming 940 m of emerged structures into submerged ones (Figure 14).

Figure 13. Permeable structure at Marina di Pisa in a *circa* 1915 postcard.

Figure 14. Marina di Pisa: on the right the old defence system with revetments and detached breakwater and on the left the new one with gravel beaches and submerged breakwaters (Photo Provincia di Livorno).

On the southern side of the town, shore protection structures expanded towards Tirrenia, not to protect the road or the houses but to maintain the beach in front of the numerous beach establishments. It is not known how many defences were unauthorized but it is accepted that they did not follow a general project. Now detached breakwaters, oblique and shore-connected structures and traditional groins limit safe access to the sea, producing low quality water and they have transformed a sandy coast into a rocky one (Figure 15). The value on this coast in terms of tourism income was strongly reduced and all this was done simply to support tourism.

Four kilometres north of the Arno River mouth (Figure 16), the five detached breakwaters were built between 1962 and 1968 to defend Gombo beach, in front of the Presidential Villa of San Rossore (now part of a Regional Park). The house was more than 300 m far from the shoreline and only two huts were present on the beach: the project was carried out exclusively to maintain beach facilities for park employers more than for the Italian President.

Figure 15. Between Marina di Pisa and Tirrenia: coastal armouring following emergency, un-coordinated interventions (Photo Provincia di Livorno).

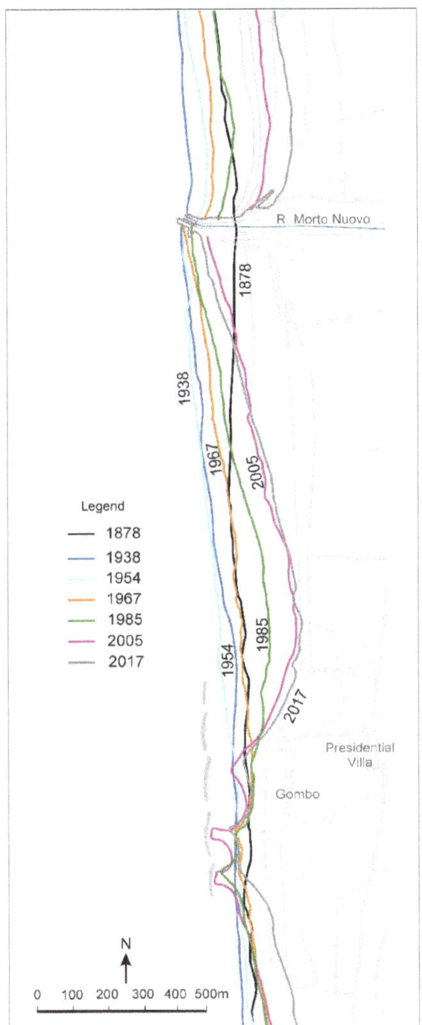

Figure 16. Shoreline evolution at Gombo from 1787 to 2017.

However, results were limited and 20 years later only smooth salients were formed. In 1984 four out of the five breakwaters were elevated and extended: resulting immediately in four tombolos, which fitted well with model predictions of beach response to structures geometry [66]. Longshore transport was interrupted by the tombolos and downdrift beach caused severe erosion; here the deepening of a spiral bay induced updrift erosion, destroying the tombolos and leaving the breakwaters too far offshore to be effective [67].

In 2009 an artificial tombolo with geo-textiles was constructed to connect the fourth breakwater and two submerged groins, also these with geo-textiles, were built to the north to limit erosion [15]. Such structures were quickly destroyed, very likely by the tree trunks brought to the sea by the R. Arno and transported northwards under the dominant winds and waves. Within the same project, the fifth breakwater was demolished, although, its elements can still be found on the sea floor. Meanwhile, a project to defend the northern lobe of the Arno River delta, retreating at a rate of 10 m/yr in the period 1997–2001, was carried out in 2001 comprising nine groins which were extended seaward with submerged segments for a total structural length of 1.3 km of structures. The total length of orthogonal structures on the study area reached 7.8 km. The need for this intervention, unusual in Italy, was not to protect settlements for tourist activities but to prevent further erosion which was connecting the interdune swales to the sea, thus causing the death of fresh water vegetation, which is one of the most relevant natural values of the Regional Park. Beach monitoring proved the effectiveness of the project [68]. This represents an interesting case because shore protection structures were built to restore a natural habitat; elsewhere, for this purpose, structure removal is proposed [69].

5. Conclusions

Shore protection structures have expanded greatly along the northern Tuscany coastline mostly following the developing tourist industry and under a policy of emergency intervention (Figure 6). The early projects mainly aimed at defending coastal roads by revetments but within a sedimentary deficit framework, this invariably resulted in beach disappearances. Several projects were addressed all limiting the effect of previous ones, each never anticipating the erosion process and constructing structures in an updrift direction.

Coastal erosion was trigged by the sediment river input reduction, commencing from 'two hot spots' that is, at the mouths of the River Magra and River Arno.

At the River Magra mouth, coastal development took place later than on the R. Arno delta southern side and no protection works were initially carried out except emplacement of a detached breakwater to protect a coastal road (1928–1938). In the following years tourist beach use exponentially increased and, accordingly the necessity of maintaining and/or enlarging beach width, resulted in artificial islands, groins and detached breakwaters being built in a downdrift sequence producing the "domino" effect [70].

On the left bank of the R. Arno mouth, at Marina di Pisa, the "new town" had to be strongly defended although tourism demanded a beach. After weak attempts of beach protection via short groins, a seawall and 10 detached breakwaters were built. Shore protection was extended to the south to satisfy beach tourist demand and ended only at Tirrenia, area of longshore transport convergence —where erosion never occurred (Figure 1).

On the right-hand side of the R. Arno mouth, nature was untouched until 1300 m of land was lost; in 2001 emerged groins with submerged extensions were built to protect the Regional Park wetlands. A completely different approach was taken a few kilometres north, at Gombo, where detached breakwaters were positioned in the 1960s to defend the privileges of a few people (Figure 16).

A further erosive hot spot was later introduced by man: Marina di Carrara harbour, which only anticipated and boosted the erosion process, expanding southwards from the River Magra mouth. Again, coastal road protection was the first reason to build defence structures, in this case revetments and later detached breakwaters. Downdrift erosion, together with beach tourism activity, brought first groin emplacement and later submerged breakwaters.

Detached breakwaters had a gradual rise and, in 1965, their total length reached that of the previous structures. Their imposition is also a consequence of a school of engineers working for the Italian Ministry of the Public Work, who deemed that only a reduction of wave energy was able to protect the coast. In the same years, several coastal segments in Italy were protected in this way, for example, detached breakwaters number more than 240 over 23 km of coast at Pescara, Adriatic Sea [15].

Groins (after the transient case of Marina di Pisa in the 1920s) were not frequent along this coast until the 1980s, being limited to those of Marina di Massa; but later they experienced an accelerated expansion and in 2011 their total length was greater than that of detached breakwaters. However, the last project at Marina di Massa, aimed at reducing the landscape impact of defensive structures and increasing beach quality, which resulted in a small decrease in their total length (Figure 6).

The main recent change in the coastal defence philosophy was the shifting from emerged to submerged structures. The latter slowly started in 1965 and today submerged detached breakwaters and groins reach 60% of the length of their emerged equivalents (7940 m vs. 13,230 m). Decision makers are highly influenced by economic stakeholders since these represent a significant part of electoral voters and the answer to beach erosion is often a compromise between the best technical solution and the most appealing one for the tourist industry. At Marina di Massa and Marina di Pisa all interventions were designed under a local perspective, without a general view of the processes at work along the littoral cell and a clear knowledge of river sediment input, which is essential in any adequate and long-term coastal management plan [71]. Under stakeholders' urging, heavy projects were executed when updrift defences had a negative influence. This study evidences as sediment bypassing could have been implemented at most important harbours and softer solutions could have been adopted starting from stable sectors and moving towards eroding areas but this requires a long-term strategy, incompatible with the life-time of political officers.

Author Contributions: E.P. coordinated the research, G.A. assessed the morphodynamic aspects of the coast, I.C. analyzed the evolution of the shoreline, M.P. focused on the historical issues and G.V. measured and classified the shore protection structures.

Funding: This research received no external funding.

Acknowledgments: Beer will never suffice to thank Allan Williams for the helpful discussion on this topic and for the critical review of the manuscript, although he can swallow barrels!

Conflicts of Interest: The authors declare no conflicts of interest.

References

1. Honey, M.; Krantz, D. *Global Trends in Coastal Tourism*; Center of Ecotourism and Sustainable Development—Stanford University: Washington, DC, USA, 2007; 140p.
2. *World Almanac*; Funk and Wagnalls Corp.: Mahwah, NJ, USA, 2012.
3. Klein, Y.L.; Osleeb, J. Determinants of Coastal Tourism: A Case Study of Florida Beach Counties. *J. Coast. Res.* **2010**, *26*, 1149–1156. [CrossRef]
4. Houston, J.R. The economic value of beaches—A 2013 update. *Shore Beach* **2013**, *81*, 3–11.
5. Fletchert, C.H.; Mullane, R.A.; Richmond, B.M. Beach loss along armored shorelines on Oahu, Hawaiian Islands. *J. Coast. Res.* **1997**, *13*, 209–215.
6. Alexandrakis, G.; Manasakis, C.; Kampanisl, N.A. Valuating the effects of beach erosion to tourism revenue. A management perspective. *Ocean Coast. Manag.* **2015**, *111*, 1–11. [CrossRef]
7. AETIB. 2017. Available online: http://www.mondobalneare.com/news/1629/imprese-balneari-uno-studio-ne-dimostra-il-grande-valore.html (accessed on 18 October 2018).
8. GNRAC. Lo stato dei litorali in Italia. *Stud. Costieri* **2006**, *10*, 3–176.
9. MondoBalneare. Available online: http://www.mondobalneare.com/news/1629/imprese-balneari-uno-studio-ne-dimostra-il-grande-valore.html (accessed on 18 October 2018).
10. Phillips, M.R.; Jones, A.L. Erosion and tourism infrastructure in the coastal zone: Problems, consequences and management. *Tourism Manag.* **2006**, *27*, 517–524. [CrossRef]

11. Kantamaneni, K.; Phillips, M.; Thomas, T.; Jenkins, R. Assessing coastal vulnerability: Development of a combined physical and economic index. *Ocean Coast. Manag.* **2018**, *58*, 164–175. [CrossRef]
12. Thinh, N.A.; Thanh, N.N.; Tuyen, L.T.; Hens, L. Tourism and beach erosion: Valuing the damage of beach erosion for tourism in the Hoi An World Heritage site, Vietnam. *J. Environ. Dev. Sustain.* **2018**. [CrossRef]
13. Sullivan Sealy, K.; Strobl, E. A hurricane loss risk assessment of coastal properties in the caribbean: Evidence from the Bahamas. *Ocean Coast. Manag.* **2017**, *149*, 42–51. [CrossRef]
14. McDougall, C. Erosion and the beaches of Negril. *Ocean Coast. Manag.* **2017**, *148*, 204–213. [CrossRef]
15. Pranzini, E. Italy. In *Coastal Erosion and Protection in Europe*; Pranzini, E., Williams, A.T., Eds.; Earthscan/Routledge: Abingdon, UK, 2013; pp. 294–323.
16. Garbrecht, G. Wasser—Vorrat, Bedarf und Nutzung in Geschichte und Gegenwart. In *Deutsches Museum Kulturgeschichte der Naturwissenschaften und Technik*; Rororo Sachbuch; Rowohlt: Berlin, Germany, 1985.
17. Jensen, J.; Schwarzer, K. Germany. In *Coastal Erosion and Protection in Europe*; Pranzini, E., Williams, A.T., Eds.; Earthscan/Routledge: Abingdon, UK, 2013; pp. 108–135.
18. Nieuwhof, A. Living in a dynamic landscape: Prehistoric and proto-historic occupation of the northern-Netherlands coastal area. In *Science for Nature Conservation and Management: The Wadden Sea Ecosystem and EU Directives, Proceedings of the 12th International ScientificWadden Sea Symposium, Wilhelmshaven, Germany, 30 March–3 April 2009*; Marencic, H., Eskildsen, K., Farke, H., Hedtkamp, S., Eds.; CommonWadden Sea Secretariat: Wilhelmshaven, Germany, 2010.
19. Grillo, S. *Venezia, le Difese a Mare*; Arsenale Editrice: Venezia, Italy, 1989.
20. Jiang, W.; Tao, C. The seawall in Quintang estuary. In *Engineeren Coasts*; Chen, J., Eisma, D., Hotta, K., Walker, H.J., Eds.; Kluwer Academic Press: Dordrecht, The Netherlands, 2002; pp. 139–150.
21. De Moor, G. *Artificial Structures and Shorelines*; Walker, H.J., Ed.; Kluver Academic Publishers: Dordrecht, Belgium, 1988; pp. 115–125.
22. Mestanza, C.; Piccardi, M.; Pranzini, E. Coastal Erosion Management at Callao (Peru) in the 17th and 18th Centuries: The First Groin Field in South America? *Water* **2018**, *10*, 891. [CrossRef]
23. Behre, K.-E. Coastal development, sea-level change and settlement history during the later Holocene in the Clay District of Lower Saxony (Niedersachsen), northern Germany. *Quat. Int.* **2004**, *112*, 37–53. [CrossRef]
24. Holling, C.S. Resilience and stability of ecological systems. *Ann. Rev. Ecol. Syst.* **1973**, *4*, 1–23. [CrossRef]
25. Manno, G.; Anfuso, G.; Messina, E.; Williams, A.T.; Suffo, M.; Liguori, V. Decadal evolution of coastline armouring along the Mediterranean Andalusia littoral (South of Spain). *Ocean Coast. Manag.* **2016**, *124*, 84–99. [CrossRef]
26. Pranzini, E.; Williams, A.T. (Eds.) *Coastal Erosion and Protection in Europe*; Earthscan/Routledge: Abingdon, UK, 2013; 454p.
27. Pranzini, E. Shore protection in Italy: From hard to soft engineering . . . and back. *Ocean Coast. Manag.* **2018**, *156*, 43–57. [CrossRef]
28. Anthony, E.; Sabatier, F. *Coastal Erosion and Protection in Europe*; Pranzini, E., Williams, A.T.W., Eds.; Earthscan/Routlege: Abingdon, UK, 2013; pp. 226–253.
29. Arenas Granados, P. Gestión del litoral y política pública en España [Coastal Management and public policy in Spain]. In *Manejo Costero Integrado y Política Pública en Iberoamérica: Un diagnóstico: Necesidad de cambio*; Red IBERMAR (CYTED); Editorial Tébar Flores: Madrid, Spain, 2009; pp. 353–380.
30. UNWTO (United Nations World Tourism Organization). *Tourism Highlights*; UNWTO: Madrid, Spain, 2017; 16p.
31. Escudero-Castillo, M.; Felix-Delgado, A.; Silva, R.; Marino-Tapia, I.; Mendoza, E. Beach erosion and loss of protection environmental services in Cancun, Mexico. *Ocean Coast. Manag.* **2018**, *156*, 183–197. [CrossRef]
32. Nava Fuentes, J.C.; Arenas Granados, P.; Cardoso Martins, F. Coastal management in Mexico: Improvements after the marine and coastal policy publication. *Ocean Coast. Manag.* **2017**, *137*, 131–143. [CrossRef]
33. Rangel-Buitrago, N.; Williams, A.T.; Anfuso, G. Hard protection structures as a principal coastal erosion management strategy along the Caribbean coast of Colombia. A chronicle of pitfalls. *Ocean Coast. Manag.* **2017**, *156*, 58–75. [CrossRef]
34. Bernatchez, P.; Fraser, C. Evolution of Coastal Defence Structures and Consequences for Beach Width Trends, Québec, Canada. *J. Coast. Res.* **2012**, *28*, 1550–1566. [CrossRef]
35. Tofta, J.D.; Ogstonb, A.S.; Heerhartza, S.H.; Cordella, J.R.; Flemerb, E.E. Ecological response and physical stability of habitat enhancements along an urban armored shoreline. *Ecol. Eng.* **2013**, *57*, 97–108. [CrossRef]

36. Pranzini, E.; Jackson, N.L.; Lami, G.; Nordstrom, K.F.; Rossi, L. Reshaping beach morphology by modifying offshore breakwaters. *Ocean Coast. Manag.* **2018**, *154*, 168–177. [CrossRef]
37. Preti, M.; Zanuttigh, B.; De Nigris, N.; Martinelli, L.; Aguzzi, M.; Archetti, R.; Lamberti, A. Integrated beach monitoring at Igea Marina, Italy: Results of ten-years monitoring. *Coast. Eng. Proc.* **2011**, *1*, 199–226. [CrossRef]
38. Garcia, V.; Sánchez-Arcilla, A.; Anfuso, G. Spain. In *Coastal Erosion and Protection in Europe*; Pranzini, E., Williams, A.T., Eds.; Earthscan/Routledge: Abingdon, UK, 2013; 457p.
39. Alexandrakis, G.; Ghionis, G.; Poulos, S.E.; Kampanis, N.A. *Coastal Erosion and Protection in Europe*; Pranzini, E., Williams, A.T., Eds.; Earthscan/Routledge: Wolverhampton, UK, 2013; pp. 355–377.
40. Nordstrom, K.F.; Jackson, N.L. Removing shore protection structures to facilitate migration of landforms and habitats on the bayside of a barrier spit. *Geomorphology* **2013**, *199*, 179–191. [CrossRef]
41. Perna, M.; Brandini, C.; Cipriani, L.; Grasso, V.; Mammì, I.; Pelliccia, F.; Pranzini, E.; Vitale, G. La percezione dell'impatto dell'erosione costiera da parte degli esercenti del litorale compreso tra la foce del F. Magra e Marina di Carrara. *Stud. Costieri* **2017**, *27*, 17–34.
42. Becchi, I.; Paris, E. Il corso dell'Arno e la sua evoluzione storica. *Acqua Aria* **1989**, *6*, 645–652.
43. Cavazza, S. Regionalizzazione geomorfologica del trasporto solido in sospensione dei corsi d'acqua tra il Magra e l'Ombrone. *Atti Soc. Toscana Sci. Nat. Memorie Ser. A* **1984**, *91*, 119–132.
44. Pranzini, E. Caratteristiche morfologiche e sedimentologiche di una zona di convergenza del trasporto litoraneo (Versilia, Toscana). *Stud. Cost.* **2004**, *8*, 135–149.
45. DEAM—UNIFI. *Stima delle Profondità di Chiusura Lungo le Coste Toscane*; BEACHMED Project; University of Florence: Florence, Italy, 2007; 138p.
46. Anfuso, G.; Pranzini, E.; Vitale, G. An integrated approach to coastal erosion problems in northern Tuscany (Italy): Littoral morphological evolution and cells distribution. *Geomorphology* **2011**, *129*, 204–214. [CrossRef]
47. Cappietti, L.; Cammelli, C.; Farrell, E.; Ferri, S.; Aminti, P.L.; Pranzini, E. *A Case of Updrift Erosion Induced by a Harbour Breakwater*; Medcoast: Ravenna, Italy, 2003; pp. 1707–1718.
48. Pranzini, E. Bilancio sedimentario ed evoluzione storica delle spiagge. *Il Quat.* **1994**, *7*, 197–202.
49. Istituto Geografico Militare. *1:25,000 Scale Topographic Map*; Istituto Geografico Militare: Firenze, Italy, 1878.
50. Bartolini, C.; Cipriani, L.E.; Pranzini, E.; Sargentini, M. Caratteristiche geomorfologiche ed evoluzione della linea di riva del litorale toscano e criteri di lettura. *Coste Toscane Regione Toscana* **1989**, 33–56. Available online: https://flore.unifi.it/handle/2158/258640?mode=full.7#.W-q5xllqtPY (accessed on 12 November 2018).
51. Perna, M.; Vitale, G.; Brandini, C.; Pranzini, E.; Gozzini, B. Coastal monitoring through field and satellite data. In Proceedings of the 9th Conference on Italian Association of Remote Sensing, Riva del Garda, Italy, 3–7 June 2018; p. 57.
52. Anfuso, G.; Bowman, D.; Danese, C.; Pranzini, E. Transect Based Analysis versus Area Based Analysis to quantify shoreline displacement: Spatial resolution issues. *Environ. Monit. Assess.* **2016**, *188*, 568. [CrossRef] [PubMed]
53. Pranzini, E. A model for cuspate delta erosion. In Proceedings of the 6th Symposium on Coastal and Ocean Management/ASCE, Charleston, SC, USA, 11–14 July 1989; pp. 4345–4357.
54. Silvester, R.; Hsu, J. Coastal stabilization. *Adv. Ser. Ocean Eng.* **1997**, *14*, 596.
55. Milano, V. Studio sull'accessibilità del porto di Viareggio e sull'equilibrio della spiaggia a Nord. In Proceedings of the Convegno di idraulica e costruzioni idrauliche, Padova, Italy, 8–10 September 1986; pp. 255–264.
56. Istituto Idrografico della Marina Italiana. *1:100,000 Scale Sea Chart*; Istituto Idrografico della Marina Italiana: Genova, Italy, 1954.
57. Piccardi, M.; Pranzini, E.; Rombai, L. Historical cartography and coastal dynamics of the Apuan littoral in the modern and contemporary periods: The port of Marina di Carrara (Tuscany, Italy). *E-Perimetron* **2018**, *13*, 32–49.
58. Pranzini, E.; Anfuso, G.; Botero Saltaren, C. Nourishing tourist beaches. In *Beach Management Tools—Concepts, Methodologies and Case Studies*; Botero, C.M., Cervantes, O.D., Finkl, C.W., Eds.; Springer: Berlin, Germany, 2017; pp. 293–318.
59. Aminti, P.; Graniglia, N.; Stefani, G. Il recupero della spiaggia di Marina di Massa. In *La rete di Monitoraggio delle Spiagge, Resmar EU Project*; Editrice Taphros: Olbia, Italy, 2013; pp. 152–178.

60. Kraus, N.C.; Hans Hanson, H.; Blomgren, S.H. Modern functional design of groin systems. *Coast. Eng.* **1994**, *95*, 1327–1342.
61. Cipriani, L.E.; Ferri, S.; Iannotta, P.; Paolieri, F.; Pranzini, E. Morfologia e dinamica dei sedimenti del litorale della Toscana settentrionale. *Stud. Costieri* **2001**, *4*, 119–156.
62. Bruun, P. The Development of Downdrifterosion. *J. Coast. Res.* **1995**, *11*, 1242–1257.
63. Aminti, P.; Cammelli, C.; Cappietti, L.; Jackson, N.L.; Nordstrom, K.F.; Pranzini, E. Evaluation of beach response to submerged groin construction at Marina di Ronchi, Italy, using field data and a numerical simulation model. *J. Coast. Res.* **2004**, *33*, 99–120.
64. Kraus, C.N.; McDougal, W.G. The Effects of Seawalls on the Beach: Part I, An Updated Literature Review. *J. Coast. Res.* **1996**, *12*, 691–701.
65. Aminti, P.L.; Cipriani, L.E.; Pranzini, E. 'Back to the beach': Converting seawalls into gravel beaches. In *Soft Shore Protection, Coastal Systems and Continental Margins*; Goudas, C., Katsiaris, G., May, V., Karambas, T., Eds.; Kluwer Academic Publishers: Dordrecht, The Netherlands, 2003; Volume 7, pp. 261–274.
66. Pope, J.; Dean, J.L. Development of design criteria for segmented breakwaters. In Proceedings of the 20th International Conference on Coastal Engineering, Taipei, Taiwan, 9–14 November 1986; pp. 2144–2158.
67. Bowman, D.; Pranzini, E. Reversed response within a segmented detached breakwater—The Gombo case, Tuscany coast, Italy. *Coast. Eng.* **2003**, *49*, 263–274. [CrossRef]
68. Pranzini, E. L'intervento di stabilizzazione del litorale de Le Lame (Parco San Rossore Migliarino Massaciuccoli, Toscana). *Studi Costieri* **2008**, *14*, 29–42.
69. Nordstrom, K.F. Living with shore protection structures: A review. *Estuar. Coast. Shelf Sci.* **2014**, *150*, 11–23. [CrossRef]
70. Cooper, J.A.G.; Anfuso, G.; Del Rio, L. Bad beach management: European perspectives. *Geol. Soc Am.* **2009**, *460*, 167–179.
71. Samaras, A.G.; Koutitas, C.G. An integrated approach to quantify the impact of watershed management on coastal morphology. *Ocean Coast. Manag.* **2012**, *69*, 68–77. [CrossRef]

water

MDPI

Article

Efficiency Assessment of Existing Pumping/Hydraulic Network Systems to Mitigate Flooding in Low-Lying Coastal Regions under Different Scenarios of Sea Level Rise: The Mazzocchio Area Study Case

Francesco Cioffi *, Alessandro De Bonis Trapella and Federico Rosario Conticello

DICEA—Dipartimento di Ingegneria Civile, Edile ed Ambientale—Università di Roma 'La Sapienza', 00184 Rome, Italy; alessandro.debonistrapella@uniroma1.it (A.D.B.T.); federicorosario.conticello@uniroma1.it (F.R.C.)

* Correspondence: francesco.cioffi@uniroma1.it; Tel.: +39-064-991-2228

Received: 11 May 2018; Accepted: 18 June 2018; Published: 21 June 2018

Abstract: Rising of the sea level and/or heavy rainfall intensification significantly enhance the risk of flooding in low-lying coastal reclamation areas. Therefore, there is a necessity to assess whether channel hydraulic networks and pumping systems are still efficient and reliable in managing risks of flooding in such areas in the future. This study addresses these issues for the pumping system of the Mazzocchio area, which is the most depressed area within the Pontina plain, a large reclamation region in the south of Lazio (Italy). For this area, in order to assess climate change impact, a novel methodological approach is proposed, based on the development of a simulation–optimization model, which combines a multiobjective evolutionary algorithm and a hydraulic model. For assigned extreme rainfall events and sea levels, the model calculates sets of Pareto optimal solutions which are obtained by defining two optimality criteria: (a) to minimize the flooding surface in the considered area; (b) to minimize the pumping power necessary to mitigate the flooding. The application shows that the carrying capacity of the hydraulic network downstream of the pumping system is insufficient to cope with future sea level rise and intensification of rainfall.

Keywords: climate change; multiobjective optimization; coastal region; pumping plant; flooding

---

1. Introduction

Recent studies have shown that in Central Italy, the occurrence of torrential rainfall, exceeding 100 mm/d, has increased in the last decades [1]. Furthermore, future projections by global and regional climate models indicate an intensification of extreme precipitation events in Italy [2–4]. As a consequence of global warming, sea level rise is also expected in the next years in the Mediterranean region [5,6]. Lambeck et al. [7] have argued that in the central Tyrrhenian Sea, sea level rise will mainly impact the coasts near Rome. Some of these coasts include the southern Latium, with its mainly coastal lakes, the Voltuno littoral, and the Sele River area, with sea level rise ranging from 315 to 1400 mm, depending on the climate scenario considered. Rising sea levels and intensification of extreme precipitation significantly increases the flood risk in such low-lying coastal areas. This has some substantial consequences, as the coastal areas around Rome are densely populated, with extensive and highly developed agriculture along with a large presence of industrial activity.

Within this region, an area particularly vulnerable to flooding is in the Mazzocchio zone, the lowest lying area of the Pontina plain (see Figure 1a). The majority of this zone has a soil surface elevation equal to or lower than the mean sea level. Historically a swamp, this zone was recovered in the years 1926–1937 by a large reclamation work covering an area of 20,000 ha.

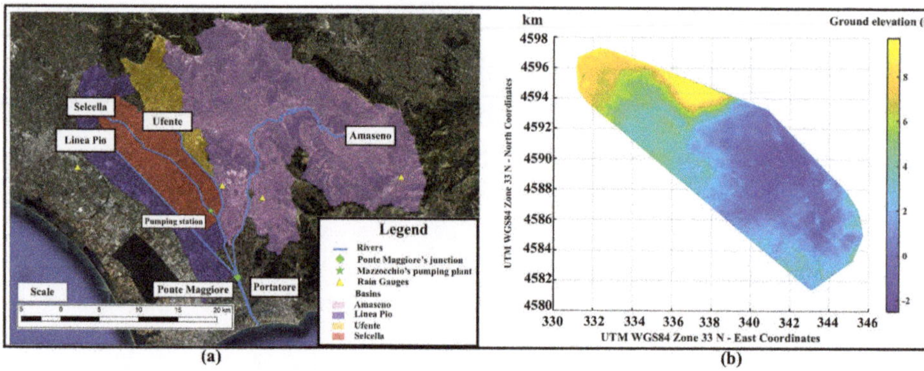

Figure 1. (**a**) Hydraulic network and basins of the study site. (**b**) Mazzocchio's basin ground elevation.

As seen in Figure 1a, the Mazzocchio zone is enclosed by two rivers: Ufente in the north and Linea Pio in the south. To ensure dry soil, water runoff is intercepted within and outside the zone. To intercept the runoff outside the zone, the two rivers act as barriers, and within the zone, water runoff—caused by rainfall over the zone—is collected by a dense channel network in the Selcella River. Downstream of the Selcella River, there is a pumping station which lifts the water from the Mazzocchio area into the Ufente River, with a maximum capacity of 36 m^3/s. Figure 1a shows the water coming from the upstream basins of the Ufente, Linea Pio, and Amaseno Rivers, together with the pumping water from the Mazzocchio area, flowing through the Portatore Channel towards the sea.

On 7 November 2014, a heavy rainfall event (approximately 100 mm/d) caused a serious crisis to the entire channel network, with flooding in the Mazzocchio area and overflowing and subsequent collapse of a large portion of the levees along the Ufente River. Most of the damage occurred along the Ufente River stretch closer to the Mazzocchio pumping station (downstream from the confluence with the Selcella collector). High rainfall amount and sea level rise due to storm surge at the outlet of the Portatore channel were the two main factors which caused the crisis of hydraulic system.

As mentioned before, the expected intensification and increment in frequency of such extreme rainfall events, as well as the sea level rise, causes serious concerns with regard to the capability of the hydraulic infrastructures to cope with similar or more intense events in the future. Therefore, methodological approaches need to be developed to assess the efficiency and reliability of existing hydraulic infrastructures. As a number of authors suggest, such methodologies must evolve to address "change" from climate variability at the global scale to local human impacts [8,9]. Recently, rainfall downscaling models have been constructed to perform projections of rainfall occurrence and amount at the basin level, under different global warming scenarios simulated by global or regional circulation models (GCMs and RCMs) [10,11]. Therefore, such models can be used to provide the hydrological inputs necessary to run hydraulic models to assess the reliability of existing hydraulic infrastructures, and eventually the design of new ones, to manage future flooding risk at the local scale. Preliminary to such assessment, it is, however, necessary to perform an analysis on the capability of the existing infrastructure to manage the risk of flooding due to extreme rainfall and high tidal sea level events. This is very useful to identify the elements of the hydraulic network which are more vulnerable.

In this context, we propose a methodology and related models to assess the reliability of hydraulic infrastructures in control flooding events and apply it for the case of the reclamation region of Mazzocchio. The first question that arises in developing such a methodology is how to assess the reliability and efficiency of existing pumping–hydraulic network systems to mitigate flooding in the Mazzocchio basin under different hydrological inputs. This question arises because different pumping schedules can be hypothesized to manage extreme stream flows. In other words, as a consequence of adopting alternative pumping schedules, for the same set of hydrological inputs, a number of different configurations of hydraulic systems may potentially exist. In order to compare the effects of the different hydrologic inputs on the hydraulic system, we should identify a particular set of pumping schedules. This restricts the analysis to a few arbitrarily chosen cases of pumping schedules. To perform a more general and less restrictive analysis, in order to compare the possible different configurations of a hydraulic system under different hydrological inputs, this paper proposes to use sets of Pareto optimal solutions as calculated by a multiobjective optimization approach, in which the switching on/off levels of the pump system are assumed as decision variables. Given two or more optimality criteria, the Pareto set identifies not a unique optimal solution, but an ensemble of nondominant configurations of the system that belong to the Pareto front. Such a set of non-dominant solutions is chosen as optimal, if no objective can be improved without sacrificing at least one other objective. Therefore, for given hydrological inputs, the set of Pareto optimal solutions is unique. This solution can then be used to compare the possible states of the hydraulic system (as identified by the free surface levels and flow rates along the rivers and channel networks) forced by different hydrological inputs—rainfall amount and sea level rise—and depending on the optimal pumping schedules associated to the solutions lying in the Pareto front. In this paper, the sets of Pareto optimal solutions are calculated by a simulation–optimization model, which combines a multiobjective evolutionary algorithm (the non-dominating sorting genetic algorithm, NSGA2) and a hydraulic model. While a number of optimization methods exist [12], we use a genetic algorithm due to their reliability in solving nonlinear, nonconvex, multimodal, and discrete problems, unlike classical optimization methods [8]. Since genetic algorithms are independent of derivative information, they also allow a less restricted formalization of the objective functions and constraints. Even though a number of genetic algorithms have been proposed in the past, we adhered to NSGA2, since a number of studies have proved the reliability and robustness of such an algorithm [13]. The use of a simulation–optimization model, in which a multiobjective optimization model and hydraulic model are combined, is not novel in the literature. For instance, Cioffi and Gallerano [13], proposed a multiobjective programming model including output from 2D hydraulic simulation for habitat assessment to optimize power production and fish habitat suitability as a Pareto set. In the past, a number of simulation–optimization models specifically aimed to find the optimal schedule of pumping systems have been proposed for urban drainage systems [14–16], irrigation pumping stations [17], water supply systems [18], and water resource management [19]. However, the above-cited studies were mainly focused on the optimal control and operation of such systems. Some researchers have proposed criteria and methodologies that use simulation–optimization models to assess how climate change and global warming affect the hydrologic cycle and its effects on the performance of water resource systems. Most of these studies are addressed to assessing the climate change impacts on hydropower production by reservoirs [20–22]. Direct application of multiobjective optimization to flood risk management under climate change is very rare in the literature [23]. Most of the papers focus on cost–benefit analysis [24]. For instance, Woodward et al. [25] identify a set of Pareto optimal solutions using NSGA2, in which costs and benefits of flood risk intervention strategies are compared, taking into account the uncertainty in the future projected sea level rise. In such studies, flooding simulations by hydraulic models are carried out separately from the optimization process; the output from hydraulic simulations are used to assess the costs and benefits related to the specific flood risk intervention strategy hypothesized. Such approaches, however, seem difficult to apply in the cases in which pumping systems are part of flooding control hydraulic infrastructures. In this paper, the hydraulic simulations are integrated in the multiobjective optimization algorithm and the objective functions are not defined on the basis of economic variables, but directly in terms of flooding surface and pumping power. The main reason of such a choice is due to the uncertainty of the estimation of damage from flooding, since agriculture is the main activity over the basin, and productivity and damage depend on the period of the year, as well as the type and state of growth of crops. To calculate the sets of Pareto optimal solutions, two optimality criteria are defined: (a) to minimize the maximum flooding surface over the Mazzocchio basin; (b) to minimize the pumping power necessary to limit the flooding over the Mazzocchio basin. In formalizing the multiobjective problem, the state variables are the free surface levels and flow rates along the hydraulic network as well as the surface of the flooding areas, and the decisional variables are the levels switching on and off the pumps in the Mazzocchio station. Temporal and spatial distribution of the rainfall amount and the sea level are the input variables. State and decisional variables range within the constraints imposed by the flow carrying capacity of the channel network downstream from the pumping station. The hydraulic state variables are simulated by a hydraulic model. In order to limit the calculation time, due to the large number of simulations necessary to solve the multiobjective problem by the genetic algorithm, a simplified version of the hydraulic model has been constructed to represent the flow river network and the rainfall–runoff and flooding processes over the river basins. The calculation along the river network is performed numerically solving the 1D de Saint-Venant equations, while the basins are represented by storage areas connected to the river network by linear channels. The paper is organized as follows: In Section 2, a description of the study areas and hydrological data is provided; in Section 2 also, the hydraulic and multiobjective optimization models are described. In Section 3, the procedure of calibration and validation for the hydraulic model, the construction of the design hydrograph, and finally the set of Pareto optimal solutions for the Mazzocchio zone obtained by solving the multiobjective problem are discussed. Comparing the different Pareto sets, the reliability of the existing pumping systems and of the hydraulic channel network is inferred.

2. Materials and Methods

2.1. Materials

2.1.1. The Mazzocchio Reclamation Area and the Drainage Hydraulic Network

The Mazzocchio basin, with a surface about equal to 103 km^2, occupies the largest depression of the Pontina Plain, which lies between the Appia and the Ufente River. The drainage of the basin is provided by the pumping plant of Mazzocchio (see Figure 1a) located at the downstream end of the Selcella river. The Selcella river is 18.9 km long with an average slope (h/L) of 0.00035. The terrain of the Mazzocchio basin is shown in Figure 1b; the ground height above mean sea level ranges from −2 m in the central stretch of the Selcella river to +8 m in the upstream region. From Figure 1b, we can observe the presence of a large area with about the same or lower height than the mean sea level. Furthermore, a dense channel network hierarchically structured, can be observed in Figure 1a collecting water to the Selcella River. All the water that flows to the Mazzocchio plant is lifted and through a short channel, conveyed into the Ufente river. The Mazzocchio pumping system (Figure 2) at the downstream end of the river Selcella consists of six pump groups of 6 m^3/s each with a total capacity of about 36 m^3/s. The pumping system was built on 20 July 1934 and inaugurated on 19 December of the same year; it is equipped with six engines and six water-immersion pumps with a modern conception for the era, each with a capacity of 6000 L per second and powered by an electric motor of about 600 hp. During the Second World War, German troops sabotaged the plant, taking away the engines. Flooding of the surrounding land meant hindering the entry by the allies. In 1948, near the Brenner railway station, located in Northern Italy, the engines were found and immediately reinstalled.

Figure 2. Hydraulic pumps in Mazzocchio's pumping station.

The starting and stopping of each pump is ruled by floating switches, which depend on the prefixed free surface levels in the collection pond immediately upstream from the pumping station. The depth of the pond close to the floating switches is −4 m.a.s.l. The pump starting and stopping levels are reported in Table 1.

Table 1. Starting and stopping levels for each pump (data provided by Agro Pontino Reclamation Consortium (APRC)).

	Pump no. 1	Pump no. 2	Pump no. 3	Pump no. 4	Pump no. 5	Pump no. 6
Starting level (Meters above sea level)	−2.15	−2.05	−1.95	−1.85	−1.7	−1.6
Stopping level (Meters above sea level)	−2.5	−2.4	−2.3	−2.2	−2.0	−1.95

Downstream of the pumping system of Mazzocchio, the waters flow into the Ufente River, which flows into the hydraulic node of Ponte Maggiore. Ponte Maggiore also receives water from the Amaseno River and the Pio Line Channel flow. From the Ponte Maggiore node, through the Portatore Channel, all the water flows into the sea (see Figure 1a). The main characteristics of the hydraulic network downstream from the pumping station are reported in Table 2.

Table 2. Hydraulic characteristics of channels, rivers, and basins (data provided by Agro Pontino Reclamation Consortium (APRC)).

Rivers	Slope ‰ (m for 1 km)	Size of Cross Section: Max Width–Max Height (m)	Shape of Cross Section	Length (m)	Basin Extension (Km^2)
Selcella	0.1	45–5	Symmetrical Trapezoidal	11,250	97.659
Ufente	0.16	30–4	Symmetrical Trapezoidal	14,750	75.1
Linea Pio	0.15	37.2–4.5	Symmetrical Trapezoidal	16,000	99.561
Amaseno	0.45	109–11.75	Symmetrical Trapezoidal	80,940	628.53
Portatore	0.1	30–4.85	Symmetrical Trapezoidal	5700	34.864

The geometric and morphological characteristic of the entire hydraulic network (river slope, size and shape of cross-sections, etc.) as well as maintenance and operation of the pumping systems described above were provided by the Agro Pontino Reclamation Consortium (APRC).

2.1.2. Precipitation and Pumping Rate Data

The closest station to the study area is the Borgo San Michele station (Lat. 41°25′12″ N Lon. 12°58′12″ E), from which 56 years (1950–2005) of hourly rainfall data has been collected. From this record, intensity and duration curves, with return periods equal to 10 and 100 years are shown in Figure 3. Hourly data of pump operation for the period 1950–2016, recorded in the archive of the Mazzocchio pumping station and provided by APRC, are also collected. From the time series of hourly rainfall amount, a number of heavy rainfall events that occurred in the past are identified, and the corresponding temporal trends of pumping discharge during such events are reconstructed. The latter data are used to calibrate the hydraulic model. The data related to the ground surface level, the bed profiles of rivers, and the geometry of river cross sections were also provided by the APRC.

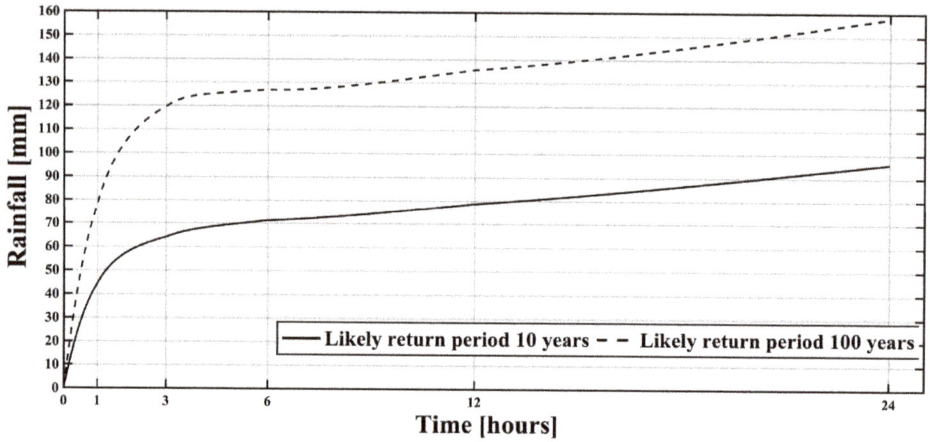

Figure 3. Intensity–duration curve from hourly rainfall records at the rain gauge located in Borgo San Michele station (Lat. 41°25′12″ N Lon. 12°58′12″ E).

2.2. Methods

The simulation–optimization model combines a hydraulic model and a multiobjective optimization model. The hydraulic model calculates the free surface level and the flowrate along rivers and channels upstream and downstream of the Mazzocchio pumping station, the flooding areas over the basins, and the pumped flowrate at the pumping station. The multiobjective optimization genetic algorithm, using the outputs from the hydraulic model, calculates the objective functions, verifies the constraints violations, and by an iterative procedure based on tournament selection of nondominant solutions and generation of new populations by crossover and mutation, identifies the set of Pareto optimal solutions. A sketch of the flowchart of the combined simulation–optimization model is shown in Figure 4.

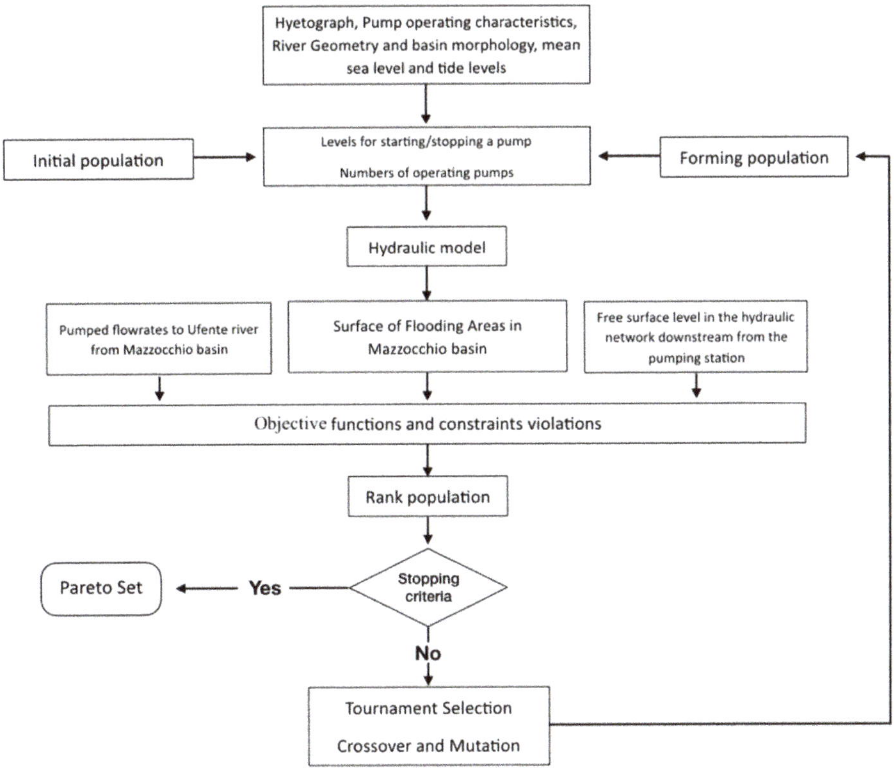

Figure 4. Optimization model's flowchart.

2.2.1. Hydraulic Model

For given input variables—rainfall amount over the basins and sea levels—as well as for given pumping switching levels in the collection pond upstream of the Mazzocchio pumping station, the hydraulic model calculates the outputs related to free surface levels and flowrates along the entire hydraulic network, flooding surface, water volumes, and depths over the basins. To simulate the temporal and spatial distribution of these quantities, 2D and 3D hydraulic models are generally used, as proposed by Cioffi and Gallerano [26] and Orton et al. [27]. Such models require a computation time which conflicts with the very high number of simulations necessary for the multiobjective optimization algorithm to calculate the Pareto front. Therefore, in this paper, a simplified hydraulic model has been constructed. This model is able to calculate the above listed hydraulic quantities after calibration. The scheme of such models, for the hydraulic networks upstream and downstream from the Mazzocchio pumping station, is shown in Figure 5.

Three different hydraulic elements may be recognized in the figure: the main river hydraulic networks upstream and downstream from the pumping station, the storage areas representing the basins, and the ideal channels connecting the storage areas to points of the rivers belonging to the hydraulic networks. As seen in the figure, the basin of Mazzocchio, upstream from the pumping station, is represented by seven storage areas connected to the river Selcella by ideal channels. Figure 6 shows how the storage areas refer to parts of the basin with homogeneous morphology and altimetry, in order to provide a sufficiently accurate representation of the spatial distribution of ground surface elevation.

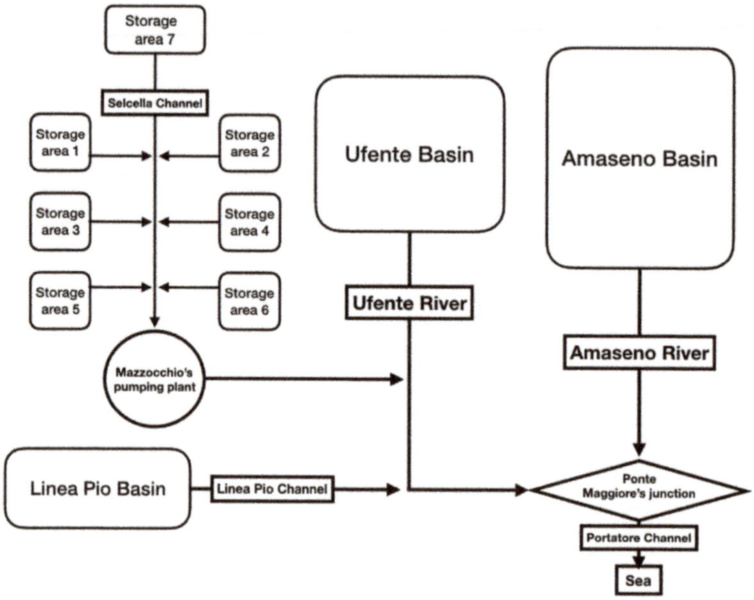

Figure 5. Sketch of hydraulic model.

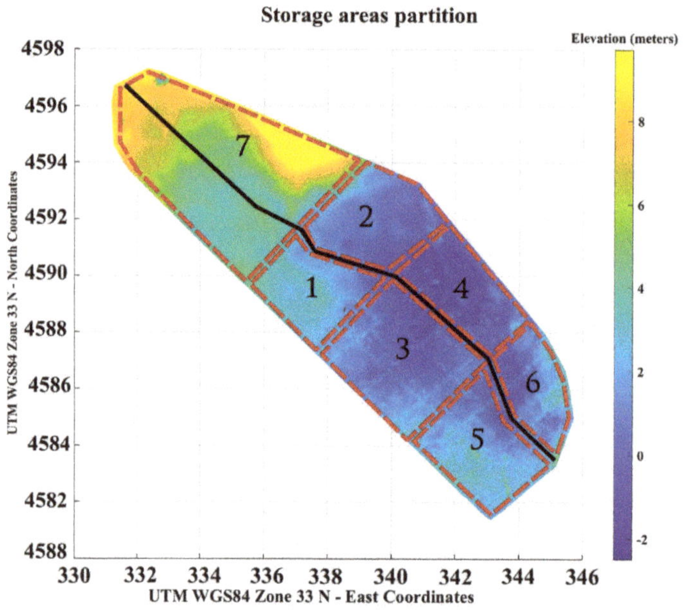

Figure 6. Storage areas partition of the Mazzocchio basin.

The ideal channels mimic the secondary drainage network collecting the rainfall water to the Selcella river. For the hydraulic network downstream from the pumping station, storage areas and related ideal channels are located upstream of the Amaseno, Ufente, and Linea Pio rivers.

The combination of storage areas and ideal channels allows for a representation of both the rainfall–runoff processes on the basins and the flooding over the riverbanks due to levee overflowing.

The 1D de Saint-Venant equations are used to simulate the flow along the river which belongs to the hydraulic networks:

$$\frac{\partial Q}{\partial x} + \frac{\partial S}{\partial t} = q_l \tag{1}$$

$$\frac{\partial Q}{\partial t} + \frac{\partial}{\partial x}\left(\beta\frac{Q^2}{S}\right) + gS\left(\frac{\partial H}{\partial x} + J\right) = \gamma_l \tag{2}$$

where Q is the flow rate, S the cross-section area, H the free surface elevation from a reference plane, and q_l represents the lateral inflow, γ_l represents the sum of the forces applied, and J is a dimensionless number, representing the average rate of energy dissipation:

$$J = \frac{Q^2}{K_m^2 S^2 R^{4/3}} \tag{3}$$

where K_m is the Strickler's roughness coefficient and R the hydraulic radius.

A number of river junctions (nodes) are present in the hydraulic network downstream from the pumping station (Figure 5). The equations of the junctions are derived from the equality of the elevations and the conservation of the discharges at the junction following the approach of 1D–2D coupling proposed by Goutal et al. [28]. The temporal trend of water level over the storage areas is calculated by the continuity equation, which is a function of the flows entering or going out from the ideal channels and of the rainfall amount directly falling over the area. It is assumed that the water free surface of the storage areas remains horizontal while moving vertically. The link between the storage area and the river is seen in Figure 7.

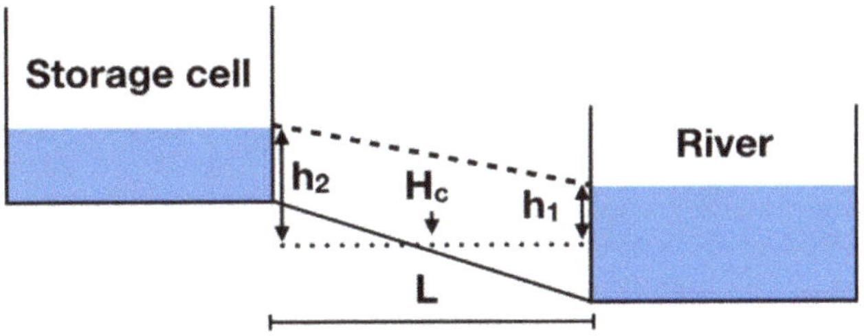

Figure 7. Sketch of the link between a river and a storage cell by an ideal channel. H_c is the bottom elevation of the ideal channel, h_1 and h_2 are free surface elevation in the river and in the storage cell respectively.

The flow rate through the ideal channels is calculated by the Manning–Stricker uniform free surface flow formula and depends on the difference of the free surface levels in the storage areas and in the river according to the following equations:

$$Q = Klh^{5/3}\sqrt{\frac{\Delta H}{L}} \tag{4}$$

where, as shown in Figure 7, ΔH is the difference between the free surface elevation h_2 in the storage cell and h_1 in the river, L is the length of the channel, l is the channel width, and h is the water depth in the channel, which is equal to the average between h_2 and h_1 (see Figure 7). Both h_2 and h_1 depend on the bottom elevation H_c of the ideal channel.

2.2.2. Multiobjective Optimization Problem Formalization

In formalizing the multiobjective problem, we assume the levels switching on and off the pumps in the Mazzocchio station (H^s) as decisional variables. Temporal and spatial distribution of the rainfall amount (I) and the sea levels (Sl) are the input variables. The state variables are the free surface levels $\left(H^d\right)$, the flow rates (Q^d) along the hydraulic networks, and the surface of the flooding areas (A_i^m). State and decisional variables range within imposed constraints. Specifically, we impose that the free surface level in the hydraulic network downstream from the pumping station cannot be more than the height of the river levees $\left(H_{max}^d\right)$. The multiobjective optimization problem is formalized defining two optimality criteria: (a) to minimize the maximum surface of flooding areas (A^m) over the Mazzocchio basin during a flood event caused by heavy precipitation; (b) to minimize the pumping power necessary for flooding control over the Mazzocchio basin. The pumping power here is defined as the energy consumed to lift water from the beginning of the event to the moment in which the maximum flooding surface over the Mazzocchio basin is reached. The second criterion is aimed to identify pumping schedules that limit, as much as possible, the power used to reduce the flooding surface. Such a criterion also responds to the need to limit the number of pumps working simultaneously in order to allow a more efficient and robust maintenance and operation of the plant.

The multiobjective algorithm is aimed to identify the levels of switching on and off the pumps in the Mazzocchio station which satisfy the optimality criteria described above. The mathematical formulation of the multiobjective optimization problem and the Pareto set calculation algorithms are as follows. The objective functions may be expressed as:

$$Of1 = (max(A_1^m,\ A_2^m,\ \ldots, A_n^m)) \tag{5}$$

$$Of2 = \left(\sum_{i=1}^{T_{Amax}} \gamma Q_i^p \left(H_i^m - H_i^p\right)\Delta t\right) \quad (i = 1, n) \tag{6}$$

where A_i^m, Q_i^p, H_i^m, and H_i^p at each time step i (i = 1, n) are the surface of the flooding area over the Mazzocchio basin, the pumping flowrate at the Mazzocchio pumping station, the free surface levels (from the reference level) in the pond of the Mazzocchio pumping station, and the free surface levels in the basin downstream from the pumping station, respectively; T_{Amax} is the temporal step at which the flooding surface in Mazzocchio reaches the maximum value; n is the number of time steps in which the period of hydraulic simulation T_s is divided; and Δt the temporal step.

The objective functions $Of1$ and $Of2$ are constrained by:

$$Q_i^p = n_k^p \cdot \Delta Q \ \ if \ \ H_i^m \geq H_k^s \ \ k = 1, n_p \ \ i = 1, n \tag{7}$$

$$H_{i,j}^d \leq H_{max,j}^d \ \ j = 1, n_c \ \ i = 1, n \tag{8}$$

where n_k^p is the number of pumps working simultaneously, ΔQ is the flowrate of each single pump, H_k^s is the level in the collecting pond upstream of the pumping station switching on the kth pump, and $H_{i,j}^d$ and $H_{max,j}^d$ are the free surface level and the levee top level, respectively, in the jth point along the river stretches of the hydraulic network downstream from the Mazzocchio pumping station.

Equation (7) imposes a constraint on the maximum number of pumps that may work simultaneously and the pumping switching on/off levels. Equation (8) constraints mean that the free surface levels along the hydraulic network downstream from the pumping station must be lower or equal to the top level of the levees of the Ufente, Amaseno, and Linea Pio Rivers and in the Portatore Channel.

The variables A_i^m, Q_i^p, H_i^m, H_i^p and $H_{i,j}^d$ are calculated by the hydraulic model as a function of the decision variable H_k^s and of the input variables I_i and Sl_i, that is, the temporal trend of the rainfall over the storage areas and the sea level; the latter varies in time with the tidal cycle.

All the hydrodynamic variables of the river flows, that is, free surface levels and flow rates along the two hydraulic networks, are subject to the constraints imposed by the continuity and momentum equations previously described (see Equation (1)).

For given inputs I_i and Sl_i, the resolution of the multiobjective problem requires the minimization of the two objectives $Of1$ and $Of2$ in Equations (5) and (6), within the constraints imposed by Equations (7) and (8), as a function of the decision vector $\mathrm{H}_\mathrm{k}^\mathrm{s}$ with $k = 1, n_p$.

Given two decision vectors $H_{k,1}^s$ and $H_{k,2}^s$ with $k = 1, n_p$, solutions of the multiobjective problem, the identification of the Pareto front is obtained applying Pareto dominance criteria.

In accordance with such criteria, an objective vector $\vec{y_1} \equiv \{Of1^2, Of2^2\}$ is said to dominate another objective vector $\vec{y_2} \equiv \{Of1^2, Of2^2\}$ (i.e., $\vec{y_1} < \vec{y_2}$), if no component of $\vec{y_1}$ is greater than the corresponding components of $\vec{y_2}$ and at least one component of $\vec{y_2}$ is greater; consequently, the solution $H_{k,1}^s$ dominates $H_{k,2}^s$. The nondominant solutions are optimal solutions of the problem whose ensemble identifies the set of Pareto optimal solutions.

A genetic algorithm is applied to solve the multiobjective optimization problem formalized above. Genetic algorithms are based on Darwin's theory of natural selection, which involves the language of microbiology, and in developing new potential solutions, mimics genetic operations. A population represents a group of solution points, and a generation represents the algorithm iteration, while a chromosome is equivalent to a component of the design vector. In accordance with these definitions, a genetic algorithm deals with a population of points, and hence multiple Pareto optimal solutions can be obtained from a population in a single run. Random numbers and information from previous iterations are combined to evaluate and improve a population of points, and then to select nondominant solutions. In this paper, the nondominant-sorting genetic algorithm II described by Deb et al. [29], NSGA2, is used, which has been applied successfully to many optimization problems. This algorithm uses tournament selection [30], simulated binary crossover (SBX) [31], a mutation operator, and crowding distance for diversity preservation.

3. Results

3.1. Hydraulic Model Calibration

As mentioned in Section 2.2.1, in order to limit the computation time needed to run the combined optimization–hydraulic models, the simplified hydraulic model previously described has been constructed. Such a model has a number of parameters that have to be identified: in particular, the length L, average heights h_1 and h_2, and width l of the ideal channels linking the rivers to the storage areas, in which the different basins have been divided, as well as the runoff coefficient ϕ of the storage areas and the Strickler's roughness coefficient K_m in the de Saint-Venant equations.

To identify such parameters, a calibration procedure was applied. Such a procedure consists of carrying out a single-objective optimization, through which the parameter values that minimize the sum of squares of the difference between the simulated and observed hydraulic data are identified.

Since the hydraulic network upstream of the Mazzocchio pumping station is hydraulically disconnected by the pumping from the downstream network, calibration of the upstream hydraulic network can be performed separately. For this network, two datasets were selected: the first is the precipitation trend recorded at the Borgo San Michele station, and the second is the pumping discharge trend at the Mazzocchio pumping station. Collected data from five different extreme events that occurred in the past were used to calibrate the hydraulic model. Then, the set of hydraulic parameters in the Mazzocchio area were identified, which minimizes the sum of squares of the difference between the simulated and the recorded pumping discharges during that period. The values of the identified parameters are shown in Table 3. For an extreme event, during the time period 16–20 March 2011, recorded pumping discharge and simulated pumping discharge was compared to validate the model. Figure 8a,b shows the precipitation trend used as input as well as the comparison between the simulated and recorded pumping discharges. Model accuracy was tested using the root mean square error (RMSE) between observed and simulated data, giving ~70% of accuracy.

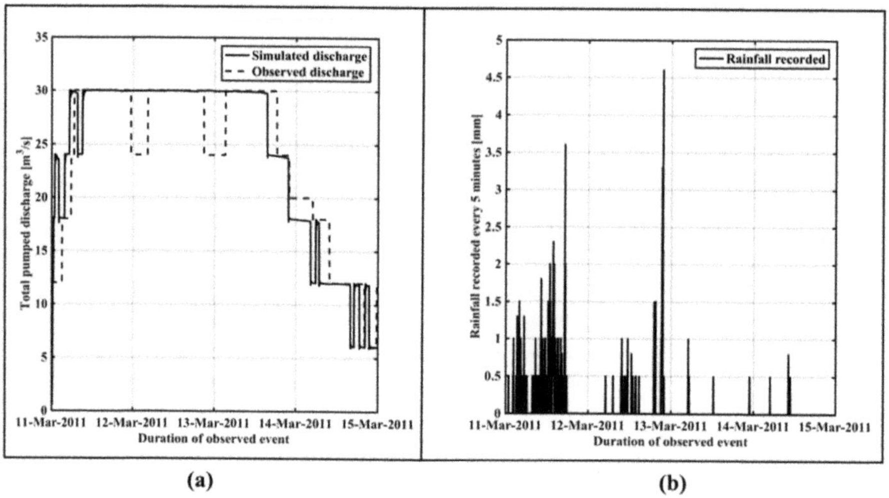

Figure 8. (**a**) Comparison between simulated and observed pumping discharges. (**b**) Rainfall intensity trend during the extreme event in March 2011.

Table 3. Hydraulic parameters after the calibration of the model upstream from the Mazzocchio pumping system (data provided by Agro Pontino Reclamation Consortium (APRC)).

	Link no. 1	Link no. 2	Link no. 3	Link no. 4	Link no. 5	Link no. 6	Link no. 7
Channel mean bottom elevation (meters above sea level)	−2.67	−2.03	−1.04	−1.35	−1.75	−2.3	−0.58
Channel width (m)	6	6	6	6	6	6	20
Channel length	1000	1000	1400	1400	1400	1400	7000
Roughness (Strickler)	40	40	40	40	40	40	30

For the hydraulic network downstream from the pumping station, we used a different heavy rainfall event (from 12–17 December 2008), since this was the only event where there were simultaneous measurements of both free surface level at Ponte Maggiore's hydrometric station and rainfall amount trend in the Pontinia rain gauge. In this case, the optimization was performed by running the simulation model of both the hydraulic networks forced by the same rainfall amount trend. In this last case, the objective function was formalized using the measured free surface levels at the hydrometric gauge of Ponte Maggiore. The values of the parameters that minimized the objective functions and referred to the ideal channels connecting the storage areas of Linea Pio, Ufente, and Amaseno Rivers are reported in Table 4.

Table 4. Hydraulic parameters after the calibration of the model downstream from the Mazzocchio pumping system.

	Amaseno River	Linea Pio Channel	Ufente River
Channel mean bottom elevation H_c (meters above sea level)	4.5	0.0	3.28
Channel width l (m)	60	20	10
Channel length L (m)	10,600	18,000	5500
Roughness (Strickler)	30	40	30

3.2. *Optimal Pareto Set for the Event in March 2011*

An optimal Pareto curve was calculated by using the rainfall temporal distribution of the event on 16–20 March 2011 as an input (see Figure 8). The main aim is to verify whether and how much the pumping schedule applied during such an event was far from the Pareto front. Furthermore, we also test the process of convergence to the Pareto front of the multiobjective optimization algorithm, finally selecting 100 populations and 20 generations. In the analysis, the discrete levels of free surface H_s, regulating the switching on or off of the pumps, was varied within the prefixed range reported in Table 5.

Table 5. Variation range for activation levels.

	Pump no. 1	Pump no. 2	Pump no. 3	Pump no. 4	Pump no. 5	Pump no. 6
Start level (meters above sea level)	−2.15	−2.05	−1.95	−1.85	−1.7	−1.6
Variation range (meters above sea level)	Min −2.40 Max −2.10	Min −2.09 Max 2.00	Min −1.99 Max −1.90	Min −1.89 Max −1.75	Min −1.74 Max −1.50	Min −1.49 Max −1.00

In Figure 9, in the space of the objective functions, the points belonging to the optimal Pareto set as well as the points obtained from the different generations show that 20 generations were sufficient to allow for the convergence of the algorithm at the Pareto front. From Figure 9, it is evident how the two objectives are conflicting, since one objective cannot be improved without sacrificing the other one; i.e., if less pumping power is consumed during the event, a greater surface of the Mazzocchio basin is flooded.

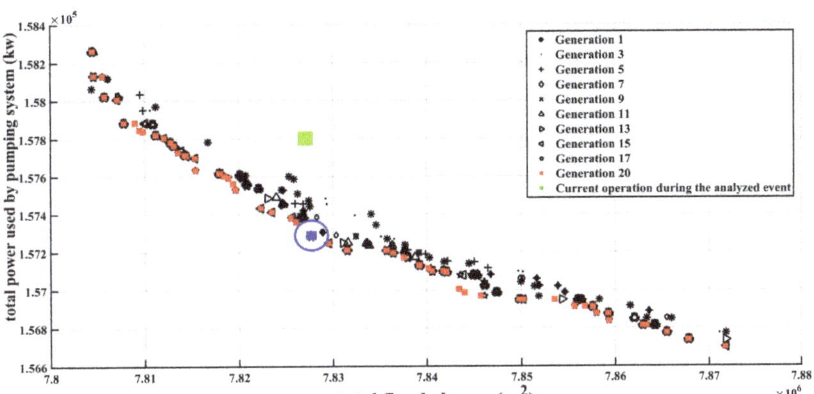

Figure 9. Pareto front obtained for the rainfall trend of the event that occurred in March 2011.

In the figure, the green-filled square refers to the values of the two objective functions obtained assuming the real operative pumping schedule applied during the rainfall event by the pumping station operators. From Figure 9, it is evident that such a point is far from the set of optimal Pareto solutions. This suggests that a more efficient pump schedule could have been identified to manage the extreme event of March 2011 between the solutions belonging to the Pareto front. As suggested by Kurek and Ostfeld [32], we used a procedure based on an utopian solution [33] to identify the optimal solution, circled in blue on Figure 9. Therefore, other comparison methods could be employed, such as game theory techniques [34]. The utopian solution has been identified as minimizing the objective functions:

$$m = \{min\{Of_1\}, min\{Of_2\}\} \tag{9}$$

Then, the span of the front ΔOf_i for each of the considered objectives is computed:

$$\Delta Of_i = max\{Of_i\} - min\{Of_i\} \quad i = 1,2 \tag{10}$$

Subsequently, the fronts are normalized using their span in the objective space; finally, a solution having the minimum distance to the utopian solution is selected for comparison:

$$\min_{p\in\Omega}\left\{\left[\left(\frac{Of_1 - m_1}{\Delta Of_1}\right)^2 + \left(\frac{Of_2 - m_2}{\Delta Of_2}\right)^2\right]^{\frac{1}{2}}\right\} \tag{11}$$

where Ω is the Pareto set resulting from solving the problem (5) or (6) and p is the selected "balanced" solution.

3.3. Sets of Pareto Optimal Solutions for Heavy Rainfall Events and Different Sea Level Rises

Running the combined simulation–optimization model, sets of Pareto optimal solutions were obtained for different design hyetographs associated to prefixed return periods of daily rainfall amount and different mean sea level. In this study, we used daily rainfall amount because previous studies [11,35] have shown that 24-h-long heavy rainfall (with a daily rainfall amount greater than 100 mm) has in the past induced serious flash flooding in the examined site. It should also be underlined that downscaling models aimed to project future changes in the precipitation regime [10] generally refer to the daily rainfall amount, and therefore it is reasonable to use the return period associated to such a quantity. In order to construct the design hyetograph, that is, the artificial rainfall temporal distribution, having a given 24-h rainfall amount return period, the dimensionless approach suggested by Kimura et al. [35], also called the modified ranking method, was used.

Such an approach can be summarized in the following steps:

1. The top ten 24-h independent rainfall events are selected, and the 24-h rainfall amounts arranged in descending order.
2. For each event, the percentage of hourly rainfalls in respect to the 24-h rainfall amount is calculated.
3. For each event, the hourly percentages are arranged in descending order.
4. The peak value among the ratios is placed at the center of the distribution, the second highest value is places in the right side of the center, and then the next highest ratio is placed at the opposite side of the edge, repeating the procedure until the rest of the ratios are all placed in the order.
5. Finally, the ratios are multiplied with the coefficient of the modified ranking method for each extreme rainfall; that coefficient is calculated by dividing the top ten 24-h rainfall amounts by the largest amount.

Using the rainfall data recorded in the Borgo San Michele rain gauge, the dimensionless hyetograph shown in Figure 10 was calculated.

Then, the hyetographs referring to the daily rainfall amount with return periods (T^R) equal to 10 and 100 years were assumed as inputs of the simulation–optimization model, and different sets of Pareto optimal solutions were calculated. While calculating the sets, we impose a further constraint on the maximum number of simultaneously working pumps. In Figure 11, two sets of Pareto optimal solutions are compared for a return period (T^R) equal to 100 years.

The first sets refer to the configuration of the system with unlimited flow carrying capacity of the hydraulic network downstream from the pumping station; that is, the Equation (6) constraint was removed. The second sets refer to the case in which the constraint of Equation (6) remains. As expected, within the admissible region defined by the constraints of Equation (6), the Pareto optimal solutions of the two sets are coinciding. The comparison in Figure 11 shows that the constraints of Equation (6) (the free surface levels must be lower than the height of levees in the downstream channel network) significantly reduces the discharge from the Mazzocchio basin that can be pumped into the river Ufente. For the reconstructed hyetograph with a return period (T^R) equal to 100 years, no more than four pumps can simultaneously work without violating the constraints. Therefore, due to the limited flow carrying capacity of the hydraulic network downstream from the pumping system, not all the potential of the pumping system to mitigate the flooding in the Mazzocchio basin can be achieved. It would be important to investigate in detail why this occurs.

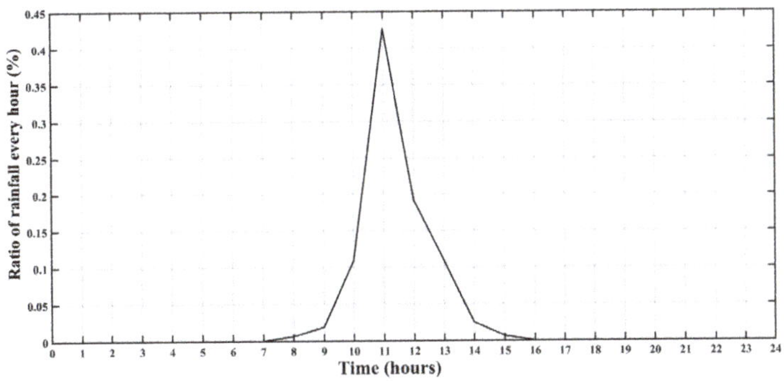

Figure 10. Design hyetograph.

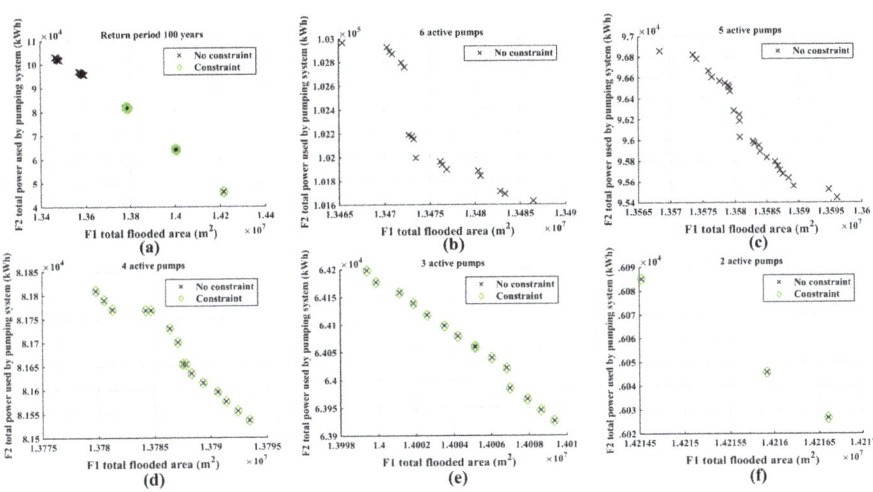

Figure 11. Pareto sets for daily rainfall amount with return period of 100 years. (**a**) Pareto set with all the pumping configurations. (**b**) Optimal Pareto set with a maximum of six pumps simultaneously working; and with (**c**) maximum of five pumps; (**d**) maximum of four pumps; (**e**) maximum of three pumps; (**f**) maximum of two pumps.

Figure 12 shows the temporal trend of the flow rates in significant cross sections of the Ufente, Amaseno, and Portatore rivers, close to the Ponte Maggiore node and the free surface level at the junction. These trends refer to the configuration of pump switching on/off corresponding to the Pareto optimal solution in the middle of the range, with a maximum number of working pumps equal to five. From Figure 12, it can be observed that a large increase in discharge, up to 250 m^3/s in the Amaseno river, causes a significant increase in the free surface level at the Ponte Maggiore node. Such an increase in the free surface level produces a backwater with a consequent inversion of the flow direction in the Ufente river (as seen by the negative values of the flow rate in the Figure 12), with a consequent blocking effect for the pumped flow rate from Mazzocchio. The consequences of this phenomenon can cause overflows and levee collapse along the Ufente River, which has already occurred in the recent past. From the results of hydraulic simulation, it is evident that the area included in a 3-km radius from Ponte Maggiore's junction, as seen in Figure 13, is at high risk of embankment overtop.

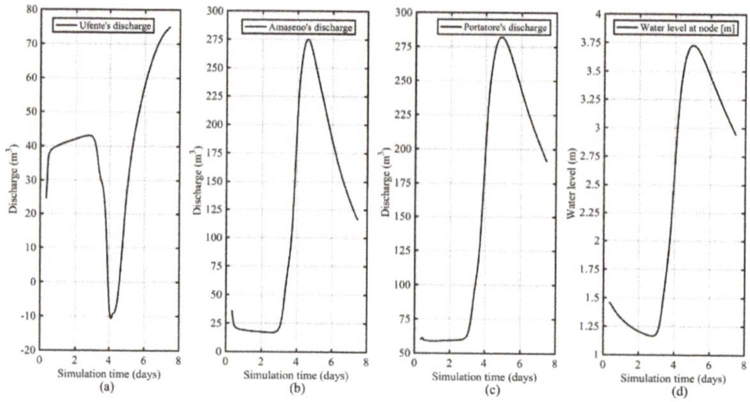

Figure 12. River discharges and free surface level trends at Ponte Maggiore's junction: rivers (**a**) Ufente, (**b**) Amaseno, and (**c**) Portatore, and (**d**) water level at node.

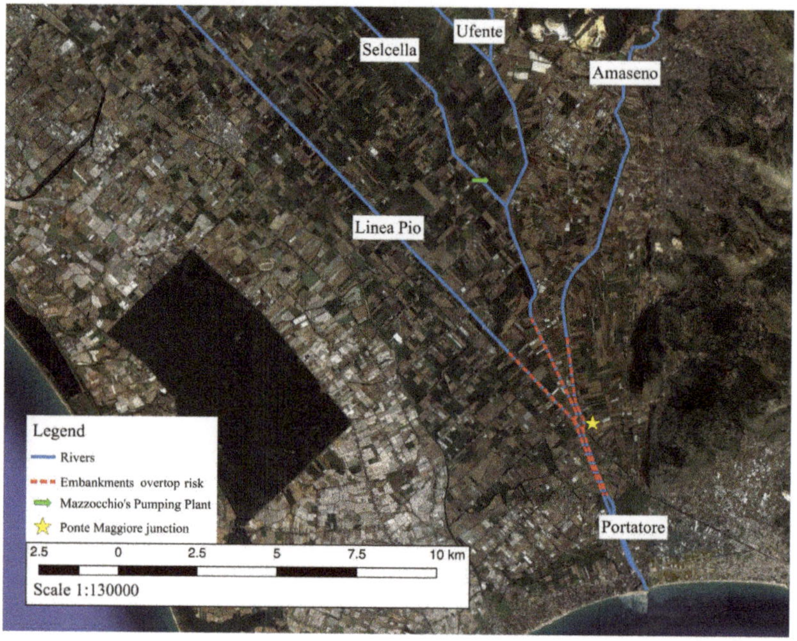

Figure 13. High risk of embankment overtop near Ponte Maggiore's junction using a 100-year return period hyetograph as the boundary condition for the hydraulic simulation.

In Figure 14, the sets of Pareto optimal solutions obtained for a hyetograph with the daily rainfall amount corresponding to a return period of T^R = 10 years and referring to three different maximum sea levels (0, +0.5, +1.0 m) at the outlet of the Portatore river are shown. These levels are typical of the shore examined and are due to the combined effects of the astronomical tide, storm, and barometric surge.

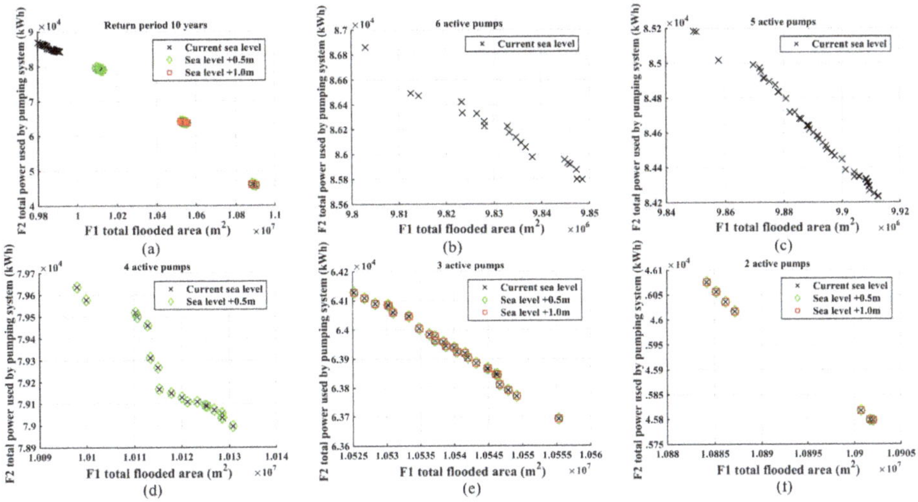

Figure 14. Pareto sets for a likely return period of 10 years. (**a**) Pareto set with all the pumping configurations. (**b**) Optimal Pareto set with a maximum of six pumps simultaneously working; and with (**c**) maximum of five pumps; (**d**) maximum of four pumps; (**e**) maximum of three pumps; and (**f**) maximum of two pumps.

As expected, the mean sea level rise modifies the boundary condition at the outlet of the network, causing an increase in free surface level in a larger part of the hydraulic network, making the violation of level constraints more likely. As a consequence of such a violation of level constraints, there is a limitation of the number of the pumps that can simultaneously work. As the figure shows, for a sea level rise equal to 0.5 m, the maximum number of pumps that may simultaneously work is equal to four, while in the case of a sea level rise equal to 1.0 m, no more than three pumps may be used simultaneously.

4. Discussion and Conclusions

This study assessed in the Mazzocchio area whether pumping systems and related channel hydraulic networks are reliable in managing the risks of flooding associated to extreme rainfall events and sea level rise. To conduct this research, we developed and proposed an approach that takes into account the different pumping schedules which are possible at parity of hydrological inputs. To compare the different states of the hydraulic system as a function of the different hydrological inputs, we referred to sets of Pareto optimal solutions calculated by a simulation–optimization model, which combined a multiobjective evolutionary algorithm (the nondominating sorting genetic algorithm, NSGA2) and a hydraulic model in order to satisfy two optimality criteria: minimize the consumption of pumping power and minimize the flooding surface over the Mazzocchio basin. The application to the study case shows how the increase of extreme rainfall amount, as well as sea level rise, affects the reliability and efficiency of the pumping system and hydraulic channel networks to mitigate the flooding in the study site. The use of the model allows identification of pumping system management solutions that are more efficient than those currently used, which are based on empirical assumptions. As seen in Figure 9, the current operating configuration of Mazzocchio's pumping plant should be more efficient in terms of energy consumption, with significant economic benefit. The Pareto set includes the optimal solutions, then through decision-making techniques, the preferred solutions can be identified. Pareto sets provide decision makers with a fundamental tool for defining and choosing the actions to be taken to reduce the hydraulic risk. For example, improving safety also means increasing costs. At present, the model can be part of a flood management system, included in a model system in which a submodel makes rainfall and sea level forecasts, starting from large-scale meteorological configurations. In this context, the developed model can be part of an optimal control system to manage both the pumping plant and the hydraulic network. The Pareto sets obtained from the optimization model (Figures 11 and 14) show that the hydraulic network will not be able to manage the increase in river flows due to increasingly intense rainfall and rising sea level, phenomena generated by climate changes. Therefore, the study suggests that in order to exploit the entire pumping capacity of the plant to mitigate the flooding in the Mazzocchio area, the flow carrying capacity of the downstream hydraulic channel network should be increased by designing new interventions; for instance, reshaping of the riverbeds and levees and designing structures for flow diversion or compensating–balancing reservoirs; all of those measures should be designed using a multiobjective optimization approach in order to minimize construction costs, but guaranteeing adequate protection from future extreme events. The identification of such technical solutions is beyond the goal of this paper, but the simulation–optimization model proposed here could still be used in order to explore such flood risk intervention strategies. This study could be further developed by taking into account the uncertainty in future projections of heavy rainfall and sea level rise, as suggested by Woodward et al. [25]. It is possible to take into account future uncertainties regarding sea level and rainfall amount and duration with the help of flexible sets of defense strategies linked to predefined situations. To assess such uncertainties, we are developing rainfall downscaling models which use General Circulation Models (GCM) ensemble simulations as inputs within a more general project aimed to assess and to identify possible strategies to preserve the coastal areas from the negative effect of climate change. Sea level rise will also be calculated using hydraulic models that will simulate storm surges caused by waves and wind, thus providing more accurate values.

Author Contributions: Conceptualization, F.C., A.D.B.T. and F.R.C.; Methodology, F.C., A.D.B.T. and F.R.C.; Software, F.C., A.D.B.T. and F.R.C.; Validation, F.C., A.D.B.T. and F.R.C.; Formal Analysis, F.C., A.D.B.T. and F.R.C.; Investigation, F.C., A.D.B.T. and F.R.C.; Resources, F.C., A.D.B.T. and F.R.C.; Data Curation, F.C., A.D.B.T. and F.R.C.; Writing-Original Draft Preparation, F.C., A.D.B.T. and F.R.C.; Writing-Review & Editing, F.C., A.D.B.T. and F.R.C.; Visualization, F.C., A.D.B.T. and F.R.C.; Supervision, F.C.; Project Administration, F.C.; Funding Acquisition, F.C.

Funding: This research was funded by Consorzio di Bonifica dell'Agro Pontino (APRC).

Conflicts of Interest: The authors declare no conflict of interest

References

1. Alpert, S.; Ben-Gai, P.; Baharad, T.; Benjamini, A.; Yekutieli, Y.; Colacino, D.; Michaelides, M. The paradoxical increase of Mediterranean extreme daily rainfall in spite of decrease in total values. *Geophys. Res. Lett.* **2002**, *29*, 1536. [CrossRef]
2. Zollo, A.L.; Rillo, V.; Bucchignani, E.; Montesarchio, M.; Mercogliano, P. Extreme temperature and precipitation events over Italy: Assessment of high-resolution simulations with COSMO-CLM and future scenarios. *Int. J. Climatol.* **2016**, *36*, 987–1004. [CrossRef]
3. Rajczak, J.; Pall, P.; Schär, C. Projections of extreme precipitation events in regional climate simulations for Europe and the Alpine Region. *J. Geophys. Res.* **2013**, *118*, 3610–3626. [CrossRef]
4. Gao, X.; Pal, J.S.; Giorgi, F. Projected changes in mean and extreme precipitation over the Mediterranean region from a high resolution double nested RCM simulation. *Geophys. Res. Lett.* **2006**, *33*, 2–5. [CrossRef]
5. Antonioli, F.; Anzidei, M.; Amorosi, A.; Lo Presti, V.; Mastronuzzi, G.; Deiana, G.; De Falco, G.; Fontana, A.; Fontolan, G.; Lisco, S.; et al. Sea-level rise and potential drowning of the Italian coastal plains: Flooding risk scenarios for 2100. *Quat. Sci. Rev.* **2017**, *158*, 29–43. [CrossRef]

6. Galassi, G.; Spada, G. Sea-level rise in the Mediterranean Sea by 2050: Roles of terrestrial ice melt, steric effects and glacial isostatic adjustment. *Glob. Planet. Chang.* **2014**, *123*, 55–66. [CrossRef]
7. Lambeck, K.; Antonioli, F.; Anzidei, M.; Ferranti, L.; Leoni, G.; Scicchitano, G.; Silenzi, S. Sea level change along the Italian coast during the Holocene and projections for the future. *Quat. Int.* **2011**, *232*, 250–257. [CrossRef]
8. Nicklow, J.; Reed, P.; Savic, D.; Dessalegne, T.; Harrell, L.; Chan-Hilton, A.; Karamouz, M.; Minsker, B.; Ostfeld, A.; Singh, A.; et al. State of the Art for Genetic Algorithms and Beyond in Water Resources Planning and Management. *J. Water Resour. Plan. Manag.* **2010**, *136*, 412–432. [CrossRef]
9. Merz, B.; Aerts, J.; Arnbjerg-Nielsen, K.; Baldi, M.; Becker, A.; Bichet, A.; Bloschl, G.; Bouwer, L.M.; Brauer, A.; Cioffi, F.; et al. Floods and climate: Emerging perspectives for flood risk assessment and management. *Nat. Hazards Earth Syst. Sci.* **2014**, *2*, 1921–1942. [CrossRef]
10. Cioffi, F.; Conticello, F.; Lall, U. Projecting changes in Tanzania rainfall for the 21st century. *Int. J. Climatol.* **2016**, *36*, 4297–4314. [CrossRef]
11. Cioffi, F.; Conticello, F.; Lall, U.; Marotta, L.; Telesca, V. Large scale climate and rainfall seasonality in a Mediterranean Area: Insights from a non-homogeneous Markov model applied to the Agro-Pontino plain. *Hydrol. Process.* **2017**, *31*, 668–686. [CrossRef]
12. Marler, R.T.; Arora, J.S. Survey of multi-objective optimization methods for engineering. *Struct. Multidiscip. Optim.* **2004**, *26*, 369–395. [CrossRef]
13. Cioffi, F.; Gallerano, F. Multi-objective analysis of dam release flows in rivers downstream from hydropower reservoirs. *Appl. Math. Model.* **2012**, *36*, 2868–2889. [CrossRef]
14. Yazdi, J.; Yoo, D.G.; Kim, J.H. Comparative study of multi-objective evolutionary algorithms for hydraulic rehabilitation of urban drainage networks. *Urban Water J.* **2017**, *14*, 483–492. [CrossRef]
15. Lin, Y.H.; Chen, Y.P.; Der Yang, M.; Su, T.C. Multiobjective Optimal Design of Sewerage Rehabilitation by Using the Nondominated Sorting Genetic Algorithm-II. *Water Resour. Manag.* **2015**, *30*, 487–503. [CrossRef]
16. Yazdi, J.; Choi, H.S.; Kim, J.H. A methodology for optimal operation of pumping stations in urban drainage systems. *J. Hydro-Environ. Res.* **2014**, *11*, 101–112. [CrossRef]
17. Moradi-Jalal, M.; Karney, B.W. Optimal design and operation of irrigation pumping stations using mathematical programming and Genetic Algorithm (GA). *J. Hydraul. Res.* **2008**, *46*, 237–246. [CrossRef]
18. Makaremi, Y.; Haghighi, A.; Ghafouri, H.R. Optimization of Pump Scheduling Program in Water Supply Systems Using a Self-Adaptive NSGA-II; a Review of Theory to Real Application. *Water Resour. Manag.* **2017**, *31*, 1283–1304. [CrossRef]
19. Lewis, A.; Randall, M. Solving multi-objective water management problems using evolutionary computation. *J. Environ. Manag.* **2017**, *204*, 179–188. [CrossRef] [PubMed]
20. Ahmadi, M.; Haddad, O.B.; Loáiciga, H.A. Adaptive Reservoir Operation Rules Under Climatic Change. *Water Resour. Manag.* **2014**, *29*, 1247–1266. [CrossRef]
21. Jahandideh-Tehrani, M.; Bozorg Haddad, O.; Loáiciga, H.A. Hydropower Reservoir Management Under Climate Change: The Karoon Reservoir System. *Water Resour. Manag.* **2014**, *29*, 749–770. [CrossRef]
22. François, B.; Hingray, B.; Creutin, J.D.; Hendrickx, F. Estimating Water System Performance Under Climate Change: Influence of the Management Strategy Modeling. *Water Resour. Manag.* **2015**, *29*, 4903–4918. [CrossRef]
23. Schardong, A.; Simonovic, S.P. *Multi-Objective Evolutionary Algorithms for Water Resource Management*; The University of Western Ontario: Ontario, ON, Canada, 2011.
24. Van der Pol, T.D.; van Ierland, E.C.; Gabbert, S.; Weikard, H.P.; Hendrix, E.M.T. Impacts of rainfall variability and expected rainfall changes on cost-effective adaptation of water systems to climate change. *J. Environ. Manag.* **2015**, *154*, 40–47. [CrossRef] [PubMed]
25. Woodward, M.; Kapelan, Z.; Gouldby, B. Adaptive flood risk management under climate change uncertainty using real options and optimization. *Risk Anal.* **2014**, *34*, 75–92. [CrossRef] [PubMed]
26. Cioffi, F.; Gallerano, F. A two-dimensional self-adaptive hydrodynamic scheme for the assessment of the effects of structures on flooding phenomena in river basins. *River Res. Appl.* **2003**, *19*, 1–26. [CrossRef]
27. MacManus, K.; Orton, P.M.; Cioffi, F.; Conticello, F.; Hall, T.M.; Lall, U.; Bloomberg, A.F.; MacManus, K. Flood hazard assessment from storm tides, rain and sea level rise for a tidal river estuary. *Nat. Hazards* **2018**, 1–29. [CrossRef]

28. Goutal, N.; Lacombe, J.; Zaoui, F. MASCARET: A 1-D open-source software for flow hydrodynamic and water quality in open channel networks. In *River Flow*; Taylor & Francis Group: London, UK, 2012; pp. 1169–1174.
29. Deb, K.; Agrawal, S.; Pratap, A.; Meyarivan, T. A fast elitist non-dominated sorting genetic algorithm for multi-objective optimization: NSGA-II. In Proceedings of the International Conference on Parallel Problem Solving From Nature 2000, Paris, France, 18–20 September 2000; pp. 849–858.
30. Goldberg, D.E.; Deb, K. A Comparative Analysis of Selection Schemes Used in Genetic Algorithms. *Found. Genet. Algorithms* **1991**, *1*, 69–93.
31. Deb, K.; Agrawal, R.B. Simulated Binary Crossover for Continuous Search Space. *Complex Syst.* **1994**, *9*, 1–34.
32. Wojciech Kurek, A.O. Multi-objective optimization of water quality, pumps operation, and storage sizing of water distribution systems. *J. Environ. Manag.* **2013**, *115*, 189–197. [CrossRef] [PubMed]
33. Miettinen, K. *Nonlinear Multiobjective Optimization*; Kluwer Academic Publishers: Boston, MA, USA, 1999.
34. Raquel, S.; Ferenc, S.; Emery, C.; Abraham, R. Application of game theory for a groundwater conflict in Mexico. *J. Environ. Manag.* **2007**, *84*, 560–571. [CrossRef] [PubMed]
35. Kimura, N.; Tai, A.; Chiang, S.; Wei, H.P.; Su, Y.F.; Cheng, C.T.; Kitoh, A. Hydrological flood simulation using a design hyetograph created from extreme weather data of a high-resolution atmospheric general circulation model. *Water* **2014**, *6*, 345–366. [CrossRef]

MDPI
St. Alban-Anlage 66
4052 Basel
Switzerland
Tel. +41 61 683 77 34
Fax +41 61 302 89 18
www.mdpi.com

Water Editorial Office
E-mail: water@mdpi.com
www.mdpi.com/journal/water

www.ingramcontent.com/pod-product-compliance
Lightning Source LLC
Chambersburg PA
CBHW061455220926
51850CB00004B/96
* 9 7 8 3 0 3 9 4 3 3 8 6 5 *